ATOMIC AND IONIC RADII OF SELECTED ELEMENTS[a]

Atomic number	Symbol	Atomic radius (nm)	Ion	Ionic radius (nm)
3	Li	0.152	Li^+	0.078
4	Be	0.114	Be^{2+}	0.054
5	B	0.097	B^{3+}	0.02
6	C	0.077	C^{4+}	<0.02
7	N	0.071	N^{5+}	0.01–0.02
8	O	0.060	O^{2-}	0.132
9	F	—	F^-	0.133
11	Na	0.186	Na^+	0.098
12	Mg	0.160	Mg^{2+}	0.078
13	Al	0.143	Al^{3+}	0.057
14	Si	0.117	Si^{4+}	0.039
15	P	0.109	P^{5+}	0.03–0.04
16	S	0.106	S^{2-}	0.174
17	Cl	0.107	Cl^-	0.181
19	K	0.231	K^+	0.133
20	Ca	0.197	Ca^{2+}	0.106
21	Sc	0.160	Sc^{2+}	0.083
22	Ti	0.147	Ti^{4+}	0.064
23	V	0.132	V^{4+}	0.061
24	Cr	0.125	Cr^{3+}	0.064
25	Mn	0.112	Mn^{2+}	0.091
26	Fe	0.124	Fe^{2+}	0.087
27	Co	0.125	Co^{2+}	0.082
28	Ni	0.125	Ni^{2+}	0.078
29	Cu	0.128	Cu^+	0.096
30	Zn	0.133	Zn^{2+}	0.083
31	Ga	0.135	Ga^{3+}	0.062
32	Ge	0.122	Ge^{4+}	0.044
35	Br	0.119	Br^-	0.196
39	Y	0.181	Y^{3+}	0.106
40	Zr	0.158	Zr^{4+}	0.087
41	Nb	0.143	Nb^{4+}	0.074
42	Mo	0.136	Mo^{4+}	0.068
46	Pd	0.137	Pd^{2+}	0.050
47	Ag	0.144	Ag^+	0.113
48	Cd	0.150	Cd^{2+}	0.103
50	Sn	0.158	Sn^{4+}	0.074
53	I	0.136	I^-	0.220
55	Cs	0.265	Cs^+	0.165
56	Ba	0.217	Ba^{2+}	0.143
74	W	0.137	W^{4+}	0.068
78	Pt	0.138	Pt^{2+}	0.052
79	Au	0.144	Au^+	0.137
80	Hg	0.150	Hg^{2+}	0.112
82	Pb	0.175	Pb^{2+}	0.132
92	U	0.138	U^{4+}	0.105

[a]For a complete listing, see Appendix 2.

INTRODUCTION to MATERIALS SCIENCE for ENGINEERS

FIFTH EDITION

JAMES F. SHACKELFORD

University of California, Davis

Prentice Hall
Upper Saddle River, New Jersey 07458

Library of Congress Cataloging-in-Publication Data

Shackelford, James F.
    Introduction to materials science for engineers / James F. Shackelford. – 5th ed.
        p.   cm.
    Includes bibliographical references and index.
    ISBN   0–13–011287–9
    1. Materials   I. Title.
TA403.S515   1999   99–41181
620.1–dc20   CIP

Editorial/production supervision: *Rose Kernan*
Editor-in-chief: *Marcia Horton*
Editor-in-chief of development: *Ray Mullaney*
Development editor: *Deena Cloud*
Copyeditor: *Pat Daly*
Interior design: *Kevin Kall*
Vice-president of production and manufacturing: *David W. Riccardi*
Creative director: *Paul Belfanti*
Associate creative director: *Amy Rosen*
Art director: *Ann France*
Cover design: *Marjory Dressler*
Manufacturing buyer: *Pat Brown*
Composition: *PreTEX, Inc.*

The cover: Weldment between Ti–6A1–4V and Ni-Cr alloys following an argon anneal. Nomarski interference contrast, with incident illumination. (Courtesy of Dr. Michael Meier, University of California, Davis)

 © 2000 by Prentice-Hall, Inc.
Upper Saddle River, New Jersey 07458

Printed in the United States of America

10 9 8 7 6 5 4 3 2 1

ISBN   0-13-011287-9

Prentice-Hall International (UK) Limited, *London*
Prentice-Hall of Australia Pty. Limited, *Sydney*
Prentice-Hall Canada Inc., *Toronto*
Prentice-Hall Hispanoamericana, S.A., *Mexico*
Prentice-Hall of India Private Limited, *New Delhi*
Prentice-Hall of Japan, Inc., *Tokyo*
Prentice-Hall (Singapore) Pte. Ltd., *Singapore*
Editora Prentice-Hall do Brasil, Ltda., *Rio de Janeiro*

Dedicated to Penelope and Scott

Contents

PART FOUR   MATERIALS IN ENGINEERING DESIGN

PREFACE

This book is designed for a first course in engineering materials. The field that covers this area of the engineering profession has come to be known as "materials science and engineering." To me, this label serves two important functions. First, it is an accurate description of the balance between scientific principles and practical engineering that is required in selecting the proper materials for modern technology. Second, it gives us a guide to organizing this book. Each word defines a distinct part. After a short introductory chapter, "science" serves as a label for Part I on "The Fundamentals." Chapters 2 through 10 cover various topics in applied physics and chemistry. These are the foundation for understanding the principles of "materials science." I assume that some students take this course at the freshman or sophomore level and may not yet have taken their required coursework in chemistry and physics. As a result, Part I is intended to be self-contained. A previous course in chemistry or physics is certainly helpful, but should not be necessary. If an entire class has finished freshman chemistry, Chapter 2 (atomic bonding) could be left as optional reading, but it is important not to overlook the role of bonding in defining the fundamental types of engineering materials. The remaining chapters in Part I are less optional, as they describe the key topics of materials science. Chapter 3 outlines the ideal, crystalline structures of important materials. Chapter 4 introduces the structural imperfections found in real, engineering materials. These structural defects are the bases of solid-state diffusion (Chapter 5) and plastic deformation in metals (Chapter 6). Chapter 6 also includes a broad range of mechanical behavior for various engineering materials. Similarly, Chapter 7 covers the thermal behavior of these materials. Subjecting materials to various mechanical and thermal processes can lead to their failure, the subject of Chapter 8. In addition, the systematic analysis of material failures can lead to the prevention of future catastrophes. Chapters 9 and 10 are especially important in providing a bridge between "materials science" and "materials engineering." Phase diagrams (Chapter 9) are an effective tool for describing the equilibrium microstructures of practical engineering materials. Instructors will note that this topic is introduced in a descriptive and empirical way. Since some students in this course may not have taken a course in thermodynamics, I avoid the use of the free-energy property. (A companion volume, described below, is available for those instructors wishing to discuss phase diagrams with a complementary introduction to thermodynamics.) Kinetics (Chapter 10) is the foundation of the heat treatment of engineering materials.

The word *materials* gives us a label for Part II of the book. We identify the four categories of *structural materials*. Metals (Chapter 11), ceramics (Chapter 12), and polymers (Chapter 13) are traditionally identified as the three types of engineering materials. I have entitled Chapter 12 "Ceramics and Glasses" to emphasize the distinctive character of the noncrystalline glasses, which are chemically similar to the crystalline ceramics. Chapter 14 adds "composites" as a fourth category that involves some combination of the three fundamental types. Fiberglass, wood, and concrete are some common examples. Advanced composites, such as the graphite/epoxy system, represent some of the most dramatic developments in structural materials. In Part II, each chapter catalogues examples of each type of structural material and describes their processing, the techniques used to produce the materials.

The word *materials* also labels Part III. Materials used primarily for *electronic, optical*, and *magnetic* applications can generally be classified in one of the categories of structural materials. But a careful inspection of electrical conduction (Chapter 15) shows that a separate category, semiconductors, can be defined. Metals are generally good electrical conductors, while ceramics and polymers are generally good insulators, and semiconductors are intermediate. The exceptional discovery of superconductivity in certain ceramic materials at relatively high temperatures augments the long-standing use of superconductivity in certain metals at very low temperatures.

Chapter 16 covers optical behavior which determines the application of many materials, from traditional glass windows to some of the latest advances in telecommunications. Chapter 17 is devoted to the important category of semiconductor materials that is the basis of the solid-state electronics industry. A wide variety of magnetic materials is discussed in Chapter 18. Traditional metallic and ceramic magnets are being supplemented by superconducting metals and ceramics, which can provide some intriguing design applications based on their magnetic behavior.

The "engineering" in "materials science and engineering" labels Part IV, "Materials in Engineering Design," which focuses on the role of materials in engineering applications. Chapter 19 (Environmental Degradation) discusses limitations imposed by the environment. Chemical degradation, radiation damage, and wear must be considered in making a final judgment on a material's application. Finally, in Chapter 20 (Materials Selection), we see that our previous discussions of properties have left us with "design parameters." Herein lies a final bridge between the principles of materials science and the use of those materials in modern engineering designs.

I hope that students and instructors alike will find what I have attempted to produce: a clear and readable textbook organized around the title of this important branch of engineering. It is also worth noting that materials play a central role across the broad spectrum of contemporary science and technology. In the National Research Council's report *Materials Science and Engineering for the 1990s: Maintaining Competitiveness in the Age of Materials*, it is estimated that approximately one-third of all employed physicists and chemists work in materials. In the report *Science: The End of the Frontier?* from the American Association for the Advancement of Science, 10 of the 26 technologies identified at the forefront of economic growth are various types of advanced materials.

In the presentation of this book I have attempted to be generous with sample problems and practice problems within each chapter, and I have tried to be even more generous with the end-of-chapter homework problems (with the level of difficulty for the homework problems clearly noted). Problems dealing with the role of materials in the engineering design process are noted with the use of a design icon **D**. One of the most enjoyable parts of writing the book was the preparation of biographical footnotes for those cases in which a person's name has become intimately associated with a basic concept in materials science and engineering. I suspect that most readers will share my fascination with these great contributors to science and engineering from the distant and not-so-distant past. In addition to a substantial set of useful data, the Appendixes provide convenient location of materials' properties, characterization tools, and key term definitions.

The various editions of this book have been produced during a period of fundamental change in the field of materials science and engineering. This change was exemplified by the change of name in the Fall of 1986 for the "American Society for Metals" to "ASM International"—a society for *materials*, as opposed to metals only. An adequate introduction to materials science can no longer be a traditional treatment of physical metallurgy with supplementary introductions to nonmetallic materials. The first edition was based on a balanced treatment of the full spectrum of engineering materials. The second edition reinforced that approach with the timely addition of new materials poised to play key roles in the economy of the twenty-first century: lightweight metal alloys, "high-tech" ceramics for power generation, engineering polymers for metal substitution, advanced composites for aerospace applications, semiconductors for increasingly sophisticated electronic devices, and nonmetallic superconducting magnets with increasingly high operating temperatures. The third edition was updated to include recent, fundamental research on grain boundary structure and quasicrystals, which have significantly broadened our understanding of the structure of materials. Similarly, advances in high-temperature superconductors, semiconductor devices, and biomaterials were included. The fourth edition was updated to include several new developments. Numerous crystal structure illustrations were created with computer simulation software, reflecting the significant advances in the computer modeling of materials. The discussions of microscopy and processing were expanded to include developments such as the scanning tunneling microscope (STM) and self-propagating high temperature synthesis (SHS), respectively. Developments in advanced composites, biomaterials, and "smart materials" were also included.

CHANGES IN THE FIFTH EDITION

The Fifth Edition represents the most extensive revision to date. After consultation with a wide spectrum of users, I have made a number of structural changes which I believe will provide an even more effective introduction to the field of materials science and engineering. There are now separate chapters on mechanical, thermal, and optical behavior, as well as diffusion and failure analysis and prevention. Discussions on the various materials types have been expanded and updated, and materials processing has been integrated into those discussions. I also provide new or expanded treatments of several topics including heat capacity, polarization curves, luminescence, photoconductors, lasers, optical fibers, liquid crystal displays, light-emitting diodes, amorphous metals for electric power distribution, honeycomb structures, and environmental issues including recycling.

Consistent with previous editions, over 150 new and revised homework problems are provided. Symbolic of the substantial nature of this revision, there are nearly 80 new Key Terms identified throughout the text and added to the glossary in Appendix 7. Finally, several one-page inserts labelled "The Material World" have been added to provide focus on some fascinating topics in the world of both engineered and natural materials.

SUPPLEMENTARY MATERIAL

An *Instructor's Manual* with fully worked-out solutions to the practice problems and homework problems is available from the publisher. This volume includes a set of *Laboratory Experiments* of general interest for an introductory course on engineering materials and a discussion of *Thermodynamics* which can be introduced just prior to Chapter 9 as a foundation for the discussions of phase diagrams and kinetics in Chapters 9 and 10.

All illustrations and tables are available for classroom use at a website devoted to this text. This site is readily accessed through

www.prenhall.com/shackelford.

The "Shackelford site" also contains a variety of other material of interest to instructors and students alike.

ACKNOWLEDGMENTS

Finally, I want to acknowledge a number of people who have been immensely helpful in making this book possible. My family has been more than the usual "patient and understanding." They are a constant reminder of the rich life beyond the material plane. Peter Gordon (first edition), David Johnstone (second and third editions), and Bill Stenquist (fourth and fifth editions) are much appreciated in their roles as editors. I am especially indebted to Prentice-Hall for providing Deena Cloud as Development Editor. Her expert guidance was exceptionally helpful and effective. Lilian Davila skillfully produced the computer-generated crystal structure images. A special appreciation is due to my colleagues at the University of California—Davis and to the many reviewers of all editions, especially D. J. Montgomery, John M. Roberts, D. R. Rossington, R. D. Daniels, R. A. Johnson, D. H. Morris, J. P. Mathers, Richard Fleming, Ralph Graff, Ian W. Hall, John J. Kramer, Enayat Mahajerin, Carolyn W. Meyers, Ernest F. Nippes, Richard L. Porter, Eric C. Skaar, E. G. Schwartz, William N. Weins, M. Robert Baren, John Botsis, D. L. Douglass, Robert W. Hendricks, J. J. Hren, Sam Hruska, I. W. Hull, David B. Knoor, Harold Koelling, John McLaughlin, Alvin H. Meyer, M. Natarajan, Jay Samuel, John R. Schlup, Theodore D. Taylor, Ronald Kander, Alan Lawley, and Joanna McKittrick.

J.F.S.
Davis, California

AUTHOR BIOGRAPHY

James F. Shackelford has B.S. and M.S. degrees in Ceramic Engineering from the University of Washington and a Ph.D. in Materials Science and Engineering from the University of California, Berkeley. He is currently a Professor in the Department of Chemical Engineering and Materials Science and the Associate Dean for Undergraduate Studies in the College of Engineering at the University of California, Davis. He teaches and conducts research in the areas of materials science, the structure of materials, nondestructive testing, and biomaterials. A member of ASM International and the American Ceramic Society, he was named a Fellow of the American Ceramic Society in 1992 and the Outstanding Educator of the American Ceramic Society in 1996. He has published over 100 archived papers and books, including *The CRC Materials Science and Engineering Handbook*, now in its third edition.

CHAPTER 1
Materials for Engineering

The modern automobile is a case study in the selection of a wide range of traditional and advanced materials. For example the large air scoops that extend from the front of this vehicle to the doors are an integral part of the fenders, which are made of a sophisticated, moldable polymer. (Courtesy of Dow Automotive Division of Dow Chemical Corporation)

1.1   THE MATERIAL WORLD

We live in a world of material possessions that largely define our social relationships and economic quality of life. Figure 1–1 dramatically illustrates the extent of this world for a family matching the current statistical average for "material wealth" in the United States.

The material possessions of our earliest ancestors were probably their tools and weapons. In fact, the most popular way of naming the era of early human civilization is in terms of the materials from which these tools and weapons were made. The **Stone Age** has been traced as far back as 2.5 million years ago when human ancestors, or hominids, chipped stones to form weapons for hunting. The **Bronze Age** roughly spanned the period from 2000 B.C. to 1000 B.C. and represents the foundation of metallurgy, in which **alloys** of copper and tin were discovered to produce superior tools and weapons. (An *alloy* is a metal composed of more than one element.)

Contemporary archaeologists note that an earlier but less well known "Copper Age" existed between roughly 4000 B.C. and 3000 B.C. in Europe, in which relatively pure copper was used before tin became available. The limited utility of those copper products provided an early lesson in the importance of proper alloy additions. The **Iron Age** defines the period from 1000 B.C. to 1 B.C. By 500 B.C. iron alloys had largely replaced bronze for tool and weapon making in Europe.

Although archaeologists do not refer to a "pottery age," the presence of domestic vessels made from baked clay has provided some of the best descriptions of human cultures for thousands of years. Similarly, glass artifacts have been traced back to 4000 B.C. in Mesopotamia.

Modern culture in the second half of the twentieth century is sometimes referred to as "plastic," a not entirely complimentary reference to the lightweight and economical polymeric materials from which so many products are made. Some observers have suggested instead that this same time frame should be labeled the "silicon age," given the pervasive impact of modern electronics largely based on silicon technology.

Figure 1-1 *The material possessions of a family matching the statistical average for the United States. (From Peter Menzel, Material World— A Global Family Portrait, Sierra Club Books, San Francisco, 1994.)*

1.2   MATERIALS SCIENCE AND ENGINEERING

In the past three decades, the term that has come to label the general branch of engineering concerned with materials is *materials science and engineering.* This label is accurate in that this field is a true blend of fundamental, scientific studies and practical engineering. It has grown to include contributions from many traditional fields, including metallurgy, ceramic engineering, polymer chemistry, condensed matter physics, and physical chemistry.

The term *materials science and engineering* will serve a special function in this introductory textbook. It will provide the basis for the text's organization. First, the word *science* describes Part I (Chapters 2 to 10), which deals with the fundamentals of structure and classification. Second, the word *materials* describes Part II (Chapters 11 to 14), which describes the four types of *structural materials*, and Part III (Chapters 15 to 18), which describes various electronic and magnetic materials, including the separate category of semiconductors. Finally, the word *engineering* describes Part IV (Chapters 19 and 20), which puts the materials to work with discussions of key aspects of the degradation and selection of materials.

1.3   TYPES OF MATERIALS

The most obvious question to be addressed by the engineering student entering an introductory course on materials is "What materials are available to me?" Various classification systems are possible for the wide-ranging answer to this question. In this book we distinguish five categories that encompass the materials available to practicing engineers: metals, ceramics and glasses, polymers, composites, and semiconductors.

METALS

If there is a "typical" material associated in the public's mind with modern engineering practice, it is structural *steel.* This versatile construction material has several properties that we consider **metallic:** (1) is strong and can be readily formed into practical shapes. (2) Its extensive, permanent deformability, or **ductility,** is an important asset in permitting small amounts of yielding to sudden and severe loads. Many Californians have been able to observe moderate earthquake activity that leaves windows (of glass, which is relatively **brittle**, i.e., lacking in ductility) cracked while steel support framing still functions normally. (3) A freshly cut steel surface has a characteristic metallic luster, and (4) a steel bar shares a fundamental characteristic with other metals: It is a good conductor of electrical current. Although structural steel is an especially common example of metals for engineering, a little thought produces numerous others (Figure 1–2).

Figure 1-2   *These examples of common metal parts, including various springs and clips, are characteristic of their wide range of engineering applications. (Courtesy of Elgiloy Company)*

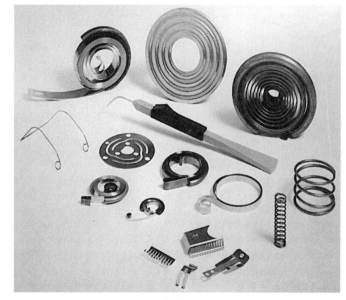

In Chapter 2, the nature of metals will be defined and placed in perspective relative to the other categories. It is useful to consider the extent of metallic behavior in the currently known range of chemical elements. Figure 1–3 highlights the chemical elements in the periodic table that are inherently metallic. This is a large family indeed. The shaded elements are the bases of the various engineering alloys,including the irons and steels (from Fe), aluminum alloys (Al), magnesium alloys (Mg), titanium alloys (Ti), nickel alloys (Ni), zinc alloys (Zn), and copper alloys (Cu) including the brasses (Cu, Zn). Figure 1–4 illustrates an example of the state of the art in metalworking, namely, parts formed by superplastic deformation, to be discussed further in Chapter 11.

Figure 1-3   *Periodic table of the elements with those elements that are inherently metallic in nature in color.*

I A													III A	IV A	V A	VI A	VII A	O
1 H	II A																	2 He
3 Li	4 Be						VIII						5 B	6 C	7 N	8 O	9 F	10 Ne
11 Na	12 Mg	III B	IV B	V B	VI B	VII B				I B	II B		13 Al	14 Si	15 P	16 S	17 Cl	18 Ar
19 K	20 Ca	21 Sc	22 Ti	23 V	24 Cr	25 Mn	26 Fe	27 Co	28 Ni	29 Cu	30 Zn		31 Ga	32 Ge	33 As	34 Se	35 Br	36 Kr
37 Rb	38 Sr	39 Y	40 Zr	41 Nb	42 Mo	43 Tc	44 Ru	45 Rh	46 Pd	47 Ag	48 Cd		49 In	50 Sn	51 Sb	52 Te	53 I	54 Xe
55 Cs	56 Ba	57 La	72 Hf	73 Ta	74 W	75 Re	76 Os	77 Ir	78 Pt	79 Au	80 Hg		81 Tl	82 Pb	83 Bi	84 Po	85 At	86 Rn
87 Fr	88 Ra	89 Ac																

58 Ce	59 Pr	60 Nd	61 Pm	62 Sm	63 Eu	64 Gd	65 Tb	66 Dy	67 Ho	68 Er	69 Tm	70 Yb	71 Lu
90 Th	91 Pa	92 U	93 Np	94 Pu	95 Am	96 Cm	97 Bk	98 Cf	99 Es	100 Fm	101 Md	102 No	103 Lw

Figure 1-4   *Various aluminum parts fabricated by superplastic deformation. The unusually high degree of deformability for these alloys is possible with a carefully controlled, fine-grained microstructure. Superplastic forming uses air pressure to stretch a "bubble" of metal sheet over a metal preform. (Courtesy of Superform USA)*

CERAMICS AND GLASSES

Aluminum (Al) is a common metal, but aluminum *oxide*, a compound of aluminum and oxygen such as Al_2O_3, is typical of a fundamentally different family of engineering materials, **ceramics.** Aluminum oxide has two principal advantages over metallic aluminum. First, Al_2O_3 is chemically stable in a wide variety of severe environments whereas metallic aluminum would be oxidized (a term discussed in detail in Chapter 19). In fact, a common reaction product in the chemical degradation of aluminum is the more chemically stable oxide. Second, the ceramic Al_2O_3 has a significantly higher melting point (2020°C) than does the metallic Al (660°C). This makes Al_2O_3 a popular **refractory,** that is, a high-temperature-resistant material of wide use in industrial furnace construction.

With its superior chemical and temperature-resistant properties, why isn't Al_2O_3 used for applications such as automotive engines in place of metallic aluminum? The answer to this question lies in the most limiting property of ceramics—brittleness. Aluminum and other metals have high ductility, a desirable property that permits them to undergo relatively severe impact loading without fracture, whereas aluminum oxide and other ceramics do not. Thus ceramics are eliminated from many structural application because they are brittle.

Recent developments in ceramic technology are expanding the utility of ceramics for structural applications, not by eliminating their inherent brittleness, but by increasing their strength to sufficiently high levels and increasing their resistance to fracture. (The important concept of *fracture toughness* will be introduced in Chapter 8.) In Chapter 6 we will explore the basis for the brittleness of ceramics as well as the promise of the new high-strength structural ceramics. An example of these new materials is silicon nitride (Si_3N_4), a primary candidate for high-temperature, energy-efficient automobile engines—an application unthinkable for traditional ceramics.

Aluminum oxide is typical of the traditional ceramics, with magnesium oxide (MgO) and **silica** (SiO_2) being other good examples. In addition, SiO_2 is the basis of a large and complex family of **silicates,** which includes clays and claylike minerals. Silicon nitride (Si_3N_4), mentioned previously, is an important nonoxide ceramic used in a variety of structural applications. The vast majority of commercially important ceramics are chemical compounds made up of at least one metallic element (see Figure 1–3) and one of five

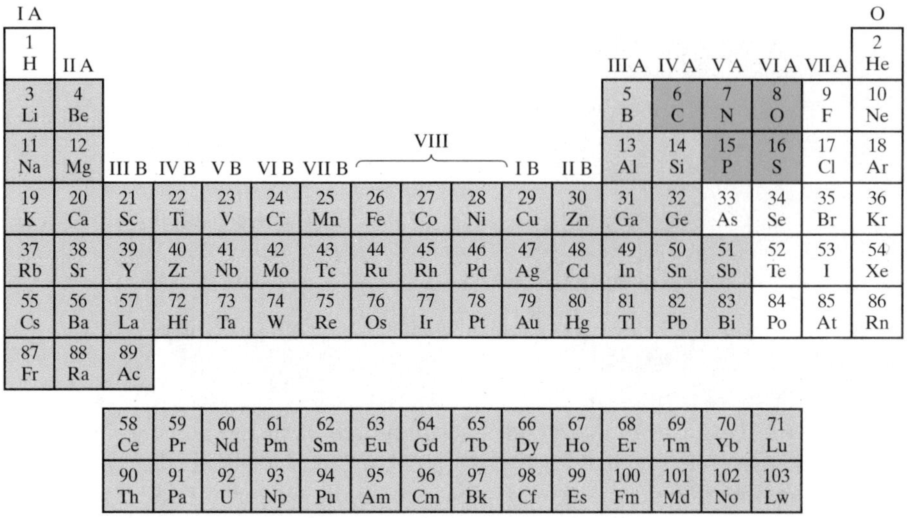

Figure 1-5 *Periodic table with ceramic compounds indicated by a combination of one or more metallic elements (in light color) with one or more nonmetallic elements (in dark color). Note that elements silicon (Si) and germanium (Ge) are included with the metals in this figure, but were not in Figure 1–3. This is because, in elemental form, Si and Ge behave as semiconductors (Figure 1–16). Elemental tin (Sn) can be either a metal or a semiconductor, depending on its crystalline structure.*

nonmetallic elements (C, N, O, P, or S). Figure 1–5 illustrates the various metals (in light color) and the five key nonmetals (in dark color) that can be combined to form an enormous range of ceramic materials. Bear in mind that many commercial ceramics include compounds and solutions of many more than two elements, just as commercial metal alloys are composed of many elements. Figure 1–6 illustrates some traditional, commercial ceramics. Figure 1–7 illustrates an example of the use of ceramics for an automobile engine.

The metals and ceramics shown in Figures 1–2, 1–4, 1–6, and 1–7 have a similar structural feature on the atomic scale: They are **crystalline,** which means that their constituent atoms are stacked together in a regular, repeating pattern. A distinction between metallic- and ceramic-type materials is that, by fairly simple processing techniques, many ceramics can be made in a **noncrystalline** form; that is, their atoms are stacked in irregular, random patterns. This is illustrated in Figure 1–8. The general term for noncrystalline solids with compositions comparable to the crystalline ceramics is **glass** (Figure 1–9). Most common glasses are silicates; ordinary window glass is approximately 72% silica (SiO_2) by weight, with the balance of the material being primarily sodium oxide (Na_2O) and calcium oxide (CaO). Glasses share the property of brittleness with crystalline ceramics. Glasses are important engineering materials because of other properties, such as their ability to transmit visible light (as well as ultraviolet and infrared radiation) and chemical inertness.

Figure 1-6 *Some common ceramics for traditional engineering applications. These miscellaneous parts with characteristic resistance to damage by high temperatures and corrosive environments are used in a variety of furnaces and chemical processing systems. (Courtesy of Duramic Products, Inc.)*

Figure 1-7 *Cutaway view of an advanced gas turbine design incorporating various ceramic components, for example, silicon carbide for turbine rotors, vanes, and flow path walls, silicon nitride for turbine rotors, and aluminum silicate for regenerator disks. (Courtesy of Allison Gas Turbine Operations, General Motors Corporation)*

A relatively recent materials development is a third category, **glass-ceramics.** Certain glass compositions (such as lithium aluminosilicates) can be fully **devitrified** (that is, transformed from the vitreous, or glassy, state to the crystalline state) by an appropriate thermal treatment. By forming the product shape during the glassy stage, complex forms can be obtained.

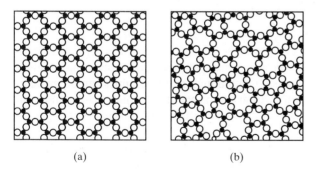

(a)   (b)

Figure 1-8 *Schematic comparison of the atomic-scale structure of (a) a ceramic (crystalline) and (b) a glass (noncrystalline). The open circles represent a nonmetallic atom, and the solid black circles represent a metal atom.*

Figure 1-9 *Some common silicate glasses for engineering applications. These materials combine the important qualities of transmitting clear visual images and resisting chemically aggressive environments. (Courtesy of Corning Glass Works)*

The high-quality microscopic-scale structure (fine grained with no porosity) gives a product with mechanical strength superior to that of many traditional crystalline ceramics. An added bonus is that lithium aluminosilicate compounds tend to have low thermal-expansion coefficients, making them resistant to fracture due to rapid temperature changes. This is an important advantage in applications such as cookware (Figure 1–10).

POLYMERS

A major impact of modern engineering technology on everyday life has been made by the class of materials known as **polymers**. An alternative name for this category is **plastics**. which describes the extensive formability of many polymers during fabrication. These synthetic, or human-made, materials are a special branch of organic chemistry. Examples of inexpensive, functional polymer products are readily available to each of us (Figure 1–11). The "mer" in a polymer is a single hydrocarbon molecule such as ethylene (C_2H_4). Polymers are long-chain molecules composed of many mers bonded together. The most common commercial polymer is **polyethylene** $(C_2H_4)_n$, where n can range from approximately 100 to 1000. Figure 1–12 shows the relatively limited portion of the periodic table that is associated with commercial polymers. Many important polymers, including polyethylene, are simply compounds of hydrogen and carbon. Others contain oxygen

Figure 1-10 *Cookware made of a glass-ceramic provides good mechanical and thermal properties. The casserole dish can withstand the thermal shock of simultaneous high temperature (the torch flame) and low temperature (the block of ice). (Courtesy of Corning Glass Works)*

Figure 1-11 *Miscellaneous internal parts of a contemporary parking meter are made of an acetal polymer. Engineered polymers are typically inexpensive and characterized by ease of formation and adequate structural properties. (Courtesy of the Du Pont Company, Engineering Polymers Division)*

I A																	O
1 H	II A											III A	IV A	V A	VI A	VII A	2 He
3 Li	4 Be											5 B	6 C	7 N	8 O	9 F	10 Ne
11 Na	12 Mg	III B	IV B	V B	VI B	VII B		VIII		I B	II B	13 Al	14 Si	15 P	16 S	17 Cl	18 Ar
19 K	20 Ca	21 Sc	22 Ti	23 V	24 Cr	25 Mn	26 Fe	27 Co	28 Ni	29 Cu	30 Zn	31 Ga	32 Ge	33 As	34 Se	35 Br	36 Kr
37 Rb	38 Sr	39 Y	40 Zr	41 Nb	42 Mo	43 Tc	44 Ru	45 Rh	46 Pd	47 Ag	48 Cd	49 In	50 Sn	51 Sb	52 Te	53 I	54 Xe
55 Cs	56 Ba	57 La	72 Hf	73 Ta	74 W	75 Re	76 Os	77 Ir	78 Pt	79 Au	80 Hg	81 Tl	82 Pb	83 Bi	84 Po	85 At	86 Rn
87 Fr	88 Ra	89 Ac															

58 Ce	59 Pr	60 Nd	61 Pm	62 Sm	63 Eu	64 Gd	65 Tb	66 Dy	67 Ho	68 Er	69 Tm	70 Yb	71 Lu
90 Th	91 Pa	92 U	93 Np	94 Pu	95 Am	96 Cm	97 Bk	98 Cf	99 Es	100 Fm	101 Md	102 No	103 Lw

Figure 1-12 *Periodic table with the elements associated with commercial polymers in color.*

Figure 1-13 *The rear quarter-panel on this sports car was a pioneering application of an engineering polymer in a traditional structural metal application. The polymer is an injection-molded nylon. (Courtesy of the Du Pont Company, Engineering Polymers Division)*

(e.g., acrylics), nitrogen (nylons), fluorine (fluoroplastics), and silicon (silicones). As the descriptive title implies, "plastics" commonly share with metals the desirable mechanical property of ductility. Unlike brittle ceramics, polymers are frequently lightweight, low-cost alternatives to metals in structural design applications. The nature of chemical bonding in polymeric materials will be explored in Chapter 2. Important bonding-related properties include lower strength compared with metals, and lower melting point and higher chemical reactivity compared with ceramics and glasses. In spite of the limitations, polymers are highly versatile and useful materials. Substantial progress has been made in the past decade in the development of engineering polymers with sufficiently high strength and stiffness to permit substitution for traditional structural metals. A good example is the automobile body panel in Figure 1–13.

COMPOSITES

The three categories of structural engineering materials we have discussed so far—metals, ceramics, and polymers—contain various elements and compounds that can be classified by their chemical bonding. Such a classification appears in Chapter 2. Another important set of materials is made up of some combinations of individual materials from the previous categories. This fourth group is **composites,** and perhaps our best example is **fiberglass.** This composite of glass fibers embedded in a polymer matrix is a relatively recent invention but has, in a few decades, become commonplace. Characteristic of good composites, fiberglass has the best properties of each component, producing a product that is superior to either of the components separately. The high strength of the small-diameter glass fibers is combined

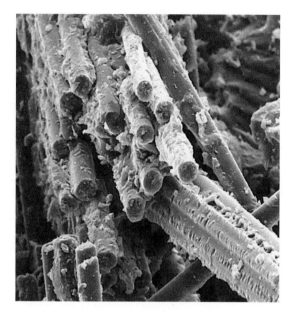

Figure 1-14 *Example of a fiberglass composite composed of microscopic-scale reinforcing glass fibers in a polymer matrix. The tremendous depth of field in this microscopic image is characteristic of the scanning electron microscope (SEM) to be discussed in Section 4.7 (Courtesy of Owens-Corning Fiberglas Corporation)*

with the ductility of the polymer matrix to produce a strong material capable of withstanding the normal loading required of a structural material.

There is no need to illustrate a region of the periodic table as characteristic of composites, since they involve virtually the entire table except for the noble gases (column O). Three main types of engineering composite structures will be discussed in detail in Chapter 14. Fiberglass is typical of many synthetic fiber-reinforced materials (Figure 1–14). *Wood* is an excellent example of a natural material with useful mechanical properties because of a fiber-reinforced structure. *Concrete* is a common example of an aggregate composite. Both rock and sand reinforce a complex silicate cement matrix. In addition to these relatively common examples, the field of composites includes some of the most advanced materials used in engineering (Figure 1–15).

SEMICONDUCTORS

Whereas polymers are highly visible engineering materials with a major impact on contemporary society, semiconductors are relatively invisible but have a comparable social impact. Technology has clearly revolutionized society, but solid-state electronics has revolutionized technology itself. A relatively small group of elements and compounds has an important electrical property, *semiconduction*, in which they are neither good electrical

Figure 1-15 *Golf club head and shaft molded of a graphite fiber-reinforced epoxy composite. Golf clubs made of this advanced composite system are stronger, stiffer, and lighter than conventional equipment, allowing the golfer to drive the ball farther with greater control. (Courtesy of Fiberite Corporation)*

I A	II A	III B	IV B	V B	VI B	VII B	VIII	VIII	VIII	I B	II B	III A	IV A	V A	VI A	VII A	O
1 H																	2 He
3 Li	4 Be											5 B	6 C	7 N	8 O	9 F	10 Ne
11 Na	12 Mg											13 Al	14 Si	15 P	16 S	17 Cl	18 Ar
19 K	20 Ca	21 Sc	22 Ti	23 V	24 Cr	25 Mn	26 Fe	27 Co	28 Ni	29 Cu	30 Zn	31 Ga	32 Ge	33 As	34 Se	35 Br	36 Kr
37 Rb	38 Sr	39 Y	40 Zr	41 Nb	42 Mo	43 Tc	44 Ru	45 Rh	46 Pd	47 Ag	48 Cd	49 In	50 Sn	51 Sb	52 Te	53 I	54 Xe
55 Cs	56 Ba	57 La	72 Hf	73 Ta	74 W	75 Re	76 Os	77 Ir	78 Pt	79 Au	80 Hg	81 Tl	82 Pb	83 Bi	84 Po	85 At	86 Rn
87 Fr	88 Ra	89 Ac															

58 Ce	59 Pr	60 Nd	61 Pm	62 Sm	63 Eu	64 Gd	65 Tb	66 Dy	67 Ho	68 Er	69 Tm	70 Yb	71 Lu
90 Th	91 Pa	92 U	93 Np	94 Pu	95 Am	96 Cm	97 Bk	98 Cf	99 Es	100 Fm	101 Md	102 No	103 Lw

Figure 1-16 *Periodic table with the elemental semiconductors in dark color and those elements that form semiconducting compounds in light color. The semiconducting compounds are composed of pairs of elements from columns III and V (e.g., GaAs) or from columns II and VI (e.g., CdS).*

conductors nor good electrical insulators. Instead, their ability to conduct electricity is intermediate. These materials are called **semiconductors,** and in general, they do not fit into any of the four structural materials categories based on atomic bonding. As discussed earlier, metals are inherently good electrical conductors. Ceramics and polymers (nonmetals) are generally poor conductors but good insulators. An important section of the periodic table is shown in dark color in Figure 1–16. These three semiconducting elements (Si, Ge, and Sn) from column IVA serve as a kind of boundary between metallic and nonmetallic elements. Silicon (Si) and germanium (Ge), widely used elemental semiconductors, are excellent examples of this class of materials. Precise control of chemical purity allows precise control of electronic properties. As techniques have been developed to produce variations in chemical purity over small regions, sophisticated electronic circuitry has been produced in exceptionally small areas (Figure 1–17). Such **microcircuitry** is the basis of the current revolution in technology.

The elements shaded in light color in Figure 1–16 form compounds that are semiconducting. Examples include gallium arsenide (GaAs), which is used as a high-temperature rectifier and a laser material, and cadmium sulfide (CdS), which is used as a relatively low-cost solar cell for conversion of solar energy to useful electrical energy. The various compounds formed by these elements show similarities to many of the ceramic compounds. With appropriate impurity additions, some of the ceramics display semiconducting behavior—for example, zinc oxide (ZnO), which is widely used as a phosphor in color television screens.

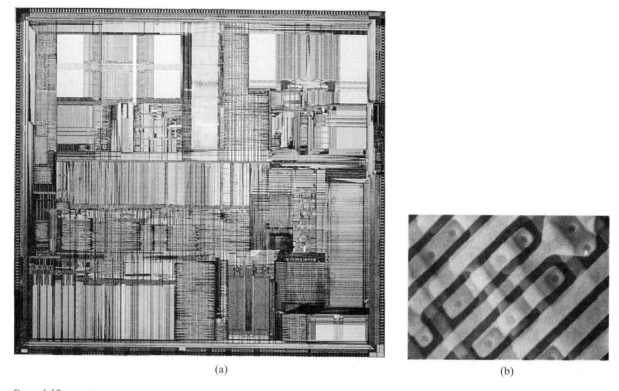

(a)   (b)

Figure 1-17 *(a) Typical microcircuit containing a complex array of semiconducting regions. (Photograph courtesy of Intel Corporation). (b) A microcircuit viewed with a scanning electron microscope. (From Metals Handbook, 9th ed., Vol. 10: Materials Characterization, American Society for Metals, Metals Park, Ohio, 1986.)*

1.4   FROM STRUCTURE TO PROPERTIES

To understand the properties or observable characteristics of engineering materials, it is necessary to understand their structure on an atomic and/or microscopic scale. Virtually every major property of the five materials' categories just outlined will be shown to result directly from mechanisms occurring at either the atomic or the microscopic level.

There is a special sort of architecture associated with these minute scales. Figure 1–8 illustrated, in a simplified way, the nature of **atomic-scale architecture** for crystalline (regular, repeating) and noncrystalline (irregular, random) arrangements of atoms. Figure 1–14 illustrated the nature of **microscopic-scale architecture,** in which the reinforcing glass fibers of a high-strength composite are contrasted against the surrounding matrix of polymer. The difference in scale between "atomic" and "microscopic" levels should be appreciated. The structure shown in Figure 1–14 represents a magnification of approximately 1000 times, whereas that in Figure 1–8 is approximately 10,000,000 times.

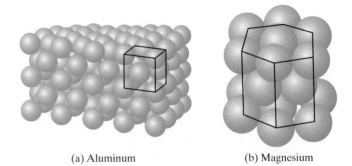

(a) Aluminum   (b) Magnesium

Figure 1-18 *Comparison of crystal structures for (a) aluminum and (b) magnesium.*

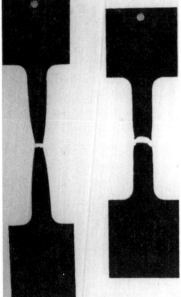

Figure 1-19 *Contrast in mechanical behavior of (a) aluminum (relatively ductile) and (b) magnesium (relatively brittle) resulting from the atomic-scale structure shown in Figure 1–18. Each sample was pulled in tension until it fractured. (Courtesy of R. S. Wortman)*

The dramatic effect that structure has on properties is well illustrated by two examples, one on the atomic scale and one on the microscopic scale. Any engineer responsible for selecting various metals for design applications must be aware that some alloys are relatively ductile, whereas others are relatively brittle. Aluminum alloys are characteristically ductile, while magnesium alloys are typically brittle. This fundamental difference relates directly to their different crystal structures (Figure 1–18). The nature of these crystal structures will be detailed in Chapter 3. For now, note only that the aluminum structure follows a cubic packing arrangement and the magnesium a hexagonal one. In Chapter 6, it will be shown that ductility depends on mechanical deformation occurring easily on the atomic scale, and that there are four times as many mechanisms for deformation in the aluminum crystal structure as in magnesium. This is equivalent to having four times the number of avenues available for ductility in aluminum-based alloys as in magnesium-based alloys. The relative brittleness of magnesium alloys is the result (Figure 1–19). In Chapters 6 and 10 we shall find that the mechanical behavior of a given type of metal alloy can also be dramatically affected by heat treatment and/or variations in alloy chemistry.

A significant achievement in materials technology in recent decades is the development of transparent ceramics, which has made possible new products and substantial improvements in others (such as commercial lighting). To make traditionally opaque ceramics, such as aluminum oxide (Al_2O_3), into optically transparent materials required a fundamental change in microscopic-scale architecture. Commercial ceramics are frequently produced by heating crystalline powders to high temperatures until a relatively strong and dense product results. Traditional ceramics made in this way contained a substantial amount of residual porosity (Figure 1–20a and b), corresponding to the open space between the original powder particles prior to high-temperature processing. The porosity leads to loss of visible light transmission, that is, a loss in transparency, by providing a light-scattering mechanism. Each Al_2O_3–air interface at a pore surface is a source of light refraction (change of direction). Only about 0.3% porosity can cause Al_2O_3 to be

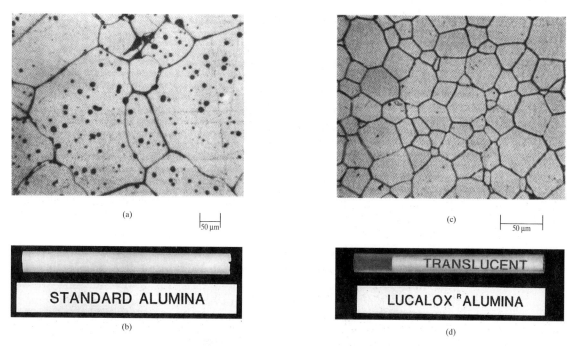

Figure 1-20 *Porous microstructure in polycrystalline* Al_2O_3 *(a) leads to an opaque material (b). Nearly pore-free microstructure in poly crystalline* Al_2O_3 *(c) leads to a translucent material (d). (Courtesy of C. E. Scott, General Electric Company)*

translucent (capable of transmitting a diffuse image), and 3% porosity can cause the material to be completely opaque (Figures 1–20a and b). The elimination of porosity resulted from a relatively simple invention[1] that involved adding a small amount of impurity (0.1 wt % MgO), which caused the high-temperature densification process for the Al_2O_3 powder to go to completion. The resulting pore-free microstructure produced a nearly transparent material (Figure 1–20c and d) with an important additional property—excellent resistance to chemical attack by high-temperature sodium vapor. Cylinders of translucent Al_2O_3 became the heart of the design of high-temperature (1000°C) sodium vapor lamps, which provide substantially higher illumination than do conventional light bulbs (100 lumens/W compared to 15 lumens/W). A commercial sodium vapor lamp is shown in Figure 1–21.

The two examples just cited show typical and important demonstrations of how properties of engineering materials follow directly from structure. Throughout this book, we shall be alert to the continuous demonstration of this interrelationship for all the materials of importance to engineers.

---

[1] R. L. Coble, U.S. Patent 3,026,210, March 20, 1962.

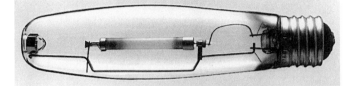

Figure 1-21 *High-temperature sodium vapor lamp made possible by use of a translucent Al_2O_3 cylinder for containing the sodium vapor. (Note that the Al_2O_3 cylinder is inside the exterior glass envelope.) (Courtesy of General Electric Company)*

1.5 PROCESSING MATERIALS

Our use of materials in modern technology ultimately depends on our ability to make those materials. Throughout Parts II and III of this text, we will discuss how each of the five types of materials are produced. The topic of materials **processing** serves two functions. First, it provides a fuller understanding of the nature of each type of material. Second, and more important, it provides an appreciation of the effects of processing history on properties.

We shall find that processing technology ranges from traditional methods such as metal casting (Figure 1–22) to the most comtemporary techniques of electronic microcircuit fabrication (Figure 1–23).

Figure 1-22 *Pouring molten iron into molds for casting. Even this traditional form of materials processing is becoming increasingly sophisticated. This pour occurred at the "Foundry of the Future" discussed in the Feature Box in Chapter 11. (Courtesy of the Casting Emission Reduction Program [CERP].)*

Figure 1-23 *The modern integrated circuit fabrication laboratory represents the state of the art in materials processing. (Courtesy of the College of Engineering, University of California, Davis)*

1.6 SELECTION OF MATERIALS

In Section 1.3 we answered the question "What materials are available to me?" In Section 1.4 we gained some appreciation of why these different materials behave as they do. In Section 1.5 we asked "How do I produce a material with optimal properties?"We must now face a new and obvious question: "Which material do I now select for a particular application?" **Materials selection** is the final, practical decision in the engineering design process and can determine that design's ultimate success or failure. (This important aspect of the engineering design process will be discussed in Chapter 20.) In fact, there are two separate decisions to be made. First, one must decide which general type of material is appropriate (metal, ceramic, etc.). Second, the best specific material within that category must be found (e.g., is a magnesium alloy preferable to aluminum or steel?).

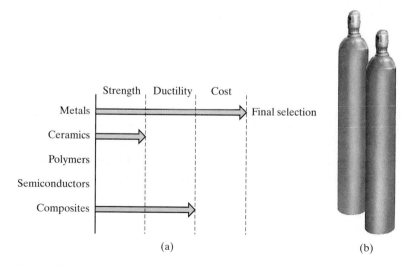

Figure 1-24 *(a) Sequence of choices leading to selection of metal as the appropriate type of material for construction of a commercial gas cylinder. (b) Commercial gas cylinders. (Courtesy of Matheson Division of Searle Medical Products)*

The choice of the appropriate type of material is sometimes simple and obvious. A solid-state electronic device requires a semiconductor component, and either conductors or insulators are entirely inappropriate in its place. Most choices are less obvious. Figure 1–24 illustrates the sequence of choices necessary to make a final selection of metal as the appropriate type of material for a commercial gas cylinder, that is, a container that must be capable of storing gases at pressures as high as 14 MPa (2000 psi) for indefinite periods.

Just as metal is an inappropriate substitute for semiconductors, the semiconductor materials cannot be considered for routine structural applications. Of the three common structural materials (metals, ceramics, and polymers), polymers must be initially rejected because of typically low strengths. Although some structural ceramics can withstand the anticipated service load, they generally fail to provide the necessary ductility to survive practical handling. The use of such a brittle material in a pressure-containing design can be extremely dangerous. Several common metals provide sufficient strength and ductility to serve as excellent candidates. It must also be noted that many fiber-reinforced composites can satisfy the design requirements. However, the third criterion, cost, eliminates them from competition. The added cost of fabricating these more sophisticated material systems is justified only if a special advantage results. Reduced weight is one such advantage that frequently does justify the cost.

Reducing the selection to metals still leaves an enormous list of candidate materials. Even consideration of commercially available, moderately priced alloys with acceptable mechanical properties can provide a substantial list of candidates. In making the final alloy selection, property comparisons must be made at each step in the path. Superior mechanical properties can dominate the selection at certain junctions in the path, but more often, cost dominates. Mechanical performance typically focuses on a trade-off between the strength of the material and its deformability.

SUMMARY

The wide range of materials available to engineers can be divided into five categories: metals, ceramics and glasses, polymers, composites, and semiconductors. The first three categories can be associated with distinct types of atomic bonding. Composites involve combinations of two or more materials from the previous three categories. The first four categories comprise the structural materials. Semiconductors comprise a separate category of electronic materials that is distinguished by its unique, intermediate electrical conductivity. Understanding the properties of these various materials requires examination of structure at either a microscopic or atomic scale. The relative ductility of certain metal alloys is related to atomic-scale architecture. Similarly, the development of transparent ceramics has required careful control of microscopic-scale architecture. Once the properties of materials are understood, the processing and selection of the appropriate material for a given application can be done. The selection of materials is done at two levels. First, there is competition among the various categories of materials. Second, there is competition within the most appropriate category for the optimum, specified material. In addition, new developments can lead to the selection of an alternate material for a given design. We now move on to the body of the text, with the term *materials science and engineering* serving to define this branch of engineering as well as providing the key words to entitle the various parts of the text: I ("science" → the fundamentals), II and III ("materials" → the structural, electronic, optical, and magnetic materials), and IV ("engineering" → materials in engineering design).

KEY TERMS

Many technical journals now list a set of "key terms" with each article. These words serve the practical purpose of information retrieval, but also provide a convenient summary of important concepts in that publication. In this spirit, a list of key terms will be given at the end of each chapter. Students can use this list as a convenient guide to the major concepts that they should be taking away from that chapter. A comprehensive glossary is provided in Appendix 7 giving definitions of the key terms from all chapters.

alloy (2)
atomic-scale architecture (13)
brittle (3)
Bronze Age (2)
ceramic (5)
composite (10)
crystalline (6)
devitrified (7)
ductility (3)
fiberglass (10)

glass (6)
glass-ceramic (7)
Iron Age (2)
materials selection (17)
metallic (3)
microcircuitry (12)
microscopic-scale architecture (13)
noncrystalline (6)
nonmetallic (6)
plastic (8)

polyethylene (8)
polymer (8)
processing (16)
refractory (5)
semiconductor (12)
silica (5)
silicate (5)
Stone Age (2)

REFERENCES

At the end of each chapter, a short list of selected references will be cited to indicate some primary sources of related information for the student wishing to do outside reading. For Chapter 1 the references are some of the general textbooks in the field of materials science and engineering.

Askeland, D. R., *The Science and Engineering of Materials*, 3rd ed., PWS-Kent, Boston, 1994.

Callister, W. D., *Materials Science and Engineering—An Introduction*, 5th ed., John Wiley & Sons, Inc., New York, 2000.

Flinn, R. A., and **P. K. Trojan,** *Engineering Materials and Their Applications*, 4th ed., Houghton Mifflin Company, Boston, 1990.

Schaffer, J. P., A. Saxena, S. D. Antolovich, T. H. Sanders, Jr., and **S. B. Warner,** *The Science and Design of Engineering Materials*, 2nd ed., McGraw-Hill Book Company, New York, 1999.

Smith, W. F., *Principles of Materials Science and Engineering*, 3rd ed., McGraw-Hill Book Company, New York, 1995.

Van Vlack, L. H., *Elements of Materials Science and Engineering*, 6th ed., Addison-Wesley Publishing Co., Inc., Reading, Mass., 1989.

PART ONE

CHAPTER 2
Atomic Bonding

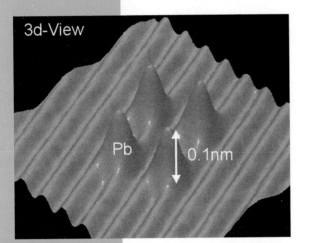

3d-View

Pb   0.1nm

The scanning tunneling microscope (Section 4.7) allows the imaging of individual atoms bonded to a material surface. In this case, the microscope was also used to manipulate the atoms into a simple pattern. Four lead atoms are shown forming a rectangle on the surface of a copper crystal. (From G. Meyer and K. H. Rieder, MRS Bulletin 23 28 [1998].)

Chapter 1 introduced the basic types of materials available to engineers. One basis of that classification system is found in the nature of atomic bonding in materials. Atomic bonding falls into two general categories. *Primary bonding* involves transfer or sharing of electrons and produces a relatively strong joining of adjacent atoms. Ionic, covalent, and metallic bonds are in this category. *Secondary bonding* involves a relatively weak attraction between atoms in which no electron transfer or sharing occurs. Van der Waals bonds are in this category. Each of the four fundamental types of engineering materials (metals, ceramics and glasses, polymers, and semiconductors) is associated with a certain type (or types) of atomic bonding. Composites, of course, are combinations of fundamental types.

2.1   ATOMIC STRUCTURE

In order to understand bonding between atoms, we must appreciate the structure within the individual atoms. For this purpose, it is sufficient to use a relatively simple planetary model of atomic structure—that is, **electrons** (the planets) orbit about a **nucleus** (the sun).

It is not necessary to consider the detailed structure of the nucleus for which physicists, in recent decades, have catalogued a vast number of elementary particles. We need consider only the number of **protons** and **neutrons** in the nucleus as the basis of the chemical identification of a given atom. Figure 2–1 is a planetary model of a carbon atom. This illustration is schematic and definitely not to scale. In reality, the nucleus is much smaller, even though it contains nearly all of the mass of the atom. Each proton and neutron has a mass of approximately 1.66×10^{-24} g. This value is referred to as an **atomic mass unit (amu).** It is convenient to express the mass of elemental materials in these units. For instance, the most common isotope of carbon, C^{12} (shown in Figure 2–1), contains in its nucleus six protons and six neutrons, for an **atomic mass** of 12 amu. It is also convenient to note that there are 0.6023×10^{24} amu per gram. This large value, known as **Avogadro's* number**, represents the number of protons or neutrons necessary to produce a mass of 1 g. Avogadro's number of atoms of a given element is termed a **gram-atom**. For a compound, the corresponding term is **mole**; that is, one mole of NaCl contains Avogadro's number of Na atoms *and* Avogadro's number of Cl atoms.

Avogadro's number of C^{12} atoms would have a mass of 12.00 g. Naturally occurring carbon actually has an atomic mass of 12.011 amu because not all carbon atoms contain six neutrons in their nuclei. Instead, some

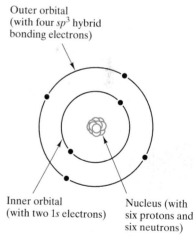

Outer orbital
(with four sp^3 hybrid
bonding electrons)

Inner orbital
(with two $1s$ electrons)

Nucleus (with
six protons and
six neutrons)

Figure 2-1 *Schematic of the planetary model of a ^{12}C atom.*

---

*   Amadeo Avogadro (1776–1856), Italian physicist, who, among other contributions, coined the word *molecule*. Unfortunately, his hypothesis that all gases (at a given temperature and pressure) contain the same number of molecules per unit volume was not generally acknowledged as correct until after his death.

I A	II A	III B	IV B	V B	VI B	VII B	VIII	VIII	VIII	I B	II B	III A	IV A	V A	VI A	VII A	0
1 H 1.008																	2 He 4.003
3 Li 6.941	4 Be 9.012											5 B 10.81	6 C 12.01	7 N 14.01	8 O 16.00	9 F 19.00	10 Ne 20.18
11 Na 22.99	12 Mg 24.31											13 Al 26.98	14 Si 28.09	15 P 30.97	16 S 32.06	17 Cl 35.45	18 Ar 39.95
19 K 39.10	20 Ca 40.08	21 Sc 44.96	22 Ti 47.90	23 V 50.94	24 Cr 52.00	25 Mn 54.94	26 Fe 55.85	27 Co 58.93	28 Ni 58.71	29 Cu 63.55	30 Zn 65.38	31 Ga 69.72	32 Ge 72.59	33 As 74.92	34 Se 78.96	35 Br 79.90	36 Kr 83.80
37 Rb 85.47	38 Sr 87.62	39 Y 88.91	40 Zr 91.22	41 Nb 92.91	42 Mo 95.94	43 Tc 98.91	44 Ru 101.07	45 Rh 102.91	46 Pd 106.4	47 Ag 107.87	48 Cd 112.4	49 In 114.82	50 Sn 118.69	51 Sb 121.75	52 Te 127.60	53 I 126.90	54 Xe 131.30
55 Cs 132.91	56 Ba 137.33	57 La 138.91	72 Hf 178.49	73 Ta 180.95	74 W 183.85	75 Re 186.2	76 Os 190.2	77 Ir 192.22	78 Pt 195.09	79 Au 196.97	80 Hg 200.59	81 Tl 204.37	82 Pb 207.2	83 Bi 208.98	84 Po (210)	85 At (210)	86 Rn (222)
87 Fr (223)	88 Ra 226.03	89 Ac (227)															

58 Ce 140.12	59 Pr 140.91	60 Nd 144.24	61 Pm (145)	62 Sm 150.4	63 Eu 151.96	64 Gd 157.25	65 Tb 158.93	66 Dy 162.50	67 Ho 164.93	68 Er 167.26	69 Tm 168.93	70 Yb 173.04	71 Lu 174.97
90 Th 232.04	91 Pa 231.04	92 U 238.03	93 Np 237.05	94 Pu (244)	95 Am (243)	96 Cm (247)	97 Bk (247)	98 Cf (251)	99 Es (254)	100 Fm (257)	101 Md (258)	102 No (259)	103 Lw (260)

Figure 2-2 *Periodic table of the elements indicating atomic number and atomic mass (in amu).*

contain seven. Different numbers of neutrons (six or seven) identify different **isotopes**—various forms of an element that differ in the number of neutrons in the nucleus. In nature, 1.1% of the carbon atoms are the isotope C^{13}. However, the nuclei of all carbon atoms contain six protons. In general, the number of protons in the nucleus is knows as the **atomic number** of the element. The well-known periodicity of chemical elements is based on this system of elemental atomic numbers and atomic masses arranged in chemically similar **groups** (vertical columns) in a **periodic table** (Figure 2–2).

While chemical identification is done relative to the nucleus, atomic bonding involves electrons and **electron orbitals.** The electron, with a mass of 0.911×10^{-27} g, makes a negligible contribution to the atomic mass of an element. However, this particle has a negative charge of 0.16×10^{-18} coulomb (C), equal in magnitude to the $+0.16 \times 10^{-18}$ C charge of each proton. (The neutron is, of course, electrically neutral.)

Electrons are excellent examples of the wave-particle duality; that is, they are atomic-scale entities exhibiting both wavelike and particlelike behavior. It is beyond the scope of this book to deal with the principles of quantum mechanics that define the nature of the electron orbitals (based on the wavelike character of electrons). However, a brief summary of the nature of electron orbitals is helpful. As shown schematically in Figure 2–1, electrons are grouped at fixed orbital positions about a nucleus. In addition, each orbital radius is characterized by an **energy level,** a fixed binding

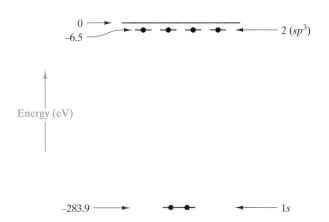

Figure 2-3 *Energy-level diagram for the orbital electrons in a ^{12}C atom. Notice the sign convention. An attractive energy is negative. The 1s electrons are closer to the nucleus (see Figure 2–1) and more strongly bound (binding energy = -283.9 eV). The outer orbital electrons have a binding energy of only -6.5 eV. The zero level of binding energy corresponds to an electron completely removed from the attractive potential of the nucleus.*

energy between the electron and its nucleus. Figure 2–3 shows an *energy-level diagram* for the electrons in a C^{12} atom. It is important to note that the electrons around a C^{12} nucleus occupy these specific energy levels, with intermediate energies forbidden. The forbidden energies correspond to unacceptable quantum mechanical conditions; that is, standing waves cannot be formed.

A detailed list of electronic configurations for the elements of the periodic table is given in Appendix 1 together with various useful data. The arrangement of the periodic table (Figure 2–2) is largely a manifestation of the systematic "filling" of the electron orbitals with electrons, as summarized in Appendix 1. The notation for labeling electron orbitals is derived from the quantum numbers of wave mechanics. These integers relate to solutions to the appropriate wave equations. We do not deal with this numbering system in detail in this book. It is sufficient to appreciate the basic labeling system. For instance, Appendix 1 tells us that there are two electrons in the 1s orbital. The 1 is a principal quantum number, identifying this as the first energy level. There are also two electrons each associated with the 2s and 2p orbitals. The *s*, *p*, and so on, notation refers to an additional set of quantum numbers. The rather cumbersome letter notation is derived from the terminology of early spectographers. The six electrons in the C^{12} atom are then described as a $1s^22s^22p^2$ distribution; that is, two electrons in the 1s orbital, two in 2s, and two in 2p. In fact, the four electrons in the outer orbital of C^{12} redistribute themselves in a more symmetrical fashion to produce the characteristic geometry of bonding between carbon atoms and adjacent atoms (generally described as $1s^22s^12p^3$). This sp^3 configuration in the second energy level of carbon, called **hybridization,** is indicated in Figures 2–1 and 2–3 and is discussed in further detail in Section 2.3. (Note especially Figure 2–19.)

The bonding of adjacent atoms is essentially an electronic process. Strong **primary bonds** are formed when outer orbital electrons are transferred or shared between atoms. Weaker **secondary bonds** result from more subtle attraction between positive and negative charges with no actual transfer or sharing of electrons. In the next section we will look at the various possibilities of bonding in a systematic way, beginning with the ionic bond.

SAMPLE PROBLEM 2.1

Chemical analysis in materials science laboratories is frequently done by means of the scanning electron microscope. As discussed in Section 4.7, an electron beam generates characteristic x-rays that can be used to identify chemical elements. This instrument samples a roughly cylindrical volume at the surface of a solid material. Calculate the number of atoms sampled in a 1-μm-diameter by 1-μm-deep cylinder in the surface of solid copper.

SOLUTION

From Appendix 1,

$$\text{density of copper} = 8.93 \text{g/cm}^3$$
$$\text{atomic mass of copper} = 63.55 \text{amu}$$

The atomic mass indicates that there are

$$\frac{63.55 \text{ g Cu}}{\text{Avogadro's number of Cu atoms}}$$

The volume sampled is

$$V_{\text{sample}} = \pi \left(\frac{1\mu\text{m}}{2} \right)^2 \times 1\mu\text{m}$$

$$= 0.785 \ \mu\text{m}^3 \times \left(\frac{1\text{cm}}{10^4 \ \mu\text{m}} \right)^3$$

$$= 0.785 \times 10^{-12} \text{cm}^3$$

Thus the number of atoms sampled is

$$N_{\text{sample}} = \frac{8.93\text{g}}{\text{cm}^3} \times 0.785 \times 10^{-12} \text{cm}^3 \times \frac{0.602 \times 10^{24} \text{atoms}}{63.55\text{g}}$$

$$= 6.64 \times 10^{10} \text{atoms}$$

SAMPLE PROBLEM 2.2

One mole of solid MgO occupies a cube 22.37 mm on a side. Calculate the density of MgO (in g/cm^3).

SOLUTION

From Appendix 1,

$$\text{mass of 1 mol of MgO} = \text{atomic mass of Mg (in g)}$$
$$+ \text{ atomic mass of O (in g)}$$
$$= 24.31g + 16.00g = 40.31g$$

$$\text{density} = \frac{\text{mass}}{\text{volume}}$$

$$= \frac{40.31g}{(22.37 \text{ mm})^3 \times 10^{-3}\text{cm}^3/\text{mm}^3}$$

$$= 3.60g/\text{cm}^3$$

SAMPLE PROBLEM 2.3

Calculate the dimensions of a cube containing 1 mol of solid magnesium.

SOLUTION

From Appendix 1,

$$\text{density of Mg} = 1.74g/\text{cm}^3$$

$$\text{atomic mass of Mg} = 24.31 \text{ amu}$$

$$\text{volume of 1 mol} = \frac{24.31g/\text{mol}}{1.74g/\text{cm}^3}$$

$$= 13.97\text{cm}^3/\text{mol}$$

$$\text{edge of cube} = (13.97)^{1/3}\text{cm}$$

$$= 2.408\text{cm} \times 10 \text{ mm/cm}$$

$$= 24.08 \text{ mm}$$

Beginning at this point, a few elementary problems, called Practice Problems, will be provided immediately following the solved Sample Problems. These exercises follow directly from the preceding solutions and are intended to provide a carefully guided journey into the first calculations in each new area.

More independent and challenging problems are provided at the conclusion of the chapter. Answers for nearly all of the Practice Problems are given following the appendixes.

....

PRACTICE PROBLEM 2.1

Calculate the number of atoms contained in a cylinder 1 μm in diameter by 1 μm deep of (**a**) magnesium and (**b**) lead. (See Sample Problem 2.1.)

PRACTICE PROBLEM 2.2

Using the density of MgO calculated in Sample Problem 2.2, calculate the mass of an MgO refractory (temperature-resistant) brick with dimensions 50 mm × 100 mm × 200 mm.

PRACTICE PROBLEM 2.3

Calculate the dimensions of (**a**) a cube containing 1 mol of copper and (**b**) a cube containing 1 mol of lead. (See Sample Problem 2.3.)

2.2   THE IONIC BOND

An **ionic bond** is the result of *electron transfer* from one atom to another. Figure 2–4 illustrates an ionic bond between sodium and chlorine. The transfer of an electron *from* sodium is favored because this produces a more stable electronic configuration; that is, the resulting Na$^+$ species has a full outer **orbital shell,** defined as a set of electrons in a given orbital. Similarly, the chlorine readily accepts the electron, producing a stable Cl$^-$ species, also with a full outer orbital shell. The charged species (Na$^+$ and Cl$^-$) are termed **ions,** giving rise to the name *ionic bond*. The positive species (Na$^+$) is a **cation,** and the negative species (Cl$^-$) is an **anion.**

It is important to note that the ionic bond is *nondirectional*. A positively charged Na$^+$ will attract any adjacent Cl$^-$ equally in all directions. Figure 2–5 shows how Na$^+$ and Cl$^-$ ions are stacked together in solid sodium chloride (rock salt). Details about this structure will be discussed in Chapter 3. For now, it is sufficient to note that this is an excellent example of an ionically bonded material, and the Na$^+$ and Cl$^-$ ions are stacked together systematically to maximize the number of oppositely charged ions adjacent to any given ion. In NaCl, six Na$^+$ surround each Cl$^-$, and six Cl$^-$ surround each Na$^+$.

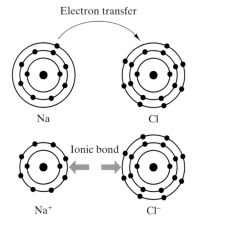

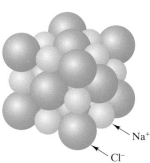

Figure 2-4 *Ionic bonding between sodium and chlorine atoms. Electron transfer from Na to Cl creates a cation (Na$^+$) and an anion (Cl$^-$). The ionic bond is due to the coulombic attraction between the ions of opposite charge.*

Figure 2-5 *Regular stacking of Na$^+$ and Cl$^-$ ions in solid NaCl. This is indicative of the nondirectional nature of ionic bonding.*

The ionic bond is the result of the **coulombic* attraction** between the oppositely charged species. It is convenient to illustrate the nature of the bonding force for the ionic bond because the coulombic attraction force follows a simple, well-known relationship,

$$F_c = \frac{-K}{a^2} \qquad (2.1)$$

where F_c is the coulombic force of attraction between two oppositely charged ions, a is the separation distance between the *centers* of the ions, and K is

$$K = k_0(Z_1 q)(Z_2 q) \qquad (2.2)$$

where Z is the *valence* of the charged ion (e.g., $+1$ for Na$^+$ and -1 for Cl$^-$), q is the charge of a single electron (0.16×10^{-18} C), and k_0 is a proportionality constant ($9 \times 10^9 V \cdot$ m/C).

---

* Charles Augustin de Coulomb (1736–1806), French physicist, was first to experimentally demonstrate the nature of Equations 2.1 and 2.2 (for large spheres, not ions). Beyond major contributions to the understanding of electricity and magnetism, Coulomb was an important pioneer in the field of applied mechanics (especially in the areas of friction and torsion).

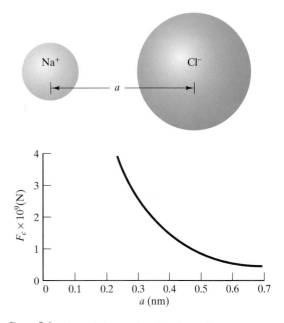

Figure 2-6 *Plot of the coulombic force (Equation 2.1) for a Na^+—Cl^- pair.*

A plot of Equation 2.1 is shown in Figure 2–6. This shows that the coulombic force of attraction increases dramatically as the separation distance between adjacent ion centers (a) decreases. This, in turn, implies that the **bond length** (a) would ideally be zero. In fact, bond lengths are most definitely not zero. This is because the attempt to move two oppositely charged ions closer together to increase coulombic attraction is counteracted by an opposing **repulsive force,** F_R, which is due to the overlapping of the similarly charged (negative) electric fields from each ion as well as the attempt to bring the two positively charged nuclei closer together. The repulsive force as a function of a follows an exponential relationship:

$$F_R = \lambda e^{-a/\rho} \tag{2.3}$$

where λ and ρ are experimentally determined constants for a given ion pair. **Bonding force** is the net force of attraction (or repulsion) as a function of the separation distance between two atoms or ions. Figure 2–7 shows the *bonding force curve* for an ion pair in which the *net* bonding force, $F(= F_c + F_R)$, is plotted against a. The *equilibrium bond length*, a_0, occurs at the point where the forces of attraction and repulsion are precisely balanced ($F_c + F_R = 0$). It should be noted that the coulombic force (Equation 2.1) dominates for larger values of a, whereas the repulsive force (Equation 2.3) dominates for small values of a. Up to this point, we have concentrated on the attractive coulombic force between two ions of opposite charge. Of course, bringing two similarly charged ions together would produce a coulombic *repulsive*

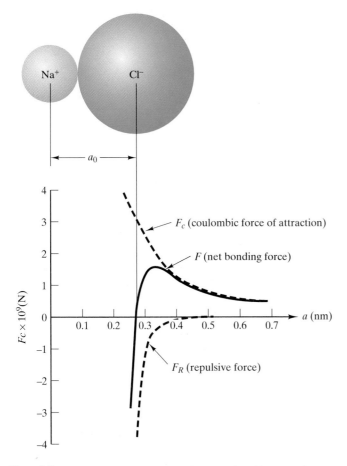

Figure 2-7 *Net bonding force curve for a* $Na^+ - Cl^-$ *pair show-ing an equilibrium bond length of* $a_0 = 0.28$ *nm.*

force (separate from the F_R term). In an ionic solid such as Figure 2–5, the similarly charged ions experience this "coulombic repulsion" force. The net cohesiveness of the solid is due to the fact that any given ion is immediately surrounded by ions of opposite sign for which the coulombic term (Equations 2.1 and 2.2) is positive. This overcomes the smaller, repulsive term due to more distant ions of like sign.

It should also be noted that an externally applied compressive force is required to push the ions closer together (i.e., closer than a_0). Similarly, an externally applied tensile force is required to pull the ions farther apart. This has implications for the mechanical behavior of solids, which is discussed in detail later (especially Chapter 6).

Bonding energy, E, is related to bonding force through the differential expression

$$F = \frac{dE}{da} \qquad (2.4)$$

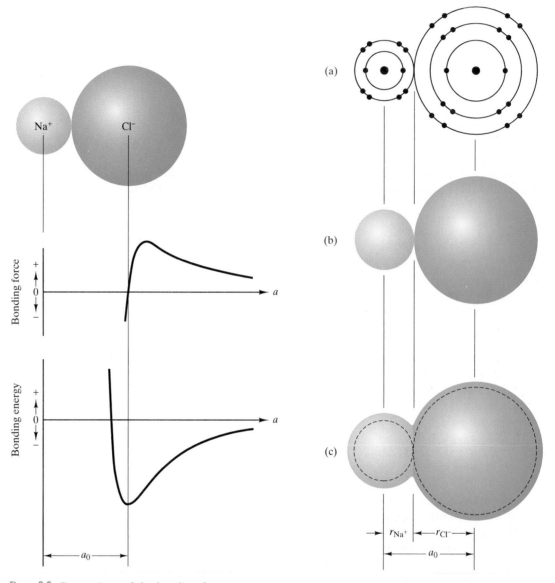

Figure 2-8 *Comparison of the bonding force curve and the bonding energy curve for a* Na^+-Cl^- *pair. Since* $F = dE/da$, *the equilibrium bond length* (a_0) *occurs where* $F = 0$ *and* E *is a minimum (see Equation 2.5).*

Figure 2-9 *Comparison of (a) a planetary model of a* Na^+-Cl^- *pair with (b) a hard-sphere model and (c) a soft-sphere model.*

In this way, the net bonding force curve in Figure 2–7 is the derivative of the bonding energy curve. This relationship is shown in Figure 2–8, which demonstrates that the equilibrium bond length, a_0, which corresponds to $F = 0$, also corresponds to a minimum in the energy curve. This is a consequence of Equation 2.5; that is, the slope of the energy curve at a minimum equals zero:

$$F = 0 = \left(\frac{dE}{da}\right)_{a=a_0} \qquad (2.5)$$

This is an important concept in materials science and will be seen again many times throughout the book. The stable ion positions correspond to an energy minimum. To move the ions from their equilibrium spacing, energy must be supplied to this system (for example, by compressive or tensile loading).

Having established that there is an equilibrium bond length, a_0, it follows that this bond length is the sum of two ionic radii; that is, for NaCl,

$$a_0 = r_{Na^+} + r_{Cl^-} \qquad (2.6)$$

This implies that the two ions are hard spheres touching at a single point. In Section 2.1 it was noted that, while electron orbitals are represented as particles orbiting at a fixed radius, electron charge is found in a range of radii. This is true for ions as well as for neutral atoms. An **ionic,** or **atomic, radius** is, then, the radius corresponding to the average electron density in the outermost electron orbital. Figure 2–9 compares three models of an Na^+—Cl^- ion pair: (a) shows a simple planetary model of the two ions; (b) shows a hard-sphere model of the pair; (c) shows the soft-sphere model in which the actual electron density in the outer orbitals of Na^+ and Cl^- extend farther out than shown for the hard sphere. The precise nature of actual bond lengths, a_0, allows us to use the hard-sphere model almost exclusively through the remainder of the book. Appendix 2 provides a detailed list of calculated ionic radii for a large number of ionic species.

Ionization has a significant effect on the effective (hard-sphere) radii for the atomic species involved. Although Figure 2–4 did not indicate this factor, the loss or gain of an electron by a neutral atom changes its radius. Figure 2–10 illustrates again the formation of an ionic bond between Na^+ and Cl^-. (Compare this with Figure 2–4.) In this case, atomic and ionic sizes are shown to correct scale. The loss of an electron by the sodium atom leaves 10 electrons to be drawn closer around the nucleus still containing 11 protons. Conversely, a gain of one electron by the chlorine atom gives 18 electrons around a nucleus with 17 protons and, therefore, a larger effective radius.

COORDINATION NUMBER

Earlier in this section, the nondirectional nature of the ionic bond was introduced. Figure 2–5 indicated a structure for NaCl in which six Na^+ surround each Cl^-, and vice versa. The **coordination number** (CN) is the number of adjacent ions (or atoms) surrounding a reference ion (or atom). For each ion in Figure 2–5, the CN is 6; that is, each has six nearest neighbors. For ionic compounds, the coordination number of the smaller ion can be calculated in

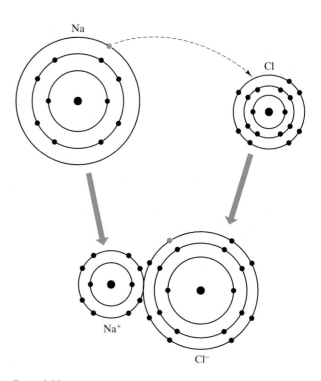

Figure 2-10 *Formation of an ionic bond between sodium and chlorine in which the effect of ionization on atomic radius is illustrated. The cation (Na^+) becomes smaller than the neutral atom (Na), while the anion (Cl^-) becomes larger than the neutral atom (Cl).*

a systematic way by considering the greatest number of larger ions (of opposite charge) that can be in contact with, or coordinate with, the smaller one. This number (CN) depends directly on the relative sizes of the oppositely charged ions. This relative size is characterized by the **radius ratio** (r/R), where r is the radius of the smaller ion and R is the radius of the larger one.

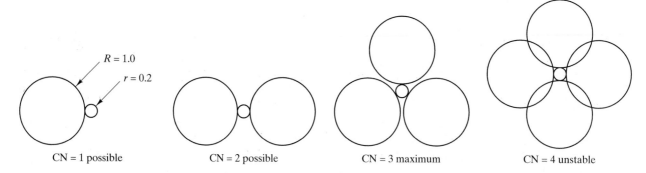

Figure 2-11 *The largest number of ions of radius R that can coordinate an atom of radius r is 3 when the radius ratio, r/R = 0.2. (Note: The instability for CN = 4 can be reduced but not eliminated by allowing a three-dimensional, rather than a coplanar, stacking of the larger ions.)*

To illustrate the dependence of CN on radius ratio, consider the case of $r/R = 0.20$. Figure 2–11 shows how the greatest number of larger ions that can coordinate the smaller one is 3. Any attempt to place four larger ions in contact with the smaller one requires the larger ions to overlap. This is a condition of great instability because of high repulsive forces. The minimum value of r/R that can produce threefold coordination ($r/R = 0.155$) is shown in Figure 2–12; that is, the larger ions are just touching the smaller ion as well as just touching each other. In the same way that fourfold coordination was unstable in Figure 2–11, an r/R value of *less* than 0.155 cannot allow threefold coordination. As r/R increases above 0.155, threefold coordination is stable (e.g., Figure 2–11 for $r/R = 0.20$) until at $r/R = 0.225$, fourfold coordination becomes possible. Table 2.1 summarizes the relationship between coordination number and radius ratio. As r/R increases to 1.0, a coordination number as high as 12 is possible. As will be noted in Sample Problem 2.8, calculations based on Table 2.1 serve as guides, not absolute predictors.

$$\cos 30^\circ = 0.866 = \frac{R}{r+R} \rightarrow \frac{r}{R} = 0.155$$

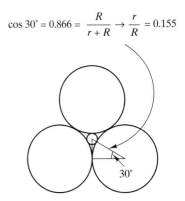

Figure 2-12 *The minimum radius ratio, r/R, that can produce three-fold coordination is 0.155.*

Table 2.1 *Coordination Numbers for Ionic Bonding*

Coordination number	Radius ratio, r/R	Coordination geometry
2	$0 < \dfrac{r}{R} < 0.155$	
3	$0.155 \leq \dfrac{r}{R} < 0.225$	
4	$0.225 \leq \dfrac{r}{R} < 0.414$	
6	$0.414 \leq \dfrac{r}{R} < 0.732$	
8	$0.732 \leq \dfrac{r}{R} < 1$	
12	1	or[a]

[a] The geometry on the left is for the hexagonal close-packed (hcp) structure and that on the right for the face-centered cubic (fcc) structure. These crystal structures are discussed in Chapter 3.

An obvious question is "Why does Table 2.1 not include radius ratios greater than 1?" Certainly, more than 12 small ions could simultaneously touch a single larger one. However, there are practical constraints in the connecting of the coordination groups of Table 2.1 together into a periodic, three-dimensional structure, and the coordination number for the larger ions tends to be less than 12. A good example is again Figure 2–5, in which the coordination number of Na^+ is 6, as predicted by the r/R value ($= 0.098nm/0.181nm = 0.54$), and the regular stacking of the six-coordinated sodiums, in turn, gives Cl^- a coordination number of 6. These structural details will be discussed further in Chapter 3. One might also inquire why coordination numbers of 5, 7, 9, 10, and 11 are absent. These cannot be integrated into the repetitive, crystalline structures described in Chapter 3.

SAMPLE PROBLEM 2.4

(a) Compare the electronic configurations for the atoms and ions in Figure 2–4.

(b) Which noble gas atoms have electronic configurations equivalent to those for the ions in Figure 2–4?

SOLUTION

(a) From Appendix 1,

$$Na: \quad 1s^2 2s^2 2p^6 3s^1$$

$$Cl: \quad 1s^2 2s^2 2p^6 3s^2 3p^5$$

Because Na loses its outer orbital ($3s$) electron in becoming Na^+,

$$Na^+: \quad 1s^2 2s^2 2p^6$$

Because Cl gains an outer orbital electron in becoming Cl^-, its $3p$ shell becomes filled:

$$Cl^-: \quad 1s^2 2s^2 2p^6 3s^2 3p^6$$

(b) From Appendix 1,

$$Ne: \quad 1s^2 2s^2 2p^6$$

equivalent to Na^+ (of course, the nuclei of Ne and Na^+ differ)

$$Ar: \quad 1s^2 2s^2 2p^6 3s^2 3p^6$$

equivalent to Cl^- (again, the nuclei differ).

SAMPLE PROBLEM 2.5

(a) Using the ionic radii data in Appendix 2, calculate the coulombic force of attraction between Na^+ and Cl^- in NaCl.

(b) What is the repulsive force in this case?

SOLUTION

(a) From Appendix 2,

$$r_{Na^+} = 0.098nm$$

$$r_{Cl^-} = 0.181nm$$

Then

$$a_0 = r_{Na^+} + r_{Cl^-} = 0.098nm + 0.181nm$$

$$= 0.278nm$$

From Equations 2.1 and 2.2,

$$F_c = -\frac{k_0(Z_1q)(Z_2q)}{a_0^2}$$

where the equilibrium bond length is used. Substituting, we get

$$F_c = -\frac{(9 \times 10^9 \text{ V} \cdot \text{m/C})(+1)(0.16 \times 10^{-18}\text{C})(-1)(0.16 \times 10^{-18}\text{C})}{(0.278 \times 10^{-9}\text{m})^2}$$

Noting that $1 \text{ V} \cdot \text{C} = 1 \text{ J}$, we obtain

$$F_c = 2.98 \times 10^{-9} \text{ N}$$

Note.   This result can be compared to Figures 2–6 and 2–7.

(b) Because $F_c + F_R = 0$,

$$F_R = -F_c = -2.98 \times 10^{-9} \text{ N}$$

SAMPLE PROBLEM 2.6

Repeat Sample Problem 2.5 for Na_2O, an oxide component in many ceramics and glasses.

SOLUTION

(a) From Appendix 2,

$$r_{Na^+} = 0.098nm$$

$$r_{O^{2-}} = 0.132nm$$

Then

$$a_0 = r_{Na^+} + r_{O^{2-}} = 0.098nm + 0.132nm$$

$$= 0.231nm$$

Again,

$$F_c = -\frac{k_0(Z_1q)(Z_2q)}{a_0^2}$$

$$= -\frac{(9 \times 10^9 \text{ V} \cdot \text{m/C})(+1)(0.16 \times 10^{-18}\text{C})(-2)(0.16 \times 10^{-18}\text{C})}{(0.231 \times 10^{-9}\text{m})^2}$$

$$= 8.64 \times 10^{-9} \text{ N}$$

(b) $F_R = -F_c = -8.64 \times 10^{-9} \text{ N}$

SAMPLE PROBLEM 2.7

Calculate the minimum radius ratio for a coordination number of 8.

SOLUTION

From Table 2.1 it is apparent that the ions are touching along a body diagonal. If the cube edge length is termed l,

$$2R + 2r = \sqrt{3}l$$

For the minimum radius ratio coordination, the large ions are also touching each other (along a cube edge), giving

$$2R = l$$

Combining gives us

$$2R + 2r = \sqrt{3}(2R)$$

Then

$$2r = 2R(\sqrt{3} - 1)$$

and

$$\frac{r}{R} = \sqrt{3} - 1 = 1.732 - 1$$

$$= 0.732$$

Note.   There is no shortcut to visualizing three-dimensional structures of this type. It might be helpful to sketch slices through the cube of Table 2.1 with the ions drawn full scale. Many more exercises of this type will be given in Chapter 3.

SAMPLE PROBLEM 2.8

Estimate the coordination number for the cation in each of these ceramic oxides: Al_2O_3, B_2O_3, CaO, MgO, SiO_2, and TiO_2.

SOLUTION

From Appendix 2, $r_{Al^{3+}} = 0.057$ nm, $r_{B^{3+}} = 0.02$ nm, $r_{Ca^{2+}} = 0.106$ nm, $r_{Mg^{2+}} = 0.078$ nm, $r_{Si^{4+}} = 0.039$ nm, $r_{Ti^{4+}} = 0.064$ nm, and $r_{O^{2-}} = 0.132$ nm.
    For Al_2O_3,

$$\frac{r}{R} = \frac{0.057\text{nm}}{0.132\text{nm}} = 0.43$$

for which Table 2.1 gives

$$CN = 6$$

For B_2O_3,

$$\frac{r}{R} = \frac{0.02\text{nm}}{0.132\text{nm}} = 0.15 \quad \text{giving CN} = 2^*$$

---

*   The actual CN for B_2O_3 is 3 and for CaO is 6. Discrepancies are due to a combination of uncertainty in the estimation of ionic radii and bond directionality due to partially covalent character.

For CaO,

$$\frac{r}{R} = \frac{0.106\text{nm}}{0.132\text{nm}} = 0.80 \quad \text{giving CN} = 8$$

For MgO,

$$\frac{r}{R} = \frac{0.078\text{nm}}{0.132\text{nm}} = 0.59 \quad \text{giving CN} = 6$$

For SiO$_2$,

$$\frac{r}{R} = \frac{0.039\text{nm}}{0.132\text{nm}} = 0.30 \quad \text{giving CN} = 4$$

For TiO$_2$,

$$\frac{r}{R} = \frac{0.064\text{nm}}{0.132\text{nm}} = 0.48 \quad \text{giving CN} = 6$$

PRACTICE PROBLEM 2.4

(a) Make a sketch similar to Figure 2–4 illustrating Mg and O atoms and ions in MgO. (b) Compare the electronic configurations for the atoms and ions illustrated in part (a). (c) Show which noble gas atoms have electronic configurations equivalent to those illustrated in part (a). (See Sample Problem 2.4.)

PRACTICE PROBLEM 2.5

(a) Using the ionic radii data in Appendix 2, calculate the coulombic force of attraction between the $Mg^{2+} - O^{2-}$ ion pair. (b) What is the repulsive force in this case? (See Sample Problems 2.5 and 2.6.)

PRACTICE PROBLEM 2.6

Calculate the minimum radius ratio for a coordination number of (a) 4 and (b) 6. (See Sample Problem 2.7.)

PRACTICE PROBLEM 2.7

In the next chapter we shall see that MgO, CaO, FeO, and NiO all share the NaCl crystal structure. As a result, in each case the metal ions will have the same coordination number (6). The case of MgO and CaO is treated in Sample Problem 2.8. Use the radius ratio calculation to see if it estimates the CN = 6 for FeO and NiO.

2.3 THE COVALENT BOND

While the ionic bond is nondirectional, the **covalent bond** is highly directional. The name *covalent* derives from the cooperative sharing of valence electrons between two adjacent atoms. **Valence electrons** are those outer orbital electrons that take part in bonding.* Figure 2–13 illustrates the covalent bond in a molecule of chlorine gas (Cl_2) with (a) a planetary model compared with (b) the actual **electron density,** which is clearly concentrated along a straight line between the two Cl nuclei. Common shorthand notations of electron dots and a bond line are shown in parts (c) and (d), respectively.

Figure 2–14a shows a bond-line representation of another covalent molecule, ethylene (C_2H_4). The double line between the two carbons signifies a *double bond*, or covalent sharing of two pairs of valence electrons. By converting the double bond to two single bonds, adjacent ethylene molecules can be covalently bonded together, leading to a long-chain molecule of *polyethylene* (Figure 2–14b). Such **polymeric molecules** (each C_2H_4 unit is a "mer") are the structural basis of polymers. In Chapter 13, these materials will be discussed in detail. For now, it is sufficient to re-

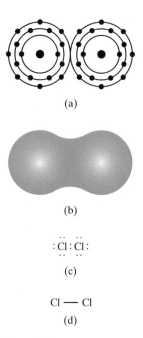

(a)

(b)

$: \overset{..}{\underset{..}{Cl}} : \overset{..}{\underset{..}{Cl}} :$

(c)

Cl —— Cl

(d)

Figure 2-13 *The covalent bond in a molecule of chlorine gas, Cl_2, is illustrated with (a) a planetary model compared with (b) the actual electron density and (c) an "electron-dot" schematic and (d) a "bond-line" schematic.*

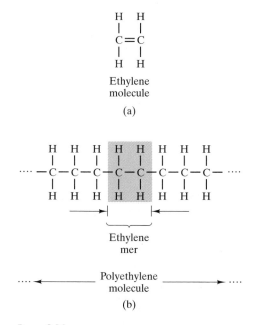

Figure 2-14 *(a) An ethylene molecule (C_2H_4) is compared with (b) a polyethylene molecule* $+C_2H_4+_n$ *that results from the conversion of the C=C double bond into two C—C single bonds.*

---

* Remember that, in ionic bonding, the valence of Na^+ was +1 because one electron had been transferred to an anion.

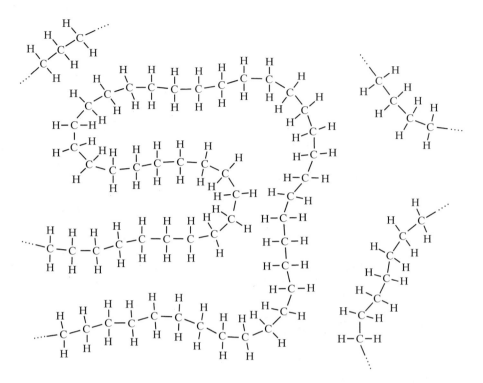

Figure 2-15 *Two-dimensional schematic representation of the "spaghettilike" structure of solid polyethylene.*

alize that long-chain molecules of this type have sufficient flexibility to fill three-dimensional space by a complex coiling structure. Figure 2–15 is a two-dimensional schematic of such a "spaghetti like" structure. The straight lines between C and C and between C and H represent strong, covalent bonds. Only weak, secondary bonding occurs between adjacent sections of the long molecular chains. It is this secondary bonding that is the "weak link" that leads to the low strengths and low melting points for traditional polymers. By contrast, diamond, with exceptionally high hardness and a melting point of greater than 3500°C, has covalent bonding between each adjacent pair of C atoms (Figure 2–16).

It is important to note that covalent bonding can produce coordination numbers substantially smaller than predicted by the radius ratio considerations of ionic bonding. For diamond, the radius ratio for the equally sized carbon atoms is $r/R = 1.0$, but Figure 2–16 shows that the coordination number is 4 rather than 12, as predicted by Table 2.1. In this case, the coordination number for carbon is determined by its characteristic sp^3 hybridization bonding in which the four outer-shell electrons of the carbon atom are shared with adjacent atoms in equally spaced directions (see Section 2.1).

In some cases, the efficient packing considerations of Table 2.1 are in agreement with covalent bonding geometry. For example, the basic structural unit in silicate minerals and in many commercial ceramics and glasses is the SiO_4^{4-} tetrahedron shown in Figure 2–17. Silicon resides just be-

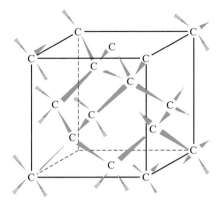

Figure 2-16 *Three-dimensional structure of bonding in the covalent solid,
carbon (diamond). Each carbon
atom (C) has four covalent bonds
to four other carbon atoms. (This
geometry can be compared with the
"diamond cubic" structure of Figure
3–23.) In this illustration, the "bond-
line" schematic of covalent bonding
is given a perspective view to emphasize the spatial arrangement of
bonded carbon atoms.*

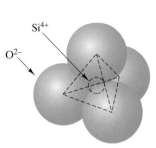

Figure 2-17 *The SiO_4^{4-} tetrahedron represented as a
cluster of ions. In fact, the
Si—O bond exhibits both
ionic and covalent character.*

low carbon in group IVA of the periodic table and exhibits similar chemical behavior. Silicon forms many compounds with fourfold coordination.
The SiO_4^{4-} unit maintains this bonding configuration but, simultaneously,
has strong ionic character, including agreement with Table 2.1. The radius ratio ($r_{Si^{4+}}/r_{O^{2-}} = 0.039nm/0.132nm = 0.295$) is in the correct range
($0.225 < r/R < 0.414$) to produce maximum efficiency of ionic coordination with CN = 4. In fact, the Si—O bond is roughly one-half ionic (electron
transfer) and one-half covalent (electron sharing) in nature.

The bonding force and bonding energy curves for covalent bonding
look similar to those shown in Figure 2–8 for ionic bonding. The different
nature of the two types of bonding implies, of course, that the ionic force
equations (2.1 and 2.2) do not apply. Nonetheless, the general terminology
of bond energy and bond length apply in both cases (Figure 2–18). Table
2.2 summarizes values of bond energy and bond length for major covalent
bonds.

Another important characteristic of covalent solids is the **bond angle,**
determined by the directional nature of valence electron sharing. Figure
2–19 illustrates the bond angle for a typical carbon atom, which tends to form
four equally spaced bonds. This tetrahedral configuration (see Figure 2–17)
gives a bond angle of 109.5°. The bond angle can vary slightly depending on
the species to which the bond is linked, double bonds, and so on. In general,
bond angles involving carbon are close to the ideal 109.5°.

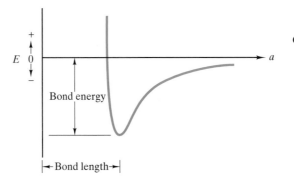

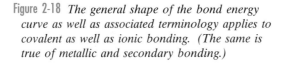

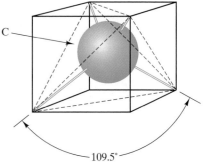

Figure 2-18 *The general shape of the bond energy curve as well as associated terminology applies to covalent as well as ionic bonding. (The same is true of metallic and secondary bonding.)*

Figure 2-19 *Tetrahedral configuration of covalent bonds with carbon. The bond angle is 109.5°.*

Table 2.2 *Bond Energies and Bond Lengths for Representative Covalent Bonds*

Bond	Bond energy[a]		Bond length,
	kcal/mol	kJ/mol	nm
C—C	88[b]	370	0.154
C—C	162	680	0.13
C—C	213	890	0.12
C—H	104	435	0.11
C—N	73	305	0.15
C—O	86	360	0.14
C—O	128	535	0.12
C—F	108	450	0.14
C—Cl	81	340	0.18
O—H	119	500	0.10
O—O	52	220	0.15
O—Si	90	375	0.16
N—H	103	430	0.10
N—O	60	250	0.12
F—F	38	160	0.14
H—H	104	435	0.074

Source: L. H. Van Vlack, *Elements of Materials Science and Engineering,* 4th ed., Addison–Wesley Publishing Co., Inc., Reading, Mass., 1980.

[a] Approximate. The values vary with the type of neighboring bonds. For example, methane (CH_4) has the value shown above for its C—H bond; however, the C—H bond energy is about 5% less in CH_3Cl and 15% less in $CHCl_3$.

[b] All values are negative for forming bonds (energy is released) and positive for breaking bonds (energy is required).

SAMPLE PROBLEM 2.9

Sketch the polymerization process for polyvinyl chloride, PVC. The vinyl chloride molecule is C_2H_3Cl.

SOLUTION

Similar to Figure 2–14, the vinyl chloride molecule would appear as

Polymerization would occur when several adjacent vinyl chloride molecules connect, transforming their double bonds to single bonds:

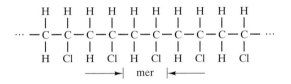

SAMPLE PROBLEM 2.10

Calculate the reaction energy for the polymerization of polyvinyl chloride in Sample Problem 2.9.

SOLUTION

In general, each C—C bond is broken to form two single C—C bonds:

$$C = C \rightarrow 2C - C$$

By using data from Table 2.2, the energy associated with this reaction is

$$680 \text{ kJ/mol} \rightarrow 2(370 \text{ kJ/mol}) = 740 \text{ kJ/mol}$$

The reaction energy is then

$$(740 - 680) \text{ kJ/mol} = 60 \text{ kJ/mol}$$

Note.   As stated in the footnote in Table 2.2, the reaction energy is released during polymerization, making this a spontaneous reaction in which the product, polyvinyl chloride, is stable relative to individual vinyl chloride molecules. Since carbon atoms in the backbone of the polymeric molecule are involved rather than side members, this reaction energy also applies for polyethylene (Figure 2–14) and other "vinyl"-type polymers.

SAMPLE PROBLEM 2.11

Calculate the length of a polyethylene molecule, $+C_2H_4+_n$, where $n = 500$.

SOLUTION

Looking only at the carbon atoms in the backbone of the polymeric chain, we must acknowledge the characteristic bond angle of 109.5°:

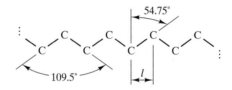

This produces an effective bond length, l, of

$$l = (C—C \text{ bond length}) \times \sin 54.75°$$

Using Table 2.2, we obtain

$$l = (0.154 \text{nm})(\sin 54.75°)$$

$$= 0.126 \text{nm}$$

With two bond lengths per mer and 500 mers, the total molecule length, L, is

$$L = 500 \times 2 \times 0.126 \text{nm}$$

$$= 126 \text{nm}$$

$$= 0.126 \ \mu\text{m}$$

Note.   In Chapter 13 we shall calculate the degree of coiling of these long, linear molecules.

PRACTICE PROBLEM 2.8

In Figure 2–14 we see the polymerization of polyethylene $-(C_2H_4)_n$ illustrated. Sample Problem 2.9 illustrates polymerization for polyvinyl chloride $-(C_2H_3Cl)_n$. Make a similar sketch to illustrate the polymerization of polypropylene $-(C_2H_3R)_n$, where R is a CH_3 group.

PRACTICE PROBLEM 2.9

Use a sketch to illustrate the polymerization of polystyrene $-(C_2H_3R)_n$, where R is a benzene group, C_6H_5.

PRACTICE PROBLEM 2.10

Calculate the reaction energy for polymerization of **(a)** propylene (see Practice Problem 2.8) and **(b)** styrene (see Practice Problem 2.9).

PRACTICE PROBLEM 2.11

The length of an average polyethylene molecule in a commercial clear plastic wrap is 0.2 μm. What is the average degree of polymerization (n) for this material? (See Sample Problem 2.11.)

2.4   THE METALLIC BOND

The ionic bond involves electron transfer and is nondirectional. The covalent bond involves electron sharing and is directional. The third type of primary bond, the **metallic bond,** involves electron sharing and is nondirectional. In this case, the valence electrons are said to be *delocalized*, that is, they have an equal probability of being associated with any of a large number of adjacent atoms. In typical metals, this delocalization is associated with the entire material, leading to an electron cloud, or electron gas (Figure 2–20). This mobile "gas" is the basis for the high electrical conductivity in metals. (The role of electronic structure in producing conduction electrons in metals is discussed in Section 15.2.)

Again, the concept of a bonding *energy "well,"* or *"trough,"* as shown in Figure 2–18, applies. As with ionic bonding, bond angles and coordination numbers are determined primarily by efficient packing considerations, so coordination numbers tend to be high (8 and 12). Relative to the bonding energy curve, a detailed list of atomic radii for the elements is given in Appendix 2, which includes the important elemental metals. Appendix 2 also includes a list of ionic radii. Some of these ionic species are found in the important ceramics and glasses. Inspection of Appendix 2 shows that

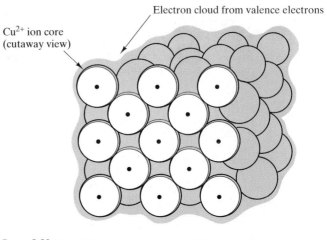

Electron cloud from valence electrons

Cu²⁺ ion core
(cutaway view)

Figure 2-20 *Metallic bond consisting of an electron cloud, or gas. An imaginary slice is shown through the front face of the crystal structure of copper, revealing Cu²⁺ ion cores bonded by the delocalized valence electrons.*

the radius of the metal ion core involved in metallic bonding (Figure 2–20) differs substantially from the radius of a metal ion from which valence electrons have been transferred.

Instead of a list of bond energies for metals and ceramics, similar to those included for covalent bonds in Table 2.2, data that represent the energetics associated with the bulk solid rather than isolated atom (or ion) pairs are more useful. For example, Table 2.3 lists the heats of sublimation of some common metals and their oxides (some of the common ceramic compounds). The heat of sublimation represents the amount of thermal energy necessary to turn 1 mol of solid directly into vapor at a fixed temperature. It is a good indication of the relative strength of bonding in the solid. However, caution must be used in making direct comparisons to the bond energies in Table 2.2,

Table 2.3 *Heats of Sublimation (at 25° C) of Some Metals and Their Oxides*

	Heat of sublimation			Heat of sublimation	
Metal	**kcal/mol**	**kJ/mol**	**Metal oxide**	**kcal/mol**	**kJ/mol**
Al	78	326			
Cu	81	338			
Fe	100	416	FeO	122	509
Mg	35	148	MgO	145	605
Ti	113	473	α-TiO	143	597
			TiO$_2$ (rutile)	153	639

Source: Data from *JANAF Thermochemical Tables,* 2nd ed., National Standard Reference Data Series, Natl. Bur. Std. (U.S.), *37* (1971), and Supplement in *J. Phys. Chem. Ref. Data 4* (1), 1–175 (1975).

IA	IIA	IIIB	IVB	VB	VIB	VIIB	VIII	VIII	VIII	IB	IIB	IIIA	IVA	VA	VIA	VIIA	0
1 H 2.1																	2 He –
3 Li 1.0	4 Be 1.5											5 B 2.0	6 C 2.5	7 N 3.0	8 O 3.5	9 F 4.0	10 Ne –
11 Na 0.9	12 Mg 1.2											13 Al 1.5	14 Si 1.8	15 P 2.1	16 S 2.5	17 Cl 3.0	18 Ar –
19 K 0.8	20 Ca 1.0	21 Sc 1.3	22 Ti 1.5	23 V 1.6	24 Cr 1.6	25 Mn 1.5	26 Fe 1.8	27 Co 1.8	28 Ni 1.8	29 Cu 1.9	30 Zn 1.6	31 Ga 1.6	32 Ge 1.8	33 As 2.0	34 Se 2.4	35 Br 2.8	36 Kr –
37 Rb 0.8	38 Sr 1.0	39 Y 1.2	40 Zr 1.4	41 Nb 1.6	42 Mo 1.8	43 Tc 1.9	44 Ru 2.2	45 Rh 2.2	46 Pd 2.2	47 Ag 1.9	48 Cd 1.7	49 In 1.7	50 Sn 1.8	51 Sb 1.9	52 Te 2.1	53 I 2.5	54 Xe –
55 Cs 0.7	56 Ba 0.9	57-71 La-Lu 1.1-1.2	72 Hf 1.3	73 Ta 1.5	74 W 1.7	75 Re 1.9	76 Os 2.2	77 Ir 2.2	78 Pt 2.2	79 Au 2.4	80 Hg 1.9	81 Tl 1.8	82 Pb 1.8	83 Bi 1.9	84 Po 2.0	85 At 2.2	86 Rn –
87 Fr 0.7	88 Ra 0.9	89-102 Ac-No 1.1-1.7															

Figure 2-21 *The electronegativities of the elements. (After Linus Pauling, The Nature of the Chemical Bond and the Structure of Molecules and Crystals; An Introduction to Modern Structural Chemistry, 3rd ed., Cornell University Press, Ithaca, New York, 1960)*

which corresponds to specific atom pairs. Nonetheless, the magnitudes of energies in Tables 2.2 and 2.3 are comparable in range.

In this chapter, we have seen that the nature of the chemical bonds between atoms of the same element and atoms of different elements depends on the transfer or sharing of electrons between adjacent atoms. The American chemist Linus Pauling systematically defined **electronegativity** as the ability of an atom to attract electrons to itself. Figure 2–21 summarizes Pauling's electronegativity values for the elements in the periodic table. We can recall from Chapter 1 that the majority of elements in the periodic table are metallic in nature (Figure 1–3). In general, the values of electronegativities increase from the left to the right side of the periodic table, with cesium and francium (in group IA) having the lowest value (0.7) and flourine (in group VIIA) having the highest (4.0). Clearly, the metallic elements tend to have the lower values of electronegativity and the nonmetallic elements have the higher values. Although Pauling specifically based his electronegativities on thermochemical data for molecules, we shall see in Section 4.1 that the data of Figure 2–21 are useful for predicting the nature of metallic alloys.

SAMPLE PROBLEM 2.12

Several metals, such as α-Fe, have a body-centered cubic crystal structure in which the atoms have a coordination number of 8. Discuss this in light of the prediction of Table 2.1 that nondirectional bonding of equal-sized spheres should have a coordination number of 12.

SOLUTION

The presence of some covalent character in these predominantly metallic materials can reduce the coordination number below the predicted value. (See Sample Problem 2.8.)

....

PRACTICE PROBLEM 2.12

Discuss the low coordination number (= 4) for the diamond cubic structure found for some elemental solids, such as silicon. (See Sample Problem 2.12.)

2.5   THE SECONDARY, OR VAN DER WAALS, BOND

The major source of cohesion in a given engineering material is one or more of the three primary bonds just covered. As seen in Table 2.2, typical primary bond energies range from 200 to 700 kJ/mol ($\approx$ 50 to 170 kcal/mol). It is possible to obtain some atomic bonding (with substantially smaller bonding energies) without electron transfer or sharing. This is known as *secondary bonding*, or **van der Waals* bonding.** The mechanism of secondary bonding is somewhat similar to ionic bonding, that is, the attraction of opposite charges. The key difference is that no electrons are transferred.[†] Attraction depends on asymmetrical distributions of positive and negative charge within each atom or molecular unit being bonded. Such charge asymmetry is referred to as a **dipole.** Secondary bonding can be of two types depending on whether the dipoles are (1) temporary or (2) permanent.

Figure 2–22 illustrates how two neutral atoms can develop a weak bonding force between them by a slight distortion of their charge distributions. The example is argon, a noble gas, which does not tend to form primary bonds because it has a stable, filled outer orbital shell. An isolated argon atom has a perfectly spherical distribution of negative electrical charge surrounding its positive nucleus. However, when another argon atom is brought nearby, the negative charge is drawn slightly toward the positive nucleus of the adjacent atom. This slight distortion of charge distribution occurs simultaneously in both atoms. The result is an *induced dipole*. Because the degree of charge distortion related to an induced dipole is small, the magnitude of the resulting dipole is small, leading to a relatively small bond energy (0.99 kJ/mol or 0.24 kcal/mol).

---

[*]   Johannes Diderik van der Waals (1837–1923), Dutch physicist, improved the equations of state for gases by taking into account the effect of secondary bonding forces. His brilliant research was first published as a thesis dissertation arising from his part-time studies of physics. The immediate acclaim for the work led to his transition from a job as headmaster of a secondary school to a professorship at the University of Amsterdam.

[†]   Primary bonds are sometimes referred to as *chemical bonds*, with secondary bonds being *physical bonds*.

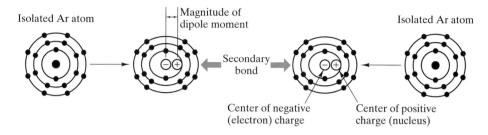

Figure 2-22 *Development of induced dipoles in adjacent argon atoms leading to a weak, secondary bond. The degree of charge distortion shown here is greatly exaggerated.*

Secondary bonding energies are somewhat greater when molecular units containing *permanent dipoles* are involved. Perhaps the best example of this is the *hydrogen bridge*, which connects adjacent molecules of water, H_2O (Figure 2–23). Because of the directional nature of electron sharing in the covalent O—H bonds, the H atoms become positive centers and O atoms become negative centers for the H_2O molecules. The greater charge separation possible in such a **polar molecule,** a molecule with a permanent separation of charge, gives a larger **dipole moment** (product of charge and separation distance between centers of positive and negative charge) and therefore a greater bond energy (21 kJ/mol or 5 kcal/mol). The secondary bonding between adjacent polymeric chains in polymers such as polyethylene is of this type.

Note that one of the important properties of water derives from the hydrogen bridge. The expansion of water upon freezing is due to the regular and repeating alignment of adjacent H_2O molecules, as seen in Figure 2–23. This leads to a relatively open structure. Upon melting, the adjacent H_2O molecules, while retaining the hydrogen bridge, pack together in a more random and more dense arrangement.

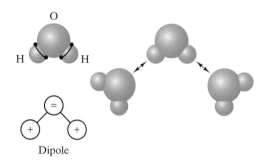

Figure 2-23 *"Hydrogen bridge." This secondary bond is formed between two permanent dipoles in adjacent water molecules. (From W. G. Moffatt, G. W. Pearsall, and J. Wulff, The Structure and Properties of Materials, Vol. 1: Structures, John Wiley & Sons, Inc., New York, 1964.)*

SAMPLE PROBLEM 2.13

A common way to describe the bonding energy curve (Figure 2–18) for secondary bonding is the "6–12" potential, which states that

$$E = -\frac{K_A}{a^6} + \frac{K_R}{a^{12}}$$

where K_A and K_R are constants for attraction and repulsion, respectively. This relatively simple form is a quantum mechanical result for this relatively simple bond type. Given $K_A = 10.37 \times 10^{-78}$ J · m^6 and $K_R = 16.16 \times 10^{-135}$ J · m^{12}, calculate the bond energy and bond length for argon.

SOLUTION

The (equilibrium) bond length occurs at $dE/da = 0$:

$$\left(\frac{dE}{da}\right)_{a=a_0} = 0 = \frac{6K_A}{a_0^7} - \frac{12K_R}{a_0^{13}}$$

Rearranging gives us

$$a_0 = \left(2\frac{K_R}{K_A}\right)^{1/6}$$

$$= \left(2 \times \frac{16.16 \times 10^{-135}}{10.37 \times 10^{-78}}\right)^{1/6} \text{m}$$

$$= 0.382 \times 10^{-9}\text{m} = 0.382\text{nm}$$

Note that bond energy $= E(a_0)$ yields

$$E(0.382\text{nm}) = -\frac{K_A}{(0.382\text{nm})^6} + \frac{K_R}{(0.382\text{nm})^{12}}$$

$$= -\frac{(10.37 \times 10^{-78}\text{J} \cdot \text{m}^6)}{(0.382 \times 10^{-9}\text{m})^6} + \frac{(16.16 \times 10^{-135} \text{ J} \cdot \text{m}^{12})}{(0.382 \times 10^{-9}\text{m})^{12}}$$

$$= -1.66 \times 10^{-21} \text{ J}$$

For 1 mol of Ar,

$$E_{\text{bonding}} = -1.66 \times 10^{-21} \text{ J/bond} \times 0.602 \times 10^{24} \frac{\text{bonds}}{\text{mole}}$$

$$= -0.999 \times 10^3 \text{ J/mol}$$

$$= -0.999\text{kJ/mol}$$

> Note. This bond energy is less than 1% of the magnitude of any of the primary (covalent) bonds listed in Table 2.2. It should also be noted that the footnote in Table 2.2 indicates a consistent sign convention (bond energy is negative).

PRACTICE PROBLEM 2.13

The bond energy and bond length for argon are calculated (assuming a "6–12" potential) in Sample Problem 2.13. Plot E as a function of a over the range 0.33 to 0.80 nm.

PRACTICE PROBLEM 2.14

Using the information from Sample Problem 2.13, plot the van der Waals bonding force curve for argon, that is, F versus a over the same range covered in Practice Problem 2.13.

2.6   MATERIALS—THE BONDING CLASSIFICATION

A dramatic representation of the relative bond energies of the various bond types of this chapter is obtained by comparison of melting points. The **melting point** of a solid indicates the temperature to which the material must be subjected to provide sufficient thermal energy to break its cohesive bonds. Table 2.4 shows representative examples used in this chapter. A special note must be made for polyethylene, which is of mixed-bond character. As discussed in Section 2.3, the secondary bonding is a weak link that causes the material to lose structural rigidity above approximately 120°C. This is not a precise melting point but a temperature above which the material softens rapidly with increasing temperature. The irregularity of the polymeric structure (Figure 2–15) produces variable secondary bond lengths and, therefore, variable bond energies. More important than the variation in bond energy

Table 2.4 *Comparison of Melting Points for Some of the Representative Materials of Chapter 2*

Material	Bonding type	Melting point (°C)
NaCl	Ionic	801
C (diamond)	Covalent	$\sim$ 3550
$(C_2H_4)_n$	Covalent and secondary	$\sim$ 120[a]
Cu	Metallic	1084.87
Ar	Secondary (induced dipole)	-189
H_2O	Secondary (permanent dipole)	0

Source: [a]Because of the irregularity of the polymeric structure of polyethylene, it does not have a precise melting point. Instead, it softens with increasing temperature above 120°C. In this case, the 120°C value is a "service temperature" rather than a true melting point.

Table 2.5 *Bonding Character of the Four Fundamental Types of Engineering Materials*

Material type	Bonding character	Example
Metal	Metallic	Iron (Fe) and the ferrous alloys
Ceramics and glasses	Ionic/covalent	Silica (SiO_2): crystalline and noncrystalline
Polymers	Covalent and secondary	Polyethylene $+(C_2H_4)_n$
Semiconductors	Covalent or covalent/ionic	Silicon (Si) or cadmium sulfide (CdS)

is the average magnitude, which is relatively small. Even though polyethylene and diamond each have similar C—C covalent bonds, the absence of secondary-bond weak links allows diamond to retain its structural rigidity more than 3000°C beyond polyethylene.

We have now seen four major types of atomic bonding consisting of three primary bonds (ionic, covalent, and metallic) and secondary bonding. It has been traditional to distinguish the three fundamental structural materials (metals, ceramics, and polymers) as being directly associated with the three types of primary bonds (metallic, ionic, and covalent, respectively). This is a useful concept, but we have already seen in Sections 2.3 and 2.5 that polymers owe their behavior to both covalent and secondary bonding. We also noted in Section 2.3 that some of the most important ceramics have strong covalent as well as ionic character. Table 2.5 summarizes the bonding character associated with the four fundamental types of engineering materials together with some representative examples. Remember that the mixed-bond character for ceramics referred to both ionic and covalent nature for a given bond (e.g., Si—O), whereas the mixed-bond character for polymers referred to different bonds being covalent (e.g., C—H) and secondary (e.g., between chains). The relative contribution of different bond types can be graphically displayed in the form of a tetrahedron of bond types (Figure 2–24) in which each apex of the tetrahedron represents a pure bonding type. In Chapter 14 we shall add another perspective on materials classification—electrical conductivity. This will follow directly from the nature of bonding and is especially helpful in defining the unique character of semiconductors.

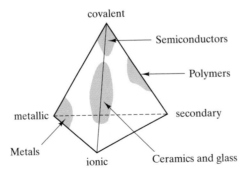

Figure 2-24 *Tetrahedron representing the relative contribution of different bond types to the four fundamental categories of engineering materials (the three structural types plus semiconductors).*

SUMMARY

One basis for the classification of engineering materials is atomic bonding. While the chemical identity of each atom is determined by the number of protons and neutrons within its nucleus, the nature of atomic bonding is determined by the behavior of the electrons that orbit the nucleus.

There are three kinds of strong, or primary, bonds responsible for the cohesion of solids: (1) The ionic bond involves electron transfer and is nondirectional. The electron transfer creates a pair of ions with opposite charge. The attractive force between ions is coulombic in nature. An equilibrium ionic spacing is established due to the strong repulsive forces associated with attempting to overlap the two atomic cores. The nondirectional nature of the ionic bond allows ionic coordination numbers to be determined by strictly geometrical packing efficiency (as indicated by the radius ratio). (2) The covalent bond involves electron sharing and is highly directional. This can lead to relatively low coordination numbers and more open atomic structures. (3) The metallic bond involves sharing of delocalized electrons, producing a nondirectional bond. The resulting electron cloud or gas results in high electrical conductivity. The nondirectional nature results in relatively high coordination numbers, as in ionic bonding. In the absence of electron transfer or sharing, a weaker form of bonding is possible. This secondary bonding is the result of attraction between either temporary or permanent electrical dipoles.

The classification of engineering materials acknowledges a particular bonding type or combination of types for each category. Metals involve metallic bonding. Ceramics and glasses involve ionic bonding but usually in conjunction with a strong covalent character. Polymers typically involve strong covalent bonds along polymeric chains but have weaker secondary bonding between adjacent chains. The secondary bonding acts as a weak link in the structure, giving characteristically low strengths and melting points. Semiconductors are predominantly covalent in nature, with some semiconducting compounds having a significant ionic character. These four categories of engineering materials are, then, the fundamental types. Composites are combinations of the first three fundamental types and have bonding characteristics appropriate to their constituents.

KEY TERMS

anion (26)	bond length (28)	electron (21)
atomic mass (21)	cation (26)	electron density (39)
atomic mass unit (21)	coordination number (31)	electron orbital (22)
atomic number (22)	coulombic attraction (26)	electronegativity (47)
atomic radius (31)	covalent bond (39)	energy level (22)
Avogadro's number (21)	delocalized electron (45)	energy trough (45)
bond angle (41)	dipole (48)	energy well (45)
bonding energy (29)	dipole moment (49)	gram-atom (21)
bonding force (28)	double bond (39)	group (22)

hard sphere (31)
hybridization (23)
hydrogen bridge (49)
ion (26)
ionic bond (26)
ionic radius (31)
isotope (22)
melting point (51)
metallic bond (45)

mole (21)
neutron (21)
nucleus (21)
orbital shell (26)
periodic table (22)
polar molecule (49)
polymeric molecule (39)
primary bond (23)
proton (21)

radius ratio (32)
repulsive force (28)
secondary bond (23)
soft sphere (31)
valence (27)
valence electron (39)
van der Waals bond (48)

REFERENCES

Virtually any introductory textbook on college-level chemistry will be useful background for this chapter. Good examples are the following

Brown, T. L., H. E. LeMay, Jr., and **B. E. Bursten,** *Chemistry— The Central Science*, 7th ed., Prentice Hall, Upper Saddle River, New Jersey, 1997.

McQuarrie, D. A., and **P. A. Rock,** *General Chemistry,* 3rd ed., W. H. Freeman & Co., New York, 1991.

Oxtoby, D. W., and **N. H. Nachtrieb,** *Principles of Modern Chemistry,* 3rd ed., Saunders College Publishing, Philadelphia, 1996.

PROBLEMS

Beginning with this chapter, a set of problems will be provided at the conclusion of each chapter of the book. Instructors may note that there are few of the subjective, discussion-type problems that are so often used in materials textbooks. I strongly feel that such problems are generally frustrating to students being introduced to materials science and engineering. I shall concentrate on objective problems. For this reason, no problems were given in the general, introductory Chapter 1.

A few points about the organization of problems should be noted. All problems are clearly related to the appropriate chapter section. Also, some Practice Problems for each section were already given following the solved Sample Problems within that section. These are intended to provide a carefully guided journey into the first calculations in each new area and could be used by students for self-study. Answers are given for nearly all of the Practice Problems following the appendixes. The following problems are increasingly challenging. Problems not marked with a bullet are relatively straightforward but are not explicitly connected to a sample problem. Those problems marked with a bullet ($\bullet$) are intended to be relatively challenging. Answers to odd-numbered problems are given following the appendixes.

Section 2.1 • Atomic Structure

2.1. A gold O-ring is used to form a gastight seal in a high-vacuum chamber. The ring is formed from a 100-mm length of 1.5-mm-diameter wire. Calculate the number of gold atoms in the O-ring.

2.2. Common aluminum foil for household use is nearly pure aluminum. A box of this product at a local supermarket is advertised as giving 75 ft^2 of material (in a roll 304 mm wide by 22.8 m long). If the foil is 0.5 mil (12.7 μm) thick, calculate the number of atoms of aluminum in the roll.

2.3. In a metal-oxide-semiconductor (MOS) device, a thin layer of SiO_2 (density $= 2.20$ Mg/m^3) is grown on a single crystal chip of silicon. How many Si atoms *and* how many O atoms are present per square millimeter of the oxide layer? Assume that the layer thickness is 150 nm.

2.4. A box of clear plastic wrap for household use is polyethylene, $\text{+C}_2\text{H}_4\text{+}_n$, with density $= 0.910$ Mg/m^3. A box of this product contains 100 ft^2 of material (in a roll 304 mm wide by 30.5 m long). If the wrap is 0.5 mil (12.7 μm) thick, calculate the number of carbon atoms *and* the number of hydrogen atoms in this roll.

2.5. An Al_2O_3 *whisker* is a small single crystal used to reinforce metal-matrix composites. Given a cylindrical shape, calculate the number of Al atoms *and* the number of O atoms in a whisker with a diameter of 1 μm and a length of 25 μm. (The density of Al_2O_3 is 3.97 Mg/m^3.)

2.6. An optical fiber for telecommunication is made of SiO_2 glass (density = 2.20 Mg/m^3). How many Si atoms *and* how many O atoms are present per millimeter of length of a fiber 10 μm in diameter?

2.7. Thirty grams of magnesium filings are to be oxidized in a laboratory demonstration. **(a)** How many O_2 molecules would be consumed in this demonstration? **(b)** How many moles of O_2 does this represent?

2.8. Naturally occurring copper has an atomic weight of 63.55. Its principal isotopes are Cu^{63} and Cu^{65}. What is the abundance (in atomic percent) of each isotope?

2.9. A copper penny has a mass of 2.60 g. Assuming pure copper, how much of this mass is contributed by **(a)** the neutrons in the copper nuclei? and **(b)** electrons?

2.10. The orbital electrons of an atom can be ejected by exposure to a beam of electromagnetic radiation. Specifically, an electron can be ejected by a photon with energy greater than or equal to the electron's binding energy. Given that the photon energy (E) is equal to hc/λ, where h is Planck's constant, c the speed of light, and λ the wavelength, calculate the maximum wavelength of radiation (corresponding to the minimum energy) necessary to eject a $1s$ electron from a C^{12} atom. (See Figure 2–3.)

2.11. Once the $1s$ electron is ejected from a ^{12}C atom, as described in Problem 2.10, there is a tendency for one of the $2(sp^3)$ electrons to drop into the $1s$ level. The result is the emission of a photon with an energy precisely equal to the energy change associated with the electron transition. Calculate the wavelength of the photon that would be emitted from a ^{12}C atom. (You will note various examples of this concept throughout the text in relation to the chemical analysis of engineering materials.)

2.12. The mechanism for producing a photon of specific energy is outlined in Problem 2.11. The magnitude of photon energy increases with the atomic number of the atom from which emission occurs. (This is due to the stronger binding forces between the negative electrons and the positive nucleus as the numbers of protons and electrons increase with atomic number.) As noted in Problem 2.10, $E = hc/\lambda$, which means that a higher-energy photon will have a shorter wavelength. Verify that higher atomic number materials will emit higher-energy, shorter-wavelength photons by calculating E and λ for emission from iron (atomic number 26 compared to 6 for carbon), given that the energy levels for the first two electron orbitals in iron are at $-7,112$ eV and -708 eV.

Section 2.2 • The Ionic Bond

2.13. Make an accurate plot of F_c versus a (comparable to Figure 2–6) for an Mg^{2+}—O^{2-} pair. Consider the range of a from 0.2 to 0.7 nm.

2.14. Make an accurate plot of F_c versus a for an Na^+—O^{2-} pair.

2.15. So far, we have concentrated on the coulombic force of attraction between ions. But like ions repel each other. A nearest-neighbor pair of Na^+ ions in Figure 2–5 are separated by a distance of $\sqrt{2}a_0$, where a_0 is defined in Figure 2–7. Calculate the coulombic force of *repulsion* between such a pair of like ions.

2.16. Calculate the coulombic force of attraction between Ca^{2+} and O^{2-} in CaO, which has the NaCl-type structure.

2.17. Calculate the coulombic force of repulsion between nearest-neighbor Ca^{2+} ions in CaO. (Note Problems 2.15 and 2.16.)

2.18. Calculate the coulombic force of repulsion between nearest-neighbor O^{2-} ions in CaO. (Note Problems 2.15, 2.16, and 2.17.)

2.19. Calculate the coulombic force of repulsion between nearest-neighbor Ni^{2+} ions in NiO, which has the NaCl-type structure. (Note Problem 2.17.)

2.20. Calculate the coulombic force of repulsion between nearest-neighbor O^{2-} ions in NiO. (Note Problems 2.18 and 2.19.)

2.21. SiO_2 is known as a "glass former" because of the tendency of SiO_4^{4-} tetrahedra (Figure 2–17) to link together in a noncrystalline network. Al_2O_3 is known as an intermediate glass former due to the ability of Al^{3+} to substitute for Si^{4+} in the glass network, although Al_2O_3 does not by itself tend to be noncrystalline. Discuss the substitution of Al^{3+} for Si^{4+} in terms of the radius ratio.

2.22. Repeat Problem 2.21 for TiO_2, which, like Al_2O_3, is an intermediate glass former.

2.23. The coloration of glass by certain ions is often sensitive to the coordination of the cation by oxygen ions. For example, Co^{2+} gives a blue-purple color when in the fourfold coordination characteristic of the silica network (see Problem 2.21) and gives a pink color when in a sixfold coordination. Which color from Co^{2+} is predicted by the radius ratio?

2.24. One of the first nonoxide materials to be produced as a glass was BeF_2. As such, it was found to be similar to SiO_2 in many ways. Calculate the radius ratio for Be^{2+} and F^- and comment.

2.25. A common feature in high-temperature ceramic superconductors is a Cu–O sheet that serves as a superconducting plane. Calculate the coulombic force of attraction between a Cu^{2+} and an O^{2-} within one of these sheets.

2.26. In contrast to the calculation for the superconducting Cu—O sheets discussed in Problem 2.25, calculate the coulombic force of attraction between a Cu^+ and an O^{2-}.

• 2.27. For an ionic crystal, such as NaCl, the net coulombic bonding force is a simple multiple of the force of attraction between an adjacent ion pair. To demonstrate this, consider the hypothetical, one-dimensional "crystal" shown:

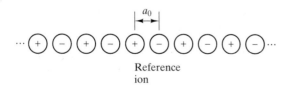

Reference
ion

(a) Show that the net coulombic force of attraction between the reference ion and all other ions in the crystal is

$$F = AF_c$$

where F_c is the force of attraction between an adjacent ion pair (see Equation 2.1) and A is a series expansion.

(b) Determine the value of A.

2.28. In Problem 2.27, a value for A was calculated for the simple one-dimensional case. For the three-dimensional NaCl structure, A has been calculated to be 1.748. Calculate the net coulombic force of attraction, F, for this case.

Section 2.3 • The Covalent Bond

2.29. Calculate the total reaction energy for polymerization required to produce the roll of clear plastic wrap described in Problem 2.4.

2.30. Natural rubber is polyisoprene. The polymerization reaction can be illustrated as

$$n\begin{pmatrix} \overset{\displaystyle H}{|} & \overset{\displaystyle H}{|} & \overset{\displaystyle CH_3}{|} & \overset{\displaystyle H}{|} \\ C = C & - & C = C \\ | & & & | \\ H & & & H \end{pmatrix} \rightarrow \begin{pmatrix} \overset{\displaystyle H}{|} & \overset{\displaystyle H}{|} & \overset{\displaystyle CH_3}{|} & \overset{\displaystyle H}{|} \\ -C & - & C = C & - & C- \\ | & & & | \\ H & & & H \end{pmatrix}_n$$

Calculate the reaction energy (per mole) for polymerization.

2.31. Neoprene is a synthetic rubber, polychloroprene, with a chemical structure similar to natural rubber (see Problem 2.30) except that it contains a Cl atom in place of the CH_3 group of the isoprene molecule. **(a)** Sketch the polymerization reaction for neoprene, and **(b)** calculate the reaction energy (per mole) for this polymerization. **(c)** Calculate the total energy released during the polymerization of 1 kg of chloroprene.

2.32. Acetal polymers, which are widely used for engineering applications, can be represented by the following reaction, the polymerization of formaldehyde:

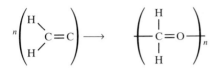

Calculate the reaction energy for this polymerization.

2.33. The first step in the formation of phenolformaldehyde, a common phenolic polymer, is shown in Figure 13–6. Calculate the net reaction energy (per mole) for this step in the overall polymerization reaction.

2.34. Calculate the molecular weight of a polyethylene molecule with $n = 500$.

2.35. The monomer upon which a common acrylic polymer, polymethyl methacrylate, is based is given in Table 13.1. Calculate the molecular weight of a polymethyl methacrylate molecule with $n = 600$.

2.36. Bone "cement" used by orthopedic surgeons to set artificial hip implants in place is methyl methacrylate polymerized during the surgery. The resulting polymer has a relatively wide range of molecular weights. Calculate the resulting range of molecular weights if $200 < n < 700$.

2.37. Orthopedic surgeons notice a substantial amount of heat evolution from polymethyl methacrylate bone cement during surgery. Calculate the reaction energy if a surgeon uses 20 g of polymethyl methacrylate to set a given hip implant.

2.38. The monomer for the common fluoroplastic, polytetrafluoroethylene, is

$$
\begin{array}{cc}
F & F \\
| & | \\
C & = C \\
| & | \\
F & F
\end{array}
$$

(a) Sketch the polymerization of polytetrafluoroethylene.

(b) Calculate the reaction energy (per mole) for this polymerization. **(c)** Calculate the molecular weight of a molecule with $n = 500$.

2.39. Repeat Problem 2.38 for polyvinylidene fluoride, an ingredient in various commercial fluoroplastics, having the monomer

2.40. Repeat Problem 2.38 for polyhexafluoropropylene, an ingredient in various commercial fluoroplastics, having the monomer:

Section 2.4 • The Metallic Bond

2.41. In Table 2.3, the heat of sublimation was used to indicate the magnitude of the energy of the metallic bond. A significant range of energy values is indicated by the data. The melting point data in Appendix 1 are another, more indirect indication of bond strength. Plot heat of sublimation versus melting point for the five metals of Table 2.3 and comment on the correlation.

2.42. In order to explore a trend within the periodic table, plot the bond length of the group IIA metals (Be to Ba) as a function of atomic number. (Refer to Appendix 2 for necessary data.)

2.43. Superimpose on the plot generated for Problem 2.42 the metal-oxide bond lengths for the same range of elements.

2.44. To explore another trend within the periodic table, plot the bond length of the metals in the row Na to Si as a function of atomic numbers. (For this purpose, Si is treated as a semimetal.)

2.45. Superimpose on the plot generated for Problem 2.44 the metal-oxide bond lengths for the same range of elements.

2.46. Plot the bond length of the metals in the long row of metallic elements (K to Ga).

2.47. Superimpose on the plot generated for Problem 2.46 the metal–oxide bond lengths for the same range of elements.

• **2.48.** The heat of sublimation of a metal, introduced in Table 2.3, is related to the ionic bonding energy of a metallic compound discussed in Section 2.2. Specifically, these and related reaction energies are summarized in the Born–Haber cycle, illustrated next. For the simple example of NaCl

$$\text{Na (solid)} + \tfrac{1}{2}\text{Cl}_2\text{ (g)} \longrightarrow \text{Na (g)} + \text{Cl (g)}$$
$$\downarrow \Delta H_f^\circ \qquad\qquad \downarrow \qquad \downarrow$$
$$\text{NaCl (solid)} \longleftarrow \text{Na}^+\text{ (g)} + \text{Cl}^-\text{ (g)}$$

Given the heat of sublimation to be 100 kJ/mol for sodium, calculate the ionic bonding energy of sodium chloride. (Additional data: ionization energies for sodium and chlor ine = 496 kJ/mol and −361 kJ/mol, respectively; dissociation energy for diatomic chlorine gas = 243 kJ/mol; heat of formation, ΔH_f°, of NaCl = −411 kJ/mol.)

Section 2.5 • The Secondary, or van der Waals, Bond

2.49. The secondary bonding of gas molecules to a solid surface is a common mechanism for measuring the surface area of porous materials. By lowering the temperature of a solid well below room temperature, a measured volume of gas will condense to form a monolayer coating of molecules on the porous surface. For a 100-g sample of fused copper catalyst, a volume of $9 \times 10^3\text{mm}^3$ of nitrogen (measured at standard temperature and pressure, $0°\text{C}$ and 1 atm) is required to form a monolayer upon condensation. Calculate the surface area of the catalyst in units of m^2/kg. (Take the area covered by a nitrogen molecule as 0.162 nm^2 and recall that, for an ideal gas, $pV = nRT$, where n is the number of moles of the gas.)

2.50. Repeat Problem 2.49 for a highly porous silica gel that has a volume of $1.16 \times 10^7\text{ mm}^3$ of N_2 gas (at STP or Standard Temperature and Pressure) condensed to form a monolayer.

2.51. Small-diameter noble gas atoms, such as helium, can dissolve in the relatively open network structure of silicate glasses. (See Figure 1–8b for a schematic of glass structure.) The secondary bonding of helium in vitreous silica is represented by a heat of solution, ΔH_s, of −3.96 kJ/mol. The relationship between solubility, S, and the heat of solution is

$$S = S_0 e^{-\Delta H_s/(RT)}$$

where S_0 is a constant, R the gas constant, and T the absolute temperature (in K). If the solubility of helium in vitreous silica is 5.51×10^{23} atoms/($\text{m}^3 \cdot$ atm) at 25°C, calculate the solubility at 250°C.

2.52. Due to its larger atomic diameter, neon has a higher heat of solution in vitreous silica than helium. If the heat of solution of neon in vitreous silica is −6.70 kJ/mol and the solubility at 25°C is 9.07×10^{23} atoms/($\text{m}^3 \cdot$ atm), calculate the solubility at 250°C. (See Problem 2.51.)

CHAPTER 3
Crystalline Structure–Perfection

113 pm

The transmission electron microscope (Section 4.7) can be used to image the regular arrangement of atoms in a crystalline structure. This atomic-resolution view is along individual columns of gallium and nitrogen atoms in gallium nitride. The distance marker is 113 picometers or 0.113 nm. (Courtesy of C. Kisielowski, C. Song, and E. C. Nelson, National Center for Electron Microscopy, Berkeley, California.)

With the categories of engineering materials firmly established, we can now begin characterizing these materials. We will begin with atomic-scale structure, which for most engineering materials, is crystalline; that is, the atoms of the material are arranged in a regular and repeating manner.

Common to all crystalline materials are the fundamentals of crystal geometry. We must identify the seven crystal systems and the 14 crystal lattices. Each of the thousands of crystal structures found in natural and synthetic materials can be placed within these few systems and lattices.

The crystalline structures of most metals belong to one of three relatively simple types. Ceramic compounds, which have a wide variety of chemical compositions, exhibit a similarly wide variety of crystalline structures. Some are relatively simple, but many, such as the silicates, are quite complex. Glass is noncrystalline, and its structure and the nature of noncrystalline materials are discussed in Chapter 4. Polymers share two features with ceramics and glasses. First, their crystalline structures are relatively complex. Second, because of this complexity, the material is not easily crystallized, and common polymers may have as much as 50% to 100% of their volume noncrystalline. Elemental semiconductors, such as silicon, exhibit a characteristic structure (diamond cubic), whereas semiconducting compounds have structures similar to some of the simpler ceramic compounds.

Within a given structure, we must know how to describe atom positions, crystal directions, and crystal planes. With these quantitative "ground rules" in hand, we conclude this chapter with a brief introduction to x-ray diffraction, the standard experimental tool for determining crystal structure.

3.1   SEVEN SYSTEMS AND FOURTEEN LATTICES

The central feature of crystalline structure is that it is regular and repeating. This repetition is apparent from inspection of a typical model of a crystalline arrangement of atoms (see Figure 1–18). In order to quantify this repetition, we must determine which structural unit is being repeated. Actually, any crystalline structure could be described as a pattern formed by repeating various structural units (Figure 3–1). As a practical matter, there will generally be a simplest choice to serve as a representative structural unit. Such a choice is referred to as a **unit cell.** The geometry of a general unit cell is shown in Figure 3–2. The length of unit cell edges and the angles between crystallographic axes are referred to as **lattice constants,** or **lattice parameters.** The key feature of the unit cell is that it contains a full description of the structure as a whole because the complete structure can be generated by the repeated stacking of adjacent unit cells face to face throughout three-dimensional space.

The description of crystal structures by means of unit cells has an important advantage. All possible structures reduce to a small number of basic unit cell geometries. This is demonstrated in two ways. First, there are only seven, unique unit cell shapes that can be stacked together to fill three-dimensional space. These are the seven **crystal systems** defined and illustrated in Table 3.1. Second, we must consider how atoms (viewed as

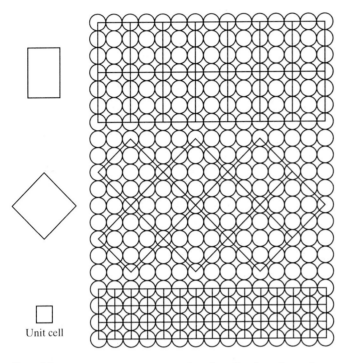

Figure 3-1 *Various structural units that describe the schematic crystalline structure. The simplest structural unit is the unit cell.*

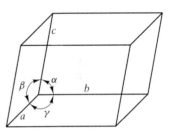

Figure 3-2 *Geometry of a general unit cell.*

hard spheres) can be stacked together within a given unit cell. To do this in a general way, we begin by considering **lattice points,** theoretical points arranged periodically in three-dimensional space, rather than actual atoms or spheres. Again, there is a limited number of possibilities, referred to as the 14 **Bravais* lattices,** defined in Table 3.2. Periodic stacking of unit cells from Table 3.2 generates **point lattices,** arrays of points with identical surroundings in three-dimensional space. These lattices are skeletons upon which crystal structures are built by placing atoms or groups of atoms on or near the lattice points. Figure 3–3 shows the simplest possibility with one atom centered on each lattice point. Some of the simple metal structures are of this type. However, a very large number of actual crystal structures is known to exist. Most of these result from having more than one atom associated with a given lattice point. We shall find many examples of these in the crystal structures of common ceramics and polymers (Sections 3.3 and 3.4).

---

*   Auguste Bravais (1811–1863), French crystallographer, was productive in an unusually broad range of areas, including botany, astronomy, and physics. However, it is his derivation of the 14 possible arrangements of points in space that is best remembered. This achievement provided the foundation for our current understanding of the atomic structure of crystals.

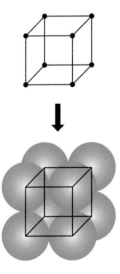

Figure 3-3 *The simple cubic lattice becomes the simple cubic crystal structure when an atom is placed on each lattice point.*

Table 3.1 *The Seven Crystal Systems*

System	Axial lengths and angles[a]	Unit cell geometry
Cubic	$a = b = c, \alpha = \beta = \gamma = 90°$	
Tetragonal	$a = b \neq c, \alpha = \beta = \gamma = 90°$	
Orthorhombic	$a \neq b \neq c, \alpha = \beta = \gamma = 90°$	
Rhombohedral	$a = b = c, \alpha = \beta = \gamma \neq 90°$	
Hexagonal	$a = b \neq c, \alpha = \beta = 90°, \gamma = 120°$	
Monoclinic	$a \neq b \neq c, \alpha = \gamma = 90° \neq \beta$	
Triclinic	$a \neq b \neq c, \alpha \neq \beta \neq \gamma \neq 90°$	

[a] The lattice parameters a, b, and c are unit cell edge lengths. The lattice parameters α, β, and γ are angles between adjacent unit cell axes where α is the angle viewed *along* the a axis (i.e., the angle *between* the b and c axes). The inequality sign ($\neq$) means that equality is not required. Accidental equality occasionally occurs in some structures.

Table 3.2 *The Fourteen Crystal (Bravais) Lattices*

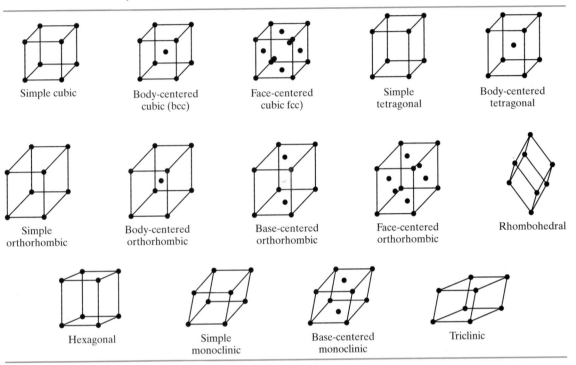

| Simple cubic | Body-centered cubic (bcc) | Face-centered cubic fcc) | Simple tetragonal | Body-centered tetragonal |

| Simple orthorhombic | Body-centered orthorhombic | Base-centered orthorhombic | Face-centered orthorhombic | Rhombohedral |

| Hexagonal | Simple monoclinic | Base-centered monoclinic | Triclinic |

SAMPLE PROBLEM 3.1

Sketch the five point lattices for two-dimensional crystal structures.

SOLUTION

Unit cell geometries are

   i. Simple square

  ii. Simple rectangle

 iii. Area-centered rectangle (or rhombus)

  iv. Parallelogram

   v. Area-centered hexagon

Note. It is a useful exercise to construct other possible geometries that must be equivalent to these five basic types. For example, an area-centered square can be resolved into a simple square lattice (inclined at 45°).

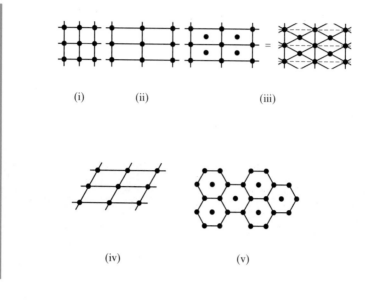

(i)   (ii)   (iii)

(iv)   (v)

PRACTICE PROBLEM 3.1

The note in Sample Problem 3.1 states that an area-centered square lattice can be resolved into a simple square lattice. Sketch this equivalence.

3.2   METAL STRUCTURES

With the structural ground rules behind us, we can now list the main crystal structures associated with important engineering materials. For our first group, the metals, this listing is fairly simple. As we see from an inspection of Appendix 1, most elemental metals at room temperature are found in one of three crystal structures.

Figure 3–4 shows the **body-centered cubic (bcc)** structure. This is the body-centered cubic Bravais lattice with one atom centered on each lattice point. There is one atom at the center of the unit cell and one-eighth atom at each of eight unit cell corners. (Each corner atom is shared by eight adjacent unit cells.) Thus there are two atoms in each bcc unit cell. The **atomic packing factor (APF)** for this structure is 0.68 and represents the fraction of the unit cell volume occupied by the two atoms. Typical metals with this structure include α-Fe (the form stable at room temperature), V, Cr, Mo, and W. An alloy in which one of these metals is the predominant constituent will tend to have this structure also. However, the presence of alloying elements diminishes crystalline perfection and will be discussed in Chapter 4.

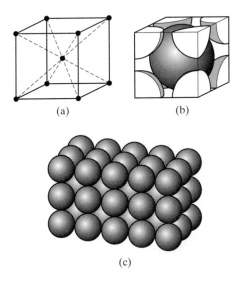

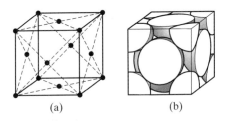

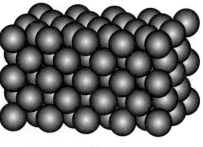

Structure: body-centered cubic (bcc)
Bravais lattice: bcc
Atoms/unit cell: $1 + 8 \times \frac{1}{8} = 2$
Typical metals: α-Fe, V, Cr, Mo, and W

Figure 3-4 *Body-centered cubic (bcc) structure for metals showing (a) the arrangement of lattice points for a unit cell; (b) the actual packing of atoms (represented as hard spheres) within the unit cell; and (c) the repeating bcc structure, equivalent to many adjacent unit cells (Part (c) courtesy of Molecular Simulations, Inc.).*

Structure: face-centered cubic (fcc)
Bravais lattice: fcc
Atoms/unit cell: $6 \times \frac{1}{2} + 8 \times \frac{1}{8} = 4$
Typical metals: γ-Fe, Al, Ni, Cu, Ag, Pt, and Au

Figure 3-5 *Face-centered cubic (fcc) structure for metals showing (a) the arrangement of lattice points for a unit cell; (b) the actual packing of atoms within the unit cell; and (c) the repeating fcc structure, equivalent to many adjacent unit cells (Part (c) courtesy of Molecular Simulations, Inc.).*

Figure 3–5 shows the **face-centered cubic (fcc)** structure, which is the fcc Bravais lattice with one atom per lattice point. There is one-half atom (i.e., one atom shared between two unit cells) in the center of each unit cell face and one-eighth atom at each unit cell corner, for a total of four atoms in each fcc unit cell. The atomic packing factor for this structure is 0.74, a value slightly higher than the 0.68 found for bcc metals. In fact, an APF of 0.74 is the highest value possible for filling space by stacking equal-sized hard spheres. For this reason, the fcc structure is sometimes referred to as **cubic close packed (ccp).** Typical metals with the fcc structure include γ-Fe (stable from 912 to 1394°C), Al, Ni, Cu, Ag, Pt, and Au.

The **hexagonal close packed (hcp)** structure (Figure 3–6) is our first encounter with a structure more complicated than its Bravais lattice (hexagonal). There are two atoms associated with each Bravais lattice point. There is one atom centered within the unit cell and various fractional atoms at the unit cell corners (four $\frac{1}{6}$ atoms and four $\frac{1}{12}$ atoms), for a total of two atoms

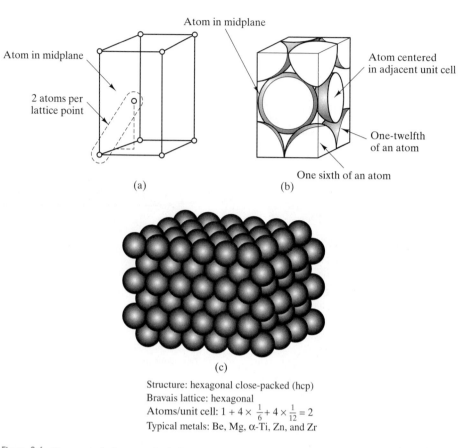

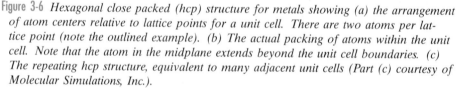

Structure: hexagonal close-packed (hcp)
Bravais lattice: hexagonal
Atoms/unit cell: $1 + 4 \times \frac{1}{6} + 4 \times \frac{1}{12} = 2$
Typical metals: Be, Mg, α-Ti, Zn, and Zr

Figure 3-6 *Hexagonal close packed (hcp) structure for metals showing (a) the arrangement of atom centers relative to lattice points for a unit cell. There are two atoms per lattice point (note the outlined example). (b) The actual packing of atoms within the unit cell. Note that the atom in the midplane extends beyond the unit cell boundaries. (c) The repeating hcp structure, equivalent to many adjacent unit cells (Part (c) courtesy of Molecular Simulations, Inc.).*

per unit cell. As the "close packed" name implies, this structure is as efficient in packing spheres as is the fcc structure. Both hcp and fcc structures have atomic packing factors of 0.74. This raises two questions: (1) In what other ways are the fcc and hcp structures alike? and (2) How do they differ? The answers to both questions can be found in Figure 3–7. The two structures are each regular stackings of close-packed planes. The difference lies in the sequence of packing of these layers. The fcc arrangement is such that the fourth close-packed layer lies precisely above the first one. In the hcp structure, the third close-packed layer lies precisely above the first. The fcc stacking is referred to as an ABCABC ... sequence, and the hcp stacking is referred to as an ABAB ... sequence. This subtle difference can lead to significant differences in material properties, as we have already discussed in Section 1.4. Typical metals with the hcp structure include Be, Mg, α-Ti, Zn, and Zr.

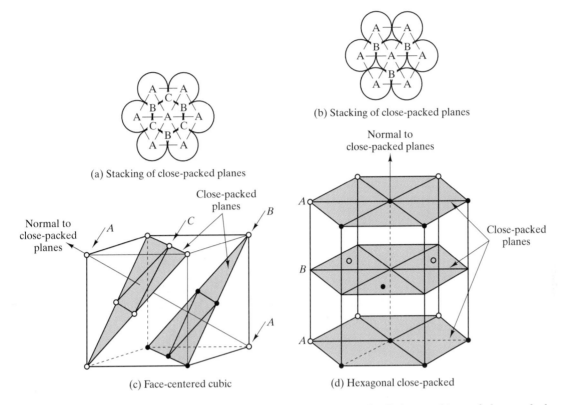

(b) Stacking of close-packed planes

(a) Stacking of close-packed planes

(c) Face-centered cubic

(d) Hexagonal close-packed

Figure 3-7 *Comparison of the fcc and the hcp structures. They are each efficient stackings of close-packed planes. The difference between the two structures is the different stacking sequences. (After B. D. Cullity,* Elements of X-Ray Diffraction, *2nd ed., Addison-Wesley Publishing Co., Inc., Reading, Mass., 1978.)*

Although the majority of elemental metals falls within one of the three structural groups just discussed, several display less common structures. We shall not dwell on these cases, which can be found from a careful inspection of Appendix 1.

In the course of analyzing the metallic structures introduced in this section, we shall frequently encounter the useful relationships between unit cell size and atomic radius given in Table 3.3. Our initial discovery of the utility of these relationships is found in the following Sample Problems and Practice Problems.

Table 3.3 *Relationship between unit cell size (edge length) and atomic radius for the common metallic structures*

Crystal structure	Relationship between edge length, a, and atomic radius, r
Body-centered cubic (bcc)	$a = 4r/\sqrt{3}$
Face-centered cubic (fcc)	$a = 4r/\sqrt{2}$
Hexagonal close packed (hcp)	$a = 2r$

SAMPLE PROBLEM 3.2

Using the data of Appendixes 1 and 2, calculate the density of copper.

SOLUTION

Appendix 1 shows this to be an fcc metal. The length, l, of a face diagonal in the unit cell (Figure 3–5) is

$$l = 4r_{\text{Cu atom}} = \sqrt{2}a$$

or

$$a = \frac{4}{\sqrt{2}}r_{\text{Cu atom}}$$

as given in Table 3.3. From the data of Appendix 2,

$$a = \frac{4}{\sqrt{2}}(0.128\,\text{nm}) = 0.362\,\text{nm}$$

The density of the unit cell (containing four atoms) is

$$\rho = \frac{4\,\text{atoms}}{(0.362\,\text{nm})^3} \times \frac{63.55\,\text{g}}{0.6023 \times 10^{24}\,\text{atoms}} \times \left(\frac{10^7\,\text{nm}}{\text{cm}}\right)^3$$

$$= 8.89\,\text{g/cm}^3$$

This can be compared with the tabulated value of $8.93\,\text{g/cm}^3$ in Appendix 1. The difference would be eliminated if a more precise value of $r_{\text{Cu atom}}$ were used, that is, with at least one more significant figure.

....

PRACTICE PROBLEM 3.2

In Sample Problem 3.2 the relationship between lattice parameter, a, and atomic radius, r, for an fcc metal is found to be $a = (4/\sqrt{2})r$, as given in Table 3.3. Derive the similar relationships in Table 3.3 for **(a)** a bcc metal and **(b)** an hcp metal.

PRACTICE PROBLEM 3.3

Calculate the density of α-Fe, which is a bcc metal. (*Caution:* A different relationship between lattice parameter, a, and atomic radius, r, applies to this different crystal structure. See Practice Problem 3.2 and Table 3.3.)

3.3   CERAMIC STRUCTURES

The wide variety of chemical compositions of ceramics is reflected in their crystalline structures. We cannot begin to give an exhaustive list of ceramic structures, but we can give a systematic list of some of the most important and representative ones. Even this list becomes rather long, so most structures will be described briefly. It is worth noting that many of these ceramic structures also describe intermetallic compounds. Also, we can define an **ionic packing factor (IPF)** for these ceramic structures, similar to our definition of the atomic packing factor (APF) for metallic structures. The IPF is the fraction of the unit cell volume occupied by the various cations and anions.

Let us begin with the ceramics with the simplest chemical formula, MX, where M is a metallic element and X is a nonmetallic element. Our first example is the **cesium chloride (CsCl) structure** shown in Figure 3–8. At first glance, we might want to call this a body-centered structure because of its similarity in appearance to Figure 3–4. In fact, the CsCl structure is built on the simple cubic Bravais lattice with two ions (one Cs^+ and one Cl^-) associated with each lattice point. There are two ions (one Cs^+ and one Cl^-) per unit cell.

Although CsCl is a useful example of a compound structure, it does not represent any commercially important ceramics. By contrast, the **sodium chloride (NaCl) structure** shown in Figure 3–9 is shared by many important ceramic materials. This can be viewed as the intertwining of two fcc structures, one of sodium ions and one of chlorine ions. Consistent with our treatment of the hcp and CsCl structures, the NaCl structure can be described as having an fcc Bravais lattice with two ions (one Na^+ and one Cl^-) associated

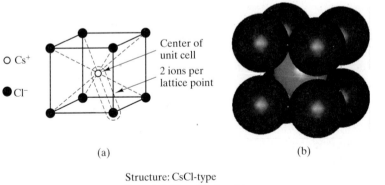

O Cs^+

Center of
unit cell
2 ions per
lattice point

● Cl^-

(a)

(b)

Structure: CsCl-type
Bravais lattice: simple cubic
Ions/unit cell: $1Cs^+ + 1Cl^-$

Figure 3-8 *Cesium chloride (CsCl) unit cell showing (a) ion positions and the 2 ions per lattice point, and (b) full-size ions. Note that the $Cs^+ - Cl^-$ pair associated with a given lattice point is not a molecule because the ionic bonding is nondirectional and a given Cs^+ is equally bonded to eight adjacent Cl^-, and vice versa. (Part (b) courtesy of Molecular Simulations, Inc.)*

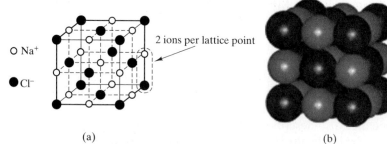

(a)

(b)

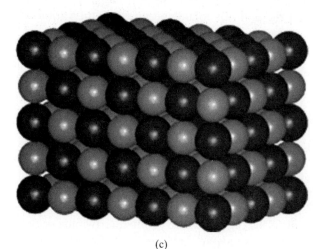

(c)

Structure: NaCl-type
Bravais lattice: fcc
Ions/unit cell: 4Na$^+$ + 4Cl$^-$
Typical ceramics: MgO, CaO, FeO, and NiO

Figure 3-9 *Sodium chloride (NaCl) structure showing (a) ion positions in a unit cell, (b) full-size ions, and (c) many adjacent unit cells. (Parts (b) and (c) courtesy of Molecular Simulations, Inc.)*

with each lattice point. There are eight ions (four Na$^+$ plus four Cl$^-$) per unit cell. Some of the important ceramic oxides with this structure are MgO, CaO, FeO, and NiO.

The chemical formula MX$_2$ includes a number of important ceramic structures. Figure 3–10 shows the **fluorite** (CaF$_2$) **structure.** This is built on an fcc Bravais lattice with three ions (one Ca^{2+} and two F$^-$) associated with each lattice point. There are 12 ions (four Ca^{2+} and eight F$^-$) per unit cell. Typical ceramics with this structure are UO$_2$, ThO$_2$, and TeO$_2$. There is an unoccupied volume near the center of the fluorite unit cell that plays an important role in nuclear materials technology. Uranium dioxide (UO$_2$) is a reactor fuel that can accommodate fission products such as helium gas without troublesome "swelling." The helium atoms are accommodated in the open regions of the fluorite unit cells.

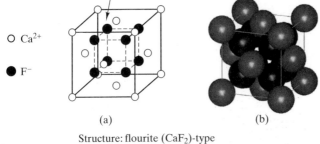

F⁻ ions located at corners
of a cube(at one-quarter
of the distance along the
body diagonal)

O Ca²⁺

● F⁻

(a)   (b)

Structure: flourite (CaF₂)-type
Bravais lattice: fcc
Ions/unit cell: 4Ca²⁺ + 8F⁻
Typical ceramics: UO₂, ThO₂, and TeO₂

Figure 3-10   *Fluorite (CaF₂) unit cell show-ing (a) ion positions and (b) full-size ions. (Part (b) courtesy of Molecular Simulations, Inc.)*

Included in the MX₂ category is perhaps the most important ceramic compound, **silica** (SiO₂), which is widely available in raw materials in the earth's crust. Silica, alone and in chemical combination with other ceramic oxides (forming silicates), represents a large fraction of the ceramic materi-als available to engineers. For this reason, the structure of SiO₂ is important. Unfortunately, this structure is not simple. In fact, there is not a single struc-ture to describe, but many (under different conditions of temperature and pressure). For a representative example, Figure 3–11 shows the **cristobalite**

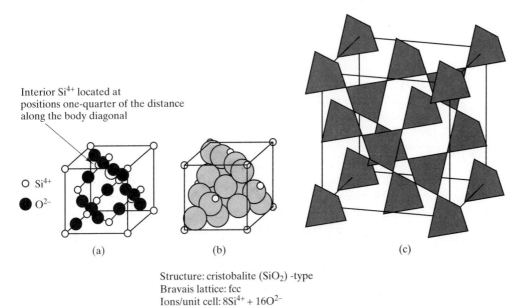

Interior Si⁴⁺ located at
positions one-quarter of the distance
along the body diagonal

O Si⁴⁺

● O²⁻

(a)   (b)   (c)

Structure: cristobalite (SiO₂) -type
Bravais lattice: fcc
Ions/unit cell: 8Si⁴⁺ + 16O²⁻

Figure 3-11 *The cristobalite (SiO₂) unit cell showing (a) ion positions, (b) full-size ions, and (c) the connectivity of SiO₄⁴⁻ tetrahedra. In the schematic, each tetrahedron has a Si⁴⁺ at its cen-ter. In addition, an O²⁻ would be at each corner of each tetrahedron and is shared with an adjacent tetrahedron. (Part (c) courtesy of Molecular Simulations, Inc.)*

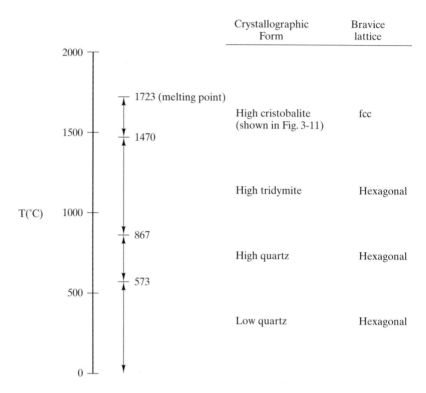

Figure 3-12 *Many crystallographic forms of SiO$_2$ are stable as they are heated from room temperature to the melting temperature. Each form represents a different way to connect adjacent SiO$_4^{4-}$ tetrahedra.*

(SiO$_2$) **structure.** Cristobalite is built on an fcc Bravais lattice with six ions (two Si^{4+} and four O^{2-}) associated with each lattice point. There are 24 ions (eight Si^{4+} plus 16 O^{2-}) per unit cell. In spite of the large unit cell needed to describe this structure, it is perhaps the simplest of the various crystallographic forms of SiO$_2$. The general feature of all SiO$_2$ structures is the same—a continuously connected network of SiO$_4^{4-}$ tetrahedra (see Section 2.3). The sharing of O^{2-} ions by adjacent tetrahedra gives the overall chemical formula SiO$_2$.

We have already noticed (in Section 3.2) that iron, Fe, had different crystal structures stable in different temperature ranges. The same is true for silica, SiO$_2$. Although the basic SiO$_4^{4-}$ tetrahedra are present in all SiO$_2$ crystal structures, the arrangement of connected tetrahedra changes. The equilibrium structures of SiO$_2$ from room temperature to its melting point are summarized in Figure 3–12. Caution must always be exercised in using materials with transformations of these types. Even the relatively subtle "low" to "high" quartz transformation can cause catastrophic structural damage when a silica ceramic is heated or cooled through the vicinity of 573°C.

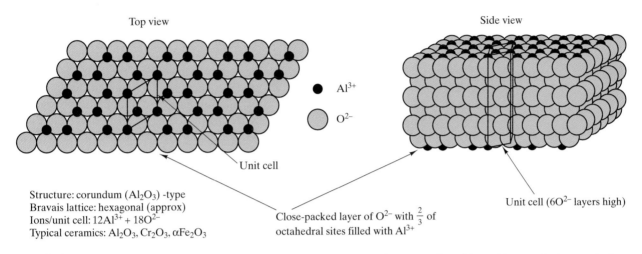

Top view

Side view

● Al^{3+}

○ O^{2-}

Unit cell

Structure: corundum (Al_2O_3) -type
Bravais lattice: hexagonal (approx)
Ions/unit cell: $12Al^{3+} + 18O^{2-}$
Typical ceramics: Al_2O_3, Cr_2O_3, αFe_2O_3

Close-packed layer of O^{2-} with $\frac{2}{3}$ of
octahedral sites filled with Al^{3+}

Unit cell ($6O^{2-}$ layers high)

Figure 3-13 *The corundum (Al_2O_3) unit cell is shown superimposed on the repeated stacking of layers of close-packed O^{2-} ions. The Al^{3+} ions fill two-thirds of the small (octahedral) interstices between adjacent layers.*

The chemical formula M_2X_3 includes the important **corundum** (Al_2O_3) **structure** shown in Figure 3–13. This is in a rhombohedral Bravais lattice, but closely approximates a hexagonal lattice. There are 30 ions per lattice site (and per unit cell). The Al_2O_3 formula requires that these 30 ions be divided as 12 Al^{3+} and 18 O^{2-}. One can visualize this seemingly complicated structure as being similar to the hcp structure of Section 3.2. The Al_2O_3 structure closely approximates close-packed O^{2-} sheets with two-thirds of the small interstices between sheets filled with Al^{3+}. Both Cr_2O_3 and α-Fe_2O_3 have the corundum structure.

Moving to ceramics with three atomic species, we find that the $M'M''X_3$ formula includes an important family of electronic ceramics with the **perovskite** ($CaTiO_3$) **structure** shown in Figure 3–14. At first glance, the

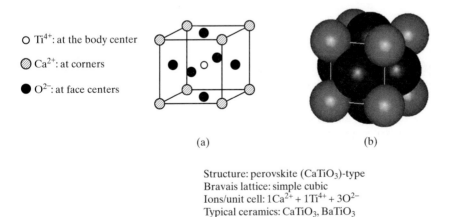

○ Ti^{4+}: at the body center

◍ Ca^{2+}: at corners

● O^{2-}: at face centers

(a)

(b)

Structure: perovskite ($CaTiO_3$)-type
Bravais lattice: simple cubic
Ions/unit cell: $1Ca^{2+} + 1Ti^{4+} + 3O^{2-}$
Typical ceramics: $CaTiO_3$, $BaTiO_3$

Figure 3-14 *Perovskite ($CaTiO_3$) unit cell showing (a) ion positions and (b) full-size ions. (Part (b) courtesy of Molecular Simulations, Inc.)*

perovskite structure appears to be a combination of simple cubic, bcc, and fcc structures. But closer inspection indicates that different atoms occupy the corner (Ca^{2+}), body-centered (Ti^{4+}), and face-centered (O^{2-}) positions. As a result, this structure is another example of a simple cubic Bravais lattice. There are five ions (one Ca^{2+}, one Ti^{4+}, and three O^{2-}) per lattice point and per unit cell. In Section 15.4 we shall find that perovskite materials such as $BaTiO_3$ have important ferroelectric and piezoelectric properties (related to the relative positions of cations and anions as a function of temperature). In Section 15.3 we shall see that the high-temperature superconductors resulted from fundamental research on variations in the structure of perovskite-type ceramics.

The $M'M_2''X_4$ formula includes an important family of magnetic ceramics based on the **spinel** ($MgAl_2O_4$) **structure** shown in Figure 3–15. This is built on an fcc Bravais lattice with 14 ions (two Mg^{2+}, four Al^{3+}, and eight O^{2-}) associated with each lattice point. There are 56 ions in the unit cell (8 Mg^{2+}, 16 Al^{3+}, and 32 O^{2-}). Typical materials sharing this structure include $NiAl_2O_4$, $ZnAl_2O_4$, and $ZnFe_2O_4$. It is noted in Figure 3–15 that the Mg^{2+} ions are in *tetrahedral* positions; that is, they are coordinated by *four* oxygens (O^{2-}), and the Al^{3+} ions are in *octahedral* positions. (Recall the discussion of coordination numbers in Section 2.2.) The Al^{3+} ions are coordinated by *six* oxygens. The "octa" prefix, of course, refers to eight, not six. This reference is to the eight-sided figure created by the six oxygens. Commercially important ceramic magnets (see Section 18.5) are actually based on a slightly modified version of the spinel structure, the *inverse spinel structure*, in which the octahedral sites are occupied by the M^{2+} and one-half of the M^{3+} ions. The remaining M^{3+} ions occupy the tetrahedral sites. These materials can be described with the formula $M''(M'M'')X_4$, where M' has a 2+ valence and M'' has a 3+ valence. Examples include $FeMgFeO_4$, $FeFe_2O_4$ ($= Fe_3O_4$

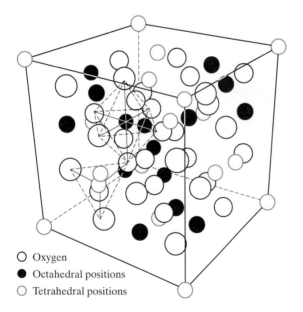

Figure 3-15   *Ion positions in the spinel ($MgAl_2O_4$) unit cell. The circles in color represent Mg^{2+} ions (in tetrahedral or four-coordinated positions), and the black circles represent Al^{3+} ions (in octahedral or six-coordinated positions). (From F. G. Brockman,* Bull. Am. Ceram. Soc. 47, *186 (1967).)*

○ Oxygen
● Octahedral positions
○ Tetrahedral positions

or magnetite), $FeNiFeO_4$, and many other commercially important *ferrites,* or ferrimagnetic ceramics.

In discussing the complexity of the SiO_2 structures, we mentioned the importance of the many silicate materials resulting from the chemical reaction of SiO_2 with other ceramic oxides. The general nature of silicate structures is that the additional oxides tend to break up the continuity of the SiO_4^{4-} tetrahedra connections. The remaining connectedness of tetrahedra may be in the form of silicate chains or sheets. One relatively simple example is illustrated in Figure 3–16, which shows the **kaolinite structure**. Kaolinite $[2(OH)_4Al_2Si_2O_5]$ is a hydrated aluminosilicate and a good example of a clay mineral. The structure is typical of sheet silicates. It is built on the

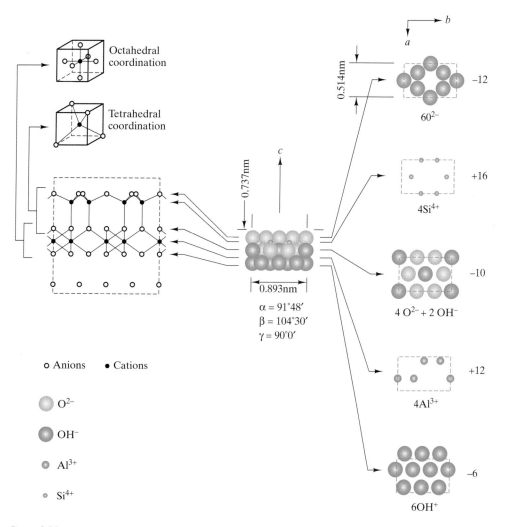

Figure 3-16 *Exploded view of the kaolinite unit cell, $2(OH)_4Al_2Si_2O_5$. (After F. H. Norton,* Elements of Ceramics, *2nd ed., Addison-Wesley Publishing Co., Inc., Reading, Mass., 1974.)*

triclinic Bravais lattice with two kaolinite "molecules" per unit cell. On a microscopic scale, we observe many clay minerals to have a platelike or flaky structure (see Figure 3–17), a direct manifestation of crystal structures such as Figure 3–16.

In this section we have surveyed crystal structures for various ceramic compounds. The structures have generally been increasingly complex as we considered increasingly complex chemistry. The contrast between CsCl (Figure 3–8) and kaolinite (Figure 3–16) is striking.

Before leaving ceramics, it is appropriate to look at some important materials that are exceptions to our general description of ceramics as compounds. First, Figure 3–18 shows the layered crystal structure of graphite, the stable room-temperature form of carbon. Although monatomic, graphite is much more ceramiclike than metallic. The hexagonal rings of carbon atoms are strongly bonded by covalent bonds. The bonds between layers are, however, of the van der Waals type (Section 2.5), accounting for graphite's friable nature and application as a useful "dry" lubricant. It is interesting to contrast the graphite structure with the high-pressure stabilized form, diamond cubic, which plays such an important role in solid-state technology because semiconductor silicon has this structure (see Figure 3–23).

Even more intriguing is a comparison of both graphite and diamond structures with an alternate form of carbon that was recently discovered as a byproduct of research in astrochemistry. Figure 3–19a illustrates the structure of a C_{60} molecule. This unique structure was discovered during experiments on the laser vaporization of carbon in a carrier gas such as helium. The experiments had intended to simulate the synthesis of carbon chains in carbon stars. The result, however, was a molecular-scale version of the geodesic dome, leading to this material's being named

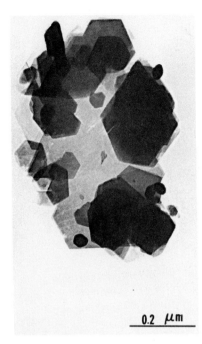

Figure 3-17 *Transmission electron micrograph (see Section 4.7) of the structure of clay platelets. This microscopic-scale structure is a manifestation of the layered crystal structure shown in Figure 3–16. (Courtesy of I. A. Aksay)*

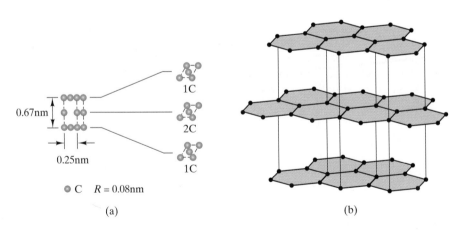

(a)   (b)

Figure 3-18 *(a) An exploded view of the graphite (C) unit cell. (From F. H. Norton,* Elements of Ceramics, *2nd ed., Addison-Wesley Publishing Co., Inc., Reading, Mass., 1974.) (b) A schematic of the nature of graphite's layered structure. (From W. D. Kingery, H. K. Bowen, and D. R. Uhlmann,* Introduction to Ceramics, *2nd ed., John Wiley & Sons, Inc., New York, 1976.)*

buckminsterfullerene,[*] or **fullerene,** in honor of the inventor of that architectural structure. A close inspection of the structure of Figure 3–19a indicates that the nearly spherical molecule is, in fact, a polyhedron composed of 5- and 6-sided faces.

The uniform distribution of 12 pentagons among 20 hexagons is precisely the form of a soccer ball, leading to the nickname for the structure as a *buckyball*. It is the presence of 5-membered rings that gives the positive curvature to the surface of the buckyball, in contrast to the flat, sheetlike structure of 6-membered rings in graphite (Figure 3–18b). Subsequent research has led to the synthesis of a wide variety of structures for a wide range of fullerenes. Buckyballs have been synthesized with the formula C_n, where n can take on various large, even values, such as 240 and 540. In each case, the structure consists of 12 uniformly distributed pentagons connecting an array of hexagons. Although pentagons are necessary to give the approximately spherical curvature of the buckyballs, extensive research on these unique materials led to the realization that cylindrical curvature can result from simply rolling the hexagonal graphite sheets. A resulting *buckytube* is shown in Figure 3–19b.

Although in a preliminary research phase, these materials have stimulated enormous interest in the fields of chemistry and physics, as well as materials science and engineering. Their molecular structure is clearly interesting, but in addition, they have unique chemical and physical properties—for example, individual C_n buckyballs are unique, passive surfaces on an nm-scale. Similarly, buckytubes hold the theoretical promise of being the highest-strength reinforcing fibers available for the advanced composites discussed in Chapter 14. Finally, we must acknowledge that the discrete molecular structures of Figure 3–19 are interesting but are not associated with long-range crystallographic structure. Such repetitive structures are, in fact, being observed, again with intriguing properties. For example, by capturing metal ions within C_n cages, fcc stackings of such buckyballs have become the latest family of superconductors, joining the metallic superconductors discussed in Sections 15.3 and 18.4 and the ceramic-oxide superconductors discussed in Sections 15.3 and 18.5. The initial research on buckyballs and buckytubes clearly indicates an intriguing set of atomic-scale structures that could lead to potentially important applications in materials technology.

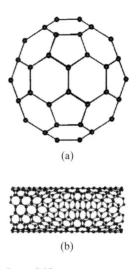

(a)

(b)

Figure 3-19 *(a) C_{60} molecule, or buckyball. (b) Cylindrical array of hexagonal rings of carbon atoms, or buckytube. (Courtesy of Molecular Simulations, Inc.)*

SAMPLE PROBLEM 3.3

Calculate the ionic packing factor (IPF) of MgO, which has the NaCl structure (Figure 3–9).

---

[*] Richard Buckminster Fuller (1895–1983), American architect and inventor, was one of the most colorful and famous personalities of the twentieth century. His creative discussions of a wide range of topics from the arts to the sciences (including frequent references to future trends) helped to establish his fame. In fact, his charismatic personality became as celebrated as his unique inventions of various architectural forms and engineering designs.

SOLUTION

Taking $a = 2r_{Mg^{2+}} + 2r_{O^{2-}}$ and the data of Appendix 2, we have

$$a = 2(0.078\,nm) + 2(0.132\,nm) = 0.420\,nm$$

Then

$$V_{unitcell} = a^3 = (0.420\,nm)^3 = 0.0741\,nm^3$$

There are four Mg^{2+} ions and four O^{2-} ions per unit cell, giving a total ionic volume of

$$4 \times \frac{4}{3}\pi r_{Mg^{2+}}^3 + 4 \times \frac{4}{3}\pi r_{O^{2-}}^3$$

$$= \frac{16\pi}{3}[(0.078\,nm)^3 + (0.132\,nm)^3]$$

$$= 0.0465\,nm^3$$

The ionic packing factor is then

$$IPF = \frac{0.0465\,nm^3}{0.0741\,nm^3} = 0.627$$

SAMPLE PROBLEM 3.4

Using data from Appendixes 1 and 2, calculate the density of MgO.

SOLUTION

From Sample Problem 3.3, $a = 0.420$ nm. This gave a unit cell volume of $0.0741\,nm^3$. The density of the unit cell is

$$\rho = \frac{[4(24.31\,g) + 4(16.00\,g)]/(0.6023 \times 10^{24})}{0.0741\,nm^3} \times \left(\frac{10^7\,nm}{cm}\right)^3$$

$$= 3.61\,g/cm^3$$

....

PRACTICE PROBLEM 3.4

Calculate the ionic packing factor of **(a)** CaO, **(b)** FeO, and **(c)** NiO. All of these compounds share the NaCl-type structure. **(d)** Is there a unique IPF value for the NaCl-type structure? Explain. (See Sample Problem 3.3.)

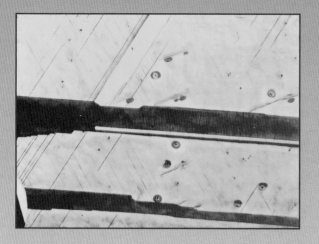

◀ Color Plate 1

The 1998 Jacquet-Lucas Award of the International Metallographic Society went to an intriguing study of cerium, unique among the elements in that it undergoes a polymorphic transformation directly from one face-centered cubic (fcc) structure to another fcc structure with a different density. Shown here is a micrograph of annealing twin bands in the starting material, the gamma phase stable at room temperature. *(Courtesy of Ramiro Pereyra, Los Alamos National Laboratory)*

Color Plate 2 ▶

In this micrograph, the same area of cerium as in Color Plate 1 is shown after one cycle of pressurization from one atmosphere to 8 kilobars and back to one atmosphere. The original twin bands have been deformed due to the complete transformation to the fcc alpha phase (at high pressure) and back to gamma (upon depressurization). *(Courtesy of Ramiro Pereyra, Los Alamos National Laboratory)*

◀ Color Plate 3

In this micrograph, cerium exhibits a more conventional phase transformation. Upon slow cooling from 440°C down to -196°C, about two-thirds of the fcc gamma phase is transformed to an hexagonal beta phase. *(Courtesy of Ramiro Pereyra, Los Alamos National Laboratory)*

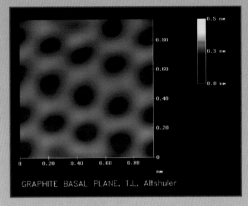

GRAPHITE BASAL PLANE, T.L. Altshuler

◀ Color Plate 4

The atomic-scale resolution of the scanning tunneling microscope permits the direct imaging of the carbon atom rings which result from highly directional covalent bonding in graphite. *(Courtesy of T.L. Altshuler, Advanced Materials Laboratory, Concord, Massachusetts)*

Color Plate 5 ▶

The C_{60} molecule (or "Buckyball") represents a breakthrough in carbon chemistry. The presence of 5-membered carbon rings leads to the spherical geometry of the molecule, in contrast to the planar 6-membered rings of graphite in Color Plate 4. *(Courtesy of Molecular Simulations, Inc.)*

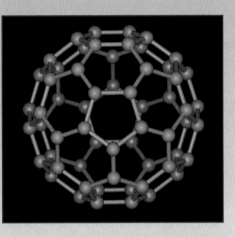

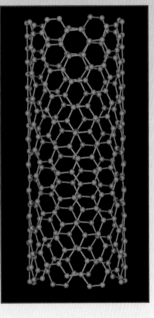

◀ Color Plate 6

Along with Color Plate 5, the "Buckytube" is another recent breakthrough in carbon chemistry. As with the graphite in color Plate 4, only 6-membered rings are involved. In this case, however, they are rolled into a cylindrical tube rather than lying in a flat plane. *(Courtesy of Molecular Simulations, Inc.)*

Color Plate 7 ▶

The linear molecular structure of polyethylene, C_2H_4, is seen in relation to a unit cell, the crystalline material's basic building block. The carbon atoms are green and the hydrogen atoms are gray. *(Courtesy of Molecular Simulations, Inc.)*

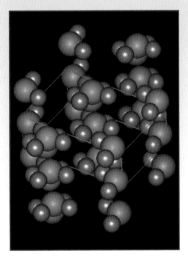

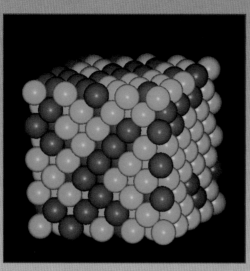

Color Plate 9 ▶

The modern metal casting operation includes consideration of minimizing hazardous air pollutants. *(Courtesy of the Casting Emission Reduction Program [CERP])*

◀ **Color Plate 10**

This heat exchanger for the chemical processing industry contains over 17 kilometers of zirconium tubing. Zirconium is the material selected for many such applications due to its strong corrosion resistance in a variety of acidic and basic chemical environments. *(Courtesy of Teledyne Wah Chang, Albany, Oregon)*

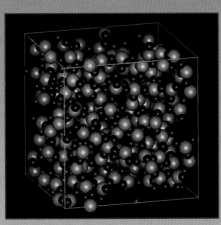

◄ Color Plate 11

The random network structure of vitreous SiO_2 is based on a linkage of SiO_4^{4-} tetrahedra. The Si^{4+} ions are yellow, and the O^{2-} ions are red. The relatively large size of the Si^{4+} in this model is due to the consideration of a significant covalent bonding character. *(Courtesy of Molecular Simulations, Inc.)*

Color Plate 12 ▶

The random linkage of SiO_4^{4-} tetrahedra in Plate 11 is emphasized in this schematic representation. *(Courtesy of Molecular Simulations, Inc.)*

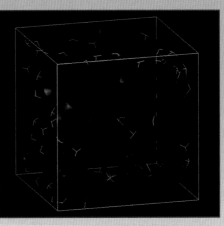

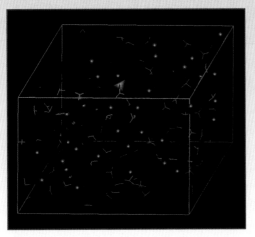

◄ Color Plate 13

In this schematic representation of the tetrahedral linkage in Na_2O-SiO_2 glass, the role of Na^+ as a "network modifier" is emphasized. The Na^+ ions break up the linkage of the SiO_4^{4-} tetrahedra in Color Plate 12. *(Courtesy of Molecular Simulations, Inc.)*

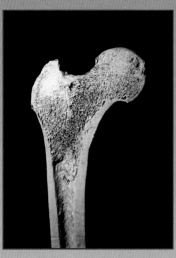

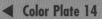

Color Plate 14

This cross section of a human femur represents a ceramic material "engineered" by the body to play a key role in our skeletal structure. Hydroxyapatite (HA), with the chemical formula $Ca_{10}(HPO_4)_6(OH)_2$ is the primary mineral content of bone, comprising 43% of overall bone weight. *(Courtesy of R.B. Martin, Orthopaedic Research Laboratories, University of California, Davis Medical Center, Sacramento, CA)*

Color Plate 15 ▶

The small white strips in the tray are examples of an engineered composite designed to fill large bone defects, such as fractures. In this system, millimeter-scale ceramic particles (including hydroxyapatite) are embedded in a matrix of collagen (a natural polymeric protein). The composite's performance is enhanced by adding bone marrow from the patient to the commercial material, thereby imitating a third component (viscous fluids) of natural bone. *(Courtesy of Zimmer Corporation)*

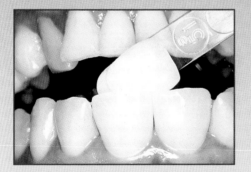

◀ Color Plate 16

Glass-ceramic crowns can provide excellent color match, along with essential mechanical performance for this dental application. *(Courtesy of David Grossman)*

Color Plate 17 ▶

Biomimetic processing, introduced in Chapter 12, was used to produce a thin coating of octacalcium phosphate (a precursor to hydroxyapatite formation) on this porous titanium implant. *(After B.C. Bunker, et al., Science, 264 48 [1994])*

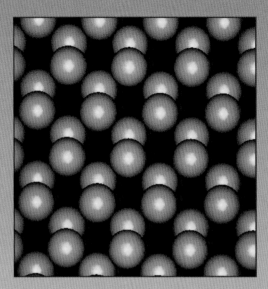

◀ Color Plate 18
The regular arrangement of silicon atoms in the diamond cubic crystal structure is seen in this view looking along the [111] crystal direction. *(Courtesy of Molecular Simulations, Inc.)*

Color Plate 19 ▶
A wafer of single crystal silicon contains a large number of microcircuit "chips." *(Courtesy of R.D. Pashley, Intel Corporation)*

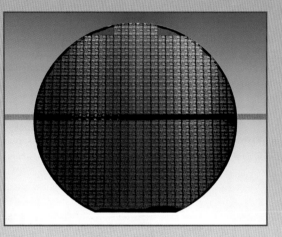

◀ Color Plate 20
A molded engineering polymer serves as a lightweight and cost-effective air-intake manifold for automotive applications. *(Courtesy of Solvay Automotive, Inc., Troy, Michigan)*

Color Plate 21 ▶
A summary of the extensive use of advanced composite materials in a commercial jet aircraft.
(Courtesy of the Boeing Airplane Company)

PRACTICE PROBLEM 3.5

Calculate the density of CaO. (See Sample Problem 3.4.)

3.4   POLYMERIC STRUCTURES

In Chapters 1 and 2 we defined the polymers category of materials by the chainlike structure of long polymeric molecules (e.g., Figure 2–15). Compared to the stacking of individual atoms and ions in metals and ceramics, the arrangement of these long molecules into a regular and repeating pattern is difficult. As a result, most commercial plastics are to a large degree noncrystalline. In those regions of the microstructure that are crystalline, the structure tends to be quite complex. The complexity of the unit cells of common polymers is generally beyond the scope of this text, but two relatively simple examples will be shown.

Polyethylene, $+C_2H_4+_n$, is chemically quite simple. However, the relatively elaborate way in which the long-chain molecule folds back and forth on itself is illustrated in Figures 3–20 and 3–21. Figure 3–20 is an orthorhombic unit cell, a common crystal system for polymeric crystals. For metals and ceramics, knowledge of the unit cell structure implies knowledge of the crystal structure over a large volume. For polymers, we must be more cautious. Single crystals of polyethylene are difficult to grow. When produced (by cooling a dilute solution), they tend to be thin platelets, about

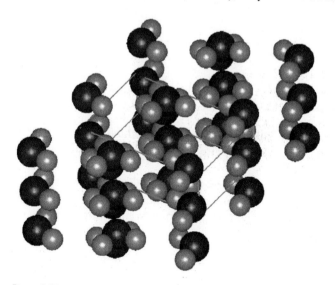

Figure 3-20 *Arrangement of polymeric chains in the unit cell of polyethylene. The dark spheres are carbon atoms, and the light spheres are hydrogen atoms. The unit cell dimensions are 0.255 nm × 0.494 nm × 0.741 nm. (Courtesy of Molecular Simulations, Inc.)*

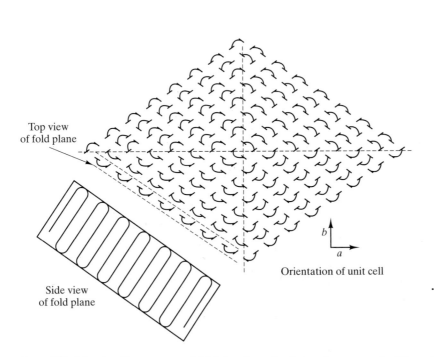

Figure 3-21 *Weaving-like pattern of folded polymeric chains that occurs in thin crystal platelets of polyethylene. (From D. J. Williams, Polymer Science and Engineering, Prentice Hall, Inc., Englewood Cliffs, N.J., 1971.)*

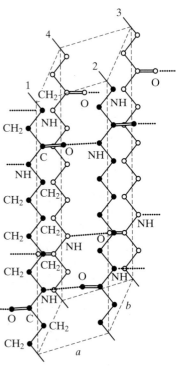

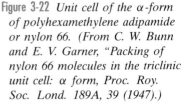

Figure 3-22 *Unit cell of the α-form of polyhexamethylene adipamide or nylon 66. (From C. W. Bunn and E. V. Garner, "Packing of nylon 66 molecules in the triclinic unit cell: α form, Proc. Roy. Soc. Lond. 189A, 39 (1947).)*

10 nm thick. Since polymer chains are generally several hundred nanometers long, the chains must be folded back and forth in a sort of atomic-scale weaving (as illustrated in Figure 3–21).

Figure 3–22 shows the triclinic unit cell for polyhexamethylene adipamide, or nylon 66. The crystal structure of other polyamides and some polymethanes is similar to this. Up to approximately 50% of the volume of these materials would be of this crystalline form, with the balance noncrystalline.

SAMPLE PROBLEM 3.5

Calculate the number of C and H atoms in the polyethylene unit cell (Figure 3–20), given a density of 0.9979 g/cm^3.

SOLUTION

Figure 3–20 gives unit cell dimensions that allow calculation of volume:

$$V = (0.741 \, \text{nm})(0.494 \, \text{nm})(0.255 \, \text{nm}) = 0.0933 \, \text{nm}^3$$

There will be some multiple (n) of C_2H_4 units in the unit cell with atomic mass:

$$m = \frac{n[2(12.01) + 4(1.008)] \, \text{g}}{0.6023 \times 10^{24}} = (4.66 \times 10^{-23}n) \, \text{g}$$

Therefore, the unit cell density is

$$\rho = \frac{(4.66 \times 10^{-23}n) \, \text{g}}{0.0933 \, \text{nm}^3} \times \left(\frac{10^7 \, \text{nm}}{\text{cm}}\right)^3 = 0.9979 \frac{\text{g}}{\text{cm}^3}$$

Solving for n gives

$$n = 2.00$$

As a result, there are

$$4 \, (= 2n) \, \text{C atoms} + 8 \, (= 4n) \, \text{H atoms per unit cell}$$

....

PRACTICE PROBLEM 3.6

How many unit cells are contained in 1 kg of commercial polyethylene that is 50 vol % crystalline (balance amorphous) and has an overall product density of 0.940 Mg/m^3? (See Sample Problem 3.5.)

3.5 SEMICONDUCTOR STRUCTURES

The technology developed by the semiconductor industry for growing single crystals has led to crystals of phenomenally high degrees of perfection. All crystal structures shown in this chapter imply structural perfection. However, all structures are subject to various imperfections, which will be discussed in Chapter 4. The "perfect" structures described in this section are approached in real materials more closely than in any other category.

A single structure dominates the semiconductor industry. The elemental semiconductors (Si, Ge, and gray Sn) share the **diamond cubic structure** shown in Figure 3–23. This is built on an fcc Bravais lattice with two atoms associated with each lattice point and eight atoms per unit cell. A key feature of this structure is that it accommodates the tetrahedral bonding configuration of these group IVA elements.

The Material World:
Growing a (Nearly) Perfect Crystal

The technological and cultural revolution created by the many products based on the modern integrated circuit begins with single crystals of exceptionally high chemical purity and structural perfection. More than any other commercially produced materials, these crystals represent the ideal of Chapter 3. The vast majority of integrated circuits are produced on thin slices (wafers) of silicon single crystals (see photo which shows a state-of-the-art 300 mm [12 in.] diameter wafer). The systematic procedures for producing fine-scale electrical circuits on these wafers are discussed in some detail in Chapter 17.

As shown in the drawing, a large crystal of silicon is produced by pulling a small "seed" crystal up from a crucible containing molten silicon. The high melting point of silicon ($T_m = 1414°C$) calls for a crucible made of high-purity SiO_2 glass. Heat is provided by radio-frequency (RF) inductive heating coils. The seed crystal is inserted into the melt and

Courtesy of SEAMATECH.

slowly withdrawn. Crystal growth occurs as the liquid silicon freezes next to the seed crystal, with individual atoms stacking against those in the seed. Successive layers of atomic planes are added at the liquid-solid interface. The overall growth rate is approximately 10 μm/s. The resulting, large crystals are sometimes called "ingots" or "boules". This overall process is generally called the Czochralski or Teal-Little technique. As noted in Chapter 17, the economics of a circuit fabrication on the single crystal wafers drives crystal growers to produce as large a diameter crystal as possible. The industry standard for crystal- and corresponding wafer-diameters has increased over the years to 200 mm and may soon approach 300 mm.

The basic science and technology involved in growing single crystals as illustrated here have been joined by considerable "art" in adjusting the various specifics of the crystal-growing apparatus and procedures. The highly specialized process of growing these crystals is largely the focus of companies separate from those that fabricate circuits. The structural perfection provided by the Czochralski technique is joined with the ability to chemically purify the resulting crystal by the zone refining technique, as discussed in Chapter 17.

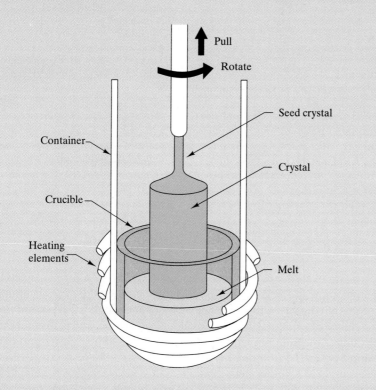

Schematic of growth of single crystals using the Czochralski technique. (After J. W. Mayer and S. S. Lau, *Electronic Materials Science: For Integrated Circuits in Si and GaAs*, Macmillan Publishing Company, New York, 1990.)

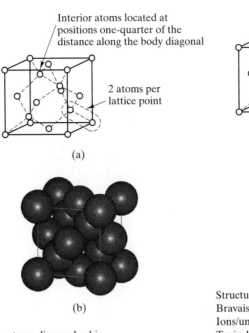

Interior atoms located at positions one-quarter of the distance along the body diagonal

2 atoms per lattice point

(a)

○ Zn^{2+} ● S^{2-}

Two ions per lattice point

(a)

(b)

(b)

Structure: diamond cubic
Bravais lattice: fcc
Atoms/unit cell: $4 + 6 \times \frac{1}{2} + 8 \times \frac{1}{8} = 8$
Typical semiconductors: Si, Ge, and gray Sn

Figure 3-23 *Diamond cubic unit cell showing (a) atom positions. There are two atoms per lattice point (note the outlined example). Each atom is tetrahedrally coordinated. (b) The actual packing of full-size atoms associated with the unit cell. (Part (b) courtesy of Molecular Simulations, Inc.)*

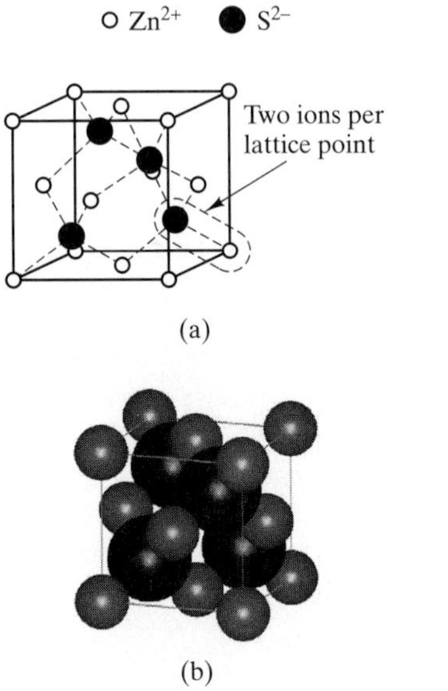

Structure: Zinc blende (ZnS)-type
Bravais lattice: fcc
Ions/unit cell: $4Zn^{2+} + 4S^{2-}$
Typical semiconductors:
  GaAs, AlP, InSb (III-V compounds),
  ZnS, ZnSe, CdS, HgTe (II-VI compounds)

Figure 3-24 *Zinc blende (ZnS) unit cell showing (a) ion positions. There are two ions per lattice point (note the outlined example). Compare this with the diamond cubic structure (Figure 3–23a). (b) The actual packing of full-size ions associated with the unit cell. (Part (b) courtesy of Molecular Simulations, Inc.)*

A small cluster of elements adjacent to group IVA forms semiconducting compounds. These tend to be MX-type compounds with combinations of atoms having an average valence of 4+. For example, GaAs combines the 3+ valence of gallium with the 5+ valence of arsenic, and CdS combines the 2+ valence of cadmium with the 6+ valence of sulfur. GaAs and CdS are examples of a *III-V compound* and a *II-VI compound*, respectively. Many of these simple MX compounds crystallize in a structure closely related to the diamond cubic. Figure 3–24 shows the **zinc blende** (ZnS) **structure,** which is essentially the diamond cubic structure with Zn^{2+} and S^{2-} ions alternating in the atom positions. This is again the fcc Bravais lattice but with two oppositely charged ions associated with each lattice site rather than two like atoms. There are eight ions (four Zn^{2+} and four S^{2-}) per unit cell. This

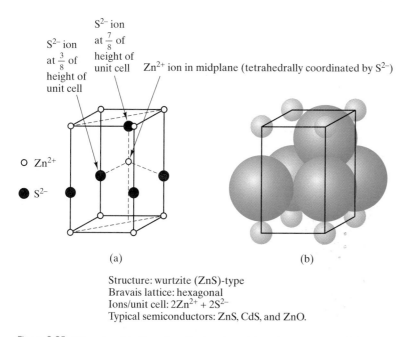

S^{2-} ion at $\frac{3}{8}$ of height of unit cell

S^{2-} ion at $\frac{7}{8}$ of height of unit cell

Zn^{2+} ion in midplane (tetrahedrally coordinated by S^{2-})

○ Zn^{2+}

● S^{2-}

(a)   (b)

Structure: wurtzite (ZnS)-type
Bravais lattice: hexagonal
Ions/unit cell: $2Zn^{2+} + 2S^{2-}$
Typical semiconductors: ZnS, CdS, and ZnO.

Figure 3-25 *Wurtzite (ZnS) unit cell showing (a) ion positions and (b) full-size ions.*

structure is shared by both III-V compounds (e.g., GaAs, AlP, and InSb) and II-VI compounds (e.g., ZnSe, CdS, and HgTe).

Depending on the details of the crystallization process, zinc sulfide also crystallizes in another crystal structure that is energetically very close to the stability of zinc blende. This alternative, the **wurtzite (ZnS) structure** shown in Figure 3–25, is built on a hexagonal Bravais lattice with four ions (two Zn^{2+} and two S^{2-}) per lattice site and per unit cell. As with ZnS, CdS can also be found with this structure. This is the characteristic structure of ZnO.

SAMPLE PROBLEM 3.6

Calculate the atomic packing factor (APF) for the diamond cubic structure (Figure 3–23).

SOLUTION

Because of the tetrahedral bonding geometry of the diamond cubic structure, the atoms lie along body diagonals. Inspection of Figure 3–23 indicates that this orientation of atoms leads to the equality

$$2r_{Si} = \frac{1}{4}(\text{body diagonal}) = \frac{\sqrt{3}}{4}a$$

or

$$a = \frac{8}{\sqrt{3}} r_{Si}$$

The unit cell volume is, then,

$$V_{unit\ cell} = a^3 = (4.62)^3 r_{Si}^3 = 98.5 r_{Si}^3$$

The volume of the eight Si atoms in the unit cell is

$$V_{atoms} = 8 \times \frac{4}{3} \pi r_{Si}^3 = 33.5 r_{Si}^3$$

This gives an atomic packing factor of

$$APF = \frac{33.5 r_{Si}^3}{98.5 r_{Si}^3} = 0.340$$

Note.   This result represents a very open structure compared to the tightly packed metals' structures of Section 3.2 (e.g., APF = 0.74 for fcc and hcp metals).

SAMPLE PROBLEM 3.7

Using the data of Appendixes 1 and 2, calculate the density of silicon.

SOLUTION

From Sample Problem 3.6,

$$V_{unit\ cell} = 98.5 r_{Si}^3 = 98.5\ (0.117\ nm)^3$$
$$= 0.158\ nm^3$$

giving a density of

$$\rho = \frac{8\ atoms}{0.158\ nm^3} \times \frac{28.09\ g}{0.6023 \times 10^{24}\ atoms} \times \left(\frac{10^7\ nm}{cm}\right)^3$$
$$= 2.36\ g/cm^3$$

As for previous calculations, a slight discrepancy of this result with Appendix 1 data (e.g., $\rho_{Si} = 2.33\ g/cm^3$) is the result of not having another significant figure with the atomic radius data of Appendix 2.

PRACTICE PROBLEM 3.7

In Sample Problem 3.6 we find the atomic packing factor for silicon to be quite low compared to the common metal structures. Comment on the relationship between this characteristic and the nature of bonding in semiconductor silicon.

PRACTICE PROBLEM 3.8

Calculate the density of germanium, using data from Appendixes 1 and 2. (See Sample Problem 3.7.)

3.6 LATTICE POSITIONS, DIRECTIONS, AND PLANES

There are a few basic rules for describing geometry in and around a unit cell. These rules and associated notation are used uniformly by crystallographers, geologists, materials scientists, and others who must deal with crystalline materials. What we are about to learn is, then, a vocabulary that allows us to communicate efficiently about crystalline structure. This will prove to be most useful when we begin to deal with structure-sensitive properties later in the book.

Figure 3–26 illustrates the notation for describing **lattice positions** expressed as fractions or multiples of unit cell dimensions. For example, the body-centered position in the unit cell projects midway along each of the three unit cell edges and is designated the $\frac{1}{2}\frac{1}{2}\frac{1}{2}$ position. One aspect of the nature of crystalline structure is that a given lattice position in a given unit cell is structurally equivalent to the same position in any other unit cell of the same structure. These equivalent positions are connected by **lattice translations,** consisting of integral multiples of lattice constants along directions parallel to crystallographic axes (Figure 3–27).

Figure 3–28 illustrates the notation for describing **lattice directions.** These directions are always expressed as sets of integers, which are obtained by identifying the *smallest integer positions* intercepted by the line from the origin of the crystallographic axes. To distinguish the notation for a direction from that of a position, the direction integers are enclosed in square brackets. The use of square brackets is important and is the standard designation for specific lattice directions. Other symbols are used to designate other geometrical features. Returning to Figure 3–28, we note that the line from the origin of the crystallographic axes through the $\frac{1}{2}\frac{1}{2}\frac{1}{2}$ body-centered position can be extended to intercept the 111 unit cell corner position. Although further extension of the line will lead to interception of other integer sets (222, 333, etc.), the 111 set is the smallest. As a result, that direction is referred to as the [111].

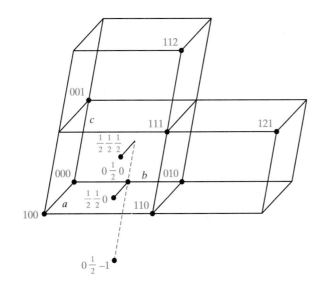

Figure 3-26 *Notation for lattice positions.*

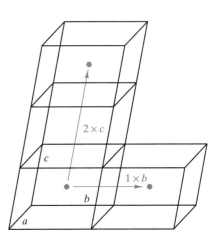

Figure 3-27 *Lattice translations connect structurally equivalent positions (e.g., the body center) in various unit cells.*

When a direction moves along a negative axis, the notation must indicate this. For example, the bar above the final integer in the $[11\bar{1}]$ direction in Figure 3–28 designates that the line from the origin has penetrated the 11–1 position. Note that the directions [111] and $[11\bar{1}]$ are structurally very similar. Both are body diagonals through identical unit cells. In fact, if you look at *all* body diagonals associated with the cubic crystal system, it is apparent that they are structurally identical, differing only in their orientation in space (Figure 3–29). In other words, the $[11\bar{1}]$ direction would become the [111] direction if we made a different choice of crystallographic axes orientations. Such a set of directions, which are structurally equivalent, is called a **family of directions** and is designated by angular brackets, $\langle \rangle$. An example of body diagonals in the cubic system is

$$\langle 111 \rangle = [111], [\bar{1}11], [1\bar{1}1], [11\bar{1}], [\overline{11}1], [1\overline{11}], [\bar{1}1\bar{1}], [\overline{111}] \qquad (3.1)$$

In future chapters, especially when dealing with calculations of mechanical properties, it will be useful to know the angle between directions. In general, the angles between directions can be determined by careful visualization and trigonometric calculation. In the frequently encountered cubic system, the angle can be determined from the relatively simple calculation of a dot product of two vectors. Taking directions $[uvw]$ and $[u'v'w']$ as vectors $\mathbf{D} = u\mathbf{a} + v\mathbf{b} + w\mathbf{c}$ and $\mathbf{D}' = u'\mathbf{a} + v'\mathbf{b} + w'\mathbf{c}$, you can determine the angle, δ, between these two directions by

$$\mathbf{D} \cdot \mathbf{D}' = |\mathbf{D}||\mathbf{D}'| \cos \delta \qquad (3.2)$$

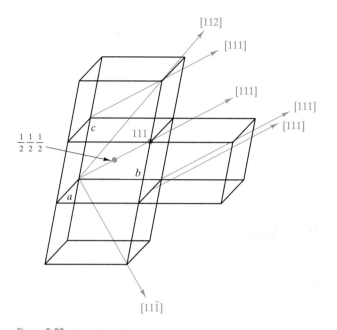

Figure 3-28 *Notation for lattice directions. Note that parallel [uvw] directions (e.g., [111]) share the same notation because only the origin is shifted.*

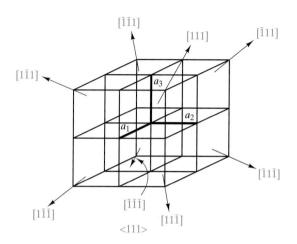

Figure 3-29 *Family of directions, ⟨111⟩, representing all body diagonals for adjacent unit cells in the cubic system.*

or

$$\cos\delta = \frac{\mathbf{D} \cdot \mathbf{D}'}{|\mathbf{D}||\mathbf{D}'|} = \frac{uu' + vv' + ww'}{\sqrt{u^2 + v^2 + w^2}\sqrt{(u')^2 + (v')^2 + (w')^2}} \qquad (3.3)$$

It is important to remember that Equations 3.2 and 3.3 apply to the cubic system only.

Another quantity of interest in future calculations is the *linear density of atoms* along a given direction. Again, a general approach to such a calculation is careful visualization and trigonometric calculation. A streamlined approach in the case where atoms are uniformly spaced along a given direction is to determine the repeat distance, r, between adjacent atoms. The linear density is simply the inverse, r^{-1}. In making linear density calculations for the first time, it is important to keep in mind that we are counting only those atoms whose centers lie directly on the direction line and not any that might intersect that line off-center.

Figure 3–30 illustrates the notation for describing **lattice planes,** that is, planes in a crystallographic lattice. As for directions, these planes are expressed as a set of integers, known as **Miller*** **indices.** Obtaining these

---

*   William Hallowes Miller (1801–1880), British crystallographer, was a major contributor along with Bravais to nineteenth-century crystallography. His efficient system of labeling crystallographic planes was but one of many achievements.

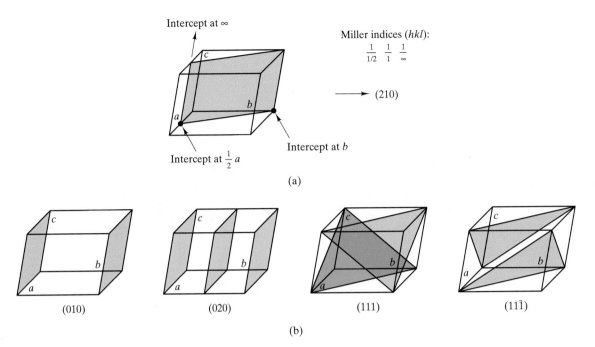

Figure 3-30 *Notation for lattice planes. (a) The (210) plane illustrates Miller indices (hkl). (b) Additional examples.*

integers is a more elaborate process than was required for directions. The integers represent the inverse of axial intercepts. For example, consider the plane (210) in Figure 3–30a. As with the square brackets of direction notation, the parentheses serve as standard notation for planes. The (210) plane intercepts the a-axis at $\frac{1}{2}a$, the b-axis at b, and is parallel to the c-axis (in effect, intercepting it at ∞). The inverses of the axial intercepts are $1/\frac{1}{2}, 1/1$, and $1/\infty$, respectively. These inverse intercepts give the 2, 1, and 0 integers leading to the (210) notation. At first, the use of these Miller indices seems like extra work. In fact, they provide an efficient labeling system for crystal planes and play an important role in equations dealing with diffraction measurements (Section 3.7). The general notation for Miller indices is (hkl), and it can be used for any of the seven crystal systems. Because the hexagonal system can be conveniently represented by four axes, a four-digit set of **Miller-Bravais indices** $(hkil)$ can be defined as shown in Figure 3–31. Since only three axes are necessary to define the three-dimensional geometry of a crystal, one of the integers in the Miller-Bravais system is redundant. Once a plane intersects any two axes in the basal plane at the bottom of the unit cell (which contains axes a_1, a_2, and a_3 in Figure 3–31), the intersection with the third basal plane axis is determined. As a result, it can be shown that $h+k = -i$ for any plane in the hexagonal system. This also permits any such hexagonal system plane to be designated by Miller-Bravais indices $(hkil)$ or by Miller indices (hkl). For the plane in Figure 3–31, the designation can be $(01\bar{1}0)$ or (010).

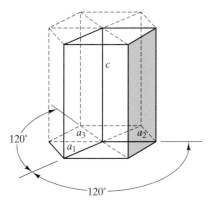

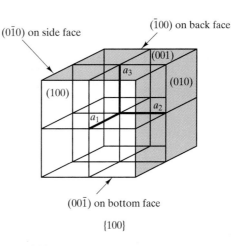

Miller-Bravais indices $(hkil)$: $\frac{1}{\infty}, \frac{1}{1}, \frac{1}{-1}, \frac{1}{\infty} \to (01\bar{1}0)$
Note: $h + k = -i$

Figure 3-31 *Miller-Bravais indices, (hkil), for the hexagonal system.*

Figure 3-32 *Family of planes, {100}, representing all faces of unit cells in the cubic system.*

As for structurally equivalent directions, we can group structurally equivalent planes as a **family of planes** with Miller or Miller-Bravais indices enclosed in braces, $\{hkl\}$ or $\{hkil\}$. Figure 3–32 illustrates that the faces of a unit cell in the cubic system are of the $\{100\}$ family with

$$\{100\} = (100), (010), (001), (\bar{1}00), (0\bar{1}0), (00\bar{1}) \qquad (3.4)$$

Future chapters will require calculation of **planar densities** of atoms (number per unit area) analogous to linear densities mentioned before. As with linear densities, only those atoms centered on the plane of interest are counted.

SAMPLE PROBLEM 3.8

Using Table 3.2, list the face-centered lattice point positions for **(a)** the *face-centered cubic* (fcc) Bravais lattice, and **(b)** the *face-centered orthorhombic* (fco) lattice.

SOLUTION

(a) For the face-centered positions, $\frac{1}{2}\frac{1}{2}0$, $\frac{1}{2}0\frac{1}{2}$, $0\frac{1}{2}\frac{1}{2}$, $\frac{1}{2}\frac{1}{2}1$, $\frac{1}{2}1\frac{1}{2}$, $1\frac{1}{2}\frac{1}{2}$

(b) Same answer as part (a). Lattice parameters do not appear in the notation for lattice positions.

SAMPLE PROBLEM 3.9

Which lattice points lie on the [110] direction in the fcc and fco unit cells of Table 3.2?

SOLUTION

Sketching this case gives

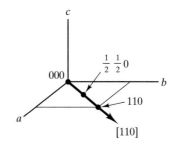

The lattice points are 000, $\frac{1}{2}\frac{1}{2}0$, and 110. This holds for either system, fcc or fco.

SAMPLE PROBLEM 3.10

List the members of the ⟨110⟩ family of directions in the cubic system.

SOLUTION

This constitutes all face diagonals of the unit cell, with two such diagonals on each face for a total of 12 members:

$$\langle 110 \rangle = [110], [1\bar{1}0], [\bar{1}10], [\bar{1}\bar{1}0], [101], [10\bar{1}], [\bar{1}01],$$
$$[\bar{1}0\bar{1}], [011], [01\bar{1}], [0\bar{1}1], [0\bar{1}\bar{1}]$$

SAMPLE PROBLEM 3.11

What is the angle between the [110] and [111] directions in the cubic system?

SOLUTION

From Equation 3.3,

$$\delta = \arccos \frac{uu' + vv' + ww'}{\sqrt{u^2 + v^2 + w^2}\sqrt{(u')^2 + (v')^2 + (w')^2}}$$

$$= \arccos \frac{1 + 1 + 0}{\sqrt{2}\sqrt{3}}$$

$$= \arccos 0.816$$

$$= 35.3°$$

SAMPLE PROBLEM 3.12

Identify the axial intercepts for the $(3\bar{1}1)$ plane.

SOLUTION

For the a axis, intercept $= \frac{1}{3}a$; for the b axis, intercept $= \frac{1}{-1}b = -b$; for the c axis, intercept $= \frac{1}{1}c = c$

SAMPLE PROBLEM 3.13

List the members of the {110} family of planes in the cubic system.

SOLUTION

$$\{110\} = (110),\ (1\bar{1}0),\ (\bar{1}10),\ (\bar{1}\bar{1}0),\ (101),\ 10\bar{1}),\ (\bar{1}01),$$

$$(\bar{1}0\bar{1}),\ (011),\ (01\bar{1}),\ (0\bar{1}1),\ (0\bar{1}\bar{1})$$

(Compare this answer with that for Sample Problem 3.10.)

SAMPLE PROBLEM 3.14

Calculate the linear density of atoms along the [111] direction in **(a)** bcc tungsten and **(b)** fcc aluminum.

SOLUTION

(a) For a bcc structure (Figure 3–4), atoms touch along the [111] direction (a body diagonal). Therefore, the repeat distance is equal to one atomic diameter. Taking data from Appendix 2, we find that the repeat distance is

$$r = d_{W \text{ atom}} = 2r_{W \text{ atom}}$$

$$= 2(0.137 \text{ nm}) = 0.274 \text{ nm}$$

Therefore,

$$r^{-1} = \frac{1}{0.274 \text{ nm}} = 3.65 \text{ atoms/nm}$$

(b) For an fcc structure, only one atom is intercepted along the body diagonal of a unit cell. To determine the length of the body diagonal, we can note that two atomic diameters equal the length of a face diagonal (see Figure 3–5). Using data from Appendix 2, we have

$$\text{face diagonal length} = 2d_{Al \text{ atom}}$$

$$= 4r_{Al \text{ atom}} = \sqrt{2}a$$

or the lattice parameter is

$$a = \frac{4}{\sqrt{2}} r_{Al \text{ atom}} \quad \text{(see also Table 3.3)}$$

$$= \frac{4}{\sqrt{2}} (0.143 \text{ nm}) = 0.404 \text{ nm}$$

The repeat distance is

$$r = \text{body diagonal length} = \sqrt{3}a$$

$$= \sqrt{3}(0.404 \text{ nm})$$

$$= 0.701 \text{ nm}$$

which gives a linear density of

$$r^{-1} = \frac{1}{0.701 \text{ nm}} = 1.43 \text{ atoms/nm}$$

SAMPLE PROBLEM 3.15

Calculate the planar density of atoms in the (111) plane of **(a)** bcc tung-sten and **(b)** fcc aluminum.

SOLUTION

(a) For the bcc structure (Figure 3–4), the (111) plane intersects only corner atoms in the unit cell:

Following the calculations of Sample Problem 3.14a, we have

$$\sqrt{3}a = 4r_{\text{W atom}}$$

or

$$a = \frac{4}{\sqrt{3}}r_{\text{W atom}} = \frac{4}{\sqrt{3}}(0.137\text{nm}) = 0.316\,\text{nm}$$

Face diagonal length, l, is then

$$l = \sqrt{2}a = \sqrt{2}(0.316\,\text{nm}) = 0.447\,\text{nm}$$

The area of the (111) plane within the unit cell is

$$A = \frac{1}{2}bh = \frac{1}{2}(0.447\,\text{nm})\left(\frac{\sqrt{3}}{2} \times 0.447\,\text{nm}\right)$$

$$= 0.0867\,\text{nm}^2$$

There is $\frac{1}{6}$ atom (i.e., $\frac{1}{6}$ of the circumference of a circle) at each corner of the equilateral triangle formed by the (111) plane in the unit cell. Therefore,

$$\text{atomic density} = \frac{3 \times \frac{1}{6}\,\text{atom}}{A}$$

$$= \frac{0.5\,\text{atom}}{0.0867\,\text{nm}^2} = 5.77\frac{\text{atoms}}{\text{nm}^2}$$

(b) For the fcc structure (Figure 3–5), the (111) plane intersects three corner atoms plus three face-centered atoms in the unit cell:

Following the calculations of Sample Problem 3.14b, we obtain the face diagonal length

$$l = \sqrt{2}a = \sqrt{2}(0.404 \text{ nm}) = 0.572 \text{ nm}$$

The area of the (111) plane within the unit cell is

$$A = \frac{1}{2}bh = \frac{1}{2}(0.572 \text{ nm})\left(\frac{\sqrt{3}}{2}0.572 \text{ nm}\right)$$

$$= 0.142 \text{ nm}^2$$

There are $3 \times \frac{1}{6}$ corner atoms plus $3 \times \frac{1}{2}$ face-centered atoms within this area, giving

$$\text{atomic density} = \frac{3 \times \frac{1}{6} + 3 \times \frac{1}{2} \text{ atoms}}{0.142 \text{ nm}^2} = \frac{2 \text{ atoms}}{0.142 \text{ nm}^2}$$

$$= 14.1 \text{ atoms/nm}^2$$

SAMPLE PROBLEM 3.16

Calculate the linear density of ions in the [111] direction of MgO.

SOLUTION

Figure 3–9 shows that the body diagonal of the unit cell intersects one Mg^{2+} and one O^{2-}. Following the calculations of Sample Problem 3.3, we find that the length of the body diagonal is

$$l = \sqrt{3}a = \sqrt{3}(0.420 \text{ nm}) = 0.727 \text{ nm}$$

The ionic linear densities are, then,

$$\frac{1\ Mg^{2+}}{0.727\ nm} = 1.37\ Mg^{2+}/nm$$

and similarly,

$$1.37\ O^{2-}/nm$$

giving

$$(1.37\ Mg^{2+} + 1.37\ O^{2-})/nm$$

SAMPLE PROBLEM 3.17

Calculate the planar density of ions in the (111) plane of MgO.

SOLUTION

There are really two separate answers to this problem. Using the unit cell of Figure 3–9, we see an arrangement comparable to an fcc metal:

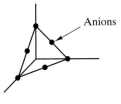

Anions

However, we could similarly define a unit cell with its origin on a cation site (rather than on an anion site as shown in Figure 3–9). In this case, the (111) plane would have a comparable arrangement of cations:

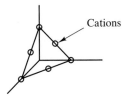

Cations

In either case, there are two ions per (111) "triangle." From Sample Problem 3.3, we know that $a = 0.420$ nm. The length of each (111) triangle side (i.e., a unit cell face diagonal) is

$$l = \sqrt{2}a = \sqrt{2}(0.420 \text{ nm}) = 0.594 \text{ nm}$$

The planar area is, then,

$$A = \frac{1}{2}bh = \frac{1}{2}(0.594 \text{ nm})\left(\frac{\sqrt{3}}{2}0.594 \text{ nm}\right) = 0.153 \text{ nm}^2$$

This gives

$$\text{ionic density} = \frac{2 \text{ ions}}{0.153 \text{ nm}^2} = 13.1 \text{ nm}^{-2}$$

or

$$13.1(\text{Mg}^{2+} \text{ or } \text{O}^{2-})/\text{nm}^2$$

SAMPLE PROBLEM 3.18

Calculate the linear density of atoms along the [111] direction in silicon.

SOLUTION

We must use some caution in this problem. Inspection of Figure 3–23 indicates that atoms along the [111] direction (a body diagonal) are *not* uniformly spaced. Therefore, the r^{-1} calculations of Sample Problem 3.14 are not appropriate.

Referring to the comments of Sample Problem 3.6, we can see that two atoms are centered along a given body diagonal (e.g., $\frac{1}{2}$ atom at 000, 1 atom at $\frac{1}{4}\frac{1}{4}\frac{1}{4}$, and $\frac{1}{2}$ atom at 111). If we take the body diagonal length in a unit cell as l,

$$2r_{\text{Si}} = \frac{1}{4}l$$

or

$$l = 8r_{\text{Si}}$$

From Appendix 2,

$$l = 8(0.117 \, \text{nm}) = 0.936 \, \text{nm}$$

Therefore, the linear density is

$$\text{linear density} = \frac{2 \, \text{atoms}}{0.936 \, \text{nm}} = 2.14 \frac{\text{atoms}}{\text{nm}}$$

SAMPLE PROBLEM 3.19

Calculate the planar density of atoms in the (111) plane of silicon.

SOLUTION

Close observation of Figure 3–23 shows that the four interior atoms in the diamond cubic structure do not lie on the (111) plane. The result is that the atom arrangement in this plane is precisely that for the metallic fcc structure (see Sample Problem 3.15b). Of course, the atoms along [110]-type directions in the diamond cubic structure do not touch as in fcc metals.

As calculated in Sample Problem 3.15b, there are two atoms in the equilateral triangle bounded by sides of length $\sqrt{2}a$. From Sample Problem 3.6 and Appendix 2, we see that

$$a = \frac{8}{\sqrt{3}}(0.117 \, \text{nm}) = 0.540 \, \text{nm}$$

$$\sqrt{2}a = 0.764 \, \text{nm}$$

giving a triangle area of

$$A = \frac{1}{2}bh = \frac{1}{2}(0.764 \, \text{nm})\left(\frac{\sqrt{3}}{2}0.764 \, \text{nm}\right)$$

$$= 0.253 \, \text{nm}^2$$

and a planar density of

$$\frac{2 \, \text{atoms}}{0.253 \, \text{nm}^2} = 7.91 \frac{\text{atoms}}{\text{nm}^2}$$

PRACTICE PROBLEM 3.9

From Table 3.2, list the body-centered lattice point positions for **(a)** the bcc Bravais lattice, **(b)** the body-centered tetragonal lattice, and **(c)** the body-centered orthorhombic lattice. (See Sample Problem 3.8.)

PRACTICE PROBLEM 3.10

Use a sketch to determine which lattice points lie along the [111] direction in the **(a)** bcc, **(b)** body-centered tetragonal, and **(c)** body-centered orthorhombic unit cells of Table 3.2. (See Sample Problem 3.9.)

PRACTICE PROBLEM 3.11

Sketch the 12 members of the ⟨110⟩ family determined in Sample Problem 3.10. (You may want to use more than one sketch.)

PRACTICE PROBLEM 3.12

(a) Determine the ⟨100⟩ family of directions in the cubic system, and **(b)** sketch the members of this family. (See Sample Problem 3.10 and Practice Problem 3.11.)

PRACTICE PROBLEM 3.13

Calculate the angles between **(a)** the [100] and [110] and **(b)** the [100] and [111] directions in the cubic system. (See Sample Problem 3.11.)

PRACTICE PROBLEM 3.14

Sketch the $(3\bar{1}1)$ plane and its intercepts. (See Sample Problem 3.12 and Figure 3–30.)

PRACTICE PROBLEM 3.15

Sketch the 12 members of the {110} family determined in Sample Problem 3.13. (To simplify matters, you will probably want to use more than one sketch.)

PRACTICE PROBLEM 3.16

Calculate the linear density of atoms along the [111] direction in **(a)** bcc iron and **(b)** fcc nickel. (See Sample Problem 3.14.)

PRACTICE PROBLEM 3.17

Calculate the planar density of atoms in the (111) plane of **(a)** bcc iron and **(b)** fcc nickel. (See Sample Problem 3.15.)

PRACTICE PROBLEM 3.18

Calculate the linear density of ions along the [111] direction for CaO. (See Sample Problem 3.16.)

PRACTICE PROBLEM 3.19

Calculate the planar density of ions in the (111) plane for CaO. (See Sample Problem 3.17.)

PRACTICE PROBLEM 3.20

Find the linear density of atoms along the [111] direction for germanium. (See Sample Problem 3.18.)

PRACTICE PROBLEM 3.21

Find the planar density of atoms in the (111) plane for germanium. (See Sample Problem 3.19.)

3.7   X-RAY DIFFRACTION

This chapter has introduced a large variety of crystal structures. We now end with a brief description of x-ray diffraction, a powerful experimental tool.

There are many ways in which x-ray diffraction is used to measure the crystal structure of engineering materials. It can be used to determine the structure of a new material, or the known structure of a common material can be used as a source of chemical identification.

Diffraction is the result of radiation's being scattered by a regular array of scattering centers whose spacing is about the same as the wavelength of the radiation. For example, parallel scratch lines spaced repeatedly about 1 μm apart cause diffraction of visible light (electromagnetic radiation with a wavelength just under 1 μm). This *diffraction grating* causes the light to be scattered with a strong intensity in a few specific directions (Figure 3–33). The precise direction of observed scattering is a function of the exact spacing between scratch lines in the diffraction grating, relative to the wavelength of the incident light. Appendix 2 shows that atoms and ions are on the order of 0.1 nm in size, so we can think of crystal structures as being diffraction gratings on a subnanometer scale. As shown in Figure 3–34, the portion of the electromagnetic spectrum with a wavelength in this range is **x-radiation** (compared to the 1000-nm range for the wavelength of visible light). As a result, x-ray diffraction is capable of characterizing crystalline structure.

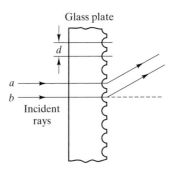

Figure 3-33 *Diffraction grating for visible light. Scratch lines in the glass plate serve as light-scattering centers. (After D. Halliday and R. Resnick,* Physics, *John Wiley & Sons, Inc., New York, 1962.)*

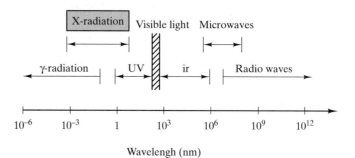

Figure 3-34 *Electromagnetic radiation spectrum. X-radiation represents that portion with wavelengths around 0.1 nm.*

For x-rays, atoms are the scattering centers. The specific mechanism of scattering is the interaction of a photon of electromagnetic radiation with an orbital electron in the atom. A crystal acts as a three-dimensional diffraction grating. Repeated stacking of crystal planes serves the same function as the parallel scratch lines in Figure 3–33. For a simple crystal lattice, the condition for diffraction is shown in Figure 3–35. For diffraction to occur, x-ray beams scattered off adjacent crystal planes must be in phase. Otherwise, destructive interference of waves occurs and essentially no scattering intensity is observed. At the precise geometry for constructive interference (scattered waves in phase), the difference in path length between the adjacent x-ray beams is some integral number (n) of radiation wavelengths (λ). The relationship that demonstrates this condition is the **Bragg* equation**

$$n\lambda = 2d \sin \theta \qquad (3.5)$$

where d is the spacing between adjacent crystal planes and θ is the angle of scattering as defined in Figure 3–35. The angle θ is usually referred to as the **Bragg angle** and the angle 2θ is referred to as the **diffraction angle** because that is the angle measured experimentally (Figure 3–36).

The magnitude of **interplanar spacing** (d in Equation 3.5) is a direct function of the Miller indices for the plane. For a cubic system, the relationship is fairly simple. The spacing between adjacent hkl planes is

$$d_{hkl} = \frac{a}{\sqrt{h^2 + k^2 + l^2}} \qquad (3.6)$$

---

* William Henry Bragg (1862–1942) and William Lawrence Bragg (1890–1971), English physicists, were a gifted father-and-son team. They were the first to demonstrate the power of Equation 3.5 by using x-ray diffraction to determine the crystal structures of several alkali halides, such as NaCl. Since this achievement in 1912, the crystal structures of over 70,000 materials have been catalogued (see the footnote on p. 106).

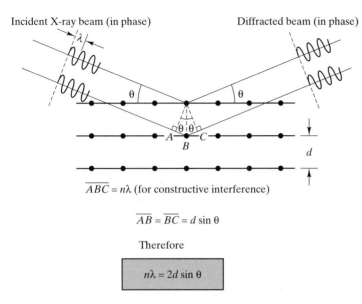

Incident X-ray beam (in phase)   Diffracted beam (in phase)

$\overline{ABC} = n\lambda$ (for constructive interference)

$\overline{AB} = \overline{BC} = d \sin \theta$

Therefore

$$n\lambda = 2d \sin \theta$$

Figure 3-35 *Geometry for diffraction of x-radiation. The crystal structure is a three-dimensional diffraction grating. Bragg's law ($n\lambda = 2d \sin \theta$) describes the diffraction condition.*

where a is the lattice parameter (edge length of the unit cell). For more complex unit cell shapes, the relationship is more complex. For a *hexagonal system,*

$$d_{hkl} = \cfrac{a}{\sqrt{\frac{4}{3}(h^2 + hk + k^2) + l^2(a^2/c^2)}} \qquad (3.7)$$

where a and c are the lattice parameters.

Bragg's law (Equation 3.5) is a necessary but not sufficient condition for diffraction. It defines the diffraction condition for **primitive unit cells,** that is, those Bravais lattices with lattice points only at unit cell corners, such as simple cubic and simple tetragonal. Crystal structures with **nonprimitive unit cells** have atoms at additional lattice sites located along a unit cell edge, within a unit cell face, or in the interior of the unit cell. The extra scattering

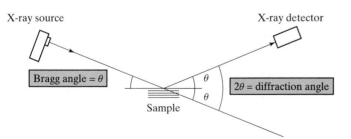

X-ray source   X-ray detector

Bragg angle = θ 2θ = diffraction angle

Sample

Figure 3-36 *Relationship of the Bragg angle (θ) and the experimentally measured diffraction angle (2θ).*

Table 3.4 *Reflection Rules of X-Ray Diffraction for the Common Metal Structures*

Crystal structure	Diffraction does not occur when	Diffraction occurs when
Body-centered cubic (bcc)	$h + k + l =$ odd number	$h + k + l =$ even number
Face-centered cubic (fcc)	h, k, l mixed (i.e., both even and odd numbers)	h, k, l unmixed (i.e., are all even numbers or all odd numbers)
Hexagonal close packed (hcp)	$(h + 2k) = 3n, l$ odd (n is an integer)	All other cases

centers can cause out-of-phase scattering to occur at certain Bragg angles. The result is that some of the diffraction predicted by Equation 3.5 does not occur. An example of this effect is given in Table 3.4, which gives the **reflection rules** for the common metal structures. This shows which sets of Miller indices do not produce diffraction as predicted by Bragg's law. Keep in mind that "reflection" here is a casual term in that diffraction rather than true reflection is being described.

Figure 3–37 shows an x-ray diffraction pattern for a single crystal of MgO. Each spot on the film is a solution to Bragg's law and represents the diffraction of an x-ray beam (with some wavelength, λ) from a crystal plane (hkl) oriented at some angle (θ). A wide range of x-radiation wavelengths is used to provide diffraction conditions for the many crystal plane orientations in the single-crystal specimen. This experiment is done in a **Laue* camera,**

Figure 3-37 *Diffraction pattern of a single crystal of MgO (with the NaCl structure of Figure 3–9). Each spot on the film represents diffraction of the x-ray beam from a crystal plane (hkl).*

---

* Max von Laue (1879–1960), German physicist, correctly suggested that atoms in a crystal would be a diffraction grating for x-rays. In 1912, he experimentally verified this fact using a single crystal of copper sulfate, and thereby laid the groundwork for the first structure determinations by the Braggs.

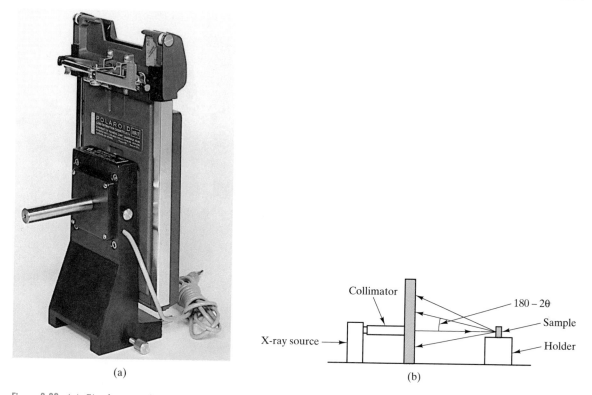

Figure 3-38 *(a) Single-crystal diffraction camera (or Laue camera). (Courtesy of Blake Industries, Inc.) (b) Schematic of the experiment.*

as shown in Figure 3–38. This method is used in the electronics industry to determine the orientation of single crystals so they can be sliced along certain desired crystal planes.

A diffraction pattern for a specimen of aluminum powder is shown in Figure 3–39. Each peak represents a solution to Bragg's law. Because the powder consists of many small crystal grains oriented randomly, a single wavelength of radiation is used to keep the number of diffraction peaks

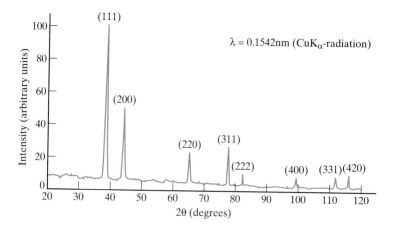

Figure 3-39 *Diffraction pattern of aluminum powder. Each peak (in the plot of x-ray intensity versus diffraction angle, 2θ) represents diffraction of the x-ray beam by a set of parallel crystal planes (hkl) in various powder particles.*

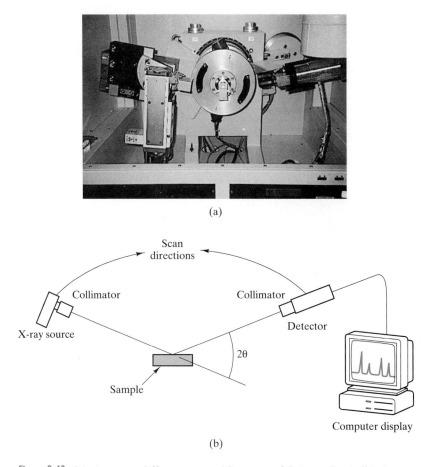

Figure 3-40 *(a) An x-ray diffractometer. (Courtesy of Scintag, Inc.) (b) A schematic of the experiment.*

in the pattern to a small, workable number. The experiment is done in a **diffractometer** (Figure 3–40), an electromechanical scanning system. The diffracted beam intensity is monitored electronically by a mechanically driven scanning radiation detector. "Powder patterns" such as Figure 3–39 are routinely used by materials engineers for comparison against a large collection of known diffraction patterns.* The comparison of an experimental diffraction pattern such as Figure 3–39 with the database of known diffraction patterns can be done in a few seconds with "search/match" computer software, an integral part of modern diffractometers such as that shown in Figure 3–40. The unique relationship between such patterns and crystal structures provides a powerful tool for chemical identification of powders and polycrystalline materials. (Typical polycrystalline grain structure was shown in Figure 1–20 and will be discussed in detail in Section 4.4.)

---

* *Powder Diffraction File,* over 70,000 powder diffraction patterns catalogued by the International Centre for Diffraction Data (ICDD), Newtown Square, Pa.

Standard procedure for analyzing the diffraction patterns of powder samples or polycrystalline solids involves the use of $n = 1$ in Equation 3.5. This is justified in that the nth-order diffraction of any (hkl) plane occurs at an angle identical to the first-order diffraction of the $(nh\ nk\ nl)$ plane [which is, by the way, parallel to (hkl)]. As a result, we can use an even simpler version of Bragg's law for powder diffraction:

$$\lambda = 2d \sin \theta \qquad\qquad (3.8)$$

SAMPLE PROBLEM 3.20

(a) A (111) diffraction spot from an MgO single crystal is produced with a Laue camera. It occurs 1 cm from the film center. Calculate the diffraction angle (2θ) and the Bragg angle (θ). Assume that the sample is 3 cm from the film. **(b)** Calculate the x-ray wavelength (λ) that would produce first-, second-, and third-order diffraction (i.e., $n = 1, 2,$ and 3).

SOLUTION

(a) The geometry is

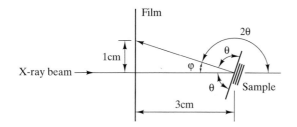

By inspection,

$$\varphi = \arctan \left(\frac{1\,\text{cm}}{3\,\text{cm}} \right) = 18.4°$$

and

$$2\theta = 180° - \varphi = 180° - 18.4° = 161.6°$$

or

$$\theta = 80.8°$$

(b) To obtain λ requires Bragg's law (Equation 3.5),

$$n\lambda = 2d \sin \theta$$

or

$$\lambda = \frac{2d}{n} \sin \theta$$

To obtain d, we can use Equation 3.6 and the value for a calculated in Sample Problem 3.3:

$$d = \frac{a}{\sqrt{h^2 + k^2 + l^2}} = \frac{0.420 \, \text{nm}}{\sqrt{1 + 1 + 1}} = \frac{0.420 \, \text{nm}}{\sqrt{3}}$$
$$= 0.242 \, \text{nm}$$

Substituting to obtain λ for $n = 1$, we obtain the following. For $n = 1$,

$$\lambda_{n=1} = 2(0.242 \, \text{nm}) \sin 80.8° = 0.479 \, \text{nm}$$

For $n = 2$,

$$\lambda_{n=2} = \frac{2(0.242 \, \text{nm}) \sin 80.8°}{2} = 0.239 \, \text{nm}$$

For $n = 3$,

$$\lambda_{n=3} = \frac{2(0.242 \, \text{nm}) \sin 80.8°}{3} = 0.160 \, \text{nm}$$

SAMPLE PROBLEM 3.21

Using Bragg's law, calculate the diffraction angles (2θ) for the first three peaks in the aluminum powder pattern of Figure 3–39.

SOLUTION

Figure 3–39 indicates that the first three (i.e., lowest-angle) peaks are for (111), (200), and (220). From Sample Problem 3.14b, we note that $a = 0.404$ nm. Therefore, Equation 3.6 yields

$$d_{111} = \frac{0.404 \, \text{nm}}{\sqrt{1 + 1 + 1}} = \frac{0.404 \, \text{nm}}{\sqrt{3}} = 0.234 \, \text{nm}$$

$$d_{200} = \frac{0.404 \, \text{nm}}{\sqrt{2^2 + 0 + 0}} = \frac{0.404 \, \text{nm}}{2} = 0.202 \, \text{nm}$$

$$d_{220} = \frac{0.404 \, \text{nm}}{\sqrt{2^2 + 2^2 + 0}} = \frac{0.404 \, \text{nm}}{\sqrt{8}} = 0.143 \, \text{nm}$$

Noting in Figure 3–39 that $\lambda = 0.1542$ nm, Equation 3.8 gives

$$\theta = \arcsin \frac{\lambda}{2d}$$

or

$$\theta_{111} = \arcsin \frac{0.1542 \text{ nm}}{2 \times 0.234 \text{ nm}} = 19.3°$$

$$\text{or} \quad (2\theta)_{111} = 38.6°$$

$$\theta_{200} = \arcsin \frac{0.1542 \text{ nm}}{2 \times 0.202 \text{ nm}} = 22.4°$$

$$\text{or} \quad (2\theta)_{200} = 44.8°$$

$$\theta_{220} = \arcsin \frac{0.1542 \text{ nm}}{2 \times 0.143 \text{ nm}} = 32.6°$$

$$\text{or} \quad (2\theta)_{220} = 65.3°$$

PRACTICE PROBLEM 3.22

In Sample Problem 3.20 we characterize the geometry for diffraction by (111) planes in MgO. Suppose the crystal is tilted slightly so that the (111) diffraction spot is shifted to a position 0.5 cm from the film center. What wavelength (λ) would produce first-order diffraction in this case?

PRACTICE PROBLEM 3.23

The diffraction angles for the first three peaks in Figure 3–39 are calculated in Sample Problem 3.21. Calculate the diffraction angles for the remainder of the peaks in Figure 3–39.

SUMMARY

Most materials used by engineers are crystalline in nature; that is, their atomic-scale structure is regular and repeating. This regularity allows the structure to be defined in terms of a fundamental structural unit, the unit cell. There are seven crystal systems, which correspond to the possible unit cell shapes. Based on these crystal systems, there are 14 Bravais lattices that represent the possible arrangements of points through three-dimensional space. These lattices are the "skeletons" on which the large number of crystalline atomic structures is based.

There are three primary crystal structures observed for common metals: the body-centered cubic (bcc), the face-centered cubic (fcc), and the hexagonal close packed (hcp). These are relatively simple structures, with the fcc and hcp forms representing optimum efficiency in packing equal-sized spheres (i.e., metal atoms). The fcc and hcp structures differ only in the pattern of stacking of close-packed atomic planes.

Chemically more complex than metals, ceramic compounds exhibit a wide variety of crystalline structures. Some, such as the NaCl structure, are similar to the simpler metal structures sharing a common Bravais lattice but with more than one ion associated with each lattice point. Silica, SiO_2, and the silicates exhibit a wide array of relatively complex arrangements of silica tetrahedra (SiO_4^{4-}). In this chapter several representative ceramic structures are displayed, including some for important electronic and magnetic ceramics.

Polymers are characterized by long-chain polymeric structures. The elaborate way in which these chains must be folded to form a repetitive pattern produces two effects: (1) the resulting crystal structures are relatively complex and (2) most commercial polymers are only partially crystalline. The unit cell structures of polyethylene and nylon 66 are illustrated in this chapter.

High-quality single crystals are an important part of *semiconductor* technology. This is possible largely because most semiconductors can be produced in a few relatively simple crystal structures. Elemental semiconductors, such as silicon, have the diamond cubic structure, a modification of the fcc Bravais lattice with two atoms associated with each lattice point. Many compound semiconductors are found in the closely related zinc blende (ZnS) structure, in which the diamond cubic atom positions are retained but with Zn^{2+} and S^{2-} ions alternating on those sites. Some semiconducting compounds are found in the energetically similar but slightly more complex wurtzite (ZnS) structure.

There are standard methods for describing the geometry of crystalline structures. These give an efficient and systematic notation for lattice positions, directions, and planes.

X-ray diffraction is the standard experimental tool for analyzing crystal structures. The regular atomic arrangement of crystals serves as a sub-nanometer diffraction grating for *x-radiation* (with a subnanometer wavelength). The use of Bragg's law in conjunction with the reflection rules permits a precise measurement of interplanar spacings in the crystal structure. Both single-crystal and polycrystalline (or powdered) materials can be analyzed in this way.

KEY TERMS

General
atomic packing factor (APF) (64)
crystal (Bravais) lattice (61)
crystal system (60)
family of directions (88)
family of planes (91)
ionic packing factor (69)

lattice constant (60)
lattice direction (87)
lattice parameter (60)
lattice plane (89)
lattice point (61)
lattice position (87)
lattice translation (87)

linear density (89)
Miller-Bravais indices (90)
Miller indices (89)
octahedral position (74)
planar density (91)
point lattice (61)
tetrahedral position (74)

REFERENCES

Barrett, C. S., and **T. B. Massalski,** *Structure of Metals,* 3rd revised ed., Pergamon Press, New York, 1980. This text includes substantial coverage of x-ray diffraction techniques.

Chiang, Y., D. P. Birnie III, and **W. D. Kingery,** *Physical Ceramics*, John Wiley & Sons, Inc., New York, 1997.

Cullity, B. D., *Elements of X-Ray Diffraction,* 2nd ed., Addison-Wesley Publishing Co., Inc., Reading, Mass., 1978. An especially clear discussion of the principles and applications of x-ray diffraction.

Molecular Simulations, Inc., San Diego, Ca. Computer-generated crystal structures of a wide range of materials available on CD-ROM for display on graphics workstations. Updated annually.

Williams, D. J., *Polymer Science and Engineering,* Prentice Hall, Inc., Englewood Cliffs, N.J., 1971.

Wyckoff, R. W. G., ed., *Crystal Structures,* 2nd ed., Vols. 1–5 and Vol. 6, Parts 1 and 2, John Wiley & Sons, Inc., New York, 1963–1971. An encyclopedic collection of crystal structure data.

PROBLEMS

Section 3.1 • Seven Systems and Fourteen Lattices

3.1. Why is the simple hexagon not a two-dimensional point lattice?

3.2. What would be an equivalent two-dimensional point lattice for the area-centered hexagon?

3.3. Why is there no base-centered cubic lattice in Table 3.2? (Use a sketch to answer.)

• **3.4.** **(a)** Which two-dimensional point lattice corresponds to the crystalline ceramic illustrated in Figure 1–8a? **(b)** Sketch the unit cell.

3.5. Under what conditions does the triclinic system reduce to the hexagonal system?

3.6. Under what conditions does the monoclinic system reduce to the orthorhombic system?

Section 3.2 • Metal Structures

3.7. Calculate the density of Mg, an hcp metal. (Note Problem 3.11 for the ideal c/a ratio.)

3.8. Calculate the atomic packing factor (APF) of 0.68 for the bcc metal structure.

3.9. Calculate the atomic packing factor (APF) of 0.74 for fcc metals.

3.10. Calculate the atomic packing factor (APF) of 0.74 for hcp metals.

3.11. **(a)** Show that the c/a ratio (height of unit cell divided by its edge length) is 1.633 for the ideal hcp structure. **(b)** Comment on the fact that real hcp metals display c/a ratios varying from 1.58 (for Be) to 1.89 (for Cd).

Section 3.3 • Ceramic Structures

3.12. Calculate the ionic packing factor for UO_2, which has the CaF_2 structure (Figure 3–10).

3.13. In Section 3.3 the open nature of the CaF_2 structure was given credit for the ability of UO_2 to absorb He gas atoms and, thereby, resist swelling. Confirm that an He atom (diameter ≈ 0.2 nm) can fit in the center of the UO_2 unit cell (see Figure 3–10 for the CaF_2 structure).

3.14. Calculate the ionic packing factor for $CaTiO_3$ (Figure 3–14).

3.15. Show that the unit cell in Figure 3–16 gives the chemical formula $2(OH)_4 Al_2 Si_2 O_5$.

3.16. Calculate the density of UO_2.

3.17. Calculate the density of $CaTiO_3$.

• **3.18.** **(a)** Derive a general relationship between the ionic packing factor of the NaCl-type structure and the radius ratio (r/R). **(b)** Over what r/R range is this relationship reasonable?

• **3.19.** Calculate the ionic packing factor for cristobalite (Figure 3–11).

• **3.20.** Calculate the ionic packing factor for corundum (Figure 3–13).

Section 3.4 • Polymeric Structures

3.21. Calculate the reaction energy involved in forming a single unit cell of polyethylene.

3.22. How many unit cells are contained in the thickness of a 10-nm-thick polyethylene platelet (Figure 3–21)?

3.23. Calculate the atomic packing factor for polyethylene.

Section 3.5 • Semiconductor Structures

3.24. Calculate the ionic packing factor for the zinc blende structure (Figure 3–24).

3.25. Calculate the density of zinc blende, using data from Appendixes 1 and 2.

• **3.26.** **(a)** Derive a general relationship between the ionic packing factor of the zinc blende structure and the radius ratio (r/R). **(b)** What is a primary limitation

of such IPF calculations for these compound semiconductors?

• **3.27.** Calculate the density of wurtzite (Figure 3–25), using data from Appendixes 1 and 2.

Section 3.6 • Lattice Positions, Directions, and Planes

3.28. **(a)** Sketch, in a cubic unit cell, a [111] and a [112] lattice direction. **(b)** Use a trigonometric calculation to determine the angle between these two directions. **(c)** Use Equation 3.3 to determine the angle between these two directions.

3.29. List the lattice point positions for the corners of the unit cell in **(a)** the base-centered orthorhombic lattice and **(b)** the triclinic lattice.

3.30. **(a)** Sketch, in a cubic unit cell, [100] and [210] directions. **(b)** Use a trigonometric calculation to determine the angle between these directions. **(c)** Use Equation 3.3 to determine this angle.

3.31. List the body-centered and base-centered lattice point positions for **(a)** the body-centered orthorhombic lattice and **(b)** the base-centered monoclinic lattice, respectively.

• **3.32.** What polyhedron is formed by "connecting the dots" between a corner atom in the fcc lattice and the three adjacent face-centered positions? Illustrate your answer with a sketch.

• **3.33.** Repeat Problem 3.32 for the six face-centered positions on the surface of the fcc unit cell.

3.34. What $[hkl]$ direction connects the adjacent face-centered positions $\frac{1}{2}\frac{1}{2}0$ and $\frac{1}{2}0\frac{1}{2}$? Illustrate your answer with a sketch.

3.35. A useful rule of thumb for the cubic system is that a given $[hkl]$ direction is the normal to the (hkl) plane. Using this rule and Equation 3.3, determine which members of the $\langle 110 \rangle$ family of directions lie within the (111) plane. (*Hint:* The dot product of two perpendicular vectors is zero.)

3.36. Which members of the $\langle 111 \rangle$ family of directions lie within the (110) plane? (See the comments in Problem 3.35.)

3.37. Repeat Problem 3.35 for the $(\bar{1}11)$ plane.

3.38. Repeat Problem 3.35 for the $(11\bar{1})$ plane.

3.39. Repeat Problem 3.36 for the (101) plane.

3.40. Repeat Problem 3.36 for the $(10\bar{1})$ plane.

3.41. Repeat Problem 3.36 for the $(\bar{1}01)$ plane.

3.42. Sketch the basal plane for the hexagonal unit cell having the Miller-Bravais indices (0001) (see Figure 3–31).

3.43. List the members of the family of prismatic planes for the hexagonal unit cell $\{01\bar{1}0\}$ (see Figure 3–31).

3.44. The four-digit notation system (Miller-Bravais indices) introduced for planes in the hexagonal system can also be used for describing crystal directions. In a hexagonal unit cell sketch **(a)** the [0001] direction, and **(b)** the $[11\bar{2}0]$ direction.

3.45. The family of directions described in Practice Problem 3.12 contains six members. The size of this family will be diminished for noncubic unit cells. List the members of the $\langle 100 \rangle$ family for **(a)** the tetragonal system, and **(b)** the orthorhombic system.

3.46. The comment in Problem 3.45 about families of directions also applies to families of planes. Figure 3–32 illustrates the six members of the $\{100\}$ family of planes for the cubic system. List the members of the $\{100\}$ family for **(a)** the tetragonal system, and **(b)** the orthorhombic system.

3.47. **(a)** List the first three lattice points (including the 000 point) lying on the [112] direction in the fcc lattice. **(b)** Illustrate your answer to part (a) with a sketch.

3.48. Repeat Problem 3.47 for the bcc lattice.

3.49. Repeat Problem 3.47 for the bct lattice.

3.50. Repeat Problem 3.47 for the base-centered orthorhombic lattice.

3.51. In the cubic system, which of the $\langle 110 \rangle$ family of directions represents the line of intersection between the (111) and $(11\bar{1})$ planes? (Note the comment in Problem 3.35.)

3.52. Sketch the directions and planar intersection described in Problem 3.51.

3.53. Sketch the members of the $\{100\}$ family of planes in the triclinic system.

• **3.54.** The first eight planes that give x-ray diffraction peaks for aluminum are indicated in Figure 3–39. Sketch each plane and its intercepts relative to a cubic unit cell. (To avoid confusion, use a separate sketch for each plane.)

• **3.55.** **(a)** List the $\langle 112 \rangle$ family of directions in the cubic system. **(b)** Sketch this family. (You will want to use more than one sketch.)

3.56. In Figures 3–4b and 3–5b we show atoms and fractional atoms making up a unit cell. An alternative convention is to describe the unit cell in terms of "equivalent points." For example, the two atoms in the bcc unit cell can be considered to be one corner atom at 000 and one body-centered atom at $\frac{1}{2}\frac{1}{2}\frac{1}{2}$. The one corner atom is equivalent to the eight $\frac{1}{8}$ atoms shown in Figure 3–4b. In a similar way, identify the four atoms associated with equivalent points in the fcc structure.

3.57. Identify the atoms associated with equivalent points in the hcp structure. (See Problem 3.56.)

3.58. Repeat Problem 3.57 for the body-centered orthorhombic lattice.

3.59. Repeat Problem 3.57 for the base-centered orthorhombic lattice.

3.60. Sketch the $[1\bar{1}0]$ direction within the (111) plane relative to an fcc unit cell. Include all atom center positions within the plane of interest.

3.61. Sketch the $[1\bar{1}1]$ direction within the (110) plane relative to a bcc unit cell. Include all atom center positions within the plane of interest.

3.62. Sketch the $[11\bar{2}0]$ direction within the (0001) plane relative to an hcp unit cell. Include all atom center positions within the plane of interest.

• **3.63.** The $\frac{1}{4}\frac{1}{4}\frac{1}{4}$ position in the fcc structure is a "tetrahedral site," an interstice with fourfold atomic coordination. The $\frac{1}{2}\frac{1}{2}\frac{1}{2}$ position is an "octahedral site," an interstice with sixfold atomic coordination. How many tetrahedral and octahedral sites are there per fcc unit cell? Use a sketch to illustrate your answer.

• **3.64.** The first eight planes that give x-ray diffraction peaks for aluminum are indicated in Figure 3–39. Sketch each plane relative to the fcc unit cell (Figure 3–5a) and emphasize atom positions within the planes. (Note Problem 3.54 and use a separate sketch for each plane.)

3.65. Calculate the linear density of ions along the [111] direction in UO_2, which has the CaF_2 structure (Figure 3–10).

3.66. Calculate the linear density of ions along the [111] direction in $CaTiO_3$ (Figure 3–14).

3.67. Identify the ions associated with equivalent points in the NaCl structure. (Note Problem 3.56).

3.68. Identify the ions associated with equivalent points in the $CaTiO_3$ structure. (Note Problem 3.56.)

3.69. Calculate the planar density of ions in the (111) plane of $CaTiO_3$.

• **3.70.** Sketch the ion positions in a (111) plane through the cristobalite unit cell (Figure 3–11).

• **3.71.** Sketch the ion positions in a (101) plane through the cristobalite unit cell (Figure 3–11).

3.72. Calculate the linear density of ions along the [111] direction in zinc blende (Figure 3–24).

3.73. Calculate the planar density of ions along the (111) plane in zinc blende (Figure 3–24).

3.74. Identify the ions associated with equivalent points in the diamond cubic structure. (Note Problem 3.56.)

3.75. Identify the ions associated with equivalent points in the zinc blende structure. (Note Problem 3.56.)

3.76. Identify the ions associated with equivalent points in the wurtzite structure. (Note Problem 3.56.)

• **3.77.** Calculate the ionic packing factor for the wurtzite structure (Figure 3–25).

Section 3.7 • X-Ray Diffraction

3.78. The diffraction peaks labeled in Figure 3–39 correspond to the reflection rules for an fcc metal (h, k, l unmixed, as shown in Table 3.4). What would be the (hkl) indices for the three lowest diffraction angle peaks for a bcc metal?

3.79. Using the result of Problem 3.78, calculate the diffraction angles (2θ) for the first three peaks in the diffraction pattern of α-Fe powder using CuK$_\alpha$-radiation ($\lambda = 0.1542$ nm).

3.80. Repeat Problem 3.79 using CrK$_\alpha$-radiation ($\lambda = 0.2291$ nm).

3.81. Repeat Problem 3.78 for the *next* three lowest diffraction angle peaks for a bcc metal.

3.82. Repeat Problem 3.79 for the *next* three lowest diffraction angle peaks for α-Fe powder using CuK$_\alpha$-radiation.

3.83. Assuming the relative peak heights would be the same for given (hkl) planes, sketch a diffraction pattern similar to Figure 3–39 for copper powder using CuK$_\alpha$-radiation. Cover the range of $20° < 2\theta < 90°$.

3.84. Repeat Problem 3.83 for lead powder.

• **3.85.** What would be the (hkl) indices for the three lowest diffraction angle peaks for an hcp metal?

• **3.86.** Using the result of Problem 3.85, calculate the diffraction angles (2θ) for the first three peaks in the diffraction pattern of magnesium powder using CuK$_\alpha$-radiation ($\lambda = 0.1542$ nm). Note that the c/a ratio for Mg is 1.62.

• **3.87.** Repeat Problem 3.86 for CrK$_\alpha$-radiation ($\lambda = 0.2291$ nm).

3.88. Calculate the first six diffraction peak positions for MgO powder using CuK$_\alpha$-radiation. (This ceramic structure based on the fcc lattice shares the reflection rules of the fcc metals.)

3.89. Repeat Problem 3.88 for CrK$_\alpha$-radiation ($\lambda = 0.2291$ nm).

• **3.90.** The first three diffraction peaks of a metal powder are $2\theta = 44.4°, 64.6°$, and $81.7°$ using CuK$_\alpha$-radiation. Is this a bcc or an fcc metal?

• **3.91.** More specifically, is the metal powder in Problem 3.90 Cr, Ni, Ag, or W?

3.92. What would the positions of the first three diffraction peaks in Problem 3.90 have been using CrK$_\alpha$-radiation ($\lambda = 0.2291$ nm)?

CHAPTER 4

Crystal Defects and Noncrystalline Structure— Imperfection

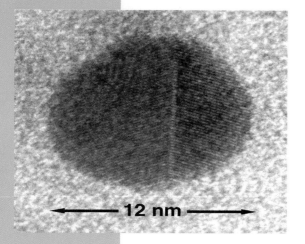

← 12 nm →

As with the chapter-opening photograph for Chapter 3, this transmission electron micrograph provides an atomic-resolution image of a crystalline compound, viz. a small crystal of zinc selenide embedded in a glass matrix. By viewing individual crystal lattice planes in ZnSe, we can see a distinctive image of a vertical twin boundary (shown schematically in Figure 14–15). This ZnSe "quantum dot" is the basis of a blue light laser. (Courtesy of V. J. Leppert and S. H. Risbud, University of California, Davis and M. J. Fendorf, National Center for Electron Microscopy, Berkeley, California.)

115

In Chapter 3 we looked at a wide variety of atomic-scale structures characteristic of important engineering materials. The primary limitation of Chapter 3 was that it dealt only with the perfectly repetitive crystalline structures. As you have learned long before this first course in engineering materials, nothing in our world is quite perfect. No crystalline material exists that does not have at least a few structural flaws. In this chapter we will systematically survey these imperfections.

Our first consideration is that no material can be prepared without some degree of chemical impurity. The impurity atoms or ions in the resulting *solid solution* serve to alter the structural regularity of the ideally pure material.

Independent of impurities, there are numerous structural flaws that represent a loss of crystalline perfection. The simplest type of flaw is the *point defect,* for example, a missing atom (vacancy). This type of flaw is the inevitable result of the normal thermal vibration of atoms in any solid at a temperature above absolute zero. Linear defects, or *dislocations,* follow an extended and sometimes complex path through the crystal structure. *Planar defects* represent the boundary between a nearly perfect crystalline region and its surroundings. Some materials are completely lacking in crystalline order. Common window glass is such a *noncrystalline solid.* Traditional concepts of order and disorder are being challenged by the study of *quasicrystals,* which expand our understanding of structural geometry at the atomic level.

We conclude this chapter with a brief introduction to microscopy, a set of powerful tools for inspecting structural order and disorder.

4.1   THE SOLID SOLUTION—CHEMICAL IMPERFECTION

It is not possible to avoid some contamination of practical materials. Even high-purity semiconductor products have some measurable level of impurity atoms. Many engineering materials contain significant amounts of several different components. Commercial metal alloys are examples. As a result, all materials that the engineer deals with on a daily basis are actually **solid solutions.** At first, the concept of a solid solution may be difficult to grasp. In fact, it is essentially equivalent to the more familiar liquid solution, such as the water–alcohol system shown in Figure 4–1. The complete solubility of alcohol in water is the result of complete molecular mixing. A similar result is seen in Figure 4–2, which shows a solid solution of copper and nickel atoms sharing the fcc crystal structure. Nickel acts as a *solute* dissolving in the copper *solvent.* This particular configuration is referred to as a *substitutional solid solution* because the nickel atoms are substituting for copper atoms on the fcc atom sites. This configuration will tend to occur when the atoms do not differ greatly in size. The water–alcohol system shown in Figure 4–1 represents two liquids completely soluble in each other in all proportions.

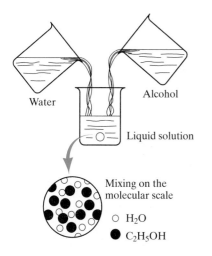

Figure 4-1 *Forming a liquid solution of water and alcohol. Mixing occurs on the molecular scale.*

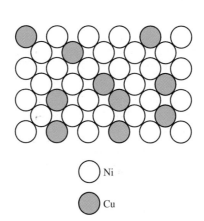

Figure 4-2 *Solid solution of copper in nickel shown along a (100) plane. This is a* substitutional solid solution *with nickel atoms substituting for copper atoms on fcc atom sites.*

For this complete miscibility to occur in metallic solid solutions, the two metals must be quite similar, as defined by the **Hume-Rothery*** **rules:**

1. Less than about 15% difference in atomic radii
2. The same crystal structure
3. Similar electronegativities (the ability of the atom to attract an electron)
4. The same valence

If one or more of the Hume-Rothery rules are violated, only partial solubility is possible. For example, less than 2 at % (atomic percent) silicon is soluble in aluminum. Inspection of Appendixes 1 and 2 shows that Al and Si violate rules 1, 2, and 4. Regarding rule 3, Figure 2–21 shows that the electronegativities of Al and Si are quite different, despite their adjacent positions on the periodic table.

Figure 4–2 shows a **random solid solution.** By contrast, some systems form **ordered solid solutions.** A good example is the alloy $AuCu_3$, shown in Figure 4–3. At high temperatures (above 390°C), thermal agitation keeps a random distribution of the Au and Cu atoms among the fcc sites. Below approximately 390°C, the Cu atoms preferentially occupy the face-centered

---

*   William Hume-Rothery (1899–1968), British metallurgist, made major contributions to theoretical and experimental metallurgy as well as metallurgical education. His empirical rules of solid solution formation have been a practical guide to alloy design for over 50 years.

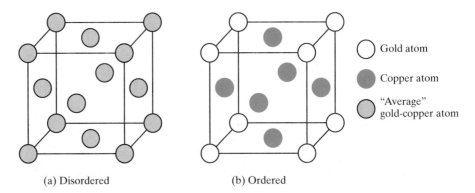

(a) Disordered   (b) Ordered

Figure 4-3 *Ordering of the solid solution in the AuCu₃ alloy system. (a) Above ~ 390° C, there is a random distribution of the Au and Cu atoms among the fcc sites. (b) Below ~ 390° C, the Au atoms preferentially occupy the corner positions in the unit cell, giving a simple cubic Bravais lattice. (From B. D. Cullity,* Elements of X-Ray Diffraction, *2nd ed., Addison-Wesley Publishing Co., Inc., Reading, Mass., 1978.)*

positions, and the Au atoms preferentially occupy corner positions in the unit cell. Ordering may produce a new crystal structure similar to some of the ceramic compound structures. For AuCu₃ at low temperatures, the compoundlike structure is based on a simple cubic Bravais lattice.

When atom sizes differ greatly, substitution of the smaller atom on a crystal structure site may be energetically unstable. In this case, it is more stable for the smaller atom simply to fit into one of the spaces, or interstices, among adjacent atoms in the crystal structure. Such an **interstitial solid solution** is shown in Figure 4–4, which shows carbon dissolved interstitially in α-Fe. This interstitial solution is a dominant phase in steels. Although more stable than a substitutional configuration of C atoms on Fe lattice sites, the interstitial structure of Figure 4–4 produces considerable strain locally to the α-Fe crystal structure, and less than 0.1 at % C is soluble in α-Fe.

To this point, we have looked at solid-solution formation in which a pure metal or semiconductor solvent dissolves some solute atoms either substitutionally or interstitially. The principles of substitutional solid-solution

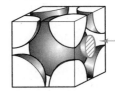

C atom dissolved interstitially at a $\frac{1}{2} 0 \frac{1}{2}$-type position in the bcc structure of α-Fe

Figure 4-4 *Interstitial solid solution of carbon in α-iron. The carbon atom is small enough to fit with some strain in the interstice (or opening) among adjacent Fe atoms in this structure of importance to the steel industry. [This unit cell structure can be compared with Figure 3–4b.]*

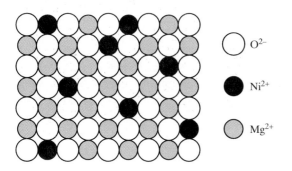

Figure 4-5 *Random, substitutional solid solution of NiO in MgO. The O^{2-} arrangement is unaffected. The substitution occurs among Ni^{2+} and Mg^{2+} ions.*

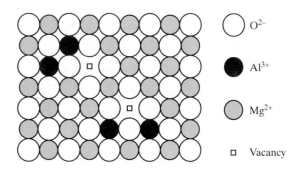

Figure 4-6 *A substitutional solid solution of Al_2O_3 in MgO is not as simple as the case of NiO in MgO (Figure 4–5). The requirement of charge neutrality in the overall compound permits only two Al^{3+} ions to fill every three Mg^{2+} vacant sites, leaving one Mg^{2+} vacancy.*

formation in these elemental systems also apply to compounds. For example, Figure 4–5 shows a random, substitutional solid solution of NiO in MgO. Here the O^{2-} arrangement is unaffected. The substitution occurs between Ni^{2+} and Mg^{2+}. The example of Figure 4–5 is a relatively simple one. In general, the charged state for ions in a compound affects the nature of the substitution. In other words, one could not indiscriminately replace all of the Ni^{2+} ions in Figure 4–5 with Al^{3+} ions. This would be equivalent to forming a solid solution of Al_2O_3 in MgO, each having distinctly different formulas and crystal structures. The higher valence of Al^{3+} would give a net positive charge to the oxide compound, creating a highly unstable condition. As a result, an additional ground rule in forming compound solid solutions is the maintenance of charge neutrality.

Figure 4–6 shows how charge neutrality is maintained in a dilute solution of Al^{3+} in MgO by having only two Al^{3+} ions fill every three Mg^{2+} sites. This leaves one Mg^{2+} site vacancy for each two Al^{3+} substitutions. This type of vacancy and several other point defects will be discussed further in Section 4.2. This example of a defect compound suggests the possibility of an even more subtle type of solid solution. Figure 4–7 shows a **nonstoichiomet-**

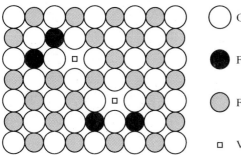

Figure 4-7 *Iron oxide, $Fe_{1-x}O$ with $x \approx 0.05$, is an example of a nonstoichiometric compound. Similar to the case of Figure 4–6, both Fe^{2+} and Fe^{3+} ions occupy the cation sites with one Fe^{2+} vacancy occurring for every two Fe^{3+} ions present.*

ric compound, $Fe_{1-x}O$, in which x is ~0.05. An ideally stoichiometric FeO would be identical to MgO with a NaCl-type crystal structure consisting of equal numbers of Fe^{2+} and O^{2-} ions. However, ideal FeO is never found in nature due to the multivalent nature of iron. Some Fe^{3+} ions are always present. As a result, these Fe^{3+} ions play the same role in the $Fe_{1-x}O$ structure as Al^{3+} in the Al_2O_3 in MgO solid solution of Figure 4–6. One Fe^{2+} site vacancy is required to compensate for the presence of every two Fe^{3+} ions in order to maintain charge neutrality.

SAMPLE PROBLEM 4.1

Do Cu and Ni satisfy Hume-Rothery's first rule for complete solid solubility?

SOLUTION

From Appendix 2,

$$r_{Cu} = 0.128 \text{ nm}$$

$$r_{Ni} = 0.125 \text{ nm}$$

$$\%\text{difference} = \frac{(0.128 - 0.125) \text{ nm}}{0.128 \text{ nm}} \times 100$$

$$= 2.3\% (< 15\%)$$

Therefore, yes.
    In fact, all four rules are satisfied by these two neighbors from the periodic table (in agreement with the observation that they are completely soluble in all proportions).

SAMPLE PROBLEM 4.2

How much "oversize" is the C atom in α-Fe? (See Figure 4–4.)

SOLUTION

By inspection of Figure 4–4, it is apparent that an ideal interstitial atom centered at $\frac{1}{2}0\frac{1}{2}$ would just touch the surface of the iron atom in the center of the unit cell cube. The radius of such an ideal interstitial would be

$$r_{\text{interstitial}} = \frac{1}{2}a - R$$

where a is the length of the unit cell edge and R is the radius of an iron atom.

Remembering Figure 3–4, we note that

length of unit cell body diagonal $= 4\,R$

$$= \sqrt{3}a$$

or

$$a = \frac{4}{\sqrt{3}}R$$

as given in Table 3.3. Then

$$r_{\text{interstitial}} = \frac{1}{2}\left(\frac{4}{\sqrt{3}}R\right) - R = 0.1547\,R$$

From Appendix 2, $R = 0.124$ nm, giving

$$r_{\text{interstitial}} = 0.1547(0.124 \text{ nm}) = 0.0192 \text{ nm}$$

But Appendix 2 gives $r_{\text{carbon}} = 0.077$ nm, or

$$\frac{r_{\text{carbon}}}{r_{\text{interstitial}}} = \frac{0.077 \text{ nm}}{0.0192 \text{ nm}} = 4.01$$

Therefore, the carbon atom is roughly *four times too large* to fit next to the adjacent iron atoms without strain. The severe local distortion required for this accommodation leads to the low solubility of C in α-Fe (< 0.1 at %).

....

PRACTICE PROBLEM 4.1

Copper and nickel (which are completely soluble in each other) satisfy the first Hume-Rothery rule of solid solubility, as shown in Sample Problem 4.1. Aluminum and silicon are soluble in each other to only a limited degree. Do they satisfy the first Hume-Rothery rule?

The interstitial site for dissolving a carbon atom in α-Fe was shown in Figure 4–4. Sample Problem 4.2 shows that a carbon atom is more than four times too large for the site and, consequently, carbon solubility in α-Fe is quite low. Consider now the case for interstitial solution of carbon in the high-temperature (fcc) structure of γ-Fe. The largest interstitial site for a carbon atom is a $\frac{1}{2}01$ type. **(a)** Sketch this interstitial solution in a manner similar to Figure 4–4. **(b)** Determine by how much the C atom in γ-Fe is oversize. (Note that the atomic radius for fcc iron is 0.127 nm.)

4.2   POINT DEFECTS—ZERO-DIMENSIONAL IMPERFECTIONS

Structural defects exist in real materials independently of chemical impurities. Imperfections associated with the crystalline point lattice are called **point defects.** Figure 4–8 illustrates the two common types of point defects associated with elemental solids: (1) The **vacancy** is simply an unoccupied atom site in the crystal structure. (2) The **interstitial,** or **interstitialcy,** is an atom occupying an interstitial site not normally occupied by an atom in the perfect crystal structure or an extra atom inserted into the perfect crystal structure such that two atoms occupy positions close to a singly occupied atomic site in the perfect structure. In the preceding section, we saw how vacancies can be produced in compounds as a response to chemical impurities and nonstoichiometric compositions. Such vacancies can also occur independently of these chemical factors (e.g., by the thermal vibration of atoms in a solid above a temperature of absolute zero).

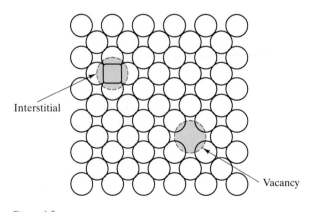

Figure 4-8 *Two common point defects in metal or elemental semiconductor structures are the vacancy and the interstitial.*

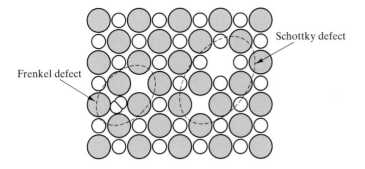

Figure 4-9 *Two common point defect structures in compound structures are the Schottky defect and the Frenkel defect. Note their similarity to the structures of Figure 4–8.*

Figure 4–9 illustrates the two analogs of the vacancy and interstitialcy for compounds. The **Schottky* defect** is a pair of oppositely charged ion vacancies. This pairing is required in order to maintain local charge neutrality in the compound's crystal structure. The **Frenkel defect**[†] is a vacancy–interstitialcy combination. Most of the compound crystal structures described in Chapter 3 were too "tight" to allow Frenkel defect formation. However, the relatively open CaF_2-type structure can accommodate cation interstitials without excessive lattice strain. Defect structures in compounds can be further complicated by charging due to "electron trapping" or "electron hole trapping" at these lattice imperfections. We shall not dwell on these more complex systems now, but they can have important implications for optical properties (Chapter 16).

SAMPLE PROBLEM 4.3

The fraction of vacant lattice sites in a crystal is typically small. For example, the fraction of aluminum sites vacant at 400°C is 2.29×10^{-5}. Calculate the density of these sites (in units of m^{-3}).

---

* Walter Hans Schottky (1886–1976), German physicist, was the son of a prominent mathematician. Besides identifying the *Schottky defect,* he invented the screen-grid tube (in 1915) and discovered the *Schottky effect* of thermionic emission, that is, the current of electrons leaving a heated metal surface increases when an external electrical field is applied.

† Yakov Ilyich Frenkel (1894–1954), Russian physicist, made significant contributions to a wide range of areas, including solid-state physics, electrodynamics, and geophysics. Although his name is best remembered in conjunction with defect structure, he was an especially strong contributor to the understanding of ferromagnetism (to be discussed in Chapter 18).

SOLUTION

From Appendix 1, we find the density of aluminum to be $2.70\,\text{Mg/m}^3$ and its atomic mass to be 26.98 amu. The corresponding density of aluminum atoms is then

$$\text{at. density} = \frac{\rho}{\text{at. mass}} = \frac{2.70 \times 10^6 \,\text{g/m}^3}{26.98 \,\text{g}/0.602 \times 10^{24}\text{atoms}}$$

$$= 6.02 \times 10^{28} \,\text{atoms·m}^{-3}$$

Then the density of vacant sites will be

$$\text{vac. density} = 2.29 \times 10^{-5} \,\text{atom}^{-1} \times 6.02 \times 10^{28} \,\text{atoms·m}^{-3}$$

$$= 1.38 \times 10^{24} \,\text{m}^{-3}$$

PRACTICE PROBLEM 4.3

Calculate the density of vacant sites (in m^{-3}) for aluminum at $660°\text{C}$ (just below its melting point) where the fraction of vacant lattice sites is 8.82×10^{-4}. (See Sample Problem 4.3.)

4.3   LINEAR DEFECTS, OR DISLOCATIONS—ONE-DIMENSIONAL IMPERFECTIONS

We have seen that point (zero-dimensional) defects are structural imperfections resulting from thermal agitation. **Linear defects,** which are one-dimensional, are associated primarily with mechanical deformation. Linear defects are also known as **dislocations.** An especially simple example is shown in Figure 4–10. The linear defect is commonly designated by the "in-

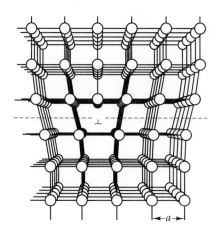

Figure 4-10   *Edge dislocation. The linear defect is represented by the edge of an extra half-plane of atoms. (From A. G. Guy,* Elements of Physical Metallurgy, *Addison-Wesley Publishing Co., Inc., Reading, Mass., 1959.)*

verted T" symbol (⊥), which represents the edge of an *extra half-plane of atoms.* Such a configuration lends itself to a simple quantitative designation, the **Burgers*   vector, b.** This parameter is simply the displacement vector necessary to close a stepwise loop around the defect. In the perfect crystal (Figure 4–11a), an $m \times n$ atomic step loop closes at the starting point. In the region of a dislocation (Figure 4–11b), the same loop fails to close. The closure vector **(b)** represents the magnitude of the structural defect. In Chapter 6, we shall see that the magnitude of **b** for the common metal structures (bcc, fcc, and hcp) is simply the repeat distance along the highest atomic density direction (the direction in which atoms are touching).

Figure 4–10 represents a specific type of linear defect, the **edge dislocation,** so named because the defect, or *dislocation line,* runs along the edge of the extra row of atoms. *For the edge dislocation, the Burgers vector is perpendicular to the dislocation line.* Figure 4–12 shows a fundamentally different type of linear defect, the **screw dislocation,** which derives its name from the spiral stacking of crystal planes around the dislocation line. *For the screw dislocation, the Burgers vector is parallel to the dislocation line.* The edge and screw dislocations can be considered the pure extremes of linear defect structure. Most linear defects in actual materials will be mixed, as shown in Figure 4–13. In this general case, the **mixed dislocation** has both edge and

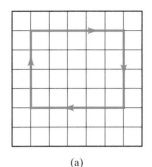

(a)

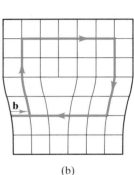

(b)

Figure 4-11 *Definition of the Burgers vector,* **b,** *relative to an edge dislocation. (a) In the perfect crystal, an $m \times n$ atomic step loop closes at the starting point. (b) In the region of a dislocation, the same loop does not close, and the closure vector* **(b)** *represents the magnitude of the structural defect. For the edge dislocation, the Burgers vector is* perpendicular *to the dislocation line.*

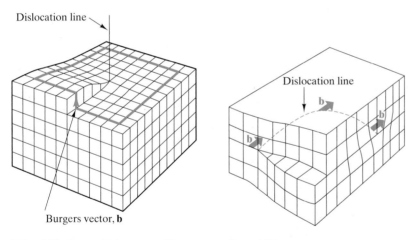

Dislocation line

Burgers vector, **b**

Figure 4-12 *Screw dislocation. The spiral stacking of crystal planes leads to the Burgers vector being parallel to the dislocation line.*

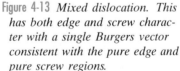

Dislocation line

Figure 4-13 *Mixed dislocation. This has both edge and screw character with a single Burgers vector consistent with the pure edge and pure screw regions.*

----

*   Johannes Martinus Burgers (1895–1981), Dutch-American fluid mechanician. Although his highly productive career centered on aero- and hydrodynamics, a brief investigation of dislocation structure around 1940 has made Burgers's name one of the best known in materials science. He was first to identify the convenience and utility of the closure vector for characterizing a dislocation.

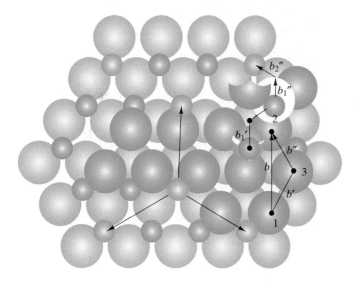

Figure 4-14 *Burgers vector for the aluminum oxide structure. The large repeat distance in this relatively complex structure causes the Burgers vector to be broken up into two (for O^{2-}) or four (for Al^{3+}) partial dislocations, each representing a smaller slip step. This complexity is associated with the brittleness of ceramics compared to metals. (From W. D. Kingery, H. K. Bowen, and D. R. Uhlmann,* Introduction to Ceramics, *2nd ed., John Wiley & Sons, Inc., New York, 1976.)*

screw character. The Burgers vector for the mixed dislocation is neither perpendicular nor parallel to the dislocation line but retains a fixed orientation in space consistent with the previous definitions for the pure edge and pure screw regions. The local atomic structure around a mixed dislocation is difficult to visualize, but the Burgers vector provides a convenient and simple description. In compound structures, even the basic Burgers vector designation can be relatively complicated. Figure 4–14 shows the Burgers vector for the aluminum oxide structure (Section 3.3). The complication arises from the relatively large repeat distance in this crystal structure, which causes the total dislocation designated by the Burgers vector to be broken up into two (for O^{2-}) or four (for Al^{3+}) partial dislocations. In Chapter 6 we will see that the complexity of dislocation structures has a good deal to do with the basic mechanical behavior of the material.

SAMPLE PROBLEM 4.4

Calculate the magnitude of the Burgers vector for **(a)** α-Fe, **(b)** Al, and **(c)** Al_2O_3.

SOLUTION

(a) As noted in the opening of this section, |**b**| is merely the repeat distance between adjacent atoms along the highest atomic density direction. For α-Fe, a bcc metal, this tends to be along the body diagonal of a unit cell. We saw in Figure 3–4 that Fe atoms are in contact along the body diagonal. As a result, the atomic repeat distance is

$$r = 2R_{Fe}$$

Using Appendix 2, we can then calculate in a simple way,

$$|\mathbf{b}| = r = 2(0.124 \text{ nm}) = 0.248 \text{ nm}$$

(b) Similarly, the highest atomic denisty direction in fcc metals such as Al tends to be along the face diagonal of a unit cell. As shown in Figure 3–5, this is also a line of contact for atoms in an fcc structure. Again,

$$|\mathbf{b}| = r = 2R_{Al} = 2(0.143 \text{ nm})$$
$$= 0.286 \text{ nm}$$

(c) Figure 4–14 shows how things are more complex for ceramics. The total slip vector connects two O^{2-} ions (labeled 1 and 2):

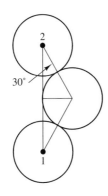

Thus

$$|\mathbf{b}| = (2)(2R_{O^{2-}})(\cos\ 30°)$$

Using Appendix 2 gives us

$$|\mathbf{b}| = (2)(2 \times 0.132 \text{ nm})(\cos\ 30°)$$
$$= 0.457 \text{ nm}$$

PRACTICE PROBLEM 4.4

Calculate the magnitude of the Burgers vector for an hcp metal, Mg. (See Sample Problem 4.4.)

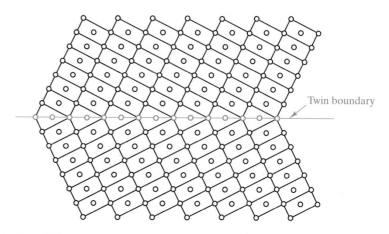

Figure 4-15 *A twin boundary separates two crystalline regions that are, structurally, mirror images of each other.*

4.4 PLANAR DEFECTS—TWO-DIMENSIONAL IMPERFECTIONS

Point defects and linear defects are acknowledgments that crystalline materials cannot be made flaw-free. These imperfections exist in the interior of each of these materials. But we must also consider that we are limited to a finite amount of any material, and it is contained within some boundary surface. This surface is, in itself, a disruption of the atomic stacking arrangement of the crystal. There are various forms of planar defects. We shall briefly list them beginning with the simplest geometrically.

Figure 4–15 illustrates a **twin boundary,** which separates two crystalline regions that are, structurally, mirror images of each other. This highly symmetrical discontinuity in structure can be produced by deformation (e.g., in bcc and hcp metals) and by annealing (e.g., in fcc metals).

All crystalline materials do not exhibit twin boundaries, but all must have a *surface.* A simple view of the crystalline surface is given in Figure 4–16. This is little more than an abrupt end to the regular atomic stacking arrangement. One should note that this schematic illustration indicates that the surface atoms are somehow different from interior (or "bulk") atoms. This is the result of different coordination numbers for the surface atoms leading to different bonding strengths and some asymmetry. A more detailed picture of atomic-scale surface geometry is shown in Figure 4–17. This **Hirth-Pound* model** of a crystal surface has elaborate ledge systems rather than atomically smooth planes.

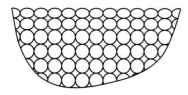

Figure 4-16 *Simple view of the surface of a crystalline material.*

---

* John Price Hirth (1930–) and Guy Marshall Pound (1920–1988), American metallurgists, formulated their model of crystal surfaces in the late 1950s after careful analysis of the kinetics of vaporization.

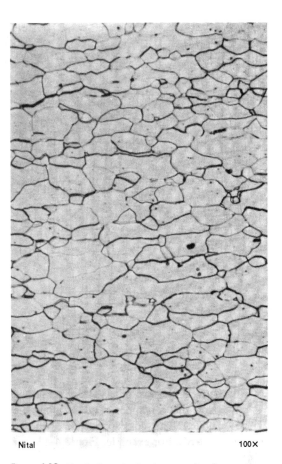

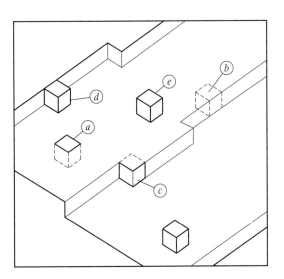

Figure 4-17 *A more detailed model of the elabo-rate ledgelike structure of the surface of a crys-talline material. Each cube represents a single atom. [From J. P. Hirth and G. M. Pound,* J. Chem. Phys. *26, 1216 (1957).]*

Figure 4-18 *Typical optical micrograph of a grain structure, 100×. The material is a low-carbon steel. The grain boundaries have been lightly etched with a chemical solution so that they re-flect light differently from the polished grains, thereby giving a distinctive contrast. (From* Met-als Handbook, *8th ed., Vol. 7:* Atlas of Mi-crostructures of Industrial Alloys, *American Society for Metals, Metals Park, Ohio, 1972.)*

The most important planar defect for our consideration in this intro-ductory course occurs at the **grain boundary,** the region between two ad-jacent single crystals, or **grains.** In the most common planar defect, the grains meeting at the boundary have different orientations. Aside from the electronics industry, most practical engineering materials are polycrystalline rather than in the form of single crystals. The predominant microstructural feature (i.e., microscopic-scale architecture as discussed in Section 1.4) of many engineering materials is the grain structure (Figure 4–18). Many ma-terials' properties are highly sensitive to such grain structures. What, then,

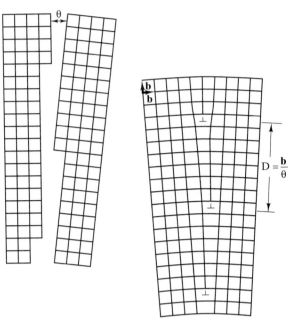

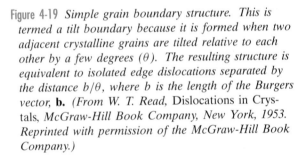

Figure 4-19 *Simple grain boundary structure. This is termed a tilt boundary because it is formed when two adjacent crystalline grains are tilted relative to each other by a few degrees (θ). The resulting structure is equivalent to isolated edge dislocations separated by the distance b/θ, where b is the length of the Burgers vector,* **b.** *(From W. T. Read,* Dislocations in Crystals, *McGraw-Hill Book Company, New York, 1953. Reprinted with permission of the McGraw-Hill Book Company.)*

is the structure of a grain boundary on the atomic scale? The answer to this depends greatly on the relative orientations of the adjacent grains.

Figure 4–19 illustrates an unusually simple grain boundary produced when two adjacent grains are tilted only a few degrees relative to each other. This **tilt boundary** is accommodated by a few isolated edge dislocations (see Section 4.3). Most grain boundaries involve adjacent grains at some arbitrary and rather large misorientation angle. The grain boundary structure in this general case is considerably more complex than that shown in Figure 4–19. However, considerable progress has been made in the past two decades in understanding the nature of the structure of the general, high-angle grain boundary. Advances in both electron microscopy and computer modeling techniques have played primary roles in this improved understanding. A central component, now, in the analysis of grain boundary structure is the concept of the **coincident site lattice** (CSL), illustrated in Figure 4–20. A high-angle tilt boundary ($\theta = 36.9°$) between two simple square lattices is shown in Figure 4–20a. This specific tilt angle has been found to occur

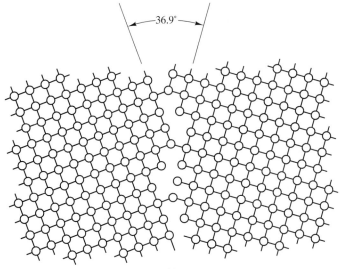

(a)

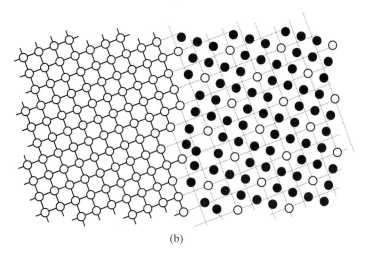

(b)

Figure 4-20 *(a) A high-angle ($\theta = 36.9°$) grain boundary between two square lattice grains can be represented by a coincidence site lattice, as shown in (b). As one in five of the atoms in the grain on the right is coincident with the lattice of the grain on the left, the boundary is said to have $\Sigma^{-1} = 1/5$, or $\Sigma = 5$.*

frequently in grain boundary structures in real materials. The reason for its stability is an especially high degree of registry between the two adjacent crystal lattices in the vicinity of the boundary region. (Note that a number of atoms along the boundary are common to each adjacent lattice.) This correspondence at the boundary has been quantified in terms of the CSL. For example, Figure 4–20b shows that, by extending the lattice grid of the crystalline grain on the left, one in five of the atoms of the grain on the right is coincident with that lattice. The fraction of coincident sites in the adjacent

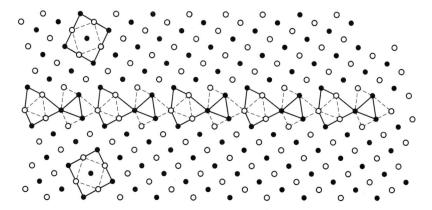

Figure 4-21 *A Σ5 boundary for an fcc metal, in which the [100] directions of two adjacent fcc grains are oriented at 36.9° to each other. (See Figure 4–20 also.) This is a three-dimensional projection with the open circles and closed circles representing atoms on two different, adjacent planes (each parallel to the plane of this page). Polyhedra formed by drawing straight lines between adjacent atoms in the grain boundary are irregular in shape due to the misorientation angle, 36.9°, but reappear at regular intervals due to the crystallinity of each grain. The crystalline grains can be considered to be composed completely of tetrahedra and octahedra.* [Reprinted with permission from M. F. Ashby, F. Spaepen, and S. Williams, Acta Metall. 26, 1647 (1978), Copyright 1978, Pergamon Press, Ltd.]

grain can be represented by the symbol $\Sigma^{-1} = 1/5$ or $\Sigma = 5$, leading to the label for the structure in Figure 4–20 as a "Σ5 boundary." The geometry of the overlap of the two lattices also indicates why the particular angle of $\theta = 36.9°$ arises. One can demonstrate that $\theta = 2 \tan^{-1}(1/3)$.

Another indication of the regularity of certain high-angle grain boundary structures is given in Figure 4–21, which illustrates a Σ5 boundary in an fcc metal. Polyhedra formed by drawing straight lines between adjacent atoms in the grain boundary region are irregular in shape due to the misorientation angle but reappear at regular intervals due to the crystallinity of each grain.

The theoretical and experimental studies of high-angle boundaries mentioned earlier have indicated that the simple, low-angle model of Figure 4–19 serves as a useful analogy for the high-angle case. Specifically, a grain boundary between two grains at some arbitrary, high angle will tend to consist of regions of good correspondence (with local boundary rotation to form a Σn structure, where n is a relatively low number) separated by **grain boundary dislocations** (GBD), linear defects within the boundary plane. The GBD associated with high-angle boundaries tend to be *secondary* in that they have Burgers vectors different from those found in the bulk material (*primary* dislocations).

With atomic-scale structure in mind, we can return to the microstructural view of grain structures (e.g., Figure 4–18). In describing microstructures, it is useful to have a simple index of *grain size*. A frequently used

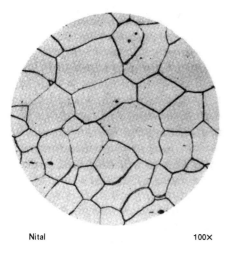

Figure 4-22 *Specimen for the calculation of the grain-size number, G, 100×. The material is a low-carbon steel similar to that shown in Figure 4–18.* (*From* Metals Handbook, *8th ed., Vol. 7:* Atlas of Microstructures of Industrial Alloys, *American Society for Metals, Metals Park, Ohio, 1972.*)

Nital   100×

parameter standardized by the American Society for Testing and Materials (ASTM) is the **grain-size number,** G, defined by

$$N = 2^{G-1} \qquad (4.1)$$

where N is the number of grains observed in an area of 1 in.^{2} ($= 645$ mm^2) on a photomicrograph taken at a magnification of 100 times (100×), as shown in Figure 4–22. The calculation of G follows.

There are 21 grains within the field of view and 22 grains cut by the circumference, giving

$$21 + \frac{22}{2} = 32 \text{ grains}$$

in a circular area with diameter $= 2.25$ in. The area density of grains is

$$N = \frac{32 \text{ grains}}{\pi(2.25/2)^2 \text{ in.}^2} = 8.04 \, \frac{\text{grains}}{\text{in.}^2}$$

From Equation 4.1,

$$N = 2^{(G-1)}$$

or

$$
\begin{aligned}
G &= \frac{\ln N}{\ln 2} + 1 \\
&= \frac{\ln(8.04)}{\ln 2} + 1 \\
&= 4.01
\end{aligned}
$$

Although the grain-size number is a useful indicator of average grain size, it has the disadvantage of being somewhat indirect. It would be useful to obtain an average value of *grain diameter* from a microstructural section. A simple indicator is to count the number of grains intersected per unit length, n_L, of a random line drawn across a micrograph. The average grain size is roughly indicated by the inverse of n_L, corrected for the magnification, M, of the micrograph. Of course, one must consider that the random line cutting across the micrograph (in itself, a random plane cutting through the microstructure) will not tend, on average, to go along the maximum diameter of a given grain. Even for a microstructure of uniform size grains, a given planar slice (micrograph) will show various size grain sections (e.g., Figure 4–22), and a random line would indicate a range of segment lengths defined by grain boundary intersections. In general, then, the true average grain diameter, d, is given by

$$d = \frac{C}{n_L M} \qquad (4.2)$$

where C is some constant greater than 1. Extensive analysis of the statistics of grain structures has led to various theoretical values for the constant, C. For typical microstructures, a value of $C = 1.5$ is adequate.

SAMPLE PROBLEM 4.5

Calculate the separation distance of dislocations in a low-angle ($\theta = 2°$) tilt boundary in aluminum.

SOLUTION

As calculated in Sample Problem 4.4b,

$$|\mathbf{b}| = 0.286 \text{ nm}$$

From Figure 4–19, we see that

$$D = \frac{|\mathbf{b}|}{\theta}$$

$$= \frac{0.286 \text{ nm}}{2° \times (1 \text{ rad}/57.3°)} = 8.19 \text{ nm}$$

SAMPLE PROBLEM 4.6

Find the grain-size number, G, for the microstructure in Figure 4–22 if the micrograph represents a magnification of 300× rather than 100×.

SOLUTION

There would still be $21 + 11 = 32$ grains in the 3.98-in.^{2} region. But to scale this grain density to $100\times$, we must note that the 3.98-in.^{2} area at $300\times$ would be comparable to an area at $100\times$ of

$$A_{100\mathrm{x}} = 3.98 \text{ in.}^2 \times \left(\frac{100}{300}\right)^2 = 0.442 \text{ in.}^2$$

Then the grain density becomes

$$N = \frac{32 \text{ grains}}{0.442 \text{ in.}^2} = 72.4 \text{ grains/in.}^2$$

Applying Equation 4.1 gives us

$$N = 2^{(G-1)}$$

or

$$\ln N = (G - 1)\ln 2$$

giving

$$G - 1 = \frac{\ln N}{\ln 2}$$

and, finally,

$$G = \frac{\ln N}{\ln 2} + 1$$

$$= \frac{\ln (72.4)}{\ln 2} + 1 = 7.18$$

or

$$G = 7+$$

....

PRACTICE PROBLEM 4.5

In Sample Problem 4.5 we find the separation distance between dislocations for a $2°$ tilt boundary in aluminum. Repeat this calculation for **(a)** $\theta = 1°$ and **(b)** $\theta = 5°$. **(c)** Plot the overall trend of D versus θ over the range $\theta = 0$ to $5°$.

PRACTICE PROBLEM 4.6

Figure 4–22 gives a sample calculation of grain-size number, G. Sample Problem 4.6 recalculates G assuming a magnification of $300\times$ rather than $100\times$. Repeat this process, assuming that the micrograph in Figure 4–22 is at $50\times$ rather than $100\times$.

4.5 NONCRYSTALLINE SOLIDS—THREE-DIMENSIONAL IMPERFECTIONS

Some engineering materials lack the repetitive, crystalline structure. These **noncrystalline,** or amorphous, **solids** are imperfect in three dimensions. The two-dimensional schematic of Figure 4–23a shows the repetitive structure of a hypothetical crystalline oxide. Figure 4–23b shows a noncrystalline version of this material. The latter structure is referred to as the **Zachariasen* model** and, in a simple way, it illustrates the important features of **oxide glass** structures. (Remember from Chapter 1 that glass generally refers to a non-crystalline material with chemical composition comparable to a ceramic.) The building block of the crystal (the AO_3^{3-} "triangle") is retained in the glass; that is, **short-range order** (SRO) is retained. But **long-range order** (LRO)—that is, crystallinity—is lost in the glass. The Zachariasen model is the visual definition of the **random network theory** of glass structure. This is the analog of the point lattice associated with crystal structure.

Our first example of a noncrystalline solid was the traditional oxide glass because many oxides (especially the silicates) are easy to form in a noncrystalline state. This is the direct result of the complexity of the oxide crystal structures. Rapidly cooling a liquid silicate or allowing a silicate vapor to condense on a cool substrate effectively "freezes in" the random stacking of silicate building blocks (SiO_4^{4-} tetrahedra). Since many silicate

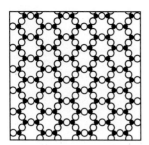

(a)

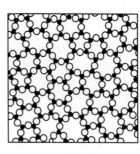

(b)

Figure 4-23 *Two-dimensional schematics give a comparison of (a) a crystalline oxide and (b) a noncrystalline oxide. The noncrystalline material retains short-range order (the triangularly coordinated building block) but loses long-range order (crystallinity). This illustration was also used to define glass in Chapter 1 (Figure 1–8).*

---

* William Houlder Zachariasen (1906–1980), Norwegian-American physicist, spent most of his career working in x-ray crystallography. But his description of glass structure in the early 1930s became a standard definition for the structure of this noncrystalline material.

glasses are made by rapidly cooling liquids, the term *supercooled liquid* is often used synonymously with *glass*. In fact, there is a distinction. The supercooled liquid is the material cooled just below the melting point, where it still behaves like a liquid (e.g., deforming by a viscous flow mechanism). The glass is the same material cooled to a sufficiently low temperature that it has become a truly rigid solid (e.g., deforming by an elastic mechanism). The relationship of these various terms is illustrated by Figure 6–40. The atomic mobility of the material at these low temperatures is insufficient for the theoretically more stable crystalline structures to form. Those semiconductors with structures similar to some ceramics can be made in amorphous forms also. There is an economic advantage to **amorphous semiconductors** compared to preparing high-quality single crystals. A disadvantage is the greater complexity of the electronic properties. As discussed in Section 3.4, the complex polymeric structure of plastics causes a substantial fraction of their volume to be noncrystalline.

Perhaps the most intriguing noncrystalline solids are the newest members of the class, **amorphous metals,** also known as *metallic glasses*. Because metallic crystal structures are typically simple in nature, they can be formed quite easily. It is necessary for liquid metals to be cooled very rapidly to prevent crystallization. Cooling rates of 1°C per microsecond are required in typical cases. This is an expensive process but potentially worthwhile due to the unique properties of these materials. For example, the uniformity of the noncrystalline structure eliminates the grain boundary structures associated with typical polycrystalline metals. This results in unusually high strengths and excellent corrosion resistance. Figure 4–24 illustrates a useful method for visualizing an amorphous metal structure: the **Bernal* model,** which is produced by drawing lines between the centers of adjacent atoms. The resulting polyhedra are comparable to those illustrating grain boundary structure in Figure 4–21. In the totally noncrystalline solid, the polyhedra are again irregular in shape but, of course, lack any repetitive stacking arrangement.

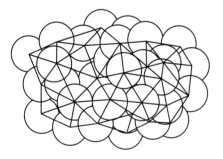

Figure 4-24 *Bernal model of an amorphous metal structure. The irregular stacking of atoms is represented as a connected set of polyhedra. Each polyhedron is produced by drawing lines between the centers of adjacent atoms. Such polyhedra are equivalent to those used to model grain boundary structure in Figure 4–21. In the noncrystalline solid, the polyhedra are not repetitive.*

---

* John Desmond Bernal (1901–1971), British physicist, was one of the pioneers in x-ray crystallography but is perhaps best remembered for his systematic descriptions of the irregular structure of liquids.

At this point it may be unfair to continue to use the term *imperfect* as a general description of noncrystalline solids. The Zachariasen structure (Figure 4–23b) is uniformly and "perfectly" random. Imperfections such as chemical impurities, however, can be defined relative to the uniformly noncrystalline structure as shown in Figure 4–25. Addition of Na^+ ions to silicate glass substantially increases formability of the material in the super-cooled liquid state (i.e., viscosity is reduced).

Finally, the state-of-the-art in our understanding of the structure of noncrystalline solids is represented by Figure 4–26, which shows the non-random arrangement of Ca^{2+} modifier ions in a CaO-SiO_2 glass. What we see in Figure 4–26 is, in fact, adjacent octahedra rather than Ca^{2+} ions. Each Ca^{2+} ion is coordinated by six O^{2-} ions in a perfect octahedral pattern. In turn, the octahedra tend to be arranged in a regular, edge-sharing fashion. This is in sharp contrast to the random distribution of Na^+ ions implied in Figure 4–25. The evidence for **medium-range order** in the study repre-sented by Figure 4–26 confirms long-standing theories of a tendency for some

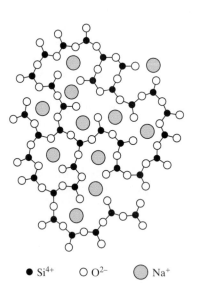

● Si^{4+} ○ O^{2-} ⬤ Na^+

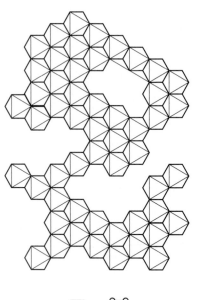

CaO_6

Figure 4-25 *A chemical impurity such as Na^+ is a glass modifier, breaking up the random network and leaving nonbridging oxy-gen ions. [From B. E. Warren, J. Am. Ceram. Soc. 24, 256 (1941).]*

Figure 4-26 *Schematic illustration of medium-range ordering in a CaO-SiO_2 glass. Edge-sharing CaO_6 octahedra have been identified by neutron diffraction experiments. [From P. H. Gaskell, et al., Na-ture 350, 675 (1991).]*

structural order to occur in the medium range of a few nanometers, between the well-known short-range order of the silica tetrahedra and the long-range randomness of the irregular linkage of those tetrahedra. As a practical matter, the random network model of Figure 4–23b is an adequate description of vitreous SiO_2. Medium-range order such as that in Figure 4–26 is, however, likely to be present in common glasses containing significant amounts of modifiers, such as Na_2O and CaO.

SAMPLE PROBLEM 4.7

Randomization of atomic packing in amorphous metals (e.g., Figure 4–24) generally causes no more than a 1% drop in density compared to the crystalline structure of the same composition. Calculate the atomic packing factor of an amorphous, thin film of nickel whose density is 8.84 g/cm^3.

SOLUTION

Appendix 1 indicates that the normal density for nickel (which would be in the crystalline state) is 8.91 g/cm^3. The atomic packing factor for the fcc metal structure is 0.74 (see Section 3.2). Therefore, the atomic packing factor for this amorphous nickel would be

$$APF = (0.74) \times \frac{8.84}{8.91} = 0.734$$

PRACTICE PROBLEM 4.7

Estimate the atomic packing factor of amorphous silicon if its density is reduced by 1% relative to the crystalline state. (Review Sample Problem 3.6.)

4.6 QUASICRYSTALS

In this chapter, we have moved methodically through discussions of increasingly disordered atomic structures, finally reaching Section 4.5, which describes materials completely lacking in long-range order. These noncrystalline solids are the antithesis of crystals. As a result, materials engineers had tended to think of materials as either crystalline or noncrystalline. A remarkable discovery by Dany Schechtman on April 8, 1982 challenged these simple categories, suggesting an intermediate structural state now referred to as a **quasicrystal.** The novel structural concepts generated to describe quasicrystals have improved our understanding of the nature of traditional crystals and glasses.

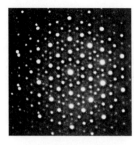

Figure 4-27 *Electron diffraction pattern of a rapidly cooled Al₆Mn alloy showing fivefold symmetry; that is, the pattern is identical with each rotation of 360°/5, or 72°, about its center. Such symmetry is impossible in traditional crystallography. [After D. Schechtman et al.,* Phys. Rev. Letters 53, *1951 (1984).]*

The Schechtman discovery that led to the idea of quasicrystals was an electron diffraction pattern of a micrometer-size crystallite of a rapidly cooled Al₆Mn alloy exhibiting *fivefold symmetry* (Figure 4–27). As will be discussed in Section 4.7, electrons can be diffracted from a crystal in much the same way as the x-rays described in Section 3.7. The symmetry of the diffraction pattern is a manifestation of the symmetry of the crystal itself. The startling nature of Figure 4–27 is that traditional crystals cannot have fivefold symmetry. You can appreciate this by trying to make a two-dimensional "crystal" out of pentagon-shaped unit cells. This impossible task leads us to tile a kitchen floor with squares or hexagons, not pentagons. How, then, can the "forbidden" diffraction pattern of Figure 4–27 be possible?

The initial clue about the structure of quasicrystals was provided by **Penrose* tilings,** intriguing two-dimensional patterns illustrated by the example in Figure 4–28. Penrose tilings were developed as an alternative to the impossible task of filling (or tiling) two-dimensional space with pentagons. The use of two rhombuses (one "skinny" and one "fat"), as shown in Figure 4–28a, allows two-dimensional space to be tiled, as shown in Figure 4–28b, a

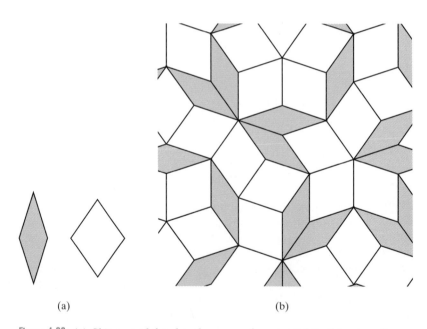

(a)   (b)

Figure 4-28 *(a) Skinny and fat rhombuses can be repeated in (b) a two-dimensional stacking to produce a space-filled pattern with fivefold symmetry. This Penrose tiling provides a schematic explanation for the diffraction pattern of Figure 4–27.*

---

*   Roger Penrose (1931–   ), British mathematician and physicist. Penrose's distinguished career in theoretical and mathematical physics has included important contributions to the study of black holes, general relativity, and quantum mechanics. His novel discovery of the fivefold symmetry tiling pattern was the result of "recreational mathematics" outside of his normal research areas.

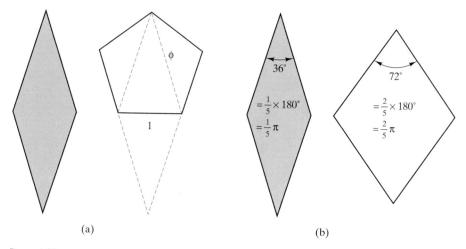

(a)

(b)

Figure 4-29 *(a) The relation of the skinny rhombus of Figure 4–28 to the geometry of a regular pentagon. The edge length of the rhombus equals a diagonal of the pentagon, with φ being the golden ratio = (√5 + 1)/2 = 1.618. (b) The acute angle in the skinny rhombus is (1/5)π and, in the fat rhombus, is (2/5)π. These angles assure the fivefold symmetry of the Penrose tiling.*

pattern with fivefold symmetry. The relationship between the Penrose tiles, which can fill two-dimensional space, and the regular pentagon, which cannot, is shown in Figure 4–29. In Figure 4–29a, we can see that the skinny rhombus is directly related to the key dimensions of the pentagon, namely, the edge and the diagonal. Note that the ratio of the length of the pentagon's diagonal to its edge is ϕ, an important irrational number equal to $(\sqrt{5} + 1)/2 = 1.618$. The number ϕ is sometimes called the **golden ratio** because of its fundamental role in numerous shapes in the natural world, as well as in the proportions of much of the architecture of the ancient world. It is also given by the limiting value of the ratio of two consecutive terms in the **Fibonacci series**,[*] in which each term is the sum of the two previous terms. The pervasive role of the golden ratio in Figure 4–28 is also shown by the fact that the ratio of the number of fat rhombuses to the number of skinny rhombuses is ϕ. The direct relation of the rhombuses to fivefold symmetry is further illustrated in Figure 4–29b, which indicates that the acute angle in the skinny rhombus is $(1/5)\pi$ and in the fat rhombus is $(2/5)\pi$. This guarantees that fivefold symmetry will be produced at the junction formed when three or more rhombuses share common vertices. To illustrate this, Figure 4–30 shows the Penrose tiling of Figure 4–28b decorated with pentagons located at each of these junctions. Each pentagon is oriented so that as many of its vertices as possible lie along a boundary between adjacent rhombuses.

---

[*] Leonardo Fibonacci (ca. 1170–ca. 1240), Italian mathematician. Also known as Leonardo de Pisa, Fibonacci was an early giant of Western mathematics. His best known work, *Liber Abaci (Book of Computation)*, put an end to the old Roman system of numerical notation by providing a systematic explanation of the advantages of Hindu and Arabic numerals.

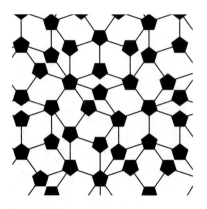

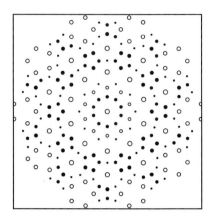

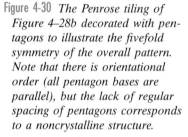

Figure 4-30 *The Penrose tiling of Figure 4–28b decorated with pentagons to illustrate the fivefold symmetry of the overall pattern. Note that there is orientational order (all pentagon bases are parallel), but the lack of regular spacing of pentagons corresponds to a noncrystalline structure.*

Figure 4-31 *A theoretical diffraction pattern for a three-dimensional Penrose tiling directly matching the experimental pattern of Figure 4–27. [After D. Levine and P. Steinhardt, Phys. Rev. Letters 53, 2477 (1984).]*

Close inspection of Figure 4–30 reveals the remarkable fact that the bases of all the pentagons are parallel. Some pentagons "point up" while others "point down," but all have a common orientation. The orientational order is combined with a lack of regular spacing in the plane. This combination of regular orientation but irregular spacing is precisely what is needed to account for the forbidden diffraction pattern of Figure 4–27.

Of course, the diffraction pattern of Figure 4–27 is produced by a three-dimensional solid and not by a two-dimensional pattern. The three-dimensional analog of the Penrose tiling is a space-filling stacking of prolate (skinny) and oblate (fat) rhombohedra. Figure 4–31 shows a theoretical diffraction pattern, based on such a three-dimensional Penrose tiling, which directly matches the experimental result of Figure 4–27. As the two-dimensional Penrose tiling produced the fivefold symmetry of the pentagon, the three-dimensional Penrose tiling produces the fivefold symmetry of the three-dimensional icosahedron. As shown in Figure 4–32, the icosahedron, composed of 20 identical equilateral triangular faces, has a fivefold symmetry (Figure 4–32a) where five triangles join at a vertex, as well as threefold symmetry (Figure 4–32b) and twofold symmetry (Figure 4–32c). All three symmetries have been seen in the diffraction patterns of Al_6Mn and other quasicrystalline materials. As a result, they are also referred to as *icosahedral phases*. Continuing research on quasicrystals may produce more than an

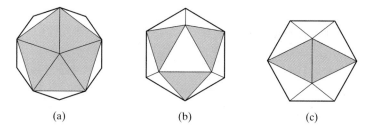

(a)   (b)   (c)

Figure 4-32 *Three views of an icosahedron showing (a) fivefold symmetry, (b) threefold symmetry, and (c) twofold symmetry.*

expanded philosophy of the structure of materials. Materials engineers are speculating that the novel structures associated with these materials could lead to novel mechanical and electrical properties. To date, numerous alloy systems have exhibited quasicrystalline structure. Some in the aluminum-lithium-copper, aluminum-cobalt-copper, and aluminum-cobalt-nickel systems are sufficiently stable as to permit the slow growth of relatively large, single crystals.

SAMPLE PROBLEM 4.8

Show that the ratio of fat to skinny rhombuses in the Penrose tiling of Figure 4–28 is given by the golden ratio, ϕ.

SOLUTION

By inspection of Figure 4–28, we find there are 23 fat rhombuses either contained wholly within the figure or, if along the edge, at least one-half within the field of view. In a similar way, we find 14 skinny rhombuses. The ratio is

$$23/14 = 1.64 \simeq 1.62 = \phi$$

Note.   An increasingly large field of view would give an increasingly close approximation.

PRACTICE PROBLEM 4.8

In Sample Problem 4.8, we show one way in which the golden ratio, ϕ, appears in a Penrose tiling. To demonstrate that this tiling is isotropic, show that the decorative pentagons in Figure 4–30 are equally probable to be oriented "up" or "down."

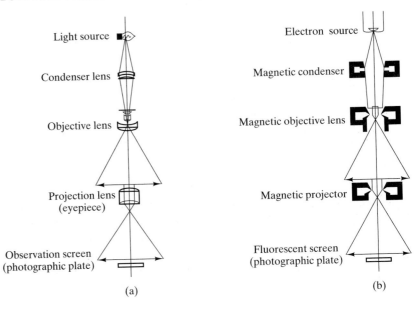

Light source

Condenser lens

Objective lens

Projection lens
(eyepiece)

Observation screen
(photographic plate)

(a)

Electron source

Magnetic condenser

Magnetic objective lens

Magnetic projector

Fluorescent screen
(photographic plate)

(b)

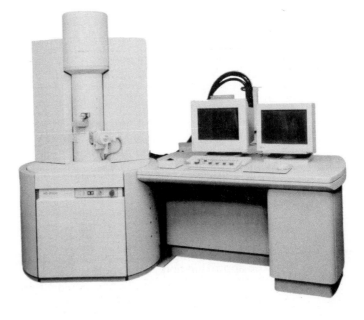

(c)

Figure 4-33 *Similarity in design between (a) an optical microscope and (b) a transmission electron microscope. The electron microscope uses solenoid coils to produce a magnetic lens in place of the glass lens in the optical microscope. (From G. Thomas,* Transmission Electron Microscopy of Metals, *John Wiley & Sons, Inc., New York, 1962.) (c) A commercial TEM. (Courtesy of Hitachi Scientific Instruments.)*

4.7 MICROSCOPY

Figure 4–18 showed an example of a common and important experimental inspection of an engineering material, a photograph of grain structure taken with an **optical microscope**. In fact, the first such inspection made in 1863 by H. C. Sorby is generally acknowledged as the beginning of the science of metallurgy and, indirectly then, as the origin of the field of materials science and engineering. The optical microscope is familiar to engineering students from any number of pre-college-level studies. Less familiar is the electron microscope. In Section 3.7, x-ray diffraction was described as a standard tool for measuring ideal, crystalline structure. Now, we shall see that electron microscopes, as well as optical microscopes, are standard tools for characterizing the microstructural features introduced in this chapter. We shall begin our discussion of electron microscopes with the two main types, the transmission and scanning designs.

The **transmission electron microscope** (TEM) is similar in design to a conventional optical microscope except that instead of a beam of light focused by glass lenses, there is a beam of electrons focused by electromagnets (Figure 4–33). This is possible due to the wavelike nature of the electron (see Section 2.1). For a typical TEM operating at a constant voltage of 100 keV, the electron beam has a monochromatic wavelength, λ, of 3.7×10^{-3} nm. This is five orders of magnitude smaller than the wavelength of visible light (400 to 700 nm) used in optical microscopy. The result is that substantially smaller structural details can be resolved by the TEM compared with the optical microscope. Practical magnifications of roughly $2000\times$ are possible in optical microscopy (corresponding to a resolution of structural dimensions as small as about 0.25 μm), whereas magnifications of $100,000\times$ are routinely obtained in the TEM (with a corresponding resolution of about 1 nm). The image in transmission electron microscopy is the result of *diffraction contrast* (Figure 4–34). The sample is oriented so that some of the beam is

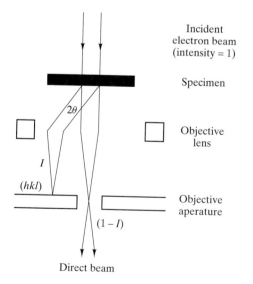

Incident electron beam (intensity = 1)

Specimen

2θ

Objective lens

I

(hkl)

Objective aperture

$(1 - I)$

Direct beam

Figure 4-34   *The basis of image formation in the TEM is diffraction contrast. Structural variations in the sample cause different fractions (I) of the incident beam to be diffracted out, giving variations in image darkness at a final viewing screen. (From G. Thomas,* Transmission Electron Microscopy of Metals, *John Wiley & Sons, Inc., New York, 1962.)*

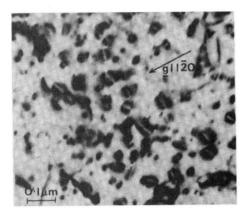

(a)

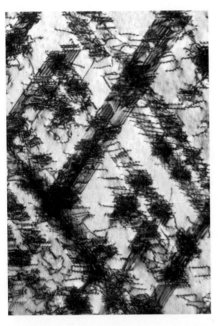

(b)

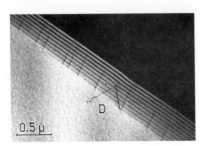

(c)

Figure 4-35   *(a) TEM image of the strain field around small dislocation loops in a zirconium alloy. These loops result from a condensation of point defects (either interstitial atoms or vacancies) after neutron irradiation. (b) Forest of dislocations in a stainless steel as seen by a TEM. (c) TEM image of a grain boundary. The parallel lines identify the boundary. A dislocation intersecting the boundary is labeled "D."* [(a) From A. Riley and P. J. Grundy, Phys. Status Solidi (a) 14, *239 (1972). (b) Courtesy of Chuck Echer, Lawrence Berkeley National Laboratory, National Center for Electron Microscopy. (c) From P. H. Pumphrey and H. Gleiter,* Philos. Mag. 30, *593 (1974).]*

transmitted and some is diffracted. Any local variation in crystalline regularity will cause a different fraction of the incident beam intensity to be "diffracted out," leading to a variation in image darkness on a viewing screen at the base of the microscope. Although it is not possible to identify single point defects, the strain field resulting around a small dislocation loop formed by a condensation of point defects (interstitial atoms or vacancies) is readily visible (Figure 4–35a). A widely used application of the transmission electron microscope is to identify various dislocation structures (e.g., Figure 4–35b). Images of grain boundary structures are also possible (Figure 4–35c).

The **scanning electron microscope** (SEM) shown in Figure 4–36 obtains structural images by an entirely different method than that used by the TEM. In the SEM, an electron beam spot $\approx 1\mu$m in diameter is scanned repeatedly over the surface of the sample. Slight variations in surface topography produce marked variations in the strength of the beam of *secondary electrons*—electrons ejected from the surface of the sample by the force of collision with *primary electrons* from the electron beam. The secondary electron beam signal is displayed on a television screen in a scanning pattern synchronized with the electron beam scan of the sample surface. The magnification possible with the SEM is limited by the beam spot size and is considerably better than that possible with the optical microscope but less than that possible with the TEM. The important feature of an SEM image is that it looks like a visual image of a large-scale piece. For instance, a small piece of lunar rock (Figure 4–37) is clearly spherical in shape. The SEM is especially useful for convenient inspections of grain structures. Figure 4–38 reveals such structure in a fractured metal surface. The depth of field of the SEM allows this irregular surface to be inspected. The optical

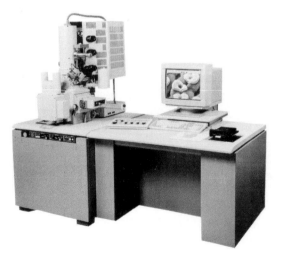

Figure 4-36 *A commercial SEM. (Courtesy of Hitachi Scientific Instruments.)*

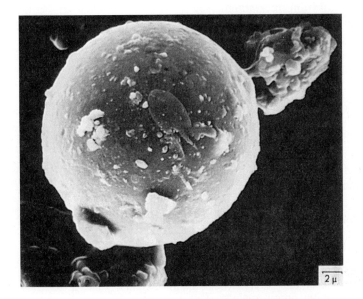

Figure 4-37 *SEM image of a 23-μm-diameter lunar rock from the SEM gives an image with "depth," in contrast to optical micrographs (e.g., Figure 4–18). The spherical shape indicates a prior melting process. (From V. A. Phillips,* Modern Metallographic Techniques and Their Applications, *John Wiley & Sons, Inc., New York, 1971.)*

Figure 4-38 *SEM image of a metal (type 304 stainless steel) fracture surface, 180×. (From* Metals Handbook, *8th ed., Vol. 9:* Fractography and Atlas of Fractographs, *American Society for Metals, Metals Park, Ohio, 1974.)*

microscope requires flat, polished surfaces (e.g., Figure 4–18). In addition to the convenience of avoiding the polishing of the sample, the irregular fracture surface can reveal information about the nature of the fracture mechanism. An additional feature of the SEM allows variations in chemistry on the microstructural scale to be monitored as shown in Figure 4–39. In addition to ejecting secondary electrons, the incident electron beam of the SEM generates characteristic wavelength x-rays that identify the elemental composition of the material under study.

The conventional transmission electron microscope is used for imaging microstructural features. Various examples have been offered, and, as noted earlier, the resolution of these instruments is about 1 nm. The most sophisticated refinements of this electron column design can improve the resolution by one order of magnitude, producing what can be more accurately described as an **atomic resolution electron microscope.** The chapter-opening micrographs for both Chapters 3 and 4 are examples of atomic resolution microscopy.

In recent years, a radically different microscope design has emerged that makes it possible to observe the packing arrangement of atoms at a solid surface. The **scanning tunneling microscope** (STM) is the first of a new family of instruments capable of providing direct images of individual atomic

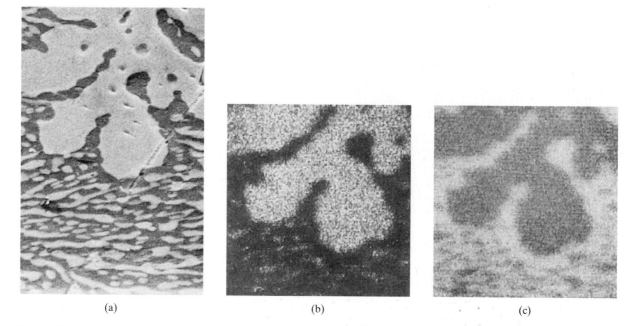

(a)   (b)   (c)

Figure 4-39 *(a) SEM image of the topography of a lead–tin solder alloy with lead-rich and tin-rich regions. (b) A map of the same area shown in (a) indicating the lead distribution (light area) in the microstructure. The light area corresponds to regions emitting characteristic lead x-rays when struck by the scanning electron beam. (c) A similar map of the tin distribution (light area) in the microstructure. [From J. B. Bindell,* Advanced Materials and Processes 143, *20 (1993).]*

packing patterns. (By contrast, atomic resolution electron microscopy images, such as the chapter-opening micrograph, represent an "average" of several adjacent atomic layers within the thickness of a thin-foil sample.) The name of the STM comes from the x-y raster (scanning) by a sharp metal tip near the surface of a conducting sample, leading to a measurable electrical current due to the quantum-mechanical *tunneling* of electrons near the surface. For gap distances around 0.5 nm, an applied bias of tens of millivolts leads to nanoampere current flows. The needle's vertical distance (z-direction) above the surface is continually adjusted to maintain a constant tunnel current. The surface topograph is the record of the trajectory of the tip (Figure 4–40).

Finally, the **atomic force microscope** (AFM), is an important derivative of the STM. The AFM is based on the concept that the atomic surface should be resolvable, using a force as well as a current. This hypothesis was confirmed by demonstrating that a small cantilever can be constructed to have a spring constant weaker than the equivalent spring between adjacent atoms (Figure 4–41). For example, the interatomic force constant is typically 1 N/m, similar in value to that for a piece of common aluminum foil 4 mm long and 1 mm wide. This mechanical equivalence permits a sharp tip to image both conducting and nonconducting materials. (The STM is limited to materials with a significant level of conductivity.)

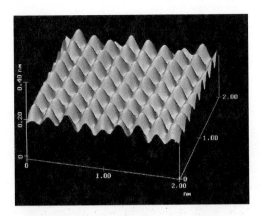

Figure 4-40 *Scanning tunneling micrograph of an interstitial atom defect on the surface of graphite. [From T. L. Altshuler,* Advanced Materials and Processes *140, 18 (1991).]*

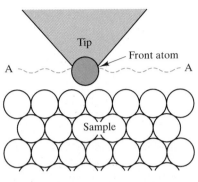

Figure 4-41   *Schematic of the principle by which the probe tip of either a scanning tunneling microscope (STM) or an atomic force microscope (AFM) operates. The sharp tip follows the contour A-A as it maintains either a constant tunneling current (in the STM) or a constant force (in the AFM). The STM requires a conductive sample while the AFM can also inspect insulators.*

SAMPLE PROBLEM 4.9

Image contrast in the transmission electron microscope is the result of electron diffraction. What is the diffraction angle for 100-keV electrons ($\lambda = 3.7 \times 10^{-3}$ nm) being diffracted from (111) planes in an aluminum sample?

SOLUTION

Turning to Bragg's law from Section 3.7, we obtain

$$n\lambda = 2d \, \sin \, \theta$$

For $n = 1$ (i.e., considering first-order diffraction),

$$\theta = \arcsin \frac{\lambda}{2d}$$

From Sample Problem 3.21,

$$d_{111} = \frac{0.404 \text{ nm}}{\sqrt{1^2 + 1^2 + 1^2}} = 0.234 \text{ nm}$$

which gives

$$\theta = \arcsin \frac{3.7 \times 10^{-3} \text{ nm}}{2 \times 0.234 \text{ nm}} = 0.453°$$

The diffraction angle (2θ) as defined in Figure 3–36 is then

$$2\theta = 2(0.453°) = 0.906°$$

Note. This characteristically small angle for electron diffraction can be compared with the characteristically large angle (38.6°) for x-ray diffraction of the same system as that shown in Figure 3–39.

....

PRACTICE PROBLEM 4.9

In Sample Problem 4.9 we calculate the diffraction angle (2θ) for 100-keV electrons diffracted from (111) planes in aluminum. What would be the diffraction angle from **(a)** the (200) planes and **(b)** the (220) planes?

SUMMARY

No real material used in engineering is as perfect as the structural descriptions of Chapter 3 would imply. There is always some contamination in the form of solid solution. When the impurity, or solute, atoms are similar to the solvent atoms, substitutional solution takes place in which impurity atoms rest on crystal lattice sites. Interstitial solution takes place when a solute atom is small enough to occupy open spaces among adjacent atoms in the crystal structure. Solid solution in ionic compounds must account for charge neutrality of the material as a whole.

Point defects can be missing atoms or ions (vacancies) or extra atoms or ions (interstitialcies). Charge neutrality must be maintained locally for point defect structures in ionic compounds.

Linear defects, or dislocations, correspond to an extra half-plane of atoms in an otherwise perfect crystal. Although dislocation structures can be complex, they can also be characterized with a simple parameter, the Burgers vector.

Planar defects include any boundary surface surrounding a crystalline structure. Twin boundaries divide two mirror-image regions. The exterior surface has a characteristic structure involving an elaborate ledge system. The predominant microstructural feature for many engineering materials is grain structure, where each grain is a region with a characteristic crystal structure orientation. A grain-size number (G) is used to quantify this microstructure. The structure of the region of mismatch between adjacent grains (i.e., the grain boundary) depends on the relative orientation of the grains.

Noncrystalline solids, on the atomic scale, are lacking in any long-range order (LRO) but may exhibit short-range order (SRO) associated with structural building blocks such as SiO_4^{4-} tetrahedra. Relative to a perfectly random structure, one can define solid solution, just as was done relative to perfectly crystalline structures. Recently, medium-range order has been found for the distribution of modifier ions such as Na^+ and Ca^{2+} in silicate glasses.

Quasicrystals represent an intermediate state between crystals and glasses. Their fivefold symmetry diffraction patterns are the result of orientational order in the absence of translational periodicity. The icosahedral phases have been described as three-dimensional Penrose tilings.

Optical and electron microscopy are powerful tools for observing structural order and disorder. The transmission electron microscope (TEM) uses diffraction contrast to obtain high-magnification (e.g., $100,000\times$) images of defects such as dislocations. The scanning electron microscope (SEM) produces three-dimensional-appearing images of microstructural features such as fracture surfaces. By analyzing characteristic x-ray emission, microstructural chemistry can be studied. The state-of-the-art in TEM design is represented by the atomic resolution electron microscope. A revolutionary new approach to microscope design has resulted in the scanning tunneling microscope (STM) and the atomic force microscope (AFM), which provide direct images of individual atomic stacking patterns.

KEY TERMS

amorphous metal (137)
amorphous semiconductor (137)
atomic force microscope (AFM) (149)
atomic resolution electron microscope (148)
Bernal model (137)
Burgers vector (125)
coincident site lattice (CSL) (130)
dislocation (124)
edge dislocation (125)
Fibonacci series (141)
fivefold symmetry (140)
Frenkel defect (123)
golden ratio (141)
grain (129)
grain boundary (129)
grain boundary dislocation (GBD) (132)
grain-size number (133)

Hirth-Pound model (128)
Hume-Rothery rules (117)
interstitialcy (122)
interstitial solid solution (118)
linear defect (124)
long-range order (136)
medium-range order (138)
mixed dislocation (125)
noncrystalline solid (136)
nonstoichiometric compound (120)
optical microscope (145)
ordered solid solution (117)
oxide glass (136)
Penrose tiling (140)
planar defect (128)
point defect (122)
quasicrystal (139)
random network theory (136)

random solid solution (117)
scanning electron microscope (SEM) (147)
scanning tunneling microscope (STM) (148)
Schottky defect (123)
screw dislocation (125)
short-range order (136)
solid solution (116)
solute (116)
solvent (116)
substitutional solid solution (116)
tilt boundary (130)
transmission electron microscope (TEM) (145)
twin boundary (128)
vacancy (122)
Zachariasen model (136)

REFERENCES

Chiang, Y., D. P. Birnie III, and **W. D. Kingery,** *Physical Ceramics*, John Wiley & Sons, Inc., New York, 1997.

Hull, D., and **D. J. Bacon,** *Introduction to Dislocations,* 3rd ed., Pergamon Press, Inc., Elmsford, N.Y., 1984.

Janot, C., *Quasicrystals: A Primer,* Oxford University Press, New York, 1992.

Williams, D. B., A. R. Pelton, and **R. Gronsky,** eds., *Images of Materials,* Oxford University Press, New York, 1992. A comprehensive as well as beautiful example of the microscopic tools available for materials characterization.

PROBLEMS

Section 4.1 • The Solid Solution—Chemical Imperfection

4.1. In Chapter 9 we shall find a phase diagram for the Al–Cu system that indicates that these two metals do not form a complete solid solution. Which of the Hume-Rothery rules can you identify for Al–Cu that are violated? (For electronegativity data relative to rule 3, consult Figure 2–21.)

4.2. For the Al–Mg system with a phase diagram in Chapter 9 showing incomplete solid solution, which of the Hume-Rothery rules are violated? (See Problem 4.1.)

4.3. For the Cu–Zn system with a phase diagram in Chapter 9 showing incomplete solid solution, which of the Hume-Rothery rules are violated? (See Problem 4.1.)

4.4. For the Pb–Sn system with a phase diagram in Chapter 9 showing incomplete solid solution, which of the Hume-Rothery rules are violated? (See Problem 4.1.)

4.5. Sketch the pattern of atoms in the (111) plane of the ordered $AuCu_3$ alloy shown in Figure 4–3. (Show an area at least five atoms wide by five atoms high.)

4.6. Sketch the pattern of atoms in the (110) plane of the ordered $AuCu_3$ alloy shown in Figure 4–3. (Show an area at least five atoms wide by five atoms high.)

4.7. Sketch the pattern of atoms in the (200) plane of the ordered $AuCu_3$ alloy shown in Figure 4–3. (Show an area at least five atoms wide by five atoms high.)

4.8. What are the equivalent points for ordered $AuCu_3$ (Figure 4–3)? (Note Problem 3.56.)

4.9. Although the Hume-Rothery rules apply strictly only to metals, the concept of similarity of cations corresponds to the complete solubility of NiO in MgO (Figure 4–5). Calculate the percent difference between cation sizes in this case.

4.10. Calculate the percent difference between cation sizes for Al_2O_3 in MgO (Figure 4–6), a system that does not exhibit complete solid solubility.

4.11. Calculate the number of Mg^{2+} vacancies produced by the solubility of 1 mol of Al_2O_3 in 99 mol of MgO (see Figure 4–6).

4.12. Calculate the number of Fe^{2+} vacancies in 1 mol of $Fe_{0.95}O$ (see Figure 4–7).

4.13. In Part III of the text we shall be especially interested in "doped" semiconductors, in which small levels of impurities are added to an essentially pure semiconductor in order to produce desirable electrical properties. For silicon with 4×10^{21} aluminum atoms per cubic meter in solid solution, calculate **(a)** the atomic percent of aluminum atoms and **(b)** the weight percent of aluminum atoms.

4.14. For 4×10^{21} aluminum atoms/m^3 in solid solution in germanium calculate **(a)** the atomic percent of aluminum atoms and **(b)** the weight percent of aluminum atoms.

4.15. For 4×10^{21} phosphorus atoms/m^3 in solid solution in silicon, calculate **(a)** the atomic percent of phosphorous atoms and **(b)** the weight percent of phosphorous atoms.

4.16. One way to determine a structural defect model (such as Figure 4–6 for a solid solution of Al_2O_3 in MgO) is to make careful density measurements. What would be the percent change in density for a 5 at % solution of Al_2O_3 in MgO (compared to pure, defect-free MgO)?

Section 4.2 • Point Defects—Zero-Dimensional Imperfections

4.17. Calculate the density of vacant sites (in m^{-3}) in a single crystal of silicon if the fraction of vacant lattice sites is 1×10^{-7}.

4.18. Calculate the density of vacant sites (in m^{-3}) in a single crystal of germanium if the fraction of vacant lattice sites is 1×10^{-7}.

4.19. Calculate the density of Schottky pairs (in m^{-3}) in MgO if the fraction of vacant lattice sites is 5×10^{-6}. (The density of MgO is 3.60 Mg/m^3.)

4.20. Calculate the density of Schottky pairs (in m^{-3}) in CaO if the fraction of vacant lattice sites is 5×10^{-6}. (The density of CaO is 3.45 Mg/m^3.)

Section 4.3 • Linear Defects, or Dislocations— One-Dimensional Imperfections

4.21. The energy necessary to generate a dislocation is proportional to the square of the length of the Burgers vector, $|\mathbf{b}|^2$. This means that the most stable (lowest energy) dislocations have the minimum length, $|\mathbf{b}|$. For the bcc metal structure, calculate (relative to $E_{\mathbf{b}=[111]}$) the dislocation energies for **(a)** $E_{\mathbf{b}=[110]}$ and **(b)** $E_{\mathbf{b}=[100]}$.

4.22. The comments in Problem 4.21 also apply for the fcc metal structure. Calculate (relative to $E_{\mathbf{b}=[110]}$) the dislocation energies for **(a)** $E_{\mathbf{b}=[111]}$ and **(b)** $E_{\mathbf{b}=[100]}$.

4.23. The comments in Problem 4.21 also apply for the hcp metal structure. Calculate (relative to $E_{\mathbf{b}=[11\bar{2}0]}$) the dislocation energies for **(a)** $E_{\mathbf{b}=[1\bar{1}00]}$ and **(b)** $E_{\mathbf{b}=[0001]}$.

• **4.24.** Figure 4–14 illustrates how a Burgers vector can be broken up into partials. The Burgers vector for an fcc metal can be broken up into two partials. **(a)** Sketch the partials relative to the full dislocation, and **(b)** identify the magnitude and crystallographic orientation of each partial.

Section 4.4 • Planar Defects—Two-Dimensional Imperfections

4.25. Determine the grain-size number, G, for the microstructure shown in Figure 4–18. (Keep in mind that the precise answer will depend on your choice of an area of sampling.)

4.26. Calculate the grain-size number for the microstructures in Figures 1–20a and c given that the magnifications are 160× and 330×, respectively.

4.27. Using Equation 4.2, estimate the average grain diameter of Figures 1–20a and c using "random lines" cutting across the diagonal of each figure from its lower left corner to its upper right corner. (See Problem 4.26 for magnifications.)

• **4.28.** Note in Figure 4–21 that the crystalline regions in the fcc structure are represented by a repetitive polyhedra structure. This is an alternative to our usual unit cell configuration. In other words, the fcc structure can be equally represented by a space-filling stacking of regular polyhedra (tetrahedra and octahedra in a ratio of 2:1). **(a)** Sketch a typical tetrahedron (four-sided figure) on a perspective sketch such as Figure 3–5a. **(b)** Similarly, show a typical octahedron (eight-sided figure). (Note also Problem 3.63.)

4.29. As implied in the text, demonstrate that the tilt angle for the $\Sigma 5$ boundary is defined by $\theta = 2 \tan^{-1}(1/3)$. (**Hint:** Rotate two overlapping square lattices by 36.9° about a given common point and note the direction corresponding to one-half the rotation angle.)

4.30. Show that the tilt angle for the $\Sigma 13$ grain boundary is defined by $\theta = 2 \tan^{-1}(1/5) = 22.6°$. (Note Problem 4.29.)

Section 4.5 • Noncrystalline Solids—Three-Dimensional Imperfections

4.31. Figure 4–23b is a useful schematic for simple B_2O_3 glass, composed of rings of BO_3^{3-} triangles. To appreciate the openness of this glass structure, calculate the size of the interstice (i.e., largest inscribed circle) of a regular six-membered ring of BO_3^{3-} triangles.

4.32. In amorphous silicates, a useful indication of the lack of crystallinity is the "ring statistics." For the schematic illustration in Figure 4–23b, plot a histogram of the n-membered rings of O^{2-} ions, where n = number of O^{2-} ions in a loop surrounding an open interstice in the network structure. [*Note:* In Figure 4–23a, all rings are six-membered ($n = 6$).] (**Hint:** Ignore incomplete rings at the edge of the illustration.)

• **4.33.** In Problem 4.28 a tetrahedron and octahedron were identified as the appropriate polyhedra to define an fcc structure. For the hcp structure, the tetrahedron and octahedron are also the appropriate polyhedra. (**a**) Sketch a typical tetrahedron on a perspective sketch such as Figure 3–6a. (**b**) Similarly, show a typical octahedron. (Of course, we are dealing with a crystalline solid in this example. But as Figure 4–24 shows, the noncrystalline, amorphous metal has a range of such polyhedra that fill space.)

• **4.34.** There are several polyhedra that can occur at grain boundaries, as discussed relative to Figure 4–21. The tetrahedron and octahedron treated in Problems 4.28 and 4.33 are the simplest. The next simplest is the pentagonal bipyramid, which consists of 10 equilateral triangle faces. Sketch this polyhedron as accurately as you can.

4.35. Sketch a few adjacent CaO_6 octahedra in the pattern shown in Figure 4–26. Indicate both the nearest neighbor $Ca^{2+} - Ca^{2+}$ distance, R_1, and the *next-nearest neighbor* $Ca^{2+} - Ca^{2+}$ distance, R_2.

4.36. Diffraction measurements on the CaO-SiO_2 glass represented by Figure 4–26 show that the nearest neighbor $Ca^{2+} - Ca^{2+}$ distance, R_1, is 0.375 nm. What would be the next-nearest neighbor $Ca^{2+} - Ca^{2+}$ distance, R_2? (Note the results of Problem 4.35.)

Section 4.6 • Quasicrystals

4.37. Show, by using a graph, that the ratio of two consecutive terms in the Fibonacci series $(1, 2, 3, 5, 8, \ldots)$ approaches the golden ratio, ϕ.

4.38. The Fibonacci series can also be used to describe crack branching in a material undergoing failure. Starting with a single, initial crack, sketch a branching pattern that follows the Fibonacci series. How many subcracks appear after five branching steps?

4.39. In Problems 4.28, 4.33, and 4.34, we looked at the tetrahedron, the octahedron, and the pentagonal bipyramid as polyhedra with equilateral triangle faces. These are three of the family of polyhedra that can occur in metallic grain boundaries (Figure 4–21) or in amorphous metals (Figure 4–24). The icosahedron shown in Figure 4–32 is the largest such polyhedron. The interstice in the center of the icosahedron can *almost* accommodate another metal atom. (The next largest polyhedron is large enough to accommodate another atom and, consequently, the interstice gets filled in.) If the distance from the center of an icosahedron to the outermost vertex is 0.95 times the edge length of a triangular face, how large is the interstice (described as an inscribed sphere) compared to the atomic diameter?

4.40. Beyond determining the general icosahedral nature of quasicrystals, materials scientists are concerned with the location of specific atoms within the structure. One model for Al_6Mn is the Mackay icosahedron, in which there is an inner shell in the form of an icosahedron (with an Al atom at each vertex) and an outer shell icosahedron (with an Mn atom above each of the inner shell Al atoms and an Al atom in the center of each of the edges between Mn atoms). How many total atoms appear in this overall cluster?

• **4.41.** The C_{60} buckyball introduced in Figure 3–19a is related to quasicrystals in an interesting way. The soccer ball geometry of C_{60} can be described as a truncated icosahedron. Illustrate this on a sketch of Figure 4–32.

4.42. Given the relationship identified in Problem 4.41, describe how the golden ratio appears in the geometry of the C_{60} molecule.

Section 4.7 • Microscopy

4.43. Suppose that the electron microscope in Figure 4–33c is used to make a simple diffraction spot pattern (rather than a magnified microstructural image). That is done by turning off the electromagnetic magnifying lenses. (Figure 4–27 was produced in this way.) The result is analogous to the Laue x-ray experiment described in Section 3.7 but with very small 2θ values. If the aluminum specimen described in Sample Problem 4.9 and Practice Problem 4.9 is 1 m from the photographic plate, (**a**) how far is the (111) diffraction spot from the direct (undiffracted) beam? Repeat part (a) for (**b**) the (200) spot and (**c**) the (220) spot.

4.44. Repeat Problem 4.43 for **(a)** the (110) spot, **(b)** the (200) spot, and **(c)** the (211) spot produced by replacing the aluminum specimen with one composed of α-iron.

4.45. A transmission electron microscope is used to produce a diffraction ring pattern for a thin, polycrystalline sample of copper. The (111) ring is 12 mm from the center of the film (corresponding to the undiffracted, transmitted beam). How far would the (200) ring be from the film center?

4.46. The microchemical analysis discussed relative to Figure 4–39 is based on x-rays of characteristic wavelengths. As will be discussed in Chapter 16 on optical properties, a specific wavelength x-ray is equivalent to a photon of specific energy. Characteristic x-ray photons are produced by an electron transition between two energy levels in a given atom. For tin, the electron energy levels are as follows:

Electron Shell	Electron Energy
K	−29,199 eV
L	−3,929 eV
M	−709 eV

Which electron transition produces the characteristic K_α photon with energy of 25,270 eV?

4.47. Repeat Problem 4.46, calculating the electron transition for lead in which a characteristic L_α photon with energy of 10,553 eV is used for the microanalysis. Relevant data are as follows:

Electron Shell	Electron Energy
K	−88,018 eV
L	−13,773 eV
M	−3,220 eV

4.48. **(a)** Given only the data in Problems 4.46 and 4.47, determine whether a 28,490-eV characteristic x-ray photon would be produced by tin or by lead. **(b)** Which electron transition produces the characteristic photon in (a)?

CHAPTER 5
Diffusion

In addition to superconductivity and high melting point which lead to various important industrial applications, niobium is a metal which forms oxide coatings readily by the interdiffusion of oxygen and niobium atoms near the metal surface. Jewelry manufacturers use this fact to produce colorful earring designs. (Courtesy of Teledyne Wah Chang, Albany, Oregon.)

During production and application, the chemical composition of engineering materials is often changed as a result of the movement of atoms, or *solid-state diffusion*. In some cases, atoms are redistributed within the microstructure of the material. In other cases, atoms are added from the material's environment, or atoms from the material may be discharged into the environment. Understanding the nature of the movement of atoms within the material can be critically important both in producing the material and in applying it successfully within an engineering design.

In Chapter 4, we were introduced to a variety of point defects, such as the vacancy. These defects were said to result typically from the thermal vibration of the atoms in the material. In this chapter, we will see the detailed relationship between temperature and the number of these defects. Specifically, the concentration of such defects rises exponentially with increasing temperature. The flow of atoms in engineering materials occurs by the movement of point defects, and, as a result, the rate of this solid-state diffusion increases exponentially with temperature. The mathematics of diffusion allows a precise description of the variation of the chemical composition within materials as a result of various diffusional processes. An important example is the *carburization* of steels, in which the surface is hardened by the diffusion of carbon atoms from a carbon-rich environment.

After some time, the chemical concentration profile within a material may become linear, and the corresponding mathematics for this *steady-state diffusion* is relatively simple.

Although we generally consider diffusion within the entire volume of a material, there are some cases in which the atomic transport occurs primarily along grain boundaries (by *grain boundary diffusion*) or along the surface of the material (by *surface diffusion*).

5.1   THERMALLY ACTIVATED PROCESSES

A large number of processes in materials science and engineering share a common feature—the process rate rises exponentially with temperature. The diffusivity of elements in metal alloys, the rate of creep deformation in structural materials, and the electrical conductivity of semiconductors are a few examples that will be covered in this book. The general equation that describes these various processes is of the form

$$\text{rate} = Ce^{-Q/RT} \tag{5.1}$$

where C is a **preexponential constant** (independent of temperature), Q the **activation energy,** R the universal gas constant, and T the absolute temperature. It should be noted that the universal gas constant is as important for the solid state as for the gaseous state. The term *gas constant* derives from its role in the perfect gas law ($pV = nRT$) and related gas-phase equations. In fact, R is a fundamental constant that appears frequently in this book devoted to the solid state.

Equation 5.1 is generally referred to as the **Arrhenius* equation.** Taking the logarithm of each side of Equation 5.1 gives

$$\ln(\text{rate}) = \ln C - \frac{Q}{R}\frac{1}{T} \tag{5.2}$$

By making a semilog plot of ln (rate) versus the reciprocal of absolute temperature $(1/T)$, one obtains a straight-line plot of rate data (Figure 5–1). The slope of the resulting **Arrhenius plot** is $-Q/R$. Extrapolation of the Arrhenius plot to $1/T = 0$ (or $T = \infty$) gives an intercept equal to $\ln C$.

The experimental result of Figure 5–1 is a very powerful one. Knowing the magnitudes of process rate at any two temperatures allows the rate at a third temperature (in the linear plot range) to be determined. Similarly, knowledge of a process rate at any temperature and of the activation energy, Q, allows the rate at any other temperature to be determined. A common use of the Arrhenius plot is to obtain a value of Q from measurement of the slope of the plot. This value of activation energy can indicate the mechanism of the process. In summary, Equation 5.2 contains two constants. Therefore, only two experimental observations are required to determine them.

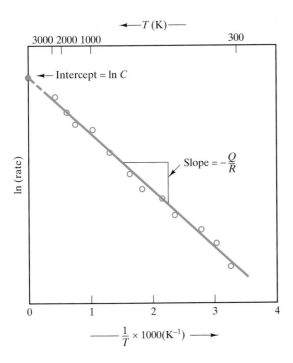

Figure 5-1 *Typical Arrhenius plot of data compared to Equation 5.2. The slope equals $-Q/R$ and the intercept (at $1/T = 0$) is $\ln C$.*

---

* Svante August Arrhenius (1859–1927), Swedish chemist, made numerous contributions to physical chemistry, including the experimental demonstration of Equation 5.1 for chemical reaction rates.

To appreciate why rate data show the characteristic behavior of Figure 5–1, we must explore the concept of the activation energy, Q. As used in Equation 5.1, Q has units of energy per mole. It is possible to rewrite this equation by dividing both Q and R by Avogadro's number (N_{AV}), giving

$$\text{rate} = Ce^{-q/kT} \tag{5.3}$$

where q ($= Q/N_{AV}$) is the activation energy per atomic scale unit (atom, electron, ion, etc.) and k ($= R/N_{AV}$) is Boltzmann's* constant ($13.8 \times 10^{-24} \, J/K$). Equation 5.3 provides for an interesting comparison with the high-energy end of the **Maxwell–Boltzmann**[†] **distribution** of molecular energies in gases:

$$P \propto e^{-\Delta E/kT} \tag{5.4}$$

where P is the probability of finding a molecule at an energy ΔE greater than the average energy characteristic of a particular temperature, T. Herein lies the clue to the nature of the activation energy. It is the energy barrier that must be overcome by **thermal activation.** Although Equation 5.4 was originally developed for gases, it applies to solids as well. As temperature increases, a larger number of atoms (or any other species involved in a given process, e.g., electrons or ions) are available to overcome a given energy barrier, q. Figure 5–2 shows a *process path* in which a single atom overcomes an energy barrier, q. Figure 5–3 shows a simple mechanical model of activation energy in which a box is moved from one position to another by going through an increase in potential energy, ΔE, analogous to the q in Figure 5–2.

In the many processes described in the text where an Arrhenius equation applies, particular values of activation energy will be found to be characteristic of process mechanisms. In each case, it is useful to remember that various possible mechanisms may be occurring simultaneously within the material, and each mechanism has a characteristic activation energy. The fact that one activation energy is representative of the experimental data means simply that a single mechanism is dominant. If the process involves several sequential steps, the slowest step will be the **rate-limiting step.** The activation energy of the rate-limiting step will, then, be the activation energy for the overall process.

---

* Ludwig Edward Boltzmann (1844–1906), Austrian physicist, is associated with many major scientific achievements of the nineteenth century (prior to the development of "modern physics"). The constant that bears his name plays a central role in the statistical statement of the second law of thermodynamics. Some ideas are difficult to put aside. His second law equation is carved on his tombstone.

[†] James Clerk Maxwell (1831–1879), Scottish mathematician and physicist, was an unusually brilliant and productive individual. His equations of electromagnetism are among the most elegant in all of science. He developed the kinetic theory of gases (including Equation 5.4) independently of his contemporary, Ludwig Edward Boltzmann.

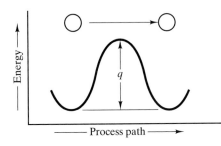

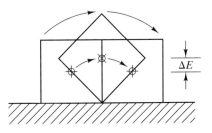

Figure 5-2 *Process path showing how an atom must overcome an activation energy, q, to move from one stable position to a similar adjacent position.*

Figure 5-3 *Simple mechanical analog of the process path of Figure 5–2. The box must overcome an increase in potential energy, ΔE, in order to move from one stable position to another.*

SAMPLE PROBLEM 5.1

The rate at which a metal alloy oxidizes in an oxygen-containing atmosphere is a typical example of the practical utility of the Arrhenius equation (Equation 5.1). For example, the rate of oxidation of a magnesium alloy is represented by a rate constant, k. The value of k at 300°C is 1.05×10^{-8} kg/(m^4· s). At 400°C, the value of k rises to 2.95×10^{-4} kg/(m^{-4}· s). Calculate the activation energy, Q, for this oxidation process (in units of kJ/mol).

SOLUTION

For this specific case, Equation 5.1 has the form

$$k = Ce^{-Q/RT}$$

Taking the ratio of rate constants at 300°C (= 573 K) and 400°C (= 673 K), we conveniently cancel out the unknown preexponential constant, C, and obtain

$$\frac{2.95 \times 10^{-4}\,\text{kg/[m}^4\cdot\text{s]}}{1.05 \times 10^{-8}\,\text{kg/[m}^4\cdot\text{s]}} = \frac{e^{-Q/(8.314\,\text{J/[mol·K]})(673\ \text{K})}}{e^{-Q/(8.314\,\text{J/[mol·K]})(573\ \text{K})}}$$

or

$$2.81 \times 10^4 = e^{\{-Q/(8.314\,\text{J/[mol·K]})\}\{1/(673\,\text{K})-1/(573\,\text{K})\}}$$

giving

$$Q = 328 \times 10^3\ \text{J/mol} = 328\,\text{kJ/mol}$$

....

Given the background provided by Sample Problem 5.1, calculate the value of the rate constant, k, for the oxidation of the magnesium alloy at 500°C.

5.2 THERMAL PRODUCTION OF POINT DEFECTS

Point defects occur as a direct result of the periodic oscillation, or **thermal vibration**, of atoms in the crystal structure. As temperature increases, the intensity of this vibration increases and, thereby, the likelihood of structural disruption and the development of point defects increases. At a given temperature, the thermal energy of a given material is fixed, but this is an average value. The thermal energy of individual atoms varies over a wide range, as indicated by the Maxwell–Boltzmann distribution. At a given temperature, a certain fraction of the atoms in the solid have sufficient thermal energy to produce point defects. An important consequence of the Maxwell–Boltzmann distribution is that this fraction increases exponentially with absolute temperature. As a result, the concentration of point defects increases exponentially with temperature; that is,

$$\frac{n_{\text{defects}}}{n_{\text{sites}}} = Ce^{-(E_{\text{defect}})/kT} \qquad (5.5)$$

where $n_{\text{defects}}/n_{\text{sites}}$ is the ratio of point defects to ideal crystal lattice sites, C is a preexponential constant, E_{defect} is the energy needed to create a single-point defect in the crystal structure, k is Boltzmann's constant, and T is the absolute temperature.

The temperature sensitivity of point defect production depends on the type of defect being considered; that is, E_{defect} for producing a vacancy in a given crystal structure is different from E_{defect} for producing an interstitialcy.

Figure 5–4 illustrates the thermal production of vacancies in aluminum. The slight difference between the thermal expansion measured by overall sample dimensions ($\Delta L/L$) and by x-ray diffraction ($\Delta a/a$) is the result of vacancies. The x-ray value is based on unit cell dimensions measured by x-ray diffraction (Section 3.7). The increasing concentration of empty lattice sites (vacancies) in the material at temperatures approaching the melting point produces a measurably greater thermal expansion as measured by overall dimensions. The concentration of vacancies (n_v/n_{sites}) follows the Arrhenius expression of Equation 5.5,

$$\frac{n_v}{n_{\text{sites}}} = Ce^{-E_v/kT} \qquad (5.6)$$

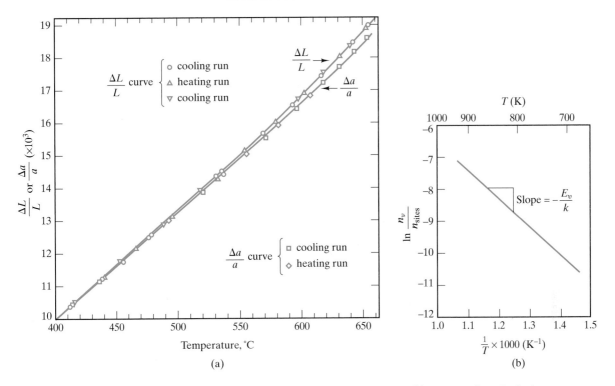

Figure 5-4 *(a) The overall thermal expansion ($\Delta L/L$) of aluminum is measurably greater than the lattice parameter expansion ($\Delta a/a$) at high temperatures because vacancies are produced by thermal agitation. (b) A semilog (Arrhenius-type) plot of ln (vacancy concentration) versus $1/T$ based on the data of part (a). The slope of the plot ($-E_v/k$) indicates that 0.76 eV of energy is required to create a single vacancy in the aluminum crystal structure. (After P. G. Shewmon,* Diffusion in Solids, *McGraw-Hill Book Company, New York, 1963.)*

where C is a preexponential constant and E_V is the energy of formation of a single vacancy. As discussed previously, this expression leads to a convenient semilog plot of data. Taking the logarithm of each side of Equation 5.6 gives

$$\ln \frac{n_v}{n_{\text{sites}}} = \ln C - \frac{E_V}{k}\frac{1}{T} \qquad (5.7)$$

Figure 5–4 shows the linear plot of ln (n_v/n_{sites}) versus $1/T$. The slope of this Arrhenius plot is $-E_V/k$. These experimental data indicate that the energy required to create one vacancy in the aluminum crystal structure is 0.76 eV.

SAMPLE PROBLEM 5.2

At 400°C, the fraction of aluminum lattice sites vacant is 2.29×10^{-5}. Calculate the fraction at 660°C (just below its melting point).

SOLUTION

From the text discussion relative to Figure 5–4, we have $E_V = 0.76$ eV. Using Equation 5.5, we have

$$\frac{n_v}{n_{sites}} = Ce^{-E_V/kT}$$

We obtain at $400°C\,(= 673\text{ K})$,

$$C = \left(\frac{n_v}{n_{sites}}\right)e^{+E_V/kT}$$

$$= (2.29 \times 10^{-5})e^{+0.76\text{eV}/(86.2\times 10^{-6}\text{eV/K})(673\text{ K})}$$

$$= 11.2$$

At $660°C\,(= 933\text{ K})$,

$$\frac{n_v}{n_{sites}} = (11.2)e^{-0.76\text{eV}/(86.2\,\times\,10^{-6}\text{eV/K})(933\text{ K})}$$

$$= 8.82 \times 10^{-4}$$

or roughly nine vacancies occur for every 10,000 lattice sites.

....

PRACTICE PROBLEM 5.2

Calculate the fraction of aluminum lattice sites vacant at **(a)** 500°C, **(b)** 200°C, and **(c)** room temperature (25°C). (See Sample Problem 5.2.)

5.3 POINT DEFECTS AND SOLID-STATE DIFFUSION

At sufficient temperatures, atoms and molecules can be quite mobile in both liquids and solids. Watching a drop of ink fall into a beaker of water and spread out until all the water is evenly colored gives a simple demonstration of **diffusion,** the movement of molecules from an area of higher concentration to an area of lower concentration. But diffusion is not restricted to different materials. At room temperature, H_2O molecules in pure water are in continuous motion and migrating through the liquid as an example of **self-diffusion.** This atomic-scale motion is relatively rapid in liquids and relatively easy to visualize. It is more difficult to visualize diffusion in rigid solids. Nonetheless, diffusion does occur in the solid state. A primary difference between solid-state and liquid-state diffusion is the low rate of diffusion in solids. Looking back at the crystal structures of Chapter 3, we can

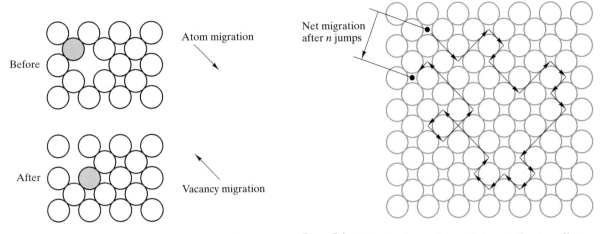

Figure 5-5 *Atomic migration occurs by a mechanism of vacancy migration. Note that the overall direction of material flow (the atom) is opposite to the direction of vacancy flow.*

Figure 5-6 *Diffusion by an interstitialcy mechanism illustrating the random walk nature of atomic migration.*

appreciate that diffusion of atoms or ions through those generally tight structures is difficult. In fact, the energy requirements to squeeze most atoms or ions through perfect crystal structures are so high as to make diffusion nearly impossible. To make solid-state diffusion practical, point defects are generally required. Figure 5–5 illustrates how atomic migration becomes possible without major crystal structure distortion by means of a **vacancy migration** mechanism. It is important to note that the overall direction of material flow is opposite to the direction of vacancy flow.

Figure 5–6 shows diffusion by an interstitialcy mechanism and illustrates effectively the **random walk** nature of atomic migration. This randomness does not preclude the net flow of material when there is an overall variation in chemical composition. This frequently occurring case is illustrated in Figures 5–7 and 5–8. Although each atom of solid A has an equal probability of randomly "walking" in any direction, the higher initial concentration of A on the left side of the system will cause such random motion to produce *interdiffusion*, a net flow of A atoms into solid B. Similarly, solid B diffuses into solid A. The formal mathematical treatment of such diffusional flow begins with an expression known as **Fick's* first law,**

$$J_x = -D\frac{\partial c}{\partial x} \tag{5.8}$$

---

* Adolf Eugen Fick (1829–1901), German physiologist. The medical sciences frequently apply principles previously developed in the fields of mathematics, physics, and chemistry. However, Fick's work in the "mechanistic" school of physiology was so excellent that it served as a guide for the physical sciences. He developed the diffusion laws as part of a study of blood flow.

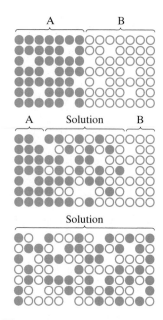

Figure 5-7 *The interdiffusion of materials A and B. Although any given A or B atom is equally likely to "walk" in any random direction (see Figure 5–6), the concentration gradients of the two materials can result in a net flow of A atoms into the B material, and vice versa. (From W. D. Kingery, H. K. Bowen, and D. R. Uhlmann,* Introduction to Ceramics, *2nd ed., John Wiley & Sons, Inc., New York, 1976.)*

where J_x is the *flux*, or flow rate, of the diffusing species in the x-direction due to a **concentration gradient** $(\partial c / \partial x)$. The proportionality coefficient, D, is called the **diffusion coefficient** or, simply, the **diffusivity.** The geometry of Equation 5.8 is illustrated in Figure 5–9. Figure 5–7 reminds us that the concentration gradient at a specific point along the diffusion path changes with time, t. This transient condition is represented by a second-order differential equation also known as **Fick's second law,**

$$\frac{\partial c_x}{\partial t} = \frac{\partial}{\partial x}\left(D\frac{\partial c_x}{\partial x}\right) \tag{5.9}$$

For many practical problems, one can assume that D is independent of c, leading to a simplified version of Equation 5.9:

$$\frac{\partial c_x}{\partial t} = D\frac{\partial^2 c_x}{\partial x^2} \tag{5.10}$$

Figure 5–10 illustrates a common application of Equation 5.10, the diffusion of material into a semi-infinite solid while the surface concentration of the diffusing species, c_s, remains constant. Two examples of this system would be the plating of metals and the saturation of materials with reactive atmospheric gases. Specifically, steel surfaces are often hardened by **carburization**, the diffusion of carbon atoms into the steel from a carbon-rich environment. The solution to this differential equation with the given boundary conditions is

$$\frac{c_x - c_0}{c_s - c_0} = 1 - \text{erf}\left(\frac{x}{2\sqrt{Dt}}\right) \tag{5.11}$$

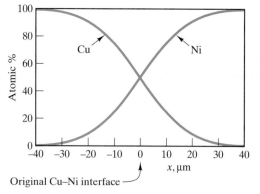

Figure 5-8 *The interdiffusion of materials on an atomic scale was illustrated in Figure 5–7. A comparable example on the microscopic scale is this interdiffusion of copper and nickel.*

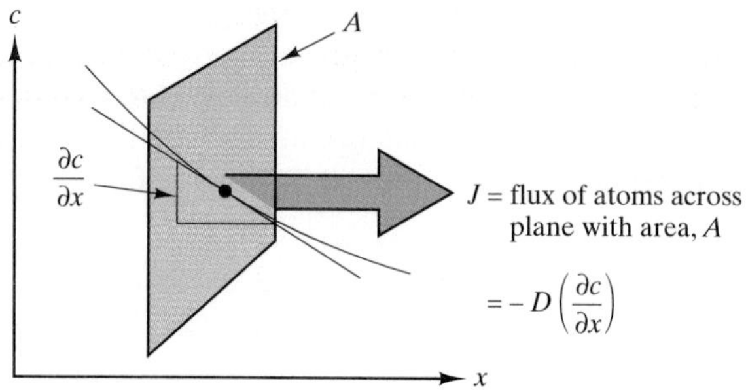

Figure 5-9 *Geometry of Fick's first law (Equation 5.8).*

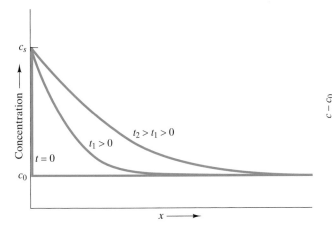

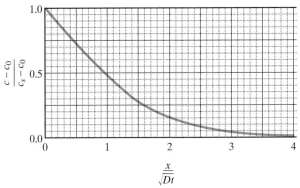

Figure 5-10 *Solution to Fick's second law (Equation 5.10) for the case of a semi-infinite solid, constant surface concentration of the diffusing species c_s, initial bulk concentration c_0, and a constant diffusion coefficient D.*

Figure 5-11 *Master plot summarizing all of the diffusion results of Figure 5–10 on a single curve.*

where c_0 is the initial bulk concentration of the diffusing species and erf refers to the **Gaussian*** **error function,** based on the integration of the "bell-shaped" curve with values readily available in mathematical tables. Representative values are given in Table 5.1. A great power of this analysis is that the result (Equation 5.11) allows all of the concentration profiles of Figure 5–10 to be redrawn on a single master plot (Figure 5–11). Such a plot permits rapid calculation of the time necessary for relative saturation of the solid as a function of x, D, and t. Figure 5–12 shows similar *saturation curves* for various geometries. It is important to keep in mind that these results are but a few of the large number of solutions that have been obtained by materials scientists for diffusion geometries in various practical processes.

The preceding mathematical analysis of diffusion implicitly assumed a fixed temperature. Our previous discussion of the dependence of diffusion on point defects causes us to expect a strong temperature dependence for diffusivity by analogy to Equation 5.5—and this is precisely the case. Diffusivity data are perhaps the best known examples of an Arrhenius equation:

$$D = D_0 e^{-q/kT} \qquad (5.12)$$

with D_0 being the preexponential constant and q the activation energy for defect motion. In general, q is not equal to the E_{defect} of Equation 5.5. E_{defect}

---

*   Johann Karl Friedrich Gauss (1777–1855), German mathematician, was one of the great geniuses in the history of mathematics. In his teens, he developed the method of least squares for curve-fitting data. Much of his work in mathematics was similarly applied to physical problems, such as astronomy and geomagnetism. His contribution to the study of magnetism led to the unit of magnetic flux density being named in his honor.

Table 5.1 *The Error Function*

z	erf(z)	z	erf(z)
0.00	0.0000	0.70	0.6778
0.01	0.0113	0.75	0.7112
0.02	0.0226	0.80	0.7421
0.03	0.0338	0.85	0.7707
0.04	0.0451	0.90	0.7969
0.05	0.0564	0.95	0.8209
0.10	0.1125	1.00	0.8427
0.15	0.1680	1.10	0.8802
0.20	0.2227	1.20	0.9103
0.25	0.2763	1.30	0.9340
0.30	0.3286	1.40	0.9523
0.35	0.3794	1.50	0.9661
0.40	0.4284	1.60	0.9763
0.45	0.4755	1.70	0.9838
0.50	0.5205	1.80	0.9891
0.55	0.5633	1.90	0.9928
0.60	0.6039	2.00	0.9953
0.65	0.6420		

Source: *Handbook of Mathematical Functions*, M. Abramowitz
and I. A. Stegun, eds., National Bureau of Standards,
Applied Mathematics Series 55, Washington, D.C., 1972.

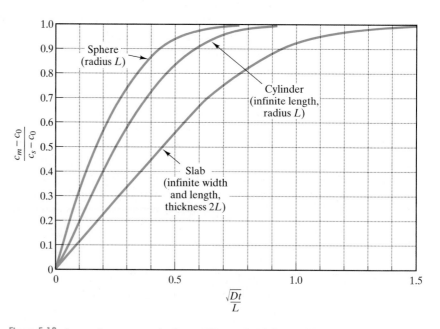

Figure 5-12 *Saturation curves similar to Figure 5–11 for various geometries. The parameter c_m is the average concentration of diffusing species within the sample. Again, the surface concentration c_s and diffusion coefficient D are assumed to be constant. (From W. D. Kingery, H. K. Bowen, and D. R. Uhlmann,* Introduction to Ceramics, *2nd ed., John Wiley & Sons, Inc., New York, 1976.)*

represents the energy required for defect formation, while q represents the energy required for movement of that defect through the crystal structure ($E_{\text{defect motion}}$) for *interstitial diffusion.* For the vacancy mechanism, vacancy formation is an integral part of the diffusional process (see Figure 5–5) and $q = E_{\text{defect}} + E_{\text{defect motion}}$.

It is more common to tabulate diffusivity data in terms of molar quantities, that is, with an activation energy, Q, per mole of diffusing species:

$$D = D_0 e^{-Q/RT} \qquad (5.13)$$

where R is the universal gas constant ($= N_{\text{AV}}k$, as discussed previously). Figure 5–13 shows an Arrhenius plot of the diffusivity of carbon in α-Fe over a range of temperatures. This is an example of an interstitialcy mechanism as sketched in Figure 5–6. Figure 5–14 collects diffusivity data for a number of metallic systems. Table 5.2 gives the Arrhenius parameters for these data. It is useful to compare different data sets. For instance, C can diffuse by an interstitialcy mechanism through bcc Fe more readily than through fcc Fe ($Q_{\text{bcc}} < Q_{\text{fcc}}$ in Table 5.2). The greater openness of the bcc structure (Section 3.2) makes this understandable. Similarly, the self-diffusion of Fe by

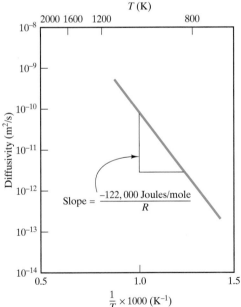

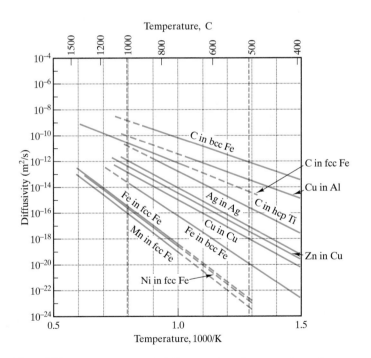

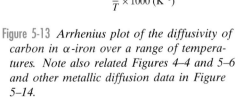

Figure 5-13 *Arrhenius plot of the diffusivity of carbon in α-iron over a range of temperatures. Note also related Figures 4–4 and 5–6 and other metallic diffusion data in Figure 5–14.*

Figure 5-14 *Arrhenius plot of diffusivity data for a number of metallic systems. (From L. H. Van Vlack,* Elements of Materials Science and Engineering, *4th ed., Addison-Wesley Publishing Co., Inc., Reading, Mass., 1980.)*

Table 5.2 *Diffusivity Data for a Number of Metallic Systems*[a]

Solute	Solvent	$D_0(m^2/s)$	Q (kJ/mol)	Q (kcal/mol)
Carbon	Fcc iron	20×10^{-6}	142	34.0
Carbon	Bcc iron	220×10^{-6}	122	29.3
Iron	Fcc iron	22×10^{-6}	268	64.0
Iron	Bcc iron	200×10^{-6}	240	57.5
Nickel	Fcc iron	77×10^{-6}	280	67.0
Manganese	Fcc iron	35×10^{-6}	282	67.5
Zinc	Copper	34×10^{-6}	191	45.6
Copper	Aluminum	15×10^{-6}	126	30.2
Copper	Copper	20×10^{-6}	197	47.1
Silver	Silver	40×10^{-6}	184	44.1
Carbon	Hcp titanium	511×10^{-6}	182	43.5

Source: Data from L. H. Van Vlack, *Elements of Materials Science and Engineering,* 4th ed., Addison-Wesley Publishing Co., Inc., Reading, Mass., 1980.
[a] See Equation 5.13.

a vacancy mechanism is greater in bcc Fe than in fcc Fe. Figure 5–15 and Table 5.3 give comparable diffusivity data for several nonmetallic systems. In many compounds, such as Al_2O_3, the smaller ionic species (e.g., Al^{3+}) diffuse much more readily through the system. The Arrhenius behavior of ionic diffusion in ceramic compounds is especially analogous to the temperature dependence of semiconductors to be discussed in Chapter 17. It is this ionic transport mechanism that is responsible for the semiconducting behavior of certain ceramics such as ZnO; that is, charged ions rather than electrons produce the measured electrical conductivity. Polymer data are not included with the other nonmetallic systems of Figure 5–15 and Table 5.3 because most commercially important diffusion mechanisms in polymers involve the liquid state or the amorphous solid state, where the point defect mechanisms of this section do not apply.

Table 5.3 *Diffusivity Data for a Number of Nonmetallic Systems*[a]

Solute	Solvent	$D_0(m^2/s)$	Q (kJ/mol)	Q (kcal/mol)
Al	Al_2O_3	2.8×10^{-3}	477	114
O	Al_2O_3	0.19	636	152
Mg	MgO	24.9×10^{-6}	330	79
O	MgO	4.3×10^{-9}	344	82.1
Ni	MgO	1.8×10^{-9}	202	48.3
Si	Si	0.18	460	110
Ge	Ge	1.08×10^{-3}	291	69.6
B	Ge	1.1×10^{3}	439	105

Source: Data from P. Kofstad, *Nonstoichiometry, Diffusion, and Electrical Conductivity in Binary Metal Oxides,* John Wiley & Sons, Inc., New York, 1972; and S. M. Hu, in *Atomic Diffusion in Semiconductors,* D. Shaw, ed., Plenum Press, New York, 1973.
[a] See Equation 5.13.

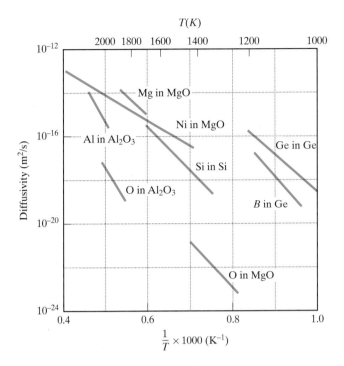

$T(K)$

Figure 5-15 *Arrhenius plot of diffusivity data for a number of nonmetallic systems. (From P. Kofstad,* Nonstoichiometry, Diffusion, and Electrical Conductivity in Binary Metal Oxides, *John Wiley & Sons, Inc., New York, 1972; and S. M. Hu in* Atomic Diffusion in Semiconductors, *D. Shaw, ed., Plenum Press, New York, 1973.)*

SAMPLE PROBLEM 5.3

Steel surfaces can be hardened by *carburization,* as discussed relative to Figure 5–10. During one such treatment at 1000°C, there is a drop in carbon concentration from 5 to 4 at % carbon between 1 and 2 mm from the surface of the steel. Estimate the flux of carbon atoms into the steel in this near-surface region. (The density of γ-Fe at 1000°C is 7.63 g/cm^3.)

SOLUTION

First, we approximate

$$\frac{\partial c}{\partial x} \simeq \frac{\Delta c}{\Delta x} = \frac{5 \text{ at } \% - 4 \text{ at } \%}{1 \text{ mm} - 2 \text{ mm}}$$

$$= -1 \text{ at } \%/\text{mm}$$

To obtain an absolute value for carbon atom concentration, we must first know the concentration of iron atoms. From the given data and Appendix 1,

$$\rho = 7.63 \frac{g}{cm^3} \times \frac{0.6023 \times 10^{24} \text{ atoms}}{55.85 \text{ g}} = 8.23 \times 10^{22} \frac{\text{atoms}}{cm^3}$$

Therefore,

$$\frac{\Delta c}{\Delta x} = -\frac{0.01(8.23 \times 10^{22} \text{ atoms/cm}^3)}{1 \text{ mm}} \times \frac{10^6 \text{ cm}^3}{m^3} \times \frac{10^3 \text{ mm}}{m}$$

$$= -8.23 \times 10^{29} \text{ atoms/m}^4$$

From Table 5.2,

$$D_{c \text{ in } \gamma-\text{Fe},1000°C} = D_0 e^{-Q/RT}$$

$$= (20 \times 10^{-6} \text{ m}^2/\text{s})e^{-(142,000 \text{ J/mol})/(8.314 \text{ J/mol/K})(1273 \text{ K})}$$

$$= 2.98 \times 10^{-11} \text{ m}^2/\text{s}$$

Using Equation 5.8 gives us

$$J_x = -D\frac{\partial c}{\partial x}$$

$$\simeq -D\frac{\Delta c}{\Delta x}$$

$$= -(2.98 \times 10^{-11} \text{ m}^2/\text{s})(-8.23 \times 10^{29} \text{ atoms/m}^4)$$

$$= 2.45 \times 10^{19} \text{ atoms/(m}^2 \cdot \text{s)}$$

SAMPLE PROBLEM 5.4

The diffusion result described by Equation 5.11 can apply to the carburization process (Sample Problem 5.3). The carbon environment (a hydrocarbon gas) is used to set the surface carbon content (c_s) at 1.0 wt %. The initial carbon content of the steel (c_0) is 0.2 wt %. Using the error function table, calculate how long it would take at 1000°C to reach a carbon content of 0.6 wt % [i.e., $(c - c_0)/(c_s - c_0) = 0.5$] at a distance of 1 mm from the surface.

SOLUTION

Using Equation 5.11, we get

$$\frac{c_x - c_0}{c_s - c_0} = 0.5 = 1 - \text{erf}\left(\frac{x}{2\sqrt{Dt}}\right)$$

or

$$\text{erf}\left(\frac{x}{2\sqrt{Dt}}\right) = 1 - 0.5 = 0.5$$

Interpolating from Table 5.1 gives

$$\frac{0.5 - 0.4755}{0.5205 - 0.4755} = \frac{z - 0.45}{0.50 - 0.45}$$

or

$$z = \frac{x}{2\sqrt{Dt}} = 0.4772$$

or

$$t = \frac{x^2}{4(0.4772)^2 D}$$

Using the diffusivity calculation from Sample Problem 5.3, we obtain

$$t = \frac{(1 \times 10^{-3} \text{ m})^2}{4(0.4772)^2 (2.98 \times 10^{-11} \text{ m}^2/\text{s})}$$

$$= 3.68 \times 10^4 \text{ s} \times \frac{1 \text{ h}}{3.6 \times 10^3 \text{ s}}$$

$$= 10.2 \text{ h}$$

SAMPLE PROBLEM 5.5

Recalculate the carburization time for the conditions of Sample Problem 5.4 using the master plot of Figure 5–11, rather than the error function table.

SOLUTION

From Figure 5–11, we see that the condition for $(c - c_0)/(c_s - c_0) = 0.5$ is

$$\frac{x}{\sqrt{Dt}} \simeq 0.95$$

or

$$t = \frac{x^2}{(0.95)^2 D}$$

Using the diffusivity calculation from Sample Problem 5.3, we obtain

$$t = \frac{(1 \times 10^{-3} \text{ m})^2}{(0.95)^2 (2.98 \times 10^{-11} \text{ m}^2/\text{s})}$$

$$= 3.72 \times 10^4 \text{ s} \times \frac{1 \text{ h}}{3.6 \times 10^3 \text{ s}}$$

$$= 10.3 \text{ h}$$

Note.   There is appropriately close agreement with the calculation of Sample Problem 5.4. Exact agreement is hindered by the need for graphical interpretation (in this problem) and tabular interpolation (in the previous one).

SAMPLE PROBLEM 5.6

For a carburization process similar to that in Sample Problem 5.5, a carbon content of 0.6 wt % is reached at 0.75 mm from the surface after 10 h. What is the carburization temperature? (Assume, as before, that $c_s = 1.0$ wt % and $c_0 = 0.2$ wt %.)

SOLUTION

As in Sample Problem 5.5,

$$\frac{x}{\sqrt{Dt}} \simeq 0.95$$

or

$$D = \frac{x^2}{(0.95)^2 t}$$

with the data given,

$$D = \frac{(0.75 \times 10^{-3} \text{ m})^2}{(0.95)^2(3.6 \times 10^4 \text{ s})} = 1.73 \times 10^{-11} \text{ m}^2/\text{s}$$

From Table 5.2, for C in γ-Fe,

$$D = (20 \times 10^{-6} \text{ m}^2/\text{s})e^{-(142,000 \text{ J/mol})/[8.314 \text{ J/(mol·}K)](T)}$$

Equating the two values for D gives

$$1.73 \times 10^{-11} \frac{\text{m}^2}{\text{s}} = 20 \times 10^{-6} \frac{\text{m}^2}{\text{s}} e^{-1.71 \times 10^4/T}$$

or

$$T^{-1} = \frac{-\ln(1.73 \times 10^{-11}/20 \times 10^{-6})}{1.71 \times 10^4}$$

or

$$T = 1225 \text{ K} = 952°\text{C}$$

··

PRACTICE PROBLEM 5.3

Suppose that the carbon concentration gradient described in Sample Problem 5.3 occurred at 1100°C rather than 1000°C. Calculate the carbon atom flux for this case.

PRACTICE PROBLEM 5.4

In Sample Problem 5.4 the time to generate a given carbon concentration profile is calculated using the error function table. The carbon content at the surface was 1.0 wt % and at 1 mm from the surface was 0.6 wt %. For this diffusion time, what is the carbon content at a distance **(a)** 0.5 mm from the surface and **(b)** 2 mm from the surface?

PRACTICE PROBLEM 5.5

Repeat Practice Problem 5.4 using the graphical method of Sample Problem 5.5.

PRACTICE PROBLEM 5.6

In Sample Problem 5.6 a carburization temperature is calculated for a given carbon concentration profile. Calculate the carburization temperature if the given profile were obtained in 8 hours rather than 10 hours, as originally stated.

5.4   STEADY-STATE DIFFUSION

The change in the concentration profile with time for processes such as carburization was shown in Figure 5–10. A similar observation for a process with slightly different boundary conditions is shown in Figure 5–16. In this case, the relatively high surface concentration of the diffusing species, c_h, is held constant with time, just as c_s was held constant in Figure 5–10, but the relatively low concentration, c_l, at x_0 is also held constant with time. As a result, the nonlinear concentration profiles at times greater than zero (e.g., at t_1 and t_2 in Figure 5–16) approach a straight line after a relatively long time (e.g., at t_3 in Figure 5–16). This *linear* concentration profile is unchanging with additional time as long as c_h and c_l remain fixed. This limiting case is an example of **steady-state diffusion** (i.e., mass transport that is unchanging with time). The concentration gradient defined by Equation 5.8 takes an especially simple form in this case:

$$\frac{\partial c}{\partial x} = \frac{\Delta c}{\Delta x} = \frac{c_h - c_l}{0 - x_0} = -\frac{c_h - c_l}{x_0} \qquad (5.14)$$

In the case of carburization represented by Figure 5–10, the surface concentration, c_s, was held fixed by maintaining a fixed carbon-source atmospheric pressure at the $x = 0$ surface. That is also how both c_h and c_l are maintained fixed in the case represented by Figure 5–16. A plate of material with a thickness of x_0 is held between two gas atmospheres: a high-pressure atmosphere on the $x = 0$ surface, which produces the fixed concentration c_h, and a low-pressure atmosphere on the $x = x_0$ surface, which produces the concentration c_l (Figure 5–17).

A common application of steady-state diffusion is the use of materials as gas purification membranes. For example, a thin sheet of palladium metal is permeable to hydrogen gas but not to common atmospheric gases such as oxygen, nitrogen, and water vapor. By introducing the "impure"

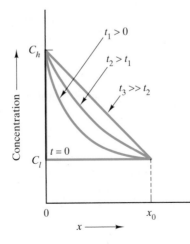

Figure 5-16 *Solution to Fick's second law (Equation 5.10) for the case of a solid of thickness x_0, constant surface concentrations of the diffusing species c_h and c_l, and a constant diffusion coefficient D. For long times, e.g., t_3, the linear concentration profile is an example of steady-state diffusion.*

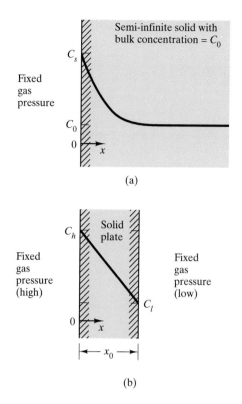

Fixed gas pressure

(a)

Fixed gas pressure (high)

Fixed gas pressure (low)

(b)

Figure 5-17 *Schematic of sample configurations in gas environments that lead, after a long time, to the diffusion profiles representative of (a) nonsteady state diffusion (Figure 5–10) and (b) steady-state diffusion (Figure 5–16).*

gas mixture on the high-pressure side of Figure 5–17b and maintaining a constant, reduced hydrogen pressure on the low-pressure side, a steady flow of purified hydrogen passes through the palladium sheet.

SAMPLE PROBLEM 5.7

A 5-mm-thick sheet of palladium with a cross-sectional area of 0.2 m² is used as a steady-state diffusional membrane for purifying hydrogen. If the hydrogen concentration on the high-pressure (impure gas) side of the sheet is 0.3 kg/m³, and the diffusion coefficient for hydrogen in Pd is 1.0×10^{-8} m²/s, calculate the mass of hydrogen being purified per hour.

SOLUTION

Fick's first law (Equation 5.8) is simplified by the steady-state concentration gradient of Equation 5.14, giving

$$J_x = -D(\partial c/\partial x) = [-(c_h - c_l)/x_0]$$
$$= -(1.0 \times 10^{-8} \, \text{m}^2/\text{s}) \left[-(1.5 \, \text{kg/m}^3 - 0.3 \, \text{kg/m}^3)/(5 \times 10^{-3} \, \text{m})\right]$$
$$= 2.4 \times 10^{-6} \, \text{kg/m}^2 \cdot \text{s} \times 3.6 \times 10^3 \, \text{s/h} = 8.64 \times 10^{-3} \, \text{kg/m}^2 \cdot \text{h}$$

The total mass of hydrogen being purified will then be this flux times the membrane area:

$$m = J_x \times A = 8.64 \times 10^{-3} \text{ kg/m}^2 \cdot \text{h} \times 0.2 \text{ m}^2 = 1.73 \times 10^{-3} \text{ kg/h}$$

....

PRACTICE PROBLEM 5.7

For the purification membrane in Sample Problem 5.7, how much hydrogen would be purified per hour if the membrane used is 3 mm thick, with all other conditions unchanged?

5.5   ALTERNATE DIFFUSION PATHS

A final word of caution is in order about using specific diffusivity data to analyze a particular material process. Figure 5–18 shows that the self-diffusion coefficients for silver vary by several orders of magnitude, depending on the route for diffusional transport. To this point, we have considered **volume diffusion,** or **bulk diffusion,** through a material's crystal structure by means of some defect mechanism. However, there can be "short circuits" associated with easier diffusion paths. As seen in Figure 5–18, diffusion is much faster (with a lower Q) along a grain boundary. As we saw in Section 4.4, this region of mismatch between adjacent crystal grains in the material's microstructure is a more open structure, allowing enhanced grain boundary diffusion. The crystal surface is an even more open region, and **surface diffusion** allows easier atom transport along the free surface less hindered by adjacent atoms. The overall result is that

$$Q_{\text{volume}} > Q_{\text{grain boundary}} > Q_{\text{surface}} \quad \text{and} \quad D_{\text{volume}} < D_{\text{grain boundary}} < D_{\text{surface}}$$

This does not mean that surface diffusion is always the important process just because D_{surface} is greatest. More important is the amount of diffusing region available. In most cases, volume diffusion dominates. For a material with a small average grain size (see Section 4.4) and therefore large grain boundary area, **grain boundary diffusion** can dominate. Similarly, in a fine-grained powder with large surface area, surface diffusion can dominate.

For a given polycrystalline microstructure, the penetration of a diffusing species will tend to be greater along grain boundaries and even greater along the free surface of the sample (Figure 5–19).

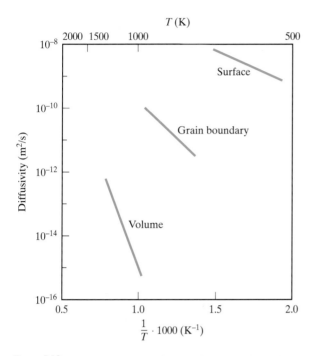

Figure 5-18 *Self-diffusion coefficients for silver depend on the diffusion path. In general, diffusivity is greater through less restrictive structural regions. (After J. H. Brophy, R. M. Rose, and J. Wulff,* The Structure and Properties of Materials, *Vol. 2:* Thermodynamics of Structure, *John Wiley & Sons, Inc., New York, 1964.)*

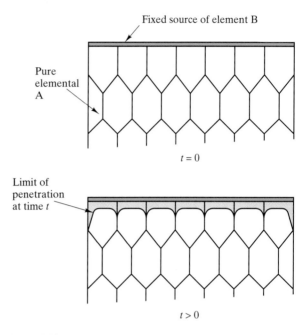

Figure 5-19 *Schematic illustration of how a coating of impurity B can penetrate more deeply into grain boundaries and even further along a free surface of polycrystalline A, consistent with the relative values of diffusion coefficients (* $D_{volume} < D_{grain\ boundary} < D_{surface}$ *).*

SAMPLE PROBLEM 5.8

We can approximate the extent of grain boundary penetration in Figure 5–19 by using the semi-infinite diffusion expression of Equation 5.11. **(a)** Taking $D_{grain\ boundary} = 1.0 \times 10^{-10}\,\text{m}^2/\text{s}$, calculate the "penetration" of B into A along the grain boundary after one hour, defined as the distance, x, at which $c_x = 0.01c_s$ (with $c_0 = 0$ for initially pure A). **(b)** For comparison, caclulate the penetration defined in the same way within the bulk grain for which $D_{volume} = 1.0 \times 10^{-14}\,\text{m}^2/\text{s}$.

SOLUTION

(a) We can simplify Equation 5.11 as

$$\frac{c_x - c_0}{c_s - c_0} = 1 - \text{erf}\left(\frac{x}{2\sqrt{Dt}}\right) = \frac{c_x - 0}{c_s - 0} = \frac{0.01c_s}{c_s} = 0.01$$

or

$$\text{erf}\left(\frac{x}{2\sqrt{Dt}}\right) = 1 - 0.01 = 0.99$$

Interpolating from Table 5.1 gives

$$\frac{0.9928 - 0.99}{0.9928 - 0.9891} = \frac{1.90 - z}{1.90 - 1.80}$$

or

$$z = \frac{x}{2\sqrt{Dt}} = 1.824$$

and then

$$x = 2(1.824)\sqrt{Dt} = 2(1.824)\sqrt{(1.0 \times 10^{-10}\,\text{m}^2/\text{s})(1\,\text{h})(3.6 \times 10^3\,\text{s/h})}$$

$$= 2.19 \times 10^{-3}\,\text{m} = 2.19\,\text{mm}$$

(b) For comparison,

$$x = 2(1.824)\sqrt{(1.0 \times 10^{-14}\,\text{m}^2/\text{s})(1\,\text{h})(3.6 \times 10^3\,\text{s/h})}$$

$$= 21.9 \times 10^{-6}\,\text{m} = 21.9\,\mu\text{m}$$

Note.   The amount of grain boundary penetration calculated in **(a)** is exaggerated, as the excess buildup of impurity, B, in the boundary leads to some of that material being "drained away" by side diffusion into the A grains, as indicated by Figure 5–19.

PRACTICE PROBLEM 5.8

In Sample Problem 5.8, we calculated the extent of impurity penetration for volume and grain boundary paths. For further comparison, calculate the penetration for surface diffusion for which $D_{\text{surface}} = 1.0 \times 10^{-8}\,\text{m}^2/\text{s}$.

SUMMARY

The point defects introduced in Chapter 4 are seen to play a central role in the movement of atoms by solid-state diffusion. Several practical problems in the production and application of engineering materials involve these diffusional processes.

Specifically, we find that point defect concentrations increase exponentially with absolute temperature following an Arrhenius expression. As solid-state diffusion in crystalline materials occurs via a point defect mechanism, the diffusivity, as defined by Fick's laws, also increases exponentially with absolute temperature in another Arrhenius expression.

The mathematics of diffusion allows a relatively precise description of the chemical concentration profiles of diffusing species. For some sample geometries, the concentration profile approaches a simple, linear form after a relatively long time. This steady-state diffusion is well illustrated by gas transport across thin membranes.

In the case of fine-grained polycrystalline materials or powders, material transport may be dominated by grain boundary diffusion or surface diffusion, respectively. This is because, in general, $D_\text{volume} < D_\text{grain boundary} < D_\text{surface}$. Another result is that, for a given polycrystalline solid, impurity penetration will be greater along grain boundaries and even greater along the free surface.

KEY TERMS

activation energy (158)
Arrhenius equation (159)
Arrhenius plot (159)
bulk diffusion (178)
carburization (166)
concentration gradient (166)
diffusion (164)
diffusion coefficient (166)

diffusivity (166)
Fick's first law (165)
Fick's second law (166)
Gaussian error function (167)
grain boundary diffusion (178)
Maxwell-Boltzmann distribution (160)
preexponential constant (158)
random walk (165)

rate-limiting step (160)
self-diffusion (164)
steady-state diffusion (176)
surface diffusion (178)
thermal activation (160)
thermal vibration (162)
vacancy migration (165)
volume diffusion (178)

REFERENCES

Chiang, Y., D. P. Birnie III, and **W. D. Kingery,** *Physical Ceramics*, John Wiley & Sons, Inc., New York, 1997.

Crank, J., *The Mathematics of Diffusion*, 2nd ed., Clarendon Press, Oxford, 1993.

Shewmon, P. G., *Diffusion in Solids*, 2nd ed., Minerals, Metals, and Materials Society, Warrendale, Pa., 1989.

PROBLEMS

Section 5.1 • Thermally Activated Processes

5.1. In steel-making furnaces, temperature-resistant ceramic bricks (refractories) are used as linings to contain the molten metal. A common byproduct of the steel-making process is a calcium aluminosilicate (slag), that is chemically corrosive to the refractories. For alumina refractories, the corrosion rate is 2.0×10^{-8} m/s at $1425°C$ and 8.95×10^{-8} m/s at $1500°C$. Caclulate the activation energy for the corrosion of these alumina refractories.

5.2. For a steel furnace similar to that described in Problem 5.1, silica refractories have corrosion rates of 2.0×10^{-7} m/s at $1345°C$ and 9.0×10^{-7} m/s at $1510°C$. Calculate the activation energy for the corrosion of these silica refractories.

5.3. Manufacturing traditional clayware ceramics typically involves driving off the water of hydration in the clay minerals. The rate constant for the dehydration of kaolinite, a common clay mineral, is $1.0 \times 10^{-4}\,s^{-1}$ at $485°C$ and $1.0 \times 10^{-3}\,s^{-1}$ at $525°C$. Calculate the activation energy for the dehydration of kaolinite.

5.4. For the thermally activated process described in Problem 5.3, calculate the rate constant at $600°C$, assuming that to be the upper temperature limit specified for the dehydration process.

Section 5.2 • Thermal Production of Point Defects

5.5. Verify that the data represented by Figure 5–4b correspond to an energy of formation of 0.76 eV for a defect in aluminum.

5.6. What type of crystal direction corresponds to the movement of interstitial carbon in α-Fe between equivalent ($\frac{1}{2}0\frac{1}{2}$-type) interstitial positions? Illustrate your answer with a sketch.

5.7. Repeat Problem 5.6 for the movement between equivalent interstices in γ-Fe. (Note Practice Problem 4.2.)

5.8. What crystallographic positions and directions are indicated by the migration shown in Figure 5–5? [Assume the atoms are in a (100) plane of an fcc metal.]

Section 5.3 • Point Defects and Solid-State Diffusion

5.9. Verify that the data represented by Figure 5–13 correspond to an activation energy of 122,000 J/mol for the diffusion of carbon in α-iron.

5.10. Carburization was described in Sample Problem 5.3. The decarburization of a steel can also be described by using the error function. Starting with Equation 5.11 and taking $c_s = 0$, derive an expression to describe the concentration profile of carbon as it diffuses out of a steel with initial concentration, c_0. (This situation can be produced by placing the steel in a vacuum at elevated temperature.)

5.11. Using the decarburization expression derived in Problem 5.10, plot the concentration profile of carbon within 1 mm of the carbon-free surface after 1 hour in vacuum at $1000°C$. Take the initial carbon content of the steel to be 0.3 wt %.

5.12. A "diffusion couple" is formed when two different materials are allowed to interdiffuse at elevated temperature (see Figure 5–8). For a block of pure metal A adjacent to a block of pure metal B, the concentration profile of A (in at %) after interdiffusion is given by

$$c_x = 50\left[1 - \mathrm{erf}\left(\frac{x}{2\sqrt{Dt}}\right)\right]$$

where x is measured from the original interface. For a diffusion couple with $D = 10^{-14}$ m²/s, plot the concentration profile of metal A over a range of 20 μm on either side of the original interface ($x = 0$) after a time of 1 hour. [Note that erf $(-z) = -\mathrm{erf}(z)$.]

5.13. Use the information from Problem 5.12 to plot the progress of interdiffusion of two metals, X and Y, with $D = 10^{-12}$ m²/s. Plot the concentration profile of metal X over a range of 300 μm on either side of the original interface after times of 1, 2, and 3 h.

5.14. Using the results of Problem 5.12 and assuming that profile occurred at a temperature of $1000°C$, superimpose the concentration profile of metal A for the same diffusion couple for 1 hour but heated at $1200°C$ at which $D = 10^{-13}$ m²/s.

5.15. Given the information in Problems 5.12 and 5.14, calculate the activation energy for the interdiffusion of metals A and B.

5.16. Use the result of Problem 5.15 to calculate the diffusion coefficient for the interdiffusion of metals A and B at 1400°C.

5.17. Using data in Table 5.2, calculate the self-diffusivity for iron in bcc iron at 900°C.

5.18. Using data in Table 5.2, calculate the self-diffusivity for iron in fcc iron at 1000°C.

5.19. Using data in Table 5.2, calculate the self-diffusivity for copper in copper at 1000°C.

5.20. The diffusivity of copper in a commercial brass alloy is 10^{-20} m^2/s at 400°C. The activation energy for diffusion of copper in this system is 195 kJ/mol. Calculate the diffusivity at 600°C.

5.21. The diffusion coefficient of nickel in an austenitic (fcc structure) stainless steel is 10^{-22} m^2/s at 500°C and 10^{-15} m^2/s at 1000°C. Calculate the activation energy for the diffusion of nickel in this alloy over this temperature range.

• **5.22.** Show that the relationship between vacancy concentration and fractional dimension changes for the case shown in Figure 5–4 is approximately

$$\frac{n_v}{n_{\text{sites}}} = 3\left(\frac{\Delta L}{L} - \frac{\Delta a}{a}\right)$$

[Note that $(1 + x)^3 \simeq 1 + 3x$ for small x.]

• **5.23.** A popular use of diffusion data in materials science is to identify mechanisms for certain phenomena. This is done by comparison of activation energies. For example, consider the oxidation of an aluminum alloy. The rate-controlling mechanism is the diffusion of ions through an Al$_2$O$_3$ surface layer. This means that the rate of growth of the oxide layer thickness is directly proportional to a diffusion coefficient. We can specify whether oxidation is controlled by Al^{3+} diffusion or O^{2-} diffusion by comparing the activation energy for oxidation with

the activation energies of the two species as given in Table 5.3. Given that the rate constant for oxide growth is 4.00×10^{-8} kg/(m$^4 \cdot$s) at 500°C and 1.97×10^{-4} kg/(m$^4 \cdot$ s) at 600°C, determine whether the oxidation process is controlled by Al^{3+} diffusion or O^{2-} diffusion.

5.24. *Diffusion length*, λ, is a popular term in characterizing the production of semiconductors by the controlled diffusion of impurities into a high-purity material. The value of λ is taken as $2\sqrt{Dt}$, where λ represents the extent of diffusion for an impurity with a diffusion coefficient, D, over a period of time, t. Calculate the diffusion length for B in Ge for a total diffusion time of 30 minutes at a temperature of **(a)** 800°C and **(b)** 900°C.

Section 5.4 • Steady-State Diffusion

5.25. A differential nitrogen pressure exists across a 2-mm-thick steel furnace wall. After some time, steady-state diffusion of the nitrogen is established across the wall. Given that the nitrogen concentration on the high-pressure surface of the wall is 2 kg/m^3 and on the low-pressure surface is 0.2 kg/m^3, calculate the flow of nitrogen through the wall (in kg/m$^2 \cdot$h) if the diffusion coefficient for nitrogen in this steel is 1.0×10^{-10} m^2/s at the furnace operating temperature.

5.26. For the furnace described in Problem 5.25, design changes are made that include a thicker wall (3 mm) and a lower operating temperature that reduces the nitrogen diffusivity to 5.0×10^{-11} m^2/s. What would be the steady-state nitrogen flow across the wall in this case?

5.27. Many laboratory furnaces have small glass windows that provide visual access to samples. The leakage of furnace atmospheres through the windows can be a problem. Consider a 3-mm-thick window of vitreous silica in a furnace containing an inert, helium atmoshpere. For a window with a cross-sectional area of 600 mm^2, calculate the steady-state flow of helium gas (in atoms/s) across the window if the concentration of helium on the high-pressure (furnace) surface of the window if 6.0×10^{23} atoms/m^3 and on the low-pressure (outside) surface is essentially zero. The diffusion coefficient for helium in vitreous silica at this wall temperature is 1.0×10^{-10} m^2/s.

5.28. Figure 4–25 shows that Na_2O additions to vitreous silica tends to "tighten" the structure, as Na^+ ions fill in open voids in the silica structure. This structural feature can have a significant effect on the gas diffusion described in Problem 5.27. Consider the replacement of the vitreous silica window with a sodium silicate window (containing 30 mol. % Na_2O) of the same dimensions. For the "tighter" sodium silicate glass, the concentration of helium on the high-pressure surface is reduced to 3.0×10^{22} atoms/m^3. Similarly, the diffusion coefficient for helium in the sodium silicate glass at this wall temperature is reduced to 2.5×10^{-12} m^2/s. Calculate the steady-state flow of helium gas (in atoms/s) across this replacement window.

Section 5.5 • Alternate Diffusion Paths

5.29. The endpoints of the Arrhenius plot of $D_{\text{grain boundary}}$ in Figure 5–18 are $D_{\text{grain boundary}} = 3.2 \times 10^{-12}$ m^2/s at a temperature of 457°C and $D_{\text{grain boundary}} = 1.0 \times 10^{-10}$ m^2/s at a temperature of 689°C. Using these data, calculate the activation energy for grain boundary diffusion in silver.

5.30. The endpoints of the Arrhenius plot of D_{surface} in Figure 5–18 are $D_{\text{surface}} = 7.9 \times 10^{-10}$ m^2/s at a temperature of 245°C and $D_{\text{surface}} = 6.3 \times 10^{-9}$ m^2/s at a temperature of 398°C. Using these data, calculate the activation energy for surface diffusion in silver.

5.31. The contribution of grain boundary diffusion can sometimes be seen from diffusivity measurements made on polycrystalline samples of increasingly small grain size. As an example, plot the following data (as ln D versus ln[grain size]) for the diffusion coefficient of Ni^{2+} in NiO at 480°C, measured as a function of sample grain size.

Grain size (μm)	D(m^2/s)
1	1.0×10^{-19}
10	1.0×10^{-20}
100	1.0×10^{-21}

5.32. Using the plot from Problem 5.31, estimate the diffusion coefficient of Ni^{2+} in NiO at 480°C for a 20 μm grain size material.

CHAPTER **6**

Mechanical Behavior

Mechanical testing machines can be automated to simplify the analysis of the mechanical performance of materials in a variety of product applications. (Courtesy of MTS Systems Corporation.)

As discussed in Chapter 1, probably no materials are more closely associated with the engineering profession than metals such as structural steel. In this chapter, we shall explore some of the major mechanical properties of metals: stress versus strain, hardness, and creep. Although this chapter will provide an introduction to these properties, an appreciation of the versatility of metals will be presented in Chapters 9 and 10. Microstructural development as related to phase diagrams will be dealt with in Chapter 9. Heat treatment based on the kinetics of solid-state reactions will be covered in Chapter 10. Each of these topics deals with methods to "fine tune" the properties of given alloys within a broad range of values.

Many of the important mechanical properties of metals apply to ceramics as well, although the values of those properties may be very different for the ceramics. For example, brittle fracture and creep play important roles in the structural applications of ceramics. The liquidlike structure of glass leads to high-temperature deformation by a viscous flow mechanism. The production of fracture-resistant tempered glass depends on precise control of that viscosity.

We follow our discussions of the mechanical properties of the inorganic materials, metals and ceramics, with a similar discussion of the mechanical properties of the organic polymers. An important trend in engineering design during the past decade has been increased concentration on so-called *engineering polymers,* which have sufficiently high strength and stiffness to be able to substitute for traditional structural metals. Often, we find that polymers exhibit behavior associated with their long-chain or network molecular structure. Viscoelastic deformation is an important example.

6.1   STRESS VERSUS STRAIN

Metals are used in engineering designs for many reasons, but they generally serve as structural elements. Thus, we shall initially concentrate on the mechanical properties of metals in this chapter.

METALS

The mechanical properties discussed next are not exhaustive but are intended to cover the major factors in selecting a durable material for structural applications under a variety of service conditions. As we go through these various properties, an attempt will be made to use a consistent and comprehensive set of sample metals and alloys to demonstrate typical data and, especially, important data trends. Table 6.1 lists 15 classes of sample metals and alloys, with each class representing one of the groups to be covered in Chapter 11, on metals.

Table 6.1 *Some Typical Metal Alloys*

Alloy	UNS designation[a]	Primary constituent	Primary alloying elements (wt %)															
			C	Mn	Si	Cr	Ni	Mo	V	Cu	Mg	Al	Zn	Sn	Fe	Zr	Ag	Pd
1. Carbon steel: 1040 cold-drawn, no stress relief	G10400	Fe	0.4	0.75														
2. Low-alloy steel: 8630 cold-drawn, hot-reduced, no stress relief	G86300	Fe	0.3	0.8	0.2	0.5	0.5	0.2										
3. Stainless steels																		
a. Type 304 stainless bar stock, hot-finished and annealed	S30400	Fe	0.08	2.0	1.0	19.0	9.0											
b. Type 304 stainless subjected to longitudinal fatigue test	S30400	Fe	0.08	2.0	1.0	19.0	9.0											
c. Type 410 stainless, 595°C temper	S41000	Fe	0.15	1.0	1.0	12.0												
4. Tool steel: L2 (low-alloy, special-purpose) oil-quenched from 855°C and single-tempered at 425°C	T61202	Fe	0.7	0.5		1.0			0.2									
5. Ferrous superalloy: Type 410 stainless (see alloy 3c)																		
6. Cast irons																		
a. Ductile iron, quench and temper		Fe	3.65	0.52	2.48		0.78			0.15								
b. Ductile iron, 60-40-18 (tested in tension)	F32800	Fe	3.0		2.5													
7. Aluminum																		
a. 3003-H14	A93003	Al		1.25							1.0							
b. 2048, plate	A92048	Al		0.40						3.3	1.5							
8. Magnesium																		
a. AZ31B, hard-rolled sheet	M11311	Mg		0.2								3.0	1.0					
b. AM100A, casting alloy, temper F	M10100	Mg		0.1								10.0						

Table 6.1 *Continued*

Alloy	UNS designation[a]	Primary constituent	Primary alloying elements (wt %)															
			C	Mn	Si	Cr	Ni	Mo	V	Cu	Mg	Al	Zn	Sn	Fe	Zr	Ag	Pd
9. a. Titanium: Ti–5Al–2.5Sn, standard grade	R54520	Ti										5.0		2.5				
b. Titanium: Ti–6Al–4V, standard grade	R56400	Ti							4.0			6.0						
10. Copper: aluminum bronze, 9% cold-finished	C62300	Cu										10.0			3.0			
11. Nickel: Monel 400, hot-rolled	N04400	Ni(66.5%)								31.5								
12. Zinc: AC41A, No. 5 die-casting alloy	Z35530	Zn								1.0	0.04	4.0						
13. Lead: 50 Pb–50 Sn solder		Pb												50.				
14. Refractory metal: Nb–1 Zr, recrystallized R04261 (commercial grade)		Nb														1.0		
15. Precious metal: dental gold alloy		Au (76%)								8.0							13.0	2.0

[a] Alloy designations and associated properties cited in Tables 6.2, 6.4, and 6.11 are from *Metals Handbook*, 8th ed., Vol. 1, and 9th Ed., Vols. 1–3, American Society for Metals, Metals Park, Ohio, 1961, 1978, 1979, and 1980.

Perhaps the simplest questions that a design engineer can ask about a structural material are (1) "How strong is it?" and (2) "How much deformation must I expect given a certain load?" This basic description of the material is obtained by the *tensile test.* Figure 6–1 illustrates this simple pull test. The load necessary to produce a given elongation is monitored as the specimen is pulled in tension at a constant rate. A load-versus-elongation curve (Figure 6–2) is the immediate result of such a test. A more general statement about material characteristics is obtained by normalizing the data of Figure 6–2 for geometry. The resulting *stress-versus-strain curve* is given in Figure 6–3. Here, the **engineering stress,** σ, is defined as

$$\sigma = \frac{P}{A_0} \tag{6.1}$$

where P is the load on the sample with an original (zero stress) cross-sectional area, A_0. *Sample cross section* refers to the region near the center of the specimen's length. Specimens are machined such that the cross-sectional area in this region is uniform and smaller than at the ends gripped by the testing machine. This smallest area region, referred to as the **gage length,** experiences the largest stress concentration so that any significant deformation at higher stresses is localized there. The **engineering strain,** ϵ, is defined as

$$\epsilon = \frac{l - l_0}{l_0} = \frac{\Delta l}{l_0} \tag{6.2}$$

where l is the gage length at a given load, and l_0 is the original (zero-stress) length. Figure 6–3 is divided into two distinct regions: (1) elastic deformation and (2) plastic deformation. **Elastic deformation** is temporary

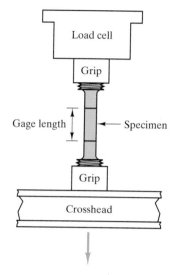

Figure 6-1 *Tensile test.*

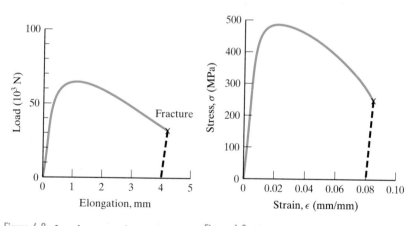

Figure 6-2 *Load-versus-elongation curve obtained in a tensile test. The specimen was aluminum 2024-T81.*

Figure 6-3 *Stress-versus-strain curve obtained by normalizing the data of Figure 6–2 for specimen geometry.*

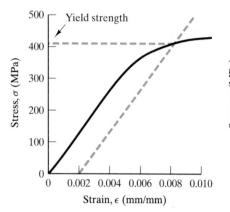

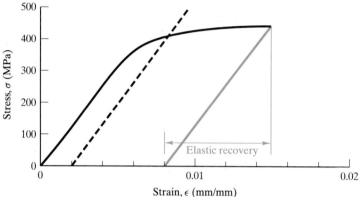

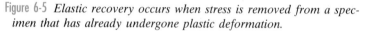

Figure 6-4 *The yield strength is defined relative to the intersection of the stress–strain curve with a "0.2% offset." This is a convenient indication of the onset of plastic deformation.*

Figure 6-5 *Elastic recovery occurs when stress is removed from a specimen that has already undergone plastic deformation.*

deformation. It is fully recovered when the load is removed. The elastic region of the stress–strain curve is the initial linear portion. **Plastic deformation** is permanent deformation. It is not recovered when the load is removed, although a small elastic component is recovered. The plastic region is the nonlinear portion generated once the total strain exceeds its elastic limit. It is often difficult to specify precisely the point at which the stress–strain curve deviates from linearity and enters the plastic region. The usual convention is to define as **yield strength** the intersection of the deformation curve with a straight-line parallel to the elastic portion and offset 0.2% on the strain axis (Figure 6–4). The yield strength represents the stress necessary to generate this small amount (0.2%) of permanent deformation. Figure 6–5 indicates the small amount of elastic recovery that occurs when a load well into the plastic region is released.

Figure 6–6 summarizes the key mechanical properties obtained from the tensile test. The slope of the stress–strain curve in the elastic region is the

Figure 6-6 *The key mechanical properties obtained from a tensile test: 1, modulus of elasticity, E; 2, yield strength, Y.S.; 3, tensile strength, T.S.; 4, ductility, $100 \times \epsilon_{failure}$ (note that elastic recovery occurs after fracture); and 5, toughness $= \int \sigma d\epsilon$ (measured under load; hence, the dashed line is vertical).*

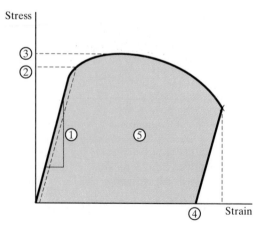

modulus of elasticity, E, also known as **Young's* modulus.** The linearity of the stress–strain plot in the elastic region is a graphical statement of **Hooke's†**
law:

$$\sigma = E\epsilon \qquad (6.3)$$

The modulus E is a highly practical piece of information. It represents the stiffness of the material, that is, its resistance to elastic strain. This manifests itself as the amount of deformation in normal use below the yield strength and the springiness of the material during forming. As with E, the yield strength has major practical significance. It shows the resistance of the metal to permanent deformation and indicates the ease with which the metal can be formed by rolling and drawing operations.

Although we are concentrating on the behavior of metals under tensile loads, the testing apparatus illustrated in Figure 6–1 is routinely used in a reversed mode producing a compressive test. The elastic modulus, in fact, tends to be the same for metal alloys tested in either tensile or compressive modes. Later, we will see the elastic behavior under shear loads is also related to the tensile modulus.

One should note that many design engineers, especially in the aerospace field, are more interested in strength-per-unit density than strength or density individually. (If two alloys each have adequate strength, the lower density one is preferred for potential fuel savings.) The strength-per-unit density is generally termed **specific strength,** or **strength-to-weight ratio,** and is discussed relative to composite properties in Section 14.5. Another term of practical engineering importance is **residual stress**, defined as the stress remaining within a structural material after all applied loads are removed. This commonly occurs following various thermomechanical treatments such as welding and machining.

As the plastic deformation represented in Figure 6–6 continues at stresses above the yield strength, the engineering stress continues to rise toward a maximum. This maximum stress is termed the *ultimate tensile strength,* or simply the **tensile strength** (T.S.). Within the region of the stress–strain curve between Y.S. and T.S., the phenomenon of increasing strength with increasing deformation is referred to as **strain hardening.** This is an important factor in shaping metals by **cold working,** that is, plastic deformation occurring well below one-half times the absolute melting point. It might appear from Figure 6–6 that plastic deformation beyond T.S. softens

---

* Thomas Young (1773–1829), English physicist and physician, was the first to define the modulus of elasticity. Although this contribution made his name a famous one in solid mechanics, his most brilliant achievements were in optics. He was largely responsible for the acceptance of the wave theory of light.

† Robert Hooke (1635–1703), English physicist, was one of the most brilliant scientists of the seventeenth century as well as one of its most cantankerous personalities. His quarrels with fellow scientists such as Sir Isaac Newton did not diminish his accomplishments, which included the elastic behavior law (Equation 6.3) and the coining of the word "cell" to describe the structural building blocks of biological systems that he discovered in optical microscopy studies.

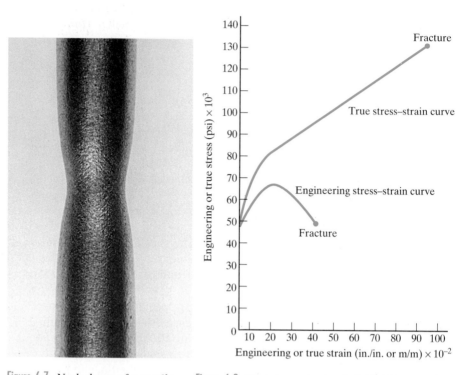

Figure 6-7 *Neck down of a tensile test specimen within its gage length after extension beyond the tensile strength. (Courtesy of R. S. Wortman)*

Figure 6-8 *True stress (= load divided by actual area in the necked-down region) continues to rise to the point of fracture, in contrast to the behavior of engineering stress. (After R. A. Flinn/P. K. Trojan:* Engineering Materials and Their Applications, *2nd Ed., Copyright © 1981, Houghton Mifflin Company, used by permission.)*

the material because the engineering stress falls. Instead, this drop in stress is simply the result of the fact that engineering stress and strain are defined relative to original sample dimensions. At the ultimate tensile strength, the sample begins to *neck down* within the gage length (Figure 6–7). The true stress ($\sigma_{\text{tr}} = P/A_{\text{actual}}$) continues to rise to the point of fracture (see Figure 6–8).

For many metals and alloys, the region of the true stress (σ_T) versus true strain (ϵ_T) curve between the onset of plastic deformation (corresponding to the yield stress in the engineering stress versus engineering strain curve) and the onset of necking (corresponding to the tensile stress in the engineering stress versus engineering strain curve) can be approximated by

$$\sigma_T = K\epsilon_T^n \tag{6.4}$$

where K and n are constants with values for a given metal or alloy depending on its thermomechanical history (e.g., degree of mechanical working, heat treatment, etc.). In other words, the true stress versus true strain curve in this region is nearly straight when plotted on logarithmic coordinates. The slope of the log-log plot is the parameter n, which is termed the **strain-hardening exponent.** For low-carbon steels used to form complex shapes, the value of n will normally be approximately 0.22. Higher values, up to 0.26, indicate an improved ability to be deformed during the shaping process without excess thinning or fracture of the piece.

The engineering stress at failure in Figure 6–6 is lower than T.S. and occasionally even lower than Y.S. Unfortunately, the complexity of the final stages of neck down causes the value of the failure stress to vary substantially from specimen to specimen. More useful is the strain at failure. **Ductility** is frequently quantified as the percent elongation at failure ($= 100 \times \epsilon_{failure}$). A less used definition is the percent reduction in area $[= (A_0 - A_{final})/A_0]$. The values for ductility from the two different definitions are not, in general, equal. It should also be noted that the value of percent elongation at failure is a function of the gage length used. Tabulated values are frequently specified for a gage length of 2 in. Ductility indicates the general ability of the metal to be plastically deformed. Practical implications of this ability include formability during fabrication and relief of locally high stresses at crack tips during structural loading (see the discussion of fracture toughness in Chapter 8).

It is also useful to know whether an alloy is both strong and ductile. A high-strength alloy that is also highly brittle may be as unusable as a deformable alloy with unacceptably low strength. Figure 6–9 compares these two extremes with an alloy with both high strength and substantial ductility. The term **toughness** is used to describe this combination of properties. Figure 6–6 shows that this is conveniently defined as the total area under the stress–strain curve. Since integrated $\sigma - \epsilon$ data are not routinely available, we shall

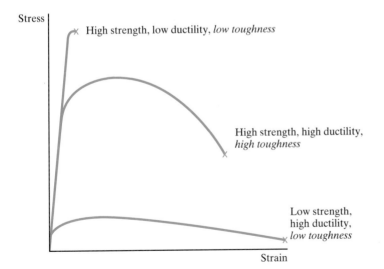

Figure 6-9 *The toughness of an alloy depends on a combination of strength and ductility.*

Table 6.2 *Tensile Test Data for the Alloys of Table 6.1*

	Alloy	E [GPa (psi)]	Y.S. [MPa (ksi)]	T.S. [MPa (ksi)]	Percent elongation at failure
1.	1040 carbon steel	200 (29 × 10⁶)	600 (87)	750 (109)	17
2.	8630 low-alloy steel		680 (99)	800 (116)	22
3.	a. 304 stainless steel	193 (28 × 10⁶)	205 (30)	515 (75)	40
	c. 410 stainless steel	200 (29 × 10⁶)	700 (102)	800 (116)	22
4.	L2 tool steel		1380 (200)	1550 (225)	12
5.	Ferrous superalloy (410)	200 (29 × 10⁶)	700 (102)	800 (116)	22
6.	a. Ductile iron, quench	165 (24 × 10⁶)	580 (84)	750 (108)	9.4
	b. Ductile iron, 60–40–18	169 (24.5 × 10⁶)	329 (48)	461 (67)	15
7.	a. 3003-H14 aluminum	70 (10.2 × 10⁶)	145 (21)	150 (22)	8–16
	b. 2048, plate aluminum	70.3 (10.2 × 10⁶)	416 (60)	457 (66)	8
8.	a. AZ31B magnesium	45 (6.5 × 10⁶)	220 (32)	290 (42)	15
	b. AM100A casting magnesium	45 (6.5 × 10⁶)	83 (12)	150 (22)	2
9.	a. Ti–5Al–2.5Sn	107–110 (15.5–16 × 10⁶)	827 (120)	862 (125)	15
	b. Ti–6Al–4V	110 (16 × 10⁶)	825 (120)	895 (130)	10
10.	Aluminum bronze, 9% (copper alloy)	110 (16.1 × 10⁶)	320 (46.4)	652 (94.5)	34
11.	Monel 400 (nickel alloy)	179 (26 × 10⁶)	283 (41)	579 (84)	39.5
12.	AC41A zinc			328 (47.6)	7
13.	50:50 solder (lead alloy)		33 (4.8)	42 (6.0)	60
14.	Nb–1 Zr (refractory metal)	68.9 (10 × 10⁶)	138 (20)	241 (35)	20
15.	Dental gold alloy (precious metal)			310–380 (45–55)	20–35

be required to monitor the relative magnitudes of strength (Y.S. and T.S.) and ductility (percent elongation at fracture).

Values of four of the five basic tensile test parameters (defined in Figure 6–6) for the alloys of Table 6.1 are given in Table 6.2. Values of the strain-hardening parameters of Equation 6.4, K and n, are given in Table 6.3.

Table 6.3 *Typical Values of Strain Hardening Parameters[a] for Various Metals and Alloys*

Alloy	K [MPa (ksi)]	n
Low-carbon steel (annealed)	530 (77)	0.26
4340 low-alloy steel (annealed)	640 (93)	0.15
304 stainless steel (annealed)	1275 (185)	0.45
Al (annealed)	180 (26)	0.20
2024 aluminum alloy (heat treated)	690 (100)	0.16
Cu (annealed)	315 (46)	0.54
Brass, 70Cu–30Zn (annealed)	895 (130)	0.49

Source: Data from S. Kalpakjian, *Manufacturing Processes for Engineering Materials*, Addison-Wesley Publishing C., Reading, Mass., 1984.
[a] Defined by Equation 6.4.

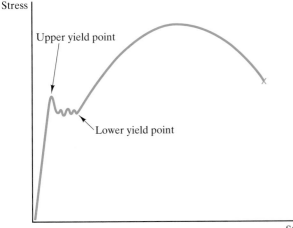

Stress

Upper yield point

Lower yield point

Strain

Figure 6-10 *For a low-carbon steel, the stress-versus-strain curve includes both an upper and lower yield point.*

The general appearance of the stress-versus-strain curve in Figure 6–3 is typical of a wide range of metal alloys. For certain alloys (especially low-carbon steels), the curve of Figure 6–10 is obtained. The obvious distinction for this latter case is a distinct break from the elastic region at a **yield point,** also termed an **upper yield point.** The distinctive ripple pattern following the yield point is associated with nonhomogeneous deformation that begins at a point of stress concentration (often near the specimen grips). A **lower yield point** is defined at the end of the ripple pattern and at the onset of general plastic deformation.

Figure 6–11 illustrates another important feature of elastic deformation, namely, a contraction perpendicular to the extension caused by a tensile stress. This effect is characterized by the **Poisson's* ratio,** ν:

$$\nu = -\frac{\epsilon_x}{\epsilon_z} \qquad (6.5)$$

where the strains in the x and z directions are defined by Figure 6–11. (There is a corresponding expansion perpendicular to the compression caused by a compressive stress.) Although the Poisson's ratio does not appear directly on the stress-versus-strain curve, it is, along with the elastic modulus, the most fundamental description of the elastic behavior of engineering materials. Table 6.4 summarizes values of ν for several common alloys. Note that values fall within the relatively narrow band of 0.26 to 0.35.

*   Simeon-Denis Poisson (1781–1840), French mathematician, succeeded Fourier in a faculty position at the Ecole Polytechnique. Although he did not generate original results in the way Fourier had, Poisson was a master of applying a diligent mathematical treatment to the unresolved questions raised by others. He is best known for the Poisson distribution dealing with probability for large number systems.

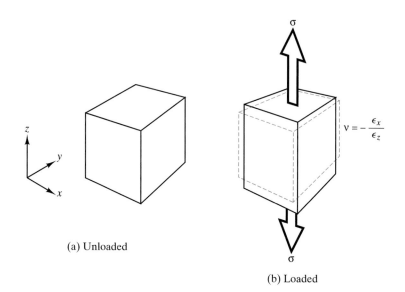

(a) Unloaded

(b) Loaded

Figure 6-11 *The Poisson's ratio (ν) characterizes the contraction per-pendicular to the extension caused by a tensile stress.*

Figure 6–12 illustrates the nature of elastic deformation in a pure shear loading. The **shear stress, τ**, is defined as

$$\tau = \frac{P_s}{A_s} \tag{6.6}$$

where P_S is the load on the sample and A_s is the area of the sample parallel (rather than perpendicular) to the applied load. The shear stress produces an angular displacement (α) with the **shear strain, γ**, being defined as

$$\gamma = \tan \alpha \tag{6.7}$$

Table 6.4 *Poisson's Ratio and Shear Modulus for the Alloys of Table 6.1*

	Alloy	ν	G (GPa)	G/E
1.	1040 carbon steel	0.3		
2.	8630 carbon steel	0.3		
3.	a. 304 stainless steel	0.29		
6.	b. Ductile iron, 60–40–18	0.29		
7.	a. 3003-H14 aluminum	0.33	25	0.36
8.	a. AZ31B magnesium	0.35	17	0.38
	b. AM100A casting magnesium	0.35		
9.	a. Ti–5Al–2.5Sn	0.35	48	0.44
	b. Ti–6Al–4V	0.33	41	0.38
10.	Aluminum bronze, 9% (copper alloy)	0.33	44	0.40
11.	Monel 400 (nickel alloy)	0.32		

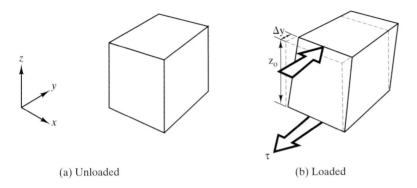

(a) Unloaded   (b) Loaded

Figure 6-12 *Elastic deformation under a shear load.*

which is equal to $\Delta y / z_0$ in Figure 6–12. The **shear modulus,** or **modulus of rigidity,** G, is defined (in a manner comparable to Equation 6.3) as

$$G = \frac{\tau}{\gamma} \qquad (6.8)$$

The shear modulus G and the elastic modulus E are related, for small strains, by Poisson's ratio; namely,

$$E = 2G(1 + \nu) \qquad (6.9)$$

Typical values of G are given in Table 6.4. As the two moduli are related by ν (Equation 6.9) and ν falls within a narrow band, the ratio of G/E is relatively fixed for most alloys at about 0.4 (see Table 6.4).

SAMPLE PROBLEM 6.1

From Figure 6–3, calculate E, Y.S., T.S., and percent elongation at failure for the aluminum 2024-T81 specimen.

SOLUTION

To obtain the modulus of elasticity, E, note that the strain at $\sigma = 300$ MPa is 0.0043 (as shown in the following figure). Then

$$E = \frac{\sigma}{\epsilon} = \frac{300 \times 10^6 \text{ Pa}}{0.0043} = 70 \text{ GPa}$$

The 0.2% offset construction gives

$$\text{Y.S.} = 410 \text{ MPa}$$

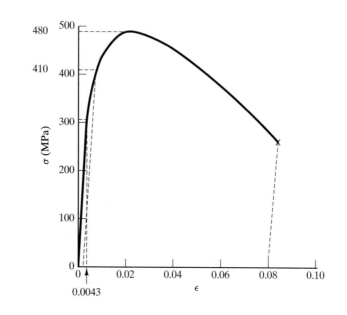

The maximum for the stress–strain curve gives

$$\text{T.S.} = 480 \text{ MPa}$$

Finally, the strain at fracture is $\epsilon_f = 0.08$, giving

$$\% \text{ elongation at failure} = 100 \times \epsilon_f = 8\%$$

SAMPLE PROBLEM 6.2

A 10-mm-diameter bar of 1040 carbon steel (see Table 6.2) is subjected to a tensile load of 50,000 N, taking it beyond its yield point. Calculate the elastic recovery that would occur upon removal of the tensile load.

SOLUTION

Using Equation 6.1 to calculate engineering stress gives

$$\sigma = \frac{P}{A_0} = \frac{50,000 \text{ N}}{\pi (5 \times 10^{-3} \text{ m})^2} = 637 \times 10^6 \frac{\text{N}}{\text{m}^2}$$
$$= 637 \text{ MPa}$$

which is between the Y.S. (600 MPa) and the T.S. (750 MPa) for this alloy (Table 6.2).

The elastic recovery can be calculated from Hooke's law (Equation 6.3) using the elastic modulus of Table 6.2:

$$\epsilon = \frac{\sigma}{E}$$

$$= \frac{637 \times 10^6 \text{ Pa}}{200 \times 10^9 \text{ Pa}}$$

$$= 3.18 \times 10^{-3}$$

SAMPLE PROBLEM 6.3

(a) A 10-mm-diameter rod of 3003-H14 aluminum alloy is subjected to a 6-kN tensile load. Calculate the resulting rod diameter.

(b) Calculate the diameter if this rod is subjected to a 6-kN compressive load.

SOLUTION

(a) From Equation 6.1, the engineering stress is

$$\sigma = \frac{P}{A_0}$$

$$= \frac{6 \times 10^3 \text{ N}}{\pi(\frac{10}{2} \times 10^{-3} \text{ m})^2} = 76.4 \times 10^6 \frac{\text{N}}{\text{m}^2} = 76.4 \text{ MPa}$$

From Table 6.2, we see that this stress is well below the yield strength (145 MPa) and, as a result, the deformation is elastic.

From Equation 6.3, we can calculate the tensile strain using the elastic modulus from Table 6.2:

$$\epsilon = \frac{\sigma}{E} = \frac{76.4 \text{ MPa}}{70 \times 10^3 \text{ MPa}} = 1.09 \times 10^{-3}$$

If we use Equation 6.5 and the value for ν from Table 6.4, the strain for the diameter can be calculated as

$$\epsilon_{\text{diameter}} = -\nu\epsilon_z = -(0.33)(1.09 \times 10^{-3})$$

$$= -3.60 \times 10^{-4}$$

The resulting diameter can then be determined (analogous to Equation 6.2) from

$$\epsilon_{\text{diameter}} = \frac{d_f - d_0}{d_0}$$

or

$$d_f = d_0(\epsilon_{\text{diameter}} + 1) = 10 \text{ mm}(-3.60 \times 10^{-4} + 1)$$

$$= 9.9964 \text{ mm}$$

(b) For a compressive stress, the diameter strain will be of equal magnitude but of opposite sign; that is,

$$\epsilon_{\text{diameter}} = +3.60 \times 10^{-4}$$

As a result, the final diameter will be

$$d_f = d_0(\epsilon_{\text{diameter}} + 1) = 10 \text{ mm}(+3.60 \times 10^{-4} + 1)$$

$$= 10.0036 \text{ mm}$$

PRACTICE PROBLEM 6.1

In Sample Problem 6.1, the basic mechanical properties of a 2024-T81 aluminum are calculated based on its stress–strain curve (Figure 6–3). Given at the top of page 201 are load-elongation data for a type 304 stainless steel similar to that presented in Figure 6–2. This steel is similar to alloy 3(a) in Table 6.2 except that it has a different thermomechanical history, giving it slightly higher strength with lower ductility. **(a)** Plot these data in a manner comparable to Figure 6–2. **(b)** Replot these data as a stress–strain curve similar to Figure 6–3. **(c)** Replot the initial strain data on an expanded scale, similar to that used for Figure 6–4. **(d)** Using the results of parts **(a)**–**(c)**, calculate **(d)** E, **(e)** Y.S., **(f)** T.S., and **(g)** percent elongation at failure for this 304 stainless steel. For parts **(d)**–**(f)**, express answers in both Pa and psi units.

PRACTICE PROBLEM 6.2

For the 304 stainless steel introduced in Practice Problem 6.1, calculate the elastic recovery for the specimen upon removal of the load of **(a)** 35,720 N and **(b)** 69,420 N. (See Sample Problem 6.2.)

Load (N)	Gage length (mm)	Load (N)	Gage length (mm)
0	50.8000	35,220	50.9778
4,890	50.8102	35,720	51.0032
9,779	50.8203	40,540	51.816
14,670	50.8305	48,390	53.340
19,560	50.8406	59,030	55.880
24,450	50.8508	65,870	58.420
27,620	50.8610	69,420	60.960
29,390	50.8711	69,670 (maximum)	61.468
32,680	50.9016	68,150	63.500
33,950	50.9270	60,810 (fracture)	66.040 (after fracture)
34,580	50.9524		

Original, specimen diameter: 12.7 mm.

PRACTICE PROBLEM 6.3

For the alloy in Sample Problem 6.3, calculate the rod diameter at the (tensile) yield stress indicated in Table 6.2.

CERAMICS AND GLASSES

Many of the mechanical properties discussed for metals are equally important to ceramics or glasses used in structural applications. In addition, the different nature of these nonmetals leads to some unique mechanical behavior.

Metal alloys generally demonstrate a significant amount of plastic deformation in a typical tensile test. In contrast, ceramics and glasses generally do not. Figure 6–13 shows characteristic results for uniaxial loading of dense,

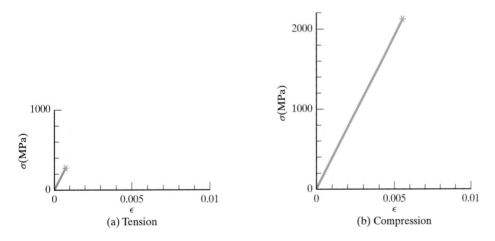

Figure 6-13 *The brittle nature of fracture in ceramics is illustrated by these stress–strain curves, which show only linear, elastic behavior. In (a), fracture occurs at a tensile stress of 280 MPa. In (b) a compressive strength of 2100 MPa is observed. The sample in both tests is a dense, polycrystalline Al_2O_3.*

polycrystalline Al_2O_3. In Figure 6–13a, failure of the sample occurred in the elastic region. This **brittle fracture** is characteristic of ceramics and glasses. An equally important characteristic is illustrated by the difference between the parts of Figure 6–13. Figure 6–13a illustrates the breaking strength in a tensile test (280 MPa), while Figure 6–13b is the same for a compressive test (2100 MPa). This is an especially dramatic example of the fact that *ceramics are relatively weak in tension but relatively strong in compression*. This behavior is shared by some cast irons (Chapter 11) and concrete (Chapter 14). Table 6.5 summarizes moduli of elasticity and strengths for several ceramics and glasses. The strength parameter is the modulus of rupture, a value calculated from data in a bending test. The **modulus of rupture (MOR)** is given by

$$\text{MOR} = \frac{3FL}{2bh^2} \qquad (6.10)$$

where F is the applied force and b, h, and L are dimensions defined in Figure 6–14. The MOR is sometimes referred to as the flexural strength (F.S.) and is similar in magnitude to the tensile strength, as the failure mode in bending is tensile (along the outermost edge of the sample). The bending test, illustrated in Figure 6–14, is frequently easier to conduct on brittle materials than the traditional tensile test. Values of Poisson's ratio are given in Table 6.6. One can note from comparing Tables 6.4 and 6.6 that ν for metals is typically $\approx \frac{1}{3}$ and for ceramics $\approx \frac{1}{4}$.

To appreciate the reason for the mechanical behavior of structural ceramics, we must consider the stress concentration at crack tips. For purely brittle materials, the simple **Griffith* crack model** is applicable. Griffith assumed that in any real material there will be numerous elliptical cracks

Figure 6-14   *The bending test that generates a modulus of rupture.   This strength parameter is similar in magnitude to a tensile strength.   Fracture occurs along the outermost sample edge, which is under a tensile load.*

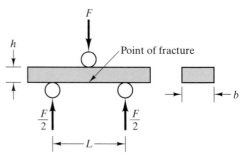

Modulus of rupture = MOR
= $3FL/(2bh^2)$

---

*   Alan Arnold Griffith (1893–1963), British engineer. Griffith's career was spent primarily in aeronautical engineering. He was one of the first to suggest that the gas turbine would be a feasible propulsion system for aircraft. In 1920, he published his research on the strength of glass fibers that was to make his name one of the best known in the field of materials engineering.

Table 6.5 *Modulus of Elasticity and Strength (Modulus of Rupture) for Some Ceramics and Glasses*

		E (MPa)	MOR (MPa)
1.	Mullite (aluminosilicate) porcelain	69×10^3	69
2.	Steatite (magnesia aluminosilicate) porcelain	69×10^3	140
3.	Superduty fireclay (aluminosilicate) brick	97×10^3	5.2
4.	Alumina (Al_2O_3) crystals	380×10^3	340–1000
5.	Sintered[a] alumina (~5% porosity)	370×10^3	210–340
6.	Alumina porcelain (90–95% alumina)	370×10^3	340
7.	Sintered[a] magnesia (~5% porosity)	210×10^3	100
8.	Magnesite (magnesia) brick	170×10^3	28
9.	Sintered[a] spinel (magnesia aluminate) (~5% porosity)	238×10^3	90
10.	Sintered[a] stabilized zirconia (~5% porosity)	150×10^3	83
11.	Sintered[a] beryllia (~5% porosity)	310×10^3	140–280
12.	Dense silicon carbide (~5% porosity)	470×10^3	170
13.	Bonded silicon carbide (~20% porosity)	340×10^3	14
14.	Hot-pressed[b] boron carbide (~5% porosity)	290×10^3	340
15.	Hot-pressed[b] boron nitride (~5% porosity)	83×10^3	48–100
16.	Silica glass	72.4×10^3	107
17.	Borosilicate glass	69×10^3	69

Source: W. D. Kingery, H. K. Bowen, and D. R. Uhlmann, *Introduction to Ceramics,* 2nd Ed., John Wiley & Sons, Inc., New York, 1976.

[a] *Sintering* refers to fabrication of the product by the bonding of powder particles by solid-state diffusion at high temperature (> one-half the absolute melting point). See Section 10.6 for a more detailed description.

[b] *Hot pressing* is sintering accompanied by high-pressure application.

Table 6.6 *Poisson's Ratio for Some Ceramics and Glasses*

		ν
1.	Al_2O_3	0.26
2.	BeO	0.26
3.	CeO_2	0.27–0.31
4.	Cordierite ($2MgO \cdot 2Al_2O_3 \cdot 5SiO_2$)	0.31
5.	Mullite ($3Al_2O_3 \cdot 2SiO_2$)	0.25
6.	SiC	0.19
7.	Si_3N_4	0.24
8.	TaC	0.24
9.	TiC	0.19
10.	TiO_2	0.28
11.	Partially stabilized ZrO_2	0.23
12.	Fully stabilized ZrO_2	0.23–0.32
13.	Glass-ceramic ($MgO–Al_2O_3–SiO_2$)	0.24
14.	Borosilicate glass	0.2
15.	Glass from cordierite	0.26

Source: Data from *Ceramic Source '86* and *Ceramic Source '87,* American Ceramic Society, Columbus, Ohio, 1985 and 1986.

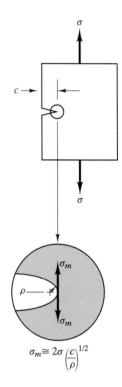

Figure 6-15 *Stress (σ_m) at the tip of a Griffith crack.*

at the surface and/or in the interior. It can be shown that the highest stress (σ_m) at the tip of such a crack is

$$\sigma_m \simeq 2\sigma \left(\frac{c}{\rho}\right)^{1/2} \tag{6.11}$$

where σ is the applied stress, c the crack length as defined in Figure 6–15, and ρ the radius of the crack tip. Since the crack tip radius can be as small as an interatomic spacing, the stress intensification can be quite large. Routine production and handling of ceramics and glasses make Griffith flaws inevitable. Hence these materials are relatively weak in tension. A compressive load tends to close, not open, the Griffith flaws and consequently does not diminish the inherent strength of the ionically and covalently bonded material.

The drawing of small-diameter glass fibers in a controlled atmosphere is one way to avoid Griffith flaws. The resulting fibers can demonstrate tensile strengths approaching the theoretical atomic bond strength of the material. This helps to make them excellent reinforcing fibers for composite systems.

SAMPLE PROBLEM 6.4

A glass plate contains an atomic-scale surface crack. (Take the crack tip radius $\simeq$ diameter of an O^{2-} ion.) Given that the crack is 1 μm long and the theoretical strength of the defect-free glass is 7.0 GPa, calculate the breaking strength of the plate.

SOLUTION

This is an application of Equation 6.11:

$$\sigma_m = 2\sigma \left(\frac{c}{\rho}\right)^{1/2}$$

Rearranging gives us

$$\sigma = \frac{1}{2}\sigma_m \left(\frac{\rho}{c}\right)^{1/2}$$

Using Appendix 2, we have

$$\rho = 2r_{O^{2-}} = 2(0.132 \text{ nm})$$
$$= 0.264 \text{ nm}$$

Then

$$\sigma = \frac{1}{2} (7.0 \times 10^9 \text{ Pa}) \left(\frac{0.264 \times 10^{-9} \text{ m}}{1 \times 10^{-6} \text{ m}} \right)^{1/2}$$

$$= 57 \text{ MPa}$$

PRACTICE PROBLEM 6.4

Calculate the breaking strength of a given glass plate containing **(a)** a 0.5-μm-long surface crack and **(b)** a 5-μm-long surface crack. Except for the length of the crack, use the conditions described in Sample Problem 6.4.

POLYMERS

As with ceramics, the mechanical properties of polymers can be described with much of the vocabulary introduced for metals. Tensile strength and modulus of elasticity are important design parameters for polymers as well as inorganic structural materials.

With the increased availability of engineering polymers for metals substitution, a greater emphasis has been placed on presenting the mechanical behavior of polymers in a format similar to that used for metals. Primary emphasis is on stress-versus-strain data. Although strength and modulus values are important parameters for these materials, design applications frequently involve a bending, rather than tensile, mode. As a result, flexural strength and flexural modulus are frequently quoted.

As noted earlier, the **flexural strength** (F.S.) is equivalent to the modulus of rupture defined for ceramics in Equation 6.10 and Figure 6–14. For the same test specimen geometry, the **flexural modulus,** or **modulus of elasticity in bending** (E_{flex}), is

$$E_{\text{flex}} = \frac{L^3 m}{4bh^3} \qquad (6.12)$$

where m is the slope of the tangent to the initial straight-line portion of the load-deflection curve and all other terms are defined relative to Equation 6.10 and Figure 6–14. An important advantage of the flexural modulus for polymers is that it describes the combined effects of compressive deformation (adjacent to the point of applied load in Figure 6–14) and tensile deformation (on the opposite side of the specimen). For metals, as noted before, tensile and compressive moduli are generally the same. For many polymers, the tensile and compressive moduli differ significantly.

Some polymers, especially the elastomers, are used in structures for the purpose of isolation and absorption of shock and vibration. For such applications, a "dynamic" elastic modulus is more useful to characterize the performance of the polymer under an oscillating mechanical load. For elastomers in general, the dynamic modulus is greater than the static modulus. For some compounds, the two moduli may differ by a factor of two. The **dynamic modulus of elasticity,** E_{dyn} (in MPa), is

$$E_{dyn} = CIf^2 \qquad (6.13)$$

where C is a constant dependent upon specific test geometry, I the moment of inertia (in kg · m^2) of the beam and weights used in the dynamic test, and f the frequency of vibration (in cycles/s) for the test. Equation 6.13 holds for both compressive and shear measurements, with the constant C having a different value in each case.

Figure 6–16 shows typical stress-versus-strain curves for an engineering polymer, polyester. Although these plots look similar to common stress-versus-strain plots for metals, there is a strong effect of temperature. On the other hand, this mechanical behavior is relatively independent of atmospheric moisture. Both polyester and acetal engineering polymers have this advantage. However, relative humidity is a design consideration for the use of nylons, as shown in Figure 6–17. Also demonstrated in Figure 6–17

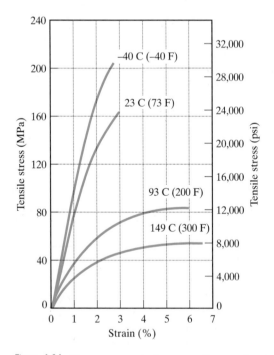

Figure 6-16 *Stress-versus-strain curves for a polyester engineering polymer. (From* Design Handbook for Du Pont Engineering Plastics, *used by permission.)*

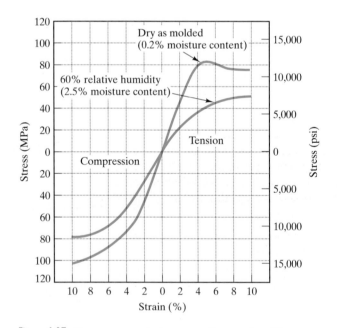

Figure 6-17 *Stress-versus-strain curves for a nylon 66 at 23° C showing the effect of relative humidity. (From* Design Handbook for Du Pont Engineering Plastics, *used by permission.)*

is the difference in elastic modulus (slope of the plots near the origin) for tensile and compressive loads. (Recall that this point was raised in the introduction of flexural modulus.) Table 6.7 gives mechanical properties of the thermoplastic polymers (those that become soft and deformable upon heating). Table 6.8 gives similar properties for the thermosetting polymers (those which become hard and rigid upon heating). Note that the dynamic modulus values in Table 6.8 are not, in general, greater than the *tensile* modulus values. The statement that the dynamic modulus of an elastomer is generally greater than the static modulus is valid for a given mode of stress application. The dynamic *shear* modulus values are, in general, greater than static *shear* modulus values.

Table 6.7 *Mechanical Properties Data for some Thermoplastic Polymers*

Polymer	$E^{\,a}$ [MPa (ksi)]	$E^{\,b}_{\text{flex}}$ [MPa (ksi)]	T.S. [MPa (ksi)]	Percent elongation at failure	Poisson's ratio v
General-use polymers					
Polyethylene					
High-density	830 (120)		28 (4)	15–100	
Low-density	170 (25)		14 (2)	90–800	
Polyvinylchloride	2800 (400)		41 (6)	2–30	
Polypropylene	1400 (200)		34 (5)	10–700	
Polystyrene	3100 (450)		48 (7)	1–2	
Polyesters	— (—)	8960 (1230)	158 (22.9)	2.7	
Acrylics (Lucite)	2900 (420)		55 (8)	5	
Polyamides (nylon 66)	2800 (410)	2830 (410)	82.7 (12.0)	60	0.41
Cellulosics	3400–28,000 (500–4000)		14–55 (2–8)	5–40	
Engineering polymers					
ABS	2100 (300)		28–48 (4–7)	20–80	
Polycarbonates	2400 (350)		62 (9)	110	
Acetals	3100 (450)	2830 (410)	69 (10)	50	0.35
Polytetrafluoroethylene (Teflon)	410 (60)		17 (2.5)	100–350	
Thermoplastic elastomers					
Polyester-type		585 (85)	46 (6.7)	400	

Source: From data collections in R. A. Flinn and P. K. Trojan, *Engineering Materials and Their Applications*, 2nd Ed., Houghton Mifflin Company, Boston, 1981; M. F. Ashby and D. R. H. Jones, *Engineering Materials*, Pergamon Press, Inc., Elmsford, N.Y., 1980; and *Design Handbook for Du Pont Engineering Plastics*.
[a] Low strain data (in tension).
[b] In shear.

Table 6.8 *Mechanical Properties Data for some Thermosetting Polymers*

Polymer	E^a [MPa (ksi)]	E^b_{Dyn} [MPa (ksi)]	T.S. [MPa (ksi)]	Percent elongation at failure
Thermosets				
Phenolics (phenolformaldehyde)	6900 (1000)	—	52 (7.5)	0
Urethanes	—	—	34 (5)	—
Urea-melamine	10,000 (1500)	—	48 (7)	0
Polyesters	6900 (1000)	—	28 (4)	0
Epoxies	6900 (1000)	—	69 (10)	0
Elastomers				
Polybutadiene/polystyrene copolymer				
Vulcanized	1.6 (0.23)	0.8 (0.12)	1.4–3.0 (0.20–0.44)	440–600
Vulcanized with 33% carbon black	3–6 (0.4–0.9)	8.7 (1.3)	17–28 (2.5–4.1)	400–600
Polyisoprene				
Vulcanized	1.3 (0.19)	0.4 (0.06)	17–25 (2.5–3.6)	750–850
Vulcanized with 33% carbon black	3.0–8.0 (0.44–1.2)	6.2 (0.90)	25–35 (3.6–5.1)	550–650
Polychloroprene				
Vulcanized	1.6 (0.23)	0.7 (0.10)	25–38 (3.6–5.5)	800–1000
Vulcanized with 33% carbon black	3–5 (0.4–0.7)	2.8 (0.41)	21–30 (3.0–4.4)	500–600
Polyisobutene/polyisoprene copolymer				
Vulcanized	1.0 (0.15)	0.4 (0.06)	18–21 (2.6–3.0)	750–950
Vulcanized with 33% carbon black	3–4 (0.4–0.6)	3.6 (0.52)	18–21 (2.6–3.0)	650–850
Silicones	—	—	7 (1)	4000
Vinylidene fluoride/ hexafluoropropylene	—	—	12.4 (1.8)	—

Source: From data collections in R. A. Flinn and P. K. Trojan, *Engineering Materials and Their Applications*, 2nd ed., Houghton Mifflin Company, Boston, 1981; M. F. Ashby and D. R. H. Jones, *Engineering Materials*, Pergamon Press, Inc., Elmsford, N.Y., 1980; and J. Brandrup and E. H. Immergut, Eds., *Polymers Handbook*, 2nd ed., John Wiley & Sons, Inc., New York, 1975.

[a] Low strain data (in tension).

[b] In shear.

SAMPLE PROBLEM 6.5

The following data are collected in a flexural test of a nylon to be used in the fabrication of lightweight gears:

> Test piece geometry: 7 mm × 13 mm × 100 mm
> Distance between supports = L = 50 mm
> Initial slope of load-deflection curve = 404×10^3 N/m

Calculate the flexural modulus for this engineering polymer.

SOLUTION

Referring to Figure 6–14 and Equation 6.12, we find that

$$E_{\text{flex}} = \frac{L^3 m}{4bh^3}$$

$$= \frac{(50 \times 10^{-3} \text{ m})^3 (404 \times 10^3 \text{ N/m})}{4(13 \times 10^{-3} \text{ m})(7 \times 10^{-3} \text{ m})^3}$$

$$= 2.83 \times 10^9 \text{ N/m}^2$$

$$= 2830 \text{ MPa}$$

SAMPLE PROBLEM 6.6

A small, uniaxial stress of 1 MPa (145 psi) is applied to a rod of high-density polyethylene.

(a) What is the resulting strain?

(b) Repeat for a rod of vulcanized isoprene.

(c) Repeat for a rod of 1040 steel.

SOLUTION

(a) For this modest stress level, we can assume Hooke's law behavior:

$$\epsilon = \frac{\sigma}{E}$$

Table 6.7 gives $E = 830$ MPa. So

$$\epsilon = \frac{1 \text{ MPa}}{830 \text{ MPa}} = 1.2 \times 10^{-3}$$

(b) Table 6.8 gives $E = 1.3$ MPa or

$$\epsilon = \frac{1 \text{ MPa}}{1.3 \text{ MPa}} = 0.77$$

(c) Table 6.2 gives $E = 200$ GPa $= 2 \times 10^5$ MPa or

$$\epsilon = \frac{1 \text{ MPa}}{2 \times 10^5 \text{ MPa}} = 5.0 \times 10^{-6}$$

Note.   The dramatic difference between elastic moduli of polymers and inorganic solids is used to advantage in composite materials (Chapter 14).

PRACTICE PROBLEM 6.5

The data in Sample Problem 6.5 permit the flexural modulus to be calculated. For the configuration described, an applied force of 680 N causes fracture of the nylon sample. Calculate the corresponding flexural strength.

PRACTICE PROBLEM 6.6

In Sample Problem 6.6 strain is calculated for various materials under a stress of 1 MPa. While the strain is relatively large for polymers, there are some high-modulus polymers with substantially lower results. Calculate the strain in a cellulosic fiber with a modulus of elasticity of 28,000 MPa (under a uniaxial stress of 1 MPa).

6.2 · ELASTIC DEFORMATION

Before leaving the discussion of stress-versus-strain behavior for materials, it is appropriate to look at the atomic-scale mechanisms involved. Figure 6–18

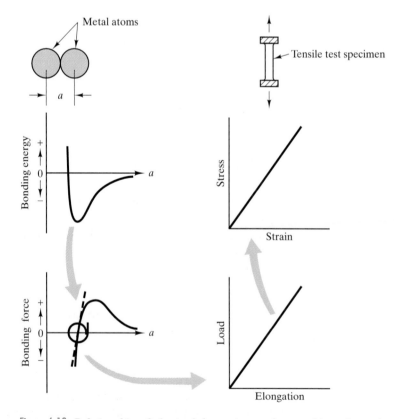

Figure 6-18 *Relationship of elastic deformation to the stretching of atomic bonds.*

shows that the fundamental mechanism of elastic deformation is the stretching of atomic bonds. The fractional deformation of the material in the initial elastic region is small so that, on the atomic scale, we are dealing only with the portion of the force–atom separation curve in the immediate vicinity of the equilibrium atom separation distance (a_0 corresponding to $F = 0$). The nearly straight-line plot of F versus a across the a axis implies that similar elastic behavior will be observed in a compressive, or push, test as well as in tension. This is often the case, especially for metals.

SAMPLE PROBLEM 6.7

In the absence of stress, the center-to-center atomic separation distance of two Fe atoms is 0.2480 nm (along a $\langle 111 \rangle$ direction). Under a tensile stress of 1000 MPa along this direction, the atomic separation distance increases to 0.2489 nm. Calculate the modulus of elasticity along the $\langle 111 \rangle$ directions.

SOLUTION

From Hooke's law (Equation 6.3),

$$E = \frac{\sigma}{\epsilon}$$

with

$$\epsilon = \frac{(0.2489 - 0.2480) \text{ nm}}{0.2480 \text{ nm}} = 0.00363$$

giving

$$E = \frac{1000 \text{ MPa}}{0.00363} = 280 \text{ GPa}$$

Note. This modulus represents the maximum value in the iron crystal structure. The minimum value of E is 125 GPa in the $\langle 100 \rangle$ direction. In polycrystalline iron with random grain orientations, an average modulus of 205 GPa occurs. This is close to the value for most steels (Table 6.2).

....

PRACTICE PROBLEM 6.7

(a) Calculate the center-to-center separation distance of two Fe atoms along the $\langle 100 \rangle$ direction in unstressed α-iron.

(b) Calculate the separation distance along that direction under a tensile stress of 1000 MPa. (See Sample Problem 6.7.)

6.3   PLASTIC DEFORMATION

The fundamental mechanism of plastic deformation is the distortion and reformation of atomic bonds. In Chapter 5 we saw that atomic diffusion in crystalline solids is extremely difficult without the presence of point defects. Similarly, the **plastic** (permanent) **deformation** of crystalline solids is difficult without dislocations, the linear defects introduced in Section 4.3. Frenkel first calculated the mechanical stress necessary to deform a perfect crystal. This would occur by sliding one plane of atoms over an adjacent plane, as shown in Figure 6–19. The shear stress associated with this sliding action can be calculated with knowledge of the periodic bonding forces along the slip plane. The result obtained by Frenkel was that the theoretical **critical shear stress** is roughly one order of magnitude less than the bulk *shear modulus, G,* for the material (see Equation 6.8). For a typical metal such as copper, the theoretical critical shear stress represents a value well over 1000 MPa. The actual stress necessary to plastically deform a sample of pure copper (i.e., slide atomic planes past each other) is at least an order of magnitude less than this. Our everyday experience with metallic alloys (opening aluminum cans or bending automobile fenders) represents deformations generally requiring stress levels of only a few hundred megapascals. What, then, is the basis of the mechanical deformation of metals, which requires only a fraction of the theoretical strength? The answer, to which we have already alluded, is the dislocation. Figure 6–20 illustrates the role a dislocation can play in the shear of a crystal along a slip plane. The key point to observe is that only a relatively small shearing force needs to operate and only in the immediate vicinity of the dislocation in order to produce a step-by-step shear that eventually yields the same overall deformation as the high stress mechanism of Figure 6–19. A perspective view of a shearing mechanism involving a more general, mixed dislocation (see Figure 4–13) is given in Figure 6–21.

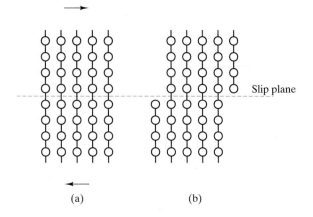

(a)   (b)

Figure 6-19 *Sliding of one plane of atoms past an adjacent one. This high-stress process is necessary to plastically (permanently) deform a perfect crystal.*

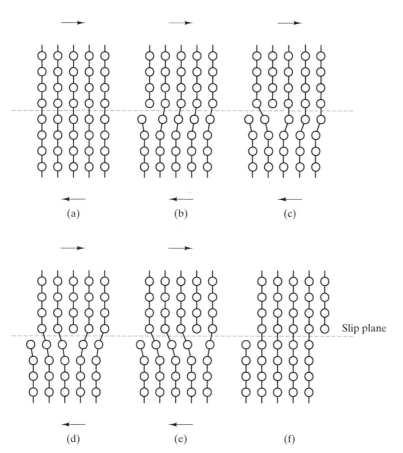

(a)

(b)

(c)

(d)

(e)

(f)

Slip plane

Figure 6-20 *A low-stress alternative for plastically deforming a crystal involves the motion of a dislocation along a slip plane.*

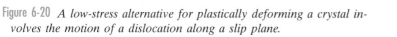

(a)

(b)

(c)

Figure 6-21 *Schematic illustration of the motion of a dislocation under the influence of a shear stress. The net effect is an increment of plastic (permanent) deformation. (Compare Figure 6–21a with Figure 4–13.)*

We can appreciate this defect mechanism of slip by considering a simple analogy. Figure 6–22 introduces Goldie the caterpillar. It is impractical to force Goldie to slide along the ground in a perfect straight line (Figure 6–22a). But Goldie "slips along" nicely by passing a "dislocation" along the length of her body (Figure 6–22b).

Reflecting on Figure 6–20, we can appreciate that the stepwise slip mechanism would tend to become more difficult as the individual atomic step distances are increased. As a result, slip is more difficult on a

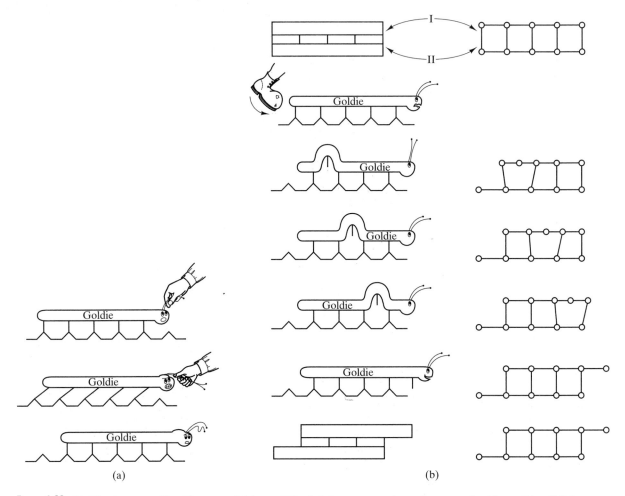

Figure 6-22 *Goldie the caterpillar illustrates (a) how difficult it is to move along the ground without (b) a "dislocation" mechanism. (From W. C. Moss, Ph.D. thesis, University of California, Davis, Calif., 1979.)*

low-atomic-density plane than on a high-atomic-density plane. Figure 6–23 shows this schematically. In general, the micromechanical mechanism of slip—dislocation motion—will occur in high-atomic-density planes and in high-atomic-density directions. A combination of families of crystallographic planes and directions corresponding to dislocation motion is referred to as a **slip system.** Figure 6–24 is similar to Figure 1–18, the difference being that we can now label the slip systems in (a) fcc aluminum and (b) hcp magnesium. As pointed out in Chapter 1, aluminum and its alloys are characteristically ductile (deformable) due to the large number (12) of high-density plane–direction combinations. Magnesium and its alloys are typically brittle (fracturing with little deformation) due to the smaller number (3) of such combinations. Table 6.9 summarizes the major slip systems in typical metal structures.

Slip plane (low atomic density)

Slip distance

(a)

Slip plane (high atomic density)

Slip distance

(b)

Figure 6-23 *Dislocation slip is more difficult along (a) a low-atomic-density plane than along (b) a high-atomic-density plane.*

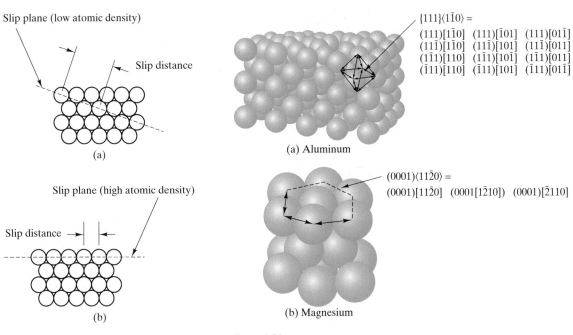

(a) Aluminum

$\{111\}\langle 1\bar{1}0\rangle =$

$(111)[1\bar{1}0]$	$(111)[\bar{1}01]$	$(111)[01\bar{1}]$
$(11\bar{1})[1\bar{1}0]$	$(11\bar{1})[101]$	$(11\bar{1})[011]$
$(1\bar{1}1)[110]$	$(1\bar{1}1)[10\bar{1}]$	$(1\bar{1}1)[011]$
$(\bar{1}11)[110]$	$(\bar{1}11)[101]$	$(\bar{1}11)[01\bar{1}]$

(b) Magnesium

$(0001)\langle 11\bar{2}0\rangle =$

$(0001)[11\bar{2}0]$ $(0001[1\bar{2}10])$ $(0001)[\bar{2}110]$

Figure 6-24 *Slip systems for (a) fcc aluminum and (b) hcp magnesium. (Compare to Figure 1–18.)*

Table 6.9 *Major Slip Systems in the Common Metal Structures*

Crystal structure	Slip plane	Slip direction	Number of slip systems	Unit cell geometry	Examples
bcc	$\{110\}$	$\langle \bar{1}11 \rangle$	$6 \times 2 = 12$	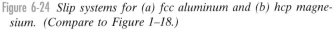	α-Fe, Mo, W
fcc	$\{111\}$	$\langle 1\bar{1}0 \rangle$	$4 \times 3 = 12$		Al, Cu, γ-Fe, Ni
hcp	(0001)	$\langle 11\bar{2}0 \rangle$	$1 \times 3 = 3$		Cd, Mg, α-Ti, Zn

Several basic concepts of the mechanical behavior of crystalline materials relate directly to simple models of dislocation motion. The *cold working* of metals involves deliberate deformation of the metal at relatively low temperatures (see Section 6.1 and Chapter 10). While forming a stronger product, an important feature of cold working is that the metal becomes more difficult to deform as the extent of deformation increases. The basic micromechanical reason for this is that a dislocation hinders the motion of another dislocation. The slip mechanism of Figure 6–20 proceeds most smoothly when the slip plane is free of obstructions. Cold working generates dislocations that serve as such obstacles. In fact, cold working generates so many dislocations that the configuration is referred to as a "forest of dislocations" (Figure 4–36). Foreign atoms can also serve as obstacles to dislocation motion. Figure 6–25 illustrates this micromechanical basis of **solution hardening** of alloys, that is, restricting plastic deformation by forming solid solutions. Hardening, or increasing strength, occurs because the elastic region is extended, producing a higher yield strength. These concepts will be discussed further in Section 6.4. Obstacles to dislocation motion harden metals, but high temperatures can help to overcome these obstacles and thereby soften the metals. An example of this is the *annealing* process, a stress-relieving heat treatment to be described in Chapter 10. The micromechanical mechanism here is rather straightforward. At sufficiently high temperatures, atomic diffusion is sufficiently great to allow highly stressed crystal grains produced by cold working to be restructured into more nearly perfect crystalline structures. The dislocation density is dramatically reduced with increasing temperature. This permits the relatively simple deformation mechanism of Figure 6–20 to occur free of the forest of dislocations. At this point, we have seen an important blending of the concepts of solid-state diffusion (from Chapter 5) and mechanical deformation. There will be many other examples of this in later chapters. In each case, a useful rule of thumb will apply: *The temperature at which atomic mobility is sufficient to affect mechanical properties is approximately one-third to one-half times the absolute melting point, T_m.*

One additional basic concept of mechanical behavior is that the more complex crystal structures correspond to relatively brittle materials. Common examples of this are intermetallic compounds (such as Ag_3Al) and ceramics (such as Al_2O_3). Relatively large Burgers vectors combined with the difficulty in creating obstacle-free slip planes create a limited opportunity for dislocation motion. The formation of *brittle intermetallics* is a common concern in high-temperature designs involving interfaces between dissimilar metals. Ceramics, as we saw in Section 6.1, are characteristically brittle materials. Inspection of Figure 4–14 confirms the statement about large Burgers vectors. An additional consideration adding to the brittleness of ceramics is that many slip systems are not possible, owing to the charged state of the ions. The sliding of ions with like charges past one another can result in high coulombic repulsive forces. As a result, even ceramic compounds with relatively simple crystal structures exhibit significant dislocation mobility only at a relatively high temperature.

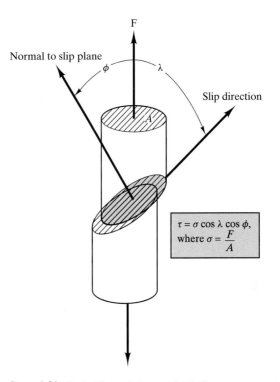

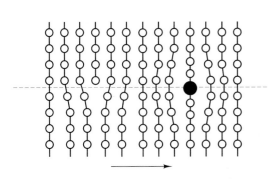

Direction of "attempted" dislocation motion

Figure 6-25 *How an impurity atom generates a strain field in a crystal lattice, causing an obstacle to dislocation motion.*

Figure 6-26 *Definition of the resolved shear stress, τ, which directly produces plastic deformation (by a shearing action) as a result of the external application of a simple tensile stress, σ.*

We close this section with a macroscopic calculation of the deformation stress for a crystalline material relative to the microscopic mechanism of a slip system. Figure 6–26 defines the **resolved shear stress, τ**, which is the actual stress operating on the slip system (in the slip plane and in the slip direction) resulting from the application of a simple tensile stress, $\sigma\,(= F/A)$, where F is the externally applied force perpendicular to the cross-sectional area (A) of the single-crystal sample. The important concept here is that the fundamental deformation mechanism is a shearing action based on the projection of the applied force onto the slip system. The component of the applied force (F) operating in the slip direction is $(F \cos \lambda)$. The projection of the sample's cross-sectional area (A) onto the slip plane gives an area of $(A/\cos \varphi)$. As a result, the resolved shear stress, τ, is

$$\tau = \frac{F \cos \lambda}{A/\cos \varphi} = \frac{F}{A} \cos \lambda \cos \varphi = \sigma \cos \lambda \cos \varphi \qquad (6.14)$$

where σ is the applied tensile stress $(= F/A)$ and λ and φ are defined in Figure 6–26. Equation 6.14 identifies the resolved shear stress, τ, resulting from a given applied stress. A value of τ great enough to produce slip by dislocation motion is called the **critical resolved shear stress** and is given by

$$\tau_c = \sigma_c \cos \lambda \cos \varphi \qquad (6.15)$$

where σ_c is, of course, the applied stress necessary to produce this deformation. In considering plastic deformation, we should always keep in mind this connection between macroscopic stress values and the micromechanical mechanism of dislocation slip.

SAMPLE PROBLEM 6.8

A zinc single crystal is being pulled in tension with the normal to its basal plane (0001) at 60° to the tensile axis and the slip direction [11$\overline{2}$0] at 40° to the tensile axis.

(a) What is the resolved shear stress, τ, acting in the slip direction when a tensile stress of 0.690 MPa (100 psi) is applied?

(b) What tensile stress is necessary to reach the critical resolved shear stress, τ_c, of 0.94 MPa (136 psi)?

SOLUTION

(a) From Equation 6.14,

$$\tau = \sigma \cos \lambda \cos \varphi$$
$$= (0.690 \text{ MPa}) \cos 40° \cos 60°$$
$$= 0.264 \text{ MPa (38.3 psi)}$$

(b) From Equation 6.15,

$$\tau_c = \sigma_c \cos \lambda \cos \varphi$$

or

$$\sigma_c = \frac{\tau_c}{\cos \lambda \cos \varphi}$$
$$= \frac{0.94 \text{ MPa}}{\cos 40° \cos 60°}$$
$$= 2.45 \text{ MPa (356 psi)}$$

....

PRACTICE PROBLEM 6.8

Repeat Sample Problem 6.8, assuming that the two directions are 45° rather than 60° and 40°.

6.4   HARDNESS

The *hardness test* (Figure 6–27) is available as a relatively simple alternative to the tensile test of Figure 6–1. The resistance of the material to indentation is a qualitative indication of its strength. The indenter can be either rounded or pointed and is made of a material much harder than the test piece, for example, hardened steel, tungsten carbide, or diamond. Table 6.10 summarizes the common types of hardness tests with their characteristic indenter geometries. Empirical *hardness numbers* are calculated from appropriate formulas using indentation geometry measurements. Microhardness measurements are made using a high-power microscope. **Rockwell* hardness** is widely used with many scales (Rockwell A, Rockwell B, etc.) available for different hardness ranges. Correlating hardness with depth of penetration allows the hardness number to be conveniently shown on a dial or digital display. In this chapter, we shall often quote **Brinell[†] hardness numbers** (BHN) because a single scale covers a wide range of material hardness and a fairly linear correlation with strength can be found. Table 6.11 gives BHN values for the alloys of Table 6.1. Figure 6–28a shows a clear trend of BHN

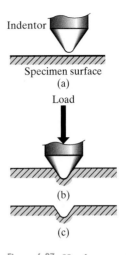

Figure 6-27 *Hardness test. The analysis of indentation geometry is summarized in Table 6.10.*

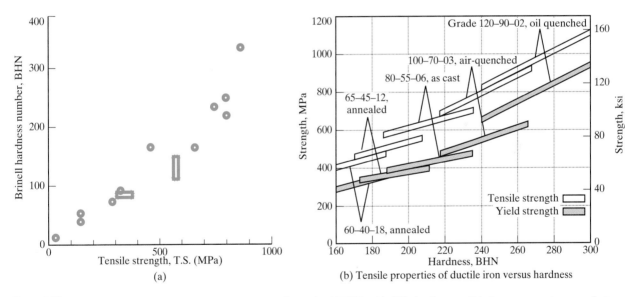

Figure 6-28 *(a) Plot of data from Table 6.11. A general trend of BHN with T.S. is shown. (b) A more precise correlation of BHN with T.S. (or Y.S.) is obtained for given families of alloys. [Part (b) from* Metals Handbook, *9th Ed., Vol. 1, American Society for Metals, Metals Park, Ohio, 1978.]*

---

*   The Rockwell hardness tester was invented in 1919 by Stanley P. Rockwell, an American metallurgist. The word *Rockwell* as applied to the tester and reference standards is a registered trademark in several countries, including the United States.

[†]   Johan August Brinell (1849–1925), Swedish metallurgist, was an important contributor to the metallurgy of steels. His apparatus for hardness testing was first displayed in 1900 at the Paris Exposition. Current "Brinell testers" are essentially unchanged in design.

Table 6.10 *Common Types of Hardness Test Geometries*

Test	Indenter	Shape of Indentation		Load	Formula for hardness number
		Side view	Top view		
Brinell	10 mm sphere of steel or tungsten carbide			P	$\mathrm{BHN} = \dfrac{2P}{\pi D\left[D - \sqrt{D^2 - d^2}\right]}$
Vickers	Diamond pyramid	136	d_1 d_1	P	$\mathrm{VHN} = 1.72P/d_1^2$
Knoop microhardness	Diamond pyramid	$l/b = 7.11$ $b/t = 4.00$	b l	P	$\mathrm{KHN} = 14.2P/l^2$
Rockwell					
A C D	Diamond cone	120		60 kg 150 kg 100 kg	$\left.\begin{array}{l} R_A = \\ R_C = \\ R_D = \end{array}\right\} 100 - 500t$
B F G E H	$\frac{1}{16}$ in. diameter steel sphere $\frac{1}{8}$ in. diameter steel sphere			100 kg 60 kg 150 kg 100 kg 60 kg	$\left.\begin{array}{l} R_B = \\ R_F = \\ R_G = \\ R_E = \\ R_H = \end{array}\right\} 130 - 500t$

Source: H. W. Hayden. W. G. Moffatt, and J. Wulff, *The Structure and Properties of Materials*, Vol. 3: *Mechanical Behavior*, John Wiley & Sons, Inc., New York, 1965.

Table 6.12 *Hardness Data for Various Polymers*

Polymer	Rockwell hardness R scale[a]
Thermoplastic polymers	
General-use polymers	
Polyethylene	
High-density	40
Low-density	10
Polyvinylchloride	110
Polypropylene	90
Polystyrene	75
Polyesters	120
Acrylics (Lucite)	130
Polyamides (nylon 66)	121
Cellulosics	50 to 115
Engineering polymers	
ABS	95
Polycarbonates	118
Acetals	120
Polytetrafluoroethylene (Teflon)	70
Thermosetting Polymers	
Phenolics (phenolformaldehyde)	125
Urea-melamine	115
Polyesters	100
Epoxies	90

Source: From data collections in R. A. Flinn and P. K. Trojan, *Engineering Materials and Their Applications,* 2nd ed., Houghton Mifflin Company, Boston, 1981; M. F. Ashby and D. R. H. Jones, *Engineering Materials,* Pergamon Press, Inc., Elmsford, N.Y., 1980; and *Design Handbook for Du Pont Engineering Plastics.*
[a] For relatively soft materials: indenter radius $\frac{1}{2}$ in. and load of 60 kg.

Table 6.11 *Comparison of Brinell Hardness Numbers (BHN) with Tensile Strength (T.S.) for the Alloys of Table 6.1*

	Alloy	BHN	T.S. (MPa)
1.	1040 carbon steel	235	750
2.	8630 low-alloy steel	220	800
3.	c. 410 stainless steel	250	800
5.	Ferrous superalloy (410)	250	800
6.	b. Ductile iron, 60–40–18	167	461
7.	a. 3003-H14 aluminum	40	150
8.	a. AZ31B magnesium	73	290
	b. AM100A casting magnesium	53	150
9.	a. Ti–5Al–2.5Sn	335	862
10.	Aluminum bronze, 9% (copper alloy)	165	652
11.	Monel 400 (nickel alloy)	110–150	579
12.	AC41A zinc	91	328
13.	50:50 solder (lead alloy)	14.5	42
15.	Dental gold alloy (precious metal)	80–90	310–380

with tensile strength for these alloys. Figure 6–28b shows that the correlation is more precise for given families of alloys. The tensile strength is generally used for this correlation rather than yield strength because the hardness test includes a substantial component of plastic deformation. There will be further discussions of hardness in relation to heat treatments in Chapter 10. Typical hardness values for a variety of polymers are given in Table 6.12.

SAMPLE PROBLEM 6.9

(a) A Brinell hardness measurement is made on a ductile iron (100–70–03, air-quenched) using a 10-mm-diameter sphere of tungsten carbide. A load of 3000 kg produces a 3.91-mm-diameter impression in the iron surface. Calculate the Brinell hardness number of this alloy. (The correct units for the Brinell equation of Table 6.10 are kilograms for load and millimeters for the diameters.)

(b) Use Figure 6–28b to predict the tensile strength of this ductile iron.

SOLUTION

(a) From Table 6.10,

$$\text{BHN} = \frac{2P}{\pi D \left(D - \sqrt{D^2 - d^2} \right)}$$

$$= \frac{2(3000)}{\pi (10) \left(10 - \sqrt{10^2 - 3.91^2} \right)}$$

$$= 240$$

(b) From Figure 6–28b,

$$(\text{T.S.})_{\text{BHN}=240} = 800 \text{ MPa}$$

PRACTICE PROBLEM 6.9

Suppose that a ductile iron (100–70–03, air-quenched) has a tensile strength of 700 MPa. What diameter impression would you expect the 3000-kg load to produce with the 10-mm-diameter ball? (See Sample Problem 6.9.)

6.5   CREEP AND STRESS RELAXATION

The tensile test alone cannot predict the behavior of a structural material used at elevated temperatures. The strain induced in a typical metal bar loaded below its yield strength at room temperature can be calculated from Hooke's law (Equation 6.3). This strain will not generally change with time under a fixed load (Figure 6–29). Repeating this experiment at a "high" temperature (T greater than one-third to one-half times the melting point on an absolute temperature scale) produces dramatically different results. Figure 6–30 shows a typical test design, and Figure 6–31 shows a typical **creep curve** in which the strain, ϵ, gradually increases with time after the initial elastic loading. **Creep** can be defined as plastic (permanent) deformation occurring at high temperature under constant load over a long time period.

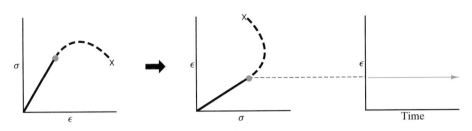

Figure 6-29 *Elastic strain induced in an alloy at room temperature is independent of time.*

After the initial elastic deformation at $t \simeq 0$, Figure 6–31 shows three stages of creep deformation. The *primary stage* is characterized by a decreasing strain rate (slope of the ϵ vs. t curve). The relatively rapid increase in length induced during this early time period is the direct result of enhanced deformation mechanisms. A commom example for metal alloys is **dislocation climb** as illustrated in Figure 6–32. As discussed in Section 6.3, this enhanced deformation comes from thermally activated atom mobility, giving dislocations additional slip planes in which to move. The *secondary stage* of creep deformation is characterized by straight-line, constant-strain-rate data (Figure 6–31). In this region, the increased ease of slip due to

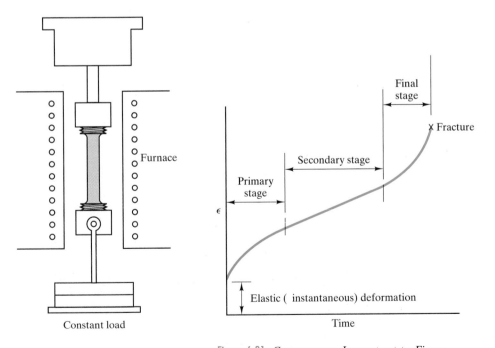

Figure 6-30 *Typical creep test.*

Figure 6-31 *Creep curve. In contrast to Figure 6–29, plastic strain occurs over time for a material stressed at high temperatures (above about one-half the absolute melting point).*

Figure 6-32 *Mechanism of dislocation climb. Obviously, many adjacent atom movements are required to produce climb of an entire dislocation line.*

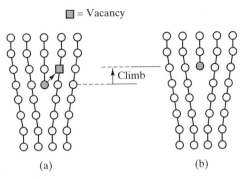

high-temperature mobility is balanced by increasing resistance to slip due to the buildup of dislocations and other microstructural barriers. In the final *tertiary stage,* strain rate increases due to an increase in true stress resulting from cross-sectional area reduction due to necking or internal cracking. In some cases, fracture occurs in the secondary stage, eliminating this final stage.

Figure 6–33 shows how the characteristic creep curve varies with changes in applied stress or environmental temperature. The thermally activated nature of creep makes this process another example of Arrhenius behavior, as discussed in Section 5.1. A demonstration of this is an Arrhenius plot of the logarithm of the steady-state creep rate ($\dot{\epsilon}$) from the secondary stage against the inverse of absolute temperature (Figure 6–34). As with other thermally activated processes, the slope of the Arrhenius plot is important in that it provides an activation energy, Q, for the creep mechanism from the Arrhenius expression

$$\dot{\epsilon} = Ce^{-Q/RT} \tag{6.16}$$

where C is the preexponential constant, R the universal gas constant, and T the absolute temperature. Another powerful aspect of the Arrhenius behavior is its predictive power. The dashed line in Figure 6–34 shows how high-temperature strain-rate data, which can be gathered in short-time laboratory experiments, can be extrapolated to predict long-term creep behavior at lower service temperatures. This extrapolation is valid as long as the same creep mechanism operates over the entire temperature range. Many elaborate semiempirical plots have been developed, based on this principle, to guide design engineers in material selection.

A shorthand characterization of creep behavior is given by the secondary-stage strain rate ($\dot{\epsilon}$) and the time to creep rupture (t) as shown in Figure 6–35. Plots of these parameters, together with applied stress (σ) and temperature (T), provide another convenient data set for design engineers responsible for selecting materials for high-temperature service (e.g., Figure 6–36).

Creep is probably more important in ceramics than in metals because high-temperature applications are so widespread. The role of diffusion mechanisms in the creep of ceramics is more complex than in the case of metals because diffusion, in general, is more complex in ceramics. The requirement of charge neutrality and different diffusivities for cations and anions contribute to this complexity. As a result, grain boundaries frequently

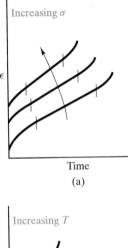

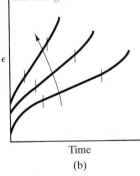

Figure 6-33 *Variation of the creep curve with (a) stress or (b) temperature. Note how the steady-state creep rate ($\dot{\epsilon}$) in the secondary stage rises sharply with temperature (see also Figure 6–34).*

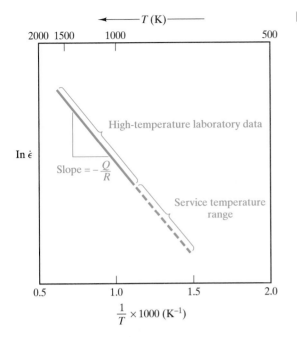

Figure 6-34 *Arrhenius plot of* $\ln \dot{\epsilon}$ *versus* $1/T$, *where* $\dot{\epsilon}$ *is the secondary-stage creep rate and* T *is the absolute temperature. The slope gives the activation energy for the creep mechanism. Extension of high-temperature, short-term data permits prediction of long-term creep behavior at lower service temperatures.*

play a dominant role in the creep of ceramics. Sliding of adjacent grains along these boundaries provides for microstructural rearrangement during creep deformation. In some relatively impure refractory ceramics, a substantial layer of glassy phase may be present at the grain boundaries. In that case,

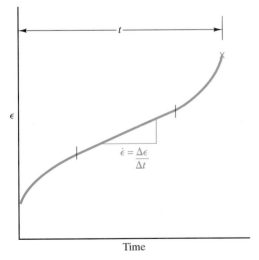

Figure 6-35 *Simple characterization of creep behavior is obtained from the secondary-stage strain rate* ($\dot{\epsilon}$) *and the time to creep rupture* (t).

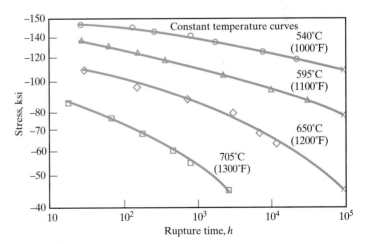

Figure 6-36 *Creep rupture data for the nickel-based superalloy Inconel 718. (From* Metals Handbook, *9th Ed., Vol. 3, American Society for Metals, Metals Park, Ohio, 1980.)*

Table 6.13 *Creep-Rate Data for Various Polycrystalline Ceramics*

Material	$\dot{\epsilon}$ at 1300°C, 1800 psi (12.4 MPa) [mm/(mm · h) × 10^6]
Al_2O_3	1.3
BeO	300
MgO (slip cast)	330
MgO (hydrostatic pressed)	33
$MgAl_2O_4$ (2–5 μm)	263
$MgAl_2O_4$ (1–3 mm)	1
ThO_2	1000
ZrO_2 (stabilized)	30

Source: W. D. Kingery, H. K. Bowen, and D. R. Uhlmann, *Introduction to Ceramics,* 2nd Ed., John Wiley & Sons, Inc., New York, 1976.

creep can again occur by the mechanism of grain boundary sliding due to the viscous deformation of the glassy phase. This "easy" sliding mechanism is generally undesirable due to the resulting weakness at high temperatures. In fact, the term *creep* is not applied to bulk glasses themselves. The subject of viscous deformation of glasses is discussed separately below.

Creep-rate data for some common ceramics at a fixed temperature are given in Table 6.13. An Arrhenius-type plot of creep-rate data at various temperatures (load fixed) is given in Figure 6–37.

For metals and ceramics, we have found creep deformation to be an important phenomenon at high temperatures (greater than one-half the absolute melting point). Creep is a significant design factor for polymers given their relatively low melting points. Figure 6–38 shows creep data for nylon 66 at moderate temperature and load. A related phenomenon, termed **stress relaxation,** is also an important design consideration for polymers. A familiar example is the rubber band, under stress for a long period of time, which does not snap back to its original size upon stress removal.

Creep deformation involves increasing strain with time for materials under constant stresses. By contrast, stress relaxation involves decreasing stress with time for polymers under constant strains. The mechanism of stress relaxation is viscous flow; that is, molecules gradually sliding past each other over an extended period of time. Viscous flow converts some of the fixed elastic strain into nonrecoverable plastic deformation. Stress relaxation is characterized by a **relaxation time,** τ, defined as the time necessary for the stress (σ) to fall to 0.37 ($= 1/e$) of the initial stress (σ_0). The exponential decay of stress with time (t) is given by

$$\sigma = \sigma_0 e^{-t/\tau} \tag{6.17}$$

In general, stress relaxation is an Arrhenius phenomenon, as was creep for metals and ceramics. The form of the Arrhenius equation for stress relaxation is

$$\frac{1}{\tau} = Ce^{-Q/RT} \tag{6.18}$$

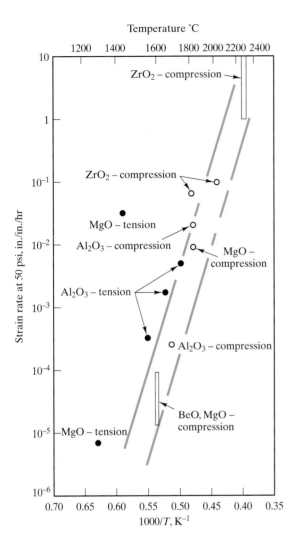

Figure 6-37 *Arrhenius-type plot of creep-rate data for several polycrystalline oxides under an applied stress of 50 psi (345 × 10³ Pa). Note that the inverse temperature scale is reversed (i.e., temperature increases to the right). (From W. D. Kingery, H. K. Bowen, and D. R. Uhlmann,* Introduction to Ceramics, *2nd Ed., John Wiley & Sons, Inc., New York, 1976.)*

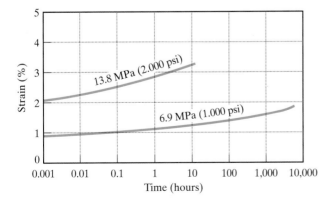

Figure 6-38 *Creep data for a nylon 66 at 60° C and 50% relative humidity. (From* Design Handbook for Du Pont Engineering Plastics, *used by permission.)*

where C is a preexponential constant, Q the activation energy (per mole) for viscous flow, R the universal gas constant, and T the absolute temperature.

SAMPLE PROBLEM 6.10

In a laboratory creep experiment at 1000°C, a steady-state reep rate of 5×10^{-1} % per hour is obtained in a metal alloy. The creep mechanism for this alloy is known to be dislocation climb with an activation energy of 200 kJ/mol. Predict the creep rate at a service temperature of 600°C. (Assume that the laboratory experiment duplicated the service stress.)

SOLUTION

Using the laboratory experiment to determine the preexponential constant in Equation 6.16, we obtain

$$C = \dot{\epsilon}e^{+Q/RT}$$

$$= (5 \times 10^{-1} \text{ % per hour})e^{+(2\times10^5\text{J/mol})/[8.314\text{J/mol·K}](1273\text{K})}$$

$$= 80.5 \times 10^6 \text{ % per hour}$$

Applying this to the service temperature yields

$$\dot{\epsilon} = (80.5 \times 10^6 \text{ % per hour})e^{-(2\times10^5)/(8.314)(873)}$$

$$= 8.68 \times 10^{-5} \text{ % per hour}$$

Note.   We have assumed that the creep mechanism remains the same between 1000 and 600°C.

Starting at this point and continuing throughout the remainder of the book, problems that deal with materials in the engineering design process will be identified with a "design icon" **D**.

D   SAMPLE PROBLEM 6.11

In designing a pressure vessel for the petrochemical industry, an engineer must estimate the temperature to which Inconel 718 could be subjected and still provide a service life of 10,000 h under a service stress of 690 MPa (100,000 psi) before failing by creep rupture. What is the service temperature?

SOLUTION

Using Figure 6–36, we must replot the data, noting that the failure stress for a rupture time of 10^4 h varies with temperature as follows:

σ (ksi)	T (°C)
125	540
95	595
65	650

Plotting gives

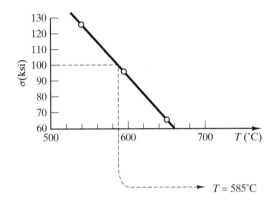

Plotting gives

$T = 585°C$

SAMPLE PROBLEM 6.12

The relaxation time for a rubber band at 25°C is 60 days.

(a) If it is stressed to 2 MPa initially, how many days will be required before the stress relaxes to 1 MPa?

(b) If the activation energy for the relaxation process is 30 kJ/mol, what is the relaxation time at 35°C?

SOLUTION

(a) From Equation 6.17,

$$\sigma = \sigma_0 e^{-t/\tau}$$

$$1 \text{ MPa} = 2 \text{ MPa} e^{-t/(60 \text{ d})}$$

Rearranging yields

$$t = -(60 \text{ days}) \left(\ln \tfrac{1}{2} \right) = 41.5 \text{ days}$$

(b) From Equation 6.18,

$$\frac{1}{\tau} = Ce^{-Q/RT}$$

or

$$\frac{1/\tau_{25°C}}{1/\tau_{35°C}} = \frac{e^{-Q/R(298 \text{ K})}}{e^{-Q/R(308 \text{ K})}}$$

or

$$\tau_{35°C} = \tau_{25°C}\exp\left[\frac{Q}{R}\left(\frac{1}{308 \text{ K}} - \frac{1}{298 \text{ K}}\right)\right]$$

giving, finally,

$$\tau_{35°C} = (60 \text{ days})\exp\left[\frac{30 \times 10^3 \text{ J/mol}}{8.314 \text{ J/(mol} \cdot \text{K)}}\left(\frac{1}{308 \text{ K}} - \frac{1}{298 \text{ K}}\right)\right]$$

$$= 40.5 \text{ days}$$

....

PRACTICE PROBLEM 6.10

Using an Arrhenius equation, we are able to predict the creep rate for a given alloy at 600°C in Sample Problem 6.10. For the same system, calculate the creep rate at **(a)** 700°C, **(b)** 800°C, and **(c)** 900°C. **(d)** Plot the results on an Arrhenius plot similar to Figure 6–34.

D PRACTICE PROBLEM 6.11

In Sample Problem 6.11, we are able to estimate a maximum service temperature for Inconel 718 in order to survive a stress of 690 MPa (100,000 psi) for 10,000 h. What is the maximum service temperature for this pressure vessel design that will allow this alloy to survive **(a)** 100,000 h and **(b)** 1000 h at the same stress?

PRACTICE PROBLEM 6.12

In Sample Problem 6.12a the time for relaxation of stress to 1 MPa at 25°C is calculated. **(a)** Calculate the time for stress to relax to 0.5 MPa at 25°C. **(b)** Repeat part (a) for 35°C using the result of Sample Problem 6.12b.

6.6   VISCOELASTIC DEFORMATION

We shall see in the next chapter on thermal behavior that materials generally expand upon heating. This thermal expansion is monitored as an incremental increase in length, ΔL, divided by its initial length, L_0. Two unique mechanical responses are found in measuring the thermal expansion of an inorganic glass or an organic polymer (Figure 6–39). First, there is a distinct break in the expansion curve at the temperature T_g. There are two different thermal expansion coefficients (slopes) above and below T_g. The thermal expansion coefficient below T_g is comparable to that of a crystalline solid of the same composition. The thermal expansion coefficient above T_g is comparable to that for a liquid. As a result, T_g is referred to as the **glass transition temperature**. Below T_g the material is a true glass (a rigid solid), and above T_g it is a supercooled liquid (see Section 4.5). In terms of mechanical behavior, elastic deformation occurs below T_g, while **viscous** (liquidlike) **deformation** occurs above T_g. Continuing to measure thermal expansion above T_g leads to a precipitous drop in the data curve at the temperature T_s. This is the **softening temperature** and marks the point where the material has become so fluid that it can no longer support the weight of the length-monitoring probe (a small refractory rod). A plot of specific volume versus temperature is given in Figure 6–40. This plot is closely related to the thermal expansion curve of Figure 6–39. The addition of data for the crystalline material (of the same composition as the glass) gives a pictorial definition of a glass in comparison to a supercooled liquid and a crystal.

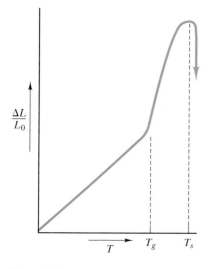

Figure 6-39 *Typical thermal expansion measurement of an inorganic glass or an organic polymer indicates a glass transition temperature, T_g, and a softening temperature, T_s.*

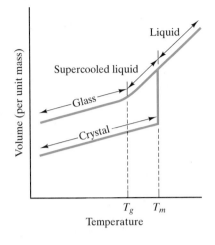

Figure 6-40   *Upon heating, a crystal undergoes modest thermal expansion up to its melting point (T_m), at which a sharp increase in specific volume occurs. Upon further heating, the liquid undergoes a greater thermal expansion. Slow cooling of the liquid would allow crystallization abruptly at T_m and a retracing of the melting plot. Rapid cooling of the liquid can suppress crystallization producing a supercooled liquid. In the vicinity of the glass transition temperature (T_g), gradual solidification occurs. A true glass is a rigid solid with thermal expansion similar to the crystal but an atomic-scale structure similar to the liquid (see Figure 4–23).*

The viscous behavior of glasses (organic or inorganic) can be described by the **viscosity,** η, which is defined as the proportionality constant between a shearing force per unit area (F/A) and velocity gradient (dv/dx):

$$\frac{F}{A} = \eta \frac{dv}{dx} \tag{6.19}$$

with the terms illustrated in Figure 6–41. The units for viscosity are traditionally the poise [$= 1$ g/(cm·s)], which is equal to 0.1 Pa·s.

INORGANIC GLASSES

The viscosity of a typical soda–lime–silica glass from room temperature to 1500°C is summarized in Figure 6–42. A good deal of useful processing information is contained in Figure 6–42 relative to the manufacture of glass products. The **melting range** is the temperature range (between about 1200 and 1500°C for soda–lime–silica glass), where η is between 50 and 500 P. This represents a very fluid material for a silicate liquid. Water and liquid metals, however, have viscosities of only about 0.01 P. The forming of product shapes is practical in the viscosity range of 10^4 to 10^8 P, the **working range**

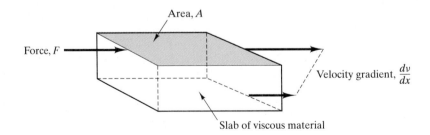

Figure 6-41   *Illustration of terms used to define viscosity, η, in Equation 6.19.*

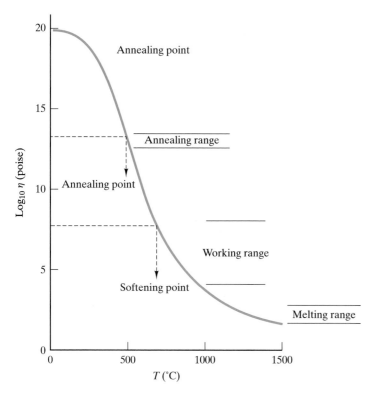

Figure 6-42 *Viscosity of a typical soda–lime–silica glass from room temperature to 1500°C. Above the glass transition temperature (∼ 450°C in this case), the viscosity decreases in the Arrhenius fashion (see Equation 6.20)*

(between about 700 and 900°C for soda–lime–silica glass). The **softening point** is formally defined at an η value of $10^{7.6}$ P (∼700°C for soda–lime–silica glass) and is at the lower temperature end of the working range. After a glass product is formed, residual stresses can be relieved by holding in the *annealing range* of η from $10^{12.5}$ to $10^{13.5}$ P. The **annealing point** is defined as the temperature at which $\eta = 10^{13.4}$ P and internal stresses can be relieved in about 15 min (∼450°C for soda–lime–silica glass). The glass transition temperature (of Figures 6–39 and 6–40) occurs around the annealing point.

Above the glass transition temperature, the viscosity data follow an Arrhenius form with

$$\eta = \eta_0 e^{+Q/RT} \tag{6.20}$$

where η_0 is the preexponential constant, Q the activation energy for viscous deformation, R the universal gas constant, and T the absolute temperature. Note that the exponential term has a positive sign rather than the usual negative sign associated with diffusivity data. This is simply the nature of the definition of viscosity, which decreases rather than increases with temperature. Fluidity, which could be defined as $1/\eta$, would, by definition, have a negative exponential sign comparable to the case for diffusivity.

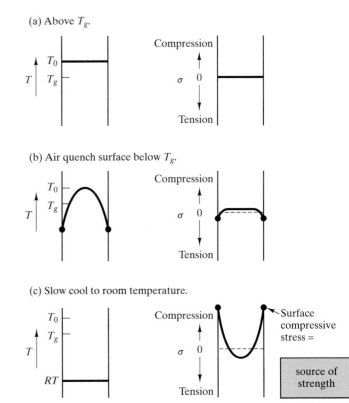

Figure 6-43 *Thermal and stress profiles occurring during the production of tempered glass. The high breaking strength of this product is due to the residual compressive stress at the material surfaces.*

A creative application of viscous deformation is **tempered glass**. Figure 6–43 shows how the glass is first equilibrated above the glass transition temperature, T_g, followed by a surface quench that forms a rigid surface "skin" at a temperature below T_g. Because the interior is still above T_g, interior compressive stresses are largely relaxed, although a modest tensile stress is present in the surface "skin." Slow cooling to room temperature allows the interior to contract considerably more than the surface, causing a net compressive residual stress on the surface balanced by a smaller tensile residual stress in the interior. This is an ideal situation for a brittle ceramic. Susceptible to surface Griffith flaws, the material must be subjected to a significant tensile load before the residual compressive load can be neutralized. An additional tensile load is necessary to fracture the material. The breaking strength becomes the normal (untempered) breaking strength plus the magnitude of the surface residual stress. A chemical rather than thermal technique to achieve the same result is to chemically exchange larger radius K^+ ions for the Na^+ ions in the surface of a sodium-containing silicate glass. The compressive stressing of the silicate network produces a product known as **chemically strengthened glass.**

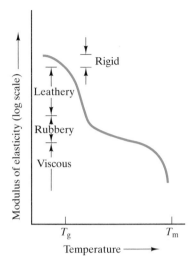

Figure 6-44 *Modulus of elasticity as a function of temperature for a typical thermoplastic polymer with 50% crystallinity. There are four distinct regions of viscoelastic behavior: (1) rigid, (2) leathery, (3) rubbery, and (4) viscous.*

ORGANIC POLYMERS

For inorganic glasses, the variation in viscosity was plotted against temperature (Figure 6–42). For organic polymers, the **modulus of elasticity** is usually plotted instead of viscosity. Figure 6–44 illustrates the drastic and complicated drop in modulus with temperature for a typical commercial thermoplastic with approximately 50% crystallinity. The magnitude of the drop is illustrated by the use of a logarithmic scale for modulus. (This was also necessary for viscosity in Figure 6–42.)

Figure 6–44 shows four distinct regions. At "low" temperatures (well below T_g), a rigid modulus occurs corresponding to mechanical behavior reminiscent of metals and ceramics. However, the substantial component of secondary bonding in the polymers causes the modulus for these materials to be substantially lower than the ones found for metals and ceramics, which were fully bonded by primary chemical bonds (metallic, ionic, and covalent). In the glass transition temperature (T_g) range, the modulus drops precipitously and the mechanical behavior is *leathery*. The polymer can be extensively deformed and slowly returns to its original shape upon stress removal. Just above T_g, a *rubbery* plateau is observed. In this region, extensive deformation is possible with rapid spring back to the original shape when stress is removed. These last two regions (leathery and rubbery) extend our understanding of elastic deformation. For metals and ceramics, elastic deformation meant a relatively small strain directly proportional to applied stress. For polymers, extensive nonlinear deformation can be fully recovered and is, by definition, elastic. This concept will be explored further when we discuss elastomers, those polymers with a predominant rubbery region. Returning to Figure 6–44, we see that, as the melting point (T_m) is approached, the modulus again drops precipitously as we enter the liquidlike viscous region. (It should be noted that, in many cases, it is more precise to define a "decomposition point" rather than a true melting point. The term *melting point* is nonetheless generally used.)

The Material World:
The Mechanical Behavior of Safety Glass

Even the most routine materials engineered for our surroundings can be the basis of health and safety concerns. Common examples include the window glass in buildings and in automobiles. Window glass is available in three basic configurations: annealed, laminated, and tempered. As discussed in this chapter in regard to the viscoelastic behavior of glass, annealing is a thermal treatment that largely removes the residual stresses of the manufacturing process. Specific glass-forming processes are described in Chapter 12. Modern windows are made largely by the float glass method of sheet glass manufacturing introduced by Pilkington Brothers, Ltd. in England in the 1950s. Annealing effectively removes processing stresses and

(Courtesy of Tamglass, Ltd.)

allows the glass plate to be cut, ground, drilled, and beveled as required. Unfortunately, annealed glass has only moderate strength and is brittle. As a result, thermal gradients, wind loads, or impact can produce characteristic dagger-shaped shards radiating out from the failure origin, as illustrated.

The obvious danger of injury from the breakage of annealed glass has led to widespread legislation requiring "safety glass" in buildings and vehicles. Laminated and tempered glasses serve this purpose. Laminated glass consists of two pieces of ordinary annealed glass with a central layer of polymer (polyvinyl butyral, PVB) sandwiched between them. As shown, the annealed glass sheets break in the same fashion as ordinary annealed glass, but the shards adhere to the PVB layer, reducing the danger of injury.

The tempering of glass is introduced in this chapter as a relatively sophisticated application of the viscoelastic nature of the material. A direct benefit of tempering is that the bending strength of tempered glass is up to five times greater than that of annealed glass. More important to its safe application, tempered glass breaks into small chunks with relatively harmless blunt shapes and dull edges. This highly desirable break pattern, as shown, is the result of the nearly instantaneous running and bifurcation of cracks initiated at the point of fracture. The energy required to propagate the break pattern comes from the strain energy associated with the residual tensile stresses in the interior of the plate. As the midplane stress rises above some threshold value, the characteristic break pattern occurs with an increasingly fine-scale fracture particle size with further increase in tensile stress.

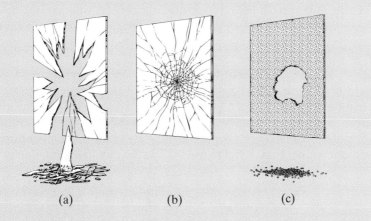

(a)   (b)   (c)

Break pattern of three states of glass used in commercial and consumer applications. (a) Annealed. (b) Laminated. (c) Tempered. (From R. A. McMaster, D. M. Shetterly, and A. G. Bueno, "Annealed and Tempered Glass," in *Engineered Materials Handbook*, Vol 4, *Ceramics and Glasses*, ASM International, Materials Park, Ohio, 1991.)

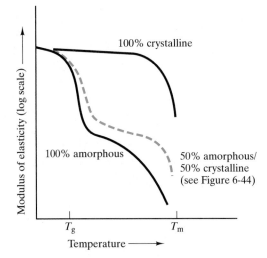

Figure 6-45 *In comparison to the plot of Figure 6–44, the behavior of the completely amorphous and completely crystalline thermoplastics falls below and above that for the 50% crystalline material. The completely crystalline material is similar to a metal or ceramic in remaining rigid up to its melting point.*

Figure 6–44 represents a linear, thermoplastic polymer with approximately 50% crystallinity. Figure 6–45 shows how that behavior lies midway between that for a fully amorphous material and a fully crystalline one. The curve for the fully amorphous polymer displays the general shape shown in Figure 6–44. The fully crystalline polymer, on the other hand, is relatively rigid up to its melting point. This is consistent with the behavior of crystalline metals and ceramics. Another structural feature that can affect mechanical behavior in polymers is the **cross-linking** of adjacent linear molecules to produce a more rigid, network structure (Figure 6–46). Figure 6–47 shows how increased cross-linking produces an effect comparable to increased crystallinity. The similarity is due to the increased rigidity of the cross-linked structure, in that cross-linked structures are generally noncrystalline.

ELASTOMERS

Figure 6–45 showed that a typical linear polymer exhibits a rubbery deformation region. For the polymers known as **elastomers,** the rubbery plateau is pronounced and establishes the normal, room-temperature behavior of these materials. (For them, the glass transition temperature is below room temperature.) Figure 6–48 shows the plot of log (modulus) versus

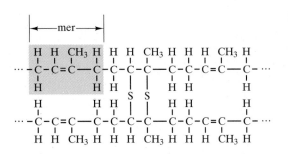

Figure 6-46 *Cross-linking produces a network structure by the formation of primary bonds between adjacent linear molecules. The classic example shown here is the* vulcanization *of rubber. Sulfur atoms form primary bonds with adjacent polyisoprene mers. This is possible because the polyisoprene chain molecule still contains double bonds after polymerization. [It should be noted that sulfur atoms can themselves bond together to form a molecule chain. Sometimes, cross-linking is by an* $(S)_n$ *chain, where n > 1.]*

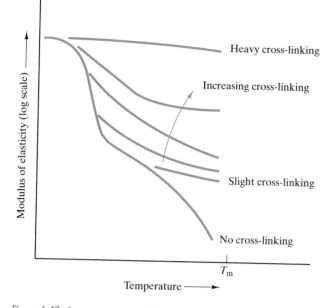

Figure 6-47 *Increased cross-linking of a thermoplastic polymer produces increased rigidity of the material.*

temperature for an elastomer. This subgroup of thermoplastic polymers includes the natural and synthetic rubbers, such as polyisoprene. These materials provide a dramatic example of the uncoiling of a linear polymer (Figure 6–49). As a practical matter, the complete uncoiling of the molecule is not achieved, but huge elastic strains do occur.

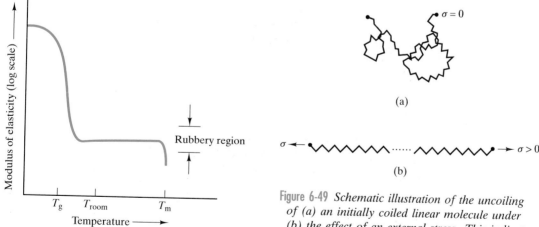

Figure 6-48 *The modulus of elasticity versus temperature plot of an elastomer has a pronounced rubbery region.*

Figure 6-49 *Schematic illustration of the uncoiling of (a) an initially coiled linear molecule under (b) the effect of an external stress. This indicates the molecular-scale mechanism for the stress versus strain behavior of an elastomer, as shown in Figure 6–50.*

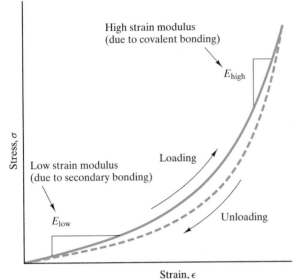

High strain modulus
(due to covalent bonding)

E_{high}

Low strain modulus
(due to secondary bonding)

E_{low}

Loading

Unloading

Stress, σ

Strain, ϵ

Figure 6-50 *The stress–strain curve for an elastomer is an example of nonlinear elasticity. The initial low-modulus (i.e., low-slope) region corresponds to the uncoiling of molecules (overcoming weak, secondary bonds), as illustrated by Figure 6–49. The high-modulus region corresponds to elongation of extended molecules (stretching primary, covalent bonds), as shown by Figure 6–49b. Elastomeric deformation exhibits hysteresis; that is, the plots during loading and unloading do not coincide.*

Figure 6–50 shows a stress–strain curve for the elastic deformation of an elastomer. This is in dramatic contrast to the stress–strain curve for a common metal (Figures 6–3 and 6–4). In that case, the elastic modulus was constant throughout the elastic region. (Stress was directly proportional to strain.) In Figure 6–50, the elastic modulus (slope of the stress-strain curve) increases with increasing strain. For low strains (up to $\approx 15\%$), the modulus is low corresponding to the small forces needed to overcome secondary bonding and to uncoil the molecules. For high strains, the modulus rises sharply, indicating the greater force needed to stretch the primary bonds along the molecular backbone. In both regions, however, there is a significant component of secondary bonding involved in the deformation mechanism, and the moduli are much lower than those for common metals and ceramics. Tabulated values of moduli for elastomers are generally for the low-strain region in which the materials are primarily used. Finally, it is important to emphasize that we are talking about elastic or temporary deformation. The uncoiled polymer molecules of an elastomer recoil to their original lengths upon removal of the stress. However, as the dashed line in Figure 6–50 indicates, the recoiling of the molecules (during unloading) has a slightly different path in the stress-versus-strain plot than does the uncoiling (during loading). The different plots for loading and unloading define **hysteresis.**

Figure 6–51 shows modulus of elasticity versus temperature for several commercial polymers. These data can be compared with the general curves in Figures 6–45 and 6–47. The deflection temperature under load (DTUL), illustrated by Figure 6–51, corresponds to the glass transition temperature.

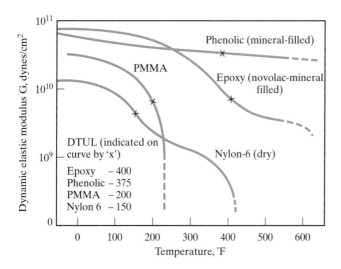

Figure 6-51 *Modulus of elasticity versus temperature for a variety of common polymers. The dynamic elastic modulus in this case was measured in a torsional pendulum (a shear mode). The DTUL is the deflection temperature under load, the load being 264 psi. This parameter is frequently associated with the glass transition temperature. (From* Modern Plastics Encyclopedia, *1981–82, Vol. 58, No. 10A, McGraw-Hill Book Company, New York, October 1981.)*

SAMPLE PROBLEM 6.13

A soda–lime–silica glass used to make lamp bulbs has an annealing point of 514°C and a softening point of 696°C. Calculate the working range and the melting range for this glass.

SOLUTION

This is an application of Equation 6.20:

$$\eta = \eta_0 e^{+Q/RT}$$

Given

$$\text{annealing point} = 514 + 273 = 787\text{K} \quad \text{for} \, \eta = 10^{13.4}\text{P}$$

$$\text{softening point} = 696 + 273 = 969\text{K} \quad \text{for} \, \eta = 10^{7.6}\text{P}$$

$$10^{13.4}\text{P} = \eta_0 e^{+Q/[8.314\text{J}/(\text{mol}\cdot\text{K})](787\text{K})}$$

$$10^{7.6}\text{P} = \eta_0 e^{+Q/[8.314\text{J}/(\text{mol}\cdot\text{K})](969\text{K})}$$

$$\frac{10^{13.4}}{10^{7.6}} = e^{+Q/[8.314\text{J}/(\text{mol}\cdot\text{K})](1/787 - 1/969)K^{-1}}$$

or

$$Q = 465 \text{ kJ/mol}$$

and

$$\eta_0 = (10^{13.4} \text{ P})e^{-(465 \times 10^3 \text{ J/mol})/[8.314 \text{ J/(mol·K)}](787 \text{ K})}$$

$$= 3.31 \times 10^{-18} \text{ P}$$

The working range is bounded by $\eta = 10^4$ P and $\eta = 10^8$ P. In general,

$$T = \frac{Q}{R \ln(\eta/\eta_0)}$$

For $\eta = 10^4$ P,

$$T = \frac{465 \times 10^3 \text{ J/mol}}{[8.314 \text{ J/(mol · K)}] \ln (10^4/3.31 \times 10^{-18})}$$

$$= 1130 \text{ K} = 858°C$$

For $\eta = 10^8$ P,

$$T = \frac{465 \times 10^3 \text{ J/mol}}{[8.314 \text{ J/(mol · K)}] \ln (10^8/3.31 \times 10^{-18})}$$

$$= 953 \text{ K} = 680°C$$

Therefore,

$$\text{working range} = 680 \text{ to } 858°C$$

For the melting range, $\eta = 50$ to 500 P. For $\eta = 50$ P,

$$T = \frac{465 \times 10^3 \text{ J/mol}}{[8.314 \text{ J/(mol · K)}] \ln (50/3.31 \times 10^{-18})}$$

$$= 1266 \text{ K} = 993°C$$

For $\eta = 500$ P,

$$T = \frac{465 \times 10^3 \text{ J/mol}}{[8.314 \text{ J/(mol · K)}] \ln (500/3.31 \times 10^{-18})}$$

$$= 1204 \text{ K} = 931°C$$

Therefore,

$$\text{melting range} = 931 \rightarrow 993°C$$

PRACTICE PROBLEM 6.13

> In Sample Problem 6.13 various viscosity ranges are characterized for a soda–lime–silica glass. For this material, calculate the annealing range (see Figure 6–42).

SUMMARY

The wide use of metals as structural elements leads us to concentrate on their mechanical properties. The tensile test gives the most basic design data, including modulus of elasticity, yield strength, tensile strength, ductility, and toughness. Closely related elastic properties are Poisson's ratio and the shear modulus. The fundamental mechanism of elastic deformation is the stretching of atomic bonds. Dislocations play a critical role in the plastic deformation of crystalline metals. They facilitate atom displacement by slipping in high-density atomic planes along high-density atomic directions. Without dislocation slip, exceptionally high stresses are required to deform these materials permanently. Many of the mechanical properties discussed in this chapter are explained in terms of the micromechanical mechanism of dislocation slip. The hardness test is a simple alternative to the tensile test that provides an indication of alloy strength. The creep test indicates that above a temperature of about one-half times the absolute melting point, an alloy has sufficient atomic mobility to deform plastically at stresses below the room-temperature yield stress.

Several mechanical properties play important roles in the structural applications and processing of ceramics and glasses. Both ceramics and glasses are characterized by brittle fracture, although they typically have compressive strengths significantly higher than their tensile strengths. Creep plays an important role in the application of ceramics in high-temperature service. Diffusional mechanisms combine with grain boundary sliding to provide the possibility of extensive deformation. Below the glass transition temperature (T_g), glasses deform by an elastic mechanism. Above T_g they deform by a viscous flow mechanism. The exponential change of viscosity with temperature provides a guideline for routine processing of glass products as well as the development of fracture-resistant tempered glass.

The major mechanical properties of polymers include many of those of importance to metals and ceramics. The wide use of polymers in design applications involving bending and shock absorption requires emphasis on the flexural modulus and the dynamic modulus, respectively. An analog of creep is stress relaxation. Due to the low melting points of polymers, these phenomena can be observed at room temperature and below. Like creep, stress relaxation is an Arrhenius process. As with glasses, viscoelastic deformation is important to polymers. There are four distinct regions of viscoelastic deformation for polymers: (1) rigid (below the glass transition temperature T_g), (2) leathery (near T_g), (3) rubbery (above T_g), and (4) viscous (near the melting temperature T_m). For typical thermosetting polymers, rigid behavior holds nearly to the melting (or decomposition) point. Polymers with a pronounced rubbery region are termed elastomers. Natural and synthetic rubbers are examples. They exhibit substantial, nonlinear elasticity.

KEY TERMS

annealing point (233)
Brinell hardness number (219)
brittle fracture (202)
chemically strengthened glass (234)
cold working (191)
creep (222)
creep curve (222)
critical resolved shear stress (218)
cirtical shear stress (212)
cross-linking (238)
dislocation climb (223)
ductility (193)
dynamic modulus of elasticity (206)
elastic deformation (189)
elastomer (238)
engineering strain (189)
engineering stress (189)
flexural modulus (205)
flexural strength (205)
gage length (189)
glass transition temperature (231)

Griffith crack model (202)
Hooke's law (191)
hysteresis (240)
lower yield point (195)
melting range (232)
modulus of elasticity (191)
modulus of elasticity in bending (205)
modulus of rigidity (197)
modulus of rupture (202)
plastic deformation (190)
Poisson's ratio (195)
relaxation time (226)
residual stress (191)
resolved shear stress (217)
Rockwell hardness (219)
shear modulus (197)
shear strain (196)
shear stress (196)
slip system (214)
softening point (233)
softening temperature (231)

solution hardening (216)
specific strength (191)
strain hardening (191)
strain-hardening exponent (193)
strength-to-weight ratio (191)
stress relaxation (226)
tempered glass (234)
tensile strength (191)
toughness (193)
upper yield point (195)
viscoelastic deformation (231)
viscosity (232)
viscous deformation (231)
working range (232)
yield point (195)
yield strength (190)
Young's modulus (191)

REFERENCES

Ashby, M. F., and **D. R. H. Jones,** *Engineering Materials—An Introduction to Their Properties and Applications,* 2nd ed., Butterworth-Heinemann, Boston, 1996.

ASM Handbook. Vols. 1 (*Properties and Selection: Irons, Steels, and High-Performance Alloys*), and 2 (*Properties and Selection: Nonferrous Alloys and Special-Purpose Metals*), ASM International, Materials Park, Ohio, 1990 and 1991.

Chiang, Y., D. P. Birnie III, and **W. D. Kingery,** *Physical Ceramics,* John Wiley & Sons, Inc., New York, 1997.

Courtney, T. H., *Mechanical Behavior of Materials,* McGraw-Hill Book Company, New York, 1990.

Davis, J. R., Ed., *Metals Handbook,* Desk Ed., 2nd ed., ASM International, Materials Park, Ohio, 1998. A one-volume summary of the extensive *Metals Handbook* series.

Engineered Materials Handbook, Desk Edition ASM International, Materials Park, Ohio, 1995.

Hull, D., and **D. J. Bacon,** *Introduction to Dislocations,* 3rd ed., Pergamon Press, Inc., Elmsford, N.Y., 1984.

Modern Plastics Encyclopedia 98, Vol. 74, No. 14, McGraw-Hill Book Company, New York, October 1997. (Revised annually)

PROBLEMS

As noted on page 228, problems that deal with materials in the engineering design process will be identified with a "design icon" [D].

Section 6.1 • Stress Versus Strain

6.1. The following three $\sigma-\epsilon$ data points are provided for a titanium alloy for aerospace applications: $\epsilon = 0.002778$ (at $\sigma = 300$ MPa), 0.005556 (600 MPa), 0.009897 (900 MPa). Calculate E for this alloy.

6.2. If the Poisson's ratio for the alloy in Problem 6.1 is 0.35, calculate **(a)** the shear modulus G, and **(b)** the shear stress τ necessary to produce an angular displacement α of 0.2865°.

6.3. In Section 6.1, the point was made that the theoretical strength (i.e., critical shear strength) of a material is roughly 0.1 G. **(a)** Use the result of Problem 6.2a to estimate the theoretical critical shear strength of the titanium alloy. **(b)** Comment on the relative value of the result in (a) compared to the apparent yield strength implied by the data given in Problem 6.1.

[D] **6.4.** Consider the 1040 carbon steel listed in Table 6.2. **(a)** A 20-mm-diameter bar of this alloy is used as a structural member in an engineering design. The unstressed length of the bar is precisely 1 m. The structural load on the bar is 1.2×10^5 N in tension. What will be the length of the bar under this structural load? **(b)** A design engineer is considering a structural change that will increase the tensile load on this member. What is the maximum tensile load that can be permitted without producing extensive plastic deformation of the bar? Give your answers in both newtons (N) and pounds force (lb_f).

[D] **6.5.** Heat treatment of the alloy in the design application of Problem 6.4 does not significantly affect the modulus of elasticity but does change strength and ductility. For a particular heat treatment, the corresponding mechanical property data are

Y.S. = 1000MPa (145ksi)

T.S. = 1380MPa (200ksi)

% elongation at failure = 12

Again considering a 20-mm-diameter by 1-m-long bar of this alloy, what is the maximum tensile load

that can be permitted without producing extensive plastic deformation of the bar?

[D] **6.6.** Repeat Problem 6.4 for a structural design using the 2024-T81 aluminum illustrated in Figure 6–3 and Sample Problem 6.1.

6.7. In normal motion, the load exerted on the hip joint is 2.5 times body weight. **(a)** Calculate the corresponding stress (in MPa) on an artificial hip implant with a cross-sectional area of 5.64 cm^2 in a patient weighing 150 lb_f. **(b)** Calculate the corresponding strain if the implant is made of Ti–6Al–4V which has an elastic modulus of 124 GPa.

6.8. Repeat Problem 6.7 for the case of an athlete who undergoes a hip implant. The same alloy is used but, because the athlete weighs 200 lb_f, a larger implant is required (with a cross-sectional area of 6.90 cm^2). Also, consider the situation in which the athlete expends his maximum effort exerting a load of five times body weight.

[D] **6.9.** Suppose that you were asked to select a material for a spherical pressure vessel to be used in an aerospace application. The stress in the vessel wall is

$$\sigma = \frac{pr}{2t}$$

where p is the internal pressure, r the outer radius of the sphere, and t the wall thickness. The mass of the vessel is

$$m = 4\pi r^2 t\rho$$

where ρ is the material density. The operating stress of the vessel will always be

$$\sigma \le \frac{\text{Y.S.}}{S}$$

where S is a safety factor. **(a)** Show that the minimum mass of the pressure vessel will be

$$m = 2S\pi pr^3 \frac{\rho}{\text{Y.S.}}$$

(b) Given Table 6.2 and the following data, select the alloy that will produce the lightest vessel.

Alloy	ρ (Mg/m³)	Cost[a] ($/kg)
1040 carbon steel	7.8	0.63
304 stainless steel	7.8	3.70
3003-H14 aluminum	2.73	3.00
Ti–5Al–2.5Sn	4.46	15.00

[a] Approximate in U.S. dollars.

(c) Given Table 6.2 and the data in the preceding table, select the alloy that will produce the minimum cost vessel.

6.10. Prepare a table comparing the tensile strength-per-unit density of the aluminum alloys of Table 6.1 with the 1040 steel in the same table. Take the densities of 1040 steel and the 2048 and 3003 alloys to be 7.85, 2.91, and 2.75 Mg/m³, respectively.

6.11. Expand on Problem 6.10 by including the magnesium alloys and the titanium alloy of Table 6.1 in the comparison of strength-per-unit density. (Take the densities of the AM100A and AZ31B alloys and the titanium alloy to be 1.84, 1.83, and 4.49 Mg/m³, respectively.)

D 6.12. (a) Select the alloy in the pressure vessel design of Problem 6.9 with the maximum tensile strength-per-unit density. (Note Problem 6.10 for a discussion of this quantity.) **(b)** Select the alloy in Problem 6.9 with the maximum (tensile strength-per-unit density)/unit cost.

• 6.13. In analyzing residual stress by x-ray diffraction, the following stress constant, K_1, is used:

$$K_1 = \frac{E \cot \theta}{2(1 + v) \sin^2 \psi}$$

where E and v are the elastic constants defined in this chapter, θ is a Bragg angle (see Section 3.7), and ψ is an angle of rotation of the sample during the x-ray diffraction experiment (generally $\psi = 45°$). To maximize experimental accuracy, one prefers to use the largest possible Bragg angle, θ. However, hardware configuration (Figure 3–40) prevents θ from being greater than 80°. **(a)** Calculate the maximum θ for a 1040 carbon steel using CrK$_\alpha$ radiation ($\lambda = 0.2291$ nm). (Note that 1040 steel is nearly pure iron, which is a bcc metal, and that the reflection rules for a bcc metal are given in Table 3.4.) **(b)** Calculate the value of the stress constant for 1040 steel.

• 6.14. Repeat Problem 6.13 for 2048 aluminum, which for purposes of the diffraction calculations can be approximated by pure aluminum. (Note that aluminum is an fcc metal and that the reflection rules for such materials are given in Table 3.4.)

6.15. (a) The following data are collected for a modulus of rupture test on an MgO refractory brick (refer to Equation 6.10 and Figure 6–14):

$$F = 7.0 \times 10^4 \text{N}$$

$$L = 175\text{mm}$$

$$b = 110\text{mm}$$

$$h = 75\text{mm}$$

Calculate the modulus of rupture. **(b)** Suppose that you are given a similar MgO refractory with the same strength and same dimensions except that its height, h, is only 64 mm. What would be the load (F) necessary to break this thinner refractory?

6.16. A single crystal Al$_2$O$_3$ rod (precisely 6 mm diameter × 60 mm long) is used to apply loads to small samples in a high-precision dilatometer (a length-measuring device). Calculate the resulting rod dimensions if the crystal is subjected to a 25-kN axial compression load.

6.17. A freshly drawn glass fiber (100 μm diameter) breaks under a tensile load of 40 N. After subsequent handling, a similar fiber breaks under a tensile load of 0.15 N. Assuming the first fiber was defect-free and that the second fiber broke due to an atomically sharp surface crack, calculate the length of that crack.

6.18. A nondestructive testing program can ensure that a given 80-μm-diameter glass fiber will have no surface cracks longer than 5 μm. Given that the theoretical strength of the fiber is 5 GPa, what can you say about the expected breaking strength of this fiber?

6.19. The following data are collected in a flexural test of a polyester to be used in the exterior trim of an automobile:

Test piece geometry: 6 mm × 15 mm × 50 mm

Distance between supports = $L = 60$ mm

Initial slope of load-deflection curve = 538×10^3 N/m

Calculate the flexural modulus of this engineering polymer.

6.20. The following data are collected in a flexural test of a polyester to be used in the fabrication of molded office furniture:

Test piece geometry: 10 mm × 30 mm × 100 mm

Distance between supports = L = 50 mm

Load at fracture = 6000 N

Calculate the flexural strength of this engineering polymer.

6.21. Figure 6–17 illustrates the effect of humidity on stress-versus-strain behavior for a nylon 66. In addition, the distinction between tensile and compressive behavior is shown. Approximating the data between 0 and 20 MPa as a straight line, calculate **(a)** the initial elastic modulus in tension and **(b)** the initial elastic modulus in compression for the nylon at 60% relative humidity.

6.22. An acetal disk precisely 5 mm thick by 25 mm diameter is used as a cover plate in a mechanical loading device. If a 20-kN load is applied to the disk, calculate the resulting dimensions.

Section 6.2 • Elastic Deformation

6.23. The maximum modulus of elasticity for a copper crystal is 195 GPa. What tensile stress is required along the corresponding crystallographic direction in order to increase the interatomic separation distance by 0.05%?

6.24. Repeat Problem 6.23 for the crystallographic direction corresponding to the minimum modulus of elasticity for copper, which is 70 GPa.

6.25. An expression for the van der Waals bonding energy as a function of interatomic distance is given in Sample Problem 2.13. Derive an expression for the slope of the force curve at the equilibrium bond length, a_0. (As shown in Figure 6–18, that slope is directly related to the elastic modulus of solid argon, which exists at cryogenic temperatures.)

6.26. Using the result of Problem 6.25 and data in Sample Problem 2.13, calculate the value of the slope of the force curve at the equilibrium bond length, a_0, for solid argon. (Note that the units will be N/m rather than MPa in that we are dealing with the slope of the force versus elongation curve rather than the stress-versus-strain curve.)

Section 6.3 • Plastic Deformation

6.27. A crystalline grain of aluminum in a metal plate is situated so that a tensile load is oriented along the [111] crystal direction. If the applied stress is 0.5 MPa (72.5 psi), what will be the resolved shear stress, τ, along the [101] direction within the $(11\bar{1})$ plane? (Review the comments in Problem 3.35.)

6.28. In Problem 6.27, what tensile stress is required to produce a critical resolved shear stress, τ_c, of 0.242 MPa?

6.29. A crystalline grain of iron in a metal plate is situated so that a tensile load is oriented along the [110] crystal direction. If the applied stress is 50 MPa (7.25×10^3 psi), what will be the resolved shear stress, τ, along the $[11\bar{1}]$ direction within the (101) plane? (Review the comments in Problem 3.35.)

6.30. In Problem 6.29, what tensile stress is required to produce a critical resolved shear stress, τ_c, of 31.1 MPa?

• 6.31. Consider the slip systems for aluminum shown in Figure 6–24. For an applied tensile stress in the [111] direction, which slip system(s) would be most likely to operate?

• 6.32. Figure 6–24 lists the slip systems for an fcc and an hcp metal. For each case, this represents all unique combinations of close-packed planes and close-packed directions (contained within the close-packed planes). Make a similar list for the 12 slip systems in the bcc structure (see Table 6.9). (*A few important hints:* It will help to first verify the list for the fcc metal. Note that each slip system involves a plane $(h_1k_1l_1)$ and a direction $[h_2k_2l_2]$ whose indices give a dot product of zero (i.e., $h_1h_2+k_1k_2+l_1l_2 = 0$). Further, all members of the $\{hkl\}$ family of planes are not listed. Because a stress involves simultaneous force application in two antiparallel directions, only nonparallel planes need to be listed. Similarly, antiparallel crystal directions are redundant. You may want to review Problems 3.35 to 3.37.)

6.33. Sketch the atomic arrangement and Burgers vector orientations in the slip plane of a bcc metal. (Note the shaded area of Table 6.9.)

6.34. Sketch the atomic arrangement and Burgers vector orientation in the slip plane of an fcc metal. (Note the shaded area of Table 6.9.)

6.35. Sketch the atomic arrangement and Burgers vector orientation in the slip plane of an hcp metal. (Note the shaded area of Table 6.9.)

• **6.36.** In some bcc metals, an alternate slip system operates, namely, the $\{211\}\langle\bar{1}11\rangle$. This system has the same Burgers vector but a lower-density slip plane, as compared to the slip system in Table 6.9. Sketch the unit cell geometry for this alternate slip system in the manner used in Table 6.9.

• **6.37.** Identify the 12 individual slip systems for the alternate system given for bcc metals in Problem 6.36. (Recall the comments in Problem 6.32.)

• **6.38.** Sketch the atomic arrangement and Burgers vector orientation in a (211) slip plane of a bcc metal. (Note Problems 6.36 and 6.37.)

Section 6.4 • Hardness

6.39. You are provided an unknown alloy with a measured Brinell hardness value of 100. Having no other information than the data of Figure 6–28a, estimate the tensile strength of the alloy. (Express your answer in the form $x \pm y$.)

6.40. Show that the data of Figure 6–28b are consistent with the plot of Figure 6–28a.

D **6.41.** A ductile iron (65–45–12, annealed) is to be used in a spherical pressure vessel. The specific alloy obtained for the vessel has a Brinell hardness number of 200. The design specifications for the vessel include a spherical outer radius of 0.30 m, wall thickness of 20 mm, and a safety factor of 2. Using the information in Figure 6–28 and Problem 6.9, calculate the maximum operating pressure p for this vessel design.

D **6.42.** Repeat Problem 6.41 for another ductile iron (grade 120–90–02, oil-quenched) with a Brinell hardness number of 280.

6.43. The simple expressions for Rockwell hardness numbers in Table 6.10 involve indentation, t, expressed in millimeters. A given steel with a BHN of 235 is also measured by a Rockwell hardness tester. Using a $\frac{1}{16}$-in.-diameter steel sphere and a load of 100 kg,

the indentation t is found to be 0.062 mm. What is the Rockwell hardness number?

6.44. An additional Rockwell hardness test is made on the steel considered in Problem 6.43. Using a diamond cone under a load of 150 kg, an indentation t of 0.157 mm is found. What is the resulting alternative Rockwell hardness value?

6.45. You are asked to measure nondestructively the yield strength and tensile strength of an annealed 65–45–12 cast iron structural member. Fortunately, a small hardness indentation in this structural design will not impair its future usefulness, which is a working definition of *nondestructive*. A 10-mm-diameter tungsten carbide sphere creates a 4.26-mm-diameter impression under a 3000-kg load. What are the yield and tensile strengths?

6.46. As in Problem 6.45, calculate the yield and tensile strengths for the case of a 4.48-mm-diameter impression under identical conditions.

6.47. The Ti–6Al–4V orthopedic implant material introduced in Problem 6.7 gives a 3.27-mm-diameter impression when a 10-mm-diameter tungsten carbide sphere is applied to the surface with a 3000-kg load. What is the Brinell hardness number of this alloy?

6.48. In Section 6.4 a useful correlation between hardness and tensile strength was demonstrated for metallic alloys. Plot hardness versus tensile strength for the data given in Table 6.12 and comment on whether a similar trend is shown for these common thermoplastic polymers. (You can compare this plot with Figure 6–28a.)

Section 6.5 • Creep and Stress Relaxation

6.49. An alloy is evaluated for potential creep deformation in a short-term laboratory experiment. The creep rate ($\dot{\epsilon}$) is found to be 1% per hour at 800°C and 5.5×10^{-2} % per hour at 700°C. **(a)** Calculate the activation energy for creep in this temperature range. **(b)** Estimate the creep rate to be expected at a service temperature of 500°C. **(c)** What important assumption underlies the validity of your answer in part (b)?

6.50. The inverse of time to reaction (t_R^{-1}) can be used to approximate a rate and, consequently, can be estimated using the Arrhenius expression (Equation 6.16). The same is true for time-to-creep rupture, as defined in Figure 6–35. If the time to rupture for a given superalloy is 2000 h at 650°C and 50 h at 700°C, calculate the activation energy for the creep mechanism.

6.51. Estimate the time to rupture at 750°C for the superalloy of Problem 6.50.

• **6.52.** Figure 6–33 indicates the dependence of creep on both stress (σ) and temperature (T). For many alloys, such dependence can be expressed in a modified form of the Arrhenius equation,

$$\dot{\epsilon} = C_1 \sigma^n e^{-Q/RT}$$

where $\dot{\epsilon}$ is the steady-state creep rate, C_1 is a constant, and n is a constant that usually lies within the range of 3 to 8. The exponential term ($e^{-Q/RT}$) is the same as in other Arrhenius expressions (see Equation 6.16). The product of $C_1\sigma^n$ is a temperature-independent term equal to the preexponential constant, C, in Equation 6.16. The presence of the σ^n term gives the name "power-law" creep to this expression. Given the power-law creep relationship with $Q = 250$ kJ/mol and $n = 4$, calculate what percentage increase in stress will be necessary to produce the same increase in $\dot{\epsilon}$ as a 10°C increase in temperature from 1000 to 1010°C.

6.53. Using Table 6.13, calculate the lifetime of **(a)** a slip cast MgO refractory at 1300°C and 12.4 MPa if 1% total strain is permissible. **(b)** Repeat the calculation for a hydrostatically pressed MgO refractory. **(c)** Comment on the effect of processing on the relative performance of these two refractories.

6.54. Assume the activation energy for the creep of Al_2O_3 is 425 kJ/mol. **(a)** Predict the creep rate, $\dot{\epsilon}$, for Al_2O_3 at 1000°C and 1800 psi applied stress. (See Table 6.13 for data at 1300°C and 1800 psi.) **(b)** Calculate the lifetime of an Al_2O_3 furnace tube at 1000°C and 1800 psi if 1% total strain is permissible.

6.55. In Problem 6.52 "power law" creep was introduced in which

$$\dot{\epsilon} = c_1 \sigma^n e^{-Q/RT}$$

(a) For a value of $n = 4$, calculate the creep rate, $\dot{\epsilon}$, for Al_2O_3 at 1300°C and 900 psi. **(b)** Calculate the lifetime of an Al_2O_3 furnace tube at 1300°C and 900 psi if 1% total strain is permissible.

• **6.56.** **(a)** The creep plot in Figure 6–37 indicates a general "band" of data roughly falling between the two parallel lines. Calculate a general activation energy for the creep of oxide ceramics using the slope indicated by those parallel lines. **(b)** Estimate the uncertainty in the answer to part (a) by considering the maximum and minimum slopes within the band between temperatures of 1400 and 2200°C.

6.57. The stress on a rubber disk is seen to relax from 0.75 to 0.5 MPa in 100 days. **(a)** What is the relaxation time, τ, for this material? **(b)** What will be the stress on the disk after (i) 50 days, (ii) 200 days, or (iii) 365 days? (Consider time $= 0$ to be at the stress level of 0.75 MPa.)

6.58. Increasing temperature from 20 to 30°C decreases the relaxation time for a polymeric fiber from 3 to 2 days. Determine the activation energy for relaxation.

6.59. Given the data in Problem 6.58, calculate the expected relaxation time at 40°C.

6.60. A spherical pressure vessel is fabricated from nylon 66 and will be used at 60°C and 50% relative humidity. The vessel dimensions are 50 mm outer radius and 2 mm wall thickness. **(a)** What internal pressure is required to produce a stress in the vessel wall of 6.9 MPa (1000 psi)? (The stress in the vessel wall is

$$\sigma = \frac{pr}{2t}$$

where p is the internal pressure, r the outer radius of the sphere, and t the wall thickness.) **(b)** Calculate the circumference of the sphere after 10,000 h at this pressure. (Note Figure 6–38.)

Section 6.6 • Viscoelastic Deformation

6.61. A borosilicate glass used for sealed-beam headlights has an annealing point of 544°C and a softening point of 780°C. Calculate **(a)** the activation energy for viscous deformation in this glass, **(b)** its working range, and **(c)** its melting range.

6.62. The following viscosity data are available on a borosilicate glass used for vacuum-tight seals:

T (°C)	η (poise)
700	4.0×10^7
1080	1.0×10^4

Determine the temperatures at which this glass should be **(a)** melted and **(b)** annealed.

6.63. For the vacuum sealing glass described in Problem 6.62, assume you have traditionally annealed the product at the viscosity of 10^{13} poise. After a cost–benefit analysis, you realize that it is more economical to anneal for a longer time at a lower temperature. If you decide to anneal at a viscosity of $10^{13.4}$

poise, how many degrees (°C) should your annealing furnace operator lower the furnace temperature?

6.64. You are asked to help design a manufacturing furnace for a new optical glass. Given that it has an annealing point of 460°C and a softening point of 647°C, calculate the temperature range in which the product shape would be formed (i.e., the working range).

CHAPTER 7
Thermal Behavior

Refractories are high-temperature resistant ceramics used in applications such as metal casting. The most effective refractories have low values of thermal expansion and thermal conductivity. (Courtesy of R. T. Vanderbilt Company, Inc.)

In the previous chapter, we surveyed a range of properties that define the mechanical behavior of materials. In a similar way, we will now survey various properties that define the thermal behavior of materials, indicating how materials respond to the application of heat.

Both *heat capacity* and *specific heat* indicate a material's ability to absorb heat from its surroundings. The energy imparted to the material from the external heat source produces an increase in the thermal vibration of the atoms in the material. Most materials increase slightly in size as they are heated. This *thermal expansion* is a direct result of the greater separation distance between the centers of adjacent atoms as the thermal vibration of the individual atoms increases with increasing temperature.

In describing the flow of heat through a material, the *thermal conductivity* is the proportionality constant between the heat flow rate and the temperature gradient, exactly analogous to the diffusivity, defined in Chapter 5 as the proportionality constant between the mass flow rate and the concentration gradient.

There can be mechanical consequences from the flow of heat in materials. *Thermal shock* refers to the fracture of a material due to a temperature change, usually a sudden cooling.

7.1 HEAT CAPACITY

As a material absorbs heat from its enviroment, its temperature rises. This commom observation can be quantified with a fundamental material property, the **heat capacity**, C, defined as the amount of heat required to raise its temperature by 1 K ($= 1°C$):

$$C = \frac{Q}{\Delta T} \tag{7.1}$$

with Q being the amount of heat producing a temperature change ΔT. It is important to note that, for incremental temperature changes, the magnitude of ΔT is the same in either the Kelvin (K) or Celcius (°C) temperature scales.

The magnitude of C will depend on the amount of material. The heat capacity is ordinarily specified for a basis of one gram-atom (for elements) or one mole (for compounds) in units of J/g-atom·K or J/mol·K. A common alternative is the **specific heat** using a basis of unit mass, such as J/kg·K. Along with the heat and the mass, the specific heat is designated by lowercase letters:

$$c = \frac{q}{m \Delta T} \tag{7.2}$$

There are two common ways in which heat capacity (or specific heat) is measured. One is while maintaining a constant volume, $C_v(c_v)$, and the other is while maintaining a constant pressure, $C_p(c_p)$. The magnitude of

C_p is always greater than C_v, but the difference is minor for most solids at room temperature or below. Because we ordinarily use mass-based amounts of engineering materials under a fixed, atmospheric pressure, we will tend to use c_p data in this book. Such values of specific heat for a variety of engineering materials are given in Table 7.1.

Fundamental studies of the relationship between atomic vibrations and heat capacity in the early part of the twentieth century led to the discovery that, at very low temperatures, C_v rises sharply from zero at 0 K as

$$C_v = AT^3 \tag{7.3}$$

where A is a temperature-independent constant. Furthermore, above a **Debye* temperature** (θ_D), the value of C_v was found to level off at approximately $3R$, where R is the universal gas constant. (As in Section 5.1, we see that R is a fundamental constant equally important to the solid state, even though it carries the label "gas constant" due to its presence in the perfect gas law.) Figure 7–1 summarizes how C_v rises to an asymptotic value of $3R$ above θ_D. As θ_D is below room temperature for many solids and $C_p \approx C_v$, we have a useful rule of thumb for the value of the heat capacity of many engineering materials.

Finally, it can be noted that there are other energy-absorbing mechanisms, besides atomic vibrations, that can contribute to the magnitude of heat capacity. Examples are the energy absorption by free electrons in metals and the randomization of electron spins in ferromagnetic materials (to be discussed in Chapter 18). On the whole, however, the general behavior summarized by Figure 7–1 and Table 7.1 will be adequate for most engineering materials in ordinary applications.

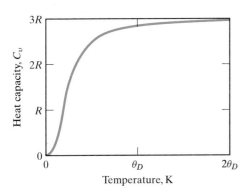

Figure 7-1 *The temperature dependence of the heat capacity at constant volume, C_v. The magnitude of C_v rises sharply with temperature near 0 K and, above the Debye temperature (θ_D), levels off at a value of approximately $3R$.*

---

* Peter Joseph Wilhelm Debye (1884–1966), Dutch-American physical chemist, developed the results displayed in Figure 7–1 as a refinement of Einstein's theory of specific heats, incorporating the then newly developed quantum theory and the elastic constants of the material. Debye made numerous contributions to the fields of physics and chemistry, including pioneering work on x-ray diffraction of powdered materials, thereby helping to lay the foundation for Section 3.7, along with von Laue and the Braggs.

Table 7.1 *Values of Specific Heat for a Variety of Materials*

Material	c_p [J/kg·K]
Metals[a]	
Aluminum	900
Copper	385
Gold	129
Iron (α)	444
Lead	159
Nickel	444
Silver	237
Titanium	523
Tungsten	133
Ceramics[a, b]	
Al_2O_3	160
MgO	457
SiC	344
Carbon (diamond)	519
Carbon (graphite)	711
Polymers[a]	
Nylon 66	1260–2090
Phenolic	1460–1670
Polyethylene (high density)	1920–2300
Polypropylene	1880
Polytetraflouroethylene (PTFE)	1050

Source: Data from [a] J. F. Shackelford, W. Alexander, and J. S. Park, *The CRC Materials Science and Engineering Handbook*, 2nd ed., CRC Press, Boca Raton, Fl., 1994, and [b] W. D. Kingery, H. K. Bowen, and D. R. Uhlmann, *Introduction to Ceramics*, 2nd ed., John Wiley & Sons, Inc., New York, 1976.

SAMPLE PROBLEM 7.1

Show that the rule of thumb that the heat capacity of a solid is approximately $3R$ is consistent with the specific heat value for aluminum in Table 7.1.

SOLUTION

From Appendix 3, we have

$$3R = 3(8.314 \, \text{J/mol·K})$$

$$= 24.94 \, \text{J/mol·K}$$

From Appendix 1, we see that, for aluminum, there are 26.98 g per g-atom, which corresponds, for this elemental solid, to a mole. Then

$$3R = (24.94 \text{ J/mol·K})(1 \text{ mol}/26.98 \text{ g})(1000 \text{ g/kg})$$

$$= 924 \text{ J/kg·K}$$

in reasonable agreement with the value of 900 J/kg·K in Table 7.1

....

PRACTICE PROBLEM 7.1

Show that a heat capacity of $3R$ is a reasonable approximation to the specific heat of copper given in Table 7.1 (See Sample Problem 7.1.)

7.2 THERMAL EXPANSION

An increase in temperature leads to greater thermal vibration of the atoms in a material and an increase in the average separation distance of adjacent atoms (Figure 7–2). In general, the overall dimension of the material in a given direction, L, will increase with increasing temperature, T. This is reflected by the **linear coefficient of thermal expansion,** α, given by

$$\alpha = \frac{dL}{L\,dT} \tag{7.4}$$

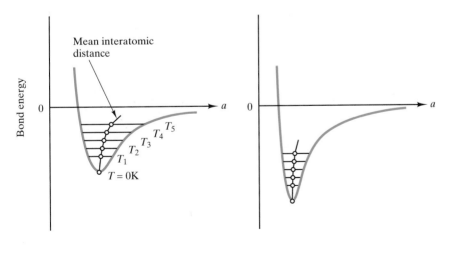

(a)   (b)

Figure 7-2 *Plot of atomic bonding energy versus interatomic distance for (a) weakly bonded solid and (b) a strongly bonded solid. Thermal expansion is the result of a greater interatomic distance with increasing temperature. The effect (represented by the coefficient of thermal expansion in Equation 7.4) is greater for the more asymmetrical energy well of the weakly bonded solid. As shown in Table 7.3, melting point and elastic modulus increase with increasing bond strength.*

Table 7.2 *Values of Linear Coefficient of Thermal Expansion for a Variety of Materials*

Material	α [mm/(mm·°C) $\times 10^6$]		
	Temperature = 27°C (300 K)	527°C (800 K)	0–1000°C
Metals[a]			
Aluminum	23.2	33.8	
Copper	16.8	20.0	
Gold	14.1	16.5	
Nickel	12.7	16.8	
Silver	19.2	23.4	
Tungsten	4.5	4.8	
Ceramics and glasses[a, b]			
Mullite ($3Al_2O_3 \cdot 2SiO_2$)			5.3
Porcelain			6.0
Fireclay refractory			5.5
Al_2O_3			8.8
Spinel ($MgO \cdot Al_2O_3$)			7.6
MgO			13.5
UO_2			10.0
ZrO_2 (stabilized)			10.0
SiC			4.7
Silica glass			0.5
Soda–lime–silica glass			9.0
Polymers[a]			
Nylon 66	30–31		
Phenolic	30–45		
Polyethylene (high-density)	149–301		
Polypropylene	68–104		
Polytetrafluoroethylene (PTFE)	99		

Source: Data from [a] J. F. Shackelford, W. Alexander, and J. S. Park, *The CRC Materials Science and Engineering Handbook*, 2nd ed., CRC Press, Boca Raton, Fl., 1994, and [b] W. D. Kingery, H. K. Bowen, and D. R. Uhlmann, *Introduction to Ceramics*, 2nd ed., John Wiley & Sons, Inc., New York, 1976.

with α having units of mm/(mm · °C). Thermal expansion data for various materials are given in Table 7.2.

Note that the thermal expansion coefficients of ceramics and glasses are generally smaller than those for metals, which are, in turn, smaller than those for polymers. The differences are related to the asymmetrical shape of the energy well in Figure 7–2. The ceramics and glasses generally have deeper wells (i.e., higher bonding energies) associated with their ionic and covalent-type bonding. The result is a more symmetrical energy well, with relatively less increase in interatomic separation with increasing temperature, as shown in Figure 7–2b.

The elastic modulus is directly related to the derivative of the bonding energy curve near the bottom of the well (Figure 6–18), and it follows that the deeper the energy well, the larger the value of that derivative and hence the greater the elastic modulus. Furthermore, the stronger bonding associated with deeper energy wells corresponds to higher melting points. These various, useful correlations with bonding strength are summarized in Table 7.3.

Table 7.3 *Correlation of Bonding Strength with Material Properties*

Weakly bonded solids	Strongly bonded solids
Low melting point	High melting point
Low elastic modulus	High elastic modulus
High thermal expansion coefficient	Low thermal expansion coefficient

The thermal expansion coefficient itself is a function of temperature. A plot showing the variation in the linear coefficient of thermal expansion of some common ceramic materials over a wide temperature range is shown in Figure 7–3.

As will be discussed in Chapter 12, crystallites of β-eucryptite are an important part of the microstructure of some glass-ceramics. The β-eucryptite ($Li_2O \cdot Al_2O_3 \cdot SiO_2$) has a *negative* coefficient of thermal expansion and helps to give the overall material a low thermal expansion coefficient and, therefore, excellent resistance to thermal shock, a problem discussed in Section 7.4. In the exceptional cases such as β-eucryptite, the overall atomic architecture "relaxes" in an accordion style as the temperature rises.

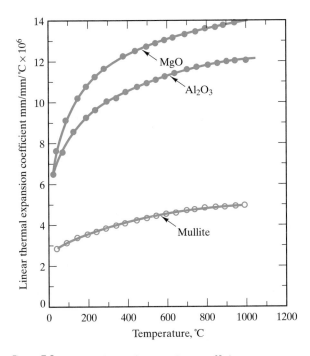

Figure 7-3 *Linear thermal expansion coefficient as a function of temperature for three ceramic oxides (mullite = $3Al_2O_3 \cdot 2SiO_2$). (From W. D. Kingery, H. K. Bowen, and D. R. Uhlmann,* Introduction to Ceramics, *2nd Ed., John Wiley & Sons, Inc., New York, 1976.)*

SAMPLE PROBLEM 7.2

A 0.1-m-long Al_2O_3 furnace tube is heated from room temperature (25°C) to 1000°C. Assuming the tube is not mechanically constrained, calculate the increase in length produced by this heating.

SOLUTION

Rearranging Equation 7.4,

$$dL = \alpha L \, dT$$

We can assume linear thermal expansion using the overall thermal expansion coefficient for this temperature range given in Table 7.2. Then

$$\Delta L = \alpha L_0 \Delta T$$

$$= [8.8 \times 10^{-6} \, \text{mm/(mm•°C)}](0.1 \, \text{m})(1000 - 25)°\text{C}$$

$$= 0.858 \times 10^{-3} \, \text{m}$$

$$= 0.858 \, \text{mm}$$

PRACTICE PROBLEM 7.2

A 0.1-m-long mullite furnace tube is heated from room temperature (25°C) to 1000°C. Assuming the tube is not mechanically constrained, calculate the increase in length produced by this heating. (See Sample Problem 7.2.)

7.3   THERMAL CONDUCTIVITY

The mathematics for the conduction of heat in solids is analogous to that for diffusion (see Section 5.3). The analog for diffusivity, D, is **thermal conductivity,** k, which is defined by **Fourier's* law:**

$$k = -\frac{dQ/dt}{A(dT/dx)} \tag{7.5}$$

---

* Jean Baptiste Joseph Fourier (1768–1830), French mathematician, left us with some of the most useful concepts in applied mathematics. His demonstration that complex waveforms can be described as a series of trigonometric functions brought him his first great fame (and the title "baron" bestowed by Napoleon). In 1822, his master work on heat flow, entitled *Analytical Theory of Heat,* was published.

where dQ/dt is the rate of heat transfer across an area, A, due to a temperature gradient dT/dx. Figure 7–4 relates the various terms of Equation 7.5 and should be compared with the illustration of Fick's first law in Figure 5–9. The units for k are J/(s · m · K). For steady-state heat conduction through a flat slab, the differentials of Equation 7.5 become average terms:

$$k = -\frac{\Delta Q / \Delta t}{A(\Delta T / \Delta x)} \qquad (7.6)$$

Equation 7.6 is appropriate for describing heat flow through refractory walls in high-temperature furnaces.

Thermal conductivity data are shown in Table 7.4. Like the thermal expansion coefficient, thermal conductivity is a function of temperature. A plot of thermal conductivity for several common ceramic materials over a wide temperature range is shown in Figure 7–5.

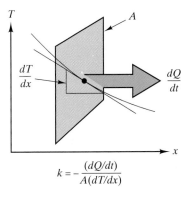

$$k = -\frac{(dQ/dt)}{A(dT/dx)}$$

Figure 7-4 *Heat transfer is defined by Fourier's law (Equation 7.5).*

Table 7.4 *Values of Thermal Conductivity for a Variety of Materials*

Material	k [J/(s·m·K)] Temperature = 27°C (300 K)	100°C	527°C (800 K)	1000°C
Metals[a]				
Aluminum	237		220	
Copper	398		371	
Gold	315		292	
Iron	80		43	
Nickel	91		67	
Silver	427		389	
Titanium	22		20	
Tungsten	178		128	
Ceramics and glasses[a, b]				
Mullite ($3Al_2O_3 \cdot 2SiO_2$)		5.9		3.8
Porcelain		1.7		1.9
Fireclay refractory		1.1		1.5
Al_2O_3		30.0		6.3
Spinel ($MgO \cdot Al_2O_3$)		15.0		5.9
MgO		38.0		7.1
ZrO_2 (stabilized)		2.0		2.3
TiC		25.0		5.9
Silica glass		2.0		2.5
Soda–lime–silica glass		1.7		—
Polymers[a]				
Nylon 66	2.9			
Phenolic	0.17–0.52			
Polythylene (high-density)	0.33			
Polypropylene	2.1–2.4			
Polytetrafluoroethylene (PTFE)	0.24			

Source: Data from [a] J. F. Shackelford, W. Alexander, and J. S. Park, *The CRC Materials Science and Engineering Handbook*, 2nd ed., CRC Press, Boca Raton, Fl., 1994, and [b] W. D. Kingery, H. K. Bowen, and D. R. Uhlmann, *Introduction to Ceramics*, 2nd ed., John Wiley & Sons, Inc., New York, 1976.

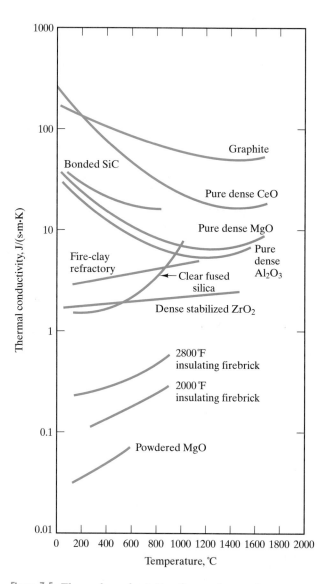

Figure 7-5 *Thermal conductivity of several ceramics over a range of temperatures. (From W. D. Kingery, H. K. Bowen, and D. R. Uhlmann,* Introduction to Ceramics, *2nd Ed., John Wiley & Sons, Inc., New York, 1976.)*

The conduction of heat in engineering materials involves two primary mechanisms, atomic vibrations and the conduction of free electrons. For poor electrical conductors such as ceramics and polymers, thermal energy is transported primarily by the vibration of atoms. For electrically conductive metals, the kinetic energy of the conducting (or "free") electrons can provide substantially more efficient conduction of heat than atomic vibrations.

In Chapter 15, we shall look in greater detail at the mechanism of electrical conduction. A general feature of this mechanism is that the electron can be viewed as a wave as well as a particle. For a wave, any structural disorder interferes with the movement of the waveform. The increasing vibration of the crystal lattice at increasing temperature, then, generally results in a decrease in thermal conductivity. Similarly, the structural disorder created by chemical impurities results in a similar decrease in thermal conductivity. As a result, metal alloys tend to have lower thermal conductivities than pure metals.

For ceramics and polymers, atomic vibrations are the predominant source of thermal conductivity, given the very small number of conducting electrons. These lattice vibrations are, however, also wavelike in nature and are similarly impeded by structural disorder. As a result, glasses will tend to have a lower thermal conductivity than crystalline ceramics of the same chemical composition. In the same way, amorphous polymers will tend to have a greater thermal conductivity than crystalline polymers of comparable compositions. Also, the thermal conductivities of ceramics and polymers will drop with increasing temperature due to the increasing disorder caused by the increasing degree of atomic vibration. For some ceramics, conductivity will eventually begin to rise with further increase in temperature due to radiant heat transfer. Significant amounts of infrared radiation can be transmitted through ceramics, that tend to be optically transparent. These issues will be discussed further in Chapter 16.

The thermal conductivity of ceramics and polymers can be further reduced by the presence of porosity. The gas in the pores has a very low thermal conductivity, giving a low net conductivity to the overall microstructure. Prominent examples include the sophisticated space shuttle tile (to be discussed in Chapter 20) and the common foamed polystyrene (Styrofoam) drinking cup.

SAMPLE PROBLEM 7.3

Calculate the steady-state heat transfer rate (in $J/m^2 \cdot s$) through a sheet of copper 10 mm thick if there is a 50°C temperature drop (from 50°C to 0°C) across the sheet.

SOLUTION

Rearranging Equation 7.6,

$$(\Delta Q / \Delta t)/A = -k(\Delta T/\Delta x)$$

Over this temperature range (average $T = 25°C = 298K$), we can use the thermal conductivity for copper at 300 K given in Table 7.4, giving

$$(\Delta Q/\Delta t)/A = -(398\ \text{J/s·m·K})([0°C - 50°C]/[10 \times 10^{-3}\ \text{m}])$$

$$= -(398\ \text{J/s·m·K})(-5 \times 10^{-3°}\ \text{C/m})$$

The K and °C units cancel given that we are dealing with an incremental change in temperature, so

$$(\Delta Q/\Delta t)/A = 1.99 \times 10^6\ \text{J/m}^2\text{·s}$$

....

PRACTICE PROBLEM 7.3

Calculate the steady-state heat transfer rate through a 10-mm-thick sheet of copper for a 50°C temperature drop, from 550°C to 500°C. (See Sample Problem 7.3.)

7.4   THERMAL SHOCK

The common use of some inherently brittle materials, especially ceramics and glasses, at high temperatures leads to a special engineering problem called **thermal shock**. This can be defined as the fracture (partial or complete) of the material as a result of a temperature change (usually a sudden cooling).

The mechanism of thermal shock can involve both thermal expansion and thermal conductivity. Thermal shock follows from these properties in one of two ways. First, a failure stress can be built up by constraint of uniform thermal expansion. Second, rapid temperature changes produce temporary temperature gradients in the material with resulting internal residual stress. Figure 7–6 shows a simple illustration of the first case. It is equivalent to

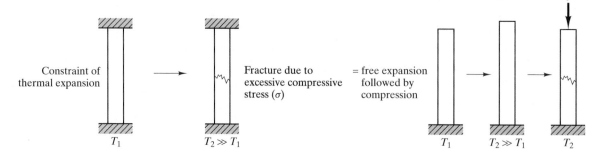

Figure 7-6 *Thermal shock resulting from constraint of uniform thermal expansion. This process is equivalent to free expansion followed by mechanical compression back to the original length.*

allowing free expansion followed by mechanical compression of the rod back to its original length. More than one furnace design has been flawed by inadequate allowance for expansion of refractory ceramics during heating. Similar consideration must be given to expansion coefficient matching of coating and substrate for glazes (glass coatings on ceramics) and enamels (glass coatings on metals).

Even without external constraint, thermal shock can occur due to the temperature gradients created because of a finite thermal conductivity. Figure 7–7 illustrates how rapid cooling of the surface of a high-temperature wall is accompanied by surface tensile stresses. The surface contracts more than the interior, which is still relatively hot. As a result, the surface "pulls" the interior into compression and is itself "pulled" into tension. With the inevitable presence of Griffith flaws at the surface, this surface tensile stress creates the clear potential for brittle fracture. The ability of a material to withstand a given temperature change depends on a complex combination of thermal expansion, thermal conductivity, overall geometry, and the inherent

Thickness of slab of material

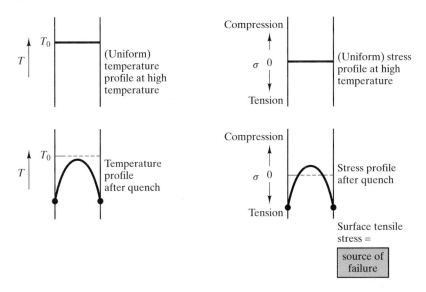

Figure 7-7 *Thermal shock resulting from temperature gradients created by a finite thermal conductivity. Rapid cooling produces surface tensile stresses.*

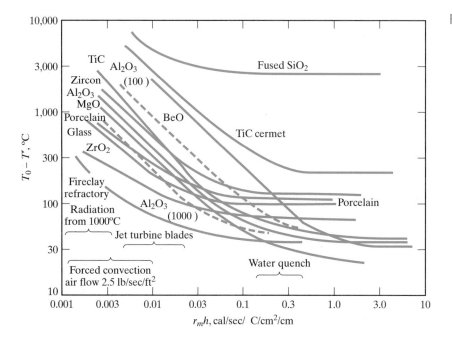

Figure 7-8 *Thermal quenches that produce failure by thermal shock are illustrated. The temperature drop necessary to produce fracture ($T_0 - T'$) is plotted against a heat transfer parameter ($r_m h$). More important than the values of $r_m h$ are the regions corresponding to given types of quench (e.g., "water quench" corresponds to an $r_m h$ around 0.2 to 0.3). (From W. D. Kingery, H. K. Bowen, and D. R. Uhlmann,* Introduction to Ceramics, *2nd Ed., John Wiley & Sons, Inc., New York, 1976.)*

brittleness of that material. Figure 7–8 shows the kinds of thermal quenches (temperature drops) necessary to fracture various ceramics and glasses by thermal shock. Our discussion of thermal shock has been independent of the contribution of phase transformations. In Chapter 9, we shall see the effect of a phase transformation on the structural failure of unstabilized zirconia (ZrO_2). In such cases, even moderate temperature changes through the transformation range can be destructive. Susceptibility to thermal shock is also a limitation of partially stabilized zirconia, which includes small grains of unstabilized phase.

SAMPLE PROBLEM 7.4

Consider an Al_2O_3 furnace tube constrained in the way illustrated in Figure 7–6. Calculate the stress that would be generated in the tube if it were heated to 1000°C.

SOLUTION

Table 7.2 gives the thermal expansion coefficient for Al_2O_3 over this range:

$$\alpha = 8.8 \times 10^{-6} \text{mm/(mm} \cdot °C)$$

If we take room temperature as 25°C, the unconstrained expansion associated with heating to 1000°C is

$$\epsilon = \alpha \, \Delta T$$
$$= [8.8 \times 10^{-6} \text{mm/(mm} \cdot ^\circ \text{C)}](1000 - 25)^\circ \text{C}$$
$$= 8.58 \times 10^{-3}$$

The compressive stress resulting from constraining that expansion is

$$\sigma = E\epsilon$$

Table 6.5 gives an E for sintered Al_2O_3 as $E = 370 \times 10^3$ MPa. Then

$$\sigma = (370 \times 10^3 \text{MPa})(8.58 \times 10^{-3})$$
$$= 3170 \text{MPa(compressive)}$$

This is substantially above the failure stress for alumina ceramics (see Figure 6–13).

◰ SAMPLE PROBLEM 7.5

Engineers should consider the possibility of an accident occuring in the design of a high-temperature furnace. If a cooling water line breaks, causing water to spray on the Al_2O_3 furnace tube at 1000°C, estimate the temperature drop that will cause the furnace tube to crack.

SOLUTION

Figure 7–8 gives the appropriate plot for Al_2O_3 at 1000°C. In the range of $r_m h$ around 0.2, a drop of

$$T_0 - T' \simeq 50^\circ \text{C}$$

will cause a thermal shock failure.

....

PRACTICE PROBLEM 7.4

In Sample Problem 7.4 the stress in an Al_2O_3 tube is calculated as a result of constrained heating to 1000°C. To what temperature could the furnace tube be heated to be stressed to an acceptable (but not necessarily desirable) compressive stress of 2100 MPa?

D   PRACTICE PROBLEM 7.5

In Sample Problem 7.5 a temperature drop of approximately 50°C caused by a water spray is seen to be sufficient to fracture an Al_2O_3 furnace tube originally at 1000°C. Approximately what temperature drop due to a 2.5-lb/(s · ft^2) airflow would cause a fracture?

SUMMARY

A variety of properties describe the way in which materials respond to the application of heat. The heat capacity indicates the amount of heat necessary to raise the temperature of a given amount of material. The term *specific heat* is used when the property is determined for a unit mass of the material. Fundamental understanding of the mechanism of heat absorption by atomic vibrations leads to a useful rule of thumb for estimating heat capacity of materials at room temperature and above ($C_p \approx C_V \approx 3R$).

The increasing vibration of atoms with increasing temperature leads to increasing interatomic separations and, generally, a positive coefficient of thermal expansion. A careful inspection of the relationship of this expansion to the atomic bonding energy curve reveals that strong bonding correlates with low thermal expansion as well as high elastic modulus and high melting point.

Heat conduction in materials can be described with a thermal conductivity, k, in the same way as mass transport was described in Chapter 5 using the diffusivity, D. The mechanism of thermal conductivity in metals is largely associated with their conductive electrons, whereas the mechanism for ceramics and polymers is largely associated with atomic vibrations. Due to the wavelike nature of both mechanisms, increasing temperature and structural disorder both tend to diminish thermal conductivity. Porosity is especially effective in diminishing thermal conductivity.

The inherent brittleness of ceramics and glasses, combined with thermal expansion mismatch or low thermal conductivities, can lead to mechanical failure by thermal shock. Sudden cooling is especially effective in creating excessive surface tensile stress and subsequent fracture.

KEY TERMS

Debye temperature (253)
Fourier's law (258)
heat capacity (252)

linear coefficient of thermal expansion (255)
specific heat (252)
thermal conductivity (258)

thermal shock (262)

REFERENCES

Bird, R. B., W. E. Stewart, and **E. N. Lightfoot,** *Transport Phenomena,* John Wiley & Sons, Inc., New York, 1960.

Chiang, Y., D. P. Birnie III, and **W. D. Kingery,** *Physical Ceramics,* John Wiley & Sons, Inc., New York, 1997.

Kubaschewski, O., C. B. Alcock, and **P. J. Spencer,** *Materials Thermochemistry,* Oxford and Pergamon Press, New York, 1993.

PROBLEMS

7.1 • Heat Capacity

7.1. Estimate the amount of heat (in J) required to raise 2 kg of (a) α-iron, (b) graphite, and (c) polypropylene from room temperature (25°C) to 100°C.

7.2. The specific heat of silicon is 702 J/kg·K. How many J of heat are required to raise the temperature of a silicon chip (of volume = 6.25×10^{-9} m^3) from room temperature (25°C) to 35°C?

7.3. A house designed for passive solar heating has a substantial amount of brickwork in its interior to serve as a heat absorber. Each brick weighs 2.0 kg and has a specific heat of 850 J/kg·K. How many bricks are needed to absorb 5.0×10^4 kJ of heat by a temperature increase of 10°C?

7.4. How many liters of water would be required to provide the same heat storage as the bricks in Problem 7.3? The specific heat of water is 1.0 cal/g·K, and its density is 1.0 Mg/m^3. (Note that 1 liter = 10^{-3} m^3.)

7.2 • Thermal Expansion

7.5. A 0.01-m-long bar of nickel is placed in a laboratory furnace and heated from room temperature (25°C) to 500°C. What will be the length of the bar at 500°C? (Take the coefficient of thermal expansion over this temperature range to be the average of the two values given in Table 7.2.)

7.6. Repeat Problem 7.5 for the case of a tungsten bar of the same length heated over the same temperature range.

7.7. At room temperature (25°C), a 5.000-mm-diameter tungsten pin is too large for a 4.999-mm-diameter hole in a nickel bar. To what temperature must these two parts be heated in order for the pin to just fit?

7.8. The thermal expansion of aluminum is plotted against temperature in Figure 5–4. Measure the plot at 800 K and see how well the result compares with the datum in Table 7.2.

7.3 • Thermal Conductivity

7.9. Calculate the rate of heat loss per square meter through the fireclay refractory wall of a furnace operated at 1000°C. The external face of the furnace wall is at 100°C, and the wall is 10 cm thick.

7.10. Repeat Problem 7.9 for a 5-cm-thick refractory wall.

7.11. Repeat Problem 7.9 for a 10-cm-thick mullite refractory wall.

7.12. Calculate the rate of heat loss per cm^2 through a stabilized zirconia lining of a high-temperature laboratory furnace operated at 1400°C. The external face of the lining is at 100°C and its thickness is 1 cm. (Assume the data for stabilized zirconia in Table 7.4 are linear with temperature and can be extrapolated to 1400°C.)

7.4 • Thermal Shock

7.13. What would be the stress developed in a mullite furnace tube constrained in the way illustrated in Figure 7–6 if it were heated to 1000°C?

7.14. Repeat Problem 7.13 for magnesia (MgO).

7.15. Repeat Problem 7.13 for silica glass.

• 7.16. A textbook on the mechanics of materials gives the following expression for the stress due to thermal expansion mismatch in a coating of (thickness a) on a substrate (of thickness b) at a temperature T:

$$\sigma = \frac{E}{1-\nu}(T_0 - T)(\alpha_c - \alpha_s)\left[1 - 3\left(\frac{a}{b}\right) + 6\left(\frac{a}{b}\right)^2\right]$$

where E and ν are the elastic modulus and Poisson's ratio of the coating, respectively; T_0 is the temperature at which the coating is applied (and the coating stress is initially zero); and α_c and α_s are the thermal expansion coefficients of the coating and the substrate, respectively. Calculate the room-temperature (25°C) stress in a thin soda–lime–silica glaze applied at 1000°C on a porcelain ceramic. (Take $E = 65 \times 10^3$ MPa and $\nu = 0.24$ and see Table 7.2 for relevant thermal expansion data.)

• **7.17.** Repeat Problem 7.16 for a special high-silica glaze with an average thermal expansion coefficient of 3×10^{-6}°C^{-1}. (Take $E = 72 \times 10^3$ MPa and $\nu = 0.24$.)

D **7.18.** **(a)** A processing engineer suggests that a fused SiO$_2$ crucible be used for a water quench from 500°C. Would you endorse this plan? Explain. **(b)** Another processing engineer suggests that a porcelain crucible be used for the water quench from 500°C. Would you endorse this plan? Again explain.

D **7.19.** In designing an automobile engine seal made of stabilized zirconia, an engineer must consider the possibility of the seal being subjected to a sudden spray of cooling oil corresponding to a heat transfer parameter $(r_m h)$ of 0.1 (see Figure 7–8). Will a temperature drop of 30°C fracture this seal?

D **7.20.** For the stabilized zirconia described in Problem 7.19, will a temperature drop of 100°C fracture the seal?

CHAPTER **8**
Failure Analysis and Prevention

The repetitive loading of engineering materials opens up additional opportunities for structural failure. Shown here is a mechanical testing machine, introduced in Chapter 6, modified to provide rapid cycling of a given level of mechanical stress. The resulting fatigue failure is a major concern for design engineers. (Courtesy of Instron Corporation)

In Chapters 6 and 7, we saw numerous examples of the failure of engineering materials. At room temperature, metal alloys and polymers stressed beyond their elastic limit eventually fracture following a period of nonlinear plastic deformation. Brittle ceramics and glasses typically break following elastic deformation, without plastic deformation. The inherent brittleness of ceramics and glasses combined with their common use at high temperatures make thermal shock a major concern. With continuous service at relatively high temperatures, any engineering material can fracture when creep deformation reaches its limit.

In this chapter, we will look at additional ways in which materials fail. For the rapid application of stress to materials with preexisting surface flaws, the measurement of the *impact energy* corresponds to the measurement of toughness, or area under the stress-versus-strain curve. Monitoring impact energy as a function of temperature reveals that, for bcc metal alloys, there is a distinctive *ductile-to-brittle transition temperature*, below which otherwise ductile materials can fail in a catastrophic, brittle fashion.

The general analysis of the failure of structural materials with preexisting flaws is termed *fracture mechanics*. The key material property coming from fracture mechanics is *fracture toughness*, which is large for materials such as pressure-vessel steels and small for brittle materials such as typical ceramics and glasses.

Under cyclic loading conditions, otherwise ductile metal alloys and engineering polymers can eventually fail in a brittle fashion, a phenomenon appropriately termed *fatigue*. Ceramics and glasses can exhibit *static fatigue* without cyclic loading due to a chemical reaction with atmospheric moisture.

Nondestructive testing, the evaluation of engineering materials without impairing their future usefulness, is an important technology for identifying microstructural flaws in engineering systems. Since such flaws, including surface and internal cracks, play a central role in the failure of materials, nondestructive testing is a critical component of failure analysis and prevention programs. *Failure analysis* can be defined as the systematic study of the nature of the various modes of material failure. The related goal of *failure prevention* is to apply the understanding provided by failure analysis to avoid future disasters.

8.1   IMPACT ENERGY

In Section 6.4, hardness was seen to be the analog of strength measured by the tensile test. **Impact energy**, the energy necessary to fracture a standard test piece under an impact load, is a similar analog of toughness. The most common laboratory measurement of impact energy is the **Charpy*   test**,

---

*   Augustin Georges Albert Charpy (1865–1945), French metallurgist. Trained as a chemist, Charpy became one of the pioneering metallurgists of France and was highly productive in this field. He developed the first platinum resistance furnace and the silicon steel routinely used in modern electrical equipment, as well as the impact test that bears his name.

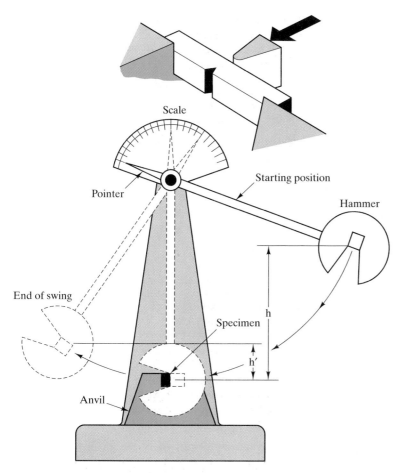

Figure 8-1 *Charpy test of impact energy. (From H. W. Hayden, W. G. Moffatt, and J. Wulff,* The Structure and Properties of Materials, *Vol. 3:* Mechanical Behavior, *John Wiley & Sons, Inc., New York, 1965.)*

illustrated in Figure 8–1. The test principle is straightforward. The energy necessary to fracture the test piece is directly calculated from the difference in initial and final heights of the swinging pendulum. To provide control over the fracture process, a stress-concentrating notch is machined into the side of the sample subjected to maximum tensile stress. The net test result is to subject the sample to elastic deformation, plastic deformation, and fracture in rapid succession. Although rapid, the deformation mechanisms involved are the same as those involved in tensile testing the same material. The load impulse must approach the ballistic range before fundamentally different mechanisms come into play.

In effect, a Charpy test takes the tensile test to completion very rapidly. The impact energy from the Charpy test correlates with the area under the total stress–strain curve (i.e., toughness). Table 8.1 gives Charpy impact en-

Table 8.1 *Impact Test (Charpy) Data for some of the Alloys of Table 6.1*

	Alloy	Impact energy [J (ft·lb)]
1.	1040 carbon steel	180 (133)
2.	8630 low-alloy steel	55 (41)
3.	c. 410 stainless steel	34 (25)
4.	L2 tool steel	26 (19)
5.	Ferrous superalloy (410)	34 (25)
6.	a. Ductile iron, quench	9 (7)
7.	b. 2048, plate aluminum	10.3 (7.6)
8.	a. AZ31B magnesium	4.3 (3.2)
	b. AM100A casting magnesium	0.8 (0.6)
9.	a. Ti–5Al–2.5Sn	23 (17)
10.	Aluminum bronze, 9% (copper alloy)	48 (35)
11.	Monel 400 (nickel alloy)	298 (220)
13.	50:50 solder (lead alloy)	21.6 (15.9)
14.	Nb–1 Zr (refractory metal)	174 (128)

ergy data for the alloys of Table 6.1. In general, we expect alloys with large values of both strength (Y.S. and T.S.) and ductility (percent elongation at fracture) to have large impact fracture energies. Although this is frequently so, the impact data are sensitive to test conditions. For instance, increasingly sharp notches can give lower impact energy values due to the stress concentration effect at the notch tip. The nature of stress concentration at notch and crack tips is explored further in the next section.

Impact energy data for a variety of polymers are given in Table 8.2. For polymers, the impact energy is typically measured with the **Izod* test** rather than the Charpy. These two standardized tests differ primarily in the configuration of the notched test specimen. Impact test temperature can also be a factor. Face-centered cubic (fcc) alloys generally show ductile fracture modes in Charpy testing, and hexagonal close-packed (hcp) alloys are generally brittle (Figure 8–2). However, body-centered cubic (bcc) alloys show a dramatic variation in fracture mode with temperature. In general, they fail in a brittle mode at relatively low temperatures and in a ductile mode at relatively high temperatures. Figure 8–3 shows this behavior for two series of low-carbon steels. The ductile-to-brittle transition for bcc alloys can be considered a manifestation of the slower dislocation mechanics for these alloys compared to that for fcc and hcp alloys. (In bcc metals, slip occurs on non-close-packed planes.) Increasing yield strength combined with decreasing dislocation velocities at decreasing temperatures eventually leads to brittle fracture. The microscopic fracture surface of the high-temperature ductile failure has a dimpled texture with many cuplike projections of deformed metal, and brittle fracture is characterized by cleavage

---

* E. G. Izod, "Testing Brittleness of Steels," *Engr. 25* (September 1903).

Table 8.2 *Impact Test (Izod) Data for Various Polymers*

Polymer	Impact energy [J (ft·lb)]
General-use polymers	
Polyethylene	
High-density	1.4–16 (1–12)
Low-density	22 (16)
Polyvinylchloride	1.4 (1)
Polypropylene	1.4–15 (1–11)
Polystyrene	0.4 (0.3)
Polyesters	1.4 (1)
Acrylics (Lucite)	0.7 (0.5)
Polyamides (nylon 66)	1.4 (1)
Cellulosics	3–11 (2–8)
Engineering polymers	
ABS	1.4–14 (1–10)
Polycarbonates	19 (14)
Acetals	3 (2)
Polytetrafluoroethylene (Teflon)	5 (4)
Thermosets	
Phenolics (phenolformaldehyde)	0.4 (0.3)
Urea-melamine	0.4 (0.3)
Polyesters	0.5 (0.4)
Epoxies	1.1 (0.8)

Source: From data collections in R. A. Flinn and P. K. Trojan, *Engineering Materials and Their Applications,* 2nd ed., Houghton Mifflin Company, Boston, 1981; M. F. Ashby and D. R. H. Jones, *Engineering Materials,* Pergamon Press, Inc., Elmsford, N.Y., 1980; and *Design Handbook for Du Pont Engineering Plastics.*

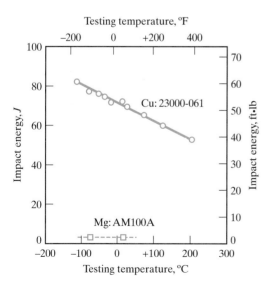

Figure 8-2 *Impact energy for a ductile fcc alloy (copper C23000–061, "red brass") is generally high over a wide temperature range. Conversely, the impact energy for a brittle hcp alloy (magnesium AM100A) is generally low over the same range. (From* Metals Handbook, *9th Ed., Vol. 2, American Society for Metals, Metals Park, Ohio, 1979.)*

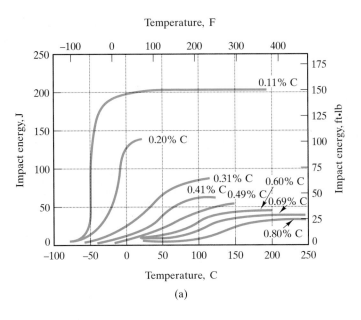

(a)

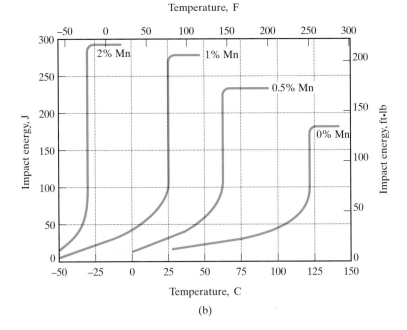

(b)

Figure 8-3 *Variation in ductile-to-brittle transition temperature with alloy composition. (a) Charpy V-notch impact energy with temperature for plain-carbon steels with various carbon levels (in weight percent). (b) Charpy V-notch impact energy with temperature for Fe–Mn–0.05C alloys with various manganese levels (in weight percent). (From* Metals Handbook, *9th Ed., Vol. 1, American Society for Metals, Metals Park, Ohio, 1978.)*

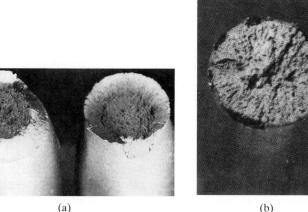

Figure 8-4 *(a) Typical "cup and cone" ductile fracture surface. Fracture originates near the center and spreads outward with a dimpled texture. Near the surface, the stress state changes from tension to shear with fracture continuing at approximately 45°. (From Metals Handbook, 9th Ed., Vol. 12, ASM International, Metals Park, Ohio, 1987.) (b) Typical cleavage texture of brittle fracture surface. (From Metals Handbook, 9th Ed., Vol. 11, American Society Metals, Metals Park, Ohio, 1986.)*

(a)   (b)

surfaces (Figure 8–4). Near the transition temperature between brittle and ductile behavior, the fracture surface exhibits a mixed texture. The **ductile-to-brittle transition temperature** is of great practical importance. The alloy that exhibits a ductile-to-brittle transition loses toughness and is susceptible to catastrophic failure below this transition temperature. Because a large fraction of the structural steels are included in the bcc alloy group, the ductile-to-brittle transition is a design criterion of great importance. The transition temperature can fall between roughly −100 and +100°C, depending on alloy composition and test conditions. Several disastrous failures of Liberty ships occurred during World War II because of this phenomenon. Some literally split in half. Low-carbon steels that were ductile in room-temperature tensile tests became brittle when exposed to lower-temperature ocean environments. Figure 8–3 shows how alloy composition can dramatically shift the transition temperature. Such data are an important guide in material selection.

D SAMPLE PROBLEM 8.1

You are required to use a furnace-cooled Fe–Mn–0.05 C alloy in a structural design that may see service temperatures as low as 0°C. Suggest an appropriate Mn content for the alloy.

SOLUTION

Figure 8–3 provides the specific guidance we need. A 1% Mn alloy is relatively brittle at 0°C, whereas a 2% Mn alloy is highly ductile. Therefore, a secure choice (based on notch toughness considerations *only*) would be

$$Mn \text{ content} = 2\%$$

🔲 PRACTICE PROBLEM 8.1

Find the necessary carbon level to ensure that a plain-carbon steel will be relatively ductile down to 0°C. (See Sample Problem 8.1.)

8.2 FRACTURE TOUGHNESS

A substantial effort has been made to quantify the nature of material failures such as the Liberty ship disasters just described. The term **fracture mechanics** has come to mean the general analysis of failure of structural materials with preexisting flaws. This is a broad field that is the focus of much active research. We shall concentrate on a material property that is the most widely used single parameter from fracture mechanics. **Fracture toughness** is represented by the symbol K_{IC} (pronounced "kay-one-cee") and is the critical value of the stress-intensity factor at a crack tip necessary to produce catastrophic failure under simple uniaxial loading. The subscript "I" stands for "mode I" (uniaxial) loading and "C" stands for "critical." A simple example of the concept of fracture toughness comes from blowing up a balloon containing a small pinhole. When the internal pressure of the balloon reaches a critical value, catastrophic failure originates at the pinhole (i.e., the balloon pops). In general, the value of fracture toughness is given by

$$K_{IC} = Y\sigma_f\sqrt{\pi a} \tag{8.1}$$

where Y is a dimensionless geometry factor on the order of 1, σ_f is the overall applied stress at failure, and a is the length of a surface crack (or one-half the length of an internal crack). Fracture toughness (K_{IC}) has units of MPa $\sqrt{m}$. Figure 8–5 shows a typical measurement of K_{IC}, and Table 8.3 gives values for various materials. It must be noted that K_{IC} is associated with so-called plane strain conditions in which the specimen thickness (Figure 8–5) is relatively large compared with the notch dimension. For thin specimens ("plane stress" conditions), fracture toughness is denoted K_C and is a sensitive function of specimen thickness. Plane strain conditions generally prevail when thickness $\geq 2.5(K_{IC}/\text{Y.S.})^2$.

The microscopic concept of toughness indicated by K_{IC} is consistent with that expressed by the macroscopic measurements of tensile and impact testing. Highly brittle materials, with little or no ability to deform plastically in the vicinity of a crack tip, have low K_{IC} values and are susceptible to catastrophic failures. By contrast, highly ductile alloys can undergo substantial plastic deformation on both a microscopic and a macroscopic scale prior to fracture. The major use of fracture mechanics in metallurgy is to characterize those alloys of intermediate ductility that can undergo catastrophic failure below their yield strength due to the stress-concentrating effect of structural flaws. In designing a pressure vessel, for example, it is convenient

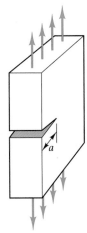

Figure 8-5 *Fracture toughness test.*

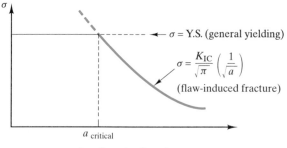

σ = Y.S. (general yielding)

$$\sigma = \frac{K_{IC}}{\sqrt{\pi}} \left(\frac{1}{\sqrt{a}} \right)$$

(flaw-induced fracture)

a critical

Log flaw size (log a)

Figure 8-6   *A design plot of stress versus flaw size for a pressure vessel material in which general yielding occurs for flaw sizes less than a critical size, a_{critical}, but catastrophic "fast fracture" occurs for flaws larger than a_{critical}.*

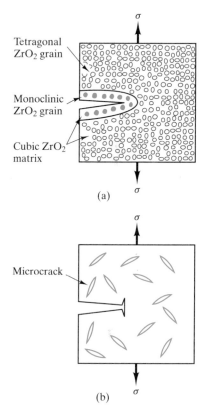

Figure 8-7 *Two mechanisms for improving fracture toughness of ceramics by crack arrest. (a) Transformation toughening of partially stabilized zirconia involves the stress-induced transformation of tetragonal grains to the monoclinic structure which has a larger specific volume. The result is a local volume expansion at the crack tip, squeezing the crack shut and producing a residual compressive stress. (b) Microcracks produced during fabrication of the ceramic can blunt the advancing crack tip.*

to plot operating stress (related to operating pressure) as a function of flaw size. (It is usually possible to ensure that flaws above a given size are not present by a careful inspection program involving nondestructive testing techniques (see Section 8.4).) **General yielding** (independent of a flaw) was covered in Section 6.1. **Flaw-induced fracture** is described by Equation 8.1. Taking Y in that equation as 1 gives the schematic design plot of Figure 8–6. An important practical point about the design plot is that failure by general yielding is preceded by observable deformation, whereas flaw-induced fracture occurs rapidly with no such warning. As a result, flaw-induced fracture is sometimes referred to as "**fast fracture.**"

In the past two decades, major progress has been made in improving the fracture toughness and, hence, the range of applications of structural ceramics. Figure 8–7 summarizes two microstructural techniques for significantly raising fracture toughness. Figure 8–7a illustrates the mechanism of **transformation toughening** in partially stabilized zirconia (PSZ). Having second-phase particles of tetragonal zirconia in a matrix of cubic zirconia is the key to improved toughness. A propagating crack creates a local stress field that induces a transformation of tetragonal zirconia particles to the monoclinic structure in that vicinity. The slightly larger specific volume of the monoclinic phase causes an effective compressive load locally and, in turn, the "squeezing" of the crack shut. Another technique of crack arrest is shown in Figure 8–7b. Microcracks purposely introduced by internal stresses during processing of the ceramic are available to blunt the tip of an advancing crack. The expression associated with Griffith cracks (Equation 6.1) indicates that the larger tip radius can dramatically reduce the local stress at the crack tip. Another technique, involving reinforcing fibers, will be discussed in Chapter 14 relative to ceramic-matrix composites.

The absence of plastic deformation in traditional ceramics and glass on the macroscopic scale (the stress–strain curve) is matched by a similar absence on the microscopic scale. This is reflected in the characteristically low fracture toughness (K_{IC}) values (≤ 5 MPa $\sqrt{m}$) for traditional ceramics and glass as shown in Table 8.3. Most K_{IC} values are lower than those of the most brittle metals listed there. Only the recently developed transformation-toughened PSZ is competitive with some of the moderate toughness metal alloys. Further improvement in toughness will be demonstrated by some ceramic-matrix composites in Chapter 14.

Table 8.3 *Typical Values of Fracture Toughness (K_{IC}) for Various Materials*

Material	K_{IC} (MPa $\sqrt{m}$)
Metal or alloy	
Mild steel	140
Medium-carbon steel	51
Rotor steels (A533; Discalloy)	204–214
Pressure-vessel steels (HY130)	170
High-strength steels (HSS)	50–154
Cast iron	6–20
Pure ductile metals (e.g., Cu, Ni, Ag, Al)	100–350
Be (brittle, hcp metal)	4
Aluminum alloys (high strength–low strength)	23–45
Titanium alloys (Ti–6Al–4V)	55–115
Ceramic or glass	
Partially stabilized zirconia	9
Electrical porcelain	1
Alumina (Al_2O_3)	3–5
Magnesia (MgO)	3
Cement/concrete, unreinforced	0.2
Silicon carbide (SiC)	3
Silicon nitride (Si_3N_4)	4–5
Soda glass ($Na_2O–SiO_2$)	0.7–0.8
Polymer	
Polyethylene	
High-density	2
Low-density	1
Polypropylene	3
Polystyrene	2
Polyesters	0.5
Polyamides (nylon 66)	3
ABS	4
Polycarbonates	1.0–2.6
Epoxy	0.3–0.5

Source: Data from M. F. Ashby and D. R. H. Jones, *Engineering Materials—An Introduction to Their Properties and Applications,* Pergamon Press, Inc., Elmsford, N.Y., 1980; GTE Laboratories, Waltham, Mass., and *Design Handbook for Dupont Engineering Plastics.*

SAMPLE PROBLEM 8.2

A high-strength steel has a yield strength of 1460 MPa and a K_{IC} of 98 MPa $\sqrt{m}$. Calculate the size of a surface crack that will lead to catastrophic failure at an applied stress of $\frac{1}{2}$ Y.S.

SOLUTION

We may use Equation 8.1 with the realization that we are assuming an ideal case of plane strain conditions. In lieu of specific geometrical information, we are forced to take $Y = 1$. Within these limitations, we can calculate

$$K_{IC} = Y\sigma_f\sqrt{\pi a}$$

With $Y = 1$ and $\sigma_f = 0.5$ Y.S.,

$$K_{IC} = 0.5\text{Y.S.}\sqrt{\pi a}$$

or

$$
\begin{aligned}
a &= \frac{1}{\pi}\frac{K_{IC}^2}{(0.5\text{Y.S.})^2} \\
&= \frac{1}{\pi}\frac{(98\,\text{MPa}\sqrt{\text{m}})^2}{[0.5(1460\,\text{MPa})]^2} \\
&= 5.74 \times 10^{-3}\,\text{m} \\
&= 5.74\,\text{mm}
\end{aligned}
$$

SAMPLE PROBLEM 8.3

Given that a quality-control inspection can ensure that a structural ceramic part will have no flaws greater than 25 μm in size, calculate the maximum service stress available with **(a)** SiC and **(b)** partially stabilized zirconia.

SOLUTION

In lieu of more specific information, we can treat this as a general fracture mechanics problem using Equation 8.1 with $Y = 1$, in which case

$$\sigma_f = \frac{K_{IC}}{\sqrt{\pi a}}$$

This assumes that the maximum service stress will be the fracture stress for a part with flaw size $= a = 25\mu$m. Values of K_{IC} are given in Table 8.3.

(a) For SiC,

$$\sigma_f = \frac{3 \text{ MPa}\sqrt{m}}{\sqrt{\pi \times 25 \times 10^{-6} \text{ m}}} = 339 \text{ MPa}$$

(b) For PSZ,

$$\sigma_f = \frac{9 \text{ MPa}\sqrt{m}}{\sqrt{\pi \times 25 \times 10^{-6} \text{ m}}} = 1020 \text{ MPa}$$

··

PRACTICE PROBLEM 8.2

What crack size is needed to produce catastrophic failure in the alloy in Sample Problem 8.2 at **(a)** $\frac{1}{3}$ Y.S. and **(b)** $\frac{3}{4}$ Y.S.?

PRACTICE PROBLEM 8.3

In Sample Problem 8.3, maximum service stress for two structural ceramics is calculated based on the assurance of no flaws greater than 25 μm in size. Repeat these calculations given that a more economical inspection program can only guarantee detection of flaws greater than 100 μm in size.

8.3 FATIGUE

Up to this point, we have characterized the mechanical behavior of metals under a single load application either slowly (e.g., the tensile test) or rapidly (e.g., the impact test). Many structural applications involve cyclic rather than static loading, and a special problem arises. **Fatigue** is the general phenomenon of material failure after several cycles of loading to a stress level *below* the ultimate tensile stress (Figure 8–8). Figure 8–9 illustrates a common laboratory test used to cycle a test piece rapidly to a predetermined stress level. A typical **fatigue curve** is shown in Figure 8–10. This plot of stress (S) versus number of cycles (N), on a logarithmic scale, at a given stress is also called the *S–N curve*. The data indicate that while the material can withstand a stress of 800 MPa (T.S.) in a single loading ($N = 1$), it fractures after 10,000 applications ($N = 10^4$) of a stress less than 600 MPa. The reason for this decay in strength is a subtle one. Figure 8–11 shows how repeated stress applications can create localized plastic deformation at the metal surface, eventually manifesting as sharp discontinuities (extrusions and intrusions). These intrusions, once formed, continue to grow into cracks, reducing the load-carrying ability of the material and serving as stress concentrators (see the preceding section).

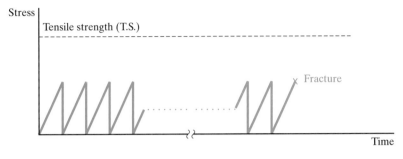

Figure 8-8 *Fatigue corresponds to the brittle fracture of an alloy after a total of N cycles to a stress below the tensile strength.*

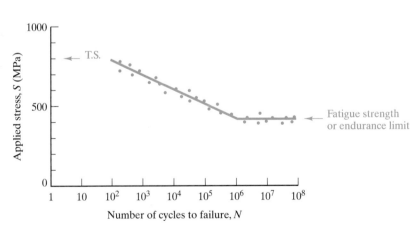

Figure 8-9 *Fatigue test. (From C. A. Keyser,* Materials Science in Engineering, *4th Ed., Charles E. Merrill Publishing Company, Columbus, Ohio, 1986.)*

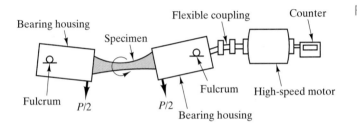

Figure 8-10 *Typical fatigue curve. (Note that a log scale is required for the horizontal axis.)*

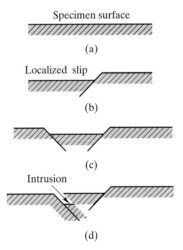

Figure 8-11 *An illustration of how repeated stress applications can generate localized plastic deformation at the alloy surface leading eventually to sharp discontinuities.*

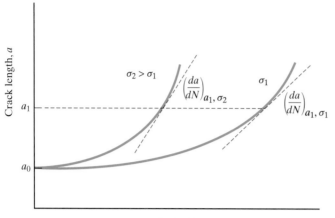

Figure 8-12 *Illustration of crack growth with number of stress cycles, N, at two different stress levels. Note that, at a given stress level, the crack growth rate, da/dN, increases with increasing crack length, and, for a given crack length such as a_1, the rate of crack growth is significantly increased with increasing magnitude of stress.*

Fracture mechanics studies of cyclic loading provide substantial, quantitative understanding of the nature of crack growth. In particular, crack growth continues until the crack length reaches the critical value as defined by Equation 8.1 and Figure 8–6.

At low stress levels or for small crack sizes, preexisting cracks do not grow during cyclic loading. Figure 8–12 shows how, once the stress exceeds some threshold value, crack length increases, as indicated by the slope of the plot (da/dN), the rate of crack growth. Figure 8–12 also shows that, at a given stress level, the crack growth rate increases with increasing crack length, and, for a given crack length, the rate of crack growth is significantly increased with increasing magnitude of stress. The overall growth of a fatigue crack as a function of the stress intensity factor, K, is illustrated in Figure 8–13. Region I in Figure 8–13 corresponds to the absence of crack growth mentioned earlier in conjunction with low stress and/or small cracks. Region II corresponds to the relationship

$$(da/dN) = A(\Delta K)^m \qquad (8.2)$$

where A and m are material parameters dependent on environment, test frequency, and the ratio of minimum and maximum stress applications, and ΔK is the stress intensity factor range at the crack tip. Relative to Equation 8.1,

$$\Delta K = K_{max} - K_{min}$$
$$= Y \Delta\sigma \sqrt{\pi a} = Y(\sigma_{max} - \sigma_{min})\sqrt{\pi a} \qquad (8.3)$$

In Equations 8.2 and 8.3, one should note that the K is the more general **stress intensity factor** rather than the more specific fracture toughness, K_{Ic}, and N is the number of cycles associated with a given crack length prior to failure rather than the total number of cycles to fatigue failure associated

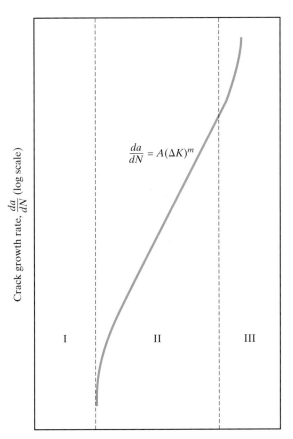

$$\frac{da}{dN} = A(\Delta K)^{m}$$

Crack growth rate, $\frac{da}{dN}$ (log scale)

I   II   III

Stress intensity factor range, ΔK (log scale)

Figure 8-13 *Illustration of logarithmic relationship between crack growth rate, da/dN, and the stress intensity factor range, ΔK. Region I corresponds to nonpropagating fatigue cracks. Region II corresponds to a linear relationship between log da/dN and log ΔK. Region III represents unstable crack growth prior to catastrophic failure.*

with an $S-N$ curve. In Region II of Figure 8–13, Equation 8.2 implies a linear relationship between the logarithm of crack growth rate, da/dN, and the stress intensity factor range, ΔK, with the slope being m. Values of m typically range between 1 and 6. Region III corresponds to accelerated crack growth just prior to fast fracture.

The fatigue fracture surface has a characteristic texture shown in Figure 8–14. The smoother portion of the surface is referred to as "clamshell" or "beachmark" texture. The concentric line pattern is a record of the slow, cyclic buildup of crack growth from a surface intrusion. The granular portion of the fracture surface identifies the rapid crack propagation at the time of catastrophic failure. Even for normally ductile materials, fatigue failure can occur by a characteristically brittle mechanism.

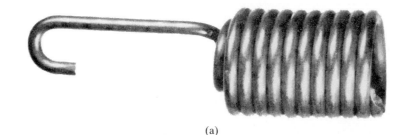

(a)

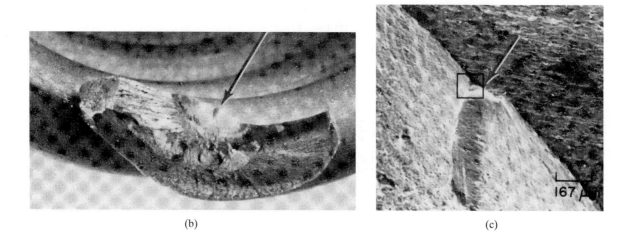

(b)   (c)

Figure 8-14 *Characteristic fatigue fracture surface. (a) Photograph of an aircraft throttle-control spring ($1\frac{1}{2}\times$) that broke in fatigue after 274 h of service. The alloy is 17–7PH stainless steel. (b) Optical micrograph ($10\times$) of the fracture origin (arrow) and the adjacent smooth region containing a concentric line pattern as a record of cyclic crack growth (an extension of the surface discontinuity shown in Figure 8–11). The granular region identifies the rapid crack propagation at the time of failure. (c) Scanning electron micrograph ($60\times$), showing a closeup of the fracture origin (arrow) and adjacent "clamshell" pattern. (From* Metals Handbook, 8th Ed., Vol. 9: Fractography and Atlas of Fractographs, *American Society for Metals, Metals Park, Ohio, 1974.)*

Figure 8–10 showed that the decay in strength with increasing numbers of cycles reaches a limit. This **fatigue strength,** or *endurance limit,* is characteristic of ferrous alloys. Nonferrous alloys tend not to have such a distinct limit, although the rate of decay decreases with N (Figure 8–15). As a practical matter, the fatigue strength of a nonferrous alloy is defined as the strength value after an arbitrarily large number of cycles (usually $N = 10^8$ as illustrated in Figure 8–15). Fatigue strength usually falls between one-fourth and one-half of the tensile strength. Table 8.1 and Figure 8–16 illustrate this for the alloys of Table 6.1. For a given alloy, the resistance to fatigue will be increased by prior mechanical deformation (cold working) or reduction of structural discontinuities (Figure 8–17).

Metal fatigue has been defined as a loss of strength created by microstructural damage generated during cyclic loading. The fatigue phenomenon is also observed for ceramics and glasses, but *without* cyclic loading. The reason is that a chemical rather than mechanical mechanism is involved.

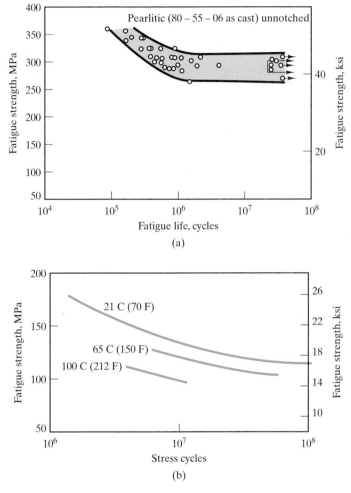

(a)

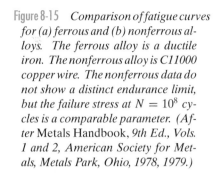

Figure 8-15 *Comparison of fatigue curves for (a) ferrous and (b) nonferrous alloys. The ferrous alloy is a ductile iron. The nonferrous alloy is C11000 copper wire. The nonferrous data do not show a distinct endurance limit, but the failure stress at $N = 10^8$ cycles is a comparable parameter. (After* Metals Handbook, *9th Ed., Vols. 1 and 2, American Society for Metals, Metals Park, Ohio, 1978, 1979.)*

(b)

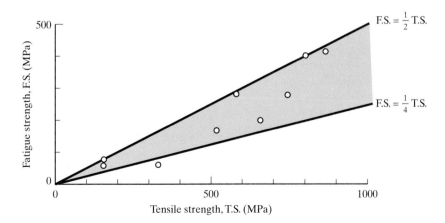

Figure 8-16 *Plot of data from Table 8.4 showing how fatigue strength is generally one-fourth to one-half of the tensile strength.*

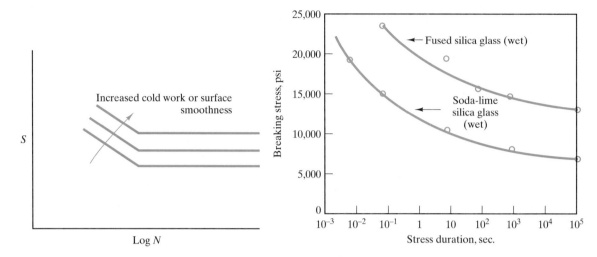

Figure 8-17 *Fatigue strength is increased by prior mechanical deformation or reduction of structural discontinuities.*

Figure 8-18 *The drop in strength of glasses with duration of load (and* without *cyclic load applications) is termed* static fatigue. *(From W. D. Kingery,* Introduction to Ceramics, *John Wiley & Sons, Inc., New York, 1960.)*

Figure 8–18 illustrates the phenomenon of **static fatigue** for common silicate glasses. Two key observations can be made about this phenomenon: (1) It occurs in water-containing environments, and (2) it occurs around room temperature. The role of water in static fatigue is shown in Figure 8–19. By chemically reacting with the silicate network, an H_2O molecule generates two Si–OH units. The hydroxyl units are not bonded to each other, leaving a break in the silicate network. When this reaction occurs at the tip of a surface crack, the crack is lengthened by one atomic-scale step.

Table 8.4 *Comparison of Fatigue Strength (F.S.) and Tensile Strength (T.S.) for some of the Alloys of Table 6.1*

	Alloy	F.S. (MPa)	T.S. (MPa)
1.	1040 carbon steel	280	750
2.	8630 low-alloy steel	400	800
3.	a. 304 stainless steel		515
3.	b. 304 stainless steel	170	
7.	a. 3003-H14 aluminum	62	150
8.	b. AM100A casting magnesium	69	150
9.	a. Ti–5Al–2.5Sn	410	862
10.	Aluminum bronze, 9% (copper alloy)	200	652
11.	Monel 400 (nickel alloy)	290	579
12.	AC41A zinc	56	328

Cyclic fatigue in metals and static fatigue in ceramics are compared in Figure 8–20. Because of the chemical nature of the mechanism in ceramics and glasses, the phenomenon is found predominantly around room temperature. At relatively high temperatures (above about 150°C), the hydroxyl reaction is so fast that the effects are difficult to monitor. At those temperatures, other factors such as viscous deformation can also contribute to static fatigue. At low temperatures (below about −100°C), the rate of hydroxyl reaction is too low to produce a significant effect in practical time periods. Analogies to static fatigue in metals would be stress corrosion cracking and hydrogen embrittlement, involving crack growth mechanisms under severe environments.

Fatigue in polymers is treated in a way similar to metal alloys. Acetal polymers are noted for good fatigue resistance. Figure 8–21 summarizes S–N curves for such a material at various temperatures. The fatigue limit for polymers is generally reported at 10^6 cycles rather than 10^8 cycles, as commonly used for nonferrous alloys (e.g., Figure 8–15).

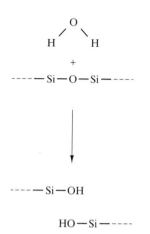

Figure 8-19 *The role of H_2O in static fatigue depends on its reaction with the silicate network. One H_2O molecule and one –Si–O–Si– segment generate two Si–OH units. This is equivalent to a break in the network.*

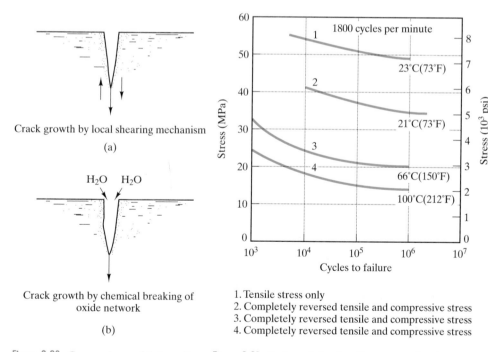

Crack growth by local shearing mechanism

(a)

Crack growth by chemical breaking of oxide network

(b)

Figure 8-20 *Comparison of (a) cyclic fatigue in metals and (b) static fatigue in ceramics.*

1. Tensile stress only
2. Completely reversed tensile and compressive stress
3. Completely reversed tensile and compressive stress
4. Completely reversed tensile and compressive stress

Figure 8-21 *Fatigue behavior for an acetal polymer at various temperatures. (From* Design Handbook for Du Pont Engineering Plastics, *used by permission.)*

SAMPLE PROBLEM 8.4

Given only that the alloy for a structural member has a tensile strength of 800 MPa, estimate a maximum permissible service stress knowing that the loading will be cyclic in nature and that a safety factor of 2 is required.

SOLUTION

If we use Figure 8–16 as a guide, a *conservative* estimate of the fatigue strength will be

$$\text{F.S.} = \tfrac{1}{4}\text{T.S.} = \tfrac{1}{4}(800 \text{ MPa}) = 200 \text{ MPa}$$

Using a safety factor of 2 will give a permissible service stress of

$$\text{service stress} = \frac{\text{F.S.}}{2} = \frac{200 \text{ MPa}}{2} = 100 \text{ MPa}$$

Note.   The safety factor helps to account for, among other things, the approximate nature of the relationship between F.S. and T.S.

SAMPLE PROBLEM 8.5

Static fatigue depends on a chemical reaction (Figure 8–19) and, as a result, is another example of Arrhenius behavior. Specifically, at a given load, the inverse of time to fracture has been shown to increase exponentially with temperature. The activation energy associated with the mechanism of Figure 8–19 is 78.6 kJ/mol. If the time to fracture for a soda–lime–silica glass is 1 s at $+50°C$ at a given load, what is the time to fracture at $-50°C$ at the same load?

SOLUTION

As stated, we can apply the Arrhenius equation (Equation 5.1). In this case,

$$t^{-1} = Ce^{-Q/RT}$$

where t is the time to fracture.
     At 50°C (323 K),

$$t^{-1} = 1\text{s}^{-1} = Ce^{-(78.6\times10^3 \text{ J/mol})/[8.314 \text{ J/mol·K}](323 \text{ K})}$$

giving

$$C = 5.15 \times 10^{12} \text{ s}^{-1}$$

Then

$$t_{-50°C}^{-1} = (5.15 \times 10^{12} \text{ s}^{-1})e^{-(78.6\times 10^3 \text{ J/mol})/[8.314 \text{ J/(mol·K)}](223 \text{ K})}$$

$$= 1.99 \times 10^{-6} \text{ s}^{-1}$$

or

$$t = 5.0 \times 10^5 \text{ s}$$

$$= 5.0 \times 10^5 \text{ s} \times \frac{1 \text{ h}}{3.6 \times 10^3 \text{ s}}$$

$$= 140 \text{ h}$$

$$= 5 \text{ days, 20 h}$$

....

PRACTICE PROBLEM 8.4

In Sample Problem 8.4, a service stress is calculated with consideration for fatigue loading. Using the same considerations, estimate a maximum permissible service stress for an 80–55–06 as-cast ductile iron with a Brinell hardness number of 200 (see Figure 6–28).

PRACTICE PROBLEM 8.5

For the system discussed in Sample Problem 8.5, what would be the time to fracture **(a)** at $0°C$ and **(b)** at room temperature, $25°C$?

8.4   NONDESTRUCTIVE TESTING

Nondestructive testing is the evaluation of engineering materials without impairing their usefulness. A central focus of many of the nondestructive testing techniques is the identification of potentially critical flaws, such as surface and internal cracks. As with fracture mechanics, nondestructive testing can serve to analyze an existing failure or it can be used to prevent future failures. The dominant techniques of this field are x-radiography and ultrasonics.

X-RADIOGRAPHY

Although diffraction allows dimensions on the order of the x-ray wavelength (typically < 1 nm) to be measured, **x-radiography** produces a "shadow-graph" of the internal structure of a part with a much coarser resolution, typically on the order of 1 mm (Figure 8–22). The medical chest x-ray is a common example. Industrial x-radiography is widely used for inspecting castings and weldments. For a given material being inspected by a given energy x-ray beam, the intensity of the beam, I, transmitted through a thickness of material, x, is given by Beer's* law:

$$I = I_0 e^{-\mu x}$$

(8.4)

where I_0 is the incident beam intensity and μ is the linear absorption coefficient for the material. The intensity is proportional to the number of photons in the beam and is distinct from the energy of photons in the beam. The absorption coefficient is a function of the beam energy and of the elemental composition of the material. Experimental values for the μ of iron as a function of energy are given in Table 8.5. There is a general drop in the magnitude of μ with increasing beam energy primarily due to mechanisms of photon absorption and scattering. The dependence of the linear

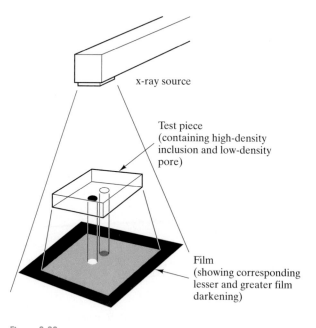

x-ray source

Test piece
(containing high-density
inclusion and low-density
pore)

Film
(showing corresponding
lesser and greater film
darkening)

Figure 8-22 *A schematic of x-radiography.*

---

* August Beer (1825–1863), German physicist. Beer graduated from the University of Bonn, where he was to remain as a teacher for the remainder of his relatively short life. He is primarily remembered for the law first stated relative to his observations on the absorption of visible light.

Table 8.5 *Linear Absorption Coefficient of Iron as a Function of X-ray Beam Energy*

Energy (MeV)	μ (mm^{-1})
0.05	1.52
0.10	0.293
0.50	0.0662
1.00	0.0417
2.00	0.0334
4.00	0.0260

Source: Selected data from Bray, D. E., and R. K. Stanley, 1989. *Nondestructive Evaluation*, McGraw-Hill Book Co., New York.

absorption coefficient on elemental composition is illustrated by the data of Table 8.6. Note that μ for a given beam energy generally increases with atomic number, causing low-atomic-number metals such as aluminum to be relatively transparent and high-atomic-number metals such as lead to be relatively opaque.

ULTRASONIC TESTING

While x-radiography is based on a portion of the electromagnetic spectrum with relatively short wavelengths in comparison to the visible region, **ultrasonic testing** is based on a portion of the acoustic spectrum (typically 1 to 25 MHz) with frequencies well above those of the audible range (20 to 20,000 Hz). An important distinction between x-radiography and ultrasonic testing is that the ultrasonic waves are mechanical in nature, requiring a transmitting medium, while electromagnetic waves can be transmitted in a vacuum. A typical ultrasonic source involving a piezoelectric transducer is shown in Figure 15–26.

Table 8.6 *Linear Absorption Coefficient of Various Elements for an X-ray Beam with Energy = 100 keV (= 0.1 MeV)*

Element	Atomic number	μ (mm^{-1})
Aluminum	13	0.0459
Titanium	22	0.124
Iron	26	0.293
Nickel	28	0.396
Copper	29	0.410
Zinc	30	0.356
Tungsten	74	8.15
Lead	82	6.20

Source: Selected data from Bray, D. E., and R. K. Stanley, 1989. *Nondestructive Evaluation*, McGraw-Hill Book Co., New York.

X-ray attenuation is a dominant factor in x-radiography, but typical engineering materials are relatively transparent to ultrasonic waves. The key factor in ultrasonic testing is the reflection of the ultrasonic waves at interfaces of dissimilar materials. The reflection coefficient, R, defined as the ratio of reflected beam intensity, I_r, to incident beam intensity, I_i, is given by

$$R = I_r/I_i = [(Z_2 - Z_1)/(Z_2 + Z_1)]^2 \qquad (8.5)$$

where Z is the acoustic impedance, defined as the product of the material's density and velocity of sound, with the subscripts 1 and 2 referring to the two dissimilar materials on either side of the interface. The high degree of reflectivity by a typical flaw, such as an internal crack, is the basis for defect inspection. Figure 8–23 illustrates a typical "pulse echo" ultrasonic

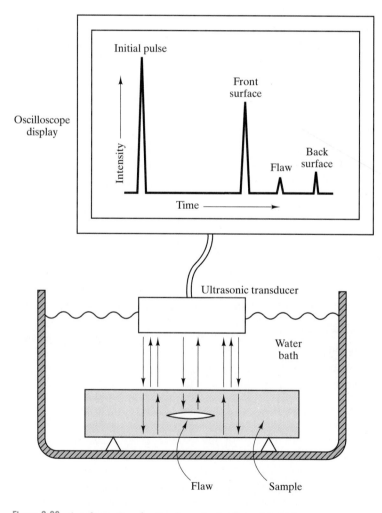

Figure 8-23 *A schematic of a "pulse echo" ultrasonic test.*

inspection. This technique is not suitable for use on complex-shaped parts, and there is a tendency for ultrasonic waves to scatter due to microstructural features such as porosity and precipitates.

OTHER NONDESTRUCTIVE TESTS

A wide spectrum of additional methods is available for nondestructive testing. Among the most widely used for failure analysis and prevention are eddy currents, magnetic particles, liquid penetrants, and acoustic emission.

1. In **eddy-current testing,** the impedance of an inspection coil is affected by the presence of an adjacent, electrically conductive test piece, in which alternating, or eddy, currents have been induced by the coil. The net impedance is a function of the composition and/or geometry of the test piece. The popularity of this test is due to its convenience, rapidity, and noncontact nature. By varying the test frequency, the method can be used for both surface and subsurface flaws. Limitations include its qualitative nature and the requirement of electrical conductivity.

2. In **magnetic-particle testing,** a fine powder of magnetic particles (Fe or Fe_3O_4) is attracted to the magnetic leakage flux around a discontinuity, such as a surface or near-surface crack in a magnetized test piece. It is a simple, traditional technique widely used because of its convenience and low cost. A primary limitation is the restriction to magnetic materials. On the other hand, an enormous volume of structural steels used in engineering is magnetic.

3. In **liquid-penetrant testing,** the capillary action of a fine powder on the surface of a sample draws out a high-visibility liquid that has previously penetrated into surface defects. Like magnetic-particle testing, it is an inexpensive and convenient technique for surface defect inspection. Liquid-penetrant testing is largely used on nonmagnetic materials for which magnetic-particle inspection is not possible. The limitations of the liquid-penetrant technique include the inability to inspect subsurface flaws and a loss of resolution on porous materials.

4. **Acoustic-emission testing** measures the ultrasonic waves produced by defects within the microstructure of a material in response to an applied stress. It has assumed a unique role in failure prevention. In addition to being able to locate defects, it can provide an early warning of impending failure due to those defects. In contrast to conventional ultrasonic testing in which a transducer provides the source of ultrasound, the material is the source of ultrasonic acoustic emissions. Transducers serve only as receivers. In general, the rate of acoustic-emission events rises sharply just prior to failure. By continuously monitoring these emissions, the structural load can be removed in time to prevent failure. A primary example of this application is in the continuous surveillance of pressure vessels.

SAMPLE PROBLEM 8.6

Calculate the fraction of x-ray beam intensity transmitted through a 10-mm-thick plate of low-carbon steel. Take the beam energy to be 100 keV. Because of the small amount of carbon and its inherently low absorption of x-rays, the steel can be approximated as elemental iron.

SOLUTION

Using Equation 8.4 and the attenuation coefficient from Table 8.6,

$$I = I_0 e^{-\mu x}$$

or

$$I/I_0 = e^{-\mu x}$$

$$= e^{-(0.293 \text{ mm}^{-1})(10 \text{ mm})}$$

$$= e^{-2.93} = 0.0534$$

SAMPLE PROBLEM 8.7

(a) Given that the velocity of sound is 6320 m/s and 5760 m/s in aluminum and 316 stainless steel, respectively, calculate the fraction of an ultrasonic pulse reflected as it travels across a bonded interface from a plate of aluminum into a plate of 316 stainless steel.

(b) Calculate the fraction reflected for the same interface but with the pulse traveling in the opposite direction, from stainless steel into aluminum. (The density of 316 stainless steel is 7.85 Mg/m^3.)

SOLUTION

(a) Using density data from the problem statement and Appendix 1, we can calculate the acoustic impedance of each medium ($Z_i = \rho_i V_i$) as

$$Z_{Al} = (2.70 \text{ Mg/m}^3)(6320 \text{ m/s})$$

$$= 17.1 \times 10^3 \text{ Mg/(m}^2 \text{ s)}$$

and

$$Z_{st} = (7.85 \text{ Mg/m}^3)(5760 \text{ m/s})$$

$$= 45.2 \times 10^3 \text{ Mg/(m}^2 \text{ s)}$$

Then Equation 8.5 gives

$$I_r/I_i = [(Z_{st} - Z_{Al})/(Z_{st} + Z_{Al})]^2$$
$$= [(45.2 - 17.1)/(45.2 + 17.1)]^2 = 0.203$$

(b) For the reverse direction of ultrasonic pulse travel,

$$I_r/I_i = [(Z_{Al} - Z_{st})/(Z_{Al} + Z_{st})]^2$$
$$= [(17.1 - 45.2)/(17.1 + 45.2)]^2 = 0.203$$

Note.   Because of the nature of Equation 8.5, the result is the same for the transmission of the ultrasonic pulse across the interface from either direction.

....

PRACTICE PROBLEM 8.6

For a 100-keV x-ray beam, calculate the fraction of beam intensity transmitted through a 10-mm-thick plate of **(a)** titanium and **(b)** lead. (See Sample Problem 8.6.)

PRACTICE PROBLEM 8.7

Given the data in Sample Problem 8.7 and $\rho_{H_2O} = 1.00$ Mg/m^3 and $V_{H_2O} = 1,483$ m/s, calculate the fraction of an ultrasonic pulse reflected from the surface of an aluminum plate in a water immersion bath.

8.5   FAILURE ANALYSIS AND PREVENTION

Failure analysis and prevention are important components of the application of materials in engineering design. There is now a well-established, systematic methodology for **failure analysis** of engineering materials. The related issue of **failure prevention** is equally important for avoiding future disasters. Ethical and legal issues are moving the field of materials science and engineering into a central role in the broader topic of engineering design.

A wide spectrum of failure modes has been identified.

1. **Ductile fracture** is observed in a large number of the failures occurring in metals due to "overload" (i.e., taking a material beyond the elastic limit and, subsequently, to fracture). The microscopic result of ductile fracture is shown in Figure 8–4a.

2. **Brittle fracture,** shown in Figure 8–4b, is characterized by rapid crack propagation without significant plastic deformation on a macroscopic scale.

The Material World:
Analysis of the Titanic Failure

We see in this chapter both the utility and the necessity of analyzing engineering failures. Such information can help prevent the recurrence of such disasters. In other cases, the analysis may serve a largely historical purpose, providing insight about the nature of a famous catastrophe of the past. An example of the latter is available for one of the most dramatic events of the twentieth century, the sinking of the Royal Mail Ship Titanic on the night of April 12, 1912 during its maiden voyage across the North Atlantic from England to New York. The wreckage was not discovered until September 1, 1985 when Robert Ballard observed it on the ocean floor at a depth of 3700 m. During an expedition to the wreckage on August 15, 1996, researchers retrieved some steel from the ship's hull, allowing metallurgical analysis at the University of Missouri–Rolla.

(Courtesy of Paramount Pictures and Twentieth Century Fox)

The general failure analysis is straightforward. The Titanic sank as a result of striking an iceberg three to six times larger than the ship itself. The six forward compartments of the ship were ruptured, leading to flooding that caused the ship to sink in less than three hours with a loss of more than 1500 lives. There is some benefit, however, from closer inspection of the material from which the ship's hull was constructed. Chemical analysis of the steel showed that it was similar to that of contemporary 1020 steel, defined in Chapter 11 as a common, low-carbon steel consisting primarily of iron with 0.20 wt % carbon. The following table, from the UM–R analysis, shows that the mechanical behavior of the Titanic hull alloy is also similar to that of 1020 steel, although the Titanic material has a slightly lower yield strength and higher elongation at failure associated with a larger average grain size (approximately 50 μm versus approximately 25 μm).

More important than the basic tensile properties, the ductile-to-brittle transition temperature measured in Charpy impact tests of the Titanic steel is significantly higher than for contemporary alloys. Modern steels in this composition range tend to have higher manganese contents and lower sulfur contents. The significantly higher Mn:S ratio reduces the ductile-to-brittle transition temperature substantially. The ductile-to-brittle transition temperature measured at an impact energy of 20 J is −27° C for a comparable, contemporary alloy and 32° C and 56° C for the Titanic plat specimens cut in directions longitudinal and transverse, respectively, to the hull configuration. Given that the sea water temperature at the time of the collision was −2° C, the materials selection clearly contributed to the failure. On the other hand, the choice was not inappropriate in the early part of the twentieth century. The Titanic steel was likely the best plain carbon ship plate available at the time. Contemporary ship passengers have, in addition to significantly better alloy selection, the benefit of superior navigational aides decreasing the probability of such collisions. Finally, it is worth noting that the Titanic's sister ship, the Olympic, had a successful career of more than 20 years with a similar hull steel but the better fortune of no encounter with a large iceberg.

Comparison of tensile properties of *Titanic* steel and SAE 1020

Property	Titanic	SAE 1020
Yield strength	193 MPa	207 MPa
Tensile strength	417 MPa	379 MPa
Elongation	29%	26%
Reduction in area	57%	50%

Source: K. Felkins, H. P. Leigh, Jr., and A. Jankovic, *Journal of Materials*, January 1998, pp. 12–18.

3. **Fatigue failure** by a mechanism of slow crack growth gives the distinctive "clamshell" fatigue fracture surface shown in Figure 8–14.

4. **Corrosion-fatigue failure** is due to the combined actions of a cyclic stress and a corrosive environment. In general, the fatigue strength of the metal will be decreased in the presence of an aggressive, chemical environment.

5. **Stress-corrosion cracking** (SCC) is another combined mechanical and chemical failure mechanism in which a noncyclic tensile stress (below the yield strength) leads to the initiation and propagation of fracture in a relatively mild chemical environment. Stress-corrosion cracks may be intergranular, transgranular, or a combination.

6. **Wear failure** is a term encompassing a broad range of relatively complex, surface-related damage phenomena. Both surface damage and wear debris can constitute "failure" of materials intended for sliding contact applications.

7. **Liquid-erosion failure** is a special form of wear damage in which a liquid is responsible for the removal of material. Liquid-erosion damage typically results in a pitted or honeycomb-like surface region.

8. **Liquid-metal embrittlement** involves the material losing some degree of ductility or fracturing below its yield stress in conjunction with its surface being wetted by a lower-melting-point liquid metal.

9. **Hydrogen embrittlement** is perhaps the most notorious form of catastrophic failure in high-strength steels. A few parts-per-million of hydrogen dissolved in these materials can produce substantial internal pressure, leading to fine, hairline cracks and a loss of ductility.

10. **Creep and stress-rupture failures** were introduced in Section 6.5. Failure of this type can occur near room temperature for many polymers and certain low-melting-point metals, such as lead, but may occur above 1000°C in many ceramics and certain high-melting-point metals, such as the superalloys.

11. **Complex failures** are those in which the failure occurs by the sequential operation of two distinct fracture mechanisms. An example would be initial cracking due to stress-corrosion cracking and, then, ultimate failure by fatigue after a cyclic load is introduced simultaneously with the removal of the corrosive environment.

A systematic sequence of procedures has been developed for the analysis of the failure of an engineering material. Although the specific methodology will vary with the specific failure, the principal components of the investigation and analysis are given in Table 8.7. In regard to failure analysis, *fracture mechanics* (Section 8.2) has provided a quantitative framework for evaluating structural reliability. It is worth noting that values of fracture toughness for various metals and alloys range between 20 and 200 MPa$\sqrt{\text{m}}$. Values of fracture toughness for ceramics and glass are typically in the range of 1 to 9 MPa$\sqrt{\text{m}}$, values for polymers are typically 1 to 4 MPa$\sqrt{\text{m}}$, and values for composites are typically 10 to 60 MPa$\sqrt{\text{m}}$.

Table 8.7 *Principal Components of Failure Analysis Methodology*

Collection of background data and samples
Preliminary examination of the failed part
Nondestructive testing
Mechanical testing
Selection, preservation, and cleaning of fracture surfaces
Macroscopic (1 to 100 ×) examination of fracture surfaces
Microscopic (> 100×) examination of fracture surfaces
Application of fracture mechanics
Simulated-service testing
Analyzing the evidence, formulating conclusions, and writing the report

Source: After 1986 *ASM Handbook, Vol. 11: Failure Analysis and Prevention*,
     ASM International, Materials Park, Ohio.

Finally, it is important to note that engineering designs can be improved by applying the concepts of failure analysis to failure prevention. An important example of this approach is to use designs without structural discontinuities that can serve as stress concentrators.

SUMMARY

In discussing the mechanical and thermal behavior of materials in Chapters 6 and 7, we found numerous examples of failure. In this chapter, we have systematically surveyed a variety of additional examples. The impact test of a notched test specimen provides a measure of impact energy, an analog of the toughness (area under the stress-versus-strain curve). Monitoring the impact energy over a range of environmental temperatures helps to identify the ductile-to-brittle transition temperature of bcc metal alloys, including common structural steels.

An introduction to fracture mechanics provides an especially useful and quantitative material parameter, the fracture toughness. For design applications of metal alloys, the fracture toughness helps to define the critical flaw size at the boundary between more desirable, general yielding and catastrophic, fast fracture. Transformation toughening techniques involve engineering design at the microstructural level to improve the fracture toughness of traditionally brittle ceramics.

Metal alloys and engineering polymers exhibit fatigue, a drop in strength as a result of cyclic loading. Fracture mechanics can provide a quantitative treatment of the approach to failure, as cracks grow in length with these repetitive stress cycles. Certain ceramics and glasses exhibit static fatigue, which is due to a chemical reaction with atmospheric moisture rather than cyclic loading.

Nondestructive testing, the evaluation of materials without impairing their usefulness, is critically important in identifying potentially critical flaws in engineering materials. Many techniques are available, but x-radiography and ultrasonic testing are primary examples.

Overall, failure analysis is the systematic methodology for determining which of a wide range of failure modes are operating for a specific material. Failure prevention utilizes the understanding provided by failure analysis to avoid future disasters.

KEY TERMS

acoustic-emission testing (293)
brittle fracture (295)
Charpy test (270)
complex failure (298)
corrosion-fatigue failure (298)
creep and stress-rupture failures (298)
ductile fracture (295)
ductile-to-brittle transition temperature (275)
eddy-current testing (293)
failure analysis (295)
failure prevention (295)

fast fracture (277)
fatigue (280)
fatigue curve (280)
fatigue failure (298)
fatigue strength (endurance limit) (284)
flaw-induced fracture (277)
fracture mechanics (276)
fracture toughness (276)
general yielding (277)
hydrogen embrittlement (298)
impact energy (270)
Izod test (272)

liquid-erosion failure (298)
liquid-metal embrittlement (298)
liquid-penetrant testing (293)
magnetic-particle testing (293)
nondestructive testing (289)
static fatigue (286)
stress-corrosion cracking (298)
stress intensity factor (282)
transformation toughening (277)
ultrasonic testing (291)
wear failure (298)
x-radiography (290)

REFERENCES

Ashby, M. F., and **D. R. H. Jones,** *Engineering Materials—An Introduction to Their Properties and Applications,* 2nd ed., Butterworth-Heinemann, Boston, 1996.

ASM Handbook, Vol. 11: Failure Analysis and Prevention, ASM International, Materials Park, Ohio, 1986, 843 pp.

ASM Handbook, Vol. 17: Nondestructive Evaluation and Quality Control, ASM International, Materials Park, Ohio, 1989, 795 pp.

Chiang, Y., D. P. Birnie III, and **W. D. Kingery,** *Physical Ceramics,* John Wiley & Sons, Inc., New York, 1997.

Engineered Materials Handbook, Desk Ed., ASM International, Materials Park, Ohio, 1995.

PROBLEMS

Section 8.1 • Impact Energy

8.1. Which of the alloys in Table 8.1 would you expect to exhibit ductile-to-brittle transition behavior? (State the basis of your selection.)

8.2. The relationship between service temperature and alloy selection in engineering design can be illustrated by the following case. **(a)** For the Fe–Mn–0.05 C alloys of Figure 8–3b, plot the ductile-to-brittle transition temperature (indicated by the sharp vertical rise in impact energy) against percentage Mn. **(b)** Using the plot from (a), estimate the percentage Mn level (to the nearest 0.1%) necessary to produce a ductile-to-brittle transition temperature of precisely 0°C.

8.3. As in Problem 8.2, estimate the percentage Mn level (to the nearest 0.1%) necessary to produce a ductile-to-brittle transition temperature of −25°C in the Fe–Mn–0.05 C alloy series of Figure 8–3b.

8.4. The following data were obtained from a series of Charpy impact tests on a particular structural steel. **(a)** Plot the data as impact energy versus temperature. **(b)** What is the ductile-to-brittle transition temperature, determined as corresponding to the average of the maximum and minimum impact energies?

Temperature (°C)	Impact energy (J)
100	84.9
60	82.8
20	81.2
0	77.9
−20	74.5
−40	67.8
−60	50.8
−80	36.0
−100	28.1
−140	25.6
−180	25.1

Section 8.2 • Fracture Toughness

8.5. Calculate the specimen thickness necessary to make the plane strain assumption used in Sample Problem 8.2 valid.

D 8.6. Generate a design plot similar to Figure 8–6 for a pressure vessel steel with Y.S. = 1000 MPa and K_{IC} = 170 MPa $\sqrt{m}$. For convenience, use the logarithmic scale for flaw size and cover a range of flaw sizes from 0.1 to 100 mm.

D 8.7. Repeat Problem 8.6 for an aluminum alloy with Y.S.= 400 MPa and K_{IC} = 25 MPa $\sqrt{m}$.

8.8. Critical flaw size corresponds to the transition between general yielding and fast fracture. If the fracture toughness of a high-strength steel can be increased by 50% (from 100 to 150 MPa $\sqrt{m}$) without changing its yield strength of 1250 MPa, by what percentage is its critical flaw size changed?

8.9. A nondestructive testing program for a design component using the 1040 steel of Table 6.2 can ensure that no flaw greater than 1 mm will exist. If this steel has a fracture toughness of 120 MPa $\sqrt{m}$, can this inspection program prevent the occurrence of fast fracture?

8.10. Would the nondestructive testing program described in Problem 8.9 be adequate for the cast iron alloy labeled 6(b) in Table 6.2, given a fracture toughness of 15 MPa $\sqrt{m}$?

8.11. A silicon nitride turbine rotor fractures at a stress level of 300 MPa. Estimate the flaw size responsible for this failure.

8.12. Estimate the flaw size responsible for the failure of a turbine rotor made from alumina that fractures at a stress level of 300 MPa.

8.13. Estimate the flaw size responsible for the failure of a turbine rotor made from partially stabilized zirconia that fractures at a stress level of 300 MPa.

8.14. Plot the breaking stress for MgO as a function of flaw size, a, on a logarithmic scale using Equation 8.1 and taking $Y = 1$. Cover a range of a from 1 to 100 mm. (See Table 8.3 for fracture toughness data and note the appearance of Figure 8–6.)

8.15. To appreciate the relatively low values of fracture toughness for traditional ceramics, plot, on a single graph, breaking stress versus flaw size, a, for an aluminum alloy with a K_{IC} of 30 MPa $\sqrt{m}$ and silicon carbide (see Table 8.3). Use Equation 8.1 and take $Y = 1$. Cover a range of a from 1 to 100 mm on a logarithmic scale. (Note also Problem 8.14.)

8.16. To appreciate the improved fracture toughness of the new generation of structural ceramics, superimpose a plot of breaking stress versus flaw size for partially stabilized zirconia on the result for Problem 8.15.

8.17. Some fracture mechanics data are given in Table 8.3. Plot the breaking stress for low-density polyethylene as a function of flaw size, a (on a logarithmic scale), using Equation 8.1 and taking $Y = 1$. Cover a range of a from 1 to 100 mm. (Note Figure 8–6.)

8.18. Superimpose a breaking stress plot for high-density polyethylene and ABS polymer on the result for Problem 8.17.

8.19. Calculate the breaking stress for a rod of ABS with a surface flaw size of 100 μm.

8.20. A nondestructive testing program can ensure that a thermoplastic polyester part will have no flaws greater than 0.1 mm in size. Calculate the maximum service stress available with this engineering polymer.

Section 8.3 • Fatigue

8.21. In Problem 6.41, a ductile iron was evaluated for a pressure vessel application. For that alloy, determine the maximum pressure to which the vessel can be repeatedly pressurized without producing a fatigue failure.

8.22. Repeat Problem 8.21 for the ductile iron of Problem 6.42.

8.23. A structural steel with a fracture toughness of 60 MPa $\sqrt{m}$ has no surface crack larger than 3 mm in length. By how much (in %) would this largest surface crack have to grow before the system would experience fast fracture under an applied stress of 500 MPa?

8.24. For the conditions given in Problem 8.23, calculate the % increase in crack size if the applied stress in the structural design is 600 MPa.

8.25. The application of a C11000 copper wire in a control circuit design will involve cyclic loading for extended periods at the elevated temperatures of a production plant. Use the data of Figure 8–15b to specify an upper temperature limit to ensure a fatigue strength of at least 100 MPa for a stress life of 10^7 cycles.

8.26. **(a)** The landing gear on a commercial aircraft experiences an impulse load upon landing. Assuming six such landings per day on average, how long would it take before the landing gear has been subjected to 10^8 load cycles? **(b)** The crankshaft in a given automobile rotates, on average, at 2000 revolutions per minute for a period of 2 h per day. How long would it take before the crankshaft has been subjected to 10^8 load cycles?

8.27. In analyzing a hip implant material for potential fatigue damage, it is important to note that an average person takes 4800 steps on an average day. How many stress cycles would this average person produce in **(a)** one year, **(b)** 10 years? (Note Problem 6.7.)

8.28. In analyzing a hip implant material for potential fatigue damage when used by an active athlete, we find that he takes a total of 10,000 steps and/or strides on an average day. How many stress cycles would this active person produce in **(a)** one year, **(b)** 10 years? (Note Problem 6.8.)

8.29. The time to fracture for a vitreous silica glass fiber at $+50°C$ is 10^4 s. What will be the time to fracture at room temperature ($25°C$)? Assume the same activation energy as given in Sample Problem 8.5.

8.30. To illustrate the very rapid nature of the water reaction with silicate glasses above $150°C$, calculate the time to fracture for the vitreous silica fiber of Problem 8.29 at $200°C$.

D **8.31.** A small pressure vessel is fabricated from an acetal polymer. The stress in the vessel wall is

$$\sigma = \frac{pr}{2t}$$

where p is the internal pressure, r the outer radius of the sphere, and t the wall thickness. For the vessel in question, $r = 30$ mm and $t = 2$ mm. What is the maximum permissible internal pressure for this design if the application is at room temperature and the wall stress is only tensile (due to internal pressurizations that will occur no more than 10^6 times)? (See Figure 8–21 for relevant data.)

D **8.32.** Calculate the maximum permissible internal pressure for the design in Problem 8.31 if all conditions are the same except that you are certain there will be no more than 10,000 pressurizations.

Section 8.4 • Nondestructive Testing

8.33. In doing x-radiography of steel, assume the film can detect a variation in radiation intensity represented by $\Delta I / I_0 = 0.001$. What thickness variation could be detected using this system for the inspection of a 12.5-mm-thick plate of steel using a 100-keV beam?

8.34. A good rule of thumb for doing x-radiographic inspections is that the test piece should be 5 to 8 times the half-value thickness ($t_{1/2}$) of the material, where $t_{1/2}$ is defined as the thickness value corresponding to an I/I_0 value of 0.5. Calculate the appropriate test piece thickness range for titanium inspected by a 100-keV beam.

8.35. Using the background of Problem 8.34, calculate the appropriate test piece thickness range for tungsten inspected by a 100-keV beam.

8.36. Using the background of Problem 8.34, calculate the appropriate test piece thickness range for iron inspected by **(a)** a 100-keV beam and **(b)** a 1-MeV beam.

8.37. Assume the geometry of the pulse echo ultrasonic test illustrated in Figure 8–23 represents an aluminum sample as follows: transducer-to-sample front surface = 25 mm, flaw depth = 10 mm, and overall sample thickness = 20 mm. Using data from Sample Problem 8.7 and Practice Problem 8.7, calculate the time delay between the initial pulse and **(a)** the front surface echo, **(b)** the flaw echo, and **(c)** the back surface echo.

8.38. For the specific test configuration given in Problem 8.37, calculate the relative intensities of the **(a)** front surface, **(b)** flaw, and **(c)** back surface echoes. (Take $I_{initial} = 100\%$, flaw area = 1/3 beam area, and flaw to be air-filled, for which you can take $Z_{air} = 0$.)

8.39. For an x-radiographic inspection of the plate in Problem 8.37 using a 100-keV beam, calculate the % change in I/I_0 between the flawed and flaw-free areas given that the effect of the flaw is to reduce the effective thickness of the plate by 10 μm.

8.40. Given the result of Problem 8.39, comment on the relative advantage of ultrasonic testing for the inspection of the flaw in that sample.

CHAPTER 9

Phase Diagrams—
Equilibrium
Microstructural
Development

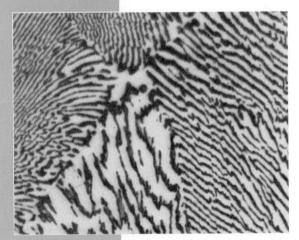

The microstructure of a slowly cooled "eutectic" soft solder ($\approx$ 38 wt % Pb — wt % Sn) consists of a lamellar structure of tin-rich solid solution (white) and lead-rich solid solution (dark), 375X. (From ASM Handbook, Vol. 3: Alloy Phase Diagrams, ASM International, Materials Park, Ohio, 1992.)

From the beginning of this book we have seen that a fundamental concept of materials science is that the properties of materials follow from their atomic and microscopic structures. The dependence of transport and mechanical properties on atomic-scale structure was seen in Chapters 5 and 6. To appreciate fully the nature of the many microstructure-sensitive properties of engineering materials, we must spend some time exploring the ways in which microstructure is developed. An important tool in this exploration is the *phase diagram,* which is a "map" that will guide us in answering the general question: What microstructure should exist at a given temperature for a given material composition? This is a question with a specific answer based in part on the equilibrium nature of the material. Closely related is the next chapter, dealing with the heat treatment of materials. Still other related questions to be addressed in Chapter 10 will be "How fast will the microstructure form at a given temperature?" and "What temperature versus time history will result in an optimal microstructure?"

The discussion of microstructural development via phase diagrams begins with the *phase rule,* which identifies the number of microscopic phases associated with a given "state condition," a set of values for temperature, pressure, and other variables that describe the nature of the material. We shall then describe the various characteristic phase diagrams for typical material systems. The *lever rule* will be used to quantify our interpretation of these phase diagrams. We shall specifically want to identify the composition and amount of each phase present. With these tools in hand, we can illustrate typical cases of microstructural development. Phase diagrams for several commercially important engineering materials are presented in this chapter. The most detailed discussion is reserved for the $Fe–Fe_3C$ diagram, which is the foundation for much of the iron and steel industry.

9.1   THE PHASE RULE

In this chapter we shall be quantifying the nature of microstructures. We begin with definitions of terms you need to understand the following discussion.

A **phase** is a chemically and structurally homogeneous portion of the microstructure. A single-phase microstructure can be polycrystalline (e.g., Figure 9–1), but each crystal grain differs only in crystalline orientation, not in chemical composition. Phase must be distinguished from **component,** which is a distinct chemical substance from which the phase is formed. For instance, we found in Section 4.1 that copper and nickel are so similar in nature that they are completely soluble in each other in any alloy proportions (e.g., Figure 4–2). For such a system, there is a single phase (a solid solution) and two components (Cu and Ni). For material systems involving compounds rather than elements, the compounds can be components.

Figure 9-1   *Single-phase microstructure of commercially pure molybdenum, 200×. Although there are many grains in this microstructure, each grain has the same, uniform composition. (From* Metals Handbook, *8th ed., Vol. 7:* Atlas of Microstructures, *American Society for Metals, Metals Park, Ohio, 1972.)*

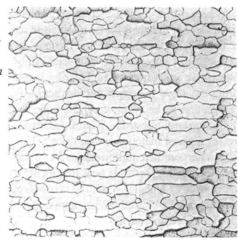

For example, MgO and NiO form solid solutions in a way similar to that for Cu and Ni (see Figure 4–5). In this case, the two components are MgO and NiO. As pointed out in Section 4.1, solid solubility is limited for many material systems. For certain compositions, the result is two phases, each richer in a different component. A classic example is the pearlite structure shown in Figure 9–2, which consists of alternating layers of ferrite and cementite. The ferrite is α-Fe with a small amount of cementite in solid solution. The cementite is nearly pure Fe_3C. The components are, then, Fe and Fe_3C.

Describing the ferrite phase as α-Fe with cementite in solid solution is appropriate in terms of our definition of the components for this system. However, on the atomic scale, the solid solution consists of carbon atoms dissolved interstitially in the α-Fe crystal lattice. The component Fe_3C does not dissolve as a discrete molecular unit. This is generally true for compounds in solid solution.

Figure 9-2   *Two-phase microstructure of pearlite found in a steel with 0.8 wt % C, 500×. This carbon content is an average of the carbon content in each of the alternating layers of ferrite (with <0.02 wt % C) and cementite (a compound, Fe_3C, which contains 6.7 wt % C). The narrower layers are the cementite phase. (From* Metals Handbook, *9th ed., Vol. 9:* Metallography and Microstructures, *American Society for Metals, Metals Park, Ohio, 1985.)*

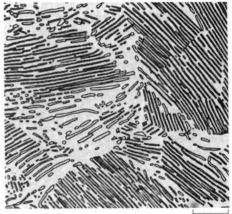

20 μm

A third term can be defined relative to "phase" and "component." The **degrees of freedom** are the number of independent variables available to the system. For example, a pure metal at precisely its melting point has no degrees of freedom. At this condition, or **state,** the metal exists in two phases in equilibrium, that is, in solid and liquid phases simultaneously. Any increase in temperature will change the state of the microstructure. (All of the solid phase will melt and become part of the liquid phase.) Similarly, even a slight reduction in temperature will completely solidify the material. The important **state variables** over which the materials engineer has control in establishing microstructure are temperature, pressure, and composition.

The general relationship between microstructure and these state variables is given by the **Gibbs* phase rule,** which, without derivation, can be stated as

$$F = C - P + 2 \tag{9.1}$$

where F is the number of degrees of freedom, C the number of components, and P the number of phases.[†] The 2 in Equation 9.1 comes from limiting the state variables to two (temperature and pressure). For most routine materials processing involving condensed systems, the effect of pressure is slight, and we can consider pressure to be fixed at 1 atm. In this case the phase rule can be rewritten to reflect one less degree of freedom:

$$F = C - P + 1 \tag{9.2}$$

For the case of the pure metal at its melting point, $C = 1$ and $P = 2$ (solid + liquid), giving $F = 1 - 2 + 1 = 0$, as we had noted previously. For a metal with a single impurity (i.e., with two components), solid and liquid phases can usually coexist over a range of temperatures (i.e., $F = 2 - 2 + 1 = 1$). The single degree of freedom means simply that we can maintain this two-phase microstructure while we vary the temperature of the material. However, we have only one independent variable ($F = 1$). By varying temperature, we indirectly vary the compositions of the individual phases. Composition is, then, a dependent variable. Such information obtainable from the Gibbs phase rule is most useful but also difficult to appreciate without the visual aid of phase diagrams. So now we proceed to introduce these fundamentally important "maps."

---

* Josiah Willard Gibbs (1839–1903), American physicist. As a professor of mathematical physics at Yale University, Gibbs was known as a quiet individual who made a profound contribution to modern science by almost single-handedly developing the field of thermodynamics. His phase rule was a cornerstone of this achievement.

[†] This derivation is an elementary one in the field of thermodynamics. Most students of this course will probably have a first course in thermodynamics one or two years hence. For now, we can take the Gibbs phase rule as a highly useful experimental fact. For those wishing to follow the derivation at this point, it is provided in the thermodynamics chapter of the *Instructor's Manual for Introduction to Materials Science for Engineers.*

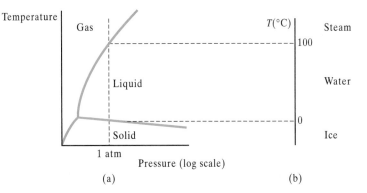

Figure 9-3 *(a) Schematic representation of the one-component phase diagram for H_2O. (b) A projection of the phase diagram information at 1 atm generates a temperature scale labeled with the familiar transformation temperatures for H_2O (melting at $0°C$ and boiling at $100°C$).*

Our first, simple example of a phase diagram is given in Figure 9–3. The one-component phase diagram (Figure 9–3a) summarizes the phases present for H_2O as a function of temperature and pressure. For the fixed pressure of 1 atm, we find a single, vertical temperature scale (Figure 9–3b) labeled with the appropriate transformation temperatures that summarize our common experience that solid H_2O (ice) transforms to liquid H_2O (water) at $0°C$ and that water transforms to gaseous H_2O (steam) at $100°C$. More relevant to materials engineering, Figure 9–4 provides a similar illustration for pure iron. As practical engineering materials are typically impure, we shall next discuss phase diagrams, in the more general sense, for the case of more than one component.

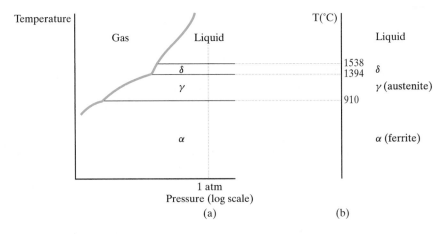

Figure 9-4 *(a) Schematic representation of the one-component phase diagram for pure iron. (b) A projection of the phase diagram information at 1 atm generates a temperature scale labeled with important transformation temperatures for iron. This projection will become one end of important binary diagrams such as Figure 9–19.*

SAMPLE PROBLEM 9.1

At 200°C, a 50:50 Pb–Sn solder alloy exists as two phases, a lead-rich solid and a tin-rich liquid. Calculate the degrees of freedom for this alloy and comment on its practical significance.

SOLUTION

Using Equation 9.2, that is, assuming a constant pressure of 1 atm above the alloy, we obtain

$$F = C - P + 1$$

There are two components (Pb and Sn) and two phases (solid and liquid), giving

$$F = 2 - 2 + 1 = 1$$

As a practical matter, we may retain this two-phase microstructure upon heating or cooling. But such a temperature change exhausts the "freedom" of the system and must be accompanied by changes in composition. (The nature of the composition change will be illustrated by the Pb–Sn phase diagram in Figure 9–16.)

....

PRACTICE PROBLEM 9.1

Calculate the degrees of freedom at a constant pressure of 1 atm for **(a)** a single-phase solid solution of Sn dissolved in the solvent Pb, **(b)** pure Pb below its melting point, and **(c)** pure Pb at its melting point. (See Sample Problem 9.1.)

9.2   THE PHASE DIAGRAM

A **phase diagram** is any graphical representation of the state variables associated with microstructures through the Gibbs phase rule. As a practical matter, phase diagrams in wide use by materials engineers are **binary diagrams,** representing two-component systems ($C = 2$ in the Gibbs phase rule), and **ternary diagrams,** representing three-component systems ($C = 3$). In this book our discussion will be restricted to the binary diagrams. There are an abundant number of important binary systems that will give us full appreciation of the power of the phase rule, and at the same time, we avoid the complexities involved in extracting quantitative information from ternary diagrams.

In the following examples, keep in mind that phase diagrams are maps. Specifically, binary diagrams are maps of the equilibrium phases associated with various combinations of temperature and composition. Our concern will be to illustrate the change in phases and associated microstructure that follows from changes in the state variables (temperature and composition).

COMPLETE SOLID SOLUTION

Probably the simplest type of phase diagram is that associated with binary systems in which the two components are completely soluble in each other in both the solid and the liquid states. There have been previous references to such completely miscible behavior for Cu and Ni and for MgO and NiO. Figure 9–5 shows a typical phase diagram for such a system. Note that the diagram shows temperature as the variable on the vertical scale and composition as the horizontal variable. The melting points of pure components A and B are indicated. For relatively high temperatures, any composition will have melted completely to give a liquid **phase field,** the region of the phase diagram corresponding to the existence of a liquid and labeled L. In other words, A and B are completely soluble in each other in the liquid state. What is unusual about this system is that A and B are also completely soluble in the solid state. The Hume-Rothery rules (see Section 4.1) state the criteria for this phenomenon in metal systems. In this book, we shall generally encounter complete miscibility in the liquid state (e.g., the L field in Figure 9–5). However, there are some systems in which liquid immiscibility occurs. Oil and water is a commonplace example. More relevant to materials engineering is the combination of various silicate liquids.

At relatively low temperatures, there is a single, solid-solution phase field labeled SS. Between the two single-phase fields is a two-phase region labeled L + SS. The upper boundary of the two-phase region is called the **liquidus,** that is, the line above which a single liquid phase will be present. The lower boundary of the two-phase region is called the **solidus** and is the line below which the system has completely solidified. At a given **state point** (a pair of temperature and composition values) within the two-phase region, an A-rich liquid exists in equilibrium with a B-rich solid solution. The composition of each phase is established as shown in Figure 9–6. The horizontal (constant-temperature) line passing through the state point cuts across both the liquidus and solidus lines. The composition of the liquid

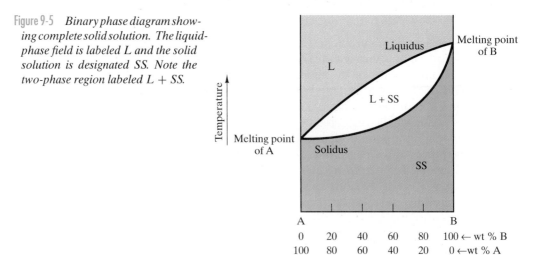

Figure 9-5 *Binary phase diagram showing complete solid solution. The liquid-phase field is labeled L and the solid solution is designated SS. Note the two-phase region labeled L + SS.*

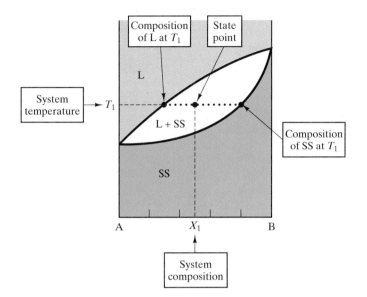

Figure 9-6 *The compositions of the phases in a two-phase region of the phase diagram are determined by a tie line (the horizontal line connecting the phase compositions at the system temperature).*

phase is given by the intersection point with the liquidus. Similarly, the solid solution composition is established by the point of intersection with the solidus. This horizontal line connecting the two phase compositions is termed a **tie line.** This construction will prove even more useful in Section 9.3 when we calculate the relative amounts of the two phases by the lever rule.

Figure 9–7 shows the application of the Gibbs phase rule (Equation 9.2) to various points in this phase diagram. The discussions in Section 9.1 can now be appreciated in terms of the graphical summary provided by the phase diagram. For example, an **invariant point** (where $F = 0$) occurs at the melting point of pure component B. At this limiting case, the material becomes a one-component system and any change of temperature changes

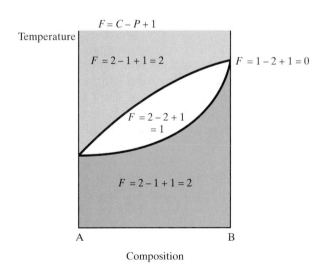

Figure 9-7 *Application of Gibbs phase rule (Equation 9.2) to various points in the phase diagram of Figure 9–5.*

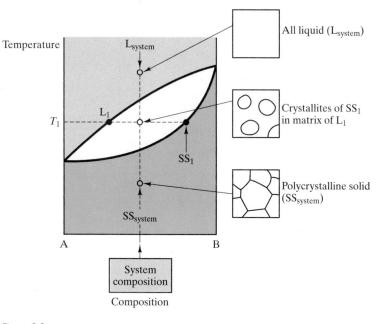

Figure 9-8 *Various microstructures characteristic of different regions in the complete solid-solution phase diagram.*

the microstructure to either all liquid (for heating) or all solid (for cooling). Within the two-phase (L + SS) region, there is one degree of freedom. A change of temperature is possible, but as Figure 9–6 indicates, the phase compositions are not independent. They will be established by the tie line associated with a given temperature. In the single-phase solid-solution region, there are two degrees of freedom; that is, both temperature and composition can be varied independently without changing the basic nature of the microstructure. Figure 9–8 summarizes microstructures characteristic of the various regions of this phase diagram.

By far the greatest use of phase diagrams in materials engineering is for the inorganic materials of importance to the metals and ceramics industries. Polymer applications generally involve single-component systems and/or nonequilibrium structures that are not amenable to presentation as phase diagrams. The common use of high-purity phases in the semiconductor industry also makes phase diagrams of limited use. The Cu–Ni system in Figure 9–9 is the classic example of a binary diagram with complete solid solution. A variety of commercial copper and nickel alloys falls within this system, including a superalloy called Monel.

The NiO–MgO system (Figure 9–10) is a ceramic system analogous to the Cu–Ni system, that is, exhibiting complete solid solution. While the Hume-Rothery rules (Section 4.1) addressed solid solution in metals, the requirement of similarity of cations is a comparable basis for solid solution in this oxide structure (see Figure 4–5).

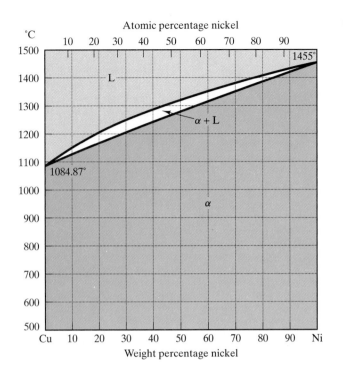

Figure 9-9 *Cu–Ni phase diagram. (After* Metals Handbook, *8th ed., Vol. 8:* Metallography, Structures, and Phase Diagrams, *American Society for Metals, Metals Park, Ohio, 1973, and* Binary Alloy Phase Diagrams, *Vol. 1, T. B. Massalski, ed., American Society for Metals, Metals Park, Ohio, 1986.)*

Note that the composition axis for the NiO–MgO phase diagram (and the other ceramic phase diagrams) is expressed in mole percent rather than weight percent. This has no effect on the validity of the lever rule calculations. The only effect on the results is that the answers obtained from such a calculation are in mole fractions rather than weight fractions.

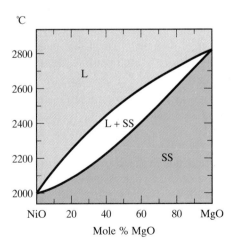

Figure 9-10 *NiO–MgO phase diagram. (After* Phase Diagrams for Ceramists, *Vol. 1, American Ceramic Society, Columbus, Ohio, 1964.)*

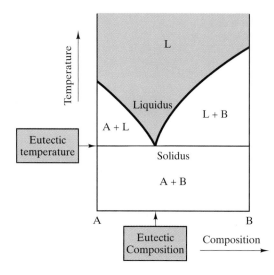

Figure 9-11 *Binary eutectic phase diagram showing no solid solution. This general appearance can be contrasted to the opposite case of complete solid solution illustrated in Figure 9–5.*

EUTECTIC DIAGRAM WITH NO SOLID SOLUTION

We now turn to a binary system that is the opposite of the one just discussed. Some components are so dissimilar that their solubility in each other is nearly negligible. Figure 9–11 illustrates the characteristic phase diagram for such a system. There are several features that distinguish this diagram from the type characteristic of complete solid solubility. First is the fact that, at relatively low temperatures, there is a two-phase field for pure solids A and B, consistent with our observation that the two components (A and B) cannot dissolve in each other. Second, the solidus is a horizontal line that corresponds to the **eutectic temperature.** This name comes from the Greek word *eutektos,* meaning "easily melted." In this case, the material with the **eutectic composition** is fully melted at the eutectic temperature. Any composition other than the eutectic will not fully melt at the eutectic temperature. Instead, such a material must be heated further through a two-phase region to the liquidus line. This situation is analogous to the two-phase region (L + SS) found in Figure 9–5. Figure 9–11 differs in that we have two such two-phase regions (A + L and B + L) in the binary eutectic diagram.

Some representative microstructures for the binary eutectic diagram are shown in Figure 9–12. The liquid and the liquid + solid microstructures are comparable to cases found in Figure 9–8. However, a fundamental difference exists in the microstructure of the fully solid system. In Figure 9–12, we find a fine-grained eutectic microstructure in which there are alternating layers of the components, pure A and pure B. A fuller discussion of solid-state microstructures will be appropriate after the lever rule has been introduced in Section 9.3. For now, we can emphasize that the sharp solidification point of the eutectic composition generally leads to the fine-grained nature of the eutectic microstructure. Even during slow cooling of the eutectic composition through the eutectic temperature, the system must transform from the

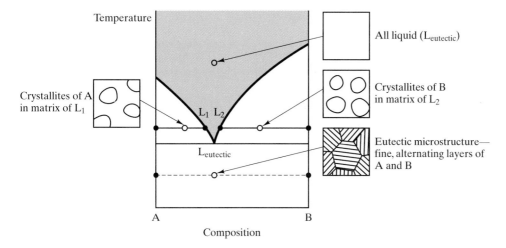

Figure 9-12 *Various microstructures characteristic of different regions in a binary eutectic phase diagram with no solid solution.*

liquid state to the solid state relatively quickly. The limited time available prevents a significant amount of diffusion (Section 5.3). The segregation of A and B atoms (which were randomly mixed in the liquid state) into separate solid phases must be done on a small scale. Various morphologies occur for various eutectic systems. But whether lamellar, nodular, or other morphologies are stable, these various eutectic microstructures are commonly fine-grained.

The simple eutectic system Al–Si (Figure 9–13) is a close approximation to Figure 9–11, although a small amount of solid solubility does exist. The aluminum-rich side of the diagram describes the behavior of some important aluminum alloys. Although we are not dwelling on semiconductor-related examples, the silicon-rich side illustrates the limit of aluminum doping in producing p-type semiconductors (see Section 17.2).

EUTECTIC DIAGRAM WITH LIMITED SOLID SOLUTION

For many binary systems, the two components are partially soluble in each other. The result is a phase diagram intermediate between the two cases we have treated so far. Figure 9–14 shows a eutectic diagram with limited solid solution. It generally looks like Figure 9–11 except for the solid-solution regions near each edge. These single-phase regions are comparable to the SS region in Figure 9–5 except for the fact that the components in Figure 9–14 do not exist in a single solid solution near the middle of the composition range. As a result, the two solid-solution phases, α and β, are distinguishable. They frequently have different crystal structures. In any case, the crystal structure of α will be that of A, and the crystal structure of β will be that of B. This is because each component serves as a solvent for the other, "impurity" component, for example, α consists of B atoms in solid solution

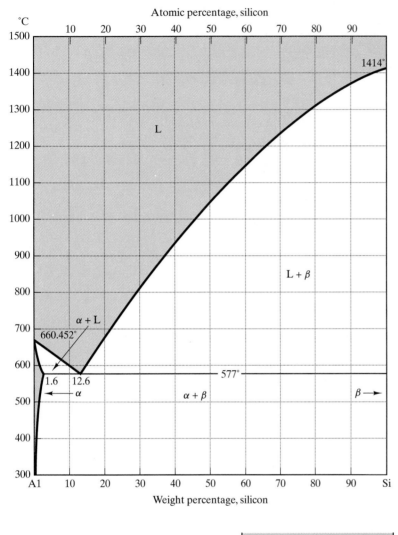

Figure 9-13 *Al–Si phase diagram. (After* Binary Alloy Phase Diagrams, *Vol. 1, T. B. Massalski, ed., American Society for Metals, Metals Park, Ohio, 1986.)*

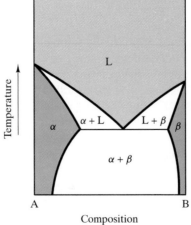

Figure 9-14 *Binary eutectic phase diagram with limited solid solution. The only difference from Figure 9–11 is the presence of solid-solution regions α and β.*

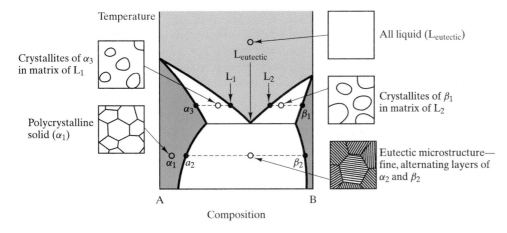

Figure 9-15 *Various microstructures characteristic of different regions in the binary eutectic phase diagram with limited solid solution. This illustration is essentially equivalent to Figure 9–12 except that the solid phases are now solid solutions (α and β) rather than pure components (A and B).*

in the crystal lattice of A. The use of tie lines to determine the compositions of α and β in the two-phase regions is identical to that illustrated in Figure 9–6, and examples are shown in Figure 9–15 together with representative microstructures.

The Pb–Sn System (Figure 9–16) is a good example of a binary eutectic with limited solid solution. Common solder alloys fall within this system. Their low melting ranges allow for joining of most metals by convenient heating methods, with low risk of damage to heat-sensitive parts. Solders with less than 5 wt % tin are used for sealing containers, coating and joining metals, and applications with service temperatures exceeding 120° C. Solders

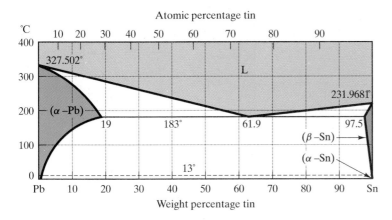

Figure 9-16 *Pb–Sn phase diagram. (After* Metals Handbook, *8th ed., Vol. 8:* Metallography, Structures, and Phase Diagrams, *American Society for Metals, Metals Park, Ohio, 1973, and* Binary Alloy Phase Diagrams, *Vol. 2, T. B. Massalski, ed., American Society for Metals, Metals Park, Ohio, 1986.)*

with between 10 and 20 wt % tin are used for sealing cellular automobile radiators and filling seams and dents in automobile bodies. General-purpose solders are generally 40 or 50 wt % tin. These solders have a characteristic pastelike consistency during application, associated with the two-phase liquid plus solid region just above the eutectic temperature. Their wide range of applications includes well-known examples from plumbing to electronics. Solders near the eutectic composition (approximately 60 wt % tin) are used for heat-sensitive electronic components requiring minimum heat application.

EUTECTOID DIAGRAM

The transformation of eutectic liquid to a relatively fine-grained microstructure of two solid phases upon cooling can be described as a special type of chemical reaction. This **eutectic reaction** can be written as

$$L(\text{eutectic}) \xrightarrow{\text{cooling}} \alpha + \beta \qquad (9.3)$$

where the notation corresponds to the phase labels from Figure 9–14. Some binary systems contain a solid-state analog of the eutectic reaction. Figure 9–17 illustrates such a case. The *eutectoid reaction* is

$$\gamma(\text{eutectoid}) \xrightarrow{\text{cooling}} \alpha + \beta \qquad (9.4)$$

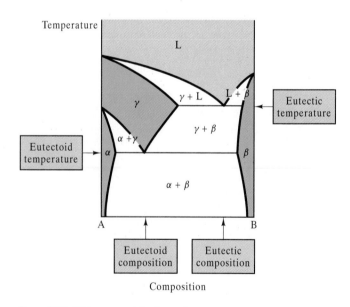

Figure 9-17 *This eutectoid phase diagram contains both a eutectic reaction (Equation 9.3) and its solid-state analog, a eutectoid reaction (Equation 9.4).*

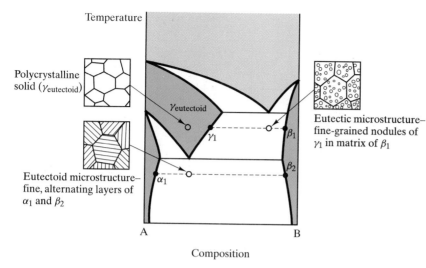

Temperature

Polycrystalline solid ($\gamma_{eutectoid}$)

$\gamma_{eutectoid}$

γ_1

β_1

β_2

α_1

Eutectoid microstructure– fine, alternating layers of α_1 and β_2

Eutectic microstructure– fine-grained nodules of γ_1 in matrix of β_1

A   B

Composition

Figure 9-18 *Representative microstructures for the eutectoid diagram of Figure 9–17.*

where *eutectoid* means "eutectic-like." Some representative microstructures are shown in Figure 9–18. The different morphologies of the eutectic and eutectoid microstructures emphasize our previous point that although the specific nature of these diffusion-limited structures will vary, they will generally be relatively fine-grained. A eutectoid reaction plays a fundamental role in the technology of steelmaking.

The Fe–Fe_3C system (Figure 9–19) is, by far, the most important commercial phase diagram we shall encounter. It provides the major scientific basis for the iron and steel industries. In Chapter 11 the boundary between "irons" and "steels" will be identified as a carbon content of 2.0 wt %. This point roughly corresponds to the carbon solubility limit in the **austenite**[*] (γ) phase of Figure 9–19. In addition, this diagram is representative of microstructural development in many related systems with three or more components (e.g., some stainless steels that include large amounts of chromium). Although Fe_3C, and not carbon, is a component in this system, the composition axis is customarily given in weight percent carbon. The important areas of interest on this diagram are around the eutectic and the eutectoid reactions. The reaction near 1500°C is of no practical consequence.

---

[*] William Chandler Roberts-Austen (1843–1902), English metallurgist. Young William Roberts set out to be a mining engineer, but his opportunities led to an appointment in 1882 as "chemist and assayer of the mint," a position he held until his death. His varied studies of the technology of coin making led to his appointment as a professor of metallurgy at the Royal School of Mines. He was a great success in both his government and academic posts. His textbook, *Introduction to the Study of Metallurgy,* was published in six editions between 1891 and 1908. In 1885, he adopted the additional surname in honor of his uncle (Nathaniel Austen).

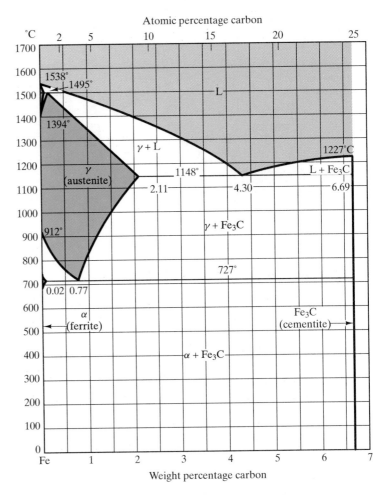

Figure 9-19 *Fe–Fe₃C phase diagram. Note that the composition axis is given in weight percent carbon even though Fe₃C, and not carbon, is a component. (After* Metals Handbook, *8th ed., Vol. 8:* Metallography, Structures, and Phase Diagrams, *American Society for Metals, Metals Park, Ohio, 1973, and* Binary Alloy Phase Diagrams, *Vol. 1, T. B. Massalski, ed., American Society for Metals, Metals Park, Ohio, 1986.)*

A final note of some irony (no pun is intended) is that the Fe–Fe₃C diagram is not a true equilibrium diagram. The Fe–C system (Figure 9–20) represents true equilibrium. Although graphite (C) is a more stable precipitate than Fe₃C, the rate of graphite precipitation is enormously slower than that of Fe₃C. The result is that in common steels (and many cast irons), the Fe₃C phase is **metastable;** that is, for all practical purposes it is stable with time and conforms to the Gibbs phase rule.

As just noted, the Fe–C system (Figure 9–20) is fundamentally more stable but less common than the Fe–Fe₃C because of slow "kinetics" (the

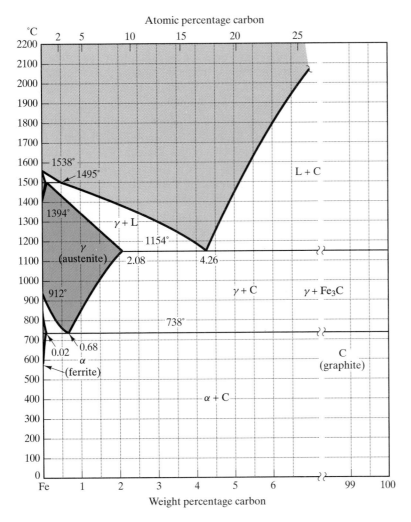

Figure 9-20 *Fe–C phase diagram. The left side of this diagram is nearly identical to that for the Fe–Fe₃C diagram (Figure 9–19). In this case, however, the intermediate compound Fe₃C does not exist. (After* Metals Handbook, *8th ed., Vol. 8:* Metallography, Structures, and Phase Diagrams, *American Society for Metals, Metals Park, Ohio, 1973, and* Binary Alloy Phase Diagrams, *Vol. 1, T. B. Massalski, ed., American Society for Metals, Metals Park, Ohio, 1986.)*

subject of Chapter 10). Extremely slow cooling rates can produce the results indicated on the Fe–C diagram. The more practical method is to promote graphite precipitation by a small addition of a third component, such as silicon. Typically, silicon additions of 2 to 3 wt % are used to stabilize the graphite precipitation. This third component is not acknowledged in Figure 9–20. The result, however, is that the figure does describe microstructural development for some practical systems. An example will be given in Section 9.4.

PERITECTIC DIAGRAM

In all of the binary diagrams inspected to this point, the pure components have had distinct melting points. In some systems, however, the components will form stable compounds that may not have such a distinct melting point. An example of this is illustrated in Figure 9–21. In this simple example, A and B form the stable compound AB, which does not melt at a single temperature as do components A and B. An additional simplification used here is to ignore the possibility of some solid solution for the components and intermediate compound. The components are said to undergo **congruent melting;** that is, the liquid formed upon melting has the same composition as the solid from which it was formed. On the other hand, compound AB (which is 50 mol % A plus 50 mol % B) is said to undergo **incongruent melting;** that is, the liquid formed upon melting has a composition other than AB. The term *peritectic* is used to describe this incongruent melting phenomenon. "Peritectic" comes from the Greek phrase meaning to "melt nearby." The **peritectic reaction** can be written as

$$AB \xrightarrow{\text{heating}} L + B \qquad (9.5)$$

where the liquid composition is noted in Figure 9–21. Some representative microstructures are shown in Figure 9–22. The Al_2O_3–SiO_2 phase diagram (9–23) is a classic example of a peritectic diagram.

The Al_2O_3–SiO_2 binary diagram is as important to the ceramic industry as the Fe–Fe_3C diagram is to the steel industry. Several important ceramics fall within this system. Refractory silica bricks are nearly pure SiO_2, with 0.2 to 1.0 wt % (0.1 to 0.6 mol %) Al_2O_3. For silica bricks required to operate at temperatures above 1600°C, it is obviously important to keep the Al_2O_3 content as low as possible (by careful raw material selection) to minimize the amount of liquid phase. A small amount of liquid

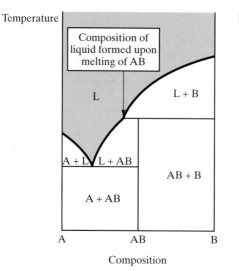

Figure 9-21 *Peritectic phase diagram showing a peritectic reaction (Equation 9.5). For simplicity, no solid solution is shown.*

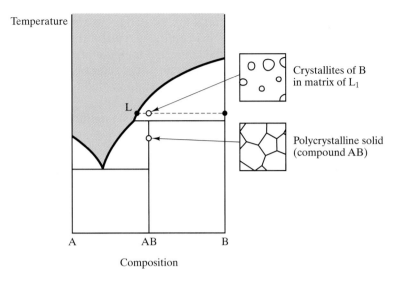

Figure 9-22 *Representative microstructures for the peritectic diagram of Figure 9–21.*

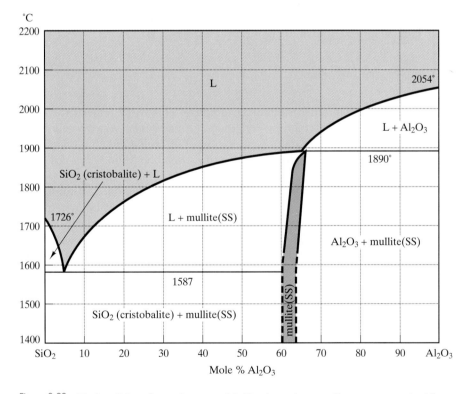

Figure 9-23 *Al_2O_3–SiO_2 phase diagram. Mullite is an intermediate compound with ideal stoichiometry $3Al_2O_3 \cdot 2SiO_2$. (After F. J. Klug, S. Prochazka, and R. H. Doremus,* J. Am. Ceram. Soc. 70, *750 (1987).)*

is tolerable. Common fireclay refractories are located in the range of 25 to 45 wt % (16 to 32 mol %) Al_2O_3. Their utility as structural elements in furnace designs is limited by the solidus (eutectic) temperature of 1587°C. A dramatic increase in "refractoriness," or temperature resistance, occurs at the composition of the incongruently melting compound mullite ($3Al_2O_3 \cdot 2SiO_2$).

A controversy over the nature of the melting for mullite has continued for several decades. For the past several years, the peritectic reaction shown in Figure 9–23 has been widely accepted. The debate about such a commercially important system illustrates a significant point. Establishing equilibrium in high-temperature ceramic systems is not easy. Silicate glasses are similar examples of this point. Phase diagrams for this text represent our best understanding at this time, but we must remain open to refined experimental results in the future.

Care is exercised in producing mullite refractories to ensure that the overall composition is greater than 72 wt % (60 mol %) Al_2O_3 to avoid the two-phase region (mullite + liquid). By so doing, the refractory remains completely solid to the peritectic temperature of 1890°C. So-called high-alumina refractories fall within the composition range 60 to 90 wt % (46 to 84 mol %) Al_2O_3. Nearly pure Al_2O_3 represents the highest refractoriness (temperature resistance) of commercial materials in the Al_2O_3–SiO_2 system. These materials are used in such demanding applications as refractories in glass manufacturing and laboratory crucibles.

GENERAL BINARY DIAGRAMS

The peritectic diagram was our first example of a binary system with an **intermediate compound,** a chemical compound formed between two components in a binary system. In fact, the formation of intermediate compounds is a relatively common occurrence. Of course, such compounds are not associated solely with the peritectic reaction. Figure 9–24a shows the case of an intermediate compound, AB, which melts congruently. An important point about this system is that it is equivalent to two adjacent binary eutectic diagrams of the type first introduced in Figure 9–11. Again, for simplicity, we are ignoring solid solubility. This is our first encounter with what can be termed a **general diagram,** a composite of two or more of the types covered in this section. The approach to analyzing these more complex systems is straightforward. Simply deal with the smallest binary system associated with the overall composition and ignore all others. This procedure is illustrated in Figure 9–24b, which shows that for an overall composition between AB and B, we can treat the diagram as a simple binary eutectic of AB and B. For all practical purposes, the A–AB binary does not exist for the overall composition shown in Figure 9–24b. Nowhere in the development of microstructure for that composition will crystals of A be found in a liquid matrix or will crystals of A and AB exist simultaneously in equilibrium. A more elaborate illustration of this point is shown in Figure 9–25.

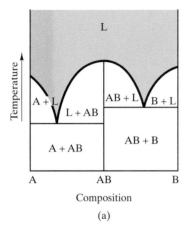

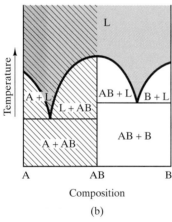

Figure 9-24 *(a) Binary phase diagram with a congruently melting intermediate compound, AB. This diagram is equivalent to two simple binary eutectic diagrams (the A–AB and AB–B systems). (b) For analysis of microstructure for an overall composition in the AB–B system, only that binary eutectic diagram need be considered.*

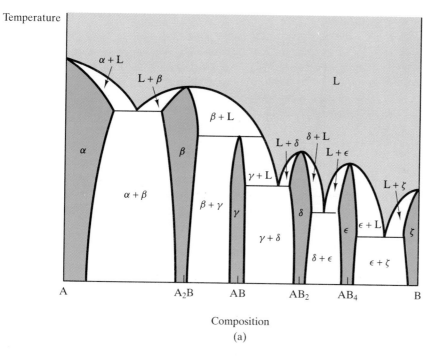

(a)

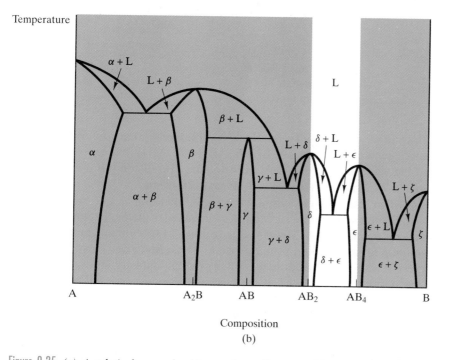

(b)

Figure 9-25 *(a) A relatively complex binary phase diagram. (b) For an overall composition between AB_2 and AB_4, only that binary eutectic diagram is needed to analyze microstructure.*

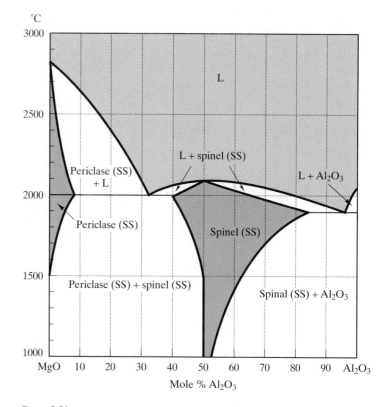

Figure 9-26 *MgO–Al$_2$O$_3$ phase diagram. Spinel is an intermediate compound with ideal stoichiometry MgO · Al$_2$O$_3$. (After* Phase Diagrams for Ceramists, *Vol. 1, American Ceramic Society, Columbus, Ohio, 1964.)*

In Figure 9–25a we find a relatively complex general diagram with four intermediate compounds (A$_2$B, AB, AB$_2$, and AB$_4$) and several examples of individual binary diagrams. But for the overall compositions shown in Figure 9–25b, only the AB$_2$–AB$_4$ binary is relevant.

The MgO–Al$_2$O$_3$ system (Figure 9–26) is similar to Figure 9–24 but with limited solid solubility. Figure 9–26 includes the important intermediate compound, spinel, MgO·Al$_2$O$_3$ or MgAl$_2$O$_4$, with an extensive solid-solution range. (A dilute solution of Al$_2$O$_3$ in MgO was illustrated in Figure 4–6.) Spinel refractories are widely used in industry. The spinel crystal structure (see Figure 3–15) is the basis of an important family of magnetic materials (see Section 18.5).

Figures 9–27 to 9–29 are good examples of general diagrams comparable to Figure 9–25. Important age-hardenable aluminum alloys are found near the κ phase boundary in the Al–Cu system (Figure 9–27). We will discuss this point in relation to Figure 9–37, and we will discuss the subtleties of precipitation hardening further in Section 10.4. The Al–Cu system is a good example of a complex diagram that can be analyzed as a simple binary eutectic in the high-aluminum region.

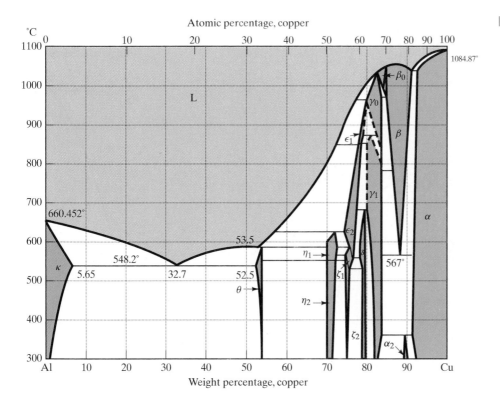

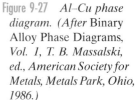

Figure 9-27 *Al–Cu phase diagram. (After* Binary Alloy Phase Diagrams, *Vol. 1, T. B. Massalski, ed., American Society for Metals, Metals Park, Ohio, 1986.)*

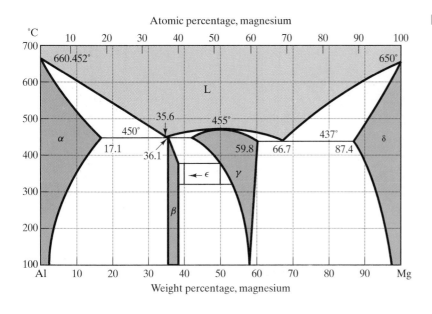

Figure 9-28 *Al–Mg phase diagram. (After* Binary Alloy Phase Diagrams, *Vol. 1, T. B. Massalski, ed., American Society for Metals, Metals Park, Ohio, 1986.)*

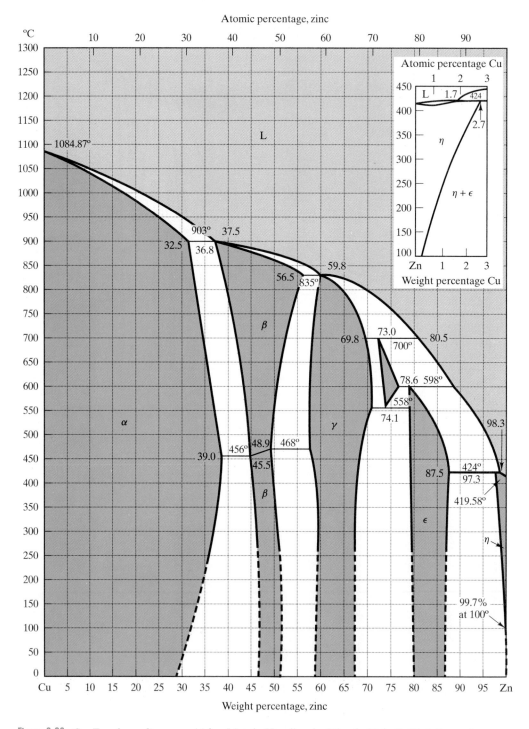

Figure 9-29 *Cu–Zn phase diagram.* (*After* Metals Handbook, *8th ed., Vol. 8:* Metallography, Structures, and Phase Diagrams, *American Society for Metals, Metals Park, Ohio, 1973, and* Binary Alloy Phase Diagrams, *Vol. 1, T. B. Massalski, ed., American Society for Metals, Metals Park, Ohio, 1986.)*

Several aluminum alloys (with small magnesium additions) and magnesium alloys (with small aluminum additions) can be described by the Al–Mg system (Figure 9–28). Like the Al–Cu system, the Cu–Zn system in Figure 9–29 is a complex diagram that is, for some practical systems, easy to analyze. For example, many commercial brass compositions lie in the single-phase α region.

The CaO–ZrO_2 phase diagram (Figure 9–30) is an example of a general diagram of a ceramic system. ZrO_2 has become an important refractory material through the use of stabilizing additions such as CaO. As seen in the phase diagram (Figure 9–30), pure ZrO_2 has a phase transformation at $1000°C$ in which the crystal structure changes from monoclinic to tetragonal upon heating. This transformation involves a substantial volume change that is structurally catastrophic to the brittle ceramic. Cycling the pure material through the transformation temperature will effectively reduce it to powder. As the phase diagram also shows, the addition of approximately 10 wt % (20 mol %) CaO produces a solid-solution phase with a cubic crystal structure from room temperature to the melting point (near $2500°C$). This "stabilized zirconia" is a practical, and obviously very refractory, structural material. Various other ceramic components, such as Y_2O_3, serve as stabilizing components and have phase diagrams with ZrO_2 that look quite similar to Figure 9–30.

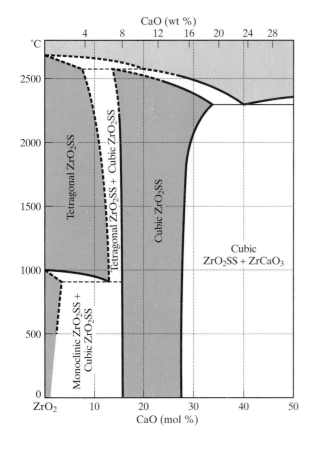

Figure 9-30 *CaO–ZrO₂ phase diagram. The dashed lines represent tentative results. (After* Phase Diagrams for Ceramists, *Vol. 1, American Ceramic Society, Columbus, Ohio, 1964.)*

SAMPLE PROBLEM 9.2

An alloy in the A–B system described by Figure 9–25 is formed by melting equal parts of A and A_2B. Qualitatively describe the microstructural development that will occur upon slow cooling of this melt.

SOLUTION

A 50:50 combination of A and A_2B will produce an overall composition midway between A and A_2B. The cooling path is illustrated as follows:

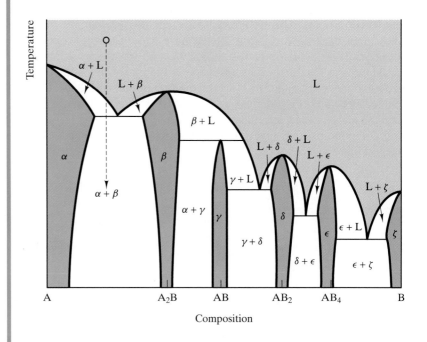

The first solid to precipitate from the liquid is the A-rich solid solution α. At the A–A_2B eutectic temperature, complete solidification occurs, giving a two-phase microstructure of solid solutions α and β.

....

PRACTICE PROBLEM 9.2

Qualitatively describe the microstructural development that will occur upon slow cooling of a melt of an alloy of equal parts of A_2B and AB.

9.3   THE LEVER RULE

In Section 9.2 we surveyed the use of phase diagrams to determine the phases present at equilibrium in a given system and their corresponding microstructure. The tie line (e.g., Figure 9–6) gives the composition of each phase in a two-phase region. We now extend this analysis to determine the amount of each phase in the two-phase region. We should first note that, for single-phase regions, the analysis is trivial. By definition, the microstructure is 100% of the single phase. In the two-phase regions, the analysis is not trivial but is, nonetheless, simple.

The relative amounts of the two phases in a microstructure are easily calculated from a mass balance. Let us consider again the case of the binary diagram for complete solid solution. Figure 9–31 is equivalent to Figure 9–6 and, again, shows a tie line giving the composition of the two phases associated with a state point in the L + SS region. In addition, the compositions of each phase and of the overall system are indicated. An overall **mass balance** requires that the sum of the two phases equal the total system. Assuming a total mass of 100 g gives an expression

$$m_L + m_{SS} = 100g \tag{9.6}$$

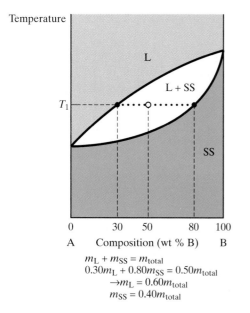

$$m_L + m_{SS} = m_{total}$$
$$0.30m_L + 0.80m_{SS} = 0.50m_{total}$$
$$\rightarrow m_L = 0.60m_{total}$$
$$m_{SS} = 0.40m_{total}$$

Figure 9-31 *A more quantitative treatment of the tie line introduced in Figure 9–6 allows the amount of each phase (L and SS) to be calculated by means of a mass balance (Equations 9.6 and 9.7).*

We can also do an independent mass balance on either of the two components. For example, the amount of B in the liquid phase plus that in solid solution must equal the total amount of B in the overall composition. Noting in Figure 9–6 that (for temperature, T_1) L contains 30% B, SS 80% B, and the overall system 50% B, we can write

$$0.30m_L + 0.80m_{SS} = 0.50(100g) = 50g \qquad (9.7)$$

Equations 9.6 and 9.7 are two equations with two unknowns, allowing us to solve for the amounts of each phase:

$$m_L = 60g$$

and

$$m_{SS} = 40g$$

This material balance calculation is convenient, but an even more streamlined version can be generated. To obtain this, we can do the mass balance in general terms. For two phases, α and β, the general mass balance will be

$$x_\alpha m_\alpha + x_\beta m_\beta = x(m_\alpha + m_\beta) \qquad (9.8)$$

where x_α and x_β are the compositions of the two phases and x is the overall composition. This expression can be rearranged to give the relative amount of each phase in terms of compositions:

$$\frac{m_\alpha}{m_\alpha + m_\beta} = \frac{x_\beta - x}{x_\beta - x_\alpha} \qquad (9.9)$$

and

$$\frac{m_\beta}{m_\alpha + m_\beta} = \frac{x - x_\alpha}{x_\beta - x_\alpha} \qquad (9.10)$$

Together, Equations 9.9 and 9.10 constitute the **lever rule.** This mechanical analogy to the mass balance calculation is illustrated in Figure 9–32. Its utility is due largely to the fact that it can be visualized so easily in terms of the phase diagram. The overall composition corresponds to the fulcrum of a lever with length corresponding to the tie line. The mass of each phase is suspended from the end of the lever corresponding to its composition. The relative amount of phase α is directly proportional to the length of the "opposite lever arm" ($= x_\beta - x$). It is this relationship that allows the relative amounts of phases to be determined by a simple visual inspection. With this final quantitative tool in hand, we can now proceed to the step-by-step analysis of microstructural development.

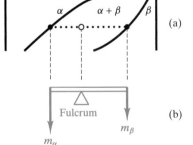

Figure 9-32 *The lever rule is a mechanical analogy to the mass balance calculation. The (a) tie line in the two-phase region is analogous to (b) a lever balanced on a fulcrum.*

SAMPLE PROBLEM 9.3

The temperature of 1 kg of the alloy shown in Figure 9–31 is lowered slowly until the liquid-solution composition is 18 wt % B and the solid-solution composition is 66 wt % B. Calculate the amount of each phase.

SOLUTION

Using Equations 9.9 and 9.10, we obtain

$$m_L = \frac{x_{SS} - x}{x_{SS} - x_L}(1 \text{ kg}) = \frac{66 - 50}{66 - 18}(1 \text{ kg})$$

$$= 0.333 \text{ kg} = 333 \text{ g}$$

$$m_{SS} = \frac{x - x_L}{x_{SS} - x_L}(1 \text{kg}) = \frac{50 - 18}{66 - 18}(1 \text{kg})$$

$$= 0.667 \text{ kg} = 667 \text{ g}$$

Note. We can also calculate m_{SS} more swiftly by simply noting that $m_{SS} = 1000 \text{ g} - m_L = (1000 - 333) \text{ g} = 667 \text{ g}$. However, we shall continue to use both Equations 9.9 and 9.10 in the sample problems in this chapter for the sake of practice and as a cross-check.

SAMPLE PROBLEM 9.4

For 1 kg of eutectoid steel at room temperature, calculate the amount of each phase (α and Fe_3C) present.

SOLUTION

Using Equations 9.9 and 9.10 and Figure 9–19, we have

$$m_\alpha = \frac{x_{Fe_3C} - x}{x_{Fe_3C} - x_\alpha}(1 \text{kg}) = \frac{6.69 - 0.77}{6.69 - 0}(1 \text{ kg})$$

$$= 0.885 \text{ kg} = 885 \text{ g}$$

$$m_{Fe_3C} = \frac{x - x_\alpha}{x_{Fe_3C} - x_\alpha}(1 \text{ kg}) = \frac{0.77 - 0}{6.69 - 0}(1 \text{ kg})$$

$$= 0.115 \text{ kg} = 115 \text{ g}$$

SAMPLE PROBLEM 9.5

A partially stabilized zirconia is composed of 4 wt % CaO. This product contains some monoclinic phase together with the cubic phase, which is the basis of fully stabilized zirconia. Estimate the mole percent of each phase present at room temperature.

SOLUTION

Noting that 4 wt % CaO = 8 mol % CaO and assuming that the solubility limits shown in Figure 9–30 do not change significantly below 500°C, we can use Equations 9.9 and 9.10:

$$\text{mol \% monoclinic} = \frac{x_{cub} - x}{x_{cub} - x_{mono}} \times 100\%$$

$$= \frac{15 - 8}{15 - 2} \times 100\% = 53.8 \text{ mol\%}$$

$$\text{mol \% cubic} = \frac{x - x_{mono}}{x_{cub} - x_{mono}} \times 100\%$$

$$= \frac{8 - 2}{15 - 2} \times 100\% = 46.2 \text{ mol\%}$$

PRACTICE PROBLEM 9.3

Suppose the alloy in Sample Problem 9.3 is reheated to a temperature at which the liquid composition is 48 wt % B and the solid-solution composition is 90 wt % B. Calculate the amount of each phase.

PRACTICE PROBLEM 9.4

In Sample Problem 9.4, we found the amount of each phase in a eutectoid steel at room temperature. Repeat this calculation for a steel with an overall composition of 1.13 wt % C.

PRACTICE PROBLEM 9.5

In Sample Problem 9.5, the phase distribution in a partially stabilized zirconia is calculated. Repeat this calculation for a zirconia with 5 wt % CaO.

9.4   MICROSTRUCTURAL DEVELOPMENT DURING SLOW COOLING

We are now in a position to follow closely the microstructural development in various binary systems. In all cases, we shall assume the common situation of cooling a given composition from a single-phase melt. Microstructure is developed in the process of solidification. We consider only the case of *slow* cooling; that is, equilibrium is essentially maintained at all points along the cooling path. The effect of more rapid temperature changes is the subject of Chapter 10, which deals with time-dependent microstructures developed during heat treatment.

Let us return to the simplest of binary diagrams, the case of complete solubility in both the liquid and solid phases. Figure 9–33 shows the gradual solidification of the 50% A–50% B composition treated previously (Figures 9–6, 9–8, and 9–31). The lever rule (Figure 9–32) is applied at three different temperatures in the two-phase (L + SS) region. It is important to appreciate that the appearance of the microstructures in Figure 9–33 corresponds directly with the relative position of the overall system composition along the tie line. At higher temperatures (e.g., T_1), the overall composition is near the liquid-phase boundary, and the microstructure is predominantly liquid. At lower temperatures (e.g., T_3), the overall composition is near the solid-

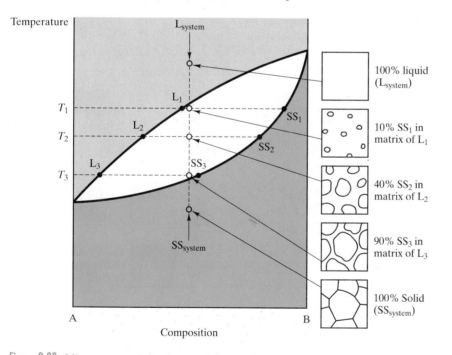

Figure 9-33 *Microstructural development during the slow cooling of a 50% A– 50% B composition in a phase diagram with complete solid solution. At each temperature, the amounts of the phases in the microstructure correspond to a lever rule calculation. The microstructure at T_2 corresponds to the calculation in Figure 9–31.*

phase boundary, and the microstructure is predominantly solid. Of course, the compositions of the liquid and solid phases change continuously during cooling through the two-phase region. At any temperature, however, the relative amounts of each phase are such that the overall composition is 50% A and 50% B. This is a direct manifestation of the lever rule as defined by the mass balance of Equation 9.8.

The understanding of microstructural development in the binary eutectic is greatly aided by the lever rule. The case for the eutectic composition, itself, is straightforward and was illustrated previously (Figures 9–12 and 9–15). Figure 9–34 repeats those cases in slightly greater detail. An additional comment is that the composition of each solid-solution phase (α and β) and their relative amounts will change slightly with temperature below the eutectic temperature. The microstructural effect (corresponding to this compositional adjustment due to solid-state diffusion) is generally minor.

Microstructural development for a noneutectic composition is more complex. Figure 9–35 illustrates the microstructural development for a **hypereutectic composition** (composition greater than that of the eutectic). The gradual growth of β crystals above the eutectic temperature is comparable to the process found in Figure 9–33 for the complete solid-solution diagram. The one difference is that, in Figure 9–35, the crystallite growth stops at the eutectic temperature with only 67% of the microstructure solidified. Final solidification occurs when the remaining liquid (with the eutectic composition) transforms suddenly to the eutectic microstructure upon cooling through the eutectic temperature. In a sense, the 33% of the microstructure that is liquid just above the eutectic temperature undergoes the eutectic re-

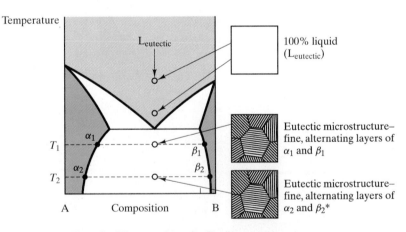

*The only differences from the T_1 microstructure are the phase compositions and the relative amounts of each phase. For example, the amount of b will be proportional to

$$\frac{x_{\text{eutectic}} - x_\alpha}{x_\beta - x_\alpha}$$

Figure 9-34 *Microstructural development during the slow cooling of a eutectic composition.*

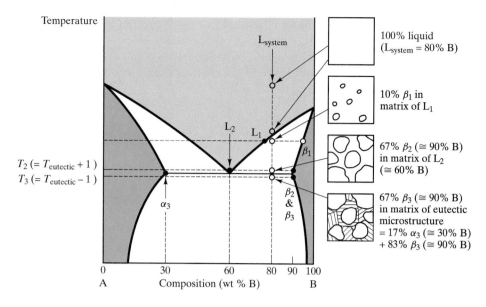

Figure 9-35 *Microstructural development during the slow cooling of a hypereutectic composition.*

action illustrated in Figure 9–34. A lever rule calculation just below the eutectic temperature (T_3 in Figure 9–35) indicates correctly that the microstructure is 17% α_3 and 83% β_3. However, following the entire cooling path has indicated that the β phase is present in two forms. The large grains produced during the slow cooling through the two-phase ($L + \beta$) region are termed **proeutectic** β; that is, they appear "before the eutectic." The finer β in the lamellar eutectic is appropriately termed *eutectic* β.

Figure 9–36 shows a similar situation that develops for a **hypoeutectic composition** (composition less than that of the eutectic). This case is

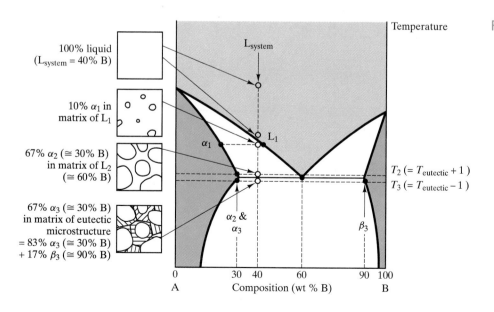

Figure 9-36 *Microstructural development during the slow cooling of a hypoeutectic composition.*

analogous to that for the hypereutectic composition. In Figure 9–36 we see the development of large grains of proeutectic α along with the eutectic microstructure of α and β layers. Two other types of microstructural development are illustrated in Figure 9–37. For an overall composition of 10%

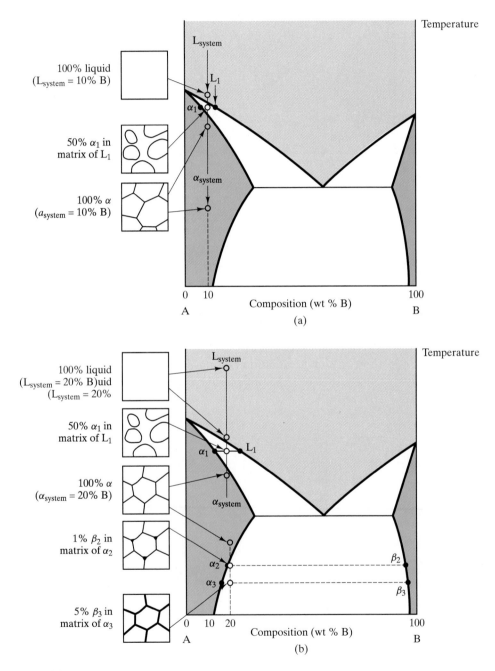

Figure 9-37 *Microstructural development for two compositions that avoid the eutectic reaction.*

B, the situation is quite similar to that for the complete solid-solution binary in Figure 9–33. The solidification leads to a single-phase solid solution that remains stable upon cooling to low temperatures. The 20% B composition behaves in a similar fashion except that, upon cooling, the α phase becomes saturated with B atoms. Further cooling leads to precipitation of a small amount of β phase. In Figure 9–37b, this precipitation is shown occurring along grain boundaries. In some systems, the second phase precipitates within grains. For a given system, the morphology of the second phase can be a sensitive function of the rate of cooling. In Section 10.4 we shall encounter such a case for the Al–Cu system, in which precipitation hardening is an important example of heat treatment.

With the variety of cases encountered in this section, we are now in a position to treat any composition in any of the binary systems presented in this chapter, including the general diagrams represented by Figure 9–25b.

The cooling path for a **white cast iron** (see also Chapter 11) is shown in Figure 9–38. The schematic microstructure can be compared with a micrograph in Figure 11–1a. The eutectoid reaction to produce pearlite is shown

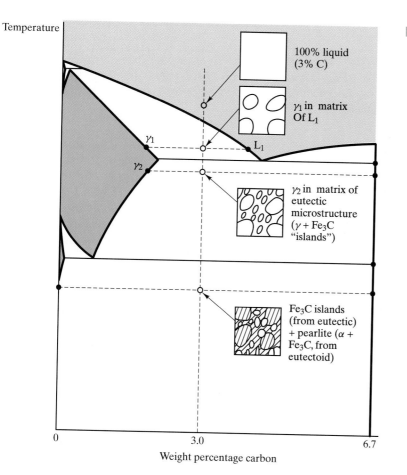

Figure 9-38 *Microstructural development for white cast iron (of composition 3.0 wt % C) shown with the aid of the Fe–Fe$_3$C phase diagram. The resulting (low-temperature) sketch can be compared with a micrograph in Figure 11–1a.*

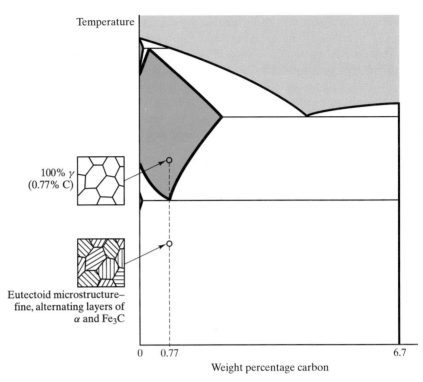

100% γ
(0.77% C)

Eutectoid microstructure–
fine, alternating layers of
α and Fe₃C

Temperature

Weight percentage carbon

0   0.77   6.7

Figure 9-39 *Microstructural development for eutectoid steel (of composition 0.77 wt % C). The resulting (low-temperature) sketch can be compared with the micrograph in Figure 9–2.*

in Figure 9–39. This composition (0.77 wt % C) is close to that for a 1080 plain-carbon steel (see Table 11.1). Many phase diagrams for the Fe–Fe₃C system give the eutectoid composition rounded off to 0.8 wt % C. As a practical matter, any composition near 0.77 wt % C will give a microstructure that is predominantly eutectoid. The actual pearlite microstructure is shown in the micrograph of Figure 9–2. A **hypereutectoid composition** (composition greater than the eutectoid composition of 0.77 wt % C) is treated in Figure 9–40. This case is similar, in many ways, to the hypereutectic path shown in Figure 9–35. A fundamental difference is that the **proeutectoid** cementite (Fe₃C) is the matrix in the final microstructure, whereas the proeutectic phase in Figure 9–35 was the isolated phase. This is because the precipitation of proeutectoid cementite is a solid-state transformation and is favored at grain boundaries. Figure 9–41 illustrates the development of microstructure for a **hypoeutectoid composition** (less than 0.77 wt % C).

The Fe–C system (Figure 9–20) provides an illustration of the development of the microstructure of **gray cast iron** (Figure 9–42). This sketch can be compared with the micrograph in Figure 11–1b.

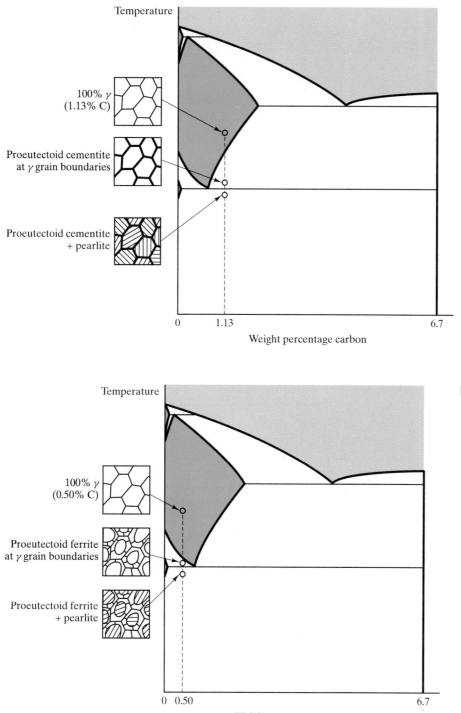

Figure 9-40 *Microstructural development for a slowly cooled hypereutectoid steel (of composition 1.13 wt % C).*

Figure 9-41 *Microstructural development for a slowly cooled hypoeutectoid steel (of composition 0.50 wt % C).*

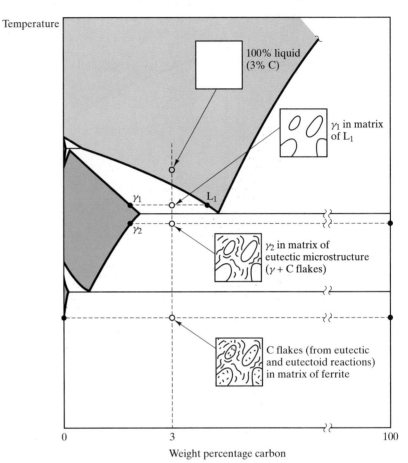

Figure 9-42 *Microstructural development for gray cast iron (of composition 3.0 wt % C) shown on the Fe–C phase diagram. The resulting low-temperature sketch can be compared with the micrograph in Figure 11–1b. A dramatic difference is that, in the actual microstructure, a substantial amount of metastable pearlite was formed at the eutectoid temperature. It is also interesting to compare this sketch with that for white cast iron in Figure 9–38. The small amount of silicon added to promote graphite precipitation is not shown in this two-component diagram.*

SAMPLE PROBLEM 9.6

Figure 9–35 shows the microstructural development for an 80 wt % B alloy. Consider instead 1 kg of a 70 wt % B alloy.

(a) Calculate the amount of β phase at T_3.

(b) Calculate what weight fraction of this β phase at T_3 is proeutectic.

SOLUTION

(a) Using Equation 9.10 gives us

$$m_{\beta,T_3} = \frac{x - x_\alpha}{x_\beta - x_\alpha}(1 \text{ kg}) = \frac{70 - 30}{90 - 30}(1 \text{ kg})$$

$$= 0.667 \text{ kg} = 667 \text{ g}$$

(b) The proeutectic β was that present in the microstructure at T_2:

$$m_{\beta,T_2} = \frac{x - x_L}{x_\beta - x_L}(1 \text{ kg}) = \frac{70 - 60}{90 - 60}(1 \text{ kg})$$

$$= 0.333 \text{ kg} = 333 \text{ g}$$

This portion of the microstructure is retained upon cooling through the eutectic temperature, giving

$$\text{fraction proeutectic} = \frac{\text{proeutectic } \beta}{\text{total } \beta}$$

$$= \frac{333 \text{ g}}{667 \text{ g}} = 0.50$$

SAMPLE PROBLEM 9.7

For 1 kg of 0.5 wt % C steel, calculate the amount of proeutectoid α at the grain boundaries.

SOLUTION

Using Figure 9–41 for illustration and Figure 9–19 for calculation, we essentially need to calculate the equilibrium amount of α at 728°C (i.e., one degree above the eutectoid temperature). Using Equation 9.9, we have

$$m_\alpha = \frac{x_\gamma - x}{x_\gamma - x_\alpha}(1 \text{ kg}) = \frac{0.77 - 0.50}{0.77 - 0.02}(1 \text{ kg})$$

$$= 0.360 \text{ kg} = 360 \text{ g}$$

Note. You might have noticed that this calculation near the eutectoid composition used a value of x_α representative of the maximum solubility of carbon in α-Fe (0.02 wt %). At room temperature (see Sample Problem 9.4), this solubility goes to nearly zero.

SAMPLE PROBLEM 9.8

For 1 kg of 3 wt % C gray iron, calculate the amount of graphite flakes present in the microstructure **(a)** at 1153°C and **(b)** at room temperature.

SOLUTION

(a) Using Figures 9–20 and 9–42, we note that 1153°C is just below the eutectic temperature. Using Equation 9.10 gives us

$$m_C = \frac{x - x_\gamma}{x_C - x_\gamma}(1 \text{ kg}) = \frac{3.00 - 2.08}{100 - 2.08}(1 \text{ kg})$$

$$= 0.00940 \text{ kg} = 9.40 \text{ g}$$

(b) At room temperature, we obtain

$$m_C = \frac{x - x_\alpha}{x_C - x_\alpha}(1 \text{ kg}) = \frac{3.00 - 0}{100 - 0}(1 \text{ kg})$$

$$= 0.030 \text{ kg} = 30.0 \text{ g}$$

Note.   This follows the ideal system of Figure 9–42 and ignores the possibility of any metastable pearlite being formed.

SAMPLE PROBLEM 9.9

Consider 1 kg of an aluminum casting alloy with 10 wt % silicon.

(a) Upon cooling, at what temperature would the first solid appear?
(b) What is the first solid phase and what is its composition?
(c) At what temperature will the alloy completely solidify?
(d) How much proeutectic phase will be found in the microstructure?
(e) How is the silicon distributed in the microstructure at 576°C?

SOLUTION

We follow this microstructural development with the aid of Figure 9–13.

(a) For this composition, the liquidus is at ~595°C.
(b) It is solid solution α with a composition of ~1 wt % Si.
(c) At the eutectic temperature, 577°C.

(d) Practically all of the proeutectic α will have developed by 578°C. Using Equation 9.9, we obtain

$$m_\alpha = \frac{x_L - x}{x_L - x_\alpha}(1 \text{ kg}) = \frac{12.6 - 10}{12.6 - 1.6}(1 \text{ kg})$$

$$= 0.236 \text{ kg} = 236 \text{ g}$$

(e) At 576°C, the overall microstructure is $\alpha + \beta$. The amounts of each are

$$m_\alpha = \frac{x_\beta - x}{x_\beta - x_\alpha}(1 \text{ kg}) = \frac{100 - 10}{100 - 1.6}(1 \text{ kg})$$

$$= 0.915 \text{ kg} = 915 \text{ g}$$

$$m_\beta = \frac{x - x_\alpha}{x_\beta - x_\alpha}(1 \text{ kg}) = \frac{10 - 1.6}{100 - 1.6}(1 \text{ kg})$$

$$= 0.085 \text{ kg} = 85 \text{ g}$$

But we found in (d) that 236 g of the α is in the form of relatively large grains of proeutectic phase, giving

$$\alpha_{\text{eutectic}} = \alpha_{\text{total}} - \alpha_{\text{proeutectic}}$$

$$= 915 \text{ g} - 236 \text{ g} = 679 \text{ g}$$

The silicon distribution is then given by multiplying its weight fraction in each microstructural region by the amount of that region:

Si in proeutectic $\alpha = (0.016)(236 \text{ g}) = 3.8 \text{ g}$

Si in eutectic $\alpha = (0.016)(679 \text{ g}) = 10.9 \text{ g}$

Si in eutectic $\beta = (1.000)(85 \text{ g}) = 85.0 \text{ g}$

Finally, note that the total mass of silicon in the three regions sums to 99.7 g rather than 100.0 g ($= 10$ wt % of the total alloy) due to round-off errors.

SAMPLE PROBLEM 9.10

The solubility of copper in aluminum drops to nearly zero at 100°C. What is the maximum amount of θ phase that will precipitate out in a 4.5 wt % copper alloy quenched and aged at 100°C? Express your answer in weight percent.

SOLUTION

As indicated in Figure 9–27, the solubility limit of the θ phase is essentially unchanging with temperature below $\sim 400°\,C$ and is near a composition of 53 wt % copper. Using Equation 9.10 gives us

$$\text{wt \% } \theta = \frac{x - x_\kappa}{x_\theta - x_\kappa} \times 100\% = \frac{4.5 - 0}{53 - 0} \times 100\%$$

$$= 8.49\%$$

SAMPLE PROBLEM 9.11

In Sample Problem 9.1, we considered a 50 : 50 Pb–Sn solder.

(a) For a temperature of 200°C, determine (i) the phases present, (ii) their compositions, and (iii) their relative amounts (expressed in weight percent).

(b) Repeat part (a) for 100°C.

SOLUTION

(a) Using Figure 9–16, we find the following results at 200°C:

  i. The phases are α and liquid.

  ii. The composition of α is $\sim$18 wt % Sn and of L is $\sim$54 wt % Sn.

  iii. Using Equations 9.9 and 9.10, we have

$$\text{wt \% } \alpha = \frac{x_L - x}{x_L - x_\alpha} \times 100\% = \frac{54 - 50}{54 - 18} \times 100\%$$

$$= 11.1\%$$

$$\text{wt \% L} = \frac{x - x_\alpha}{x_L - x_\alpha} \times 100\% = \frac{50 - 18}{54 - 18} \times 100\%$$

$$= 88.9\%$$

(b) Similarly, at 100°C, we obtain

  i. α and β.

  ii. α is $\sim$5 wt % Sn and β is $\sim$99 wt % Sn.

$$\text{iii. wt \% } \alpha = \frac{x_\beta - x}{x_\beta - x_\alpha} \times 100\% = \frac{99 - 50}{99 - 5} \times 100\% = 52.1\%$$

$$\text{wt \% } \beta = \frac{x - x_\alpha}{x_\beta - x_\alpha} \times 100\% = \frac{50 - 5}{99 - 5} \times 100\% = 47.9\%$$

SAMPLE PROBLEM 9.12

A fireclay refractory ceramic can be made by heating the raw material kaolinite, $Al_2(Si_2O_5)(OH)_4$, driving off the waters of hydration. Determine the phases present, their compositions, and their amounts for the resulting microstructure (below the eutectic temperature).

SOLUTION

A modest rearrangement of the kaolinite formula helps to clarify the production of this ceramic product:

$$Al_2(Si_2O_5)(OH)_4 = Al_2O_3 \cdot 2SiO_2 \cdot 2H_2O$$

The firing operation yields

$$Al_2O_3 \cdot 2SiO_2 \cdot 2H_2O \xrightarrow{\text{heat}} Al_2O_3 \cdot 2SiO_2 + 2H_2O \uparrow$$

The remaining solid has, then, an overall composition

$$\text{mol \% } Al_2O_3 = \frac{\text{mol } Al_2O_3}{\text{mol} Al_2O_3 + \text{mol} SiO_2} \times 100\%$$

$$= \frac{1}{1+2} \times 100\% = 33.3\%$$

Using Figure 9–23, we see that the overall composition falls in the SiO_2 + mullite two-phase region below the eutectic temperature. The SiO_2 composition is 0 mol % Al_2O_3 (i.e., 100% SiO_2). The composition of mullite is 60 mol % Al_2O_3.

Using Equations 9.9 and 9.10 yields

$$\text{mol\% } SiO_2 = \frac{x_{\text{mullite}} - x}{x_{\text{mullite}} - x_{SiO_2}} \times 100\% = \frac{60 - 33.3}{60 - 0} \times 100\%$$

$$= 44.5 \text{ mol\%}$$

$$\text{mol \% mullite} = \frac{x - x_{SiO_2}}{x_{\text{mullite}} - x_{SiO_2}} \times 100\% = \frac{33.3 - 0}{60 - 0} \times 100\%$$

$$= 55.5 \text{ mol\%}$$

Note. Because the Al_2O_3–SiO_2 phase diagram is presented in mole percent, we have made our calculations in a consistent system. It would be a minor task to convert results to weight percent using data from Appendix 1.

PRACTICE PROBLEM 9.6

In Sample Problem 9.6, we calculate microstructural information about the β phase for the 70 wt % B alloy in Figure 9–35. In a similar way, calculate **(a)** the amount of α phase at T_3 for 1 kg of a 50 wt % B alloy and **(b)** the weight fraction of this α phase at T_3, which is proeutectic. (See also Figure 9–36.)

PRACTICE PROBLEM 9.7

Calculate the amount of proeutectoid cementite at the grain boundaries in 1 kg of the 1.13 wt % C hypereutectoid steel illustrated in Figure 9–40. (See Sample Problem 9.7.)

PRACTICE PROBLEM 9.8

In Sample Problem 9.8, the amount of carbon in 1 kg of a 3 wt % C gray iron is calculated at two temperatures. Plot the amount as a function of temperature over the entire temperature range of 1135°C to room temperature.

PRACTICE PROBLEM 9.9

In Sample Problem 9.9, we monitor the microstructural development for 1 kg of a 10 wt % Si–90 wt % Al alloy. Repeat this problem for a 20 wt % Si–80 wt % Al alloy.

PRACTICE PROBLEM 9.10

In Sample Problem 9.10, we calculate the weight percent of θ phase at room temperature in a 95.5 Al–4.5 Cu alloy. Plot the weight percent of θ (as a function of temperature) that would occur upon slow cooling over a temperature range of 548°C to room temperature.

PRACTICE PROBLEM 9.11

Calculate microstructures for **(a)** a 40:60 Pb–Sn solder and **(b)** a 60:40 Pb–Sn solder at 200°C and 100°C. (See Sample Problem 9.11.)

PRACTICE PROBLEM 9.12

In the note at the end of Sample Problem 9.12, the point is made that the results can be easily converted to weight percent. Make these conversions.

SUMMARY

The development of microstructure during slow cooling of materials from the liquid state can be analyzed by using phase diagrams. These "maps" identify the amounts and compositions of phases that are stable at given temperatures. Phase diagrams can be thought of as visual displays of the Gibbs phase rule. In this chapter we restricted our discussion to binary diagrams, which represent phases present at various temperatures and compositions (with pressure fixed at 1 atm) in systems with two components; the components can be elements or compounds.

Several types of binary diagrams are commonly encountered. For very similar components, complete solid solution can occur in the solid state as well as in the liquid state. In the two-phase (liquid solution + solid solution) region, the composition of each phase is indicated by a tie line. Many binary systems exhibit a eutectic reaction in which a low melting point (eutectic) composition produces a fine-grained, two-phase microstructure. Such eutectic diagrams are associated with limited solid solution. The completely solid-state analogy to the eutectic reaction is the eutectoid reaction, in which a single solid phase transforms upon cooling to a fine-grained microstructure of two other solid phases. The peritectic reaction represents the incongruent melting of a solid compound. Upon melting, the compound transforms to a liquid and another solid, each of composition different from the original compound. Many binary diagrams include various intermediate compounds, leading to a relatively complex appearance. However, such general binary diagrams can always be reduced to a simple binary diagram associated with the overall composition of interest.

The tie line that identifies the compositions of the phases in a two-phase region can also be used to calculate the amount of each phase. This is done with the lever rule, in which the tie line is treated as a lever with its fulcrum located at the overall composition. The amounts of the two phases are such that they "balance the lever." The lever rule is, of course, a mechanical analog, but it follows directly from a mass balance for the two-phase system. The lever rule can be used to follow microstructural development as an overall composition is slowly cooled from the melt. This is especially helpful in understanding the microstructure that results in a composition near a eutectic composition. Several binary diagrams of importance to the metals and ceramics industries were given in this chapter. Special emphasis was given to the Fe–Fe$_3$C system, which provides the major scientific basis for the iron and steel industries.

KEY TERMS

austenite(319)

binary diagram(309)

complete solid solution(310)

component(305)

congruent melting(322)

degrees of freedom(307)

eutectic composition(314)

eutectic diagram(314)

eutectic reaction(318)

eutectic temperature(314)

eutectoid diagram(318)

general diagram(324)

Gibbs phase rule(307)

gray cast iron(340)

hypereutectic composition(336)

REFERENCES

ASM Handbook, Vol. 3: *Alloy Phase Diagrams,* ASM International, Materials Park, Ohio, 1992.

Binary Alloy Phase Diagrams, 2nd ed., Vols. 1–3, T. B. Massalski, et al., eds., ASM International, Materials Park, Ohio, 1990. The result of a cooperative program between ASM International and the National Institute of Standards and Technology for the critical review of 4700 phase-diagram systems.

Phase Diagrams for Ceramists, Vols. 1–12, American Ceramic Society, Columbus, Ohio, 1964–1996.

PROBLEMS

9.1 • The Phase Rule

9.1. Apply the Gibbs phase rule to the various points in the one-component H_2O phase diagram (Figure 9–3).

9.2. Apply the Gibbs phase rule to the various points in the one-component iron phase diagram (Figure 9–4).

9.3. Calculate the degrees of freedom for a 50:50 copper–nickel alloy at **(a)** 1400°C where it exists as a single, liquid phase, **(b)** 1300°C where it exists as a two-phase mixture of liquid and solid solutions, and **(c)** 1200°C where it exists as a single, solid-solution phase. Assume a constant pressure of 1 atm above the alloy in each case.

9.4. In Figure 9–7, the Gibbs phase rule was applied to a hypothetical phase diagram. In a similar way, apply the phase rule to a sketch of the Pb–Sn phase diagram (Figure 9–16).

9.5. Apply the Gibbs phase rule to a sketch of the MgO–Al_2O_3 phase diagram (Figure 9–26).

9.6. Apply the Gibbs phase rule to the various points in the Al_2O_3–SiO_2 phase diagram (Figure 9–23).

9.2 • The Phase Diagram

9.7. Describe qualitatively the microstructural development that will occur upon slow cooling of a melt of equal parts (by weight) of copper and nickel (see Figure 9–9).

9.8. Describe qualitatively the microstructural development that will occur upon slow cooling of a melt composed of 50 wt % Al and 50 wt % Si (see Figure 9–13).

9.9. Describe qualitatively the microstructural development that will occur upon slow cooling of a melt composed of 87.4 wt % Al and 12.6 wt % Si (see Figure 9–13).

9.10. Describe qualitatively the microstructural development during the slow cooling of a melt composed of **(a)** 10 wt % Pb–90 wt % Sn, **(b)** 40 wt % Pb–60 wt % Sn, and **(c)** 50 wt % Pb–50 wt % Sn (see Figure 9–16).

9.11. Repeat Problem 9.10 for a melt composed of 38.1 wt % Pb–61.9 wt % Sn.

9.12. Describe qualitatively the microstructural development that will occur upon slow cooling of an alloy with equal parts (by weight) of aluminum and θ phase (Al_2Cu) (see Figure 9–27).

9.13. Describe qualitatively the microstructural development that will occur upon slow cooling of a melt composed of **(a)** 20 wt % Mg, 80 wt % Al and **(b)** 80 wt % Mg, 20 wt % Al (see Figure 9–28).

9.14. Describe qualitatively the microstructural development during the slow cooling of a 30:70 brass (Cu with 30 wt % Zn). See Figure 9–29 for the Cu–Zn phase diagram.

9.15. Repeat Problem 9.14 for a 35 : 65 brass.

9.16. Describe qualitatively the microstructural development during the slow cooling of **(a)** a 50 mol % Al_2O_3–50 mol % SiO_2 ceramic and **(b)** an 80 mol % Al_2O_3–20 mol % SiO_2 ceramic (see Figure 9–23).

9.3 • The Lever Rule

9.17. Calculate the amount of each phase present in 1 kg of a 50 wt % Ni–50 wt % Cu alloy at **(a)** 1400°C, **(b)** 1300°C, and **(c)** 1200°C (see Figure 9–9).

9.18. Calculate the amount of each phase present in 1 kg of a 50 wt % Pb–50 wt % Sn solder alloy at **(a)** 300°C, **(b)** 200°C, **(c)** 100°C, and **(d)** 0°C (see Figure 9–16).

9.19. Repeat Problem 9.18 for a 60 wt % Pb–40 wt % Sn solder alloy.

9.20. Repeat Problem 9.18 for an 70 wt % Pb–30 wt % Sn solder alloy.

9.21. Calculate the amount of each phase present in 50 kg of a brass with composition 35 wt % Zn–65 wt % Cu at **(a)** 1000°C, **(b)** 900°C, **(c)** 800°C, **(d)** 700°C, **(e)** 100°C, and **(f)** 0°C (see Figure 9–29).

9.22. Some aluminum from a "metallization" layer on a solid-state electronic device has diffused into the silicon substrate. Near the surface, the silicon has an overall concentration of 1.0 wt % Al. In this region, what percentage of the microstructure would be composed of α-phase precipitates, assuming equilibrium? (See Figure 9–13 and assume the phase boundaries at 300°C will be essentially unchanged to room temperature.)

9.23. Calculate the amount of proeutectoid α present at the grain boundaries in 1 kg of a common 1020 structural steel (0.20 wt % C). (See Figure 9–19.)

9.24. Repeat Problem 9.23 for a 1060 structural steel (0.60 wt % C).

9.25. The ideal stoichiometry of the γ phase in the Al–Mg system is $Al_{12}Mg_{17}$. **(a)** What is the atomic percentage of excess Al in the most aluminum-rich γ composition at 450°C? **(b)** What is the atomic percentage of excess Mg in the most magnesium-rich γ composition at 437°C? (See Figure 9–28.)

• **9.26.** Suppose that you have a crucible containing 1 kg of an alloy of composition 90 wt % Sn–10 wt % Pb at a temperature of 184°C. How much Sn would you have to add to the crucible to completely solidify the alloy *without* changing the system temperature? (See Figure 9–16.)

9.27. Determine the phases present, their compositions, and their amounts (below the eutectic temperature) for a refractory made from equal molar fractions of kaolinite and mullite ($3Al_2O_3 \cdot 2SiO_2$). (See Figure 9–23.)

9.28. Repeat Problem 9.27 for a refractory made from equal molar fractions of kaolinite and silica (SiO_2).

• **9.29.** You have supplies of kaolinite, silica, and mullite as raw materials. Using kaolinite plus *either* silica *or* mullite, calculate the batch composition (in weight percent) necessary to produce a final microstructure that is equimolar in silica and mullite. (See Figure 9–23.)

9.30. Calculate the phases present, their compositions, and their amounts (in weight percent) for the microstructure at 1000°C for **(a)** a spinel (MgO · Al_2O_3) refractory with 1 wt % excess MgO (i.e., 1 g MgO per 99 g MgO · Al_2O_3), and **(b)** a spinel refractory with 1 wt % excess Al_2O_3. (See Figure 9–26.)

9.31. A partially stabilized zirconia (for a novel structural application) is desired to have an equimolar microstructure of tetragonal and cubic zirconia at an operating temperature of 1500°C. Calculate the proper CaO content (in weight percent) for this structural ceramic. (See Figure 9–30.)

9.32. Repeat Problem 9.31 for a microstructure with equal weight fractions of tetragonal and cubic zirconia.

9.33. Calculate the amount of each phase present in a 1-kg alumina refractory with composition 80 mol % Al_2O_3–20 mol % SiO_2 at **(a)** 2000°C, **(b)** 1900°C, and **(c)** 1800°C. (See Figure 9–23.)

9.34. In a test laboratory, quantitative x-ray diffraction determines that a refractory brick has 25 wt % alumina phase and 75 wt % mullite solid solution. What is the overall SiO_2 content (in wt %) of this material? (See Figure 9–23.)

9.35. An important structural ceramic is partially stabilized zirconia (PSZ), which has a composition lying in the two-phase ZrO_2-cubic ZrO_2(ss) region. Use Figure 9–30 to calculate the amount of each phase present in a 10 mol % CaO PSZ at 500°C.

• **9.36.** In a materials laboratory experiment, a student sketches a microstructure observed under an optical microscope. The sketch appears as

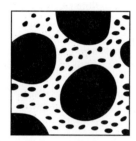

The phase diagram for this alloy system is

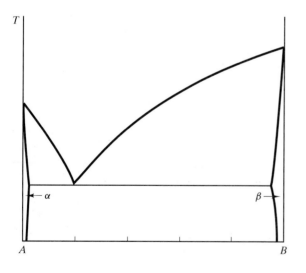

Determine **(a)** whether the black regions in the sketch represent α or β phase and **(b)** the approximate alloy composition.

9.4 • Microstructural Development During Slow Cooling

9.37. Calculate **(a)** the weight fraction of the α phase that is proeutectic in a 10 wt % Si–90 wt % Al alloy at 576°C and **(b)** the weight fraction of the β phase that is proeutectic in a 20 wt % Si–80 wt % Al alloy at 576°C.

9.38. Plot the weight percent of phases present as a function of temperature for a 10 wt % Si–90 wt % Al alloy slowly cooled from 700 to 300°C.

9.39. Plot the weight percent of phases present as a function of temperature for a 20 wt % Si–80 wt % Al alloy slowly cooled from 800 to 300°C.

9.40. Calculate the *weight* fraction of mullite that is proeutectic in a slowly cooled 30 mol % Al_2O_3–70 mol % SiO_2 refractory cooled to room temperature.

9.41. Microstructural analysis of a slowly cooled Al–Si alloy indicates there is a 5 *volume* % silicon-rich proeutectic phase. Calculate the overall alloy composition (in weight percent).

9.42. Repeat Problem 9.41 for a 10 volume % silicon-rich proeutectic phase.

9.43. Calculate the amount of proeutectic γ that has formed at 1149°C in the slow cooling of the 3.0 wt % C white cast iron illustrated in Figure 9–38. Assume a total of 100 kg of cast iron.

9.44. Plot the weight percent of phases present as a function of temperature for the 3.0 wt % C white cast iron illustrated in Figure 9–38 slowly cooled from 1400 to 0°C.

9.45. Plot the weight percent of phases present as a function of temperature from 1000 to 0°C for the 0.77 wt % C eutectoid steel illustrated in Figure 9–39.

9.46. Plot the weight percent of phases present as a function of temperature from 1000 to 0°C for the 1.13 wt % C hypereutectoid steel illustrated in Figure 9–40.

9.47. Plot the weight percent of phases present as a function of temperature from 1000 to 0°C for a common 1020 structural steel (0.20 wt % C).

9.48. Repeat Problem 9.47 for a 1040 structural steel (0.40 wt % C).

9.49. Plot the weight percent of phases present as a function of temperature from 1000 to 0°C for the 0.50 wt % C hypoeutectoid steel illustrated in Figure 9–41.

9.50. Plot the weight percent of phases present as a function of temperature from 1400 to 0°C for a white cast iron with an overall composition of 2.5 wt % C.

9.51. Plot the weight percent of all phases present as a function of temperature from 1400 to 0°C for a gray cast iron with an overall composition of 3.0 wt % C.

9.52. Repeat Problem 9.51 for a gray cast iron with an overall composition of 2.5 wt % C.

9.53. In comparing the equilibrium schematic microstructure in Figure 9–42 with the actual, room temperature microstructure shown in Figure 11–1b, it is apparent that metastable pearlite can form at the eutectoid temperature (due to insufficient time for the more stable, but slower, graphite formation). Assuming that Figures 9–20 and 9–42 are accurate for 100 kg of a gray cast iron (3.0 wt % C) down to 738°C but that pearlite forms upon cooling through the eutectoid temperature, calculate the amount of pearlite to be expected in the room temperature microstructure.

9.54. For the assumptions in Problem 9.53, calculate the amount of flake graphite in the room temperature microstructure.

9.55. Plot the weight percent of phases present as a function of temperature from 800 to 300°C for a 95 Al–5 Cu alloy.

9.56. Consider 1 kg of a brass with composition 35 wt % Zn–65 wt % Cu. **(a)** Upon cooling, at what temperature would the first solid appear? **(b)** What is the first solid phase to appear, and what is its composition? **(c)** At what temperature will the alloy completely solidify? **(d)** Over what temperature range will the microstructure be completely in the α-phase?

9.57. Repeat Problem 9.56 for 1 kg of a brass with composition 30 wt % Zn–70 wt % Cu.

9.58. Plot the weight percent of phases present as a function of temperature from 1000 to 0°C for a 35 wt % Zn–65 wt % Cu brass.

9.59. Repeat Problem 9.58 for a 30 wt % Zn–70 wt % Cu brass.

9.60. Repeat Problem 9.58 for 1 kg of brass with a composition of 15 wt % Zn–85 wt % Cu.

9.61. For a 15 wt % Zn–85 wt % Cu brass, plot the weight percent of phases present as a function of temperature from 1100°C to 0°C.

9.62. Calculate the amount of β phase that would precipitate from 1 kg of 95 wt % Al–5 wt % Mg alloy slowly cooled to 100°C.

9.63. Identify the composition ranges in the Al–Mg system for which precipitation of the type illustrated in Sample Problem 9.10 can occur (i.e., a second phase can precipitate from a single-phase microstructure upon cooling).

9.64. Plot the weight percent of phases present as a function of temperature from 700 to 100°C for a 90 Al–10 Mg alloy.

9.65. A solder batch is made by melting together 64 g of a 40:60 Pb–Sn alloy with 53 g of a 60:40 Pb–Sn alloy. Calculate the amounts of α and β phase that would be present in the overall alloy, assuming it is slowly cooled to room temperature, 25°C.

9.66. Plot the weight percent of phases present as a function of temperature from 400 to 0°C for a slowly cooled 50:50 Pb–Sn solder.

9.67. Plot the phases present (in mole percent) as a function of temperature for the heating of a refractory with the composition 60 mol % Al_2O_3–40 mol % MgO from 1000 to 2500°C.

9.68. Plot the phases present (in mole percent) as a function of temperature for the heating of a partially stabilized zirconia with 10 mol % CaO from room temperature to 2800°C.

CHAPTER 10
Kinetics—Heat Treatment

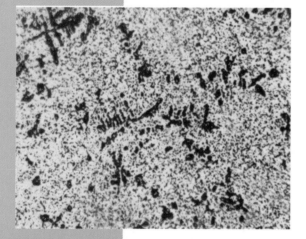

The microstructure of a rapidly cooled "eutectic" soft solder (≈ 38 wt % Pb − 62 wt % Sn) consists of globules of lead-rich solid solution (dark) in a matrix of tin-rich solid solution (white), 375X. The contrast to the slowly-cooled microstructure at the opening of Chapter 9 illustrates the effect of time on microstructural development. (From ASM Handbook, Vol. 3: Alloy Phase Diagrams, ASM International, Materials Park, Ohio, 1992.)

354

Chapter 9 introduced the powerful tool of phase diagrams for describing equilibrium microstructural development during slow cooling from the melt. Throughout that chapter, however, we were cautioned that phase diagrams represent microstructure that "should" develop, assuming that temperature is changed slowly enough to maintain equilibrium at all times. In practice, materials processing, like so much of daily life, is rushed, and time becomes an important factor. The practical aspect of this is *heat treatment*, the temperature versus time history necessary to generate a desired microstructure. The fundamental basis for heat treatment is *kinetics*, which we shall define as the science of time-dependent phase transformations.

We begin by adding a time scale to phase diagrams to show the approach to equilibrium. A systematic treatment of this kind generates a *TTT diagram*, which summarizes, for a given composition, the percentage completion of a given phase transformation on temperature–time axes (giving the three "T's" of temperature, time, and transformation). Such diagrams are maps in the same sense that phase diagrams are maps. TTT diagrams can include descriptions of transformations that involve time-dependent solid-state diffusion and of transformations that occur by a rapid, shearing mechanism, essentially independent of time. As with phase diagrams, some of our best illustrations of TTT diagrams will involve ferrous alloys. We shall explore some of the basic considerations in the heat treatment of steel. Related to this is the characterization of *hardenability*. *Precipitation hardening* is an important heat treatment illustrated by some nonferrous alloys. *Annealing* is a heat treatment leading to reduced hardness by means of successive stages of *recovery*, *recrystallization*, and *grain growth*. Heat treatment is not a topic isolated to metallurgy. To illustrate this, we conclude this chapter with a discussion of some important phase transformations in nonmetallic systems.

10.1 TIME—THE THIRD DIMENSION

Time did not appear in any quantitative way in the discussion of phase diagrams in Chapter 9. Aside from requiring temperature changes to occur relatively slowly, we did not consider time as a factor at all. Phase diagrams summarized equilibrium states and, as such, those states (and associated microstructures) should be stable and unchanging with time. However, these equilibrium structures take time to develop, and the approach to equilibrium can be mapped on a time scale.* A simple illustration of this is given in Figure 10–1, which shows a time axis perpendicular to the temperature–composition plane of a phase diagram. For component A, the phase diagram indicates that solid A should exist at any temperature below the melting point. However, Figure 10–1 indicates that the time required for the liquid phase to transform to the solid phase is a strong function of temperature.

---

* The relationship between thermodynamics and kinetics is explored in the thermodynamics chapter of the *Instructor's Manual for Introduction to Materials Science for Engineers.*

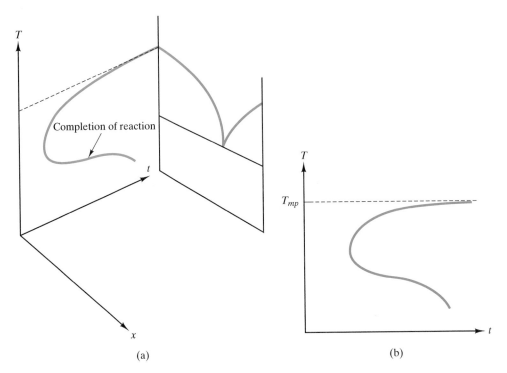

Figure 10-1 *Schematic illustration of the approach to equilibrium. (a) The time for solidification to go to completion is a strong function of temperature, with the minimum time occurring for a temperature considerably below the melting point. (b) The temperature–time plane with "transformation curve." We shall see later that the time axis is often plotted on a logarithmic scale.*

Another way of stating this is that the time necessary for the solidification reaction to go to completion varies with temperature. In order to compare the reaction times in a consistent way, Figure 10–1 represents the rather ideal case of quenching the liquid from the melting point *instantaneously* to some lower temperature and then measuring the time for solidification to go to completion at that temperature. At first glance, the nature of the plot in Figure 10–1 may seem surprising. The reaction proceeds slowly near the melting point and at relatively low temperatures. The reaction is fastest at some intermediate temperature. To understand this "knee-shaped" transformation curve, we must explore some fundamental concepts of kinetics theory.

For this discussion, we focus more closely on the precipitation of a single-phase solid within a liquid matrix (Figure 10–2). This process is an example of **homogeneous nucleation,** meaning that the precipitation occurs within a completely homogeneous medium. The more common case is **heterogeneous nucleation,** in which the precipitation occurs at some structural imperfection such as a foreign surface. The imperfection reduces the surface energy associated with forming the new phase.

Even homogeneous nucleation is rather involved. The precipitation process actually occurs in two stages. First is **nucleation**. The new phase,

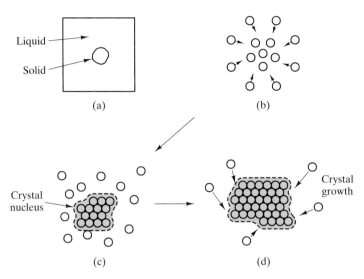

(a)   (b)   (c)   (d)

Figure 10-2   *(a) On a microscopic scale, a solid precipitate in a liquid matrix. The precipitation process is seen on the atomic scale as (b) a clustering of adjacent atoms to form (c) a crystalline nucleus followed by (d) the growth of the crystalline phase.*

which forms because it is more stable, first appears as small nuclei. These result from local atomic fluctuations and are, typically, only a few hundred atoms in size. This initial stage involves the random production of many nuclei. Only those larger than a given size are stable and can continue to grow. These critical-size nuclei must be large enough to offset the energy of formation for the solid–liquid interface. The rate of nucleation (i.e., the rate at which nuclei of critical size or larger appear) is the result of two competing factors. At the precise transformation temperature (in this case, the melting point), the solid and liquid phases are in equilibrium, and there is no net driving force for the transformation to occur. As the liquid is cooled below the transformation temperature, it becomes increasingly unstable. The classical theory of nucleation is based on an energy balance between the nucleus and its surrounding liquid. The key principle is that a small cluster of atoms (the nucleus) will be stable only if further growth reduces the net energy of the system. Taking the nucleus in Figure 10–2a as spherical, the energy balance can be illustrated by Figure 10–3, showing that the nucleus will be stable if its radius, r, is greater than a critical value, r_c.

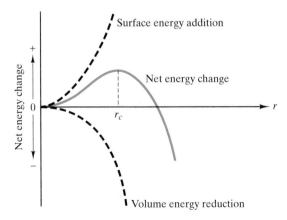

Figure 10-3   *Classical nucleation theory involves an energy balance between the nucleus and its surrounding liquid. A nucleus (cluster of atoms) as shown in Figure 10–2(c) will be stable only if further growth reduces the net energy of the system. An ideally spherical nucleus will be stable if its radius, r, is greater than a critical value, r_c.*

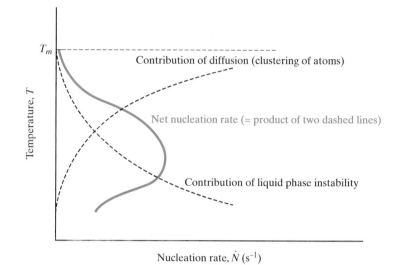

Figure 10-4 *The rate of nucleation is a product of two curves that represent two opposing factors (instability and diffusivity).*

The driving force for solidification increases with decreasing temperature, and the rate of nucleation increases sharply. This increase cannot continue indefinitely. The clustering of atoms to form a nucleus is a local-scale diffusion process. As such, this step will decrease in rate with decreasing temperature. This rate decrease is exponential in nature and another example of Arrhenius behavior (see Section 5.1). The overall nucleation rate reflects these two factors by increasing from zero at the transformation temperature (T_m) to a maximum value somewhere below T_m and then decreasing with further decreases in temperature (Figure 10–4). In a preliminary way, we now have an explanation for the shape of the curve in Figure 10–1. The time for reaction is long just below the transformation temperature because the driving force for reaction is small and the reaction rate is therefore small. The time for reaction is again long at low temperatures because the diffusion rate is small. In general, the time axis in Figure 10–1 is the inverse of the rate axis in Figure 10–4.

Our explanation of Figure 10–1 using Figure 10–4 is preliminary because we have not yet included the growth step (see Figure 10–2). This process, like the initial clustering of atoms in nucleation, is diffusional in nature. This makes the growth rate, $\dot{G}$, an Arrhenius expression:

$$\dot{G} = Ce^{-Q/RT} \tag{10.1}$$

where C is a preexponential constant, Q the activation energy for self-diffusion in this system, R the universal gas constant, and T the absolute temperature. This expression is discussed in some detail in Section 5.1. Figure 10–5 shows the nucleation rate, $\dot{N}$, and the growth rate, $\dot{G}$, together. The overall transformation rate is shown as a product of $\dot{N}$ and $\dot{G}$. This more complete picture of phase transformation shows the same general behavior

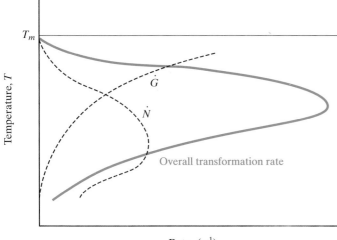

Figure 10-5 *The overall transformation rate is the product of the nucleation rate $\dot{N}$ (from Figure 10–4) and the growth rate $\dot{G}$ (given by Equation 10.1).*

as nucleation rate. The temperature corresponding to the maximum rate has shifted, but the general argument has remained the same. The maximum rate occurs in a temperature range where the driving forces for solidification *and* diffusion rates are both significant. Although this principle explains these knee-shaped curves in a qualitative way, we must acknowledge that transformation curves for many practical engineering materials frequently include additional factors, such as multiple diffusion mechanisms and mechanical strains associated with solid-state transformations.

SAMPLE PROBLEM 10.1

At 900°C, growth rate $\dot{G}$ is a dominant term in the crystallization of a copper alloy. By dropping the system temperature to 400°C, the growth rate drops six orders of magnitude and effectively reduces the crystallization rate to zero. Calculate the activation energy for self-diffusion in this alloy system.

SOLUTION

This is a direct application of Equation 10.1:

$$\dot{G} = Ce^{-Q/RT}$$

Considering two different temperatures yields

$$\frac{\dot{G}_{900°C}}{\dot{G}_{400°C}} = \frac{Ce^{-Q/R(900+273)K}}{Ce^{-Q/R(400+273)K}}$$

$$= e^{-Q/R(1/1173 - 1/673)K^{-1}}$$

This gives

$$Q = -\frac{R \ln (\dot{G}_{900°C}/\dot{G}_{400°C})}{(1/1173 - 1/673) \text{ K}^{-1}}$$

$$= -\frac{[8.314\text{J}/(\text{mol} \cdot \text{K})] \ln 10^6}{(1/1173 - 1/673) \text{ K}^{-1}} = 181 \text{ kJ/mol}$$

Note.   Because the crystallization rate is so high at elevated temperatures, it is not possible to suppress crystallization completely unless the cooling is accomplished at exceptionally high quench rates. The results in those special cases are the interesting amorphous metals (see Section 4.5).

....

PRACTICE PROBLEM 10.1

In Sample Problem 10.1, the activation energy for crystal growth in a copper alloy is calculated. Using that result, calculate the temperature at which the growth rate would have dropped three orders of magnitude relative to the rate at 900°C.

10.2   THE TTT DIAGRAM

The preceding section introduced time as an axis in monitoring microstructural development. The general term for a plot of the type shown in Figure 10–1 is **TTT diagram,** where the letters stand for temperature, time, and (percent) transformation. This plot is also known as an **isothermal transformation diagram.** In the case of Figure 10–1 the time necessary for 100% completion of transformation was plotted. Figure 10–6 shows how the progress of the transformation can be traced with a family of curves showing different percentages of completion. Using the industrially important eutectoid transformation in steels as an example, we can now discuss, in further detail, the nature of **diffusional transformations** in solids (a change in structure due to the long-range migration of atoms). In addition, we shall find that some **diffusionless transformations** play an important role in microstructural development and can be superimposed on the TTT diagrams.

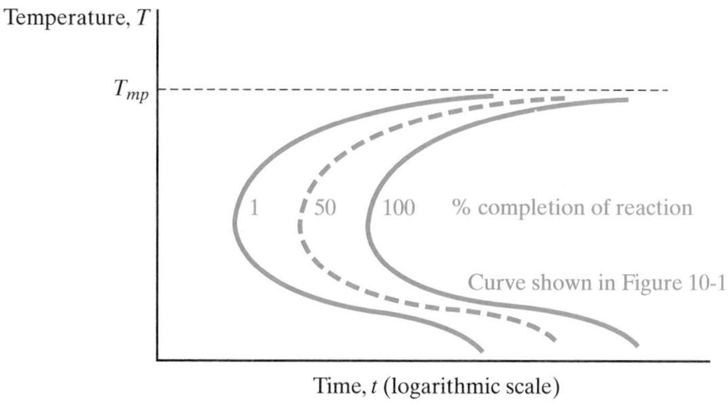

Figure 10-6 *A time–temperature–transformation diagram for the so-lidification reaction of Figure 10–1 with various percent comple-tion curves illustrated.*

DIFFUSIONAL TRANSFORMATIONS

Diffusional transformations involve a change of structure due to the long-range migration of atoms. The development of microstructure during the slow cooling of eutectoid steel (Fe with 0.77 wt % C) was shown in Figure 9–39. A TTT diagram for this composition is shown in Figure 10–7. It is quite similar to the schematic for solidification shown in Figure 10–1. The most important new information provided in Figure 10–7 is that **pearlite** is not the only microstructure that can develop from the cooling of **austenite**. In fact, various types of pearlite are noted at various transformation temperatures. The slow cooling path assumed in Chapter 9 is illustrated in Figure 10–8. Clearly, this leads to the development of a coarse pearlite. Here, all references to size are relative. In Chapter 9, we made an issue of the fact that eutectic and eutectoid structures are generally fine-grained. Figure 10–7 indicates that the pearlite produced near the eutectoid temperature is not as fine-grained as that produced at slightly lower temperatures. The reason for this trend can be appreciated from Figure 10–5. Low nucleation rates and high diffusion rates near the eutectoid temperature lead to a relatively coarse structure. The increasingly fine pearlite formed at lower tempera-tures is eventually beyond the resolution of optical microscopes (approxi-mately 0.25 μm features observable at about $2000\times$ magnification). Such fine structure can be observed with electron microscopy.

Pearlite formation is found from the eutectoid temperature ($727°$C) down to about $400°$C. Below $400°$C, the pearlite microstructure is no longer formed. The ferrite and cementite form as extremely fine needles in a mi-crostructure known as **bainite*** (Figure 10–9). This represents an even finer

---

* Edgar Collins Bain (1891–1971), American metallurgist, discovered the microstructure that now bears his name. His many achievements in the study of steels made him one of the most honored metallurgists of his generation.

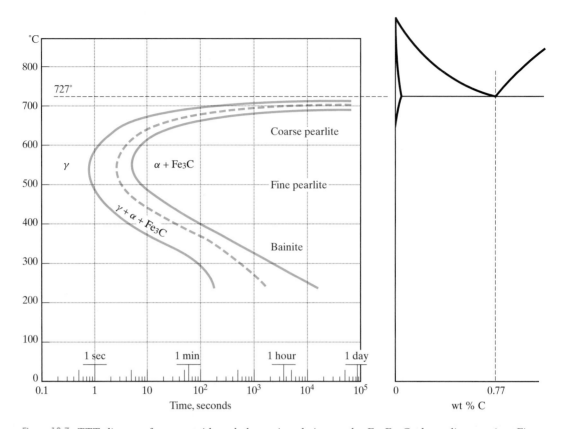

Figure 10-7 *TTT diagram for eutectoid steel shown in relation to the Fe–Fe$_3$C phase diagram (see Figure 9–39). This shows that, for certain transformation temperatures, bainite rather than pearlite is formed. In general, the transformed microstructure is increasingly fine-grained as the transformation temperature is decreased. Nucleation rate increases and diffusivity decreases as temperature decreases. The solid curve on the left represents the onset of transformation ($\sim$ 1% completion). The dashed curve represents 50% completion. The solid curve on the right represents the effective ($\sim$ 99%) completion of transformation. This convention is used in subsequent TTT diagrams. (TTT diagram after* Atlas of Isothermal Transformation and Cooling Transformation Diagrams, *American Society for Metals, Metals Park, Ohio, 1977.)*

distribution of ferrite and cementite than in fine pearlite. Although a different morphology is found in bainite, the general trend of finer structure with decreasing temperature is continued. It is important to note that the variety of morphologies that develops over the range of temperatures shown in Figure 10–7 all represent the same phase compositions and relative amounts of each phase. These terms all derive from the equilibrium calculations (using the tie line and lever rule) of Chapter 9. It is equally important to note that TTT diagrams represent specific thermal histories and are *not* state diagrams in the way that phase diagrams are. For instance, coarse pearlite is more stable than fine pearlite or bainite because it has less total interfacial boundary area (a high-energy region as discussed in Section 4.4). As a result, coarse pearlite, once formed, remains upon cooling, as illustrated in Figure 10–10.

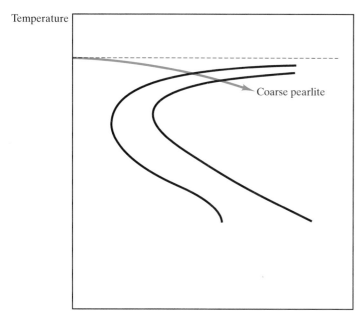

Figure 10-8 *A slow cooling path that leads to coarse pearlite formation is superimposed on the TTT diagram for eutectoid steel. This type of thermal history was assumed, in general, throughout Chapter 9.*

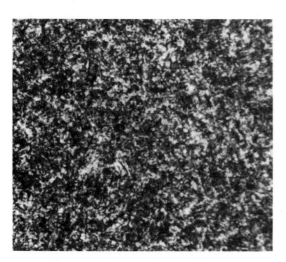

Figure 10-9 *The microstructure of bainite involves extremely fine needles of α-Fe and Fe₃C, in contrast to the lamellar structure of pearlite (see Figure 9–2), 535×. (From* Metals Handbook, *8th Ed., Vol. 7:* Atlas of Microstructures, *American Society for Metals, Metals Park, Ohio, 1972.)*

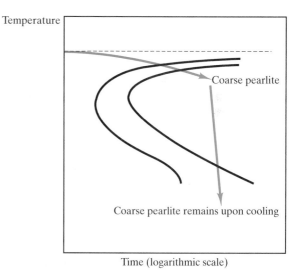

Figure 10-10 *The interpretation of TTT diagrams requires consideration of the thermal history "path." For example, coarse pearlite, once formed, remains stable upon cooling. The finer-grain structures are less stable because of the energy associated with the grain boundary area. (By contrast, phase diagrams represent equilibrium and identify stable phases independent of the path used to reach a given state point.)*

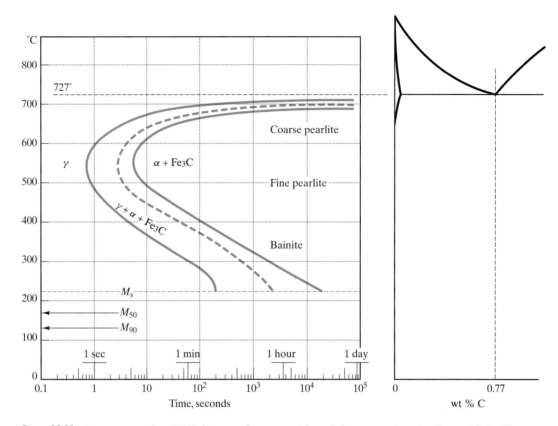

Figure 10-11 *A more complete TTT diagram for eutectoid steel than was given in Figure 10–7. The various stages of the time-independent (or diffusionless) martensitic transformation are shown as horizontal lines. M_s represents the start, M_{50} 50% transformation, and M_{90} 90% transformation. One hundred percent transformation to martensite is not complete until a final temperature (M_f) of $-46°C$.*

DIFFUSIONLESS (MARTENSITIC) TRANSFORMATIONS

The eutectoid reactions in Figure 10–7 are all diffusional in nature. But close inspection of that TTT diagram indicates that no information is given below about 250°C. Figure 10–11 shows that a very different process occurs at lower temperatures. Two horizontal lines are added to represent the occurrence of a *diffusionless* process known as the **martensitic* transformation**. This is a generic term that refers to a broad family of diffusionless transformations in metals and nonmetals alike. The most common example is the specific transformation in eutectoid steels. In this system, the product

---

*   Adolf Martens (1850–1914), German metallurgist, was originally trained as a mechanical engineer. Early in his career, he became involved in the developing field of testing materials for construction. He was a pioneer in using the microscope as a practical analytical tool for metals. Later, in an academic post he produced the highly regarded *Handbuch der Materialienkunde* (1899).

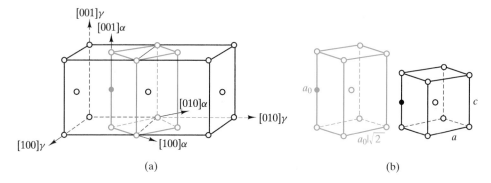

Figure 10-12 *For steels, the martensitic transformation involves the sudden reorientation of C and Fe atoms from the fcc solid solution of γ-Fe (austenite) to a body-centered tetragonal (bct) solid solution (martensite). In (a), the bct unit cell is shown relative to the fcc lattice by the $\langle 100 \rangle_\alpha$ axes. In (b), the bct unit cell is shown before (left) and after (right) the transformation. The open circles represent iron atoms. The solid circle represents an interstitially dissolved carbon atom. This illustration of the martensitic transformation was first presented by Bain in 1924, and while subsequent study has refined the details of the transformation mechanism, this remains a useful and popular schematic. (After J. W. Christian, in* Principles of Heat Treatment of Steel, *G. Krauss, Ed., American Society for Metals, Metals Park, Ohio, 1980.)*

formed from the quenched austenite is termed **martensite.** In effect, the quenching of austenite rapidly enough to bypass the pearlite "knee" at approximately 550°C allows any diffusional transformation to be suppressed. However, there is a price to pay for the avoidance of the diffusional process. The austenite phase is still unstable and is, in fact, increasingly unstable with decreasing temperature. At approximately 215°C, the instability of austenite is so great that a small fraction (less than 1%) of the material transforms spontaneously to martensite. Instead of the diffusional migration of carbon atoms to produce separate α and Fe_3C phases, the martensite transformation involves the sudden reorientation of C and Fe atoms from the fcc solid solution of γ-Fe to a body-centered tetragonal (bct) solid solution, which is martensite (Figure 10–12). The relatively complex crystal structure and the supersaturated concentration of carbon atoms in the martensite lead to a characteristically brittle nature. The start of the martensitic transformation is labeled M_s and is shown as a horizontal line (i.e., time independent) in Figure 10–11. If the quench of austenite proceeds below M_s, the austenite phase is increasingly unstable and a larger fraction of the system is transformed to martensite. Various stages of the martensitic transformation are noted in Figure 10–11. Quenching to −46°C or below leads to the complete transformation to martensite. The acicular, or needlelike, microstructure of martensite is shown in Figure 10–13. Martensite is a **metastable** phase; that is, it is stable with time but upon reheating will decompose into the even more stable phases of α and Fe_3C. The careful control of the proportions of these various phases is the subject of heat treatment, which is discussed in the next section.

Figure 10-13 *Acicular, or needlelike, microstructure of martensite 1000×. (From* Metals Handbook, *8th Ed., Vol. 7:* Atlas of Microstructures, *American Society for Metals, Metals Park, Ohio, 1972.)*

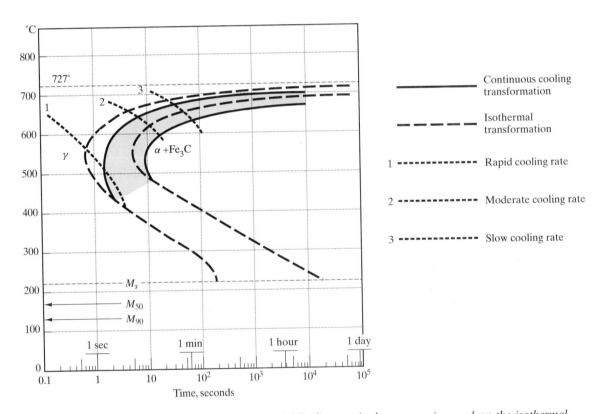

Figure 10-14 *A continuous cooling transformation (CCT) diagram is shown superimposed on the isothermal transformation diagram of Figure 10–11. The general effect of continuous cooling is to shift the transformation curves downward and toward the right. (After* Atlas of Isothermal Transformation and Cooling Transformation Diagrams, *American Society for Metals, Metals Park, Ohio, 1977.)*

As one might expect, the complex set of factors (discussed in Section 10.1) that determine transformation rates requires the TTT diagram to be defined in terms of a specific thermal history. The TTT diagrams in this chapter are generally *isothermal;* that is, the transformation time at a given temperature represents the time for transformation at the fixed temperature following an instantaneous quench. Figure 10–8 and several subsequent diagrams will superimpose cooling or heating paths on these diagrams. Such paths can affect the time at which the transformation will have occurred at a given temperature. In other words, the positions of transformation curves are shifted slightly downward and toward the right for nonisothermal conditions. Such a **continuous cooling transformation** (CCT) **diagram** is shown in Figure 10–14. For the purpose of illustration, we shall not generally make this refinement in this book. The principles demonstrated are, nonetheless, valid.

Our discussion to this point has centered on the eutectoid composition. Figure 10–15 shows the TTT diagram for the hypereutectoid composition in-

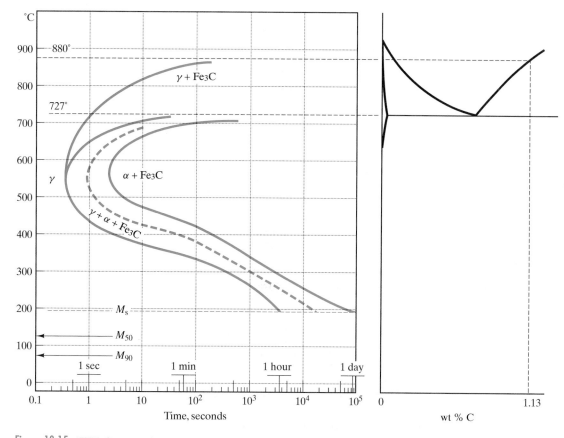

Figure 10-15 *TTT diagram for a hypereutectoid composition (1.13 wt % C) compared to the Fe–Fe₃C phase diagram. Microstructural development for the slow cooling of this alloy was shown in Figure 9–40. (TTT diagram after* Atlas of Isothermal Transformation and Cooling Transformation Diagrams, *American Society for Metals, Metals Park, Ohio, 1977.)*

troduced in Figure 9–40. The most obvious difference about this diagram relative to the eutectoid is the additional curved line extending from the pearlite "knee" to the horizontal line at 880°C. This corresponds to the additional diffusional process for the formation of proeutectoid cementite. Less obvious is the downward shift in the martensitic reaction temperatures, such as M_s. A similar TTT diagram is shown in Figure 10–16 for the hypoeutectoid composition introduced in Figure 9–41. This diagram includes the formation of proeutectoid ferrite and shows martensitic temperatures higher than those for the eutectoid steel. In general, the martensitic reaction occurs at decreasing temperatures with increasing carbon contents around the eutectoid composition region.

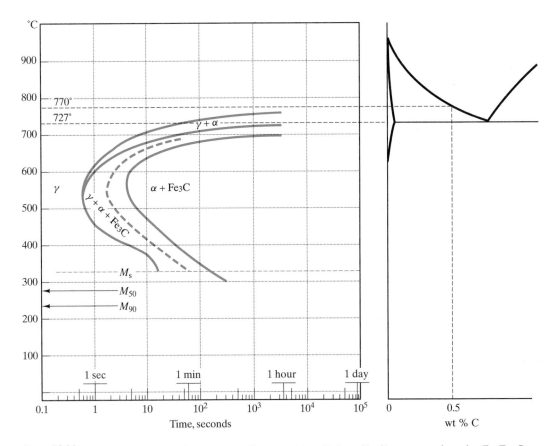

Figure 10-16 *TTT diagram for a hypoeutectoid composition (0.5 wt % C) compared to the Fe–Fe$_3$C phase diagram. Microstructural development for the slow cooling of this alloy was shown in Figure 9–41. By comparing Figures 10–11, 10–15, and 10–16, one will note that the martensitic transformation occurs at decreasing temperatures with increasing carbon content in the region of the eutectoid composition. (TTT diagrams after* Atlas of Isothermal Transformation and Cooling Transformation Diagrams, *American Society for Metals, Metals Park, Ohio, 1977.)*

HEAT TREATMENT OF STEEL

With the principles of TTT diagrams now available, we can illustrate some of the basic principles of the heat treatment of steels. This is a large field in itself with enormous, commercial significance. We can, of course, only touch on some elementary examples in this introductory textbook. For illustration, we shall select the eutectoid composition.

As discussed previously, martensite is a brittle phase. In fact, it is so brittle that a product of 100% martensite would be impractical, akin to a glass hammer. A common approach to fine tuning the mechanical properties of a steel is to first form a completely martensitic material by rapid quenching. Then, this steel can be made less brittle by a careful reheat to a temperature where transformation to the equilibrium phases of α and Fe_3C is possible. By reheating for a short time at a moderate temperature, a high-strength, low-ductility product is obtained. By reheating for longer times, greater ductility occurs (due to less martensite). Figure 10–17 shows a thermal history $[T = fn(t)]$ superimposed on a TTT diagram representing this conventional process, known as **tempering.** (It is important to recall that superimposing heating and cooling curves on an isothermal TTT diagram is a schematic illustration.) The $\alpha + Fe_3C$ microstructure produced by

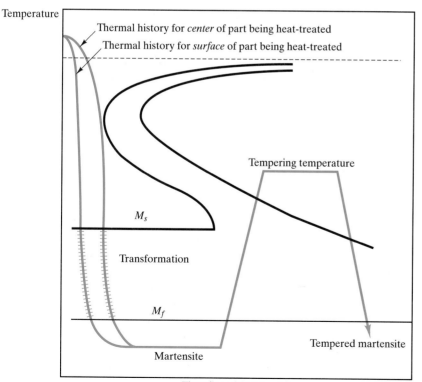

Temperature

Thermal history for *center* of part being heat-treated

Thermal history for *surface* of part being heat-treated

Tempering temperature

M_s

Transformation

M_f

Tempered martensite

Martensite

Time (logarithmic scale)

Figure 10-17 *Tempering is a thermal history $[T = fn(t)]$ in which martensite, formed by quenching austenite, is reheated. The resulting tempered martensite consists of the equilibrium phase of α-Fe and Fe_3C but in a microstructure different from both pearlite and bainite (note Figure 10–18). (After Metals Handbook, 8th Ed., Vol. 2, American Society for Metals, Metals Park, Ohio, 1964. It should be noted that the TTT diagram is, for simplicity, that of eutectoid steel. As a practical matter, tempering is generally done in steels with slower diffusional reactions permitting less severe quenches.)*

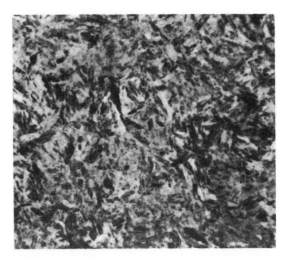

Figure 10-18 *The microstructure of tempered martensite, although an equilibrium mixture of α-Fe and Fe₃C, differs from those for pearlite (Figure 9–2) and bainite (Figure 10–9), 825×. This particular microstructure is for a 0.50 wt % C steel comparable with that described for Figure 10–16. (From* Metals Handbook, 8th Ed., Vol. 7: *Atlas of Microstructures, American Society for Metals, Metals Park, Ohio, 1972.)*

tempering is different from both pearlite and bainite. This is not surprising in light of the fundamentally different paths involved. Pearlite and bainite are formed by the cooling of austenite, a face-centered cubic solid solution. The microstructure known as **tempered martensite** (Figure 10–18) is formed by the heating of martensite, a body-centered tetragonal solid solution of Fe and C. The morphology in Figure 10–18 shows that the carbide has coalesced into isolated particles in a matrix of ferrite.

A possible problem with conventional quenching and tempering is that the part can be distorted and cracked due to uneven cooling during the quench step. The exterior will cool fastest and, therefore, transform to martensite before the interior. During the brief period of time in which the exterior and interior have different crystal structures, significant stresses can occur. The region that has the martensite structure is, of course, highly brittle and susceptible to cracking. A simple solution to this problem is a heat treatment known as **martempering** (or *marquenching*), illustrated in Figure 10–19. By stopping the quench above M_s, the entire piece can be brought to the same temperature by a brief isothermal step. Then, a slow cool allows the martensitic transformation to occur evenly through the piece. Again, ductility is produced by a final tempering step.

An alternative method to avoid the distortion and cracking of conventional tempering is the heat treatment known as **austempering**, illustrated in Figure 10–20. This has the advantage of completely avoiding the costly

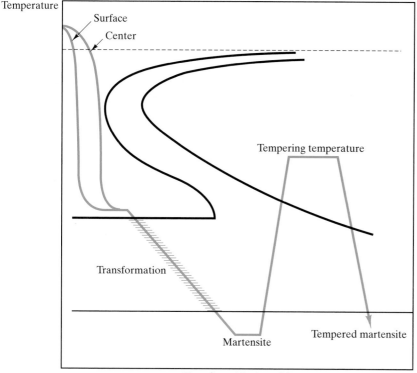

Temperature

Surface

Center

Tempering temperature

Transformation

Martensite

Tempered martensite

Time (logarithmic scale)

Figure 10-19 *In martempering, the quench is stopped just above M_s. Slow cooling through the martensitic transformation range reduces stresses associated with the crystallographic change. The final reheat step is equivalent to that in conventional tempering. (After* Metals Handbook, *8th Ed., Vol. 2, American Society for Metals, Metals Park, Ohio, 1964.)*

reheating step. As with martempering, the quench is stopped just above M_s. In austempering, the isothermal step is extended until complete transformation to bainite occurs. Since this microstructure ($\alpha + Fe_3C$) is more stable than martensite, further cooling produces no martensite. Control of hardness is obtained by careful choice of the bainite transformation temperature. Hardness increases with decreasing transformation temperature due to the increasingly fine-grained structure.

A final comment on these schematic illustrations of heat treatment is in order. The principles were adequately shown using the simple eutectoid TTT diagram. However, the various heat treatments are similarly applied to a wide variety of steel compositions that have TTT diagrams that can differ substantially from that for the eutectoid. As an example, austempering is not practical for some alloy steels because the alloy addition substantially increases the time for bainite transformation. In addition, tempering of eutectoid steel as illustrated is of limited practicality due to the high quench rate needed to avoid the pearlite "knee."

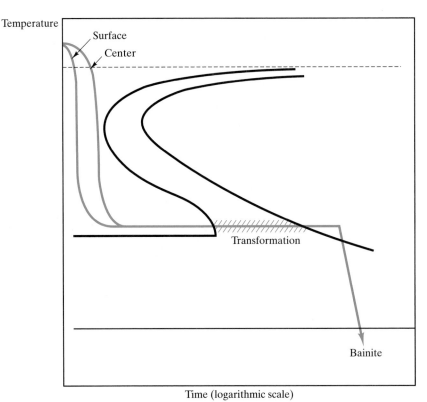

Figure 10-20 *As with martempering, austempering avoids the distortion and cracking associated with quenching through the martensitic transformation range. In this case, the alloy is held long enough just above M_s to allow full transformation to bainite. (After* Metals Handbook, *8th Ed., Vol. 2, American Society for Metals, Metals Park, Ohio, 1964.)*

SAMPLE PROBLEM 10.2

(a) How much time is required for austenite to transform to 50% pearlite at 600°C?

(b) How much time is required for austenite to transform to 50% bainite at 300°C?

SOLUTION

(a) This is a direct application of Figure 10–7. The dotted line denotes the halfway point in the $\gamma \rightarrow \alpha + Fe_3C$ transformation. At 600°C, the time to reach that line is $\sim 3\frac{1}{2}$ s.

(b) At 300°C, the time is ~ 480 s or 8 min.

SAMPLE PROBLEM 10.3

(a) Calculate the microstructure of a 0.77 wt % C steel that has the following heat treatment: (i) instantly quenched from the γ region to 500°C, (ii) held for 5 s, and (iii) quenched instantly to 250°C.

(b) What will happen if the resulting microstructure is held for 1 day at 250°C and then cooled to room temperature?

(c) What will happen if the resulting microstructure from part (a) is quenched directly to room temperature?

(d) Sketch the various thermal histories.

SOLUTION

(a) By having ideally fast quenches, we can answer this precisely in terms of Figure 10–7. The first two parts of the heat treatment lead to $\sim 70\%$ transformation to fine pearlite. The final quench will retain this state:

$$30\% \ \gamma + 70\% \text{ fine pearlite } (\alpha + Fe_3C)$$

(b) The pearlite remains stable, but the retained γ will have time to transform to bainite, giving a final state:

$$30\% \text{ bainite } (\alpha + Fe_3C) + 70\% \text{ fine pearlite } (\alpha + Fe_3C)$$

(c) Again, the pearlite remains stable but most of the retained γ will become unstable. For this case, we must consider the martensitic transformation data in Figure 10–11. The resulting microstructure will be

$$70\% \text{ fine pearlite } (\alpha + Fe_3C) + \sim 30\% \text{ martensite}$$

(Because the martensitic transformation is not complete until $-46°C$, a small amount of untransformed γ will remain at room temperature.)

(d)

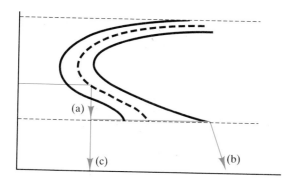

SAMPLE PROBLEM 10.4

Estimate the quench rate needed to avoid pearlite formation in

(a) 0.5 wt % C steel,

(b) 0.77 wt % C steel,

(c) 1.13 wt % C steel.

SOLUTION

In each case, we are looking at the rate of temperature drop needed to avoid the pearlite "knee":

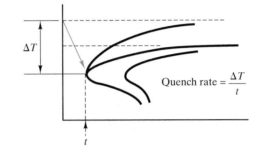

Note. This is an isothermal transformation diagram used to illustrate a continuous cooling process. Precise calculation would require a true continuous cooling transformation curve.

(a) From Figure 10–16 for a 0.5 wt % C steel, we must quench from the austenite boundary (770°C) to ∼ 520°C in ∼ 0.6 s giving

$$\frac{\Delta T}{t} = \frac{(770 - 520)°C}{0.6s} = 420°C/s$$

(b) From Figure 10–11 for a 0.77 wt % C steel, we quench from the eutectoid temperature (727°C) to ∼550°C in ∼ 0.7 s, giving

$$\frac{\Delta T}{t} = \frac{(727 - 550)°C}{0.7s} = 250°C/s$$

(c) From Figure 10–15 for a 1.13 wt % C steel, we quench from the austenite boundary (880°C) to ∼550°C in ∼ 3.5 s, giving

$$\frac{\Delta T}{t} = \frac{(880 - 550)°C}{0.35 \text{ s}} = 940°C/s$$

SAMPLE PROBLEM 10.5

Calculate the time required for austempering at 5°C above the M_s temperature for

(a) 0.5 wt % C steel,

(b) 0.77 wt % C steel,

(c) 1.13 wt % C steel.

SOLUTION

(a) Figure 10–16 for 0.5 wt % C steel indicates that complete bainite formation will have occurred 5°C above M_s by

$$\sim 180s \times 1m/60s = 3 \text{ min}$$

(b) Similarly, Figure 10–11 for 0.77 wt % C steel gives a time of

$$\sim \frac{1.9 \times 10^4 s}{3600 \text{ s/h}} = 5.3 \text{ h}$$

(c) Finally, Figure 10–15 for 1.13 wt % C steel gives an austempering time of

$$\sim 1 \text{day}$$

....

PRACTICE PROBLEM 10.2

In Sample Problem 10.2, we use Figure 10–7 to determine the time for 50% transformation to pearlite and bainite at 600 and 300°C, respectively. Repeat these calculations for **(a)** 1% transformation and **(b)** 99% transformation.

PRACTICE PROBLEM 10.3

A detailed thermal history is outlined in Sample Problem 10.3. Answer all of the questions in that problem if only one change is made in the history; namely, step (i) is an instantaneous quench to 400°C (not 500°C).

PRACTICE PROBLEM 10.4

In Sample Problem 10.4, we estimate quench rates necessary to retain austenite below the pearlite "knee." What would be the percentage of martensite formed in each of the alloys if these quenches were continued to 200°C?

PRACTICE PROBLEM 10.5

The time necessary for austempering is calculated for three alloys in Sample Problem 10.5. In order to do martempering (Figure 10–19), it is necessary to cool the alloy before bainite formation begins. How long can the alloy be held at 5° above M_s before bainite formation begins in (a) 0.5 wt % C steel, (b) 0.77 wt % C steel, and (c) 1.13 wt % C steel?

10.3   HARDENABILITY

In the remainder of this chapter, we shall encounter several heat treatments in which the primary purpose is to affect the **hardness** of a metal alloy. In Section 6.4, hardness was defined by the degree of indentation produced in a standard test. The indentation decreases with increasing hardness. An important feature of the hardness measurement is its direct correlation with strength. We shall now concentrate on heat treatments, with hardness serving to monitor the effect of the thermal history on alloy strength.

Our experience with TTT diagrams has shown a general trend. For a given steel, hardness is increased with increasing quench rates. However, a systematic comparison of the behavior of different steels must take into account the enormous range of commercial steel compositions. The relative ability of a steel to be hardened by quenching is termed **hardenability.** Fortunately, a relatively simple experiment has become standardized for industry to provide such a systematic comparison. The **Jominy* end-quench test** is illustrated in Figure 10–21. A standard-size steel bar (25 mm in diameter by 100 mm long) is taken to the austenizing temperature and then one end is subjected to a water spray. For virtually all carbon and low-alloy steels this standard quench process produces a common cooling-rate gradient along the Jominy bar. This is because the thermal properties (such as thermal conductivity) are nearly identical for these various alloys. (See Chapter 11 for other members of the steel family. The carbon and low-alloy steels are the most commonly used ones for quenching-induced hardness, for which the Jominy test is so useful.)

---

*   Walter Jominy (1893–1976), American metallurgist. A contemporary of E. C. Bain, Jominy was a similarly productive researcher in the field of ferrous metallurgy. He held important appointments in industrial, government, and university laboratories.

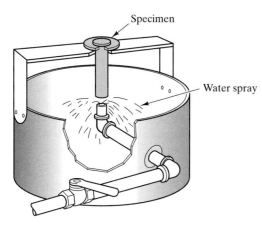

Specimen

Water spray

Figure 10-21   *Schematic illustration of the Jominy end-quench test for hardenability. (After W. T. Lankford et al., Eds.,* The Making, Shaping, and Treating of Steel, *10th Ed., United States Steel, Pittsburgh, Pa., 1985. Copyright 1985 by United States Steel Corporation.)*

Figure 10–22 shows how the cooling rate varies along the Jominy bar. An end-quench test of the type illustrated by Figure 10–21 and 10–22 is the basis of the continuous cooling diagram of Figure 10–14. Of course, the cooling rate is greatest near the end subjected to the water spray. The resulting variation in hardness along a typical steel bar is illustrated in Figure 10–23. A similar plot comparing various steels is given in Figure 10–24. Here comparisons of hardenability can be made where hardenability corresponds to the relative magnitude of hardness along the Jominy bar.

The hardenability information from the end-quench test can be used in two complementary ways. If the quench rate for a given part is known, the Jominy data can predict the hardness of that part. Conversely, hardness measurements on various areas of a large part (which may have experienced uneven cooling) can identify different quench rates.

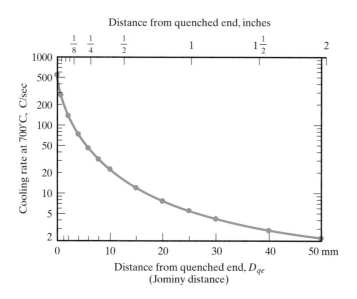

Figure 10-22   *The cooling rate for the Jominy bar (see Figure 10–21) varies along its length. This curve applies to virtually all carbon and low-alloy steels. (After L. H. Van Vlack,* Elements of Materials Science and Engineering, *4th Ed., Addison-Wesley Publishing Co., Inc., Reading, Mass., 1980.)*

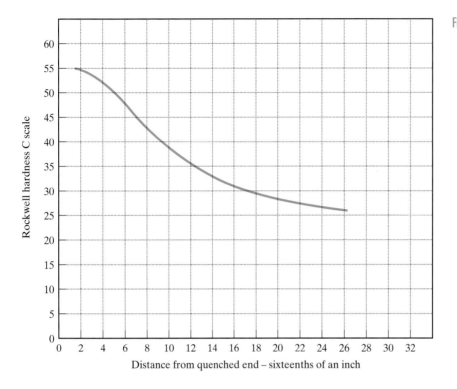

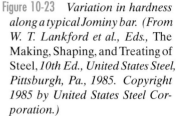

Figure 10-23 *Variation in hardness along a typical Jominy bar. (From W. T. Lankford et al., Eds.,* The Making, Shaping, and Treating of Steel, *10th Ed., United States Steel, Pittsburgh, Pa., 1985. Copyright 1985 by United States Steel Corporation.)*

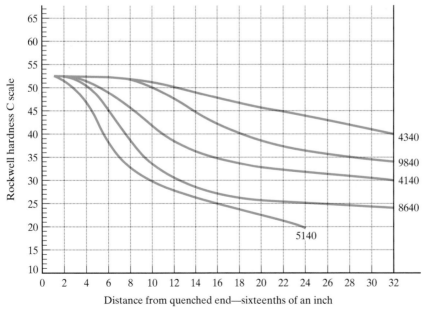

Figure 10-24 *Hardenability curves for various steels with the same carbon content (0.40 wt %) and various alloy contents. The codes designating the alloy compositions are defined in Table 11.1. (From W. T. Lankford et al., Eds.,* The Making, Shaping, and Treating of Steel, *10th Ed., United States Steel, Pittsburgh, Pa., 1985. Copyright 1985 by United States Steel Corporation.)*

SAMPLE PROBLEM 10.6

A hardness measurement is made at a critical point on a trailer-axle forging of 4340 steel. The hardness value is 45 on the Rockwell C scale. What cooling rate was experienced by the forging at the point in question?

SOLUTION

Using Figure 10–24, we see that a Jominy end-quench test on this alloy produces a hardness of Rockwell C45 at 22/16 in. from the quenched end. This is equal to

$$D_{qe} = \frac{22}{16} \text{ in.} \times 25.4 \text{ mm/in.} = 35 \text{ mm}$$

Turning to Figure 10–22, which applies to carbon and low-alloy steels, we see that the cooling rate was approximately

$$4°C/s(\text{at}700°C)$$

Note. To be more precise in answering a question such as this, it is appropriate to consult a plot of the "hardness band" from a series of Jominy tests on the alloy in question. For most alloys, there is a considerable range of hardness that can occur at a given point along D_{qe}.

SAMPLE PROBLEM 10.7

Estimate the hardness that would be found at the critical point on the axle discussed in Sample Problem 10.6 if that part were fabricated from 4140 rather than 4340 steel.

SOLUTION

This is straightforward in that Figure 10–22 shows us that the cooling behavior of various carbon and low-alloy steel is essentially the same. We can read the 4140 hardness from the plot in Figure 10–24 at the same D_{qe} as that calculated in Sample Problem 10.6 (i.e., at $\frac{22}{16}$ of an inch). The result is a hardness of Rockwell C32.5.

Note. The comment in Sample Problem 10.6 applies here also; that is, there are "uncertainty bars" associated with the Jominy data for any given alloy. However, Figure 10–24 is still very useful for indicating that the 4340 alloy is significantly more hardenable and, for a given quench rate, can be expected to yield a higher hardness part.

....

PRACTICE PROBLEM 10.6

In Sample Problem 10.6, we are able to estimate a quench rate that leads to a hardness of Rockwell C45 in a 4340 steel. What quench rate would be necessary to produce a hardness of **(a)** C50 and **(b)** C40?

PRACTICE PROBLEM 10.7

In Sample Problem 10.7, we find that the hardness of a 4140 steel is lower than that for a 4340 steel (given equal quench rates). Determine the corresponding hardness for **(a)** a 9840 steel, **(b)** an 8640 steel, and **(c)** a 5140 steel.

10.4 PRECIPITATION HARDENING

In Section 6.3, we found that small obstacles to dislocation motion can strengthen (or harden) a metal (e.g., Figure 6–25). Small, second-phase precipitates are effective in this way. In Chapter 9, we found that cooling paths for certain alloy compositions lead to second-phase precipitation (e.g., Figure 9–37b). Many alloy systems use such **precipitation hardening**. The most common illustration is found in the Al–Cu system. Figure 10–25

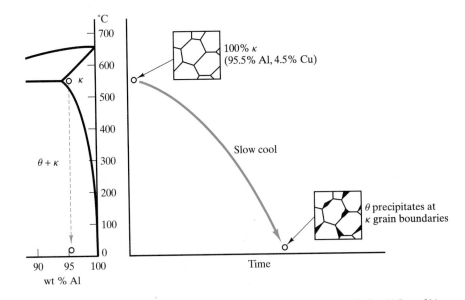

Figure 10-25 *Coarse precipitates form at grain boundaries in an Al–Cu (4.5 wt %) alloy when slowly cooled from the single-phase (κ) region of the phase diagram to the two-phase (θ+κ) region. These isolated precipitates do little to affect alloy hardness.*

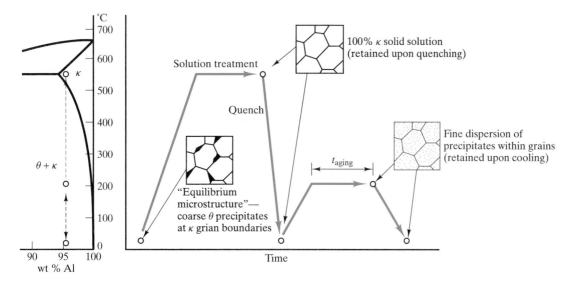

Figure 10-26 *By quenching and then reheating an Al–Cu (4.5 wt %) alloy, a fine dispersion of precipitates forms within the κ grains. These precipitates are effective in hindering dislocation motion and, consequently, increasing alloy hardness (and strength). This is known as precipitation hardening, or age hardening.*

shows the aluminum-rich end of the Al–Cu phase diagram together with the microstructure that develops upon *slow* cooling. As the precipitates are relatively coarse and isolated at grain boundaries, little hardening is produced by the presence of the second phase. A substantially different thermal history is shown in Figure 10–26. Here, the coarse microstructure is first reheated to the single-phase (κ) region. This is appropriately termed a **solution treatment**. Then, the single-phase structure is quenched to room temperature, where the precipitation is quite slow and the supersaturated solid solution remains a metastable phase. Upon reheating to some intermediate temperature, the solid-state diffusion of copper atoms in aluminum is sufficiently rapid to allow a fine dispersion of precipitates to form. These precipitates are effective dislocation barriers and lead to a substantial hardening of the alloy. Because this precipitation takes time, this process is also termed **age hardening**. Figure 10–27 illustrates **overaging**, in which the precipitation process is continued so long that the precipitates have an opportunity to coalesce into a more coarse dispersion, which is less effective as a dislocation barrier. Figure 10–28 shows the structure (formed during the early stages of precipitation), which is so effective as a dislocation barrier. These precipitates are referred to as **Guinier-Preston*** (or *G.P.*) **zones** and are distinguished by **coherent interfaces** at which the crystal structures of the matrix and precipitate maintain registry. This coherency is lost in the larger precipitates formed as overaging occurs.

---

*   Andre Guinier (1911–   ), French physicist, and George Dawson Preston (1896–1972), English physicist. The detailed atomic structure [Figure 10–28] was determined in the 1930's by these physicists using the powerful tool of x-ray diffraction (see Section 3.7).

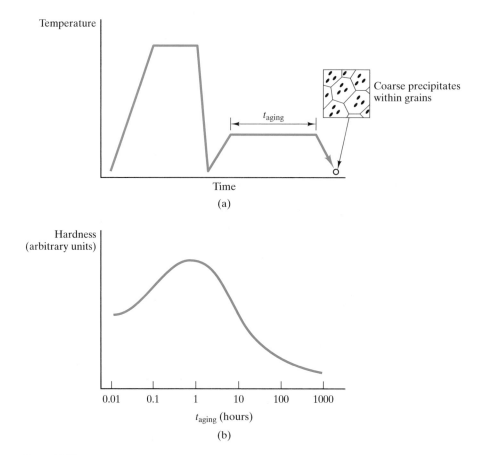

Coarse precipitates within grains

Figure 10-27 *(a) By extending the reheat step, precipitates coalesce and become less effective in hardening the alloy. The result is referred to as "overaging." (b) The variation in hardness with the length of the reheat step ("aging time").*

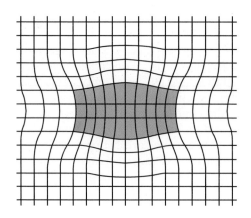

Figure 10-28 *Schematic illustration of the crystalline geometry of a Guinier–Preston (G.P.) zone. This structure is most effective for precipitation hardening, and is the structure developed at the hardness maximum in Figure 10–27b. Note the coherent interfaces lengthwise along the precipitate. The precipitate is approximately 15 nm × 150 nm. (From H. W. Hayden, W. G. Moffatt, and J. Wulff,* The Structure and Properties of Materials, *Vol. 3: Mechanical Behavior,* John Wiley & Sons, Inc., New York, *1965.)*

SAMPLE PROBLEM 10.8

(a) Calculate the amount of θ phase that would precipitate at the grain boundaries in the equilibrium microstructure shown in Figure 10–25.

(b) What is the maximum amount of Guinier-Preston zones to be expected in a 4.5 wt % Cu alloy?

SOLUTION

(a) This is an equilibrium question and returns us to the concept of phase diagrams from Chapter 9. Using the Al–Cu phase diagram (Figure 9–27) and Equation 9.10, we obtain

$$\text{wt \% } \theta = \frac{x - x_\kappa}{x_\theta - x_\kappa} \times 100\% = \frac{4.5 - 0}{53 - 0} \times 100\%$$

$$= 8.49\%$$

(b) As the G.P. zones are precursors to the equilibrium precipitation, the maximum amount would be 8.49%.

Note.   This calculation was made in a similar case treated in Sample Problem 9.10.

PRACTICE PROBLEM 10.8

The nature of precipitation in a 95.5 Al–4.5 Cu alloy is considered in Sample Problem 10.8. Repeat these calculations for a 96 Al–4 Cu alloy.

10.5   ANNEALING

One of the more important heat treatments introduced in this chapter (in Section 10.2) is tempering, in which a material (martensite) is softened by high temperature for an appropriate time. **Annealing** is a comparable heat treatment in which the hardness of a mechanically deformed microstructure is reduced at high temperatures. In order to appreciate the details of this microstructural development, we need to explore four terms: *cold work, recovery, recrystallization,* and *grain growth.*

COLD WORK

Cold work means to mechanically deform a metal at relatively low temperatures. This concept was introduced in Section 6.3 in relating dislocation motion to mechanical deformation. The amount of cold work is defined rel-

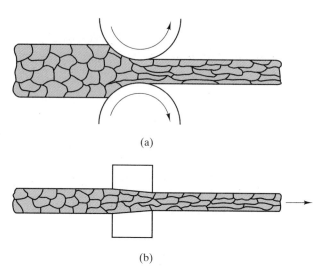

Figure 10-29 *Examples of cold-working operations: (a) cold-rolling of a bar or sheet and (b) cold-drawing a wire. Note in these schematic illustrations that the reduction in area caused by the cold-working operation is associated with a preferred orientation of the grain structure.*

ative to the reduction in cross-sectional area of the alloy by processes such as rolling or drawing (Figure 10–29). The percent cold work is given by

$$\% \text{ CW} = \frac{A_0 - A_f}{A_0} \times 100\% \tag{10.2}$$

where A_0 is the original cross-sectional area and A_f is the final cross-sectional area after cold working. The hardness and strength of alloys are increased with increasing % CW, a process termed *strain hardening*. The relationship of mechanical properties to % CW of brass is illustrated in Figure 11–3, relative to a discussion of design specifications. The mechanism for this hardening is the resistance to plastic deformation caused by the high density of dislocations produced in the cold working. (Recall the discussion in Section 6.3.) The density of dislocations can be expressed as the length of dislocation lines per unit volume (e.g., m/m^3 or net units of m^{-2}). An annealed alloy can have a dislocation density as low as 10^{10} m^{-2}, with a correspondingly low hardness. A heavily cold-worked alloy can have a dislocation density as high as 10^{16} m^{-2}, with a significantly higher hardness (and strength).

A cold-worked microstructure is shown in Figure 10–30a. The severely distorted grains are quite unstable. By taking the microstructure to higher temperatures where sufficient atom mobility is available, the material can be softened and a new microstructure can emerge.

(a)   (b)   (c)

(d)   (e)

Figure 10-30 *Annealing can involve the complete recrystallization and subsequent grain growth of a cold-worked mi-crostructure. (a) A cold-worked brass (deformed through rollers such that the cross-sectional area of the part was reduced by one-third). (b) After 3 s at 580°C, new grains appear. (c) After 4 s at 580°C, many more new grains are present. (d) After 8 s at 580°C, complete recrystallization has occurred. (e) After 1 h at 580°C, substantial grain growth has occurred. The driving force for this is the reduction of high-energy grain boundaries. The predominant reduction in hardness for this overall process had occurred by step (d). All micrographs at magnification of 75×. (Courtesy of J. E. Burke, General Electric Company, Schenectady, N.Y.)*

RECOVERY

The most subtle stage of annealing is **recovery**. No gross microstructural change occurs. However, atomic mobility is sufficient to diminish the concentration of point defects within grains and, in some cases, to allow dislocations to move to lower-energy positions. This process yields a modest decrease in hardness and can occur at temperatures just below those needed to produce significant microstructural change. Although the structural effect of recovery (primarily a reduced number of point defects) produces a

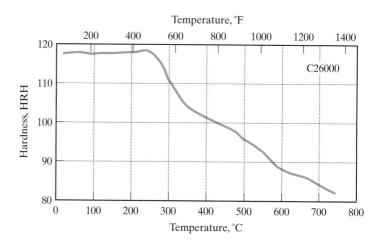

Figure 10-31 *The sharp drop in hardness identifies the recrystallization temperature as* ∼ *290° C for the alloy C26000, "cartridge brass."* (*From* Metals Handbook, *9th Ed., Vol. 4, American Society for Metals, Metals Park, Ohio, 1981.*)

modest effect on mechanical behavior, electrical conductivity does increase significantly. (The relationship between conductivity and structural regularity is explored further in Section 15.3.)

RECRYSTALLIZATION

In Section 6.3, we stated an important concept: "The temperature at which atomic mobility is sufficient to affect mechanical properties is approximately one-third to one-half times the absolute melting point, T_m." A microstructural result of exposure to such temperatures is termed **recrystallization** and is illustrated dramatically in Figures 10–30a–d. New equi-axed, stress-free grains nucleate at high-stress regions in the cold-worked microstructure (Figure 10–30b). These grains then grow together until they constitute the entire microstructure (Figure 10–30c and d). As the nucleation step occurs in order to stabilize the system, it is not surprising that the concentration of new grain nuclei increases with the degree of cold work. As a result, the grain size of the recrystallized microstructure decreases with the degree of cold work. The decrease in hardness due to annealing is substantial, as indicated by Figure 10–31. Finally, the rule of thumb quoted at the beginning of this discussion of recrystallization effectively defines the **recrystallization temperature** (Figure 10–32). For a given alloy composition, the precise recrystallization temperature will depend slightly on the percentage cold work. Higher values of % CW correspond to higher degrees of strain hardening and a correspondingly lower recrystallization temperature; that is, less thermal energy input is required to initiate the reformation of the microstructure (Figure 10–33).

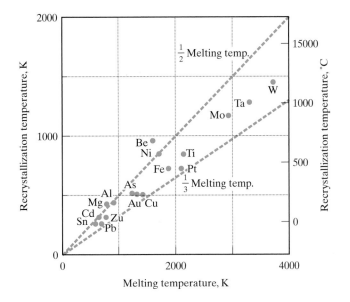

Figure 10-32 *Recrystallization temperature versus melting points for various metals. This plot is a graphic demonstration of the rule of thumb that atomic mobility is sufficient to affect mechanical properties above approximately $\frac{1}{3}$ to $\frac{1}{2} T_m$ on an* absolute *temperature scale. (From L. H. Van Vlack,* Elements of Materials Science and Engineering, *3rd Ed., Addison-Wesley Publishing Co., Inc., Reading, Mass, 1975.)*

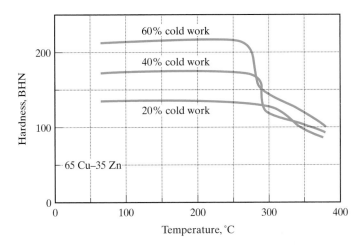

Figure 10-33 *For this cold-worked brass alloy, the recrystallization temperature drops slightly with increasing degrees of cold work. (From L. H. Van Vlack,* Elements of Materials Science and Engineering, *4th Ed., Addison-Wesley Publishing Co., Inc., Reading, Mass. 1980.)*

GRAIN GROWTH

The microstructure developed during recrystallization (Figure 10–30d) occurred spontaneously. It is stable compared with the original cold-worked
structure (Figure 10–30a). However, the recrystallized microstructure contains a large concentration of grain boundaries. We have noted frequently
since Chapter 4 that the reduction of these high-energy interfaces is a method
of stabilizing a system further. The stability of coarse pearlite (Figure 10–10)
was such an example. The coarsening of annealed microstructures by grain
growth is another. Figure 10–30e illustrates **grain growth**, which is not dissimilar to the coalescence of soap bubbles, a process similarly driven by the
reduction of surface area. Figure 10–34 shows that this grain growth stage
produces little additional softening of the alloy. That effect is associated
predominantly with recrystallization.

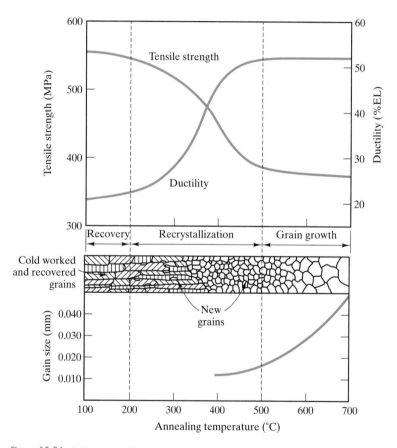

Figure 10-34 *Schematic illustration of the effect of annealing temperature
on the strength and ductility of a brass alloy shows that most of the
softening of the alloy occurs during the recrystallization stage. (After
G. Sachs and K. R. Van Horn, Practical Metallurgy: Applied Physical Metallurgy and the Industrial Processing of Ferrous and Nonferrous Metals and Alloys, American Society for Metals, Cleveland, Ohio,
1940.)*

SAMPLE PROBLEM 10.9

Cartridge brass has the approximate composition of 70 wt % Cu, 30 wt % Zn. How does this alloy compare with the trend shown in Figure 10–32?

SOLUTION

The recrystallization temperature is indicated by Figure 10–31 as $\sim 290°C$. The melting point for this composition is indicated by the Cu–Zn phase diagram (Figure 9–29) as $\sim 920°C$ (the solidus temperature). The ratio of recrystallization temperature to melting point is then

$$\frac{T_R}{T_m} = \frac{(290 + 273)\ \text{K}}{(920 + 273)\ \text{K}} = 0.47$$

which is within the range of one-third to one-half indicated by Figure 10–32.

PRACTICE PROBLEM 10.9

Noting the result of Sample Problem 10.9, plot the estimated temperature range for recrystallization of Cu–Zn alloys as a function of composition over the entire range from pure Cu to pure Zn.

10.6   THE KINETICS OF PHASE TRANSFORMATIONS FOR NONMETALS

As with phase diagrams in Chapter 9, our discussion of the kinetics of phase transformations has dwelled on metallic materials. The rates at which phase transformations occur in nonmetallic systems are, of course, also important to the processing of those materials. The crystallization of some single-component polymers is a model example of nucleation and growth kinetics illustrated in Figure 10–5. Specific data for natural rubber are given in Figure 10–35. Careful control of the rate of melting and solidification of silicon is critical to the growth and subsequent purification of large, single crystals, which are the foundation of the semiconductor industry. As with phase diagrams, ceramics (rather than polymers or semiconductors) provide the closest analogy to the treatment of kinetics for metals.

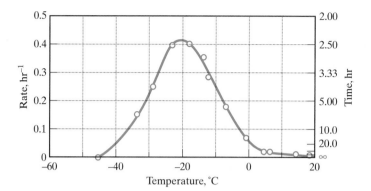

Figure 10-35 *Rate of crystallization of rubber as a function of temperature. (From L. A. Wood, in H. Mark and G. S. Whitby, Eds.,* Advances in Colloid Science, *Vol. 2, Wiley Interscience, New York, 1946, pp. 57–95.)*

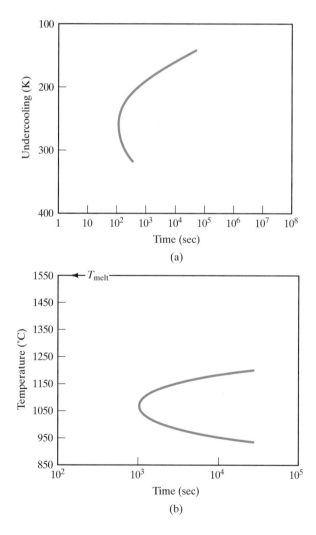

Figure 10-36 *TTT diagram for (a) the fractional crystallization (10^{-4} vol%) of a simple glass of composition $Na_2O \cdot 2SiO_2$ and (b) the fractional crystallization (10^{-1} vol%) of a glass of composition $CaO \cdot Al_2O_3 \cdot 2SiO_2$. [Part (a) from G. S. Meiling and D. R. Uhlmann,* Phys. Chem. Glasses **8**, *62 (1967) and part (b) from H. Yinnon and D. R. Uhlmann, in* Glass: Science and Technology, *Vol. 1, D. R. Uhlmann and N. J. Kreidl, Eds., Academic Press, New York, 1983, pp. 1–47.]*

TTT diagrams are not widely used for nonmetallic materials, but some examples have been generated, especially in systems where transformation rates play a critical role in processing. Examples for simple glass compositions subjected to crystallization are shown in Figure 10–36. Glass-ceramics are closely associated with nucleation and growth kinetics. These products are formed as glasses and then carefully crystallized to produce a polycrystalline product. The result can be relatively strong ceramics formed in complex shapes for modest cost. A typical temperature–time schedule for producing a glass-ceramic is shown in Figure 10–37. These interesting materials are presented in some detail in Chapter 12.

The CaO–ZrO$_2$ phase diagram was presented in Figure 9–30. The production of stabilized zirconia was shown to be the result of adding sufficient CaO (around 20 mol %) to ZrO$_2$ to form the cubic zirconia solid-solution phase. As a practical matter, partially stabilized zirconia (PSZ) with a composition in the two-phase (monoclinic zirconia + cubic zirconia) region exhibits mechanical properties superior to those of the fully stabilized material, especially thermal shock resistance (see Section 7.4). Electron microscopy has revealed that, in cooling the partially stabilized material, some monoclinic precipitates form within the cubic phase. This precipitation strengthening is the analog of precipitation hardening in metals. Electron microscopy has also revealed that the tetragonal-to-monoclinic phase transformation in pure zirconia is a martensitic-type transformation (Figure 10–38).

The subject of grain growth has played an especially important role in the development of ceramic processing in recent decades. Our first example of a microstructurally sensitive property was the transparency of a

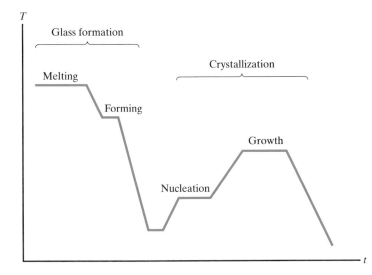

Figure 10-37 *Typical thermal history for producing a glass ceramic by the controlled nucleation and growth of crystalline grains.*

Figure 10-38 *Transmission electron micrograph of monoclinic zirconia showing a microstructure characteristic of a martensitic transformation. Included in the evidence are twins labeled T. See Figure 4–15 for an atomic-scale schematic of a twin boundary and Figure 10–13 for the microstructure of martensitic steel. (Courtesy of Arthur H. Heuer)*

polycrystalline Al_2O_3 ceramic (see Section 1.4). The material can be nearly transparent if it is essentially pore-free. However, it is formed by the densification of a powder. The bonding of powder particles occurs by solid-state diffusion. In the course of this densification stage, the pores between adjacent particles steadily shrink. This overall process is known as **sintering**, which refers to any process for forming a dense mass by heating, but without melting. This rather unusual term comes from the Greek word *sintar*, meaning "slag" or "ash." It shares this source with the more common term *cinder*. The mechanism of shrinkage is the diffusion of atoms away from the grain boundary (between adjacent particles) to the pore. In effect, the pore is filled in by diffusing material (Figure 10–39). Unfortunately, grain growth can begin long before the pore shrinkage is complete. The result is that some pores become trapped within grains. The diffusion path from grain boundary to pore is too long to allow further pore elimination (Figure 10–40). A microstructure for this case was shown in Figure 1–20a. The solution to this problem is to add a small amount (about 0.1 wt %) MgO, which severely retards grain growth and allows pore shrinkage to go to completion. The resulting microstructure was shown in Figure 1–20c. The exact mechanism for the retardation of grain growth is the subject of active research. It is probably associated with the effect of Mg^{2+} ions on the retarding of grain boundary mobility by a solid-solution pinning mechanism. In the meantime, a substantial branch of ceramic technology has been created in which nearly transparent, polycrystalline ceramics can be reliably fabricated.

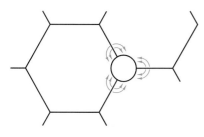

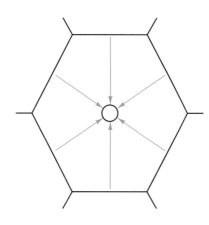

Figure 10-40 *Grain growth hinders the densification of a powder compact. The diffusion path from grain boundary to pore (now isolated within a large grain) is prohibitively long.*

Figure 10-39 *An illustration of the sintering mechanism for shrinkage of a powder compact is the diffusion of atoms away from the grain boundary to the pore, thereby "filling in" the pore. Each grain in the microstructure was originally a separate powder particle in the initial compact.*

SAMPLE PROBLEM 10.10

Calculate the maximum amount of monoclinic phase you would expect to find in the microstructure of a partially stabilized zirconia with an overall composition of 3.4 wt % CaO.

SOLUTION

This is another example of the important relationship between phase diagrams and kinetics. We can make this prediction by calculating the equilibrium concentration of monoclinic phase at room temperature in the CaO–ZrO_2 system (Figure 9–30). A composition of 3.4 wt % CaO is approximately 7 mol % CaO (as shown by the upper and lower composition scales in Figure 9–30). Extrapolating the phase boundaries in the monoclinic + cubic two-phase region to room temperature gives a monoclinic phase composition of $\sim$ 2 mol % CaO and a cubic phase composition of $\sim$ 15 mol % CaO. Using Equation 9.9, we get

$$\text{mol \% monoclinic} = \frac{x_{\text{cubic}} - x}{x_{\text{cubic}} - x_{\text{mono}}} \times 100\%$$

$$= \frac{15 - 7}{15 - 2} \times 100\% = 62 \text{ mol\%}$$

Note. We have made this calculation in terms of mole percent because of the nature of the plot in Figure 9–30. It would be a straightforward matter to convert the results to weight percent.

....

Convert the 62 mol % from Sample Problem 10.10 to weight percent.

SUMMARY

In Chapter 9, we presented phase diagrams as two-dimensional maps of temperature and composition (at a fixed pressure of 1 atm). In this chapter, we added time as a third dimension to monitor the kinetics of microstructural development. Time-dependent phase transformations begin with the nucleation of the new phase followed by its subsequent growth. The overall transformation rate is a maximum at some temperature below the equilibrium transformation temperature. This is because of a competition between the degree of instability (which increases with decreasing temperature) and atomic diffusivity (which decreases with decreasing temperature). A plot of percentage transformation on a temperature versus time plot is known as a TTT diagram. A prime example of the TTT diagram is given for the eutectoid steel composition (Fe with 0.77 wt % C). Pearlite and bainite are the products of diffusional transformation formed by cooling austenite. If the austenite is cooled in a sufficiently rapid quench, a diffusionless, or martensitic, transformation occurs. The product is a metastable, supersaturated solid solution of C in Fe known as martensite. The careful control of diffusional and diffusionless transformations is the basis of the heat treatment of steel. As an example, tempering involves a quench to produce a non-diffusional martensitic transformation followed by a reheat to produce a diffusional transformation of martensite into the equilibrium phases of α-Fe and Fe_3C. This tempered martensite is different from both pearlite and bainite. Martempering is a slightly different heat treatment in which the steel is quenched to just above the starting temperature for the martensitic transformation and then cooled slowly through the martensitic range. This reduces the stresses produced by quenching during the martensitic transformation. The reheat to produce tempered martensite is the same as before. A third type of heat treatment is austempering. Again, the quench stops short of the martensitic range, but in this case, the quenched austenite is held long enough for bainite to form. This makes the reheat step unnecessary.

The ability of steel to be hardened by quenching is termed hardenability. This characteristic is evaluated in a straightforward manner by the Jominy end-quench test. High-aluminum alloys in the Al–Cu system are excellent examples of precipitation hardening. By careful cooling from a single phase to a two-phase region of the phase diagram, it is possible to generate a fine dispersion of second-phase precipitates that are dislocation barriers and the source of hardening (and strengthening). Annealing is a heat treatment that reduces hardness in cold-worked alloys. Cold working involves the mechanical deformation of an alloy at a relatively low temperature. In combination with annealing, cold working provides for wide-ranging control of mechanical behavior. A subtle form of annealing is recovery in which a slight reduction in hardness occurs due to the effect of atomic mobility on structural defects. Elevated temperature is required to permit the atomic mobility. At slightly higher temperatures, a more dramatic reduction in hardness occurs due to recrystallization. The entire cold-worked microstructure is transformed to a set of small, stress-free grains. With longer time, grain growth occurs in the new microstructure.

Time also plays a crucial role in the development of microstructure in nonmetallic materials. Examples are given for polymers and ceramics. The production of glass ceramics is a classic example of nucleation and growth theory as applied to the crystallization of glass. Structural zirconia ceramics exhibit microstructures produced by both diffusional and diffusionless transformations. The production of transparent, polycrystalline ceramics has been the result of control of grain growth in the densification of compressed powders (sintering).

KEY TERMS

age hardening(381)
annealing(383)
austempering(370)
austenite(361)
bainite(361)
coherent interface(381)
cold work(383)
continuous cooling transformation (CCT) diagram (367)
diffusional transformation(360)
diffusionless transformation(360)
grain growth(388)
Guinier-Preston zone(381)

hardenability(376)
hardness(376)
heat treatment(355)
heterogeneous nucleation (356)
homogeneous nucleation (356)
isothermal transformation diagram(360)
Jominy end-quench test(376)
kinetics(355)
martempering(370)
martensite(365)
martensitic transformation(364)
metastable(365)
nucleation(356)

overaging(381)
pearlite(361)
precipitation hardening(380)
recovery(385)
recrystallization(386)
recrystallization temperature(386)
sintering(392)
solution treatment(381)
tempered martensite(370)
tempering(369)
TTT diagram(360)

REFERENCES

ASM Handbook, Vol. 4: *Heat Treating,* ASM International, Materials Park, Ohio, 1991.

Chiang, Y., D. P. Birnie III, and **W. D. Kingery**, *Physical Ceramics,* John Wiley & Sons, Inc., New York, 1997.

PROBLEMS

10.1 • Time—The Third Dimension

10.1. For an aluminum alloy, the activation energy for crystal growth is 120 kJ/mol. By what factor would the rate of crystal growth change by dropping the alloy temperature from 500°C to room temperature (25°C)?

10.2. Repeat Problem 10.1 for a copper alloy for which the activation energy for crystal growth is 200 kJ/mol.

10.3. Although Section 10.1 concentrates on crystal nucleation and growth from a liquid, similar kinetics laws apply to solid state transformations. For example, Equation 10.1 can be used to describe the rate of β

precipitation upon cooling supersaturated α phase in a 10 wt % Sn–90 wt % Pb alloy. Given precipitation rates of 3.77×10^3 s^{-1} and 1.40×10^3 s^{-1} at 20°C and 0°C, respectively, calculate the activation energy for this process.

10.4. Use the result of Problem 10.3 to calculate the precipitation rate at room temperature, 25°C.

10.5. Relative to Figure 10–3, derive an expression for r_c as a function of σ, the surface energy per unit area of the nucleus, and ΔG_v, the volume energy reduction per unit volume. Recall that the area of a sphere is $4\pi r^2$, and its volume is $\left(\frac{4}{3}\right)\pi r^3$.

• **10.6.** The work of formation, W, for a stable nucleus is the maximum value of the net energy change (occurring at r_c) in Figure 10–3. Derive an expression for W in terms of σ and ΔG_v (defined in Problem 10.5).

• **10.7.** A theoretical expression for the rate of pearlite growth from austenite is

$$\dot{R} = Ce^{-Q/RT}(T_E - T)^2$$

where C is a constant, Q the activation energy for carbon diffusion in austenite, R the gas constant, and T an absolute temperature below the equilibrium transformation temperature, T_E. Derive an expression for the temperature, T_M, corresponding to the maximum growth rate, that is, the "knee" of the transformation curve.

10.8. Use the result of Problem 10.7 to calculate T_M (in °C). (Recall that the activation energy is given in Chapter 5 and the transformation temperature is given in Chapter 9.)

10.2 • The TTT Diagram

10.9. (a) A 1050 steel (iron with 0.5 wt % C) is rapidly quenched to 330°C, held for 10 minutes, and then cooled to room temperature. What is the resulting microstructure? **(b)** What is a name for this heat treatment?

10.10. (a) A eutectoid steel is (i) quenched instantaneously to 500°C, (ii) held for 5 seconds, (iii) quenched instantaneously to room temperature, (iv) reheated to 300°C for 1 hour, and (v) cooled to room temperature. What is the final microstructure? **(b)** A carbon steel with 1.13 wt % C is given exactly the same heat treatment described in part (a). What is the resulting microstructure in this case?

10.11. (a) A carbon steel with 1.13 wt % C is given the following heat treatment: (i) instantaneously quenched to 200°C, (ii) held for 1 day, and (iii) cooled slowly to room temperature. What is the resulting microstructure? **(b)** What microstructure would result if a carbon steel with 0.5 wt % C were given exactly the same heat treatment?

10.12. Three different eutectoid steels are given the following heat treatments: **(a)** instantaneously quenched to 600°C, held for 2 minutes, then cooled to room temperature; **(b)** instantaneously quenched to 400°C, held for 2 minutes, then cooled to room temperature; and **(c)** instantaneously quenched to 100°C, held for 2 minutes, then cooled to room temperature. List these heat treatments in order of decreasing hardness of the final product. Briefly explain your answer.

10.13. (a) A eutectoid steel is cooled at a steady rate from 727 to 200°C in exactly 1 day. Superimpose this cooling curve on the TTT diagram of Figure 10–11. **(b)** From the result of your plot for part (a), determine at what temperature a phase transformation would first be observed. **(c)** What would be the first phase to be observed? (Recall the approximate nature of using an isothermal diagram to represent a continuous cooling process.)

10.14. Repeat Problem 10.13 for a steady rate of cooling in exactly 1 minute.

10.15. Repeat Problem 10.13 for a steady rate of cooling in exactly 1 second.

10.16. (a) Using Figures 10–11, 10–15, and 10–16 as data sources, plot M_s, the temperature at which the martensitic transformation begins, as a function of carbon content. **(b)** Repeat part (a) for M_{50}, the temperature at which the martensitic transformation is 50% complete. **(c)** Repeat part (a) for M_{90}, the temperature at which the martensitic transformation is 90% complete.

10.17. Using the trends of Figures 10–11, 10–15, and 10–16, sketch as specifically as you can a TTT diagram for a hypoeutectoid steel that has 0.6 wt % C. (Note Problem 10.16 for one specific compositional trend.)

10.18. Repeat Problem 10.17 for a hypereutectoid steel with 0.9 wt % C.

10.19. What is the final microstructure of a 0.6 wt % C hypoeutectoid steel given the following heat treatment: (i) instantaneously quenched to 500°C, (ii) held for 10 seconds, and (iii) instantaneously quenched to room temperature. (Note the TTT diagram developed in Problem 10.17.)

10.20. Repeat Problem 10.19 for a 0.9 wt % C hypereutectoid steel. (Note the TTT diagram developed in Problem 10.18.)

10.21. It is worth noting that in TTT diagrams such as Figure 10–7, the 50% completion (dashed line) curve lies roughly midway between the onset (1%) curve and completion (99%) curve. It is also worth noting that the progress of transformation is not linear but sigmoidal (s-shaped) in nature. For example, careful observation of Figure 10–7 at 500°C shows that 1%, 50%, and 99% completion occur at 0.9 s, 3.0 s, and 9.0 s, respectively. Intermediate completion data, however, can be given as follows:

% completion	$t(s)$
20	2.3
40	2.9
60	3.2
80	3.8

Plot the % completion at 500°C versus log t to illustrate the sigmoidal nature of the transformation.

10.22. **(a)** Using the result of Problem 10.21, determine t for 25% and 75% completion. **(b)** Superimpose the result of (a) on a sketch of Figure 10–7 to illustrate that the 25% and 75% completion lines are much closer to the 50% line than to either the 1% or 99% completion lines.

10.3 • Hardenability

10.23. **(a)** Specify a quench rate necessary to ensure a hardness of *at least* Rockwell C40 in a 4140 steel. **(b)** Specify a quench rate necessary to ensure a hardness of *no more than* Rockwell C40 in the same alloy.

10.24. The surface of a forging made from a 4340 steel is unexpectedly subjected to a quench rate of 100°C/s (at 700°C). The forging is specified to have a hardness between Rockwell C46 and C48. Is the forging within specifications? Briefly explain your answer.

D 10.25. A flywheel shaft is made of 8640 steel. The surface hardness is found to be Rockwell C35. By what percentage would the cooling rate at the point in question have to change in order for the hardness to be increased to a more desirable value of Rockwell C45?

D 10.26. Repeat Problem 10.25 assuming that the shaft is made of 9840 steel.

10.27. Quenching a bar of 4140 steel at 700°C into a stirred water bath produces an instantaneous quench rate at the surface of 100°C/s. Use Jominy test data to predict the surface hardness resulting from this quench. (Note: The quench described is not in the Jominy configuration. Nonetheless, Jominy data provide general information on hardness as a function of quench rate.)

10.28. Repeat Problem 10.27 for a stirred oil bath that produces a surface quench rate of 20°C/s.

10.29. A bar of steel quenched into a stirred liquid will cool much more slowly in the center than at the surface in contact with the liquid. The following limited data represent the initial quench rates at various points across the diameter of a bar of 4140 steel (initially at 700°C):

Position	Quench rate (°C/s)
center	35
15 mm from center	55
30 mm from center (at surface)	200

(a) Plot the quench rate profile across the diameter of the bar. (Assume the profile to be symmetrical.) **(b)** Use Jominy test data to plot the resulting hardness profile across the diameter of the bar. (*Note:* The quench rates described are not in the Jominy configuration. Nonetheless, Jominy data provide general information on hardness as a function of quench rate.)

10.30. Repeat Problem 10.29b for a bar of 5140 steel that would have the same quench rate profile.

D 10.31. In heat-treating a complex-shaped part made from a 5140 steel, a final quench in stirred oil leads to a hardness of Rockwell C30 at 3 mm beneath the surface. This is unacceptable, as design specifications require a hardness of Rockwell C45 at that point.

Select an alloy substitution to provide this hardness, assuming the heat treatment must remain the same.

10.32. In general, hardenability decreases with decreasing alloy additions. Illustrate this by superimposing the following data for a plain carbon 1040 steel on the plot for other xx40 steels in Figure 10–24:

Distance from quenched end (in sixteenths of an inch)	Rockwell C hardness
2	44
4	27
6	22
8	18
10	16
12	13
14	12
16	11

10.4 • Precipitation Hardening

10.33. **(a)** Calculate the maximum amount of second-phase precipitation in a 90 wt % Al–10 wt % Mg alloy at 100°C. **(b)** What is the precipitate in this case?

10.34. Repeat Problem 10.33 for stoichiometric γ phase, $Al_{12}Mg_{17}$.

10.35. Specify an aging temperature for a 95 Al–5 Cu alloy that will produce a maximum of 5 wt % θ precipitate.

10.36. Specify an aging temperature for a 95 Al–5 Mg alloy that will produce a maximum of 5 wt % β precipitate.

10.37. As second-phase precipitation is a thermally activated process, an Arrhenius expression can be used to estimate the time required to reach maximum hardness (see Figure 10–27b). As a first approximation, you can treat t_{max}^{-1} as a "rate," where t_{max} is the time to reach maximum hardness. For a given aluminum alloy, t_{max} is 40 hours at 160°C and only 4 hours at 200°C. Use Equation 5.1 to calculate the activation energy for this precipitation process.

10.38. Estimate the time to reach maximum hardness, t_{max}, at 240°C for the aluminum alloy of Problem 10.37.

10.5 • Annealing

10.39. A 90:10 Ni–Cu alloy is heavily cold-worked. It will be used in a structural design that is occasionally subjected to 200°C temperatures for as much as 1 hour. Do you expect annealing effects to occur?

10.40. Repeat Problem 10.39 for a 90:10 Cu–Ni alloy.

10.41. A 12-mm-diameter rod of steel is drawn through a 10-mm-diameter die. What is the resulting percentage cold work?

• **10.42.** An annealed copper alloy sheet is cold worked by rolling. The x-ray diffraction pattern of the original annealed sheet with an fcc crystal structure is represented schematically by Figure 3–39. Given that the rolling operation tends to produce the preferred orientation of the (220) planes parallel to the sheet surface, sketch the x-ray diffraction pattern you would expect for the rolled sheet. (To locate the 2θ positions for the alloy, assume pure copper and note the calculations for Problem 3.83.)

10.43. Recrystallization is a thermally activated process and, as such, can be characterized by the Arrhenius expression (Equation 5.1). As a first approximation, we can treat t_R^{-1} as a "rate," where t_R is the time necessary to fully recrystallize the microstructure. For a 75% cold-worked aluminum alloy, t_R is 100 hours at 250°C and only 10 hours at 280°C. Calculate the activation energy for this recrystallization process. (Note Problem 10.37 in which a similar method was applied to the case of precipitation hardening.)

10.44. Calculate the temperature at which complete recrystallization would occur for the aluminum alloy of Problem 10.43 within one hour.

10.6 • The Kinetics of Phase Transformations for Nonmetals

10.45. The sintering of ceramic powders is a thermally activated process and shares the "rule of thumb" about temperature with diffusion and recrystallization. Estimate a minimum sintering temperature for **(a)** pure Al_2O_3, **(b)** pure mullite, and **(c)** pure spinel. (See Figures 9–23 and 9–26.)

10.46. Four ceramic phase diagrams were presented in Figures 9–10, 9–23, 9–26, and 9–30. In which systems would you expect precipitation hardening to be a possible heat treatment? Briefly explain your answer.

10.47. The total sintering rate for $BaTiO_3$ increases 10 fold between 750 and 794°C. Calculate the activation energy for sintering in $BaTiO_3$.

10.48. Given the data in Problem 10.47, predict the temperature at which the initial sintering rate for $BaTiO_3$ would have increased (a) 50-fold and (b) 100-fold compared to 750°C.

10.49. At room temperature, moisture absorption will occur slowly in parts made of nylon, a common engineering polymer. Such absorption will increase dimensions and lower strength. To stabilize dimensions, nylon products are sometimes given a preliminary "moisture conditioning" by immersion in hot or boiling water. At 60°C, the time to condition a 5-mm-thick nylon part to a 3% moisture content is 24 hours. At 77°C, the time for the same size part is 8 hours. As the conditioning is diffusional in nature, the time required for other temperatures can be estimated from the Arrhenius expression (Equation 5.1). Calculate the activation energy for this moisture conditioning process. (Note the method used for similar kinetics examples in Problems 10.37 and 10.43.)

10.50. Estimate the conditioning time in boiling water (100°C) for the nylon part discussed in Problem 10.49.

CHAPTER 11

Metals

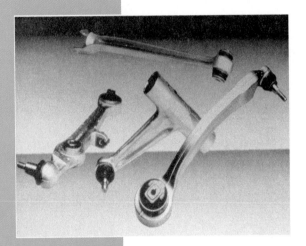

These automotive steering and suspension components are made of wrought aluminum to provide reduced weight and improved fuel economy. (Courtesy of TRW.)

As discussed in Chapter 1, probably no materials are more closely associated with the engineering profession than metals such as structural steel. In this chapter, we shall explore, in greater detail, the wide variety of engineering metals. We begin with the dominant examples: the iron-based or **ferrous alloys**. These include carbon and alloy steels and the cast irons. The **nonferrous alloys** are all other metals that do not contain iron as the major constituent. We shall look specifically at alloys based on aluminum, magnesium, titanium, copper, nickel, zinc, and lead, as well as the refractory and precious metals.

The discussion of metals in this chapter depends greatly on the fundamental concepts raised in various chapters in Part I. The key metallic crystal structures were introduced in Chapter 3. Crystallographic defects in those structures (Chapter 4) were the basis of understanding diffusional transport (Chapter 5) and mechanical behavior (Chapter 6). Combined with the understanding of thermal behavior from Chapter 7, we could then appreciate the sources of failure in metal alloys in Chapter 8.

Although this chapter will provide an introduction to the key engineering metals, an appreciation of the versatility of these materials was presented in Chapters 9 and 10. Microstructural development as related to phase diagrams was dealt with in Chapter 9. Heat treatment based on the kinetics of solid-state reactions was covered in Chapter 10. Each of these topics deals with methods to "fine tune" the properties of given alloys within a broad range of values.

Finally, we shall consider how various metal alloys are made into forms convenient for engineering applications. This topic of *processing* is the subject of more specialized courses that may be available to many students. However, even an introductory course in materials requires a brief discussion of how those materials are produced. This discussion serves two functions. First, it provides a fuller understanding of the nature of each material. Second, and more important, it provides an appreciation of the effects of processing history on properties.

11.1   FERROUS ALLOYS

More than 90% by weight of the metallic materials used by human beings are ferrous alloys. This represents an immense family of engineering materials with a wide range of microstructures and related properties. The majority of engineering designs that require structural load support or power transmission involve ferrous alloys. As a practical matter, these alloys fall into two broad categories based on the amount of carbon in the alloy composition. **Steel** generally contains between 0.05 and 2.0 wt % C. The *cast irons* generally contain between 2.0 and 4.5 wt % C. Within the steel category, we shall distinguish whether or not a significant amount of alloying elements other than carbon is used. A composition of 5 wt % total noncarbon additions will serve as an arbitrary boundary between **low-alloy** and **high-alloy steels**. These alloy additions are chosen carefully because they invariably bring with them sharply increased material costs. They are justified only by essential improvements in properties such as higher strength or improved corrosion resistance.

CARBON AND LOW-ALLOY STEELS

The majority of ferrous alloys are carbon and low-alloy steels. The reasons for this are straightforward. Such alloys are moderately priced due to the absence of large amounts of alloying elements, and they are sufficiently ductile to be readily formed. The final product is strong and durable. These eminently practical materials find applications from ball bearings to metal sheet formed into automobile bodies. A convenient designation system* for these useful alloys is given in Table 11.1. This is the AISI (American Iron and Steel Institute)–SAE (Society of Automotive Engineers) system, in which the first two numbers give a code designating the type of alloy additions and the last two or three numbers give the average carbon content in hundredths of a weight percent. As an example, a plain-carbon steel with 0.40 wt % C is a 1040 steel, whereas a steel with 1.45 wt % Cr and 1.50 wt % C is 52150 steel. One should keep in mind that chemical compositions quoted in alloy designations such as Table 11.1 are approximate and will vary slightly from product to product within acceptable limits of industrial quality control.

An interesting class of alloys known as **high-strength, low-alloy** (HSLA) **steels** has emerged in response to requirements for weight reduction of vehicles. The compositions of many commercial HSLA steels are proprietary and they are specified by mechanical properties rather than composition. A typical example, though, might contain 0.2 wt % C and about 1 wt % or less of such elements as Mn, P, Si, Cr, Ni, or Mo. The high strength of HSLA steels is the result of optimal alloy selection and carefully controlled processing, such as hot rolling (deformation at temperatures sufficiently elevated to allow some stress relief).

HIGH-ALLOY STEELS

As mentioned earlier in this section, alloy additions must be made with care and justification because they are expensive. We shall now look at three cases in which engineering design requirements justify high-alloy compositions (i.e., total noncarbon additions greater than 5 wt %). Stainless steels require alloy additions to prevent damage from a corrosive atmosphere. Tool steels require alloy additions to obtain sufficient hardness for machining applications. So-called superalloys require alloy additions to provide stability in high-temperature applications such as turbine blades.

---

* Alloy designations are convenient but arbitrary listings usually standardized by professional organizations such as AISI and SAE. These traditional designations tend to be as varied as the alloys themselves. There is a growing effort to use a unified numbering system for alloy designation. In this chapter, we shall attempt, where possible, to use the Unified Numbering System (UNS) together with the traditional designation. A student confronted with a mysterious designation not included in this introductory text might consult *Metals and Alloys in the Unified Numbering System*, 8th ed., Society of Automotive Engineers, Warrendale, Pa., and American Society for Testing and Materials, Philadelphia, Pa., 1999.

Table 11.1 *AISI–SAE Designation System for Carbon and Low-Alloy Steels*

Numerals and digits[a]	Type of steel and nominal alloy content[b]	Numerals and digits	Type of steel and nominal alloy content	Numerals and digits	Type of steel and nominal alloy content
Carbon Steels			**Nickel–Chromium–Molybdenum Steels**		**Chromium Steels**
10XX(a)	Plain carbon (Mn 1.00% max)	43XX	Ni 1.82; Cr 0.50 and 0.80; Mo 0.25	50XXX	Cr 0.50
11XX	Resulfurized			51XXX	Cr 1.02 } C 1.00 min
12XX	Resulfurized and rephosphorized	43BVXX	Ni 1.82; Cr 0.50; Mo 0.12 and 0.25; V 0.03 min	52XXX	Cr 1.45
15XX	Plain carbon (max Mn range— 1.00 to 1.65%)	47XX	Ni 1.05; Cr 0.45; Mo 0.20 and 0.35		**Chromium–Vanadium Steels**
Manganese Steels		81XX	Ni 0.30; Cr 0.40; Mo 0.12	61XX	Cr 0.60, 0.80. and 0.95; V 0.10 and 0.15 min
13XX	Mn 1.75	86XX	Ni 0.55; Cr 0.50; Mo 0.20		**Tungsten–Chromium Steel**
		87XX	Ni 0.55; Cr 0.50; Mo 0.25	72XX	W 1.75; Cr 0.75
Nickel Steels		88XX	Ni 0.55; Cr 0.50; Mo 0.35		**Silicon–Manganese Steels**
23XX	Ni 3.50	93XX	Ni 3.25; Cr 1.20; Mo 0.12	92XX	Si 1.40 and 2.00; Mn 0.65, 0.82, and 0.85; Cr 0.00 and 0.65
25XX	Ni 5.00	94XX	Ni 0.45; Cr 0.40; Mo 0.12		**High-Strength Low-Alloy Steels**
Nickel–Chromium Steels		97XX	Ni 0.55; Cr 0.20; Mo 0.20	9XX	Various SAE grades
31XX	Ni 1.25; Cr 0.65 and 0.80	98XX	Ni 1.00; Cr 0.80; Mo 0.25		**Boron Steels**
32XX	Ni 1.75; Cr 1.07		**Nickel–Molybdenum Steels**	XXBXX	B denotes boron steel
33XX	Ni 3.50; Cr 1.50 and 1.57	46XX	Ni 0.85 and 1.82; Mo 0.20 and 0.25		**Leaded Steels**
34XX	Ni 3.00; Cr 0.77	48XX	Ni 3.50; Mo 0.25	XXLXX	L denotes leaded steel
Molybdenum Steels			**Chromium Steels**		
40XX	Mo 0.20 and 0.25	50XX	Cr 0.27, 0.40, 0.50, and 0.65		
44XX	Mo 0.40 and 0.52	51XX	Cr 0.80, 0.87, 0.92, 0.95, 1.00, and 1.05		
Chromium–Molybdenum Steels					
41XX	Cr 0.50, 0.80, and 0.95; Mo 0.12, 0.20, 0.25, and 0.30				

Source: *Metals Handbook*, 9th ed., Vol. 1, American Society for Metals, Metals Park, Ohio, 1978.

[a] XX or XXX in the last two or three digits of these designations indicates that the carbon content (in hundredths of a weight percent) is to be inserted.

[b] All alloy contents are expressed in weight percent.

Table 11.2 *Alloy Designations for Some Common Stainless Steels*

		Composition (wt %)[a]							
Type	**UNS number**	**C**	**Mn**	**Si**	**Cr**	**Ni**	**Mo**	**Cu**	**Al**
Austenitic Types									
201[b]	S20100	0.15	5.5–7.5	1.00	16.0–18.0	3.5–5.5			
304	S30400	0.08	2.00	1.00	18.0–20.0	8.0–10.5			
310	S31000	0.25	2.00	1.50	24.0–26.0	19.0–22.0			
316	S31600	0.08	2.00	1.00	16.0–18.0	10.0–14.0	2.0–3.0		
347[c]	S34700	0.08	2.00	1.00	17.0–19.0	9.0–13.0			
Ferritic Types									
405	S40500	0.08	1.00	1.00	11.5–14.5				0.10–0.30
430	S43000	0.12	1.00	1.00	16.0–18.0				
Martensitic Types									
410	S41000	0.15	1.00	1.00	11.5–13.0				
501	S50100	0.10 min	1.00	1.00	4.0–6.0		0.40–0.65		
Precipitation-hardening Types									
17–4 PH[d]	S17400	0.07	1.00	1.00	15.5–17.5	3.0–5.0		3.0–5.0	
17–7 PH	S17700	0.09	1.00	1.00	16.0–18.0	6.5–7.75			0.75–1.5

Source: Data from *Metals Handbook*, 9th ed., Vol. 3, American Society for Metals, Metals Park, Ohio, 1980.

[a] Single values are maximum values unless otherwise indicated.

[b] 0.25 wt % N.

[c] 10 × %C = min.Nb + Ta(optional).

[d] 0.15 − 0.45wt%Nb+Ta.

Stainless steels are more resistant to rusting and staining than carbon and low-alloy steels, due primarily to the presence of chromium addition. The amount of chromium is at least 4 wt % and usually above 10 wt %. Levels as high as 30 wt % Cr are sometimes used. Table 11.2 summarizes the alloy designations for several of the common stainless steels in four main categories:

1. The **austenitic stainless steels** have the austenite structure retained at room temperature. As discussed in Section 3.2, γ-Fe, or austenite, has the fcc structure and is stable above 910°C. This structure can occur at room temperature when it is stabilized by an appropriate alloy addition such as nickel. While the bcc structure is energetically more stable than the fcc structure for pure iron at room temperature, the opposite is true for iron containing a significant number of nickel atoms in substitutional solid solution.

2. Without the high nickel content, the bcc structure is stable, as seen in the **ferritic stainless steels**. For many applications not requiring the high corrosion resistance of austenitic stainless steels, these lower-alloy (and less expensive) ferritic stainless steels are quite serviceable.

3. A rapid-quench heat treatment discussed in Chapter 10 allows the formation of a more complex body-centered tetragonal crystal structure called martensite. Consistent with our discussion in Section 6.3, this crystal structure yields high strength and low ductility. As a result, these **martensitic stainless steels** are excellent for applications such as cutlery and springs.

4. Precipitation hardening is another heat treatment covered in Chapter 10. Essentially, it involves producing a multiphase microstructure from a single-phase one. The result is increased resistance to dislocation motion and, thereby, greater strength or hardness. **Precipitation-hardening stainless steels** can be found in applications such as corrosion-resistant structural members. In this chapter, these four basic types of stainless steel are categorized. We leave the discussion of mechanisms of corrosion protection to Chapter 19.

Tool steels are used for cutting, forming, or otherwise shaping another material. Some of the principal types are summarized in Table 11.3. A plain-carbon steel (W1) is included at the bottom of the table. For shaping operations that are not too demanding, such a material is adequate. In fact, tool steels were historically of the plain-carbon variety until the mid-nineteenth century. Now high-alloy additions are common. Their advantage is that they can provide the necessary hardness with simpler heat treatments and retain that hardness at higher operating temperatures. The primary alloying elements used in these materials are tungsten, molybdenum, and chromium.

Superalloys include a broad class of metals with especially high strength at elevated temperatures (even above 1000°C). Table 11.4 summarizes some key examples. Many of the stainless steels from Table 11.2 serve a dual role as heat-resistant alloys. These are iron-based superalloys. However, Table 11.4 also includes cobalt- and nickel-based alloys. Most contain chromium

Table 11.3 *Alloy Designations for Some Common Tool Steels*

Designations			Composition (wt %)								
AISI	SAE	UNS	C	Mn	Si	Cr	Ni	Mo	W	V	Co
Molybdenum high-speed steels											
M1	M1	T11301	0.78–0.88	0.15–0.40	0.20–0.50	3.50–4.00	0.30 max	8.20–9.20	1.40–2.10	1.00–1.35	
Tungsten high-speed steels											
T1	T1	T12001	0.65–0.80	0.10–0.40	0.20–0.40	3.75–4.00	0.30 max		17.25–18.75	0.90–1.30	
Chromium hot-work steels											
H10		T20810	0.35–0.45	0.25–0.70	0.80–1.20	3.00–3.75	0.30 max	2.00–3.00		0.25–0.75	
Tungsten hot-work steels											
H21	H21	T20821	0.26–0.36	0.15–0.40	0.15–0.50	3.00–3.75	0.30 max		8.50–10.00	0.30–0.60	
Molybdenum hot-work steels											
H42		T20842	0.55–0.70	0.15–0.40		3.75–4.50	0.30 max	4.50–5.50	5.50–6.75	1.75–2.20	
Air-hardening medium-alloy cold-work steels											
A2	A2	T30102	0.95–1.05	1.00 max	0.50 max	4.75–5.50	0.30 max	0.90–1.40		0.15–0.50	
High-carbon, high-chromium cold-work steels											
D2	D2	T30402	1.40–1.60	0.60 max	0.60 max	11.00–13.00	0.30 max	0.70–1.20		1.10 max	1.00 max
Oil-hardening cold-work steels											
O1	O1	T31501	0.85–1.00	1.00–1.40	0.50 max	0.40–0.60	0.30 max		0.40–0.60	0.30 max	
Shock-resisting steels											
S1	S1	T41901	0.40–0.55	0.10–0.40	0.15–1.20	1.00–1.80	0.30 max	0.50 max	1.50–3.00	0.15–0.30	
Low-alloy, special-purpose tool steels											
L2		T61202	0.45–1.00	0.10–0.90	0.50 max	0.70–1.20		0.25 max		0.10–0.30	
Low-carbon mold steels											
P2		T51602	0.10 max	0.10–0.40	0.10–0.40	0.75–1.25	0.10–0.50	0.15–0.40			
Water-hardening tool steels											
W1	W108	T72301	0.70–1.50	0.10–0.40	0.10–0.40	0.15 max	0.20 max	0.10 max	0.15 max	0.10 max	
	W109										
	W110										
	W112										

Source: Data from *Metals Handbook*, 9th ed., Vol. 3, American Society for Metals, Metals Park, Ohio, 1980.

Table 11.4 *Alloy Designations for Some Common Superalloys*

Alloy	UNS number	Composition (wt %)										
		Cr	**Ni**	**Co**	**Mo**	**W**	**Nb**	**Ti**	**Al**	**Fe**	**C**	**Other**
Iron-base solid-solution alloys												
16–25–6		16.0	25.0		6.00					50.7	0.06	1.35 Mn
Cobalt-base solid-solution alloys												
Haynes 25 (L–605)	R30605	20.0	10.0	50.0		15.0				3.0	0.10	1.5 Mn
Nickel-base solid-solution alloys												
Hastelloy B	N10001	1.0 max	63.0	2.5 max	28.0					5.0	0.05 max	
Inconel 600	N06600	15.5	76.0							8.0	0.08	
Iron-base precipitation-hardening alloys												
Incoloy 903		0.1 max	38.0	15.0	0.1		3.0	1.4	0.7	41.0	0.04	
Cobalt-base precipitation-hardening alloys												
Ar–213		19.0	0.5 max	65.0		4.5			3.5	0.5 max	0.17	6.5 Ta
Nickel-base precipitation-hardening alloys												
Astroloy		15.0	56.5	15.0	5.25			3.5	4.4	<0.3	0.06	
Incoloy 901	N09901	12.5	42.5		6.0			2.7	0.2	36.2	0.10 max	
Inconel 706	N09706	16.0	41.5					1.75	0.2	37.5	0.03	2.9(Nb + Ta)
Nimonic 80A	N07080	19.5	73.0	1.0				2.25	1.4	1.5	0.05	
Rene 41	N07041	19.0	55.0	11.0	10.0			3.1	1.5	<0.3	0.09	
Rene 95		14.0	61.0	8.0	3.5	3.5	3.5	2.9	3.5	<0.3	0.16	
Udimet 500	N07500	19.0	48.0	19.0	4.0			3.0	3.0	4.0 max	0.08	
Waspaloy	N07001	19.5	57.0	13.5	4.3			3.0	1.4	2.0 max	0.07	

Source: Data from *Metals Handbook*, 9th ed., Vol. 3, American Society for Metals, Metals Park, Ohio, 1980.

additions for oxidation and corrosion resistance. These materials are expensive, in some cases extremely so, but the increasingly severe requirements of modern technology are justifying such costs. Between 1950 and 1980, for example, the use of superalloys in aircraft turbojet engines rose from 10% to 50% by weight.

At this point, our discussion of steels has taken us into closely related nonferrous alloys. Before going on to the general area of all other nonferrous alloys, we must discuss the traditional and important ferrous system, the cast irons, and the less traditional category of rapidly solidified alloys.

CAST IRONS

As stated earlier, we define **cast irons** as the ferrous alloys with greater than 2 wt % carbon. They also generally contain up to 3 wt % silicon for control of carbide formation kinetics. Cast irons have relatively low melting temperatures and liquid-phase viscosities, do not form undesirable surface films when poured, and undergo moderate shrinkage during solidification and cooling. The cast irons must balance good formability of complex shapes against inferior mechanical properties compared to wrought alloys.

A cast iron is formed into a final shape by pouring molten metal into a mold. The shape of the mold is retained by the solidified metal. Inferior mechanical properties result from a less uniform microstructure, including some porosity. **Wrought alloys** are initially cast but are rolled or forged into final, relatively simple shapes. (In fact, *wrought* simply means "worked.") A further discussion of processing is given in Section 11.3.

There are four general types of cast irons:

1. **White iron** has a characteristic white, crystalline fracture surface. Large amounts of Fe_3C are formed during casting, giving a hard, brittle material.

2. **Gray iron** has a gray fracture surface with a finely faceted structure. A significant silicon content (2 to 3 wt %) promotes graphite (C) precipitation rather than cementite (Fe_3C). The sharp, pointed graphite flakes contribute to characteristic brittleness in gray iron.

3. By adding a small amount (0.05 wt %) of magnesium to the molten metal of the gray iron composition, spheroidal graphite precipitates rather than flakes are produced. This resulting **ductile iron** derives its name from the improved mechanical properties. Ductility is increased by a factor of 20, and strength is doubled.

4. A more traditional form of cast iron with reasonable ductility is **malleable iron**, which is first cast as white iron and then heat-treated to produce nodular graphite precipitates.

Figure 11–1 shows typical microstructures of these four cast irons. Table 11.5 lists sample compositions.

Table 11.5 *Alloy Designations for Some Common Cast Irons*

Alloy	UNS number	Composition (wt %)								
		C	Mn	Si	P	S	Ni+ Cu	Cr	Mo	Mg
Low-carbon white iron, alloyed for abrasion resistance		2.2–2.8	0.2–0.6	1.0–1.6	0.15	0.15	1.5	1.0	0.5	
SAE J431 automotive gray iron for heavy-duty service, SAE Grade G2500a	F10009	3.40 min	0.60–0.90	1.60–2.10	0.12	0.12				
Ductile iron, unalloyed		3.50–3.80	0.30–1.00	2.00–2.80	0.08 max	0.02 max	0.02–0.60			0.03–0.05
Malleable iron, ferritic Grade 32510		2.30–2.70	0.25–0.55	1.00–1.75	0.05 max	0.03–0.18				

Source: Data from *Metals Handbook*, 9th ed., Vol. 1, American Society for Metals, Metals Park, Ohio, 1978.

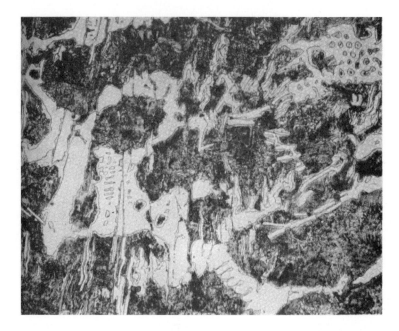

(a)

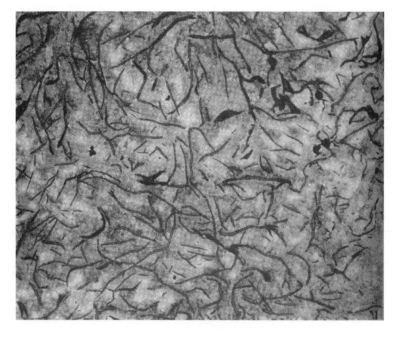

(b)

Figure 11-1 *Typical microstructures of (a) white iron (400×), eutectic carbide (light constituent) plus pearlite (dark constituent). (b) gray iron (100×), graphite flakes in a matrix of 20% free ferrite (light constituent) and 80% pearlite (dark constituent).*

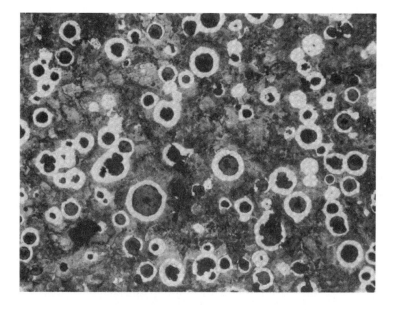

(c)

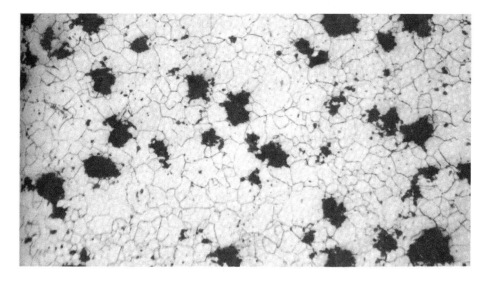

(d)

Figure 11-1 *(c) ductile iron (100×), graphite nodules (spherulites) encased in envelopes of free ferrite, all in a matrix of pearlite. (d) malleable iron (100×), graphite nodules in a matrix of ferrite. (From* Metals Handbook, *9th Ed., Vol. 1, American Society for Metals, Metals Park, Ohio, 1978.)*

RAPIDLY SOLIDIFIED FERROUS ALLOYS

In Section 4.5, the relatively new technology of amorphous metals was introduced. Various ferrous alloys in this category have been commercially produced within the past two decades. Alloy design for these systems originally involved searching for eutectic compositions, which permitted cooling to a glass transition temperature at a practical quench rate (10^5 to 10^6 °C/s). Refined alloy design has included optimizing mismatch in the size of solvent and solute atoms. Boron, rather than carbon, has been a primary alloying element for amorphous ferrous alloys. Iron–silicon alloys have been a primary example of successful commercialization of this technology. The absence of grain boundaries in these alloys helps make them among the most easily magnetized materials and especially attractive as soft-magnet transformer cores (see Sections 18.4 and 20.3). Representative amorphous ferrous alloys are given in Table 11.6.

In addition to superior magnetic properties, amorphous metals have the potential for exceptional strength, toughness, and corrosion resistance. All of these advantages can be related to the absence of microstructural inhomogeneities, especially grain boundaries. Rapid solidification methods do not, in all cases, produce a truly amorphous, or noncrystalline, product. Nonetheless, some unique and attractive materials have resulted as a by-product of the development of amorphous metals. An appropriate label for these novel materials (both crystalline and noncrystalline) is **rapidly solidified alloys**. Although rapid solidification may not produce a noncrystalline state for many alloy compositions, microstructures of rapidly solidified crystalline alloys are characteristically fine-grained (e.g., 0.5 μm compared to 50 μm for a traditional alloy). In addition, rapid solidification can produce metastable phases and novel precipitate morphologies. Correspondingly novel alloy properties are the focus of active research and development.

Table 11.6 *Some Amorphous Ferrous Alloys*

Composition (wt %)					
B	**Si**	**Cr**	**Ni**	**Mo**	**P**
20					
10	10				
28		6		6	
6			40		14

Source: Data from J. J. Gilman, "Ferrous Metallic Glasses," *Metal Progress*, July 1979.

SAMPLE PROBLEM 11.1

For every 100,000 atoms of an 8630 low-alloy steel, how many atoms of each main alloying element are present?

SOLUTION

From Table 11.1, wt % Ni $= 0.55$, wt % Cr $= 0.50$, wt % Mo $= 0.20$, and wt % C $= 0.30$.

For a 100-g alloy, there would be 0.55 g of Ni, 0.50 g of Cr, 0.20 g of Mo, and 0.30 g of C.

If we assume that the balance of the alloy is Fe, there would be

$$100 - (0.55 + 0.50 + 0.20 + 0.30) = 98.45 \text{ g Fe}$$

The number of atoms of each species (in a 100-g alloy) can be determined using the data of Appendix 1:

$$N_{\text{Fe}} = \frac{98.45 \text{ g}}{55.85 \text{ g/mol}} \times 0.6023 \times 10^{24} \text{ atoms/mol}$$

$$= 1.06 \times 10^{24} \text{ atoms}$$

Similarly,

$$N_{\text{Ni}} = \frac{0.55}{58.71} \times 0.6023 \times 10^{24} = 5.64 \times 10^{21} \text{ atoms}$$

$$N_{\text{Cr}} = \frac{0.50}{52.00} \times 0.6023 \times 10^{24} = 5.79 \times 10^{21} \text{ atoms}$$

$$N_{\text{Mo}} = \frac{0.20}{95.94} \times 0.6023 \times 10^{24} = 1.26 \times 10^{21} \text{ atoms}$$

$$N_{\text{C}} = \frac{0.30}{12.01} \times 0.6023 \times 10^{24} = 1.50 \times 10^{22} \text{ atoms}$$

So, in a 100-g alloy, there would be

$$N_{\text{total}} = N_{\text{Fe}} + N_{\text{Ni}} + N_{\text{Cr}} + N_{\text{Mo}} + N_{\text{C}}$$

$$= 1.09 \times 10^{24} \text{ atoms}$$

The atomic fraction of each alloying element is then

$$X_{\text{Ni}} = \frac{5.64 \times 10^{21}}{1.09 \times 10^{24}} = 5.19 \times 10^{-3}$$

$$X_{\text{Cr}} = \frac{5.79 \times 10^{21}}{1.09 \times 10^{24}} = 5.32 \times 10^{-3}$$

$$X_{\text{Mo}} = \frac{1.26 \times 10^{21}}{1.09 \times 10^{24}} = 1.16 \times 10^{-3}$$

$$X_{\text{C}} = \frac{1.50 \times 10^{22}}{1.09 \times 10^{24}} = 1.38 \times 10^{-2}$$

This gives, for a 100,000-atom alloy,

$$N_{Ni} = 5.19 \times 10^{-3} \times 10^5 \text{ atoms} = 519 \text{ atoms}$$

$$N_{Cr} = 5.32 \times 10^{-3} \times 10^5 \text{ atoms} = 532 \text{ atoms}$$

$$N_{Mo} = 1.16 \times 10^{-3} \times 10^5 \text{ atoms} = 116 \text{ atoms}$$

$$N_C = 1.38 \times 10^{-2} \times 10^5 \text{ atoms} = 1380 \text{ atoms}$$

PRACTICE PROBLEM 11.1

For every 100,000 atoms of an SAE J431 (F10009) gray cast iron, how many atoms of each main alloying element are present? (Use elemental compositions in the middle of the ranges given in Table 11.5.) (See Sample Problem 11.1.)

11.2 NONFERROUS ALLOYS

Although ferrous alloys are used in the majority of metallic applications in current engineering designs, nonferrous alloys play a large and indispensable role in our technology, and the list of nonferrous alloys is long and complex. We shall briefly list the major families of nonferrous alloys and their key attributes.

ALUMINUM ALLOYS

Aluminum alloys are best known for low density and corrosion resistance. Electrical conductivity, ease of fabrication, and appearance are also attractive features. Because of these, the world production of aluminum roughly doubled between the 1960s and the 1970s. During the 1980s and 1990s, demand for aluminum and other metals has tapered off due to increasing competition from ceramics, polymers, and composites. However, the importance of aluminum within the metals family has increased due to its low density, which is also a key factor in the increased popularity of nonmetallic materials. For example, the total mass of a new American automobile dropped by 16% between 1976 and 1986, from 1705 kg to 1438 kg. In large part, this was the result of a 29% *decrease* in the use of conventional steels (from 941 kg to 667 kg) and a 63% *increase* in the use of aluminum alloys (from 39 kg to 63 kg) as well as a 33% increase in the use of polymers and composites (from 74 kg to 98 kg). The total aluminum content in American automobiles further increased by 102% in the 1990s. Ore reserves for aluminum are large (representing 8% of the earth's crust), and aluminum

Table 11.7 *Alloy Designation System for Aluminum Alloys*

Numerals	Major alloying element(s)
1XXX	None ($\geq$ 99.00% Al)
2XXX	Cu
3XXX	Mn
4XXX	Si
5XXX	Mg
6XXX	Mg and Si
7XXX	Zn
8XXX	Other elements

Source: Data from *Metals Handbook*, 9th ed., Vol. 2, American Society for Metals, Metals Park, Ohio, 1979.

can be easily recycled. The alloy designation system for wrought aluminum alloys is summarized in Table 11.7.

One of the most active areas of development in aluminum metallurgy is in the 8XXX-series, involving Li as the major alloying element. The Al–Li alloys provide especially low density, as well as increased stiffness. The increased cost of Li (compared to traditional alloying elements) and controlled atmosphere processing (due to lithium's reactivity) appears justified for several advanced aircraft applications.

In Chapter 10, we discussed a wide variety of heat treatments for alloys. For some alloy systems, standard heat treatments are given code numbers and become an integral part of the alloy designations. The temper designations for aluminum alloys in Table 11.8 are good examples.

MAGNESIUM ALLOYS

Magnesium alloys have even lower density than aluminum and, as a result, appear in numerous structural applications, such as aerospace designs. Magnesium's density of 1.74 Mg/m^3 is, in fact, the lowest of any of the common structural metals. Extruded magnesium alloys have found a wide range of applications in consumer products, from tennis rackets to suitcase frames. These structural components exhibit especially high strength-to-density ratios. This is an appropriate time to recall Figure 6–24, which indicates the basis for the marked difference in characteristic mechanical behavior of fcc and hcp alloys. Aluminum is an fcc material and therefore has numerous (12) slip systems, leading to good ductility. By contrast, magnesium is hcp with only three slip systems and characteristic brittleness.

TITANIUM ALLOYS

Titanium alloys have become widely used since World War II. Before that time, a practical method of separating titanium metal from reactive oxides and nitrides was not available. Once formed, titanium's reactivity works to

Table 11.8 *Temper Designation System for Aluminum Alloys*[a]

Temper	Definition
F	As fabricated
O	Annealed
H1	Strain-hardened only
H2	Strain-hardened and partially annealed
H3	Strain-hardened and stabilized (mechanical properties stabilized by low-temperature thermal treatment)
T1	Cooled from an elevated-temperature shaping process and naturally aged to a substantially stable condition
T2	Cooled from an elevated-temperature shaping process, cold-worked, and naturally aged to a substantially stable condition
T3	Solution heat-treated, cold-worked, and naturally aged to a substantially stable condition
T4	Solution heat-treated and naturally aged to a substantially stable condition
T5	Cooled from an elevated-temperature shaping process and artificially aged
T6	Solution heat-treated and artificially aged
T7	Solution heat-treated and stabilized
T8	Solution heat-treated, cold-worked, and artificially aged
T9	Solution heat-treated, artificially aged, and cold-worked
T10	Cooled from an elevated-temperature shaping process, cold-worked, and artificially aged

Note: A more complete listing and more detailed descriptions are given on pages 24–27 of *Metals Handbook*, 9th ed., Vol. 2, American Society for Metals, Metals Park, Ohio, 1979.

[a] General alloy designation: XXXX-Temper, where XXXX is alloy numeral from Table 11.7 (e.g., 6061-T6).

its advantage. A thin, tenacious oxide coating forms on its surface, giving excellent resistance to corrosion. This "passivation" will be discussed in detail in Chapter 19. Titanium alloys, like Al and Mg, are of lower density than iron. Although more dense than Al or Mg, titanium alloys have a distinct advantage of retaining strength at moderate service temperatures (e.g., skin temperatures of high-speed aircraft), leading to numerous aerospace design applications. Titanium shares the hcp structure with magnesium, leading to characteristically low ductility. However, a high-temperature bcc structure can be stabilized at room temperature by addition of certain alloying elements, such as vanadium.

COPPER ALLOYS

Copper alloys possess a number of superior properties. Their excellent electrical conductivity makes copper alloys the leading material for electrical wiring. Their excellent thermal conductivity leads to applications for radiators and heat exchangers. Superior corrosion resistance is exhibited in marine and other corrosive environments. The fcc structure contributes to their generally high ductility and formability. Their coloration is frequently used for architectural appearance. Widespread uses of copper alloys through history have led to a somewhat confusing collection of descriptive terms. The key families of copper alloys are listed in Table 11.9 according to the primary alloying elements. The **brasses**, in which zinc is a predominant substitutional

Table 11.9 *Classification of Copper and Copper Alloys*

Family	Principal alloying element	Solid solubility (at %)[a]	UNS numbers[b]
Coppers, high-copper alloys	[c]		C10000
Brasses	Zn	37	C20000, C30000, C40000, C66400–C69800
Phosphor bronzes	Sn	9	C50000
Aluminum bronzes	Al	19	C60600–C64200
Silicon bronzes	Si	8	C64700–C66100
Copper nickels, nickel silvers	Ni	100	C70000

Source: *Metals Handbook*, 9th ed., Vol. 2, American Society for Metals, Metals Park, Ohio, 1979.
[a] At 20°C (68°F).
[b] Wrought alloys.
[c] Various elements having less than 8 at % solid solubility at 20°C (68°F).

solute, are the most common copper alloys. Examples include yellow, naval, and cartridge brass. Applications include automotive radiators, coins, cartridge casings, musical instruments, and jewelry. The **bronzes**, copper alloys involving elements such as tin, aluminum, silicon, and nickel, provide a high degree of corrosion resistance associated with brasses but a somewhat higher strength. The mechanical properties of copper alloys rival the steels in their variability. High-purity copper is an exceptionally soft material. The addition of 2 wt % beryllium followed by a heat treatment to produce CuBe precipitates is sufficient to push the tensile strength beyond 10^3 MPa.

NICKEL ALLOYS

Nickel alloys have much in common with copper alloys. We have already used the Cu–Ni system as the classic example of complete solid solubility (Section 4.1). *Monel* is the name given to commercial alloys with Ni–Cu ratios of roughly 2:1 by weight. These are good examples of **solution hardening**, in which the alloys are strengthened by the restriction of plastic deformation due to solid solution formation. (Recall the discussion relative to Figure 6–25.) Nickel is harder than copper, but Monel is harder than nickel. The effect of solute atoms on dislocation motion and plastic deformation was illustrated in Section 6.3. Nickel exhibits excellent corrosion resistance and high-temperature strength.

We have already listed some nickel alloys with the superalloys of Table 11.4. *Inconel* (nickel–chromium–iron) and *Hastelloy* (nickel–molybdenum–iron–chromium) are important examples. **Nickel–aluminum superalloy** is the object of substantial research and development. With a composition of Ni_3Al, it has the same crystal structure options shown for Cu_3Au in Figure 4–3, with Ni corresponding to Cu and Al to Au. The superalloy microstructure involves a "gamma phase" matrix with the disordered crystal structure of Figure 4–3a, and a "gamma-prime phase" precipitate with the ordered crystal structure of Figure 4–3b. This system, and related ones with modest alloy additions, have exhibited exceptional high-temperature strength and corrosion resistance. The attractive properties of Ni_3Al have led to a major research and development effort on the systematic study of intermetallic compounds for high-temperature structural use. As one example of the driving force for these investigations, the development of such alloys for jet engines allowing an increase in maximum operating temperature from the current ≈ 1400 to $\approx 1850°C$ could reduce operating costs of transoceanic commercial flights by more than 25%. A wide range of compositions, especially various titanium-based intermetallics, are under active study. The magnetic properties of various nickel alloys will be covered in Chapter 18.

ZINC, LEAD, AND OTHER ALLOYS

Zinc alloys are ideally suited for die castings due to their low melting point and lack of corrosive reaction with steel crucibles and dies. Automobile parts and hardware are typical structural applications, although the extent of use in this industry is steadily diminishing for the sake of weight savings. Zinc coatings on ferrous alloys are important means of corrosion protection. This method, termed *galvanization*, will be discussed in Chapter 19.

Lead alloys are durable and versatile materials. The lead pipes installed by the Romans at the public baths in Bath, England, nearly 2000 years ago are still in use. Lead's high density and deformability, combined with a low melting point, add to its versatility. Lead alloys find use in battery grids (alloyed with calcium or antimony), solders (alloyed with tin), radiation shielding, and sound-control structures. The toxicity of lead restricts design applications and the handling of its alloys.

The **refractory metals** include molybdenum, niobium, rhenium, tantalum, and tungsten. They are, even more than the superalloys, especially resistant to high temperatures. However, their general reactivity with oxygen requires the high-temperature service to be in a controlled atmosphere or with protective coatings.

The **precious metals** include gold, iridium, osmium, palladium, platinum, rhodium, ruthenium, and silver. Excellent corrosion resistance, combined with various inherent properties, justifies the many costly applications of these metals and alloys. Gold circuitry in the electronics industry, various dental alloys, and platinum coatings for catalytic converters are a few of the better known examples.

Rapidly solidified ferrous alloys were introduced in Section 11.1. Research and development in this area is equally active for nonferrous alloys. Various amorphous Ni-based alloys have been developed for superior magnetic properties. Rapidly solidified crystalline aluminum and titanium alloys have demonstrated superior mechanical properties at elevated temperatures. The control of fine-grained precipitates by rapid solidification is an important factor for both of these alloy systems of importance to the aerospace industry. The interesting quasicrystal structures of Section 4.6 were first produced by rapid solidification.

SAMPLE PROBLEM 11.2

In redesigning an automobile for a new model year, 25 kg of conventional steel parts are replaced by aluminum alloys of the same dimensions. Calculate the resulting mass savings for the new model, approximating the alloy densities by those for pure Fe and Al, respectively.

SOLUTION

From Appendix 1, $\rho_{Fe} = 7.87 \text{ Mg/m}^3$ and $\rho_{Al} = 2.70 \text{ Mg/m}^3$. The volume of steel parts replaced would be

$$V = \frac{m_{Fe}}{\rho_{Fe}} = \frac{25 \text{ kg}}{7.87 \text{ Mg/m}^3} \times \frac{1 \text{ Mg}}{10^3 \text{ kg}} = 3.21 \times 10^{-3} \text{ m}^3$$

The mass of new aluminum parts would be

$$m_{Al} = \rho_{Al} V_{Al} = 2.70 \text{ Mg/m}^3 \times 3.21 \times 10^{-3} \text{ m}^3 \times \frac{10^3 \text{ kg}}{1 \text{ Mg}} = 8.65 \text{ kg}$$

The resulting mass savings is then

$$m_{Fe} - m_{Al} = 25 \text{ kg} - 8.65 \text{ kg} = 16.3 \text{ kg}$$

PRACTICE PROBLEM 11.2

A common basis for selecting nonferrous alloys is their low density as compared to structural steels. Alloy density can be approximated as a weighted average of the densities of the constituent elements. In this way, calculate the densities of the aluminum alloys given in Table 6.1.

11.3   PROCESSING OF METALS

Table 11.10 summarizes some of the major **processing** techniques for metals. An especially comprehensive example is given in Figure 11–2, which summarizes the general production of steel by the **wrought process**. While the range of wrought products is large, there is a common processing history. Raw materials are combined and melted, leading eventually to a rough cast form. The casting is then worked to final product shapes. A potential problem with **casting**, as illustrated in Figure 11–3, is the presence of residual porosity (Figure 11–4). The use of mechanical deformation to form the final product shape in the wrought process largely eliminates this porosity.

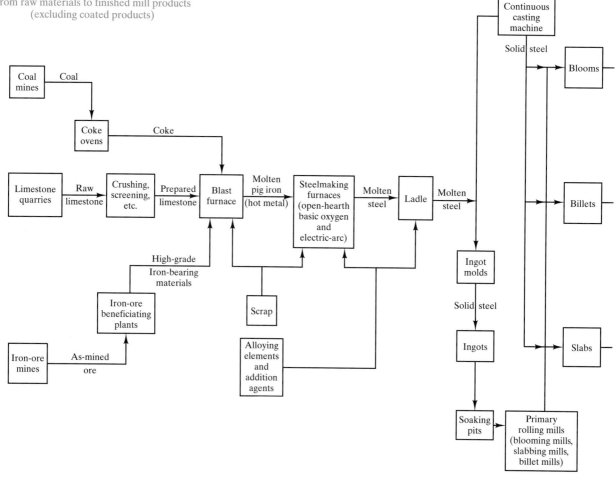

STEELMAKING
From raw materials to finished mill products
(excluding coated products)

Figure 11-2 *Schematic summary of the wrought process for producing various steel product shapes. (From W. T. Lankford et al., Eds.,* The Making, Shaping, and Treating of Steel, *10th Ed., United States Steel, Pittsburgh, Pa., 1985. Copyright 1985 by United States Steel Corporation.)*

Table 11.10 *Major Processing Methods for Metals*

Wrought processing	Joining
Rolling	Welding
Extrusion	Brazing
Forming	Soldering
Stamping	Powder metallurgy
Forging	Hot isostatic pressing
Drawing	Superplastic forming
Casting	Rapid solidification

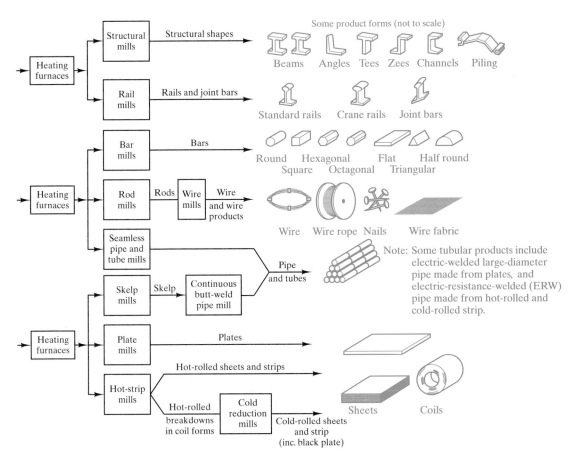

Figure 11-2 *(Continued)*

The rapid cooling of a melt during the casting process also provides examples of nonequilibrium microstructures. Figure 11–5 illustrates the development of a **cored structure** in which concentration gradients occur in individual grains. In this case, solid-state diffusion is too slow to allow the grain chemistry to remain uniform during solidification. The resulting chemical **segregation** is an example of the kinetic factors of Chapter 10 overriding the ability to maintain the equilibrium of Chapter 9. An undesirable consequence of the cored structure is preferential melting of the grain boundary region upon reheating, leading to a sudden loss of mechanical integrity. Coring can be eliminated by the homogenization of the grains in a subsequent heat treatment at a temperature below the lower (dashed) solidus line of Figure 11–5. Another example of a nonequilibrium microstructure from casting is the **dendritic structure** of a lead-tin (20 Pb–80 Sn) alloy in Figure 11–6. In this case, the growth of the cellular structure of a second phase is accompanied by side branches. At the base of the dentritic "trees" in Figure 11–6, the eutectic microstructure for the lead-tin system is formed. In general, dendritic growth is important because it can lead to casting defects such as porosity and shrinkage.

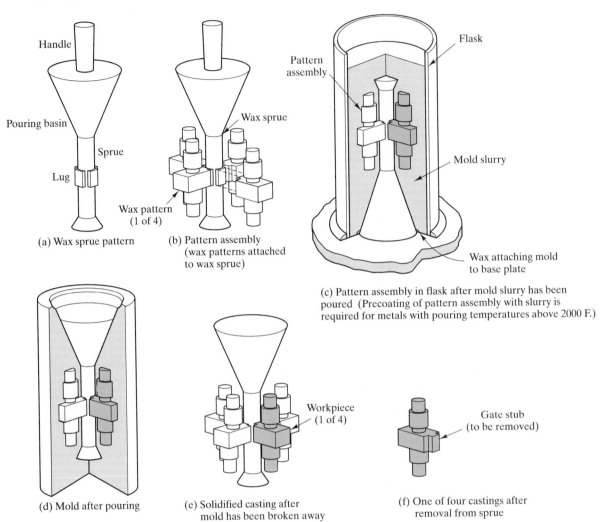

(a) Wax sprue pattern

(b) Pattern assembly (wax patterns attached to wax sprue)

(c) Pattern assembly in flask after mold slurry has been poured (Precoating of pattern assembly with slurry is required for metals with pouring temperatures above 2000 F.)

(d) Mold after pouring

(e) Solidified casting after mold has been broken away

(f) One of four castings after removal from sprue

Figure 11-3 *Schematic illustration of the casting of a metal alloy form by the "investment molding" process. (From Metals Handbook, 8th ed., Vol. 5: Forging and Casting, American Society for Metals, Metals Park, Ohio, 1970.)*

Complex structural designs are generally not fabricated in a single-step process. Instead, relatively simple forms produced by wrought or casting processes are joined together. Joining technology is a broad field in itself. Our most common example is **welding**, in which the metal parts being joined are partially melted in the vicinity of the join (Figure 11–7). Welding frequently involves a metal weld rod, which is also melted. In **brazing**, the braze metal melts, but the parts being joined may not. The bond is more often formed by the solid-state diffusion of that braze metal into the joined parts. In **soldering**, neither melting nor solid-state diffusion is required. The join is usually produced by the adhesion of the melted solder to the surface of each metal part.

Temperature

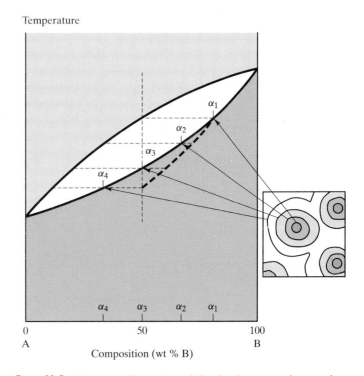

Composition (wt % B)

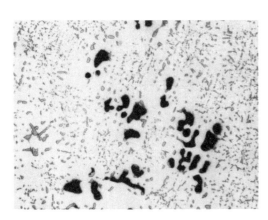

Figure 11-4 *Microstructure of a cast alloy (354-T4 aluminum), 50×. The black spots are voids, and gray particles are a silicon-rich phase. (From Metals Handbook, 9th ed., Vol. 9: Metallography and Microstructures, American Society for Metals, Metals Park, Ohio, 1985.)*

Figure 11-5 *Schematic illustration of the development of a cored structure in the nonequilibrium solidification of a 50:50 alloy in a system exhibiting complete solid solution. (This case can be contrasted with the equilibrium solidification shown in Figure 9–33.) During the rapid cooling associated with casting, the liquidus curve is unaffected given the rapid diffusion in the liquid state, but solid state diffusion may be too slow to maintain uniform grain compositions upon cooling. As a result, the solidus curve is shifted downward as indicated by the dashed line.*

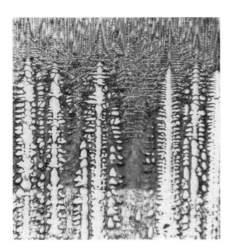

Figure 11-6 *Example of a tree-like dendritic structure in a 20 Pb–80 Sn alloy. A eutectic microstructure is seen at the base of the dendrites. (From Metals Handbook, 9th ed., Vol. 9: Casting, ASM International, Materials Park, Ohio, 1988.)*

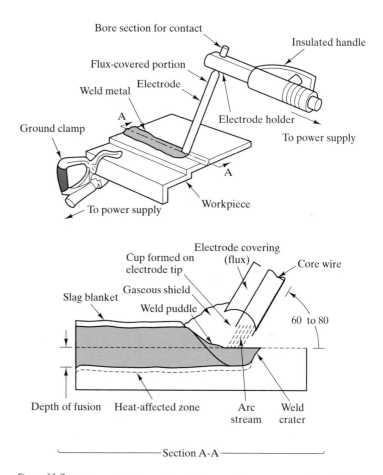

Figure 11-7 *Schematic illustration of the welding process. Specifically, "shielded metal-arc" welding is shown. (From Metals Handbook, 8th ed., Vol. 6: Welding and Brazing, American Society for Metals, Metals Park, Ohio, 1971.)*

Figure 11–8 shows a solid-state alternative to the more conventional processing techniques. **Powder metallurgy** involves the solid-state bonding of a fine-grained powder into a polycrystalline product. Each grain in the original powder roughly corresponds to a grain in the final polycrystalline microstructure. Sufficient solid-state diffusion can lead to a fully dense product, but some residual porosity is common. This processing technique is advantageous for high-melting-point alloys and products of intricate shape. The discussion of the sintering of ceramics in Section 10.6 is relevant here also. An advance in the field of powder metallurgy is the technique of **hot isostatic pressing** (HIP), shown in Figure 11–9, in which a uniform pressure is applied to the part using a high-temperature, inert gas. **Superplastic forming** was introduced in Figure 1–4 as a recent, economical technique developed

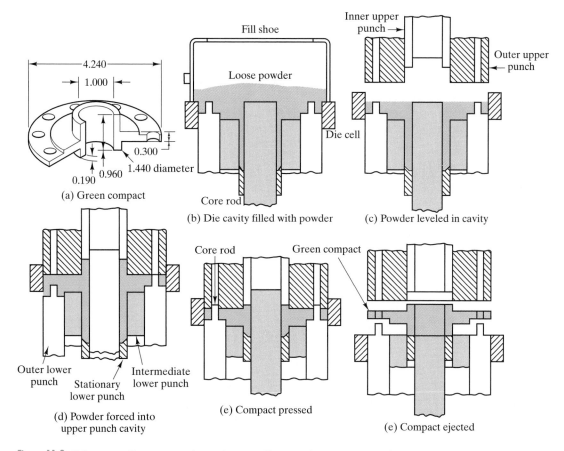

Figure 11-8 *Schematic illustration of powder metallurgy. The green, or unfired, compact is subsequently heated to a sufficiently high temperature to produce a strong piece by solid-state diffusion between the adjacent powder particles. (From* Metals Handbook, 8th Ed., Vol. 4: Forming, *American Society for Metals, Metals Park, Ohio, 1969.)*

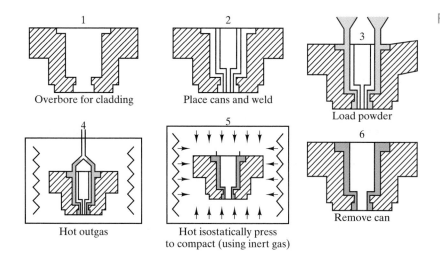

Figure 11-9 *Hot isostatic prossing (HIP) of a cladding for a complex-shaped part. (After Advanced Materials and Processes, January 1987.)*

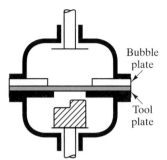

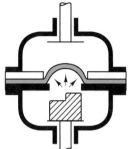

Bubble plate

Tool plate

Figure 11-10 *Superplastic forming allows deep parts to be formed with a relatively uniform wall thickness. Modest air pressure (up to 10 atmospheres) stretches a heated "bubble" of metal sheet, which then collapses over a metal former pushed up through the plane of the original sheet. (After Superform USA, Inc.)*

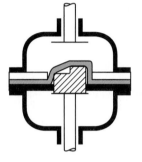

for forming complex shapes. This process, as illustrated in Figure 11–10, is closely associated with creep deformation. Certain fine-grained alloys can exhibit several thousand percent elongations, making the product shapes in Figure 1–4 possible. Figure 11–10 illustrates a typical sequence of fabrication steps. The rapid solidification of alloys was discussed earlier in this chapter in conjunction with the recent development of amorphous metals and a variety of novel crystalline microstructures. Several techniques for rapid solidification are illustrated in Figure 11–11. The quasicrystals of Section 4.6 were originally produced as a byproduct of research on rapid solidification.

Table 11.11 summarizes a few rules of thumb about the effects of metals processing on design parameters. Metals exhibit an especially wide range of behavior as a function of processing. As with any generalizations, we must be alert to exceptions. In addition to the fundamental processing topics discussed in this section, Table 11.11 refers to issues of microstructural development and heat treatment discussed in Chapters 9 and 10. A specific example is given in Figure 11–12, which shows how strength, hardness, and ductility vary with alloy composition in the Cu–Ni system. Similarly, Figure 11–13 shows how these mechanical properties vary with mechanical history for a given alloy, in this case, brass. Variations in alloy chemistry and thermomechanical history allow considerable fine-tuning of structural design parameters.

Chilling Techniques

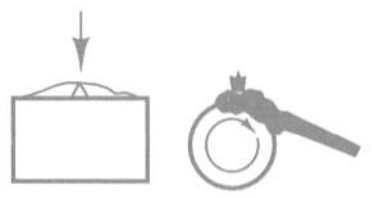	Conductive heat removal: splat cooling, planar flow casting, double roller quenching, injectin chilling, plasma spray deposition. Heat transfer coefficient, h, = 0.1–100 kW/m^{2}K
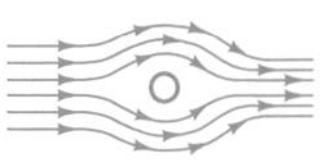	Convective heat removal: various forms of gas and water atomizers, unidirectional and centrifugal atomizers, rotating cup process, plasma spray deposition. h = 0.1– 100 kW/m^{2}K
	Radiative heat romoval: electro-hydrodynamic process, vacuum plasma process. h = 10 W/m2K
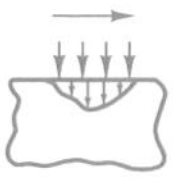	Directed and concentrated energy techniques: conductive heat removal lasers (pulsed and continuous), electron beam. h → ∞

Undercooling Techniques

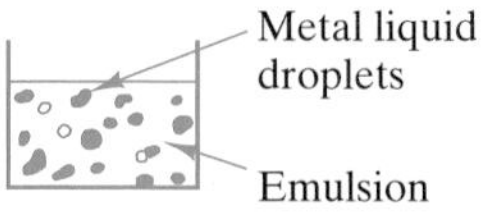 Metal liquid droplets / Emulsion	Droplet emulsion
 Liquid / Solid	Eqilibrium mushy zone
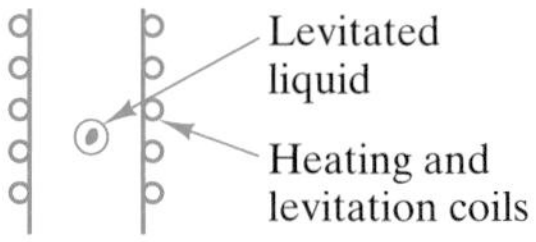 Levitated liquid / Heating and levitation coils	Levitation (gas jet or induction current)
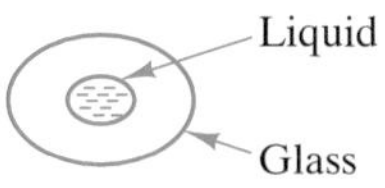 Liquid / Glass	Nucleant fluxing
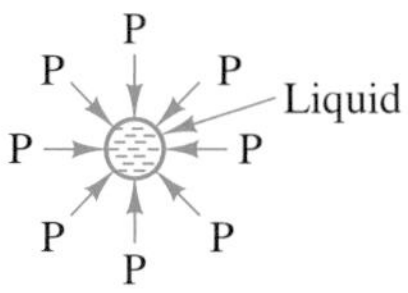 Liquid	Rapid pressure application

Figure 11-11 *Schematic summary of several techniques for the rapid solid-ification of metal alloys. (From Metal Progress, May 1986.*

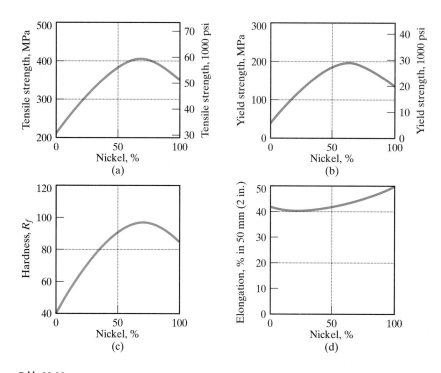

Figure 11-12 *Variation of mechanical properties of copper–nickel alloys with composition. Recall that copper and nickel form a complete solid-solution phase diagram (Figure 9–9). (From L. H. Van Vlack,* Elements of Materials Science and Engineering, *4th Ed., Addison-Wesley Publishing Co., Inc., Reading, Mass., 1980.)*

Table 11.11 *Some General Effects of Processing on the Properties of Metals*

Strengthening by	Weakening by
Cold working	Porosity (produced by casting, welding, or powder metallurgy)
Alloying (e.g., solution hardening)	Annealing
Phase transformations (e.g., martensitic)	Hot working
	Heat-affected zone (welding)
	Phase transformations (e.g., tempered martensite)

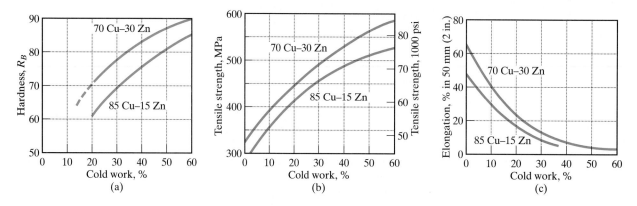

Figure 11-13 *Variation of mechanical properties of two brass alloys with degree of cold work. (From L. H. Van Vlack,* Elements of Materials Science and Engineering, *4th Ed., Addison-Wesley Publishing Co., Inc., Reading, Mass., 1980.)*

The Material World:
The Foundry of the Future

Over the past two decades, more than 2000 foundries across the United States have closed. These difficult economic decisions have often been based on increasingly stringent environmental health and safety laws. Concerns about the potential economic impact of a continuing trend of this sort has led to the Casting Emissions Reduction Program (CERP), a collaborative effort between government and private industry, located at McClellan Air Force Base in Sacramento, California. The CERP "Foundry of the Future" is a 5600-square-meter facility that can produce aluminum and gray iron castings as large as engine blocks. The CERP mission is to help the metal-casting industry remain competitive with international foundries while meeting federal clean air standards. The overall objective is to maintain manufacturing jobs while contributing to a clean environment. A specific objective is to help reduce emissions from the foundry industry, which employs over 200,000 workers in the United States alone.

(Courtesy of the Casting Emissions Reduction Program [CERP].)

The CERP foundry produces high-quality castings using environmentally sound technologies and new materials. The casting equipment within the Foundry of the Future represents the state of the art, enhanced with specialized ventilation and monitoring systems to capture air emissions. The resulting information and research data will be shared with the general foundry industry. CERP is a cooperatieve research and development agreement among McClellan AFB, the U.S. Council for Automotive Research, the U.S. Environmental Protection Agency, the California Air Resources Board, and the American Foundrymen's Society.

▣ SAMPLE PROBLEM 11.3

A copper–nickel alloy is required for a particular structural application. The alloy must have a tensile strength greater than 400 MPa and a ductility of less than 45% (in 50 mm). What is the permissible alloy composition range?

SOLUTION

Using Figure 11–12, we can determine a "window" corresponding to the given property ranges:

Tensile strength > 400MPa : 59 < %Ni < 79
Elongation < 45%   : 0 < %Ni < 79
Giving a net window of   :
  permissible alloy range : 59 < %Ni < 79

SAMPLE PROBLEM 11.4

A bar of annealed 70 Cu–30 Zn brass (10-mm diameter) is cold drawn through a die with a diameter of 8 mm. What is **(a)** the tensile strength and **(b)** the ductility of the resulting bar?

SOLUTION

The results are available from Figure 11–13 once the percent cold work is determined. This is given by

$$\% \text{ cold work} = \frac{\text{initial area} - \text{final area}}{\text{initial area}} \times 100\%$$

For the given processing history

$$\% \text{ cold work} = \frac{\pi/4(10 \text{ mm})^2 - \pi/4(8 \text{ mm})^2}{\pi/4(10 \text{ mm})^2} \times 100\%$$

$$= 36\%$$

From Figure 11–13, we see that **(a)** tensile strength = 520 MPa and **(b)** ductility (elongation) = 9%.

▣ PRACTICE PROBLEM 11.3

(a) In Sample Problem 11.3, we determine a range of copper–nickel alloy compositions that meet structural requirements for strength and ductility. Make a similar determination for the following specifications: hardness greater than 80 R_F and ductility less than 45%. **(b)** For the range of copper–nickel alloy compositions determined in part (a), which specific alloy would be preferred on a cost basis, given that the cost of copper is approximately $4.00/kg and of nickel $11.00/kg?

PRACTICE PROBLEM 11.4

In Sample Problem 11.4, we calculate the tensile strength and ductility for a cold-worked bar of 70 Cu–30 Zn brass. **(a)** What percent increase does that tensile strength represent compared to that for the annealed bar? **(b)** What percent decrease does that ductility represent compared to that for the annealed bar?

SUMMARY

Metals play a major role in engineering design, especially as structural elements. Over 90% by weight of the materials used for engineering are iron-based, or ferrous, alloys, which include the steels (containing 0.05 to 2.0 wt % C) and the cast irons (with 2.0 to 4.5 wt % C). Most steels involve a minimum of alloy additions to maintain moderate costs. These are plain-carbon or low-alloy (< 5 wt % total noncarbon additions) steels. Special care in alloy selection and processing can result in high-strength low-alloy (HSLA) steels. For demanding design specifications, high-alloy (> 5 wt % total noncarbon additions) steels are required. Chromium additions produce stainless steels for corrosion resistance. Additions such as tungsten lead to high-hardness alloys used as tool steels. Superalloys include many stainless steels that combine corrosion resistance with high strength at elevated temperatures. The cast irons exhibit a wide range of behavior depending on composition and processing history. White and gray irons are typically brittle, whereas ductile and malleable irons are characteristically ductile.

Nonferrous alloys include a wide range of materials with individual attributes. Aluminum, magnesium, and titanium alloys have found wide use as lightweight structural members. Copper and nickel alloys are especially attractive for chemical and temperature resistance and electrical and magnetic applications. Other important nonferrous alloys include the zinc and lead alloys and the refractory and precious metals.

Many applications of materials in engineering design are dependent on the processing of the material. Many common metal alloys are produced in the wrought process, in which a simple cast form is mechanically worked into a final shape. Other alloys are produced directly by casting. More complex structural shapes depend on joining techniques such as welding. An alternative to wrought and casting processes is the entirely solid-state technique of powder metallurgy. Contemporary metal forming techniques include hot isostatic pressing, superplastic forming, and rapid solidification.

KEY TERMS

austenitic stainless steel (405)
brass (416)
brazing (422)
bronze (417)
carbon steel (402)
casting (420)
cast iron (408)
cored structure (421)
dendritic structure (421)
ductile iron (408)
ferritic stainless steel (405)
ferrous alloy (401)
gray iron (408)

high-alloy steel (401)
high-strength low-alloy steel (402)
hot isostatic pressing (HIP) (424)
low-alloy steel (401)
malleable iron (408)
martensitic stainless steel (405)
nickel–aluminum superalloy (417)
nonferrous alloy (414)
powder metallurgy (424)
precious metal (418)
precipitation-hardening stainless steel (405)
processing (420)

rapidly solidified alloy (412)
refractory metal (418)
segregation (421)
soldering (422)
solution hardening (417)
steel (401)
superalloy (405)
superplastic forming (424)
tool steel (405)
welding (422)
white iron (408)
wrought alloy (408)
wrought process (420)

REFERENCES

Ashby, M. F., and **D. R. H. Jones**, *Engineering Materials 1— An Introduction to Their Properties and Applications*, 2nd ed., Butterworth-Heinemann, Boston, 1996.

Ashby, M. F., and **D. R. H. Jones**, *Engineering Materials 2: An Introduction to Microstructures, Processing and Design*, 2nd ed., Butterworth-Heinemann, Boston, 1998.

Davis, J. R., Ed., *Metals Handbook*, Desk Ed., 2nd ed., ASM International, Materials Park, Ohio, 1998. A one-volume summary of the extensive *Metals Handbook* series.

ASM Handbook. Vols. 1 (*Properties and Selection: Irons, Steels, and High-Performance Alloys*), and 2 (*Properties and Selection: Nonferrous Alloys and Special-Purpose Metals*), ASM International, Materials Park, Ohio, 1990 and 1991.

PROBLEMS

11.1 • Ferrous Alloys

11.1. **(a)** Estimate the density of 1040 carbon steel as the weighted average of the densities of constituent elements.
(b) The density of 1040 steel is what percentage of the density of pure Fe?

11.2. Repeat Problem 11.1 for the type 304 stainless steel in Table 6.1.

11.3. Use the weighted average of the densities of constituent elements to estimate the density of T1 tool steel in Table 11.3.

11.4. Use the weighted average of the densities of constituent elements to estimate the density of Incoloy 903 in Table 11.4.

11.5. Estimate the density of the 80 Fe–20 B amorphous alloy in Table 11.6 as a weighted average of the densities of constituent elements. In addition, reduce the overall, calculated density by 1% to account for the noncrystalline nature of the structure (Section 4.5).

11.6. Repeat Problem 11.5 for the 80 Fe–10 B–10 Si amorphous alloy in Table 11.6.

11.2 • Nonferrous Alloys

11.7. A prototype Al–Li alloy is being considered for replacement of a 7075 alloy in a commercial aircraft. The compositions are compared in the following table. **(a)** Assuming that the same volume of material is used, what percentage reduction in density would occur by this material substitution? **(b)** If a total mass of 70,000 kg of 7075 alloy is currently used on the aircraft, what net mass reduction would result from the substitution of the Al–Li alloy?

Primary alloying elements (wt %)						
Alloy	**Li**	**Zn**	**Cu**	**Mg**	**Cr**	**Zr**
Al–Li	2.0		3.0			0.12
7075		5.6	1.6	2.5	0.23	

11.8. Estimate the alloy densities for **(a)** the magnesium alloys of Table 6.1 and **(b)** the titanium alloys of Table 6.1.

11.9. Between 1975 and 1985, the volume of all iron and steel in a given automobile model decreased from $0.162 \ m^3$ to $0.116 \ m^3$. In the same time frame, the volume of all aluminum alloys increased from 0.012 m^3 to 0.023 m^3. Using the densities of pure Fe and Al, estimate the mass reduction resulting from this trend in materials substitution.

11.10. For an automobile design equivalent to the model described in Problem 11.9, the volume of all iron and steel is further reduced to 0.082 m^3 by the year 2000. In the same time frame, the total volume of aluminum alloys increases to 0.034 m^3. Estimate the mass reduction (compared to 1975) resulting from this materials substitution.

11.11. Estimate the density of a cobalt–chrome alloy used for artificial hip joints as a weighted average of the densities of constituent elements: 50 wt % Co, 20 wt % Cr, 15 wt % W, 15 wt % Ni.

11.12. Consider an artificial hip component made of the cobalt–chrome alloy of Problem 11.11 with a volume of $160 \times 10^{-6} \ m^3$. What mass savings would

result from substituting a Ti–6 Al–4 V alloy with the same component shape and consequently the same volume?

11.3 • Processing of Metals

At the outset of the text, we established a policy of avoiding subjective homework problems. Questions about materials processing quickly turn to the subjective. As a result, Section 11.3 contains only a few problems that can be kept in the objective style used in preceding chapters.

11.13. A bar of annealed 85 Cu–15 Zn (12-mm diameter) is cold drawn through a die with diameter of 10 mm. What are **(a)** the tensile strength and **(b)** the ductility of the resulting bar?

11.14. For the bar analyzed in Problem 11.13, **(a)** what percentage does the tensile strength represent compared to that for the annealed bar and **(b)** what percentage decrease does the ductility represent compared to that for the annealed bar?

D 11.15. You are given a 2-mm-diameter wire of 85 Cu–15 Zn brass. It must be drawn down to a diameter of 1 mm. The final product must meet specifications of tensile strength greater than 375 MPa and a ductility of greater than 20%. Describe a processing history to provide this result.

D 11.16. How would your answer to Problem 11.15 change if a 70 Cu–30 Zn brass wire were used instead of the 85 Cu–15 Zn material?

CHAPTER 12
Ceramics and Glasses

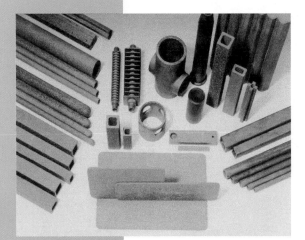

Ceramics have traditionally been used in high-temperature engineering applications. For the interior architecture of furnaces, silicon carbide provides good dimensional stability at temperatures up to 1,650° C, along with high resistance to thermal shock and corrosion and a low density. (Courtesy of Bolt Technical Ceramics.)

Ceramics and glasses represent some of the earliest and most environmentally durable materials for engineering. They also represent some of the most advanced materials being developed for the aerospace and electronics industries. In this chapter we divide this highly diverse collection of engineering materials into three main categories. *Crystalline ceramics* include the traditional silicates and the many oxide and nonoxide compounds widely used in both traditional and advanced technologies. *Glasses* are noncrystalline solids with compositions comparable to the crystalline ceramics. The absence of crystallinity, which results from specific processing techniques, gives a unique set of mechanical and optical properties. Chemically, the glasses are conveniently subdivided as silicates and nonsilicates. *Glassceramics,* the third category, are another type of crystalline ceramics that are initially formed as glasses and then crystallized in a carefully controlled way. This crystallization process will be discussed in some detail. Rather specific compositions lend themselves to this technique, with the $Li_2–Al_2O_3–SiO_2$ system being the most important commercial example.

As with the discussion of metals in Chapter 11, the treatment of ceramics in this chapter builds on concepts raised in Part I. The wide range of ceramic crystal structures was illustrated in Chapter 3. The point defects in these structures (Chapter 4) were seen to be the basis of diffusional transport in Chapter 5. The characteristic brittle nature of ceramics discussed in Chapter 6 was seen to be due to their complex dislocation structures introduced in Chapter 4. The thermal behavior of Chapter 7 was especially important for ceramics, which are often used at elevated temperatures, and a special focus of the issue of failure by thermal shock in Chapter 8. As with metals, phase equilibria (Chapter 9) and kinetics (Chapter 10) play important roles in the optimal processing of ceramic materials.

As with metals, the processing of ceramics and glasses can profoundly affect their performance as structural materials. Traditional ceramic processing methods include fusion and slip casting, sintering, and hot pressing. Traditional glass-forming methods are sometimes followed by controlled devitrification to produce glass-ceramics. More recent processing methods include sol–gel and biomimetic techniques and self-propagating high-temperature synthesis (SHS).

12.1 CERAMICS—CRYSTALLINE MATERIALS

It is appropriate to begin our discussion of crystalline ceramics by looking at the SiO_2-based **silicates**. Since silicon and oxygen together account for roughly 75% of the elements in the earth's crust (Figure 12–1), these materials are abundant and economical. Many of the traditional ceramics that we use fall into this category. One of the best tools for characterizing early civilizations is *pottery,* which is burnt clayware. This has been a commercial product since roughly 4000 B.C. Pottery is a part of the category of ceramics known as **whitewares**, which are commercial fired ceramics with a typically white and fine-grained microstructure. A well-known example is the translucent porcelain of which fine china is composed. In addition to pottery and

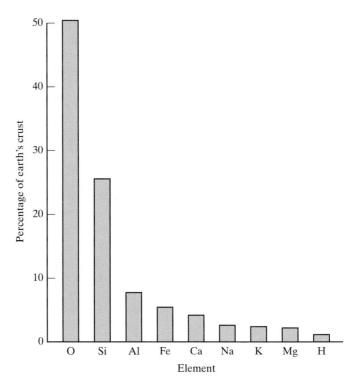

Figure 12-1 *The relative abundance of elements in the earth's crust illustrates the availability of ceramic minerals, especially the silicates.*

related whiteware ceramics, **clay** is the basis of *structural clay products,* such as brick, tile, and sewer pipe. The range of silicate ceramics reflects the diversity of the silicate minerals that are usually available to local manufacturing plants. Table 12.1 summarizes the general compositions of some common examples. This listing includes refractories based on fired clay. **Refractories** are high-temperature-resistant structural materials that play crucial roles in

Table 12.1 *Compositions[a] of Some Silicate Ceramics*

Ceramic	Composition (wt %)					
	SiO_2	Al_2O_3	K_2O	MgO	CaO	Others
Silica refractory	96					4
Fireclay refractory	50–70	45–25				5
Mullite refractory	28	72				—
Electrical porcelain	61	32	6			1
Steatite porcelain	64	5		30		1
Portland cement	25	9			64	2

[a] These are approximate compositions, indicating primary components. Impurity levels can vary significantly from product to product.

industry (e.g., in the steelmaking process). About 40% of the refractories' industry output consists of the clay-based silicates. Also listed in Table 12.1 is a representative of the cement industry. The example is portland cement, a complex mixture that can be described overall as a calcium aluminosilicate.

Table 12.2 lists several examples of **nonsilicate oxide ceramics**, which include some traditional materials such as magnesia (MgO), a refractory widely used in the steel industry. In general, however, Table 12.2 includes many of the more advanced ceramic materials. *Pure oxides* are compounds with impurity levels sometimes less than 1 wt % and, in some cases, impurity levels in the part per million (ppm) range. The expense of chemical separation and subsequent processing of these materials is a sharp contrast to the economy of silicate ceramics made from locally available and generally impure minerals. These materials find many uses in areas such as the electronics industry, where demanding specifications are required. However, many of the products in Table 12.2 with one predominant oxide compound may contain several percent oxide additions and impurities. In Table 12.2, UO_2 is our best example of a *nuclear ceramic*. This compound containing radioactive uranium is a widely used reactor fuel. **Partially stabilized zirconia** (ZrO_2) (PSZ) is a primary candidate for advanced structural applications, including many traditionally filled by metals. A key to the potential for metals-substitution is the mechanism of transformation toughening, which was discussed in Section 8.2. **Electronic ceramics**, such as $BaTiO_3$, and the **magnetic ceramics**, such as $NiFe_2O_4$ (nickel ferrite), represent the largest part of the industrial ceramics market; they are discussed in Chapters 15 and 18, respectively.

Table 12.2 *Some Nonsilicate Oxide Ceramics*

Primary composition[a]	Common product names
Al_2O_3	Alumina, alumina refractory
MgO	Magnesia, magnesia refractory, magnesite refractory, periclase refractory
$MgAl_2O_4$ ($= MgO \cdot Al_2O_3$)	Spinel
BeO	Beryllia
ThO_2	Thoria
UO_2	Uranium dioxide
ZrO_2 (stabilized[b] with CaO)	Stabilized (or partially stabilized) zirconia
$BaTiO_3$	Barium titanate
$NiFe_2O_4$	Nickel ferrite

[a] Some products, such as the industrial refractories, may have several weight percent oxide additions and impurities.

[b] Pure ZrO_2 has a phase transformation at 1000° C in which the crystal structure change produces a catastrophic volume change. The material is literally reduced to a powder. A 10 wt % CaO addition produces a cubic crystal structure stable to the melting point ($\sim$ 2500° C) making "stabilized" zirconia a highly useful refractory. Figure 9–30 illustrated this point with the CaO–ZrO_2 phase diagram. Lesser CaO additions can produce a two-phase microstructure, with cubic zirconia being one of the phases. This "partially stabilized" zirconia has even superior mechanical properties, as discussed in Section 8.2.

The Material World:
Hydroxyapatite—The Body's Own Ceramic

As we discussed ceramic materials produced by engineers for various structural applications, we can also look inward and focus on a ceramic material "engineered" by the body to play a key role in our skeletal structure. Hydroxyapatite (HA), with the chemical formula $Ca_{10}(HPO_4)_6(OH)_2$, is the primary mineral content of bone, comprising 43% of overall bone weight. Precursor calcium phosphates precipitate from fluids in bone and then undergo phase transformations to form HA. The structure and mechanical properties of HA can vary as the result of various chemical substitutions: K, Mg, Sr, and Na for Ca; carbonate for phosphate; and F for OH.

Bone is formed by cells called osteoblasts. These cells make an organic matrix that contains water in place of mineral. After about 10 days, the osteoid "matures," allowing precipitation of mineral crystals. In humans, new bone reaches approximately 70% of its possible degree of mineralization (mineral capacity) over a period of a few days. This process is called "primary" mineralization. "Secondary" mineralization occurs slowly over several months and generally reaches about 90% of mineral capacity. It appears that cells called osteocytes, which are osteoblasts that

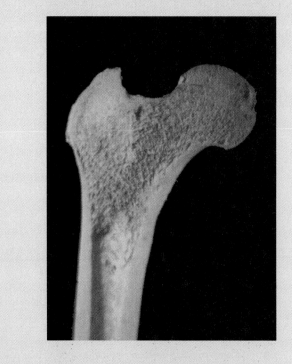

(Courtesy of R. B. Martin, Orthopaedic Research Laboratories, University of California, Davis Medical Center, Sacramento, CA)

become trapped in the bone they are forming, serve a control function by maintaining this "normal" mineral content. The death of osteocytes allows hypermineralization (100% mineral capacity) to occur, which makes the bone more brittle and less tough than normal.

The overall structure of long bones in our skeleton is similar to a bamboo shaft, with the dense outer shell called compact bone. This material has a density of 2.1 Mg/m³ and an elastic modulus of 20 GPa, a rather small value compared with most synthetic engineering materials in this book, but quite adequate for our physiological purposes. The overall geometry of the bone causes the mechanical properties to be highly directional. Compact bone pulled in tension along a direction parallel to the overall cylindrical axis has a typical strength of 135 MPa, again modest compared with many engineered materials, but generally adequate for our skeletal structure. While the mechanical properties of bone may be unimpressive compared with synthetic materials, as a living tissue, bone has the substantial advantage of being able to repair and remodel itself. Variations on the cellular mechanisms involved in bone formation are the basis of repair and remodeling.

The mechanical behavior of bone is strongly dependent on the ceramic mineral hydroxyapatite but is also affected by its complex microstructure and a significant organic phase. The primary polymeric component of bone will be the focus of a feature box in the next chapter, and the overall composite nature of bone and synthetic substitutes will be the focus of a feature box in Chapter 14.

Schematic illustration of osteon cells in a matrix of primary bone. (From R. B. Martin, "Bone as a Ceramic Composite Material," in *Bioceramics—Applications of Ceramic and Glass Materials in Medicine*, Ed. J. F. Shackelford, Trans Tech Publications, Switzerland, 1999)

Table 12.3 *Some Nonoxide Ceramics*

Primary composition[a]	Common product names
SiC	Silicon carbide
Si_3N_4	Silicon nitride
TiC	Titanium carbide
TaC	Tantalum carbide
WC	Tungsten carbide
B_4C	Boron carbide
BN	Boron nitride
C	Graphite

[a] Some products may have several weight percent additions or impurities.

Table 12.3 lists examples of **nonoxide ceramics**. Some of these, such as silicon carbide, have been common industrial materials for several decades. Silicon carbide has served as furnace heating elements and as an abrasive material. Silicon nitride and related materials (e.g., the oxygen-containing SiAlON, pronounced "sigh-a-lon") represent, along with partially stabilized zirconia, the cutting edge of ceramic technology. Substantial research and development have been devoted to these materials in the past two decades for the purpose of producing superior gas turbine components. The expanding use of structural ceramics, especially in automotive applications, is projected to continue well beyond the year 2000.

SAMPLE PROBLEM 12.1

Mullite is $3Al_2O_3 \cdot 2SiO_2$. Calculate the weight fraction of Al_2O_3 in a mullite refractory.

SOLUTION

Using data from Appendix 1, we have

$$\text{mol wt } Al_2O_3 = [2(26.98) + 3(16.00)] \text{ amu}$$
$$= 101.96 \text{ amu}$$
$$\text{mol wt } SiO_2 = [28.09 + 2(16.00)] \text{ amu}$$
$$= 60.09 \text{ amu}$$

Therefore,

$$\text{wt fraction } Al_2O_3 = \frac{3(101.96)}{3(101.96) + 2(60.09)}$$
$$= 0.718$$

PRACTICE PROBLEM 12.1

What is the weight fraction of Al_2O_3 in spinel $(MgAl_2O_4)$? (See Sample Problem 12.1.)

12.2   GLASSES—NONCRYSTALLINE MATERIALS

The concept of the noncrystalline solid was discussed in Section 4.5. As shown there, the traditional examples of this type of material are the **silicate glasses**. As with crystalline silicates, these glasses are generally moderate in cost due to the abundance of elemental Si and O in the earth's crust. For much of routine glass manufacturing, SiO_2 is readily available in local sand deposits of adequate purity. In fact, the manufacturing of various glass products accounts for a much larger tonnage than that involved in producing crystalline ceramics. Table 12.4 lists key examples of commercial silicate glassware. Table 12.5 helps to interpret the significance of the compositions in Table 12.4 by listing the oxides that are network formers, modifiers, and intermediates.

Network formers include oxides that form oxide polyhedra, with low coordination numbers. These polyhedra can connect with the network of SiO_4^{4-} tetrahedra associated with vitreous SiO_2. Alkali and alkaline earth oxides such as Na_2O and CaO do not form such polyhedra in the glass structure but, instead, tend to break up the continuity of the polymerlike

Table 12.4 *Compositions of Some Silicate Glasses*

Glass	Composition (wt %)									
	SiO_2	B_2O_3	Al_2O_3	Na_2O	CaO	MgO	K_2O	ZnO	PbO	Others
Vitreous silica	100									—
Borosilicate	76	13	4	5	1					1
Window	72		1	14	8	4				1
Container	73		2	14	10					1
Fiber (E-glass)	54	8	15		22					1
Bristol glaze	60		16		7		11	6		—
Copper enamel	34	3	4				17		42	—

Table 12.5 *Role of Oxides in Glass Formation*

Network formers	Intermediates	Network modifiers
SiO_2	Al_2O_3	Na_2O
B_2O_3	TiO_2	K_2O
GeO_2	ZrO_2	CaO
P_2O_5		MgO
		BaO
		PbO
		ZnO

SiO_2 network. One might refer back to the schematic of an alkali-silicate glass structure in Figure 4–25. The breaking of the network leads to the term **network modifier**. These modifiers make the glass article easier to form at a given temperature but increase its chemical reactivity in service environments. Some oxides such as Al_2O_3 and ZrO_2 are not, in themselves, glass formers, but the cation (Al^{3+} or Zr^{4+}) may substitute for the Si^{4+} ion in a network tetrahedron, thereby contributing to the stability of the network. Such oxides, which are neither formers nor modifiers, are referred to as **intermediates**.

Returning to Table 12.4, we can consider the nature of the major commercial silicate glasses:

1. **Vitreous silica** is high-purity SiO_2. *Vitreous* means "glassy" and is generally used interchangeably with *amorphous* and *noncrystalline*. With the absence of any significant network modifiers, vitreous silica can withstand service temperatures in excess of 1000°C. High-temperature crucibles and furnace windows are typical applications.

2. **Borosilicate glasses** involve a combination of BO_3^{3-} triangular polyhedra and SiO_4^{4-} tetrahedra in the glass-former network. About 5 wt % Na_2O provides good formability of the glassware without sacrificing the durability associated with the glass-forming oxides. The borosilicates are widely used for this durability in applications such as chemical labware and cooking ware. The great bulk of the glass industry is centered around a *soda–lime–silica* composition of approximately 15 wt % Na_2O, 10 wt % CaO, and 70 wt % SiO_2. The majority of common window glasses and glass containers can be found within a moderate range of composition.

3. The **E-glass** composition in Table 12.4 represents one of the most common glass fibers. This will be a central example of the fiber-reinforcing component of modern composite systems in Chapter 14.

4. **Glazes** are glass coatings applied to ceramics such as clayware pottery. The glaze generally provides a substantially more impervious surface than the unglazed material alone. A wide control of surface appearance is possible, as will be discussed in Chapter 16 on optical properties.

5. **Enamels** are glass coatings applied to metals. This term must be distinguished from *enamel* as applied to polymer-based paints. Frequently more important than the surface appearance provided by the enamel is the protective barrier it provides against environments corrosive to the metal. This corrosion-prevention system will be discussed further in Chapter 19. Table 12.4 lists the composition of a typical glaze and a typical enamel.

Table 12.6 lists various **nonsilicate glasses**. The nonsilicate oxide glasses such as B_2O_3 are generally of little commercial value because of their reactivity with typical environments such as water vapor. However, they can be useful additions to silicate glasses (e.g., the common borosilicate glasses). Some of

Table 12.6 *Some Nonsilicate Glasses*

B_2O_3	As_2Se_3	BeF_2
GeO_2	GeS_2	ZrF_4
P_2O_5		

the nonoxide glasses have become commercially significant. For example, chalcogenide glasses are frequently semiconductors, and are discussed in Chapter 17. The term *chalcogenide* comes from the Greek word *chalco*, meaning "copper," and is associated with compounds of S, Se, and Te. All three elements form strong compounds with copper, as well as with many other metal ions. Zirconium tetrafluoride (ZrF_4) glass fibers have proven to have superior light transmission properties in the infrared region compared to traditional silicates.

SAMPLE PROBLEM 12.2

Common soda–lime–silica glass is made by melting together Na_2CO_3, $CaCO_3$, and SiO_2. The carbonates break down, liberating CO_2 gas bubbles that help to mix the molten glass. For 1000 kg of container glass (15 wt % Na_2O, 10 wt % CaO, 75 wt % SiO_2), what is the raw material batch formula (weight percent of Na_2CO_3, $CaCO_3$, and SiO_2)?

SOLUTION

1000 kg of glass consists of 150 kg of Na_2O, 100 kg of CaO, and 750 kg of SiO_2.

Using data from Appendix 1 gives us

$$\text{mol wt } Na_2O = 2(22.99) + 16.00$$
$$= 61.98 \text{ amu}$$

$$\text{mol wt } Na_2CO_3 = 2(22.99) + 12.00 + 3(16.00)$$
$$= 105.98 \text{ amu}$$

$$\text{mol wt } CaO = 40.08 + 16.00$$
$$= 56.08 \text{ amu}$$

$$\text{mol wt } CaCO_3 = 40.08 + 12.00 + 3(16.00)$$
$$= 100.08 \text{ amu}$$

$$Na_2CO_3 \text{ required} = 150 \text{ kg} \times \frac{105.98}{61.98} = 256 \text{ kg}$$

$$CaCO_3 \text{ required} = 100 \text{ kg} \times \frac{100.08}{56.08} = 178 \text{ kg}$$

$$SiO_2 \text{ required} = 750 \text{ kg}$$

The batch formula is

$$\frac{256 \text{ kg}}{(256 + 178 + 750) \text{ kg}} \times 100 = 21.6 \text{ wt\% } Na_2CO_3$$

$$\frac{178 \text{ kg}}{(256 + 178 + 750) \text{ kg}} \times 100 = 15.0 \text{ wt\% } CaCO_3$$

$$\frac{750 \text{ kg}}{(256 + 178 + 750) \text{ kg}} \times 100 = 63.3 \text{ wt\% } SiO_2$$

....

PRACTICE PROBLEM 12.2

In Sample Problem 12.2 we calculated a batch formula for a common soda–lime–silica glass. To improve chemical resistance and working properties, Al_2O_3 is often added to the glass. This can be done by adding soda feldspar (albite), $Na(AlSi_3)O_8$, to the batch formula. Calculate the formula of the glass produced when 2000 kg of the batch formula is supplemented with 100 kg of this feldspar.

12.3 GLASS-CERAMICS

Among the most sophisticated ceramic materials are the **glass-ceramics**, which combine the nature of crystalline ceramics with glass. The result is a product with especially attractive qualities. Glass-ceramics begin as relatively ordinary glassware. A significant advantage is their ability to be formed into a product shape as economically and precisely as glasses. By a carefully controlled heat treatment, over 90% of the glassy material crystallizes (see Figure 10–37). The final crystallite grain sizes are generally between 0.1 and 1 μm. The small amount of residual glass phase effectively fills the grain boundary volume, creating a pore-free structure. The final glass-ceramic product is characterized by mechanical and thermal shock resistance far superior to conventional ceramics. In Chapter 6, the sensitivity of ceramic materials to brittle failure was discussed. The resistance of glass-ceramics to mechanical shock is largely due to the elimination of stress-concentrating pores. The resistance to thermal shock results from characteristically low thermal expansion coefficients of these materials. The significance of this was demonstrated in Section 7.4.

We have alluded to the importance of a carefully controlled heat treatment to produce the uniformly fine-grained microstructure of a glass-ceramic.

Table 12.7 *Compositions of Some Glass-Ceramics*

Glass-ceramic	Composition (wt %)							
	SiO_2	Li_2O	Al_2O_3	MgO	ZnO	B_2O_3	TiO_2^a	$P_2O_5^a$
Li_2O–Al_2O_3–SiO_2 system	74	4	16				6	
MgO–Al_2O_3–SiO_2 system	65		19	9			7	
Li_2O–MgO–SiO_2 system	73	11		7		6		3
Li_2O–ZnO–SiO_2 system	58	23			16			3

Source: Data from P. W. McMillan, *Glass-Ceramics,* 2nd ed., Academic Press, Inc., New York, 1979.
    [a] Nucleating agents.

The theory of heat treatment (the kinetics of solid-state reactions) was dealt with in Chapter 10. For now, we need to recall that the crystallization of a glass is a stabilizing process. Such a transformation begins (or is nucleated) at some impurity phase boundary. For an ordinary glass in the molten state, crystallization will tend to nucleate at a few isolated spots along the surface of the melt container. This is followed by the growth of a few large crystals. The resulting microstructure is coarse and nonuniform. Glass-ceramics differ in the presence of several weight percent of a nucleating agent such as TiO_2. A fine dispersion of small TiO_2 particles gives a nuclei density as high as 10^{12} per cubic millimeter. There is some controversy about the exact role of nucleating agents such as TiO_2. In some cases, it appears that the TiO_2 contributes to a finely dispersed second phase of TiO_2–SiO_2 glass, which is unstable and crystallizes, thereby initiating the crystallization of the entire system. For a given composition, optimum temperatures exist for nucleating and growing the small crystallites.

Table 12.7 lists the principal commercial glass-ceramics. By far the most important example is the Li_2O–Al_2O_3–SiO_2 system. Various commercial materials in this composition range exhibit excellent thermal shock resistance due to the low thermal expansion coefficient of the crystallized ceramic. Examples are Corning's Corning Ware or Pyroflam and Schott Glaswerke's Ceran or Ceradur. Contributing to the low expansion coefficient is the presence of crystallites of β-spodumene ($Li_2O \cdot Al_2O_3 \cdot 4SiO_2$), which has a characteristically small expansion coefficient, or β-eucryptite ($Li_2O \cdot Al_2O_3 \cdot SiO_2$), which actually has a negative expansion coefficient.

SAMPLE PROBLEM 12.3

What would be the composition (in weight percent) of a glass-ceramic composed entirely of β spodumene?

SOLUTION

β-spodumene is $Li_2O \cdot Al_2O_3 \cdot 4SiO_2$. Using data from Appendix 1 gives us

$$mol\ wt\ Li_2O = [2(6.94) + 16.00]\ amu$$
$$= 29.88\ amu$$
$$mol\ wt\ Al_2O_3 = [2(26.98) + 3(16.00)]\ amu$$
$$= 101.96\ amu$$
$$mol\ wt\ SiO_2 = [28.09 + 2(16.00)]\ amu$$
$$= 60.09\ amu$$

giving

$$wt\%\ Li_2O = \frac{29.88}{29.88 + 101.96 + 4(60.09)} \times 100 = 8.0\%$$

$$wt\%\ Al_2O_3 = \frac{101.96}{29.88 + 101.96 + 4(60.09)} \times 100 = 27.4\%$$

$$wt\%\ SiO_2 = \frac{4(60.09)}{29.88 + 101.96 + 4(60.09)} \times 100 = 64.6\%$$

PRACTICE PROBLEM 12.3

What would be the mole percentage of Li_2O, Al_2O_3, SiO_2, and TiO_2 in the first commercial glass-ceramic composition of Table 12.7? (See Sample Problem 12.3.)

12.4 PROCESSING OF CERAMICS AND GLASSES

Table 12.8 summarizes some of the major processing techniques for ceramics and glasses. Many of these are direct analogs of metallic processing introduced in Table 11.10. However, wrought processing does not exist per se for ceramics. The deformation forming of ceramics is limited by their inherent brittleness. Although cold working and hot working are not practical, a wider variety of casting techniques are available. **Fusion casting** refers

Table 12.8 *Some Major Processing Methods for
Ceramics and Glasses*

Fusion casting
Slip casting
Sintering
Hot isostatic pressing (HIP)
Glass forming
Controlled devitrification
Sol–gel processing
Biomimetic processing
Self-propagating high temperature synthesis (SHS)

to a process equivalent to metal casting. This technique is not a predominant one for ceramics because of their generally high melting points. Some low-porosity refractories are formed in this way, but at a relatively high cost. **Slip casting,** shown in Figure 12–2, is a more typical ceramic processing technique. Here the casting is done at room temperature. The "slip" is a powder–water mixture that is poured into a porous mold. Much of the water is absorbed by the mold, leaving a relatively rigid powder form that can be removed from the mold. To develop a strong product, the piece must be heated. Initially, the remaining absorbed water is driven off. **Firing** is done at higher temperatures, typically above 1000°C. As in powder metallurgy, much of the strength of the fired piece is due to solid-state diffusion. For many ceramics, especially clayware, additional high-temperature reactions are involved. Chemically combined water can be driven off, various phase transformations can take place, and substantial glassy phases, such as silicates, can be formed. Sintering is the direct analog of powder metallurgy. This subject was introduced in Section 10.6. The high melting points of common ceramics make sintering a widespread processing technique. As with powder metallurgy, hot isostatic pressing is finding increased applications in ceramics, especially in providing fully dense products with superior mechanical properties. Typical **glass-forming** processes are shown in Figures 12–3 and 12–4. The viscous nature of the glassy state plays a central role in this processing (see also Section 6.6). **Controlled devitrification** (i.e., crystallization) leads to the formation of glass-ceramics. This topic was raised earlier (see Sections 10.6 and 12.3). **Sol–gel processing** is among the more rapidly developing new technologies for fabricating ceramics and glasses. For ceramics, the method provides for the formation of uniform, fine particulates of high purity at relatively low temperatures. Such powders can subsequently be sintered to high density with correspondingly good mechanical properties. In such techniques, the essential feature is the formation of an organometallic solution. The dispersed phase "sol" is then converted into a rigid "gel," which, in turn, is reduced to a final composition by various thermal treatments. A key advantage of the sol–gel process is that the product formed initially through this liquid phase route can be fired at lower temperatures, compared with conventional ceramic processes. The cost savings of the lower firing temperatures can be significant.

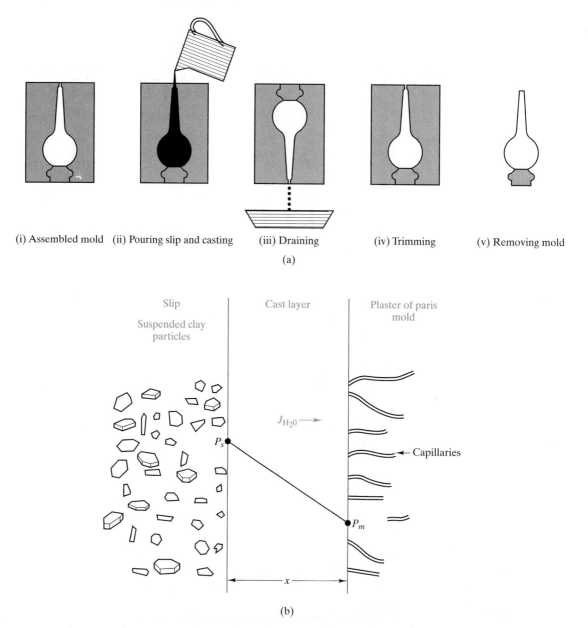

(i) Assembled mold   (ii) Pouring slip and casting   (iii) Draining   (iv) Trimming   (v) Removing mold

(a)

(b)

Figure 12-2 *(a) Schematic illustration of the slip casting of ceramics. The slip is a powder–water mixture. (After F. H. Norton,* Elements of Ceramics, *2nd Ed., Addison-Wesley Publishing Co., Inc., Reading, Mass., 1974.) (b) Much of that water is absorbed into the porous mold. The final form must be fired at elevated temperatures to produce a structurally strong piece. (From W. D. Kingery, H. K. Bowen, and D. R. Uhlmann,* Introduction to Ceramics, *2nd Ed., John Wiley & Sons, Inc., New York, 1976.)*

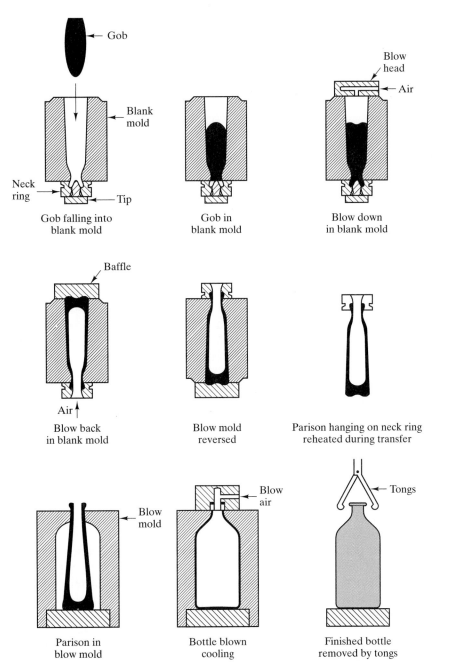

Figure 12-3 *The formation of a glass container requires careful control of the material's viscosity at various stages. (From F. H. Norton,* Elements of Ceramics, *2nd Ed., Addison-Wesley Publishing Co., Inc., Reading, Mass., 1974.)*

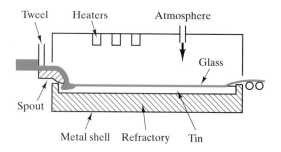

Figure 12-4 *The high degree of flatness achieved in modern architectural plate glass is the result of the float glass process in which the layer of glass is drawn across a bath of molten tin. (After Engineered Materials Handbook, Vol. 4, Ceramics and Glasses, ASM International, Materials Park, Ohio, 1991.)*

Recently, ceramic engineers have realized that certain natural ceramic fabrication processes, such as the formation of sea shells, take the liquid phase processing route to its ultimate conclusion. Figure 12–5 illustrates the formation of an abalone shell that takes place in an aqueous medium entirely at ambient temperature, with *no* firing step at all. Attractive features of this natural bioceramic, in addition to fabrication at ambient conditions from readily available materials, include a final microstructure that is fine-grained with an absence of porosity and microcracks and a resulting high strength and fracture toughness. Such bioceramics are normally produced at a slow buildup rate from a limited range of compositions, usually calcium carbonate, calcium phosphate, silica, or iron oxide.

Biomimetic processing is the name given to fabrication strategies for engineering ceramics that imitate processes such as that illustrated in Figure 12–5, viz. low-temperature, aqueous syntheses of oxides, sulfides, and other ceramics by adapting biological principles. Three key aspects of this process have been identified: (1) the occurrence within specific microenvironments (implying stimulation of crystal production at certain functional sites and inhibition of the process at other sites), (2) the production of a specific mineral with a defined crystal size and orientation, and (3) macroscopic growth by packaging many incremental units together (resulting in a unique composite structure and accommodating later stages of growth and repair). This natural process occurs for bone and dental enamel, as well as the shells. Biomimetic engineering processes have not yet been able to duplicate the sophisticated level of control exhibited in the natural materials. Nonetheless, promising results have come from initial effort in this direction. A simple example is the addition of water-soluble polymers to portland cement mixes thereby reducing freeze-thaw damage by inhibiting the growth of large ice crystals. The ceramiclike cement particles resemble biological hard tissue. The polymer addition can change hardening reactions, microstructure, and properties of cement products in the same way that extracellular biopoly-

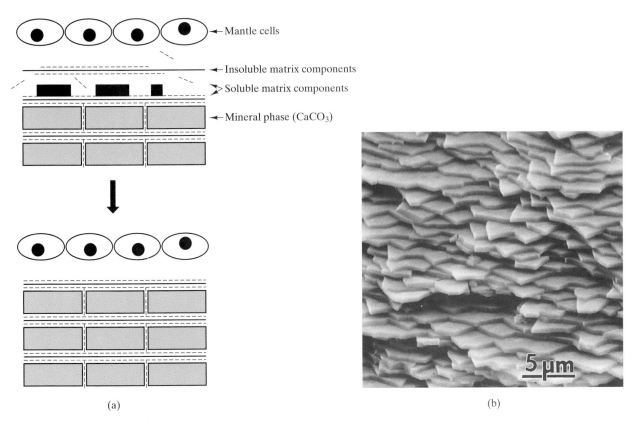

Figure 12-5 *(a) A schematic of the formation of an abalone shell. Shown is a layer of nacre, composed of platelets of* *CaCO₃ bonded together by organic molecules (proteins and sugars). The production of such fracture-resistant structures by synthetic means is known as biomimetic processing. (After A. Heuer et al.,* Science, 255 *1098–1105 (1992).)* *(b) A scanning electron micrograph of the nacre platelet structure. (Courtesy of Mehmet Sarikaya, University of Washington)*

mers contribute to the properties of bones and shells. Current research in biomimetic processing is centered on the control of crystal nucleation and growth in advanced ceramics using inorganic and organic polymers, as well as biopolymers.

A final note of importance in regard to biomimetic processing is that an additional, attractive feature of the natural formation of bones and shells is that it represents **net-shape processing** (i.e., the "product," once formed, does not require a final shaping operation). Much of the effort in recent years for both engineered metals and ceramics can be described as **near net-shape processing**, in which the goal is to minimize any final shaping operation. The superplastic forming of metal alloys and the sol–gel processing of ceramics and glass are such examples. By contrast, biominerals are formed as relatively large, dense parts in a "moving front" process in which incremental matrix-defined units are sequentially mineralized. The resulting net-shape forming of a dense material represents an exceptional level of microstructural control.

Figure 12-6 *Ignition at the top of this Ti powder pellet in an atmosphere of N_2 gas leads to a self-sustaining reaction throughout the entire sample and the complete conversion to TiN. The appropriately named self-propagating high temperature synthesis (SHS) is an example of novel techniques for processing advanced materials. (Courtesy of Zuhair Munir, University of California, Davis)*

A final type of ceramic processing, **self-propagating high-temperature synthesis** (SHS), is illustrated by Figure 12–6. This novel technique involves using the substantial heat evolved by certain chemical reactions, once initiated, to sustain the reaction and produce the final product. For example, igniting a reaction between titanium powder in a nitrogen gas atmosphere produces an initial amount of titanium nitride [$Ti(s) + 1/2 N_2(g) = TiN(s)$]. The substantial heat evolved in this highly exothermic reaction can be enough to produce a combustion wave that passes through the remaining titanium powder, sustaining the reaction and converting all of the titanium to a TiN product. Although high temperatures are involved, SHS shares an advantage of the relatively low temperature sol–gel and biomimetic processing, viz. energy savings in that much of the high temperature in SHS comes from the self-sustaining reactions. In addition, SHS is a simple process that provides relatively pure products and the possibility of simultaneous formation and densification of the product. A wide variety of materials is formed in this way. They are generally produced in powder form, although dense products can be formed by subsequent sintering or the simultaneous application of pressure or casting operations during the combustion. In addition to forming ceramics, intermetallic compounds (e.g., TiNi) and composites can be produced by SHS.

SAMPLE PROBLEM 12.4

In firing 5 kg of kaolinite, $Al_2(Si_2O_5)(OH)_4$, in a laboratory furnace to produce an aluminosilicate ceramic, how much H_2O is driven off?

SOLUTION

As in Sample Problem 9.12, note that

$$Al_2(Si_2O_5)(OH)_4 = Al_2O_3 \cdot 2SiO_2 \cdot 2H_2O$$

and

$$Al_2O_3 \cdot 2SiO_2 \cdot 2H_2O \xrightarrow{heat} Al_2O_3 \cdot 2SiO_2 + 2H_2O \uparrow$$

Then

$$
\begin{aligned}
1 \text{ mol } Al_2O_3 \cdot 2SiO_2 \cdot 2H_2O &= [2(26.98) + 3(16.00)] \text{ amu} \\
&\quad + 2[28.09 + 2(16.00)] \text{ amu} \\
&\quad + 2[2(1.008) + 16.00] \text{ amu} \\
&= 258.2 \text{ amu}
\end{aligned}
$$

and

$$2 \text{ mol } H_2O = 2[2(1.008) + 16.00] \text{ amu} = 36.03 \text{ amu}$$

As a result, the mass of H_2O driven off will be

$$m_{H_2O} = \frac{36.03 \text{ amu}}{258.2 \text{ amu}} \times 5 \text{ kg} = 0.698 \text{ kg} = 698 \text{ g}$$

SAMPLE PROBLEM 12.5

Assume the glass bottle produced in Figure 12–3 is formed at a temperature of 680°C with a viscosity of 10^7 P. If the activation energy for viscous deformation for the glass is 460 kJ/mol, calculate the annealing range for this product.

SOLUTION

Following the methods of Section 6.6, we have an application for Equation 6.20:

$$\eta = \eta_0 e^{+Q/RT}$$

For 680°C = 953 K,

$$10^7 P = \eta_0 e^{+(460 \times 10^3 \text{J/mol})/[8.314 \text{J}/(\text{mol} \cdot \text{K})](953\text{K})}$$

or

$$\eta_0 = 6.11 \times 10^{-19} P$$

For the annealing range, $\eta = 10^{12.5}$ to $10^{13.5}$P. For $\eta = 10^{12.5}$P,

$$T = \frac{460 \times 10^3 \text{J/mol}}{[8.314\text{J/(mol} \cdot \text{K)}]\ln(10^{12.5}/6.11 \times 10^{-19})}$$

$$= 782\text{K} = 509°\text{C}$$

For $\eta = 10^{13.5}$P,

$$T = \frac{460 \times 10^3 \text{J/mol}}{[8.314\text{J/(mol} \cdot \text{K)}]\ln(10^{13.5}/6.11 \times 10^{-19})}$$

$$= 758\text{K} = 485°\text{C}$$

Therefore,

$$\text{annealing range} = 485 \text{ to } 509°\text{C}$$

PRACTICE PROBLEM 12.4

In scaling up the laboratory firing operation of Sample Problem 12.4 to a production level, 6.05×10^3 kg of kaolinite are fired. How much H_2O is driven off in this case?

PRACTICE PROBLEM 12.5

For the glass bottle production described in Sample Problem 12.5, calculate the melting range for this manufacturing process.

SUMMARY

Ceramics and glasses represent a diverse family of engineering materials. The term ceramics is associated with predominantly crystalline materials. Silicates are abundant and economical examples used in numerous consumer and industrial products. Nonsilicate oxides such as MgO are widely used as refractories, nuclear fuels, and electronic materials. Partially stabilized zirconia (PSZ) is a primary candidate for high-temperature engine components. Nonoxide ceramics include silicon nitride, another candidate for engine components.

Glasses are noncrystalline solids chemically similar to the crystalline ceramics. The predominant category is the silicates, which include materials from the expensive, high-temperature vitreous silica to common soda–lime–silica window glass. Protective and decorative glass coatings on ceramics and metals are termed glazes and enamels, respectively. Many silicate glasses contain substantial amounts of other oxide components, although there is little commercial use for completely nonsilicate oxide glasses. Some nonoxide glasses have found commercial uses (e.g., chalcogenide glasses as amorphous semiconductors).

The glass-ceramics are distinctive products that are initially processed as glasses and then carefully crystallized to form a dense, fine-grained ceramic product with excellent mechanical and thermal shock resistance. Most commercial glass-ceramics are in the $Li_2O-Al_2O_3-SiO_2$ system, which is characterized by compounds with exceptionally low thermal expansion coefficients.

Ceramics can be formed by fusion casting, which is similar to metal casting. Slip casting is more common. The slip is a clay–water suspension that is fired in a process similar to powder metallurgy. This largely solid-state process is generally much less expensive than fusion casting, which, for a given material, must be done at substantially higher temperatures. Sintering and hot isostatic pressing (HIP) are directly analogous to powder metallurgy methods. Glass forming involves careful control of the viscosity of the supercooled silicate liquid. Glass-ceramics require the additional step of controlled devitrification to form a fine-grained, fully crystalline product. Sol–gel processing is rapidly developing as a method for fabricating ceramics, glasses, and glass-ceramics starting with a sol (liquid solution) route. Biomimetic processing, which imitates natural fabrication processes, furthers the use of a liquid phase route to the economical production of superior products. Self-propagating high temperature synthesis (SHS) uses the heat evolved by certain chemical reactions to produce the resulting ceramic. The energy-saving process produces economical products of high purity.

KEY TERMS

biomimetic processing (450)
borosilicate glass (442)
clay (436)
controlled devitrification (447)
crystalline ceramic (435)
E-glass (442)
electronic ceramic (437)
enamel (442)
firing (447)
fusion casting (446)
glass (441)
glass-ceramic (444)

glass forming (447)
glaze (442)
intermediate (442)
magnetic ceramic (437)
near net-shape processing (451)
net-shape processing (451)
network former (441)
network modifier (442)
nonoxide ceramic (440)
nonsilicate glass (442)
nonsilicate oxide ceramic (437)
nuclear ceramic (437)

partially stabilized zirconia (437)
pure oxide (437)
refractory (436)
self-propagating high temperature synthesis (SHS) (452)
silicate (435)
silicate glass (441)
slip casting (447)
soda–lime–silica glass (442)
sol–gel processing (447)
vitreous silica (442)
whiteware (435)

REFERENCES

Chiang, Y., D. P. Birnie III, and **W. D. Kingery,** *Physical Ceramics,* John Wiley & Sons, Inc., New York, 1997.

Doremus, R. H., *Glass Science,* 2nd ed., John Wiley & Sons, Inc., New York, 1994.

Engineered Materials Handbook, Vol. 4, *Ceramics and Glasses,* ASM International, Materials Park, Ohio, 1991.

Reed, J. S., *Principles of Ceramic Processing,* 2nd ed., John Wiley & Sons, Inc., New York, 1995.

PROBLEMS

12.1 • Ceramics—Crystalline Materials

12.1. As pointed out in the discussion relative to the Al_2O_3–SiO_2 phase diagram (Figure 9–23), an alumina-rich mullite is desirable to ensure a more refractory (temperature-resistant) product. Calculate the composition of a refractory made by adding 2.0 kg of Al_2O_3 to 100 kg of stoichiometric mullite.

12.2. A fireclay refractory of simple composition can be produced by heating the raw material kaolinite, $Al_2(Si_2O_5)(OH)_4$, driving off the waters of hydration. Calculate the composition (weight percent basis) for the resulting refractory. (Note that this process was introduced in Sample Problem 9.12 relative to the Al_2O_3–SiO_2 phase diagram.)

12.3. Using the results of Problem 12.2 and Sample Problem 9.12, calculate the weight percent of SiO_2 and mullite present in the final microstructure of a fireclay refractory made by heating kaolinite.

12.4. Estimate the density of **(a)** a partially stabilized zirconia (with 4 wt % CaO) as the weighted average of the densities of ZrO_2 ($= 5.60$ Mg/m^3) and CaO ($= 3.35$ Mg/m^3), and **(b)** a fully stabilized zirconia with 8 wt % CaO.

12.5. The primary reason for introducing ceramic components in automotive engine designs is the possibility of higher operating temperatures and, therefore, improved efficiencies. A by-product of this substitution, however, is mass reduction. For the case of 1.5 kg of cast iron (density $= 7.15$ Mg/m^3) being replaced by an equivalent volume of partially stabilized zirconia (density $= 5.50$ Mg/m^3), calculate the mass reduction.

12.6. Calculate the mass reduction attained if silicon nitride (density $= 3.18$ Mg/m^3) is used in place of 1.5 kg of cast iron (density $= 7.15$ Mg/m^3).

12.2 • Glasses—Noncrystalline Materials

12.7. A batch formula for a window glass contains 400 kg Na_2CO_3, 300 kg $CaCO_3$, and 1300 kg SiO_2. Calculate the resulting glass formula.

12.8. For the window glass in Problem 12.7, calculate the glass formula if the batch is supplemented by 75 kg of lime feldspar (anorthite), $Ca(Al_2Si_2)O_8$.

12.9. An economical substitute for vitreous silica is a high-silica glass made by leaching the B_2O_3-rich phase from a two-phase borosilicate glass. (The resulting porous microstructure is densified by heating.) A typical starting composition is 81 wt % SiO_2, 4 wt % Na_2O, 2 wt % Al_2O_3, and 13 wt % B_2O_3. A typical final composition is 96 wt % SiO_2, 1 wt % Al_2O_3, and 3 wt % B_2O_3. How much product (in kilograms) would be produced from 50 kg of starting material, assuming no SiO_2 is lost by leaching?

12.10. How much B_2O_3 (in kilograms) is removed by leaching in the glass-manufacturing process described in Problem 12.9?

12.11. A novel electronic material involves the dispersion of small silicon particles in a glass matrix. These "quantum dots" are discussed in Section 17.5. If 4.85×10^{16} particles of Si are dispersed per mm^3 of glass corresponding to a total content of 5 wt %, calculate the average particle size of the quantum dots. (Assume spherical particles and note that the density of the silicate glass matrix is 2.60 Mg/m^3.)

12.12. Calculate the average separation distance between the centers of adjacent Si particles in the quantum dot material of Problem 12.11. (For simplicity, assume a simple cubic array of dispersed particles.)

12.3 • Glass-Ceramics

12.13. Assuming the TiO_2 in a Li_2O–Al_2O_3–SiO_2 glass-ceramic is uniformly distributed with a dispersion of 10^{12} particles per cubic millimeter and a total amount of 6 wt %, what is the average particle size of the TiO_2 particles? (Assume spherical particles. The density of the glass-ceramic is 2.85 Mg/m^3 and of the TiO_2 is 4.26 Mg/m^3.)

12.14. Repeat Problem 12.13 for a 3 wt % dispersion of P_2O_5 with a concentration of 10^{12} particles per cubic millimeter. (The density of P_2O_5 is 2.39 Mg/m^3.)

12.15. What is the overall volume percent of TiO_2 in the glass-ceramic described in Problem 12.13?

12.16. What is the overall volume percent of P_2O_5 in the glass-ceramic described in Problem 12.14?

12.17. Calculate the average separation distance between the centers of adjacent TiO_2 particles in the glass-ceramic described in Problems 12.13 and 12.15. (Note Problem 12.12.)

12.18. Calculate the average separation distance between the centers of adjacent P_2O_5 particles in the glass-ceramic described in Problems 12.14 and 12.16. (Note Problem 12.12.)

12.4 • Processing of Ceramics and Glasses

12.19. For the ceramic in Sample Problem 12.4, use the Al_2O_3–SiO_2 phase diagram in Chapter 9 to determine the maximum firing temperature to prevent formation of a silica-rich liquid.

12.20. How would your answer to Problem 12.19 change if the ceramic being fired was composed of two parts Al_2O_3 in combination with one part kaolin?

12.21. For simplified processing calculations, the glass bottle shaping temperature in Figure 12–3 can be taken as the softening point (at which $\eta = 10^{7.6} P$) and the subsequent annealing temperature as the annealing point (at which $\eta = 10^{13.4} P$). If the glass bottle processing sequence of Figure 12–3 involves a shaping temperature of 690°C for a glass with an activation energy for viscous deformation of 470 kJ/mol, calculate the appropriate annealing temperature.

12.22. Using the approach of Problem 12.21, assume a change in raw material suppliers changes the glass composition thus reducing the softening point to 680°C and the activation energy to 465 kJ/mol. Recalculate the appropriate annealing temperature.

12.23. How many grams of N_2 gas are consumed in forming 100 g of TiN by self-propagating high-temperature synthesis (SHS)?

12.24. How many grams of Ti powder were initially required in the SHS process of Problem 12.23?

12.25. Using the data on modulus of elasticity and strength (modulus of rupture) in Table 6.5, select the sintered ceramic that would meet the following design specifications for a furnace refractory application:

modulus of elasticity, E: $< 350 \times 10^3$ MPa

modulus of rupture, MOR: > 125 MPa

12.26. Repeat the materials selection exercise of Problem 12.25 for any of the ceramics and glasses of Table 6.5 produced by any processing technique, including hot pressing and glass forming.

CHAPTER **13**
Polymers

A molded engineering polymer serves as a light-weight and cost-effective air-intake manifold for automotive appliations. (Courtesy of Solvay Automotive, Inc., Troy, Michigan.)

We follow our discussions of metals and ceramics with a third category of structural materials, polymers. A common synonym for polymers is *plastics,* a name derived from the deformability associated with the fabrication of most polymeric products. To some critics, *plastic* is a synonym for modern culture. Accurate or not, it represents the impact that this complex family of engineering materials has had on our society. Polymers, or plastics, are available in a wide variety of commercial forms: fibers, thin films and sheets, foams, and in bulk.

The metals, ceramics, and glasses we have considered in previous chapters are inorganic materials. The polymers discussed in this chapter are organic. Our choice to limit the discussion to organic polymers is a common one, but somewhat arbitrary. Several inorganic materials have structures composed of building blocks connected in chain and network configurations. We occasionally point out that silicate ceramics and glasses are examples. (This chapter, together with Chapter 2, will provide any of the fundamentals of organic chemistry needed to appreciate the unique nature of polymeric materials.)

We begin our discussion of polymers by investigating *polymerization.* The structural features of the resulting polymers are unique compared with inorganic materials. For many polymers, melting point and rigidity increase with the extent of polymerization and with the complexity of the molecular structure.

An important trend in engineering design during the past two decades has been increased concentration on so-called *engineering polymers,* which can substitute for traditional structural metals. Perhaps the most important examples are found in the automotive industry (e.g., Figure 1–13).

We shall find that polymers fall into one of two main categories. *Thermoplastic polymers* are materials that become less rigid upon heating, while *thermosetting polymers* become more rigid upon heating. For both categories, it is important to appreciate the roles played by *additives,* which provide important features such as improved strength and stiffness, color, and resistance to combustion.

The detailed treatment of polymeric chemistry and related molecular structure in this chapter allows us to appreciate more fully the complex mechanical behavior of polymers as discussed in Chapter 6, as well as the subsequent failure of these materials as treated in Chapter 8.

The processing of polymers follows the categories of thermoplastics and thermosets. Injection molding and extrusion molding are the predominant processes for thermoplastics. Blow molding is a third, important technique. Compression molding and transfer molding are the predominant processes for thermosets.

13.1   POLYMERIZATION

The term **polymer** simply means "many mers," where **mer** is the building block of the long-chain or network molecule. Figure 13–1 shows how a long-chain structure results from the joining together of many **monomers**

Figure 13-1 *Polymerization is the joining of individual monomers (e.g., vinyl chloride, C_2H_3Cl) to form a polymer $[(C_2H_3Cl)_n]$ consisting of many mers (again, C_2H_3Cl).*

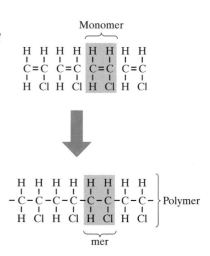

by chemical reaction. **Polymerization,** the process by which long-chain or network molecules are made from relatively small organic molecules, can take place in two distinct ways. **Chain growth,** or **addition polymerization,** involves a rapid "chain reaction" of chemically activated monomers. **Step growth,** or **condensation polymerization,** involves individual chemical reactions between pairs of reactive monomers and is a much slower process. In either case, the critical feature of a monomer, which permits it to join with similar molecules and form a polymer, is the presence of reactive sites—*double bonds* in chain growth, or reactive functional groups in step growth. As discussed in Chapter 2, each covalent bond is a pair of electrons shared between adjacent atoms. The double bond is two such pairs. The chain growth reaction in Figure 13–1 converts the double bond in the monomer to a single bond in the mer. The remaining two electrons become parts of the single bonds joining adjacent mers.

Figure 13–2 illustrates the formation of polyethylene by the process of chain growth. The overall reaction can be expressed as

$$nC_2H_4 \rightarrow +\!\!\left(C_2H_4\right)\!\!+_n \tag{13.1}$$

The process begins with an **initiator**—a hydroxyl *free radical* in this case. A free radical is a reactive atom or group of atoms containing an unpaired electron. The initiation reaction converts the double bond of one monomer into a single bond. Once completed, the one unsatisfied bonding electron (see step 1′ of Figure 13–2) is free to react with the nearest ethylene monomer, extending the molecular chain by one unit (step 2). This chain reaction can continue in rapid succession limited only by the availability of unreacted ethylene monomers. The rapid progression of steps 2 through *n* is the basis of the descriptive term *addition polymerization*. Eventually, another hydroxyl radical can act as a **terminator** (step *n*), giving a stable molecule with

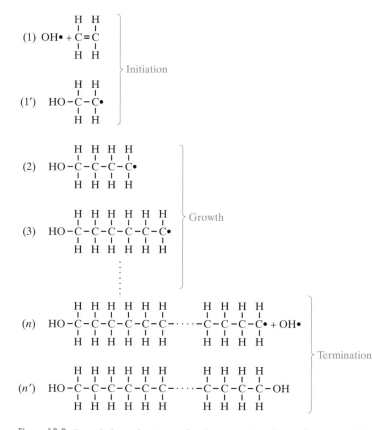

Figure 13-2 *Detailed mechanism of polymerization by a chain growth process (addition polymerization). In this case, a molecule of hydrogen peroxide, H_2O_2, provides two hydroxyl radicals, $OH\bullet$, which serve to initiate and terminate the polymerization of ethylene (C_2H_4) to polyethylene $(C_2H_4)_n$. [The large dot notation ($\bullet$) represents an unpaired electron. The joining, or pairing, of two such electrons produces a covalent bond, represented by a solid line (—).]*

n mer units (step n'). For the specific case of hydroxyl groups as initiators and terminators, hydrogen peroxide is the source of the radicals:

$$H_2O_2 \rightarrow 2OH\bullet \tag{13.2}$$

Each hydrogen peroxide molecule provides an initiator–terminator pair for each polymeric molecule. The termination step in Figure 13–2 is termed *recombination*. Although simpler to illustrate, it is not the most common mechanism of termination. Both hydrogen abstraction and disproportionation are more common termination steps than recombination. *Hydrogen abstraction* involves obtaining a hydrogen atom (with unpaired electron) from an impurity hydrocarbon group. *Disproportionation* involves the formation of a monomerlike double bond.

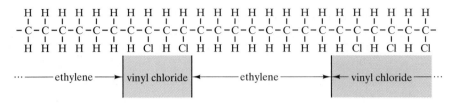

Figure 13-3 *A copolymer of ethylene and vinyl chloride is analogous to a solid-solution metal alloy.*

If an intimate solution of different types of monomers is polymerized, the result is a **copolymer** (Figure 13–3). This is analogous to the solid-solution alloy of metallic systems (Figure 4–2). Figure 13–3 represents specifically a **block copolymer;** that is, the individual polymeric components appear in "blocks" along a single carbon-bonded chain. The alternating arrangement of the different mers can be irregular (as shown in Figure 13–3) or regular. A **blend** (Figure 13–4) is another form of alloying in which different types of already formed polymeric molecules are mixed together. This is analogous to metallic alloys with limited solid solubility.

The various linear polymers illustrated in Figures 13–1 to 13–4 are based on the conversion of a carbon–carbon double bond into two carbon–carbon single bonds. It is also possible to convert the carbon–oxygen double bond in formaldehyde to single bonds. The overall reaction for this case can be expressed as

$$n \, CH_2O \rightarrow -(CH_2O)_n \qquad (13.3)$$

and is illustrated in Figure 13–5. The product is known by various names including polyformaldehyde, polyoxymethylene, and polyacetal. The important acetal group of engineering polymers is based on the reaction of Figure 13–5.

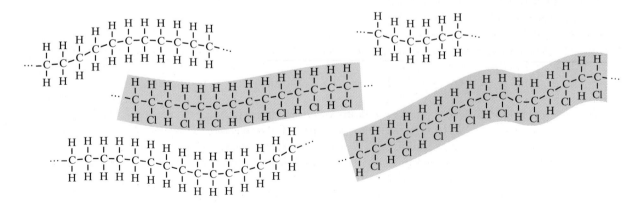

Figure 13-4 *A blend of polyethylene and polyvinyl chloride is analogous to a metal alloy with limited solid solution.*

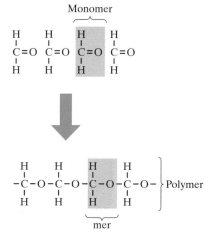

Monomer

Polymer

mer

Figure 13–6 illustrates the formation of phenol-formaldehyde by the process of step growth. Only a single step is shown. The two phenol molecules are linked by the formaldehyde molecule in a reaction in which the phenols each give up a hydrogen atom and the formaldehyde gives up an oxygen atom to produce a water molecule byproduct (condensation product). Extensive polymerization requires this three-molecule reaction to be repeated for each unit increase in molecular length. The time required for this is substantially greater than for the chain reaction of Figure 13–2. The common occurrence of condensation by-products in step growth processes provides the descriptive term *condensation polymerization*. The polyethylene mer in Figure 13–1 has two points of contact with adjacent mers and is said to be **bifunctional.** This leads to a **linear molecular structure.** On the other hand, the phenol molecule in Figure 13–6 has several potential points of contact and is termed **polyfunctional.** In practice, there is room for no more than three CH_2 connections per phenol molecule, but this is sufficient to generate a three-dimensional **network molecular structure,**

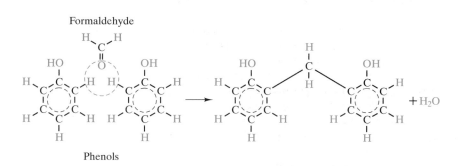

Figure 13-6 *Single, first step in the formation of phenol-formaldehyde by a step growth process (condensation polymerization). A water molecule is the condensation product.*

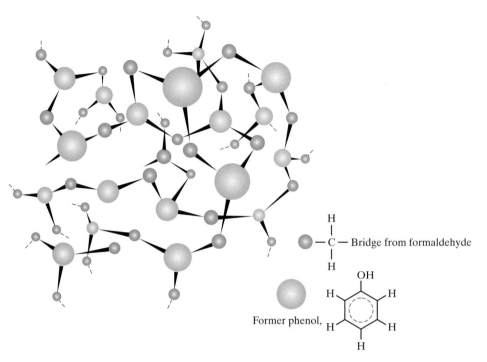

Figure 13-7 *After several reaction steps like that in Figure 13–6, polyfunctional mers form a three-dimensional network molecular structure.* (From L. H. *Van Vlack,* Elements of Materials Science and Engineering, *4th Ed., Addison-Wesley Publishing Co., Inc., Reading, Mass., 1980.)*

as opposed to the linear structure of polyethylene. Figure 13–7 illustrates this network structure. The terminology here is reminiscent of inorganic glass structure in Section 12.2. The breaking up of network arrangements of silica tetrahedra by network modifiers produced substantially "softer" glass. Similarly, linear polymers are "softer" than network polymers. A key difference between silicate "polymers" and the organic materials of this chapter is that the silicates contain predominantly primary bonds, which cause their viscous behavior to occur at substantially higher temperatures. It should be noted that a bifunctional monomer will produce a linear molecule by either chain growth or step growth processes, and a polyfunctional monomer will produce a network structure by either process.

SAMPLE PROBLEM 13.1

A sample of polyethylene is found to have an average molecular weight of 25,000 amu. What is the degree of polymerization, n, of the "average" polyethylene molecule?

SOLUTION

$$n = \frac{\text{mol wt} + C_2H_4 +_n}{\text{mol wt } C_2H_4}$$

Using the data of Appendix 1, we obtain

$$n = \frac{25{,}000 \text{ amu}}{[2(12.01) + 4(1.008)] \text{ amu}}$$

$$= 891$$

SAMPLE PROBLEM 13.2

How much H_2O_2 must be added to ethylene to yield an average degree of polymerization of 750? Assume that all H_2O_2 dissociates to OH groups that serve as terminals for the molecules and express the answer in weight percent.

SOLUTION

Referring to Figure 13–2, we note that there is one H_2O_2 molecule (= two OH groups) per polyethylene molecule.

$$\text{wt \% } H_2O_2 = \frac{\text{mol wt } H_2O_2}{750 \times (\text{mol wt } C_2H_4)} \times 100$$

Using the data of Appendix 1 yields

$$\text{wt \% } H_2O_2 = \frac{2(1.008) + 2(16.00)}{750[2(12.01) + 4(1.008)]} \times 100$$

$$= 0.162 \text{ wt \%}$$

SAMPLE PROBLEM 13.3

A regular copolymer of ethylene and vinyl chloride contains alternating mers of each type. What is the weight percent of ethylene in this copolymer?

SOLUTION

Since there is one ethylene mer for each vinyl chloride molecule, we can write

$$\text{wt \% ethylene} = \frac{\text{mol wt } C_2H_4}{\text{mol wt } C_2H_4 + \text{mol wt } C_2H_3Cl} \times 100$$

Using the data of Appendix 1, we find that

$$\text{wt \% ethylene} = \frac{[2(12.01) + 4(1.008)] \times 100}{[2(12.01) + 4(1.008)] + [2(12.01) + 3(1.008) + 35.45]}$$

$$= 31.0 \text{ wt \%}$$

The Material World: Collagen—The Body's Own Polymer

Hydroxyapatite (HA), a ceramic material "engineered" by the body for its skeletal structure, was featured in the previous chapter. Bone is, in fact, a complex composite material with an organic polymer phase called collagen comprising 36% of overall bone weight. Collagen is a protein and the most abundant structural material in the bodies of mammals (see photo). Although there are more than a dozen forms of collagen (distinguished by

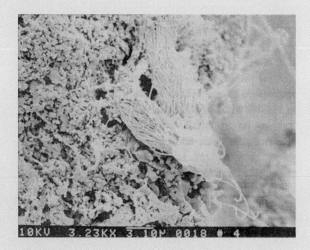

A fibrous bundle of natural collagen is shown attached to the surface of synthetic hydroxyapatite granules in a bioceramic implant. (From J.P. McIntyre, J.F. Shackelford, M.W. Chapman, and R.R. Pool, Bull. Amer. Ceram. Soc. 70 1499 (1991))

particular sequences of amino acids in the polymeric molecules), collagen in bone is Type I, the same form found in skin, tendons, and ligaments. The elaborate hierarchy of structure in Type I collagen (see illustration) begins with a triple helix molecular structure that leads to a fibril geometry and a registry of fibrils in a 64 nm banding pattern. The fibrils are not mechanically independent, but are connected by molecular cross-linking.

As the structural material bone is biologically "engineered," collagen plays a role in the formation of the ceramic phase, hydroxyapatite. It appears that the initial precipitation of mineral crystals is partially catalyzed by the elements of the collagen structure. Then the initial crystals grow in the collagen bands and subsequently spread throughout the collagen scaffold.

As noted in the discussion of hydroxyapatite in the previous chapter, the mechanical behavior of bone is not described adequately as simply that of the individual ceramic or polymer components or even their weighted average. One must always keep sight of the fact that bone is a living tissue with the ability to repair and remodel itself. The natural polymer collagen plays a central role in this function. In the next chapter, a feature box illustrates how synthetic sources of the mineral and polymer components are being combined to provide bone substitutes for orthopaedic applications.

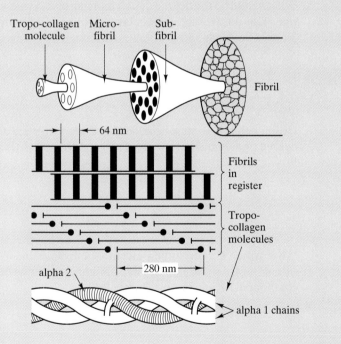

Schematic illustration of the polymeric structure of collagen in bone. (From R.B. Martin, "Bone as a Ceramic Composite Material," in Bioceramics—Applications of Ceramic and Glass Materials in Medicine, Ed. J.F. Shackelford, Trans Tech Publications, Switzerland, 1999)

SAMPLE PROBLEM 13.4

Calculate the molecular weight of a polyacetal molecule with a degree of polymerization of 500.

SOLUTION

$$\text{mol wt } (\text{CH}_2\text{O})_n = n(\text{mol wt CH}_2\text{O})$$

Using the data of Appendix 1, we obtain

$$\text{mol wt } (\text{CH}_2\text{O})_n = 500[12.01 + 2(1.008) + 16.00] \text{ amu}$$
$$= 15{,}010 \text{ amu}$$

PRACTICE PROBLEM 13.1

What would be the degree of polymerization of a polyvinyl chloride with an average molecular weight of 25,000 amu? (See Sample Problem 13.1.)

PRACTICE PROBLEM 13.2

How much H_2O_2 must be added to ethylene to yield an average degree of polymerization of **(a)** 500 and **(b)** 1000? (See Sample Problem 13.2.)

PRACTICE PROBLEM 13.3

What would be the mole percent of ethylene and vinyl chloride in an irregular copolymer that contains 50 wt % of each component? (See Sample Problem 13.3.)

PRACTICE PROBLEM 13.4

Calculate the degree of polymerization for a polyacetal molecule with a molecular weight of 25,000 amu. (See Sample Problem 13.4.)

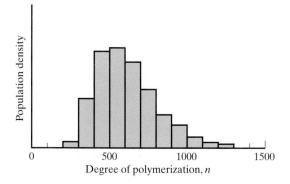

Figure 13-8 *Statistical distribution of molecular lengths in a given polymer as indicated by n, the degree of polymerization.*

13.2   STRUCTURAL FEATURES OF POLYMERS

The first aspect of polymer structure that needs to be specified is the length of the polymeric molecule. For example, how large is n in $+C_2H_4+_n$? In general, n is termed the **degree of polymerization** and is also designated $\overline{DP}$. It is usually determined from the measurement of physical properties, such as viscosity and light scattering. For typical commercial polymers, n can range from approximately 100 to 1000, but for a given polymer, the degree of polymerization represents an average. As you might suspect from the nature of both chain growth and step growth mechanisms, the extent of the molecular growth process varies from molecule to molecule. The result is a statistical distribution of molecular lengths as shown in Figure 13–8. Directly related to molecular length is molecular weight, which is simply the degree of polymerization (n) times the molecular weight of the individual mer. Less simple is the concept of molecular length. For network structures, there is, by definition, no meaningful one-dimensional measure of length. For linear structures, there are two such parameters. First is the **root-mean-square length,** $\overline{L}$, given by

$$\overline{L} = l\sqrt{m} \qquad (13.4)$$

where l is the length of a single bond in the backbone of the hydrocarbon chain and m is the number of bonds. Equation 13.4 results from the statistical analysis of a freely kinked linear chain as illustrated in Figure 13–9. Each bond angle between three adjacent C atoms is near 109.5° (as discussed in Chapter 2), but as seen in Figure 13–9, this angle can be rotated freely in space. The result is the kinked and coiled molecular configuration. The root-

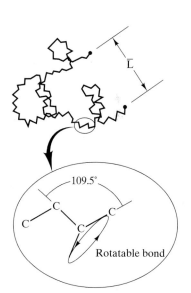

Figure 13-9 *The length of kinked molecular chain is given by Equation 13.4, due to the free rotation of the C—C—C bond angle of 109.5°.*

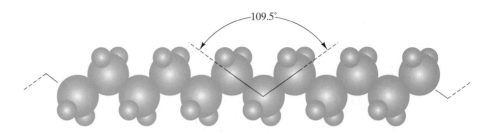

Figure 13-10 *"Sawtooth" geometry of a fully extended molecule. The relative sizes of carbon and hydrogen atoms are shown in the polyethylene configuration.*

mean-square length represents the effective length of the linear molecule as it would be present in the polymeric solid. The second length parameter is a hypothetical one in which the molecule is extended as straight as possible (without bond angle distortion),

$$L_{\text{ext}} = ml \sin \frac{109.5°}{2} \tag{13.5}$$

where L_{ext} is the **extended length.** The "sawtooth" geometry of the extended molecule is illustrated in Figure 13–10. For typical bifunctional, linear polymers such as polyethylene and polyvinyl chloride, there are two bond lengths per mer, or

$$m = 2n \tag{13.6}$$

where n is the degree of polymerization.

In general, the rigidity and melting point of polymers increase with the degree of polymerization. As with any generalization, there can be important exceptions. For example, the melting point of nylon does not change with degree of polymerization. This raises a useful rule of thumb, that is, rigidity and melting point increase as the complexity of the molecular structure increases. For example, the phenol-formaldehyde structure in Figure 13–6 produces a rigid, even brittle, polymer. By contrast, the linear polyethylene structure of Figure 13–2 produces a relatively soft material. Students of civil engineering should appreciate the rigidity of a structure with extensive cross members. The network structure has the strength of covalent bonds linking all adjacent mers. The linear structure has covalent bonding only along the backbone of the chain. Only weak secondary (van der Waals) bonding holds together adjacent molecules. Molecules are relatively free to slide past each other. The result of this was discussed in Section 6.1. Now let us explore a series of structural features that add to the complexity of linear molecules and take them closer to the nature of the network structure.

We will begin with the ideally simple hydrocarbon chain of polyethylene (Figure 13–11a). By replacing some hydrogen atoms with large side groups (R), a less symmetrical molecule results. The placement of the side

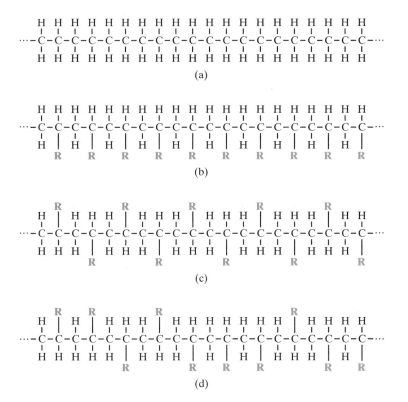

Figure 13-11 *(a) The symmetrical polyethylene molecule. (b) A less symmetrical molecule is produced by replacing one H in each mer with a large side group, R. The isotactic structure has all R along one side. (c) The syndiotactic structure has the R groups regularly alternating on opposite sides. (d) The least symmetrical structure is the atactic, in which the side groups irregularly alternate on opposite sides. Increasing irregularity decreases crystallinity while increasing rigidity and melting point. When R = CH$_3$, parts (b)–(d) illustrate various forms of polypropylene. (One might note that these schematic illustrations can be thought of as "top views" of the more pictorial representations of Figure 13–10.)*

groups can be regular and all along one side, or **isotactic** (Figure 13–11b), or the placement of the groups can be alternating along opposite sides, or **syndiotactic** (Figure 13–11c). An even less symmetrical molecule is the **atactic** form (Figure 13–11d) in which the side groups are irregularly placed. For $R = CH_3$, Figure 13–11b to d represents polypropylene. As the side groups become larger and more irregular, rigidity and melting point tend to rise for two reasons. First, the side groups serve as hindrances to molecular sliding. By contrast, the polyethylene molecules (Figure 13–11a) can slide past each other readily under an applied stress. Second, increasing size and complexity of the side group lead to greater secondary bonding forces between adjacent molecules (see Chapter 2).

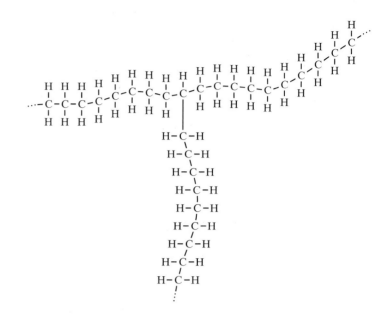

Figure 13-12 *Branching involves adding a polymeric molecule to the side of the main molecular chain.*

An extension of the concept of adding large side groups is to add a polymeric molecule to the side of the chain. This process, called *branching,* is illustrated in Figure 13–12. It can occur as a fluctuation in the chain growth process illustrated by Equation 13.1 (in which a hydrogen further back on the chain is abstracted by a free radical) or as a result of an addition agent that strips away a hydrogen, allowing chain growth to commence at that site. The complete transition from linear to network structure is produced by *cross-linking,* as shown in Figure 6–46, which illustrates **vulcanization.** Rubbers are the most common examples of cross-linking. The bifunctional isoprene mer still contains a double bond after the initial polymerization. This permits covalent bonding of a sulfur atom to two adjacent mers. The extent of cross-linking is controlled by the amount of sulfur addition. This permits control of the rubber behavior from a gummy material to a tough, elastic one and finally, a hard, brittle product as the sulfur content is increased.

In Chapters 3 and 4, we pointed out that the complexity of long-chain molecular structures leads to complex crystalline structures and a significant degree of noncrystalline structure in commercial materials. We can now comment that the degree of crystallinity will decrease with the increasing structural complexity discussed in this section. For instance, branching in polyethylene can drop the crystallinity from 90% to 40%. An isotactic polypropylene can be 90% crystalline, while atactic polypropylene is nearly all noncrystalline. Control of polymeric structure has been an essential component of the development of polymers that are competitive with metals for various engineering design applications.

SAMPLE PROBLEM 13.5

A sample of polyethylene has an average degree of polymerization of 750.

(a) What is the coiled length?

(b) Determine the extended length of an average molecule.

SOLUTION

(a) Using Equations 13.4 and 13.6, we have

$$\overline{L} = l\sqrt{2n}$$

From Table 2.2, $l = 0.154$ nm, giving

$$\overline{L} = (0.154 \text{ nm}) \sqrt{2(750)}$$

$$= 5.96 \text{ nm}$$

(b) Using Equations 13.5 and 13.6, we obtain

$$L_{\text{ext}} = 2nl \sin \frac{109.5°}{2}$$

$$= 2(750)(0.154 \text{ nm}) \sin \frac{109.5°}{2}$$

$$= 189 \text{ nm}$$

SAMPLE PROBLEM 13.6

Twenty grams of sulfur is added to 100 g of isoprene. What is the maximum fraction of cross-link sites that could be connected?

SOLUTION

As illustrated in Figure 6–46, full cross-linking involves two S atoms for two isoprene mers (i.e., 1 S:1 isoprene). The amount of sulfur needed for full cross-linking of 100 g of isoprene would be

$$m_S = \frac{\text{mol wt S}}{\text{mol wt isoprene}} \times 100\text{g}$$

Using the data of Appendix 1 yields

$$m_S = \frac{32.06}{5(12.01) + 8(1.008)} \times 100g$$

$$= 47.1$$

Assuming that all 20 g of S added in this case participates in cross-linking, we find that the maximum fraction of cross-link sites will be

$$fraction = \frac{amount \ of \ S \ added}{amount \ of \ S \ in \ fully \ cross-linked \ system}$$

$$= \frac{20g}{47.1g} = 0.425$$

PRACTICE PROBLEM 13.5

In Sample Problem 13.5, coiled and extended molecular lengths are calculated for a polyethylene with a degree of polymerization of 750. If the degree of polymerization of this material is increased by one-third (to $n = 1000$), by what percentage is **(a)** the coiled length and **(b)** the extended length increased?

PRACTICE PROBLEM 13.6

A fraction of cross-link sites is calculated in Sample Problem 13.6. What actual number of sites does this represent in the 100 g of isoprene?

13.3 THERMOPLASTIC POLYMERS

Thermoplastic polymers become soft and deformable upon heating. This is characteristic of linear polymeric molecules (including those that are branched but not cross-linked). The high-temperature plasticity is due to the ability of the molecules to slide past one another. This is another example of a thermally activated, or Arrhenius, process. In this sense, thermoplastic materials are similar to metals that gain ductility at high temperatures (e.g., creep deformation). It should be noted that, as with metals, the ductility of thermoplastic polymers is reduced upon cooling. The key distinction between thermoplastics and metals is what we mean by "high" temperatures. The secondary bonding, which must be overcome to deform thermoplastics, may allow substantial deformation around 100°C for common thermoplastics. However, metallic bonding generally restricts creep deformation to temperatures closer to 1000°C in typical alloys.

Although polymers cannot, in general, be expected to duplicate fully the mechanical behavior of traditional metal alloys, a major effort has been undertaken in recent decades to produce some polymers with sufficient strength and stiffness to be serious candidates for structural applications once dominated by metals. These are noted in Table 13.1 as **engineering polymers,** which retain good strength and stiffness up to 150–175°C. The categories are, in fact, somewhat arbitrary. The "general-use" textile fiber nylon is also a pioneering example of an engineering polymer, and it continues to be the most important. It has been estimated that industry has developed more than half a million engineering polymer part designs specifying nylon. The other members of the family of engineering polymers are part of a steadily expanding list. The importance of these materials to design engineers goes beyond their relatively small percentage of the total polymer market, as indicated by Table 13.1. Nonetheless, the bulk of that market is devoted to the materials referred to as "general-use polymers." These include the various films, fabrics, and packaging materials so much a part of everyday life. The "percentage of market" values for engineering polymers in Table 13.1 are affected by these "everyday" applications. The market shares for nylon and polyester include their major uses as textile fibers. This is the reason for polyester's larger market share even though nylon is a more common metal substitute. Table 13.1 includes some of the more familiar product trade names as well as the chemical names of the polymers.

Polyethylene, as the most common thermoplastic, is subdivided into *low-density polyethylene* (LDPE), *high-density polyethylene* (HDPE), and *ultra-high molecular-weight polyethylene* (UHMWPE). LDPE has much more chain branching than HDPE, which is essentially linear. UHMWPE has very long, linear chains. Greater chain linearity and chain length tend to increase the melting point and improve the physical and mechanical properties of the polymer due to the greater crystallinity possible within the polymer morphology. *Linear low-density polyethylene* (LLDPE) is a copolymer with α-olefins that has less chain branching and better properties than LDPE. HDPE and UHMWPE are two good examples of engineering polymers, although polyethylene overall is a general-use polymer. Note that ABS (acrylonitrile–butadiene–styrene) is an important example of a copolymer as discussed in Section 13.1. ABS is a **graft copolymer** as opposed to the block copolymer shown in Figure 13–3. Acrylonitrile and styrene chains are "grafted" onto the main polymeric chain composed of polybutadiene.

A third category of materials in Table 13.1 is that of **thermoplastic elastomers.** These are relatively recent developments in polymer technology. **Elastomers** are polymers with mechanical behavior analogous to natural rubber. Elastomeric deformation was discussed in Section 6.6. Traditional synthetic rubbers have been, upon vulcanization, thermosetting polymers as discussed in the next section. The relatively novel thermoplastic elastomers are essentially composites of rigid elastomeric domains in a relatively soft matrix of a crystalline thermoplastic polymer. A key advantage of thermoplastic elastomers is the convenience of processing by traditional thermoplastic techniques, including being recyclable.

Table 13.1 *Some Common Thermoplastic Polymers*

Name	Monomer	Typical applications	Percentage of market (weight basis)[a]
General-use polymers			
Polyethylene	$\begin{matrix} H & H \\ | & | \\ C{=}C \\ | & | \\ H & H \end{matrix}$	Clear sheet, bottles	29
Polyvinyl chloride	$\begin{matrix} H & H \\ | & | \\ C{=}C \\ | & | \\ H & Cl \end{matrix}$	Floors, fabrics, films	14
Polypropylene	$\begin{matrix} H & H \\ | & | \\ C{=}C & CH_3 \\ | & | \\ H & \end{matrix}$	Sheet, pipe, coverings	13
Polystyrene[b]	$\begin{matrix} H & \\ | & \\ C{=}C \\ | & | \\ H & H \end{matrix}$ (with phenyl group)	Containers, foams	6
Polyester, thermoplastic type [Examples: polyethylene-terephthalate (PET), Dacron[c] (fiber), Mylar[c] (film)]	$HO{-}CH_2{-}CH_2{-}OH$ and $HOOC{-}C_6H_4{-}COOH$	Magnetic tape, fibers, films	5
Nylons	$H_2N{-}(CH_2)_6{-}NH_2$ and $HOOC{-}(CH_2)_4{-}COOH$	Fabric, rope, gears, machine parts	1

Table 13.1 *Continued*

Name	Monomer	Typical applications	Percentage of market (weight basis) [a]
Acrylics (Example: polymethyl methacrylate, Lucite[c])	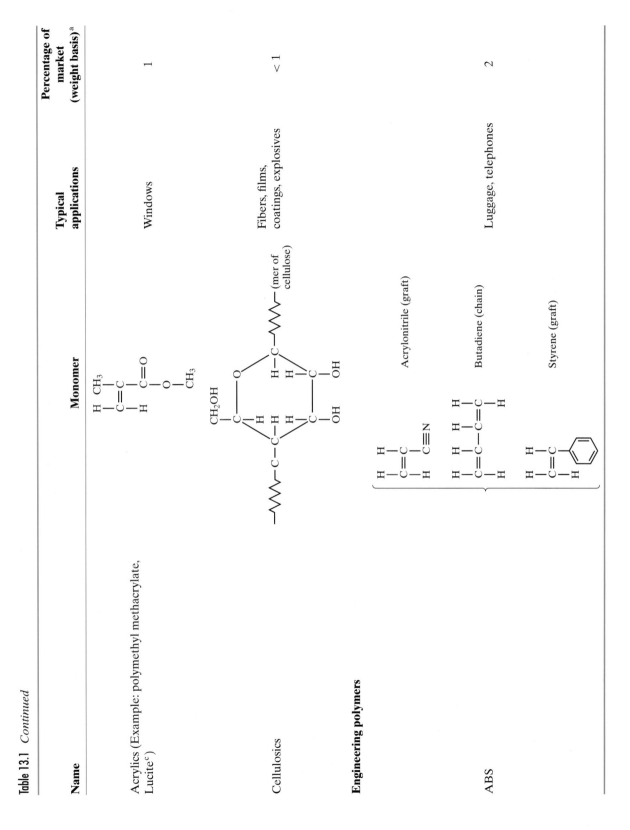	Windows	1
Cellulosics		Fibers, films, coatings, explosives	< 1
Engineering polymers			
ABS		Luggage, telephones	2

477

Table 13.1 *Continued*

Name	Monomer	Typical applications	Percentage of market (weight basis)[a]
Polycarbonates (Example: Lexan[d])	(polycarbonate mer structure with two benzene rings joined by C(CH₃)₂ and O–C(=O)–O linkage)	Machine parts, propellers	1
Acetals	(mer structure: –C(H)(H)–O–)	Hardware, gears	< 1
Fluoroplastics (Example: polytetrafluoroethylene, Teflon[c])	$F_2C=CF_2$	Chemical ware, seals, bearings, gaskets	< 1
Thermoplastic elastomers (Example: polyester-type)	$HO[(CH_2)_4O]_{n\sim14}$ + CH_3OOC—(benzene)—$COOCH_3$ + $HO(CH_2)_4OH$ + CH_3OOC—(benzene)—$COOCH_3$	Athletic footwear, couplings, tubing	1

[a] U.S. and Canada sales, from listing in *Modern Plastics*, January 1998. Industry statistics are updated annually in the January issue.

[b] (hexagon) is benzene C_6H_5

[c] Trade name, Du Pont.
[d] Trade name, General Electric.

SAMPLE PROBLEM 13.7

A copolymer of ABS contains equal weight fractions of each polymeric component. What is the mole fraction of each component?

SOLUTION

Assume that 100 g of copolymer gives 33.3 g of each component (acrylonitrile, butadiene, and styrene).

Using the information from Table 13.1 and Appendix 1 gives

$$\text{moles A} = \frac{33.3 \text{ g}}{[3(12.01) + 3(1.008) + 14.01] \text{ g/mol}}$$

$$= 0.628 \text{ mol}$$

$$\text{moles B} = \frac{33.3 \text{ g}}{[4(12.01) + 6(1.008)] \text{ g/mol}}$$

$$= 0.616 \text{ mol}$$

$$\text{moles S} = \frac{33.3 \text{ g}}{[8(12.01) + 8(1.008)] \text{ g/mol}}$$

$$= 0.320 \text{ mol}$$

Note.   There are six carbons and five hydrogens associated with the benzene ring in Table 13.1.

This gives

$$\text{mole fraction A} = \frac{0.628 \text{ mol}}{(0.628 + 0.616 + 0.320) \text{ mol}}$$

$$= 0.402$$

$$\text{mole fraction B} = \frac{0.616 \text{ mol}}{(0.628 + 0.616 + 0.320) \text{ mol}}$$

$$= 0.394$$

$$\text{mole fraction S} = \frac{0.320 \text{ mol}}{(0.628 + 0.616 + 0.320) \text{ mol}}$$

$$= 0.205$$

SAMPLE PROBLEM 13.8

An alloy of nylon and polyphenylene oxide, or PPO, produces an engineering polymer with improved toughness and high-temperature modulus compared to standard nylon. Given that PPO is

calculate the molecular weight of the PPO mer.

SOLUTION

The hexagonal symbol represents a six-carbon-atom ring. The total number of carbon atoms is, then, $6 + 2 = 8$. There is a total of eight hydrogen atoms (including two implied at the unmarked corners of the carbon ring) and, of course, only one oxygen atom. The corresponding molecular weight is

$$\text{mol wt mer} = [8(12.01) + 8(1.008) + 16.00] \text{ amu}$$

$$= 120.1 \text{ amu}$$

··

PRACTICE PROBLEM 13.7

Calculate the weight fractions for an ABS copolymer that has equal mole fractions of each component. (See Sample Problem 13.7.)

PRACTICE PROBLEM 13.8

What would be the molecular weight of a PPO polymer with a degree of polymerization of 700? (See Sample Problem 13.8.)

13.4   THERMOSETTING POLYMERS

Thermosetting polymers are the opposite of thermoplastics. They become hard and rigid upon heating. Unlike thermoplastic polymers, this phenomenon is not lost upon cooling. This is characteristic of network molecular structures formed by the step growth mechanism. The chemical reaction steps are enhanced by higher temperatures and are irreversible; that is, the polymerization remains upon cooling. Thermosetting products can be removed from the mold at the fabrication temperature (typically 200 to 300°C). By contrast, thermoplastics must be cooled in the mold to prevent distortion. Common thermosetting polymers are illustrated in Table 13.2, which is subdivided into two categories: thermosets and elastomers. In this case, thermosets refers to materials that share with the engineering polymers of Table 13.1 significant strength and stiffness so as to be common metal substitutes. However, thermosets have the disadvantages of not being recyclable and, in general, having less variable processing techniques. As noted in the previous section, traditional elastomers are thermosetting copolymers. Several key examples are listed in Table 13.2. Again, some familiar trade names are indicated. In addition to the many applications found in Table 13.1, such as films, foams, and coatings, Table 13.2 includes the important application of **adhesives**. The adhesive serves to join the surfaces of two solids (adherends) by secondary forces similar to those between molecular chains in thermoplastics. If the adhesive layer is thin and continuous, the adherend material will often fail before the adhesive.

Finally, it might also be noted that **network copolymers** can be formed similar to the block and graft copolymers already discussed for thermoplastics. The network copolymer will result from polymerization of a combination of more than one species of polyfunctional monomers.

SAMPLE PROBLEM 13.9

Metallurgical samples to be polished flat for optical microscopy are frequently mounted in a cylinder of phenol-formaldehyde, a thermosetting polymer. Because of the three-dimensional network structure, the polymer is essentially one large molecule. What would be the molecular weight of a 10-cm^3 cylinder of this polymer? (The density of the phenol-formaldehyde is 1.4 g/cm^3.)

SOLUTION

In general, the phenol molecule is trifunctional (i.e., a phenol is attached to three other phenols by three formaldehyde bridges). One such bridge is shown in Figure 13–6. A network of trifunctional bridges is shown in Figure 13–7.

Table 13.2 *Some Common Thermosetting Polymers*

Name	Monomer	Typical applications	Percentage of market (weight basis)[a]
Thermosets			
Polyurethane, also thermoplastic	$OCN-R-NCO + HO-R'-OH$ (diisocyanate) (R and R' are complex polyfunctional molecules)	Sheet, tubing, foam, elastomers, fibers	5
Phenolics (Example: phenol-formaldehyde, Bakelite[b])		Electrical equipment	4
Amino resins (Example: urea-formaldehyde)		Dishes, laminates	2
Polyesters, thermoset type		Fiberglass composite, coatings	2
Epoxies	(R and R' are complex polyfunctional molecules)	Adhesives, fiberglass composite, coatings	< 1

Table 13.2 *Continued*

Name	Monomer	Typical applications	Percentage of market (weight basis)[a]
Elastomers			
Butadiene/styrene	(butadiene; see Table 13.1 for styrene)	Tires, moldings	6
Isoprene (natural rubber)		Tires, bearings, gaskets	3
Chloroprene (Neoprene[c])		Structural bearings, fire-resistant foam, powder transmission belts	< 1
Isobutene/isoprene	(isobutene; see above for isoprene)	Tires	< 1
Silicones	(trichlorosilane) (trihydroxysilane)	Gaskets, adhesives	< 1
Vinylidene fluoride/ hexafluoropropylene (Viton[c])		Seals, O-rings, gloves	< 1

[a] U.S. sales, from listings in *Modern Plastics* and *Rubber Statistical Bulletin*.
[b] Trade name, Union Carbide. ("Bakelite" is also applied to other compounds, e.g., polyethylene.)
[c] Trade name, Du Pont.

SAMPLE PROBLEM 13.9   Solution Continued

As each formaldehyde bridge is shared by two phenols, the overall ratio of phenol to formaldehyde that must react to form the three-dimensional structure of Figure 13–6 is $1 : \frac{3}{2}$ or 1:1.5. As each formaldehyde reaction produces one H_2O molecule, we can write that

$$\text{1 phenol} + \text{1.5 formaldehyde} \rightarrow$$

$$\text{1 phenol-formaldehyde mer} + 1.5 H_2O \uparrow$$

In this way, we can calculate the mer molecular weight as

$$(\text{m.w.})_{\text{mer}} = (\text{m.w.})_{\text{phenol}} + 1.5 (\text{m.w.})_{\text{formaldehyde}}$$

$$- 1.5 (\text{m.w.})_{H_2O}$$

$$= [6(12.01) + 6(1.008) + 16.00]$$

$$+ 1.5[12.01 + 2(1.008) + 16.00]$$

$$- 1.5[2(1.008) + 16.00]$$

$$= 112.12 \text{ amu}$$

The mass of the polymer in question is

$$m = \rho V = 1.4 \frac{\text{g}}{\text{cm}^3} \times 10 \text{ cm}^3$$

$$= 14 \text{ g}$$

Therefore, the number of mers in the cylinder is

$$n = \frac{14 \text{ g}}{112.12 \text{ g}/0.6023 \times 10^{24} \text{ mers}}$$

$$= 7.52 \times 10^{22} \text{ mers}$$

This gives a molecular weight of

$$\text{mol wt} = 7.52 \times 10^{22} \text{ mers} \times 112.12 \text{ amu/mer}$$

$$= 8.43 \times 10^{24} \text{ amu}$$

SAMPLE PROBLEM 13.10

An O-ring is made from an elastomer with equimolar portions of vinylidene fluoride and hexafluoropropylene. Calculate the weight fraction of each polymer.

SOLUTION

Using the information from Table 13.2 and Appendix 1 gives

$$\text{mol wt vinylidene fluoride} = [2(12.01) + 2(1.008) + 2(19.00)] \text{ amu}$$
$$= 64.04 \text{ amu}$$

$$\text{mol wt hexafluoropropylene} = [3(12.01) + 6(19.00)] \text{ amu}$$
$$= 150.0 \text{ amu}$$

The weight fractions are then

$$\text{wt fraction vinylidene fluoride} = \frac{64.04 \text{ amu}}{(64.04 + 150.0) \text{ amu}} = 0.299$$

and

$$\text{wt fraction hexafluoropropylene} = \frac{150.0 \text{ amu}}{(64.04 + 150.0) \text{ amu}} = 0.701$$

PRACTICE PROBLEM 13.9

The molecular weight of a product of phenol-formaldehyde is calculated in Sample Problem 13.9. How much water by-product is produced in the polymerization of this product?

PRACTICE PROBLEM 13.10

For an elastomer similar to the one in Sample Problem 13.10, calculate the molecular fraction of each component if there are equal weight fractions of vinylidene fluoride and hexafluoropropylene.

13.5   ADDITIVES

Copolymers and blends were discussed in Section 13.1 as analogs of metallic alloys. There are several other alloylike *additives* that traditionally have been used in polymer technology to provide specific characteristics to the polymers.

A **plasticizer** is added to soften a polymer. This addition is essentially blending with a low-molecular-weight (approximately 300 amu) polymer. Note that a large addition of plasticizer produces a liquid. Common paint is an example. The "drying" of the paint involves the plasticizer evaporating (usually accompanied by polymerization and cross-linking by oxygen).

A **filler,** on the other hand, can strengthen a polymer by restricting chain mobility. In general, fillers are largely used for volume replacement, providing dimensional stability and reduced cost. Relatively inert materials are used. Examples include short-fiber cellulose (an organic filler) and asbestos (an inorganic filler). Roughly one-third of the typical automobile tire is a filler (i.e., carbon black). **Reinforcements** such as glass fibers are also categorized as additives. These reinforcements are widely used in the engineering polymers of Table 13.1 to enhance their strength and stiffness, thereby increasing their competitiveness as metal substitutes. Reinforcements are generally given surface treatments to ensure good interfacial bonding with the polymer and, thereby, maximum effectiveness in enhancing properties. The use of such additives up to a level of roughly 50 vol % produces a material generally still referred to as a polymer. For additions above roughly 50 vol %, the material is more properly referred to as a composite. A good example of a composite is fiberglass, which is discussed in detail in Chapter 14.

Stabilizers are additives used to reduce polymer degradation. They include a complex set of materials because of the large variety of degradation mechanisms (oxidation, thermal, and ultraviolet). As an example, polyisoprene can absorb up to 15% oxygen at room temperature, with its elastic properties being destroyed by the first 1%. Natural rubber latex contains complex phenol groups that retard the room temperature oxidation reactions. However, these naturally occurring antioxidants are not effective at elevated temperatures. Therefore, additional stabilizers (e.g., other phenols, amines, sulfur compounds, etc.) are added to rubber intended for tire applications.

Flame retardants are added to reduce the inherent combustibility of certain polymers such as polyethylene. Combustion is simply the reaction of a hydrocarbon with oxygen accompanied by substantial heat evolution. Many polymeric hydrocarbons exhibit combustibility. Others, such as polyvinyl chloride (PVC), exhibit reduced combustibility. The resistance of PVC to combustion appears to come from the evolution of the chlorine atoms from the polymeric chain. These halogens hinder the process of combustion by terminating free-radical chain reactions. Additives that provide this function for halogen-free polymers include chlorine-, bromine-, and phosphorus-containing reactants.

Colorants are additions used to provide color to a polymer where appearance is a factor in materials selection. Two types of colorants are used, pigments and dyes. **Pigments** are insoluble, colored materials added in powdered form. Typical examples are crystalline ceramics such as titanium oxide and aluminum silicate, although organic pigments are also available. **Dyes** are soluble, organic colorants that can provide transparent colors. The nature of color is discussed further in Chapter 16.

SAMPLE PROBLEM 13.11

A nylon 66 polymer is reinforced with 33 wt % glass fiber. Calculate the density of this engineering polymer. (The density of nylon 66 $= 1.14$ Mg/m^3 and the density of the reinforcing glass $= 2.54$ Mg/m^3.)

SOLUTION

For 1 kg of final product, there will be

$$0.33 \times 1 \text{ kg} = 0.33 \text{ kg glass}$$

and

$$1 \text{ kg} - 0.33 \text{ kg} = 0.67 \text{ kg nylon 66}$$

The total volume of product will be

$$V_{product} = V_{nylon} + V_{glass}$$

$$= \frac{m_{nylon}}{\rho_{nylon}} + \frac{m_{glass}}{\rho_{glass}}$$

$$= \left(\frac{0.67 \text{ kg}}{1.14 \text{ Mg/m}^3} + \frac{0.33 \text{ kg}}{2.54 \text{ Mg/m}^3} \right) \times \frac{1 \text{ Mg}}{1000 \text{ kg}} = 7.18 \times 10^{-4} \text{ m}^3$$

The overall density of the final product is then

$$\rho = \frac{1 \text{ kg}}{7.18 \times 10^{-4} \text{ m}^3} \times \frac{1 \text{ Mg}}{1000 \text{ kg}} = 1.39 \text{ Mg/m}^3$$

PRACTICE PROBLEM 13.11

Sample Problem 13.11 describes a high-strength and -stiffness engineering polymer. Strength and stiffness can be further increased by a greater "loading" of glass fibers. Calculate the density of a nylon 66 with 43 wt % glass fibers.

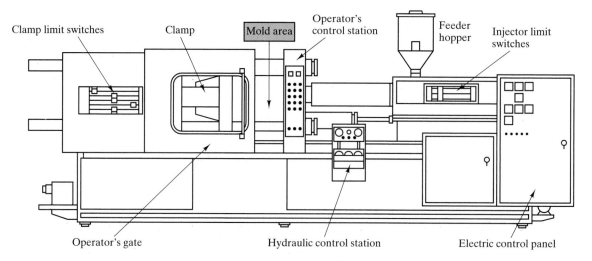

Clamp limit switches Clamp Mold area Operator's control station Feeder hopper Injector limit switches

Operator's gate Hydraulic control station Electric control panel

Figure 13-13 *Injection molding of a thermoplastic polymer. (After Modern Plastics Encyclopedia, 1981–82, Vol. 58, No. 10A, McGraw-Hill Book Company, New York, October 1981.)*

13.6   PROCESSING OF POLYMERS

Table 13.3 summarizes some of the major processing techniques for polymers. For thermoplastics, **injection molding** and **extrusion molding** are the predominant processes. Figure 13–13 illustrates the injection molding technique, and Figure 13–14 illustrates the extrusion molding technique. Injection molding involves the melting of polymer powder prior to injection. Both injection and extrusion molding are similar to metallurgical processing but are carried out at relatively low temperatures. **Blow molding** (Figure 13–15) is emerging as a third major processing technique for thermoplastics. With this technique, the specific shaping process is quite similar to the glass-forming technique of Figure 12–3, except that relatively low molding temperatures are required. As with glass container manufacturing, blow molding is often used to produce polymeric containers. In addition, various commercial products, including automobile body parts, can be economically fabricated by this method. **Compression molding** and **transfer molding** are the predominant processes for thermosetting polymers. Compression

Table 13.3 *Some Major Processing Methods for Polymers*

Thermoplastics	Thermosetting
Injection molding	Compression molding
Extrusion molding	Transfer molding
Blow molding	

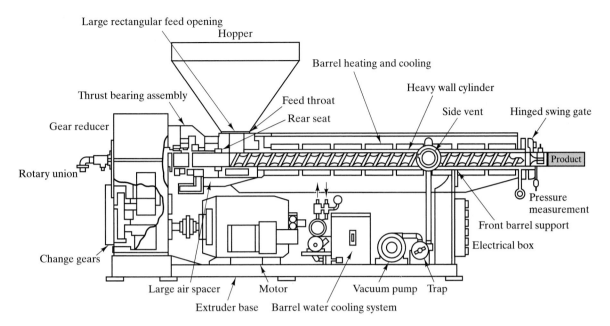

Figure 13-14 *Extrusion molding of a thermoplastic polymer. (After* Modern Plastics Encyclopedia, 1981–82, *Vol. 58, No. 10A, McGraw-Hill Book Company, New York, October 1981.)*

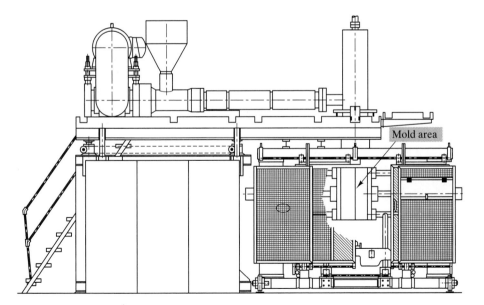

Figure 13-15 *Blow molding of a thermoplastic polymer. The specific shaping operation is similar to the glass container process of Figure 12–3. (After a Krupp-Kautex design.)*

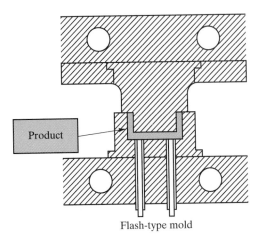

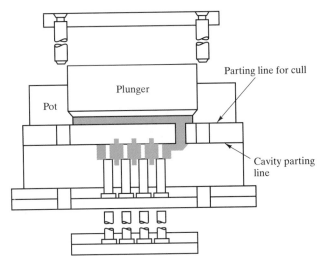

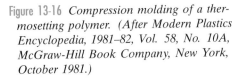

Figure 13-16 *Compression molding of a thermosetting polymer. (After Modern Plastics Encyclopedia, 1981–82, Vol. 58, No. 10A, McGraw-Hill Book Company, New York, October 1981.)*

Figure 13-17 *Transfer molding of a thermosetting polymer. (After Modern Plastics Encyclopedia, 1981–82, Vol. 58, No. 10A, McGraw-Hill Book Company, New York, October 1981.)*

molding is illustrated in Figure 13–16, and transfer molding is illustrated in Figure 13–17. Compression molding is generally impractical for thermoplastics because the mold would have to be cooled to ensure that the part would not lose its shape upon ejection from the mold. In transfer molding, a partially polymerized material is forced into a closed mold, where final cross-linking occurs at elevated temperature and pressure. Finally, Figure 13–18 summarizes the general procedures for manufacturing various rubber products.

In Chapter 6 and earlier in this chapter, there are numerous references to the effect of polymeric structure on mechanical behaviour. Table 13.4 summarizes these various relations along with corresponding references to the processing techniques leading to the various structures.

Table 13.4 *The Relationship of Processing, Molecular Structure, and Mechanical Behavior for Polymers*

Category	Processing technique	Molecular structure	Mechanical effect
Thermoplastic polymers			
	Addition agent	Branching	Increased strength and stiffness
	Vulcanization	Cross-linking	Increased strength and stiffness
	Crystallization	Increased crystallinity	Increased strength and stiffness
	Plasticizer	Decreased molecular weight	Decreased strength and stiffness
	Filler	Restricted chain mobility	Increased strength and stiffness
Thermosetting polymers			
	Setting at elevated temperatures	Network formation	Rigid (remaining upon cooling)

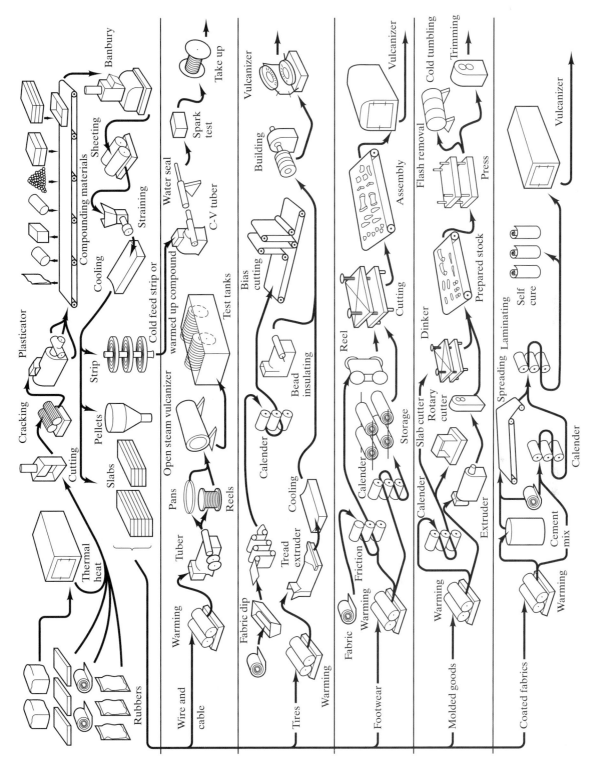

Figure 13-18 *Typical flow diagram for the manufacturing of various common rubber goods. (From the Vanderbilt Rubber Handbook, R. T. Vanderbilt Co., Norwalk, Conn., 1978.)*

SAMPLE PROBLEM 13.12

As discussed in Section 13.1, a batch of ethylene being prepared for bottle manufacturing has 0.15 wt % H_2O_2 added to establish a given degree of polymerization. **(a)** If the H_2O_2 addition is increased to 0.16 wt %, calculate the resulting % change in degree of polymerization. **(b)** Comment on the effect of this change on the bottle manufacturing process.

SOLUTION

(a) As in Sample Problem 13.2, we note there is one H_2O_2 molecule (= two OH groups) per polyethylene molecule. For 0.15 wt % H_2O_2,

$$0.15 \text{ wt \% } H_2O_2 = \frac{\text{mol wt } H_2O_2}{n \times (\text{mol wt } C_2H_4)} \times 100$$

Using Appendix 1 gives

$$0.15 \text{ wt \% } H_2O_2 = \frac{2(1.008) + 2(16.00)}{n[2(12.01) + 4(1.008)]} \times 100$$

or

$$n = 808$$

For 0.16 wt % H_2O_2,

$$0.16 \text{ wt \% } H_2O_2 = \frac{\text{mol wt } H_2O_2}{n \times (\text{mol wt } C_2H_4)} \times 100$$

Using Appendix 1 gives

$$0.16 \text{ wt \% } H_2O_2 = \frac{2(1.008) + 2(16.00)}{n[2(12.01) + 4(1.008)]} \times 100$$

or

$$n = 758$$

Therefore,

$$\% \text{ decrease} = \frac{758 - 808}{808} \times 100 = -6.2\%$$

(b) The decrease in degree of polymerization will, in general, make for a lower viscosity, more deformable material in the bottle-forming stage of manufacture.

··

PRACTICE PROBLEM 13.12

What would be the degree of polymerization for the polyethylene in Sample Problem 13.12 if the H_2O_2 addition is reduced to 0.14 wt %?

SUMMARY

Polymers, or plastics, are organic materials composed of long-chain or network organic molecules formed from small molecules (monomers) by polymerization reactions. Polymerization occurs by chain growth (addition polymerization) or step growth (condensation polymerization). Copolymers and blends are analogs of metallic alloys. Individual mers (polymer building blocks comparable to monomers) produce linear molecular structure when they are bifunctional (have two points of contact with adjacent mers). Polyfunctional mers (those having more than two points of contact) produce a network molecular structure.

The number of mers attached together to form a polymeric molecule is termed the degree of polymerization. There is a statistical distribution of both molecular weights and molecular lengths in a given polymer. Also, the length of a linear molecule can be characterized by both coiled and extended configurations. For many polymers, rigidity and melting point increase with increasing molecular length and complexity. This complexity is increased by structural irregularity, branching, and cross-linking.

Thermoplastic polymers become softer upon heating due to thermal agitation of weak, secondary bonds between adjacent linear molecules. An important trend in the past decade has been the increased development of engineering polymers for metals substitution and thermoplastic elastomers, rubberlike materials with the processing convenience of traditional thermoplastics. Thermosetting polymers are network structures that form upon heating, resulting in greater rigidity, and include the traditional vulcanized elastomers.

Additives are materials added to polymers to provide specific characteristics. Like copolymers and blends, polymers with additives are analogs of metallic alloys. Common additives include plasticizers, fillers, reinforcements, stabilizers, flame retardants, and colorants.

Thermoplastic polymers are generally processed by injection molding, extrusion molding, or blow molding. Thermosetting polymers are generally formed by compression molding or transfer molding.

KEY TERMS

addition polymerization (460)
adhesive (481)
atactic (471)
bifunctional (463)
blend (462)
block copolymer (462)
blow molding (488)
chain growth (460)
colorant (487)
compression molding (488)
condensation polymerization (460)
copolymer (462)
degree of polymerization (469)
dye (487)
elastomer (475)
engineering polymer (475)

extended length (470)
extrusion molding (488)
fillers (486)
flame retardent (486)
graft copolymer (475)
initiator (460)
injection molding (488)
isotactic (471)
linear molecular structure (463)
mer (459)
monomer (459)
network copolymer (481)
network molecular structure (463)
pigment (487)
plasticizer (486)
plastics (459)

polyfunctional (463)
polymer (459)
polymerization (460)
reinforcement (486)
root-mean-square length (469)
stabilizer (486)
step growth (460)
syndiotactic (471)
terminator (460)
thermoplastic polymers (474)
thermoplastic elastomer (475)
thermosetting polymers (481)
transfer molding (488)
vulcanization (472)

REFERENCES

Brandrup, J., and **E. H. Immergut**, Eds., *Polymer Handbook,* 3rd ed., John Wiley & Sons, Inc., New York, 1989.

Engineered Materials Handbook, Vol. 2, *Engineering Plastics,* ASM International, Metals Park, Ohio, 1988.

Mark, H. F. et al., Eds., *Encyclopedia of Polymer Science and*

Engineering, 2nd ed., Vols. 1–17, Index Vol., Supplementary Vol., John Wiley & Sons, Inc., New York, 1985–1989.

Modern Plastics Encyclopedia 98, Vol. 74, No. 13, McGraw-Hill Book Company, New York, November 1997. (Revised annually)

PROBLEMS

13.1 • Polymerization

13.1. What is the average molecular weight of a polypropylene with a degree of polymerization of 500? (Note Table 13.1.)

13.2. What is the average molecular weight of a polystyrene with a degree of polymerization of 500?

13.3. How many grams of H_2O_2 would be needed to yield 1 kg of a polypropylene, $(C_3H_6)_n$, with an average degree of polymerization of 600? (Use the same assumptions given in Sample Problem 13.2.)

13.4. A blend of polyethylene and polyvinyl chloride (see Figure 13–4) contains 20 wt % polyvinyl chloride. What is the molecular percentage of polyvinyl chloride?

13.5. A blend of polyethylene and polyvinyl chloride (see Figure 13–4) contains 20 mol % polyvinyl chloride. What is the weight percentage of polyvinyl chloride?

13.6. Calculate the degree of polymerization for **(a)** a low-density polyethylene with a molecular weight of 20,000 amu, **(b)** a high-density polyethylene with a molecular weight of 300,000 amu, and **(c)** an ultra-high molecular weight polyethylene with a molecular weight of 4,000,000 amu.

13.7. A simplified mer formula for natural rubber (isoprene) is C_5H_8. (See Table 13.2 for a more detailed illustration.) Calculate the molecular weight for a molecule of isoprene with a degree of polymerization of 500.

13.8. Calculate the molecular weight for a molecule of chloroprene (a common synthetic rubber) with a degree of polymerization of 500. (See Table 13.2.)

13.2 • Structural Features of Polymers

13.9. The data given in Figure 13–8 can be represented in tabular form as follows:

n **Range**	n_i **(Midvalue)**	**Population fraction**
1–100	50	—
101–200	150	—
201–300	250	0.01
301–400	350	0.10
401–500	450	0.21
501–600	550	0.22
601–700	650	0.18
701–800	750	0.12
801–900	850	0.07
901–1000	950	0.05
1001–1100	1050	0.02
1101–1200	1150	0.01
1201–1300	1250	0.01
		$\sum = 1.00$

Calculate the average degree of polymerization for this system.

13.10. If the polymer evaluated in Problem 13.9 is polypropylene, what would be the **(a)** coiled length and **(b)** extended length of the average molecule?

13.11. What would be the maximum fraction of cross-link sites that would be connected in 1 kg of chloroprene with the addition of 200 g of sulfur?

13.12. Calculate the average molecular length (extended) for a polyethylene with a molecular weight of 25,000 amu.

13.13. Calculate the average molecular length (extended) for polyvinyl chloride with a molecular weight of 25,000 amu.

13.14. If 0.2 g of H_2O_2 is added to 100 g of ethylene to establish the degree of polymerization, what would be the resulting average molecular length (coiled)? (Use the assumptions of Sample Problem 13.2.)

13.15. What would be the extended length of the average molecule described in Problem 13.14?

• **13.16.** The acetal polymer in Figure 13–5 contains, of course, C—O bonds rather than C—C bonds along its molecular chain backbone. As a result, there are two types of bond angles to consider. The O—C—O bond angle is approximately the same as the C—C—C bond angle (109.5°) because of the tetrahedral bonding configuration in carbon (see Figure 2–19). However, the C—O—C bond is a flexible one with a possible bond angle ranging up to 180°. **(a)** Make a sketch similar to Figure 13–10 for a fully extended polyacetal molecule. **(b)** Calculate the extended length of a molecule with a degree of polymerization of 500. (Refer to Table 2.2 for bond length data.) **(c)** Calculate the coiled length of the molecule in part (b).

13.3 • Thermoplastic Polymers

13.17. Calculate **(a)** the molecular weight, **(b)** coiled molecular length, and **(c)** extended molecular length for a polytetrafluoroethylene polymer with a degree of polymerization of 500.

13.18. Repeat Problem 13.17 for a polypropylene polymer with a degree of polymerization of 750.

13.19. Calculate the degree of polymerization of a polycarbonate polymer with a molecular weight of 100,000 amu.

13.20. Calculate the molecular weight for a polymethyl methacrylate polymer with a degree of polymerization of 500.

• **13.21.** The reaction of two molecules to form a nylon monomer is shown in Table 13.1. (The H and OH units enclosed by a dashed line become an H_2O reaction byproduct and are replaced by a C—N bond in the middle of the monomer.) **(a)** Sketch the reaction of nylon monomers to form a nylon polymer. (This occurs when one H and one OH at each end of the monomer are removed and become a reaction byproduct.) **(b)** Calculate the molecular weight of the nylon mer.

• **13.22.** A high-toughness alloy of nylon and PPO contains 10 wt % PPO. Calculate the mole fraction of PPO in this alloy. (Note Sample Problem 13.8 and Problem 13.21.)

13.23. In Problem 11.9, the mass reduction in an automobile design was calculated based on trends in metal alloy selection. A more complete picture is obtained by noting that, in the same 1975 to 1985 period, the volume of polymers used increased from $0.064 \, \text{m}^3$ to $0.100 \, \text{m}^3$. Estimate the mass reduction (compared to 1975) including this additional polymer data. (Approximate the polymer density as $1 \, \text{Mg/m}^3$.)

13.24. Repeat Problem 13.23 for the mass reduction in the year 2000 given the data in Problem 11.10 and the fact that the total volume of polymer used at that time is $0.122 \, \text{m}^3$.

13.4 • Thermosetting Polymers

13.25. What would be the molecular weight of a 40,000 mm^3 plate made from urea-formaldehyde? (The density of the urea-formaldehyde is $1.50 \, \text{Mg/m}^3$.)

13.26. How much water byproduct would be produced in the polymerization of the urea-formaldehyde product in Problem 13.25?

13.27. Polyisoprene loses its elastic properties with 1 wt % O_2 addition. If we assume that this is due to a cross-linking mechanism similar to that for sulfur, what fraction of the cross-link sites are occupied in this case?

13.28. Repeat the calculation of Problem 13.27 for the case of the oxidation of polychloroprene by 1 wt % O_2.

13.5 • Additives

13.29. An epoxy (density $= 1.1 \, \text{Mg/m}^3$) is reinforced with 30 vol % E-glass fibers (density $= 2.54 \, \text{Mg/m}^3$). Calculate **(a)** the weight percent of E-glass fibers and **(b)** the density of the reinforced polymer.

13.30. Calculate the % mass savings that would occur if the reinforced polymer described in Problem 13.29 were used to replace a steel gear. (Assume the gear volume is the same for both materials and approximate the density of steel by pure iron.)

13.31. Repeat Problem 13.30 if the polymer replaces an aluminum alloy. (Again, assume the gear volume is the same and approximate the density of the alloy by pure aluminum.)

13.32. Some injection moldable nylon 66 contains 40 wt % glass spheres as a filler. Improved mechanical properties are the result. If the average glass sphere diameter is 100 μm, estimate the density of such particles per cubic millimeter.

13.33. Calculate the average separation distance between the centers of adjacent glass spheres in the nylon described in Problem 13.32. (Assume a simple cubic array of dispersed particles as in Problem 12.12.)

13.34. Repeat Problem 13.33 for the case of the same wt % spheres but an average glass sphere diameter of 60 μm.

13.35. Bearings and other parts requiring exceptionally low friction and wear can be fabricated from an acetal polymer with an addition of polytetrafluoroethylene (PTFE) fibers. The densities of acetal and PTFE are $1.42 \, \text{Mg/m}^3$ and $2.15 \, \text{Mg/m}^3$, respectively. If the density of the polymer with additive is $1.54 \, \text{Mg/m}^3$, calculate the weight percent of the PTFE addition.

13.36. Repeat Problem 13.35 for the case of a polymer density of $1.48 \, \text{Mg/m}^3$.

13.6 • Processing of Polymers

13.37. The heat evolved in the manufacturing of polyethylene sheet was evaluated in Problem 2.29. In a similar way, calculate the heat evolution occurring during a 24-hour period in which 864 km of a sheet 1 mil ($25.4 \, \mu$m) thick by 300 mm wide are manufactured. (The density of the sheet is $0.910 \, \text{Mg/m}^3$.)

13.38. What would be the total heat evolution occurring in the manufacturing of the polyethylene sheet in Problem 13.37 if that production rate could be maintained for a full year?

D 13.39. Given data on modulus of elasticity (in tension) and tensile strength for various thermoplastic polymers in Table 6.7, select the polymers that would meet the following design specifications for a mechanical gear application:

modulus of elasticity, E: 2000 MPa $< E <$ 3000 MPa

tensile strength, T.S.: 50 MPa

D 13.40. Using the design specifications given for Problem 13.39, consider the various thermosetting polymers of Table 6.8. How many of these polymers would meet the design specifications?

CHAPTER 14
Composites

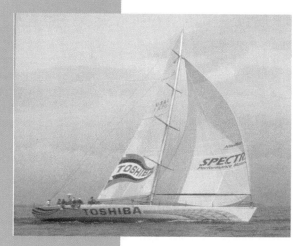

The hull of this sailing yacht is a sandwich structure, with the outer skin made of an epoxy resin reinforced with Kevlar fibers. The core is an expanded polyvinyl chloride foam. This composite material system is light weight while providing high impact strength and tear resistance. In addition, the sail is not mere cloth but a fiber-reinforced Mylar film. (Courtesy of Allied Signal.)

497

Our final category of structural engineering materials is that of composites. These materials involve some combination of two or more components from the fundamental material types covered in the preceding three chapters. A key philosophy in selecting composite materials is that they should provide the "best of both worlds," that is, attractive properties from each component. A classic example is fiberglass. The strength of small-diameter glass fibers is combined with the ductility of the polymeric matrix. The combination of these two components provides a product superior to either component alone. Many composites, such as fiberglass, involve combinations that cross over the boundaries of the preceding three chapters. Others, such as concrete, involve different components from within a single material type. In general, we shall use a fairly narrow definition of composites. We shall consider only those materials that combine different components on the microscopic (rather than macroscopic) scale. We shall not include multiphase alloys and ceramics, which are the result of routine processing discussed in Chapters 9 and 10. Similarly, the microcircuits to be discussed in Chapter 17 are not included because each component retains its distinctive character in these material systems. In spite of these restrictions, we shall find this category to include a tremendously diverse collection of materials, from the common to some of the most sophisticated. Fiberglass, wood, and concrete are among our most common construction materials. The aerospace industry has driven much of the development of our most sophisticated composite systems (e.g., "stealth" aircraft with high-performance, nonmetallic materials). Increasingly, these advanced materials are being used in civilian applications, such as improved strength-to-weight ratio bridges and more fuel-efficient automobiles.

We shall consider two broad categories of composite materials. Conveniently, these categories are demonstrated by two of our most common structural materials: fiberglass and concrete. Fiberglass, or glass fiber-reinforced polymer, is an excellent example of a synthetic fiber-reinforced composite. The fiber reinforcement is generally found in one of three primary configurations: aligned in a single direction, randomly chopped, or woven in a fabric that is laminated with the matrix. Wood is a structural analog of fiberglass, that is, a natural fiber-reinforced composite. The fibers of wood are elongated, made of dead biological cells. The matrix corresponds to lignin and hemicellulose deposits. Concrete is our best example of an aggregate composite, in which particles rather than fibers reinforce a matrix. While concrete has been a construction material for centuries, there are numerous composites developed in recent decades that use a similar particulate-reinforcement concept.

The concept of property averaging is central to understanding the utility of composite materials. An important example is the elastic modulus of a composite. The modulus is a sensitive function of the geometry of the reinforcing component. Similarly important is the strength of the interface between the reinforcing component and the matrix. Because of the wide use of composites as structural materials, we shall concentrate on these mechanical properties. So-called advanced composites have provided some unusually attractive features, such as high strength-to-weight ratios. Some care is required in citing these properties, as they can be highly directional in nature.

The processing of composites reflects the wide range of chemistry and microstructures involved in this highly diverse category of materials. We shall focus on a few, common examples. For example, the processing of polymer-matrix composites largely follows from the processing of polymers in Chapter 13.

14.1   FIBER-REINFORCED COMPOSITES

The most common examples of synthetic composite materials are those with micron-scale reinforcing fibers. Within this category are two distinct sub-groups: (1) fiberglass generally using glass fibers with moderately high values of elastic modulus, and (2) advanced composites with even higher moduli fibers. We shall also compare these synthetic materials with an important natural, fiber-reinforced composite—wood.

CONVENTIONAL FIBERGLASS

Fiberglass, is a classic example of a modern composite system. Reinforcing fibers are shown in Figure 14–1. A typical fracture surface of a composite (Figure 14–2) shows such fibers embedded in the polymeric matrix.

Figure 14-1 *Glass fibers to be used for reinforcement in a fiberglass composite. (Courtesy of Owens-Corning Fiberglas Corporation)*

Figure 14-2 *The glass fiber reinforcement in a fiberglass composite is clearly seen in a scanning electron microscope image of a fracture surface. (Courtesy of Owens-Corning Fiberglas Corporation)*

Table 14.1 *Compositions of Glass Reinforcing Fibers*

Designation	Characteristic	Composition[a] (wt %)								
		SiO_2	$(Al_2O_3 + Fe_2O_3)$	CaO	MgO	Na_2O	K_2O	B_2O_3	TiO_2	ZrO_2
A-glass	Common soda–lime silica	72	<1	10		14				
AR-glass	Alkali resistant (for concrete reinforcement)	61	<1	5	<1	14	3		7	10
C-glass	Chemical corrosion resistant	65	4	13	3	8	2	5		
E-glass	Electrical composition	54	15	17	5	<1	<1	8		
S-glass	High strength and modulus	65	25		10					

Source: Data from J. G. Mohr and W. P. Rowe, *Fiber Glass,* Van Nostrand Reinhold Company, Inc., New York, 1978.

[a] Approximate and not representing various impurities.

Table 14.1 lists some common glass compositions used for fiber reinforcement. Each is the result of substantial development that has led to optimal suitability for specific applications. For example, the most generally used glass fiber composition is **E-glass,** in which E stands for "electrical type." The low sodium content of E-glass is responsible for its especially low electrical conductivity and its attractiveness as a dielectric. Its popularity in structural composites is related to the chemical durability of the borosilicate composition. Table 14.2 lists some of the common polymeric matrix materials. Three common fiber configurations are illustrated in Figure 14–3. Parts (a) and (b) show the use of **continuous fibers** and **discrete (chopped) fibers,** respectively. Part (c) shows the **woven fabric** configuration, which is layered with the matrix polymer to form a **laminate.** The implications of these various geometries on mechanical properties will be covered in Sections 14.3 and 14.4. For now, we note that optimal strength is achieved by the aligned, continuous fiber reinforcement. Caution is necessary, however, in citing this strength because it is maximal only in the direction parallel to the fiber axes. In other words, the strength is highly **anisotropic**—it varies with direction.

ADVANCED COMPOSITES

Advanced composites include those systems in which reinforcing fibers have moduli higher than that of E-glass. For example, fiberglass used in most U.S. helicopter blades contains high modulus S-glass fibers (see Table 14.1). Advanced composites, however, generally involve fibers other than glass. Table 14.3 lists a variety of advanced composite systems. These include some of the most sophisticated materials developed for some of the most demanding engineering applications. The growth of the advanced composites industry began with the materials advances of World War II and accelerated rapidly with the space race of the 1960s and subsequent growth in demand for commercial aviation and high-performance leisure-time products (e.g., Figure 1–15).

Table 14.2 *Polymeric Matrix Materials for Fiberglass*

Polymer[a]	Characteristics and applications
Thermosetting	
Epoxies	High strength (for filament-wound vessels)
Polyesters	For general structures (usually fabric-reinforced)
Phenolics	High-temperature applications
Silicones	Electrical applications (e.g., printed-circuit panels)
Thermoplastic	
Nylon 66	
Polycarbonate	Less common, especially good ductility
Polystyrene	

Source: Data from L. J. Broutman and R. H. Krock, Eds. *Modern Composite Materials,*
   Addison-Wesley Publishing Co., Inc., Reading, Mass., 1967, Chapter 13.
   [a] See Tables 13.1 and 13.2 for chemistry.

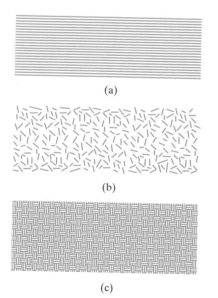

(a)

(b)

(c)

Figure 14-3 *Three common fiber configurations for composite reinforcement are (a) continuous fibers, (b) discrete (or chopped) fibers, and (c) woven fabric, which is used to make a laminated structure.*

Table 14.3 *Advanced Composite Systems Other Than Fiberglass*

Class	Fiber/Matrix
Polymer matrix	
	Para-aramid (Kevlar[a])/epoxy
	Para-aramid (Kevlar[a])/polyester
	C(graphite)/epoxy
	C(graphite)/polyester
	C(graphite)/polyetheretherketone (PEEK)
	C(graphite)/polyphenylene sulfide (PPS)
Metal matrix	
	B/Al
	C/Al
	Al_2O_3/Al
	Al_2O_3/Mg
	SiC/Al
	SiC/Ti (alloys)
Ceramic matrix	
	$Nb/MoSi_2$
	C/C
	C/SiC
	SiC/Al_2O_3
	SiC/SiC
	SiC/Si_3N_4
	SiC/Li–Al–silicate (glass-ceramic)

Source: Data from K. K. Chawla, New Mexico Institute of Mining and Technology, A. K. Dhingra, the Du Pont Company, and A. J. Klein, ASM International.
[a] Trade name, Du Pont.

Carbon and Kevlar fiber reinforcements represent advances over traditional glass fibers for **polymer–matrix composites.** Carbon fibers typically range in diameter between 4 and 10 μm, with the carbon being a combination of crystalline graphite and noncrystalline regions. Kevlar is a Du Pont trade name for poly *p*-phenyleneterephthalamide (PPD-T), a para-aramid with the formula

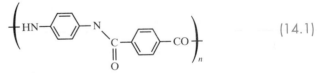

$$(14.1)$$

Epoxies and polyesters (thermosetting polymers) are traditional matrices. Substantial progress has been made in developing thermoplastic polymer matrices, such as polyetheretherketone (PEEK) and polyphenylene sulfide (PPS). These materials have the advantages of increased toughness and being recyclable. Carbon- and Kevlar-reinforced polymers are used in pressure vessels, and Kevlar reinforcement is widely used in tires. Carbon-reinforced PEEK and PPS demonstrate good temperature resistance and are, as a result, attractive for aerospace applications.

Metal–matrix composites have been developed for use in temperature, conductivity, and load conditions beyond the capability of polymer–matrix systems. For example, boron-reinforced aluminum is used in the Space Shuttle *Orbiter*, and carbon-reinforced aluminum is used in the Hubble Telescope. Alumina-reinforced aluminum is used in automobile engine components.

A dramatic entrant in the composite field is the family of **ceramic–matrix composites.** A primary driving force for their development is superior high-temperature resistance. These composites, as opposed to traditional ceramics, represent the greatest promise to obtain the requisite toughness for structural applications such as high-efficiency engine designs. An especially advanced composite system in this category is the **carbon–carbon composite**. This high-modulus and high-strength material is also quite expensive. The expense is significantly increased by the process of forming the large carbon chain molecules of the matrix by pyrolysis (heating in an inert atmosphere) of a polymeric hydrocarbon. Carbon–carbon composites are currently being used in high-performance automobiles as friction-resistant materials and in a variety of aerospace appliations, such as ablative shields for reentry vehicles.

Metal fibers are frequently small-diameter wires. Especially high strength reinforcement comes from **whiskers,** small, single-crystal fibers that can be grown with a nearly perfect crystalline structure. Unfortunately, whiskers cannot be grown as continuous filaments in the manner of glass fibers or metal wires. Figure 14–4 contrasts the wide range of cross-sectional geometries associated with reinforcing fibers.

During the 1980s, production of advanced composites in the United States doubled every five years. In the first three years of the 1990s, however, production suddenly declined by 20% due to the end of the Cold War and the resulting effect on defense budgets. Various trends have emerged in the advanced composites field in response to these changes. Emerging product applications include the marine market (e.g., high-performance power boats), improved strength-to-weight ratio civil engineering structures, and electric car development. A fundamental challenge to the wider use of advanced composites in the general automotive industry is the need for reduced costs, which cannot occur until greater production capacity is in place. In turn, production capacity cannot be increased without greater demand in the automotive field.

Specific technological developments are occurring in response to the new trend toward nondefense applications. A major thrust is reduced production costs. For this reason, nonautoclave curing of thermosetting resins is being developed for bridge construction. Similarly, resin transfer molding (RTM) involving textile preforms substantially reduces cure times. (The related technique of transfer molding for polymers is illustrated in Figure 13–17.) Also, automated fiber-placement equipment produces a more rapid fabrication process. The addition of thermoplastics or elastomeric microspheres to thermosetting resins is among the various techniques being used to improve fracture toughness and the resulting reduction of delamination and impact damage. Bismaleimide (BMI) resins are an advance over epoxies for heat resistance (over 300°C).

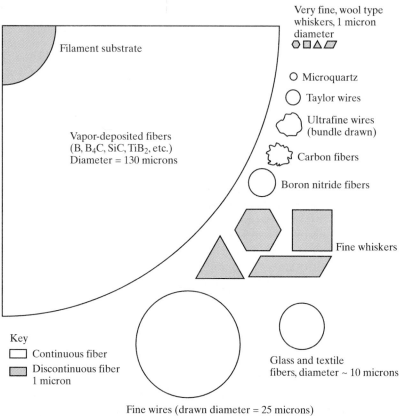

Figure 14-4 *Relative cross-sectional areas and shapes of a wide variety of reinforcing fibers. (After L. J. Broutman and R. H. Krock, Eds.,* Modern Composite Materials, *Addison-Wesley Publishing Co., Inc., Reading, Mass., 1967, Chapter 14.)*

A single composite may contain various types of reinforcing fibers. **Hybrids** are woven fabrics consisting of two or more types of reinforcing fibers (e.g., carbon and glass or carbon and aramid). The combination is a design approach to optimize composite performance. For example, high-strength, noncarbon fibers can be added to carbon fibers to improve the impact resistance of the overall composite.

WOOD—A NATURAL FIBER-REINFORCED COMPOSITE

The composites listed in Table 14.3 represent some of the most creative achievements of materials engineers. But like so many accomplishments of human beings, those fiber-reinforced composites imitate nature. Common **wood** is such a composite, which serves as an excellent structural material. In fact, the weight of wood used each year in the United States exceeds the combined total for steel and concrete. Some common woods are listed in Table 14.4. We find two categories: softwoods and hardwoods. These are relative

Table 14.4 *Some Common Woods*

Softwoods	Hardwoods
Cedar	Ash
Douglas fir	Birch
Hemlock	Hickory
Pine	Maple
Redwood	Oak
Spruce	

terms, although softwoods generally have lower strengths. The fundamental difference between the categories is their seasonal nature. Softwoods are "evergreens" with needlelike leaves and exposed seeds. Hardwoods are deciduous (i.e., lose their leaves annually) and have covered seeds (e.g., nuts).

The microstructure of wood illustrates its commonality with the synthetic composites of the preceding section. Figure 14–5 shows a dramatic view of the microstructure of southern pine, an important softwood. The dominant feature of the microstructure is the large number of tubelike cells oriented vertically. These longitudinal cells are aligned with the vertical axis of the tree. You may recall, from Chapter 6, our introduction to the brilliant Robert Hooke, who gave us Hooke's law. It was he who coined the

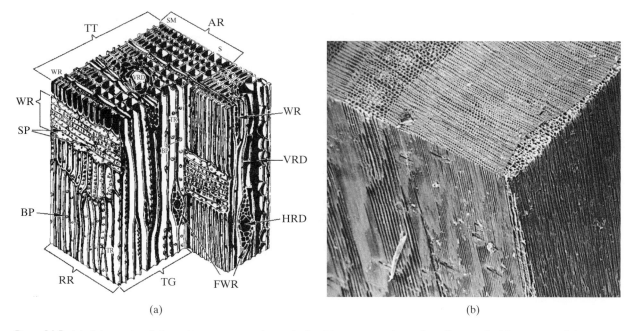

(a)   (b)

Figure 14-5 *(a) Schematic of the microstructure of wood. In this case, a softwood is illustrated. The structural features are TT, cross-sectional face; RR, radial face; TG, tangential face; AR, annual ring; S, early (spring) wood; SM, late (summer) wood; WR, wood ray; FWR, fusiform wood ray; VRD, vertical resin duct; HRD, horizontal resin duct; BP, bordered pit; SP, simple pit; and TR, tracheids. (b) A scanning electron micrograph showing the microstructure of southern pine (at 45×). (Courtesy of U.S. Dept. of Agriculture, Forest Service, Forest Products Laboratory, Madison, Wis.)*

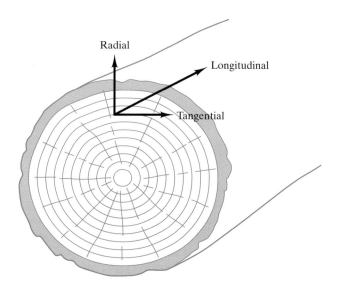

Figure 14-6 *Anisotropic macrostructure of wood.*

term cell for these biological building blocks. There are some radial cells perpendicular to the longitudinal ones. As the name implies, the radial cells extend from the center of the tree trunk out radially toward the surface. The longitudinal cells form tubes that carry sap and other fluids needed by the plant cells. Early-season cells are of larger diameter than later-season cells. This growth pattern leads to the characteristic ring structure, which indicates the tree's age. The radial cells store food for the growing tree. The cell walls are composed of cellulose (see Table 13.1). These tubular cells serve the reinforcing role played by glass fibers in fiberglass composites. The strength of the cells in the longitudinal direction is a function of fiber alignment in that direction. The cells are held together by a matrix of **lignin** and **hemicellulose.** Lignin is a phenol-propane network polymer, and hemicellulose is polymeric cellulose with a relatively low degree of polymerization (~ 200).

The complex chemistry and microstructure of wood are manifest as a highly anisotropic macrostructure as shown in Figure 14–6. Related to this, the dimensions as well as the properties of wood vary significantly with atmospheric moisture levels. Care will be required in Section 14.4 in specifying the atmospheric conditions for which mechanical property data apply.

SAMPLE PROBLEM 14.1

A fiberglass composite contains 70 vol % E-glass fibers in an epoxy matrix.

(a) Calculate the weight percent glass fibers in the composite.

(b) Determine the density of the composite. The density of E-glass is 2.54 Mg/m^3 ($= \text{g/cm}^3$) and for epoxy is 1.1 Mg/m^3.

SOLUTION

(a) For 1 m^3 composite, we would have 0.70 m^3 of E-glass and $(1.00 - 0.70) \text{ m}^3 = 0.30 \text{ m}^3$ of epoxy.
   The mass of each component will be

$$m_{\text{E-glass}} = \frac{2.54 \text{ Mg}}{\text{m}^3} \times 0.70 \text{ m}^3 = 1.77 \text{ Mg}$$

$$m_{\text{epoxy}} = \frac{1.1 \text{ Mg}}{\text{m}^3} \times 0.30 \text{ m}^3 = 0.33 \text{ Mg}$$

giving

$$\text{wt \% glass} = \frac{1.77 \text{ Mg}}{(1.77 + 0.33) \text{ Mg}} \times 100 = 84.3\%$$

(b) The density will be given by

$$\rho = \frac{m}{V} = \frac{(1.77 + 0.33) \text{ Mg}}{\text{m}^3} = 2.10 \text{ Mg/m}^3$$

SAMPLE PROBLEM 14.2

Hemicellulose is an important component in wood. Calculate its molecular weight for a degree of polymerization of 200.

SOLUTION

The mer for cellulose is given in Table 13.1 and can be described in compact form as

$$C_6H_{10}O_5$$

We recall from Chapter 13 that the molecular weight of a polymer is simply the degree of polymerization times the molecular weight of its mer. Using data from Appendix 1, we can write

$$\text{mol wt} = (200)(6 \times 12.01 + 10 \times 1.008 + 5 \times 16.00) \text{ g/mol}$$

$$= 32,430 \text{ g/mol}$$

PRACTICE PROBLEM 14.1

In Sample Problem 14.1 we found the density of a typical fiberglass composite. Repeat the calculations for **(a)** 50 vol % and **(b)** 75 vol % E-glass fibers in an epoxy matrix.

PRACTICE PROBLEM 14.2

Calculate the molecular weight of a hemicellulose molecule with a degree of polymerization **(a)** $n = 150$ and **(b)** $n = 250$. (See Sample Problem 14.2.)

The Material World:
A Novel Composite for Synthetic Bone

Previous feature boxes in Chapters 12 and 13 introduced a ceramic (hydroxyapatite) and a polymer (collagen) as major components of natural bone. With this microscopic-scale combination (43 wt % hydroxyapatite and 36 wt % collagen), bone is also a good example of a natural composite material. The balance of the composition of bone is comprised of viscous liquids. Altogether, this composite can be considered one of nature's premier building materials.

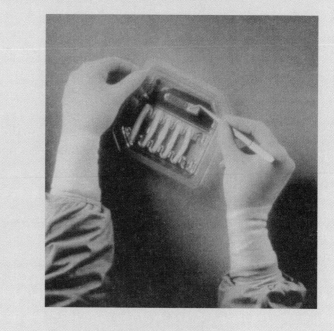

(Courtesy of Zimmer Corporation)

Bone's adequate mechanical properties combined with its ability to repair and remodel itself make animal skeletons impressive structural systems. Occasional trauma, however, extends bone beyond its ability for complete self-repair. Orthopaedic surgeons use the term large defects to describe centimeter-scale gaps in the skeletal system. Historically, such defects were repaired by harvesting bone from another part of the body (autogenous bone grafting, or autografts) or by using cadaver bone (allografts). Each of these surgeries presents problems. Autografts have significant morbidity and cost. Allografts have significant risk of immunologic reaction and disease transmission.

An example of an engineered composite to fill these large gaps is the product Collagraft (see photo). In this system, millimeter-scale ceramic particles are embedded in a matrix of collagen (see micrograph). Each ceramic particle is, in fact, a fine mixture of micrometer-scale grains of hydroxyapatite (HA) and tricalcium phosphate (TCP), a material chemically similar to HA but that reacts much more rapidly with the physiological environment. This combination is effective in that TCP is resorbed by the body in a few days, allowing rapid attachment of the implant to natural bone, while HA retains its structural integrity for a few months while new bone is growing into and filling the defect region. Collagen appears to facilitate new bone formation in a role similar to that in normal bone growth. The composite performance is maximized by adding bone marrow from the patient to the commercial material, thereby imitating the third component (viscous fluids) of natural bone.

Scanning electron micoscope image of the aggregate composite composed of ceramic (hydroxyapatite plus tricalcium phosphate) granules in a polymeric (collagen) matrix. The image shows collagen as a darker gray. (From J.P. McIntyre, J.F. Shackelford, M.W. Chapman, and R.R. Pool, Bull. Amer. Ceram. Soc. 70 1499 (1991))

14.2 AGGREGATE COMPOSITES

Fiberglass was a convenient and familiar example of fiber-reinforced composites. Similarly, **concrete** is an excellent example of an **aggregate composite,** one in which particles reinforce a matrix. Common concrete is rock and sand in a calcium aluminosilicate (cement) matrix. As with wood, this common construction material is used in staggering quantities. The weight of concrete used annually exceeds that of all metals combined.

For concrete, the **aggregate** is a combination of sand (fine aggregate) and gravel (coarse aggregate). This component of concrete is a natural material in the same sense as wood. Ordinarily, these materials are chosen for their relatively high density and strength. A table of aggregate compositions would be complex and largely meaningless. In general, aggregate materials are geological silicates chosen from locally available deposits. As such, they are complex and relatively impure examples of the crystalline silicates introduced in Section 12.1. Igneous rocks are common examples. *Igneous* means "solidified from a molten state." For quickly cooled igneous rocks, some fraction of the resulting material may be noncrystalline, corresponding to the glassy silicates introduced in Section 12.2. The relative particle sizes of sand and gravel are measured and controlled by passing these materials through standard screens, or sieves. The sizes of openings in the screen mesh are indicated in Table 14.5. A typical particle size distribution of both fine and coarse aggregate is illustrated in Figure 14–7. The reason for a combination of fine and coarse aggregate in a given concrete mix can be appreciated by inspection of Figure 14–8, which shows that space is more efficiently filled

Table 14.5 *Sizes of Openings in Standard Sieves*[a]

Sieve designation	Opening (mm)
Coarse Aggregate (Gravel)	
6	152.4
3	76.2
$1\frac{1}{2}$	38.1
$\frac{3}{4}$	19.1
$\frac{3}{8}$	9.5
Fine Aggregate (Sand)	
4	4.75
8	2.36
16	1.18
30	0.6
50	0.3
100	0.15

[a] The sieve designations for the coarse range correspond to the opening dimension in inches. For the fine range, the designation gives the number of openings per lineal inch.

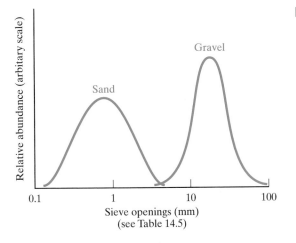

Figure 14-7 *Typical particle size distribution for aggregate in concrete. (Note the logarithmic scale for particle sizes screened through the sieve openings.)*

by a range of particle sizes. The fine particles fill the interstices between larger particles. The combination of fine and coarse aggregate accounts for 60–75% of the total volume of the final concrete.

The matrix that encloses the aggregate is **cement,** which, as the name implies, bonds the aggregate particles into a rigid solid. Modern concrete uses **portland cement,** (named for the Isle of Portland in England where a local limestone closely resembles the synthetic product) a calcium aluminosilicate. There are, in fact, five common types of portland cement, as shown in Table 14.6. They vary in the relative concentrations of four calcium-containing minerals. The resulting variations in character are noted in Table 14.6. The matrix is formed by the addition of water to the appropriate cement powder. The particle sizes for the cement powders are relatively small

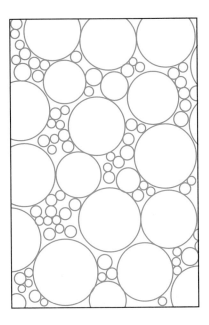

Figure 14-8 *Filling the volume of concrete with aggregate is aided by a wide particle size distribution. The smaller particles fill spaces between larger ones. This view is, of course, a two-dimensional schematic.*

Table 14.6 *Compositions of Portland Cements*

ASTM[a] Type	Characteristics	Composition[b] (wt %)				
		C_3S	C_2S	C_3A	C_4AF	Others[c]
I	Standard	45	27	11	8	9
II	Reduced heat of hydration and increased sulfate resistance	44	31	5	13	7
III	High early strength (coupled with high heat of hydration)	53	19	11	9	8
IV	Low heat of hydration (lower than II and especially good for massive structures)	28	49	4	12	7
V	Sulfate resistance (better than II and especially good for marine structures)	38	43	4	9	6

Source: Data from R. Nicholls, *Composite Construction Materials Handbook,* Prentice Hall, Inc., Englewood Cliffs, N.J., 1976.
[a] American Society for Testing and Materials.
[b] A shorthand notation is used in cement technology: $C_3S = 3CaO \cdot SiO_2$, $C_2S = 2CaO \cdot SiO_2$, $C_3A = 3CaO \cdot Al_2O_3$, $C_4AF = 4CaO \cdot Al_2O_3 \cdot Fe_2O_3$.
[c] Primarily simple oxides (MgO, CaO, alkali oxides) and $CaSO_4$.

compared to the finest of the aggregates. Variation in cement particle size can strongly affect the rate at which the cement hydrates. It is this hydration reaction that hardens the cement and produces the chemical bonding of the matrix to the aggregate particles. As you might expect from inspecting the complex compositions of portland cement (Table 14.6), the chemistry of the hydration process is equally complex. The principal hydration reactions and associated end products are shown in Table 14.7.

Concrete is frequently reinforced with steel bars (rebar) embedded prior to the setting of the cement. This technique is especially effective when the bars are held under a high tensile force until the concrete hardens. Subsequent release of the applied tensile force causes the rebar to contract and place the concrete under a residual, compressive stress. The brittle, ceramiclike concrete is more crack-resistant as a result. The **prestressed concrete** cannot be fractured until an applied tensile stress exceeds at least the magnitude of the residual, compressive stress. Prestressed concrete is routinely used in bridge construction.

In polymer technology, we noted several additives that provide certain desirable features to the end product. In cement technology, there are **admixtures,** which are additions that provide certain features. Any component of concrete other than aggregate, cement, or water is, by definition, an admixture (Table 14.8). One of the listed admixtures is an *air entrainer,* which reminds us that air can be thought of as a fourth component of concrete. Virtually all structural concrete contains some entrapped air. The air entrainer admixture increases the concentration of entrapped air bubbles, usually for the purpose of workability during forming and increased resistance to freeze–thaw cycles.

Table 14.7 *Principal Hydration Reactions[a] of Portland Cement*

(1)	$2C_3S + 6H \rightarrow 3Ch + C_3S_2H_3$ (tobermorite)
(2)	$2C_2S + 4H \rightarrow Ch + C_3S_2H_3$
(3)	$C_3A + 10H + CsH_2 \rightarrow C_3ACSH_{12}$ (calcium alumino monosulfate hydrate)
(4)	$C_3A + 12H + Ch \rightarrow C_3AChH_{12}$ (tetracalcium aluminate hydrate)
(5)	$C_4AF + 10H + 2Ch \rightarrow C_6AFH_{12}$ (calcium aluminoferrite hydrate)

Source: Data from R. Nicholls, *Composite Construction Materials Handbook,* Prentice Hall, Inc., Englewood Cliffs, N.J., 1976.
[a] The shorthand notation from cement technology is $C_3S = 3CaO \cdot SiO_2$, $C_2S = 2CaO \cdot SiO_2$, $C_3A = 3CaO \cdot Al_2O_3$, $C_4AF = 4CaO \cdot Al_2O_3 \cdot Fe_2O_3$, $H = H_2O$, $Ch = Ca(OH)_2$, $Cs = CaSO_4$.

While concrete is an important engineering material, a large number of other composite systems are based on particle reinforcement. Examples are listed in Table 14.9. As with Table 14.3, these include some of our most sophisticated engineering materials. Two groups of modern composites are identified in Table 14.9. **Particulate composites** refer specifically to systems in which the dispersed particles are relatively large (at least several micrometers in diameter) and are present in relatively high concentrations (greater than 25 vol % and frequently between 60 and 90 vol %). We earlier encountered a material system that can be included in this category—the polymers containing fillers in Section 13.5. Remember that automobile tires are a rubber with roughly one-third carbon black particles.

Table 14.8 *Admixtures*

Type	Characteristics	Example
Accelerators	Give early strength and curing	$CaCl_2$
Air entrainers	Reduce air–water interfacial tension to form entrapped air bubbles and increase workability and freeze–thaw durability	Sodium lauryl sulfate
Bonding admixtures	Bond fresh to hardened concrete	Fine iron particles plus chloride
Coloring agents	Provide surface colors	Inorganic pigments[a]
Expansion admixtures	Reduce shrinkage due to formation of an expanding rust	Fine iron particles plus chloride
Gas formers	React with hydroxides to produce H_2 bubbles and resulting porous (cellular) structure	Aluminum powder
Pozzolans	Silica reacts with free CaO to produce additional C_2S hydrate, which reduces the heat of hydration	Volcanic ash
Retarders	Retard curing and prevent bonding between hardened and fresh concrete	Lignosulfonate salts
Surface hardeners	Produce abrasion-resistant surface	Fused alumina particles
Water reducers	Increase workability	Lignosulfonate salts

Source: R. Nicholls, *Composite Construction Materials Handbook,* Prentice Hall, Inc., Englewood Cliffs, N.J., 1976.
[a] See Table 16.3 for typical coloring ions.

Table 14.9 *Aggregate Composite Systems Other Than Concrete*

Particulate composites
Thermoplastic elastomer (elastomer in thermoplastic polymer)
SiC in Al
W in Cu
Mo in Cu
WC in Co
W in NiFe
Dispersion-strengthened metals
Al_2O_3 in Al
Al_2O_3 in Cu
Al_2O_3 in Fe
ThO_2 in Ni

Source: Data from L. J. Broutman and R. H. Krock, Eds., *Modern Composite Materials,* Addison-Wesley Publishing Co., Inc., Reading, Mass., 1967, Chapters 16 and 17; and K. K. Chawla, New Mexico Institute of Mining and Technology.

A good example of a particulate composite is WC/Co, an excellent cutting tool material. A high-hardness carbide in a ductile metal matrix is an important example of a **cermet,** a ceramic–metal composite. The carbide is capable of cutting hardened steel but needs the toughness provided by the ductile matrix, which also prevents crack propagation that would be caused by particle-to-particle contact of the brittle carbide phase. As both the ceramic and metal phases are relatively refractory, they can both withstand the high temperatures generated by the machining process. The **dispersion-strengthened metals** contain fairly small concentrations (less than 15 vol %) of small-diameter oxide particles (0.01 to 0.1 μm in diameter). The oxide particles strengthen the metal by serving as obstacles to dislocation motion. This can be appreciated from the discussion in Section 6.3 and Figure 6–25. In Section 14.4, we will find that a 10 vol % dispersion of Al_2O_3 in aluminum can increase tensile strength by as much as a factor of 4.

SAMPLE PROBLEM 14.3

We have referred to portland cement as a calcium aluminosilicate. Calculate the total weight percent of $CaO + Al_2O_3 + SiO_2$ in type I portland cement. (Ignore any pure oxides such as CaO that might contribute to the "Others" column of Table 14.6.)

SOLUTION

If we consider 100 kg of type I cement as described in Table 14.6, it will contain 45 kg of C_3S, 27 kg of C_2S, 11 kg of C_3A, and 8 kg of C_4AF, where the notation of cement technology is used (C = CaO, etc.).

Using data from Appendix 1, we can determine the weight fractions from each compound:

$$\text{wt frac CaO in } C_3S = \frac{3(\text{mol wt CaO})}{3(\text{mol wt CaO}) + (\text{mol wt SiO}_2)}$$

$$= \frac{3(40.08 + 16.00)}{3(40.08 + 16.00) + (28.09 + 2 \times 16.00)}$$

$$= 0.737$$

$$\text{wt frac SiO}_2 \text{ in } C_3S = 1.000 - 0.737 = 0.263$$

Similarly,

$$\text{wt frac CaO in } C_2S = \frac{2(40.08 + 16.00)}{2(40.08 + 16.00) + (28.09 + 2 \times 16.00)}$$

$$= 0.651$$

$$\text{wt frac SiO}_2 \text{ in } C_2S = 1.000 - 0.651 = 0.349$$

$$\text{wt frac CaO in } C_3A = \frac{3(40.08 + 16.00)}{3(40.08 + 16.00) + (2 \times 26.98 + 3 \times 16.00)}$$

$$= 0.623$$

$$\text{wt frac Al}_2O_3 \text{ in } C_3A = 1.000 - 0.623 = 0.377$$

$$\text{wt frac CaO in } C_4AF = \frac{4(40.08 + 16.00)}{4(40.08 + 16.00) + (2 \times 26.98 + 3 \times 16.00)}$$
$$+ (2 \times 55.85 + 3 \times 16.00)$$

$$= 0.462$$

$$\text{wt frac Al}_2O_3 \text{ in } C_4AF = \frac{2 \times 26.98 + 3 \times 16.00}{4(40.08 + 16.00) + (2 \times 26.98 + 3 \times 16.00)}$$
$$+ (2 \times 55.85 + 3 \times 16.00)$$

$$= 0.210$$

Total mass of CaO:

$$m_{CaO} = (x_{CaO/C_3S})(m_{C_3S}) + (x_{CaO/C_2S})(m_{C_2S})$$
$$+ (x_{CaO/C_3A})(m_{C_3A}) + (x_{CaO/C_4AF})(m_{C_4AF})$$
$$= (0.737)(45 \text{ kg}) + (0.651)(27 \text{ kg}) + (0.623)(11 \text{ kg})$$
$$+ (0.462)(8 \text{ kg}) = 61.3 \text{ kg}$$

Similarly,

$$m_{Al_2O_3} = (0.377)(11 \text{ kg}) + (0.210)(8 \text{ kg}) = 5.8 \text{ kg}$$

$$m_{SiO_2} = (0.263)(45 \text{ kg}) + (0.349)(27 \text{ kg}) = 21.3 \text{ kg}$$

Because we have dealt with 100 kg of cement, these masses are numerically equal to weight percents, giving

$$\text{total wt } \% \ (CaO + Al_2O_3 + SiO_2) = (61.3 + 5.8 + 21.3)\%$$

$$= 88.4\%$$

Note. Exclusive of any single oxides, type I portland cement is roughly 90% by weight a calcium aluminosilicate.

SAMPLE PROBLEM 14.4

A dispersion-strengthened aluminum contains 10 vol % Al_2O_3. Assuming that the metal phase is essentially pure aluminum, calculate the density of the composite. (The density of Al_2O_3 is 3.97 Mg/m^3.)

SOLUTION

From Appendix 1, we see that

$$\rho_{Al} = 2.70 \text{ Mg/m}^3$$

For 1 m^3 of composite, we shall have 0.1 m^3 of Al_2O_3 and $1.0 - 0.1 = 0.9 \text{ m}^3$ of Al. The mass of each component will be

$$m_{Al_2O_3} = 3.97 \ \frac{Mg}{m^3} \times 0.1 \text{ m}^3 = 0.40 \text{ Mg}$$

$$m_{Al} = 2.70 \ \frac{Mg}{m^3} \times 0.9 \text{ m}^3 = 2.43 \text{ Mg}$$

giving

$$\rho_{composite} = \frac{m}{V} = \frac{(0.40 + 2.43) \text{ Mg}}{1 \text{ m}^3}$$

$$= 2.83 \text{ Mg/m}^3$$

PRACTICE PROBLEM 14.3

Calculate the weight percent of $CaO + Al_2O_3 + SiO_2$ in type III portland cement. (See Sample Problem 14.3.)

PRACTICE PROBLEM 14.4

Calculate the density of a particulate composite containing 50 vol % W particles in a copper matrix. (See Sample Problem 14.4.)

14.3   PROPERTY AVERAGING

It is obvious that the properties of composites must, in some way, represent an average of the properties of their individual components. However, the precise nature of the "average" is a sensitive function of microstructural geometry. Because of the wide variety of such geometries in modern composites, we must be cautious of generalities. But we will identify a few of the important examples.

Figure 14–9 illustrates three idealized geometries: (a) a direction parallel to continuous fibers in a matrix (phases in parallel), (b) a direction perpendicular to the direction of the continuous fibers (phases in series), and (c) a direction relative to a uniformly dispersed aggregate composite. The first two cases represent extremes in the highly anisotropic nature of fibrous composites such as fiberglass and wood. The third case represents an idealized model of the relatively **isotropic** nature of concrete (i.e., its properties tend not to vary with direction). We shall now consider these cases individually. Each time, we will use the modulus of elasticity to illustrate how a property is averaged. This is consistent with our emphasis on the structural applications of composites.

LOADING PARALLEL TO REINFORCING FIBERS—ISOSTRAIN

The uniaxial stressing of the geometry of Figure 14–9a is shown in Figure 14–10. If the matrix is intimately bonded to the reinforcing fibers, the strain of both the matrix and the fibers must be the same. This **isostrain** condition is true even though the elastic moduli of the components will tend to be quite different. In other words,

$$\epsilon_c = \frac{\sigma_c}{E_c} = \epsilon_m = \frac{\sigma_m}{E_m} = \epsilon_f = \frac{\sigma_f}{E_f}, \qquad (14.2)$$

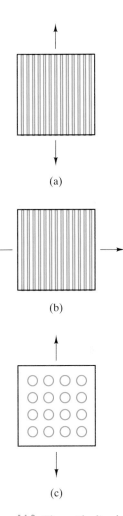

(a)

(b)

(c)

Figure 14-9 *Three idealized composite geometries: (a) a direction parallel to continuous fibers in a matrix, (b) a direction perpendicular to continuous fibers in a matrix, and (c) a direction relative to a uniformly dispersed aggregate composite.*

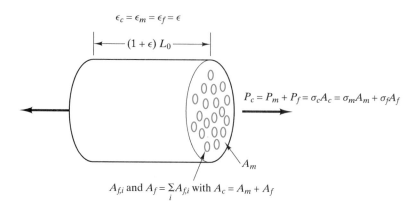

$$\epsilon_c = \epsilon_m = \epsilon_f = \epsilon$$

$$P_c = P_m + P_f = \sigma_c A_c = \sigma_m A_m + \sigma_f A_f$$

A_m

$A_{f,i}$ and $A_f = \sum_i A_{f,i}$ with $A_c = A_m + A_f$

Figure 14-10 *Uniaxial stressing of a composite with continuous fiber rein-forcement. The load is parallel to the reinforcing fibers. The terms in Equations 14.2 to 14.4 are illustrated.*

where all terms are defined in Figure 14–10. It is also apparent that the load carried by the composite, P_c, is the simple sum of loads carried by each component:

$$P_c = P_m + P_f \tag{14.3}$$

Each load is equal, by definition, to a stress times an area; that is,

$$\sigma_c A_c = \sigma_m A_m + \sigma_f A_f \tag{14.4}$$

where, again, terms are illustrated in Figure 14–10. Combining Equations 14.2 and 14.4 gives

$$E_c \epsilon_c A_c = E_m \epsilon_m A_m + E_f \epsilon_f A_f \tag{14.5}$$

Let us note that $\epsilon_c = \epsilon_m = \epsilon_f$ and divide both sides of Equation 14.4 by A_c:

$$E_c = E_m \frac{A_m}{A_c} + E_f \frac{A_f}{A_c} \tag{14.6}$$

Because of the cylindrical geometry of Figure 14–10, the area fraction is also the volume fraction, or

$$E_c = v_m E_m + v_f E_f \tag{14.7}$$

where v_m and v_f are the volume fractions of matrix and fibers, respectively. In this case, of course, $v_m + v_f$ must equal 1. Equation 14.7 is an important result. It identifies the modulus of a fibrous composite loaded axially as a simple, weighted average of the moduli of its components. Figure 14–11 shows the modulus as the slope of a stress–strain curve for a composite with 70 vol % reinforcing fibers. In this typical fiberglass (E-glass-reinforced

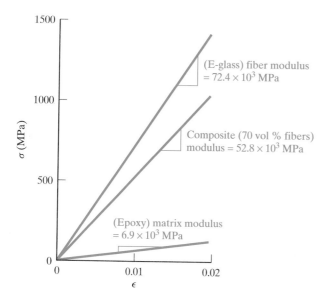

Figure 14-11   *Simple stress–strain plots for a composite and its fiber and matrix components. The slope of each plot gives the modulus of elasticity. The composite modulus is given by Equation 14.7.*

epoxy), the glass fiber modulus (72.4×10^3 MPa) is roughly 10 times that of the polymeric matrix modulus (6.9×10^3 MPa). The composite modulus, although not equal to that for glass, is substantially higher than that for the matrix.

Equally significant to the relative contribution of the glass fibers to the composite modulus is the fraction of the total composite load, P_c, in Equation 14.3, carried by the axially loaded fibers. From Equation 14.3 we note that

$$\frac{P_f}{P_c} = \frac{\sigma_f A_f}{\sigma_c A_c} = \frac{E_f \epsilon_f A_f}{E_c \epsilon_c A_c} = \frac{E_f}{E_c} v_f \tag{14.8}$$

For the fiberglass example under discussion, $P_f/P_c = 0.96$; that is, nearly the entire uniaxial load is carried by the 70 vol % of high-modulus fibers. This geometry is an ideal application of a composite. The high modulus and strength of the fibers are effectively transmitted to the composite as a whole. At the same time, the ductility of the matrix is available to produce a substantially less brittle material than glass by itself.

The result of Equation 14.7 is not unique to the modulus of elasticity. A number of important properties exhibit this behavior. This is especially true of transport properties. In general, we can write

$$X_c = v_m X_m + v_f X_f \tag{14.9}$$

where X can be diffusivity, D (see Section 4.3); thermal conductivity, k (see Section 7.3); or electrical conductivity, σ (see Section 15.1). The Poisson's ratio for loading parallel to reinforcing fibers can also be predicted from Equation 14.9.

SAMPLE PROBLEM 14.5

Calculate the composite modulus for polyester reinforced with 60 vol % E-glass under isostrain conditions.

SOLUTION

This is a direct application of Equation 14.7:

$$E_c = v_m E_m + v_f E_f$$

Data for elastic moduli of the components can be found in Tables 14.10 and 14.11:

$$E_{\text{polyester}} = 6.9 \times 10^3 \text{MPa}$$

$$E_{\text{E-glass}} = 72.4 \times 10^3 \text{MPa}$$

giving

$$E_c = (0.4)(6.9 \times 10^3 \text{MPa}) + (0.6)(72.4 \times 10^3 \text{MPa})$$

$$= 46.2 \times 10^3 \text{MPa}$$

SAMPLE PROBLEM 14.6

Calculate the thermal conductivity parallel to continuous, reinforcing fibers in a composite with 60 vol % E-glass in a matrix of polyester. [The thermal conductivity of E-glass is 0.97 W/(m · K) and for polyester is 0.17 W/(m · K). Both of these values are for room temperature.]

SOLUTION

This is the analog of our calculation in Sample Problem 14.5. We use Equation 14.9, with X being the thermal conductivity k:

$$k_c = v_m k_m + v_f k_f$$

$$= (0.4)[0.17 \text{W}/(\text{m} \cdot \text{K})] + (0.6)[0.97 \text{W}/(\text{m} \cdot \text{K})]$$

$$= 0.65 \text{W}/(\text{m} \cdot \text{K})$$

PRACTICE PROBLEM 14.5

Calculate the composite modulus for a composite with 50 vol % E-glass in a polyester matrix. (See Sample Problem 14.5.)

PRACTICE PROBLEM 14.6

The thermal conductivity of a particular fiberglass composite is calculated in Sample Problem 14.6. Repeat this calculation for a composite with 50 vol % E-glass in a polyester matrix.

LOADING PERPENDICULAR TO REINFORCING FIBERS— ISOSTRESS

The isostress condition with perpendicular loading of the reinforcing fibers produces a substantially different result from the isostrain condition with parallel loading. This **isostress** condition is defined by

$$\sigma_c = \sigma_m = \sigma_f \qquad (14.10)$$

This loading condition can be represented by a simple model illustrated in Figure 14–12. In this case, the total elongation of the composite in the direction of stress application (ΔL_c) is the sum of the elongations of matrix and fiber components:

$$\Delta L_c = \Delta L_m + \Delta L_f \qquad (14.11)$$

Dividing by total composite length (L_c) in the stress direction gives

$$\frac{\Delta L_c}{L_c} = \frac{\Delta L_m}{L_c} + \frac{\Delta L_f}{L_c} \qquad (14.12)$$

Because of the geometry of Figure 14–12, the length of each component in the stress direction is proportional to its area fraction, that is,

$$L_m = A_m L_c \qquad (14.13)$$

and

$$L_f = A_f L_c \qquad (14.14)$$

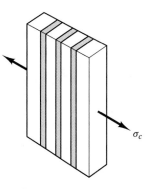

Figure 14-12 *Uniaxial loading of a composite perpendicular to the fiber reinforcement can be simply represented by this slab.*

(Some caution is required here. The length [and strain] measurement is now referred to the direction of stress application. However, the area terms still refer to the cross-sectional area as previously defined in Figure 14–10.) Combining Equations 14.13 and 14.14 with Equation 14.12 gives

$$\frac{\Delta L_c}{L_c} = \frac{A_m \Delta L_m}{L_m} + \frac{A_f \Delta L_f}{L_f} \qquad (14.15)$$

Noting the definition of strain as $\epsilon = \Delta L / L$ and that the area fraction for a component in Figure 14–12 is equal to its volume fraction, we can rewrite Equation 14.15 as

$$\epsilon_c = v_m \epsilon_m + v_f \epsilon_f \qquad (14.16)$$

Note the similarity (and difference) of this result to the stress equation for isostrain loading (Equation 14.4).

Since the isostress condition requires that $\sigma = E_c \epsilon_c = E_m \epsilon_m = E_f \epsilon_f$, we can rewrite Equation 14.16 as

$$\frac{\sigma}{E_c} = v_m \frac{\sigma}{E_m} + v_f \frac{\sigma}{E_f} \qquad (14.17)$$

Dividing out the σ term leaves

$$\frac{1}{E_c} = \frac{v_m}{E_m} + \frac{v_f}{E_f} \qquad (14.18)$$

which can be rearranged to our final result for loading perpendicular to fibers:

$$E_c = \frac{E_m E_f}{v_m E_f + v_f E_m} \qquad (14.19)$$

This result can be contrasted to Equation 14.7, which gave the composite modulus for isostrain loading. Students familiar with elementary electric circuit theory will note that the forms of Equations 14.7 and 14.19 are the reverse of those for the resistance (or resistivity) equations for parallel and series circuits. (See comment following Equation 14.20.)

The practical consequence of Equation 14.19 is a less effective use of the high modulus of the reinforcing fibers. This is illustrated by Figure 14–13, which shows that the matrix modulus dominates the composite modulus except with very high concentrations of fibers. For the example considered in the isostrain section ($v_f = 0.7$), the composite modulus under isostress loading is 18.8×10^3 MPa, substantially less than the 52.8×10^3 MPa value under isostrain.

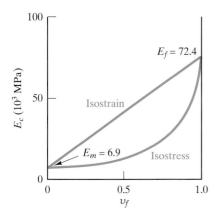

Figure 14-13 *The composite modulus, E_c, is a weighted average of the moduli of its components (E_m = matrix modulus and E_f = fiber modulus). For the isostrain case of parallel loading (Equation 14.7), the fibers make a greater contribution to E_c than for isostress (perpendicular) loading (Equation 14.19). The plot is for the specific case of E-glass-reinforced epoxy (see Figure 14–11).*

As with the similarity between Equations 14.7 and 14.9, we find that the modulus is not the only property following a form such as Equation 14.19. In general, we can write

$$X_c = \frac{X_m X_f}{v_m X_f + v_f X_m} \tag{14.20}$$

where, again, X can be diffusivity, D, thermal conductivity, k, or electrical conductivity, σ. (Note that the series equation for electrical resistivity, ρ, is of the form of Equations 14.7 and 14.9 because $\rho = 1/\sigma$.)

SAMPLE PROBLEM 14.7

Calculate the elastic modulus and thermal conductivity perpendicular to continuous, reinforcing fibers in an E-glass (60 vol %)/polyester composite.

SOLUTION

This is the isostress problem, comparable to the isostrain problem treated in Sample Problems 14.5 and 14.6. For both parameters (E and k) a series equation holds. For the elastic modulus, this is Equation 14.19:

$$E_c = \frac{E_m E_f}{v_m E_f + v_f E_m}$$

Using the data collected for Sample Problem 14.5 yields

$$E_c = \frac{(6.9 \times 10^3 \text{MPa})(72.4 \times 10^3 \text{MPa})}{(0.4)(72.4 \times 10^3 \text{MPa}) + (0.6)(6.9 \times 10^3 \text{MPa})}$$

$$= 15.1 \times 10^3 \text{MPa}$$

For thermal conductivity, we apply Equation 14.20 and the data given in Sample Problem 14.6:

$$k_c = \frac{k_m k_f}{v_m k_f + v_f k_m}$$

$$= \frac{[0.17 \text{W}/(\text{m} \cdot \text{K})][0.97 \text{W}/(\text{m} \cdot \text{K})]}{(0.4)[0.97 \text{W}/(\text{m} \cdot \text{K})] + (0.6)[0.17 \text{W}/(\text{m} \cdot \text{K})]}$$

$$= 0.34 \text{W}/(\text{m} \cdot \text{K})$$

....

PRACTICE PROBLEM 14.7

Calculate the elastic modulus and thermal conductivity perpendicular to continuous reinforcing fibers for a composite with 50 vol % E-glass in a polyester matrix. (See Sample Problem 14.7.)

LOADING A UNIFORMLY DISPERSED AGGREGATE COMPOSITE

The uniaxial stressing of the isotropic geometry of Figure 14–9c is shown in Figure 14–14. A rigorous treatment of this system can become quite complex depending on the specific nature of the dispersed and continuous phases. Fortunately, the results of the two previous cases (isostrain and isostress

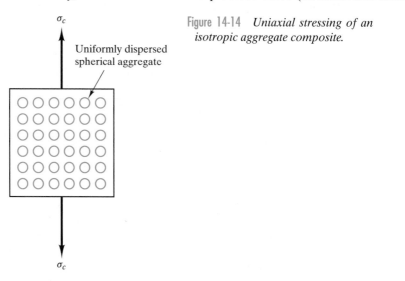

Figure 14-14   *Uniaxial stressing of an isotropic aggregate composite.*

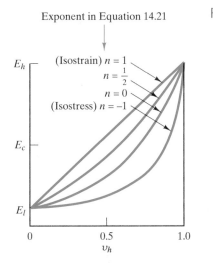

Figure 14-15 *The dependence of composite modulus, E_c, on the volume fraction of a high-modulus phase, v_h, for an aggregate composite is generally between the extremes of isostrain and isostress conditions. Two simple examples are given by Equation 14.21 for $n = 0$ and $\frac{1}{2}$. Decreasing n from a $+1$ to -1 represents a trend from a relatively low-modulus aggregate in a relatively high-modulus matrix to the reverse case of a high-modulus aggregate in a low-modulus matrix.*

fiber composites) serve as upper and lower bounds for the aggregate case. A simple approximation to the results for aggregate composites is given by noting that Equations 14.7 and 14.19 can both be written in the general form

$$E_c^n = v_l E_l^n + v_h E_h^n \qquad (14.21)$$

where the subscript l refers to the low-modulus phase and h refers to the high-modulus phase (fibers in the previous examples); n is 1 for the isostrain case (Equation 14.7) and -1 for the isostress case (Equation 14.19). For a first approximation, a higher-modulus aggregate in a lower-modulus matrix can be represented by $n = 0$ in Equation 14.21. Similarly, a lower-modulus aggregate in a higher-modulus matrix can be represented by $n = \frac{1}{2}$. Figure 14–15 summarizes these cases. The isostrain case ($n = 1$) for dispersed aggregates can be pictured as applying to an extreme case of rubber balls in a steel matrix and isostress ($n = -1$) applying to steel balls in a rubber matrix. This requires some care in using Figure 14–15 in that the high-modulus phase can be the aggregate or the matrix, depending on the nature of the stress state. For normal concrete, the modulus of aggregate is only slightly greater than for the cement matrix. Once again, the modulus results can be applied to the various conductivity parameters (D, k, and σ).

SAMPLE PROBLEM 14.8

The general expression for the modulus of an aggregate composite was given by Equation 14.21. The composite modulus is 366×10^3 MPa for 50 vol % WC aggregate in a Co matrix. The modulus for the WC phase is 704×10^3 MPa, and that for Co is 207×10^3 MPa. Calculate the value of n for this composite to be used in a cutting tool.

SOLUTION

The value of n that will satisfy Equation 14.21,

$$E_c^n = v_l E_l^n + v_h E_h^n$$

can be evaluated by trial and error. Using the given data (in units of 10^3 MPa), we can write

$$(366)^n = 0.5(207)^n + 0.5(704)^n$$

In the text we see that typical values of n include $+1$, $+\frac{1}{2}$, 0, and -1. In fact, the case of a higher-modulus aggregate is typically represented by $n = 0$. However, using $n = 0$ in Equation 14.21 gives the trivial result $1 = 1$. To overcome this, we can consider n values approaching zero, such as $n = \pm 0.01$.

Now, let us set up a table testing the equality in Equation 14.21:

n	$(366)^n = A$	$0.5(207)^n + 0.5(704)^n = B$	B/A
$+1$	366	455.5	1.24
$+\frac{1}{2}$	19.1	20.5	1.07
$+0.01$	1.06	1.06	1.00
-0.01	0.943	0.942	0.999
-1	2.73×10^{-3}	3.13×10^{-3}	1.15

A plot of B/A clearly shows that it approaches 1 (i.e., n solves Equation 14.21) as n approaches zero:

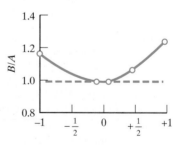

Therefore,

$$n \simeq 0$$

Therefore,

$$n \simeq 0$$

PRACTICE PROBLEM 14.8

> In Sample Problem 14.8 the case of a modulus equation with $n = 0$ is treated. Estimate the composite modulus for a reciprocal case in which 50 vol % Co aggregate is dispersed in a WC matrix. For this case, the value of n can be taken as $\frac{1}{2}$.

INTERFACIAL STRENGTH

The averaging of properties in a useful composite material can be represented by the typical examples just discussed. But before we leave this section, we must note an important consideration so far taken for granted. That is, the interface between the matrix and discontinuous phase must be strong enough to transmit the stress or strain due to a mechanical load from one phase to the other. Without this strength, the dispersed phase can fail to "communicate" with the matrix. Rather than have the "best of both worlds" as implied in the introduction to this chapter, we may obtain the worst behavior of each component. Reinforcing fibers easily slipping out of a matrix can be an example. Figure 14–16 illustrates the contrasting microstructures of (a) poorly bonded and (b) well-bonded interfaces in a fiberglass composite. Substantial effort has been devoted to controlling interfacial strength. Surface treatment, chemistry, and temperature are a few considerations in the "art and science" of interfacial bonding. To summarize, some interfacial strength is required in all composites to ensure that property averaging is available at relatively low stress levels.

Under isostrain conditions, the axial loading of a reinforcing fiber of finite length leads to a constant shear stress, τ, along the fiber surface, which, in turn, leads to the buildup of tensile stress near the ends of the fiber. Figure 14–17 shows how the tensile stress, σ, varies along the fiber. A "long" fiber (Figure 14–17b) is one with length greater than a critical length, l_c, in which the middle of the fiber reaches a maximum, constant value corresponding to fiber failure. For maximum efficiency in reinforcement, the fiber length should be much greater than l_c to ensure that the average tensile strength in the fiber is near σ_{critical}.

Two fundamentally different philosophies are applied relative to the behavior of fiber composites at relatively high stress levels. For polymer–matrix and metal–matrix composites, failure originates in or along the reinforcing fibers. As a result, a high interfacial strength is desirable to maximize the overall composite strength (Figure 14–16). In ceramic–matrix composites, failure generally originates in the matrix phase. To maximize the fracture toughness for these materials, it is desirable to have a relatively weak interfacial bond allowing fibers to pull out. As a result, a crack initiated in the matrix is deflected along the fiber–matrix interface. This increased crack path length significantly improves fracture toughness. The mechanism of fiber pull-out for improving fracture toughness is shown in Figure 14–18 and can be compared with the two mechanisms for unreinforced ceramics illustrated in Figure 8–7. In general, ceramic composites have achieved substantially higher fracture toughness levels than the unreinforced ceramics, with values between 20 and 30 MPa $\sqrt{\text{m}}$ being common.

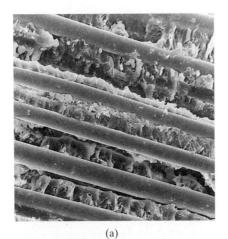

(a)

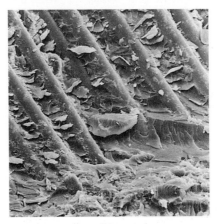

(b)

Figure 14-16 *The utility of a reinforcing phase in this polymer–matrix composite depends on the strength of the interfacial bond between the reinforcement and the matrix. These scanning electron micrographs contrast (a) poor bonding with (b) a well-bonded interface. In metal–matrix composites, high interfacial strength is also desirable to ensure high overall composite strength. (Courtesy of Owens-Corning Fiberglas Corporation)*

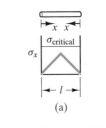

(a)

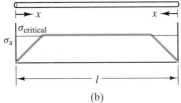

(b)

Figure 14-17 *(a) Plot of the tensile stress along a "short" fiber in which the build-up of stress near the fiber ends never exceeds the critical stress associated with fiber failure. (b) A similar plot for the case of a "long" fiber in which the stress in the middle of the fiber reaches the critical value.*

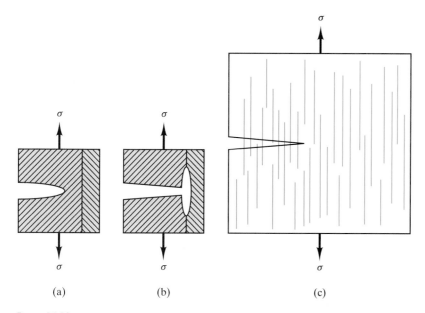

(a)   (b)   (c)

Figure 14-18 *For ceramic–matrix composites, low interfacial strength is desirable (in contrast to the case for ductile–matrix composites, such as in Figure 14–16). We see that (a) a matrix crack approaching a fiber is (b) deflected along the fiber–matrix interface. For the overall composite (c), the increased crack path length due to fiber pull-out significantly improves fracture toughness. (Two toughening mechanisms for unreinforced ceramics are illustrated in Figure 8–7.)*

14.4   MECHANICAL PROPERTIES OF COMPOSITES

It is obvious from Section 14.3 that citing a single number for a given mechanical property of a given composite material is potentially misleading. The concentration and geometry of the discontinuous phase play an important role. Unless otherwise stated, one can assume that composite properties cited in this chapter correspond to optimal conditions (e.g., loading parallel to reinforcing fibers). It is also useful to have information about the component materials separate from the composite. Table 14.10 gives some key mechanical properties for some of the common matrix materials. Table 14.11 gives a similar list for some common dispersed-phase materials. Table 14.12 gives properties for various composite systems. In the preceding tables, the mechanical properties listed are those first defined in Section 6.1 for metals. Comparing Table 14.12 with the data for metals, ceramics and glasses, and polymers in Tables 14.10 and 14.11 will give us some appreciation for the relative mechanical behavior of composites. The dramatic improvement in fracture toughness for ceramic–matrix composites compared with unreinforced ceramics is illustrated by Figure 14–19. A more dramatic comparison, which emphasizes the importance of composites in the aerospace field, is given by the list of **specific strength** (strength/density) values given in

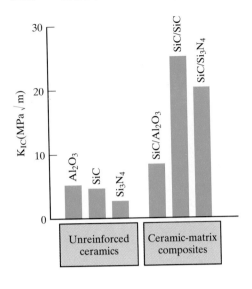

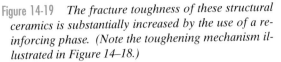

Figure 14-19 *The fracture toughness of these structural ceramics is substantially increased by the use of a reinforcing phase. (Note the toughening mechanism illustrated in Figure 14–18.)*

Table 14.13. The specific strength is sometimes referred to as the **strength-to-weight ratio.** The key point is that the substantial cost associated with many "advanced composites" is justified not so much by their absolute strength but by the fact that they can provide adequate strength in some very low-density configurations. The savings in fuel costs alone can frequently justify the higher material costs. Figure 14–20 illustrates (with the data of Table 14.13) the distinct advantage of advanced composite systems in this regard. Finally, note that the higher material costs of advanced composites can be offset by reduced assembly costs (e.g., one-piece automobile frames), as well as by high specific strength values.

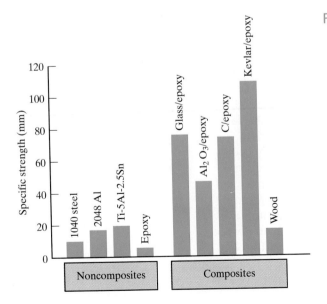

Figure 14-20 *A bar graph plot of the data of Table 14.13 illustrates the substantial increase in specific strength possible with composites.*

Table 14.10 *Mechanical Properties of Common Matrix Materials*

Class	Example	E [MPa (ksi)]	T.S. [MPa (ksi)]	Flexural strength [MPa (ksi)]	Compressive strength (After 28 days) [MPa (ksi)]	Percent elongation at failure	K_{IC} (MPa $\sqrt{m}$)
Polymer[a]							
	Epoxy	6900 (1000)	69 (10)	—	—	0	0.3–0.5
	Polyester	6900 (1000)	28 (4)	—	—	0	—
Metal[b]							
	Al	$69 \times 10^3 (10 \times 10^3)$	76 (11)	—	—	—	—
	Cu	$115 \times 10^3 (17 \times 10^3)$	170 (25)	—	—	—	—
Ceramic[c]							
	Al_2O_3	—	—	550 (80)	—	—	4-5
	SiC	—	—	500 (73)	—	—	4.0
	Si_3N_4 (reaction bonded)	—	—	260 (38)	—	—	2-3
Portland cements[d]							
	Type I	—	2.4 (0.35)	—	24 (3.5)	—	—
	Type II	—	2.3 (0.33)	—	24 (3.5)	—	—
	Type III	—	2.6 (0.38)	—	21 (3.0)	—	—
	Type IV	—	2.1 (0.30)	—	14 (2.0)	—	—
	Type V	—	2.3 (0.33)	—	21 (3.0)	—	—

[a] From Tables 6.8 and 8.3.
[b] For high-purity alloys with no significant cold working from *Metals Handbook*, 9th ed., Vol. 2. American Society for Metals, Metals Park, Ohio, 1979.
[c] Source: Data from A. J. Klein, *Advanced Materials and Processes*, 2, 26 (1986).
[d] Source: Data from R. Nicholls, *Composite Construction Materials Handbook*, Prentice Hall, Inc., Englewood Cliffs, N.J., 1976.

Table 14.11 *Mechanical Properties of Common dispersed-Phase Materials*

Group	Dispersed phase	E [MPa (ksi)]	T.S. [MPa (ksi)]	Compressive strength [MPa (ksi)]	Percent elongation at failure
Glass fiber[a]	C-glass	$69 \times 10^3 (10 \times 10^3)$	3100 (450)	—	4.5
	E-glass	$72.4 \times 10^3 (10.5 \times 10^3)$	3400 (500)	—	4.8
	S-glass	$85.5 \times 10^3 (12.4 \times 10^3)$	4800 (700)	—	5.6
Ceramic fiber[a]	C (graphite)	$340 - 380 \times 10^3 (49 - 55 \times 10^3)$	2200–2400 (320–350)	—	—
	SiC	$430 \times 10^3 (62 \times 10^3)$	2400 (350)	—	—
Ceramic whisker[a]	Al_2O_3	$430 \times 10^3 (62 \times 10^3)$	$21 \times 10^3 (3000)$	—	—
Polymer fiber[b]	Kevlar[c]	$131 \times 10^3 (19 \times 10^3)$	3800 (550)	—	2.8
Metal filament[a]	Boron	$410 \times 10^3 (60 \times 10^3)$	3400 (500)	—	—
Concrete aggregate[d]	Crushed stone and sand	$34 - 69 \times 10^3 (5 - 10 \times 10^3)$	1.4–14 (0.2–2)	69–340 (10–50)	—

[a] Source: Data from L. J. Broutman and R. H. Krock, Eds., *Modern Composite Materials*, Addison-Wesley Publishing Co., Inc., Reading, Mass., 1967.

[b] Source: Data from A. K. Dhingra, Du Pont Company.

[c] Trade name, Du Pont.

[d] Source: Data from R. Nicholls, *Composite Construction Materials Handbook*, Prentice Hall, Inc., Englewood Cliffs, N.J., 1976.

Table 14.12 *Mechanical Properties of Common Composite Systems*

Class	E [MPa (ksi)]	T.S. [MPa (ksi)]	Flexural strength [MPa (ksi)]	Compressive strength [MPa (ksi)]	Percent elongation at failure	K_{IC}^a (MPa $\sqrt{m}$)
Polymer–matrix						
E-glass (73.3 vol %) in epoxy (parallel loading of continuous fibers)[b]	56×10^3 (8.1×10^3)	1640 (238)	—	—	2.9	42–60
Al$_2$O$_3$ whiskers (14 vol %) in epoxy[b]	41×10^3 (6×10^3)	779 (113)	—	—	—	—
C (67 vol %) in epoxy (parallel loading)[c]	221×10^3 (32×10^3)	1206 (175)	—	—	—	—
Kevlar[d] (82 vol %) in epoxy (parallel loading)[c]	86×10^3 (12×10^3)	1517 (220)	—	—	—	—
B (70 vol %) in epoxy (parallel loading of continuous filaments)[b]	$210 - 280 \times 10^3$ ($30 - 40 \times 10^3$)[c]	1400 – 2100 (200–300)[c]	—	—	—	46
Metal matrix						
Al$_2$O$_3$ (10 vol %) dispersion-strengthened aluminum[b]	—	330 (48)	—	—	—	—
W (50 vol %) in copper (parallel loading of continuous filaments)[b]	260×10^3 (38×10^3)	1100 (160)	—	—	—	—
W particles (50 vol %) in copper[b]	190×10^3 (27×10^3)	380 (55)	—	—	—	—
Ceramic–matrix						
SiC whiskers in Al$_2$O$_3$[e]			800 (116)			8.7
SiC fibers in SiC[e]			750 (109)	—	—	25.0
SiC whiskers in reaction-bonded Si$_3$N$_4$[e]			900 (131)	—	—	20.0
Wood						
Douglas fir, kiln-dried at 12% moisture (loaded parallel to grain)[d]	13.4×10^3 (1.95×10^3)[f]	85.5 (12.4)[f]	—	49.9 (7.24)	—	11–13
Douglas fir, kiln-dried at 12% moisture (loaded perpendicular to grain)[d]	—			5.5 (0.80)		0.5–1
Concrete						
Standard concrete, water/cement ratio of 4 (after 28 days)[g]	—	—	—	41 (6.0)	—	0.2
Standard concrete, water/cement ratio of 4 (after 28 days) with air entrainer[g]	—	—	—	33 (4.8)	—	—

[a] *Source:* Data from M. F. Ashby and D. R. H. Jones, *Engineering Materials—An Introduction to Their Properties and Applications*, Pergamon Press, Inc., Elmsford, N.Y., 1980.
[b] L. J. Broutman and R. H. Krock, Eds., *Modern Composite Materials*, Addison-Wesley Publishing Co., Inc., Reading, Mass. 1967.
[c] A. K. Dhingra, Du Pont Company.
[d] Trade name, Du Pont.
[e] A. J. Klein. *Advanced Materials and Processes*, 2, 26 (1986).
[f] Measurement in bending. See Figure 6–14.
[g] R. Nicholls, *Composite Construction Materials Handbook*, Prentice Hall, Englewood Cliffs, N.J., 1976.

Table 14.13 *Specific Strengths (Strength/Density)*

Group	Material	Specific strength [mm (in.)]
Noncomposites	1040 steel[a]	$9.9 \times 10^6 (0.39 \times 10^6)$
	2048 plate aluminum[a]	$16.9 \times 10^6 (0.67 \times 10^6)$
	Ti–5Al–2.5Sn[a]	$19.7 \times 10^6 (0.78 \times 10^6)$
	Epoxy[b]	$6.4 \times 10^6 (0.25 \times 10^6)$
Composites	E-glass (73.3 vol %) in epoxy (parallel loading of continuous fibers)[c]	$77.2 \times 10^6 (3.04 \times 10^6)$
	Al_2O_3 whiskers (14 vol %) in epoxy[c]	$48.8 \times 10^6 (1.92 \times 10^6)$
	C (67 vol %) in epoxy (parallel loading)[d]	$76.9 \times 10^6 (3.03 \times 10^6)$
	Kevlar[e] (82 vol %) in epoxy (parallel loading)[d]	$112 \times 10^6 (4.42 \times 10^6)$
	Douglas fir, kiln-dried to 12% moisture (loaded in bending)[c]	$18.3 \times 10^6 (0.72 \times 10^6)$

[a] Sources: Data from *Metals Handbook*, 9th ed., Vols. 1–3, and 8th ed., Vol. 1, American Society for Metals, Metals Park, Ohio, 1978, 1979, 1980, and 1961.
[b] R. A. Flinn and P. K. Trojan, *Engineering Materials and Their Applications*, 2nd ed., Houghton Mifflin Company, Boston, 1981; and M. F. Ashby and D. R. H. Jones, *Engineering Materials*, Pergamon Press, Inc., Elmsford, N.Y., 1980.
[c] R. A. Flinn and P. K. Trojan, *Engineering Materials and Their Applications*, 2nd ed., Houghton Mifflin Company, Boston, 1981.
[d] A. K. Dhingra, Du Pont Company.
[e] Trade name, Du Pont.

SAMPLE PROBLEM 14.9

Calculate the isostrain modulus of epoxy reinforced with 73.3 vol % E-glass fibers and compare your result with the measured value given in Table 14.12.

SOLUTION

Using Equation 14.7 and data from Tables 14.10 and 14.11, we obtain

$$E_c = v_m E_m + v_f E_f$$

$$= (1.000 - 0.733)(6.9 \times 10^3 \text{ MPa}) + (0.733)(72.4 \times 10^3 \text{ MPa})$$

$$= 54.9 \times 10^3 \text{ MPa}$$

Table 14.12 gives for this case $E_c = 56 \times 10^3$ MPa, or

$$\% \text{ error} = \frac{56 - 54.9}{56} \times 100 = 2.0\%$$

The calculated value comes within 2% of the measured one.

SAMPLE PROBLEM 14.10

The tensile strength of the dispersion-strengthened aluminum in Sample Problem 14.4 is 350 MPa. The tensile strength of the pure aluminum is 175 MPa. Calculate the specific strengths of these two materials.

SOLUTION

The specific strength is simply

$$\text{sp. str.} = \frac{\text{T.S.}}{\rho}$$

Using the strengths given above and densities from Sample Problem 14.4, we obtain

$$\text{sp. str., Al} = \frac{(175 \text{ MPa}) \times (1.02 \times 10^{-1} \text{ kg/mm}^2)/\text{MPa}}{(2.70 \text{ Mg/m}^3)(10^3 \text{ kg/Mg})(1 \text{ m}^3/10^9 \text{ mm}^3)}$$

$$= 6.61 \times 10^6 \text{ mm}$$

$$\text{sp. str., Al/10 vol \% Al}_2\text{O}_3 = \frac{(350)(1.02 \times 10^{-1})}{(2.83)(10^{-6})} \text{ mm}$$

$$= 12.6 \times 10^6 \text{ mm}$$

Note.   In following this example, you may have been disturbed by the rather casual use of units. We canceled kg in the strength term (numerator) with the kg in the density term (denominator). This is, of course, not rigorously correct because the strength term uses kg force and the density term uses kg mass. However, this convention is commonly used and was, in fact, the basis of the numbers in Table 14.13. The important thing about specific strength is not the absolute number but the relative values for competitive structural materials.

· ·

PRACTICE PROBLEM 14.9

In Sample Problem 14.9 the isostrain modulus for a fiberglass composite is shown to be close to a calculated value. Repeat this comparison for the isostrain modulus of B (70 vol %)/epoxy composite given in Table 14.12.

PRACTICE PROBLEM 14.10

In Sample Problem 14.10 we find that dispersion-strengthened aluminum has a substantially higher specific strength than pure aluminum. In a similar way, calculate the specific strength of the E-glass/epoxy composite of Table 14.12 compared to the pure epoxy of Table 14.10. For density information, refer to Sample Problem 14.1. (You may wish to compare your calculations with the values in Table 14.13.)

14.5 PROCESSING OF COMPOSITES

Composites represent such a wide range of structural materials that a brief list of processing techniques cannot do justice to the full field. Table 14.14 is restricted to the key examples of composites from earlier in this chapter. Even these few materials represent a diverse set of processing techniques. Figure 14–21 illustrates the fabrication of typical fiberglass configurations. These are often standard polymer processing methods with glass fibers added at an appropriate point in the procedure. A major factor affecting properties is the orientation of the fibers. The issue of anistropy of properties was discussed in Section 14.3. Note that the open-mold processes include *pultrusion*, which is especially well suited for producing complex cross-section products continuously.

Sawing configurations can affect wood structure and, therefore, the nature of the product. There is also a variation in density that occurs upon equilibration with various humidity levels in the atmosphere. Mechanical properties (see Table 14.12) are commonly specified at a "standard state," such as kiln-dried to 12% moisture. The processing of portland cement is a complex manufacturing process. The final stage of concrete production is done in the familiar cement mixer in which the portland cement is combined with aggregate and water. Primary considerations at this final stage include water/cement ratio and the extent of entrained air (porosity). Figure 14–22 summarizes the variation in strength with water/cement ratio for some typical concretes.

Table 14.14 *Major Processing Methods for Three Representative Composites*

Composite	Processing methods
Fiberglass	Open mold
	Preforming
	Closed mold
Wood	Sawing
	Kiln drying
Concrete	Manufacturing of portland cement
	Mix design (mixing of cement, aggregate, and water)
	Reinforcement (with steel bars, etc.)

Open mold processes

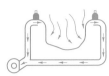

Contact Molding

Resin is in contact with air. Lay-up normally cures at room temperature. Heat may accelerate cure. A smoother exposed side may be achieved by wiping on cellophane.

Vacuum Bag

Cellophane or polyvinyl acetate is placed over lay-up. Joints are sealed with plastic; vacuum is drawn. Resultant atmospheric pressure eliminates voids and forces out entrapped air and excess resin.

Pressure Bag

Tailored bag–normally rubber sheeting–is placed against lay-up. Air or steam pressure up to 50 psi is applied between pressure plate and bag.

Autoclave

Modification of the pressure bag method: after lay-up, entire assembly is placed in steam autoclave at 50 to 100 psi. Additional pressure achieves higher glass loadings and improved removal of air.

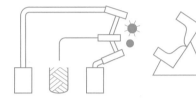

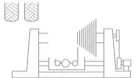

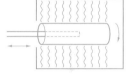

Spray-up

Roving is fed through a chopper and ejected into a resin stream, which is directed at the mold by either of two spray systems: (1) A gun carries resin premixed with catalyst, another gun carries resin premixed with accelerator. (2) Ingredients are fed into a single run mixing chamber ahead of the spray nozzle. By either method the resin mix precoats the strands and the merged spray is directed into the mold by the operator. The glass-resin mix is rolled by hand to remove air, lay down the fibers, and smooth the surface. Curing is similar to hand lay-up.

Filament Winding

Uses continuous reinforcement to achieve efficient utilization of glass fiber strength. Roving or single strands are fed from a creel through a bath of resin and wound on a mandrel. Preimpregnated roving is also used. Special lathes lay down glass in a predetermined pattern to give max. strength in the directions required. When the right number of layers have been applied, the wound mandrel is cured at room temperature or in an oven.

Centrifugal Casting

Round objects such as pipe can be formed using the centrifugal casting process Chopped strand mat is positioned inside a hollow mandrel. The assembly is then placed in an oven and rotated. Resin mix is distributed uniformly throughout the glass reinforcement. Centrifugal action forces glass and resin against walls of rotating mandrel prior to and during the cure. To accelerate cure, hot air is passed through the oven.

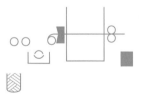

Encapsulation

Short chopped strands are combined with catalyzed resin and poured into open molds. Cure is at room temperature. A post-cure of 30 minutes at 200 F is normal.

Continuous Pultrusion

Continuous strand–roving or other forms of reinforcement–is impregnated in a resin bath and drawn through a die which sets the shape of the stock and controls the resin content. Final cure is effected in an oven through which the stock is drawn by a suitable pulling device.

(a)

Figure 14-21 *Summary of the diverse methods of processing fiberglass products: (a) open-mold processes.*

Preforming methods

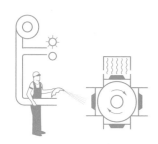 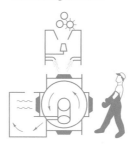 

Directed Fiber

Roving is cut into 1 to 2 inch lengths of chopped strand which are blown through a flexible hose onto a rotating preform screen. Suction holds them in place while a binder is sprayed on the preform and cured in an oven. The operator controls both deposition of chopped strands and binder.

Plenum Chamber

Roving is fed into a cutter on top of plenum chamber. Chopped strands are directed onto a spinning fiber distributor to separate chopped strands and distribute strands uniformly in plenum chamber. Falling strands are sucked onto preform screen. Resinous binder is sprayed on. Preform is positioned in a curing oven. New screen is indexed in plenum chamber for repeat cycle.

Water Slurry

Chopped strands are pre-impregnated with pigmented polyester resin and blended with cellulosic fiber in a water slurry. Water is exhausted through a contoured, perforated screen and glass fibers and cellulosic material are deposited on the surface. The wet preform is transferred to an oven where hot air is sucked through the preform. When dry, the preform is sufficiently strong to be handled and molded.

(b)

Closed mold processes

 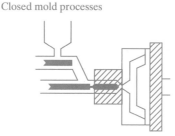 

Premix/Molding Compound

Prior to molding, glass reinforcement, usually chopped spun roving, is thoroughly mixed with resin, pigment, filler, and catalyst. The premixed material can be extruded into a rope-like form for easy handling or may be used in bulk form.

The premix is formed into accurately weighed charges and placed in the mold cavity under heat and pressure. Amount of pressure varies from 100 to 1500 psi. Length of cycle depends on cure temperature, resin, and wall thickness. Cure temperatures range from 225 F to 300 F. Time varies from 30 seconds to 5 minutes.

Injection Molding

For use with thermoplastic materials. The glass and resin molding compound is introduced into a heating chamber where it softens. This mass is then injected into a mold cavity that is kept at a temperature below the softening point of the resin. The part then cools and solidifies.

Continuous Laminating

Fabric or mat is passed through a resin dip and brought together between cellophane covering sheets; the lay-up is passed through a heating zone and the resin is cured. Laminate thickness and resin content are controlled by squeeze rolls as the various plies are brought together.

(c)

Figure 14-21 *(Continued) (b) preforming methods, (c) closed-mold processes. (After illustrations from Owens-Corning Fiberglas Corporation as abstracted in R. Nicholls,* Composite Construction Materials Handbook, *Prentice Hall, Inc., Englewood Cliffs, N.J., 1976.)*

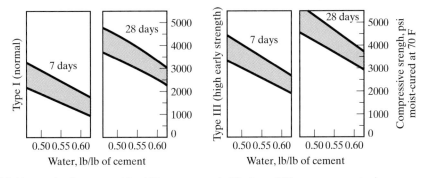

(a) Air-entrained concrete: Air within recommended limits and 2 in. max aggregate size

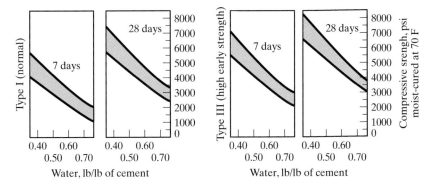

(b) Non-air-entrained concrete

Figure 14-22 *Variation in compressive strength for typical concretes (of different cement types, cure times, and air entrainment) as a function of water/cement ratio. (From* Design and Control of Concrete Mixtures, *11th Ed., Portland Cement Association, Skokie, Ill., 1968.)*

SAMPLE PROBLEM 14.11

The variation in strength of an air-entrained concrete with its water/cement ratio is given in Figure 14–22a. For Type I cement, what % increase in compressive strength would you expect 28 days after pouring the concrete if you used a water/cement ratio of 0.60 rather than 0.50? (Use midrange values to make your estimate.)

SOLUTION

Inspecting the 28-day graph on the left side of Figure 14–22a, we find the midrange strengths to be as follows:

Water/cement ratio	Strength (psi)
0.50	4100
0.60	3100

Therefore, the % increase in strength is

$$\frac{4100 - 3100}{3100} \times 100 = 32.3\%$$

....

PRACTICE PROBLEM 14.11

For a non-air-entrained concrete using Type I cement, what % increase in compressive strength would you expect 28 days after pouring if you used a water/cement ratio of 0.60 rather than 0.50? (See Sample Problem 14.11.)

SUMMARY

Composites bring together in a single material the benefits of various components discussed in previous chapters. Fiberglass typifies synthetic fiber-reinforced composites. Glass fibers provide high strength and modulus in a polymeric matrix that provides ductility. Various fiber geometries are commonly used. In any case, properties tend to reflect the highly anisotropic geometry of the composite microstructure. To produce structural materials that have properties beyond the capability of polymer–matrix composites, substantial effort is under way in the development of new advanced composites, such as metal–matrix and ceramic–matrix composites. Wood is a natural fiber-reinforced composite. Both softwoods and hardwoods exhibit similar microstructures of tubelike cells (which are the structural analogs of glass fibers) in a matrix of lignin and hemicellulose. Concrete is an important example of an aggregate composite. Aggregate in concrete specifically refers to common sand and gravel. These geological silicates are held in a matrix of portland cement, which is a synthetic calcium aluminosilicate. The hardening of cement in the fabrication of concrete forms is a complex process involving several hydration reactions. Certain admixtures are added to the cement to provide specific behavior. As an example, air entrainers provide a high concentration of entrapped air bubbles.

The property averaging that results from combining more than one component in a composite material is highly dependent on the microstructural geometry of the composite. Three representative examples are treated in this chapter. Loading parallel to reinforcing fibers produces an isostrain condition. The elastic modulus (and several transport properties) are simple, volume-weighted averages of values for each component. The result is analogous to a parallel circuit equation. The analog to a series circuit is the isostress condition produced by loading perpendicular to reinforcing fibers. This is a substantially less effective use of the reinforcing fiber modulus. The result of loading a uniformly dispersed aggregate composite is intermediate between the isostrain and isostress limiting cases. For ductile–matrix composites, the effective use of property averaging depends on good interfacial bonding between the matrix and dispersed phase and corresponding high interfacial strength. For brittle ceramic–matrix composites, low interfacial strength is desirable in order to provide high fracture toughness by a mechanism of fiber pull-out. The major mechanical properties important to structural materials are summarized for various composites. An additional parameter of importance to aerospace applications (among others) is the specific strength or strength-to-weight ratio, which is characteristically large for many advanced composite systems.

The processing of composites involves a wide range of methods, representing the especially diverse nature of this family of materials. Only a few representative examples for fiberglass, wood, and concrete have been discussed in this chapter.

KEY TERMS

admixture (512)
advanced composites (500)
aggregate (510)
aggregate composite (510)
anisotropic (500)
carbon–carbon composite (503)
cement (511)
ceramic–matrix composite (503)
cermet (514)
concrete (510)
continuous fiber (500)
discrete (chopped) fiber (500)
dispersion-strengthened metal (514)
E-glass (500)

fiberglass (499)
fiber-reinforced composite (499)
hardwood (504)
hemicellulose (506)
hybrid (504)
interfacial strength (527)
isostrain (517)
isostress (521)
isotropic (517)
laminate (500)
lignin (506)
matrix (498)
metal–matrix composite (503)
particulate composite (513)

polymer–matrix composite (502)
portland cement (511)
prestressed concrete (512)
property averaging (517)
softwood (504)
specific strength (529)
strength-to-weight ratio (530)
whisker (503)
wood (504)
woven fabric (500)

REFERENCES

Agarwal, B. D., and **L. J. Broutman,** *Analysis and Performance of Fiber Composites,* 2nd ed., John Wiley & Sons, Inc., New York, 1990.

Chawla, K. K., *Composite Materials: Science and Engineering,* 2nd ed., Springer-Verlag, New York, 1998.

Engineered Materials Handbook, Vol. 1, *Composites,* ASM International, Metals Park, Ohio, 1987.

Jones, R. M., *Mechanics of Composite Materials,* 2nd ed., Taylor and Francis, Philadelphia, 1999.

Nicholls, R., *Composite Construction Materials Handbook,* Prentice Hall, Inc., Englewood Cliffs, N.J., 1976.

PROBLEMS

Section 14.1 • Fiber-Reinforced Composites

14.1. Calculate the density of a fiber-reinforced composite composed of 14 vol % Al_2O_3 whiskers in a matrix of epoxy. The density of Al_2O_3 is 3.97 Mg/m^3 and of epoxy is 1.1 Mg/m^3.

14.2. Calculate the density of a boron filament reinforced epoxy composite containing 70 vol % filaments. The density of epoxy is 1.1 Mg/m^3.

14.3. Using the information in Equation 14.1, calculate the molecular weight of an aramid polymer with an average degree of polymerization of 500.

14.4. Calculate the density of the Kevlar fiber-reinforced epoxy composite in Table 14.12. The density of Kevlar is 1.44 Mg/m^3 and of epoxy is 1.1 Mg/m^3.

14.5. In a contemporary commercial aircraft, a total of 0.28 m^3 of its exterior surface is constructed of a Kevlar/epoxy composite, rather than a conventional aluminum alloy. Calculate the mass savings using the density calculated in Problem 14.4 and approximating the alloy density by that of pure aluminum.

14.6. What would be the mass savings (relative to the aluminum alloy) if a carbon/epoxy composite with a density of 1.5 Mg/m^3 is used rather than the Kevlar/epoxy composite of Problem 14.5?

14.7. Calculate the degree of polymerization of a cellulose molecule in the cell wall of a wood. The average molecular weight is 95,000 amu.

14.8. A small industrial building can be constructed of steel or wood framing. A decision is made to use steel for fire resistance. If 0.60 m^3 of steel is required for the same structural support as 1.25 m^3 of wood, how much more mass will the building have as a result of this decision? (Approximate the density of steel by that of pure iron. The density of wood can be taken to be 0.42 Mg/m^3.)

Section 14.2 • Aggregate Composites

14.9. Calculate the combined weight percent of CaO, Al_2O_3, and SiO_2 in type II portland cement.

14.10. A reinforced concrete beam has a cross-sectional area of 0.0350 m^2, including four steel reinforcing bars 19 mm in diameter. What is the overall density of this support structure, assuming a uniform cross-section? (The density of the nonreinforced concrete is 2.30 Mg/m^3. The density of the steel can be approximated by that of pure iron.)

14.11. Suppose larger reinforcing bars are specified for the concrete structure in Problem 14.10. If 31.8-mm diameter reinforcing bars are used, what would be the overall density of the structure?

14.12. Calculate the density of a particulate composite composed of 50 vol % Mo in a copper matrix.

14.13. Calculate the density of a dispersion-strengthened copper with 10 vol % Al_2O_3.

14.14. Calculate the density of a WC/Co cutting tool material with 60 vol % WC in a Co matrix. (The density of WC is 15.7 Mg/m^3.)

Section 14.3 • Property Averaging

14.15. Calculate the composite modulus for epoxy reinforced with 70 vol % boron filaments under isostrain conditions.

14.16. Calculate the composite modulus for aluminum reinforced with 50 vol % boron filaments under isostrain conditions.

14.17. Calculate the modulus of elasticity of a metal–matrix composite under isostrain conditions. Assume an aluminum matrix is reinforced by 60 vol % SiC fibers.

14.18. Calculate the modulus of elasticity of a ceramic–matrix composite under isostrain conditions. Assume an Al_2O_3 matrix is reinforced by 60 vol % SiC fibers.

14.19. On a plot similar to Figure 14–11, show the composite modulus for **(a)** 60 vol % fibers (the result of Sample Problem 14.5) and **(b)** 50 vol % fibers (the result of Practice Problem 14.5). Include the individual glass and polymer plots.

14.20. On a plot similar to Figure 14–11, show the composite modulus for an epoxy reinforced with 70 vol % carbon fibers under isostrain conditions. (Use the midrange value for carbon modulus in Table 14.11. Epoxy data are given in Table 14.10. For effectiveness of comparison with the case in Figure 14–11, use the same stress and strain scales. Include the individual matrix and fiber plots.)

14.21. Calculate the composite modulus for polyester reinforced with 10 vol % Al_2O_3 whiskers under isostrain conditions. (See Tables 14.10 and 14.11 for appropriate moduli.)

14.22. Calculate the composite modulus for epoxy reinforced with 70 vol % boron filaments under isostress conditions.

14.23. Calculate the modulus of elasticity of a metal–matrix composite under isostress conditions. Assume an aluminum matrix is reinforced by 50 vol % SiC fibers.

14.24. Calculate the modulus of elasticity of a ceramic–matrix composite under isostress conditions. Assume an Al_2O_3 matrix is reinforced by 50 vol % SiC fibers.

14.25. On a plot similar to Figure 14–11, show the isostress composite modulus for 50 vol % E-glass fibers in a polyester matrix (the result of Problem 14.19). It is interesting to compare the appearance of the resulting plot with that from Problem 14.19b.

14.26. Repeat Problem 14.25 for an epoxy reinforced with 70 vol % carbon fibers and compare with the isostrain results from Problem 14.20.

14.27. Plot Poisson's ratio as a function of reinforcing fiber content for an SiC fiber-reinforced Si_3N_4 composite system loaded parallel to the fiber direction and SiC contents between 50 and 75 vol %. (Note the discussion relative to Equation 14.9 and data in Table 6.6.)

14.28. Calculate the composite modulus of polyester reinforced with 10 vol % Al_2O_3 whiskers under isostress conditions. (Refer to Problem 14.21.)

14.29. Generate a plot similar to Figure 14–13 for the case of epoxy reinforced with Al_2O_3 whiskers. (Refer to Problems 14.21 and 14.28.)

14.30. Calculate the modulus of elasticity of a metal–matrix composite composed of 50 vol % SiC whiskers in a matrix of aluminum. Assume that the modulus lies exactly midway between the isostress and isostrain values.

14.31. Calculate the composite modulus for 20 vol % SiC whiskers in an Al_2O_3 matrix. Assume that the modulus lies exactly midway between the isostress and isostrain values.

14.32. Generate a plot similar to Figure 14–15 for the case of Co–WC composites. For Co being the matrix, the value of n in Equation 14.20 is zero (see Sample Problem 14.8). For WC being the matrix, the value of n is $\frac{1}{2}$ (see Practice Problem 14.8). (The extreme cases of $n = 1$ and $n = -1$ should not be plotted for this system with components having relatively similar modulus values.)

• **14.33.** Consider further the discussion of interfacial strength relative to Figure 14–17. **(a)** Taking the tensile stress in the fiber (with radius r) at a distance x from either end of the fiber to be σ_x, use a force balance between tensile and shear components to derive an expression for σ_x in terms of the fiber geometry and the interfacial shear stress, τ (which is uniform along the entire interface). **(b)** Show how this produces the plot of the tensile stress in a short fiber (in which σ_x is always less than $\sigma_{critical}$, the failure stress of the fiber), as shown in Figure 14–17a.

• **14.34.** **(a)** Referring to Problem 14.33, show how this produces the sketch of the tensile stress distribution in a long fiber (in which the stress in the middle portion of the fiber reaches a maximum, constant value, corresponding to fiber failure), as shown in Figure 14–17b. **(b)** Using the result of Problem 14.33a, derive an expression for the critical stress transfer length, l_c, the minimum fiber length that must be exceeded if fiber failure is to occur; that is, if σ_x is to reach $\sigma_{critical}$.

Section 14.4 • Mechanical Properties of Composites

14.35. Compare the calculated value of the isostrain modulus of a W fiber (50 vol %)/copper composite with that given in Table 14.12. The modulus of tungsten is 407×10^3 MPa.

14.36. Determine the error made in Problem 14.15 in calculating the isostrain modulus of the B/epoxy composite of Table 14.12.

14.37. Calculate the error in assuming the isostrain modulus of an epoxy reinforced with 67 vol % C fibers is given by Equation 14.7. (Note Table 14.12 for experimental data.)

14.38. **(a)** Calculate the error in assuming that the modulus for the Al_2O_3 whiskers (14 vol %)/epoxy composite in Table 14.12 is represented by isostrain conditions. **(b)** Calculate the error in assuming that the composite represents isostress conditions. **(c)** Comment on the nature of the agreement or disagreement indicated by your answers in parts (a) and (b).

• **14.39.** Determine the appropriate value of n in Equation 14.21 to describe the modulus of the W particles (50 vol %)/copper composite given in Table 14.12. (The modulus of tungsten is 407×10^3 MPa.)

14.40. Calculate the percent increase in composite modulus that would occur due to changing from a dispersed aggregate to an isostrain fiber loading of 50 vol % W in a copper matrix.

14.41. Calculate the fraction of the composite load carried by the W fibers in the isostrain case of Problem 14.40.

14.42. Calculate the specific strength of the Kevlar/epoxy composite in Table 14.12. (Note Problem 14.4.)

14.43. Calculate the specific strength of the B/epoxy composite in Table 14.12. (Note Problem 14.2.)

14.44. Calculate the specific strength of the W particles (50 vol %)/copper composite listed in Table 14.12. (See Practice Problem 14.4.)

14.45. Calculate the specific strength for the W fibers (50 vol %)/copper composite listed in Table 14.12.

14.46. Calculate the (flexural) specific strength of the SiC/Al_2O_3 ceramic–matrix composite in Table 14.12, assuming 50 vol % whiskers. (The density of SiC is 3.21 Mg/m^3. The density of Al_2O_3 is 3.97 Mg/m^3.)

14.47. Calculate the (tensile) specific strength of the Douglas fir (loaded parallel to the grain) in Table 14.12. (The density of this wood is 0.42 Mg/m^3.)

14.48. Calculate the (compressive) specific strength of the standard concrete (without air entrainer) in Table 14.12. (The density of this concrete is 2.30 Mg/m^3.)

14.49. To appreciate the relative toughness of (i) traditional ceramics, (ii) high-toughness, unreinforced ceramics, and (iii) ceramic–matrix composites, plot the breaking stress versus flaw size, a, for **(a)** silicon carbide, **(b)** partially stabilized zirconia, and **(c)** silicon carbide reinforced with SiC fibers. Use Equation 8.1 and take $Y = 1$. Cover a range of a from 1 to 100 mm on a logarithmic scale. (See Tables 8.3 and 14.12 for data.)

D **14.50.** In Problem 6.9 a competition among various metallic pressure vessel materials was illustrated. We can expand the selection process by including some composites, as listed in the following table:

Material	ρ (Mg/m^3)	Cost ($/kg)	Y.S. (MPa)
1040 carbon steel	7.8	0.63	
304 stainless steel	7.8	3.70	
3003-H14 aluminum	2.73	3.00	
Ti–5Al–2.5Sn	4.46	15.00	
Reinforced concrete	2.5	0.40	200
Fiberglass	1.8	3.30	200
Carbon fiber-reinforced polymer	1.5	70.00	600

(a) From this expanded list, select the material that will produce the lightest vessel. **(b)** Select the material that will produce the minimum-cost vessel.

Section 14.5 • Processing of Composites

14.51. In producing an air-entrained concrete, estimate the compressive strength 28 days after pouring if you mix 18,000 kg water with 30,000 kg Type I cement.

14.52. In producing a non-air-entrained concrete, estimate the compressive strength 28 days after pouring if you mix 18,000 kg water with 30,000 kg Type I cement.

D **14.53.** Consider the injection molding of low-cost casings using a polyethylene–clay particle composite system. The modulus of elasticity of the composite increases and the tensile strength of the composite decreases with volume fraction of clay as follows:

Volume fraction clay	Modulus of elasticity (MPa)	Tensile strength (MPa)
0.3	830	24.0
0.6	2070	3.4

Assuming both modulus and strength change lin-

early with volume fraction of clay, determine the allowable composition range that ensures a product with a modulus of at least 1000 MPa and a strength of at least 10 MPa.

D 14.54. For the injection molding process described in Problem 14.53, what specific composition would be preferred given that the cost per kg of polyethylene is ten times that of clay?

PART THREE

CHAPTER **15**

Electrical Behavior

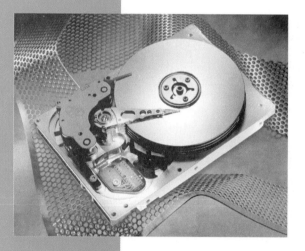

Electrical behavior is often a critical factor in materials selection. An example is the electrical flex connector seen in the lower left-hand corner of this computer hard disk drive assembly. The metal plate in the disk drive spins at 7200 rpm, generating a temperature between 260 and 315° C. A polyphenylene sulfide (PPS) polymer was chosen for the connector due to its unique combination of good electrical insulation and creep resistance. (Courtesy of Seagate Technology.)

In Chapter 2 we found that an understanding of atomic bonding can lead to a useful classification system for engineering materials. In this chapter we turn to a specific material property, *electrical conduction*, to reinforce our classification. This commonality should not be surprising in light of the electronic nature of bonding. Electrical conduction is the result of the motion of charge carriers (such as electrons) within the material. Once again, we find a manifestation of the concept that structure leads to properties. In Chapter 6, atomic and microscopic structure were found to lead to various mechanical properties. Electrical properties follow from electronic structure.

The ease or difficulty of electrical conduction in a material can be understood by returning to the concept of energy levels introduced in Chapter 2. In solid materials, discrete energy levels give way to energy bands. It is the relative spacing of these bands (on an energy scale) that determines the magnitude of conductivity. Metals, with large values of conductivity, are termed *conductors*. Ceramics, glasses, and polymers, with small values of conductivity, are termed *insulators*. *Semiconductors*, with intermediate values of conductivity, are best defined by the unique nature of their electrical conduction.

15.1 CHARGE CARRIERS AND CONDUCTION

The conduction of electricity in materials is by means of individual, atomic-scale species called **charge carriers.** The simplest example of a charge carrier is the *electron*, a particle with 0.16×10^{-18} C of negative charge (see Section 2.1). A more abstract concept is the **electron hole**, which is a missing electron in an electron cloud. The absence of the negatively charged electron gives the electron hole an effective positive charge of 0.16×10^{-18} C relative to its environment. Electron holes play a central role in the behavior of semiconductors and will be discussed in detail in Section 15.5. In ionic materials, anions can serve as negative charge carriers and cations as positive carriers. As seen in Section 2.2, the valence of each ion indicates positive or negative charge in multiples of 0.16×10^{-18} C.

A simple method for measurement of electrical conduction is shown in Figure 15–1. The magnitude of **current** flow, I, through the circuit with a given **resistance**, R, and **voltage**, V, is related by **Ohm's* law**,

$$V = IR \qquad (15.1)$$

where V is in units of volts,[†] I is in amperes[‡] (1 A = 1 C/s), and R is

---

[*] Georg Simon Ohm (1787–1854), German physicist, who first published the statement of Equation 15.1. His definition of resistance led to the unit of resistance being named in his honor.

[†] Alessandro Giuseppe Antonio Anastasio Volta (1745–1827), Italian physicist, made major contributions to the development of understanding of electricity, including the first battery, or "voltage" source.

[‡] André Marie Ampère (1775–1836), French mathematician and physicist, was another major contributor to the field of "electrodynamics" (a term he introduced).

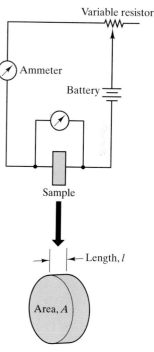

Figure 15-1 *Schematic of a circuit for measuring electrical conductivity. Sample dimensions relate to Equation 15.2.*

in ohms. The resistance value depends on the specific sample geometry; R increases with sample length, l, and decreases with sample area, A. As a result, a property more characteristic of a given material and independent of its geometry is **resistivity**, ρ, defined as

$$\rho = \frac{RA}{l} \tag{15.2}$$

The units for resistivity are $\Omega \cdot m$. An equally useful material property is the reciprocal of resistivity, **conductivity**, σ, where

$$\sigma = \frac{1}{\rho} \tag{15.3}$$

with units of $\Omega^{-1} \cdot m^{-1}$. Conductivity will be our most convenient parameter for establishing an electrical classification system for materials (Section 15.7).

Conductivity is the product of the density of charge carriers, n, the charge carried by each, q, and the mobility of each carrier, μ:

$$\sigma = nq\mu \tag{15.4}$$

The units for n are m^{-3}, for q are coulombs, and for μ are $m^2/(V \cdot s)$. The mobility is the average carrier velocity, or **drift velocity**, $\bar{v}$, divided by electrical field strength, E:

$$\mu = \frac{\bar{v}}{E} \tag{15.5}$$

The drift velocity is in units of m/s, and the **electric field strength** ($E = V/l$) in units of V/m.

When both positive and negative charge carriers are contributing to conduction, Equation 15.4 must be expanded to account for both contributions:

$$\sigma = n_n q_n \mu_n + n_p q_p \mu_p \tag{15.6}$$

The subscripts n and p refer to the negative and positive carriers, respectively. For electrons, electron holes, and monovalent ions, the magnitude of q is 0.16×10^{-18} C. For multivalent ions, the magnitude of q is $|Z_i| \times (0.16 \times 10^{-18}$ C), where $|Z_i|$ is the magnitude of the valence (e.g., 2 for O^{2-}).

Table 15.1 lists values of conductivity for a wide variety of engineering materials. It is apparent that the magnitude of conductivity produces distinctive categories of materials consistent with the types outlined in Chapters 1 and 2. We shall discuss this electrical classification system in detail at the end of this chapter, but first, we must look at the nature of electrical conduction in order to understand why conductivity varies by more than 20 orders of magnitude among common engineering materials.

Table 15.1 *Electrical Conductivities of Some Materials at Room Temperature*

Conducting range	Material	Conductivity, σ ($\Omega^{-1} \cdot m^{-1}$)
Conductors	Aluminum (annealed)	35.36×10^6
	Copper (annealed standard)	58.00×10^6
	Iron (99.99 + %)	10.30×10^6
	Steel (wire)	$5.71\text{-}9.35 \times 10^6$
Semiconductors	Germanium (high purity)	2.0
	Silicon (high purity)	0.40×10^{-3}
	Lead sulfide (high purity)	38.4
Insulators	Aluminum oxide	$10^{-10}\text{-}10^{-12}$
	Borosilicate glass	10^{-13}
	Polyethylene	$10^{-13}\text{-}10^{-15}$
	Nylon 66	$10^{-12}\text{-}10^{-13}$

Source: Data from C. A. Harper, Ed., *Handbook of Materials and Processes for Electronics*, McGraw-Hill Book Company, New York, 1970; and J. K. Stanley, *Electrical and Magnetic Properties of Metals*, American Society for Metals, Metals Park, Ohio, 1963.

SAMPLE PROBLEM 15.1

A wire sample (1 mm in diameter by 1 m in length) of an aluminum alloy (containing 1.2% Mn) is placed in an electrical circuit such as that shown in Figure 15–1. A voltage drop of 432 mV is measured across the length of the wire as it carries a 10-A current. Calculate the conductivity of this alloy.

SOLUTION

From Equation 15.1,

$$R = \frac{V}{I}$$

$$= \frac{432 \times 10^{-3} \text{ V}}{10 \text{ A}} = 43.2 \times 10^{-3} \Omega$$

From Equation 15.2,

$$\rho = \frac{RA}{l}$$

$$= \frac{(43.2 \times 10^{-3} \Omega)[\pi (0.5 \times 10^{-3} \text{ m})^2]}{1 \text{ m}}$$

$$= 33.9 \times 10^{-9} \Omega \cdot m$$

From Equation 15.3,

$$\sigma = \frac{1}{\rho}$$

$$= \frac{1}{33.9 \times 10^{-9} \Omega \cdot m}$$

$$= 29.5 \times 10^6 \Omega^{-1} \cdot m^{-1}$$

SAMPLE PROBLEM 15.2

Assuming that the conductivity for copper in Table 15.1 is entirely due to free electrons [with a mobility of 3.5×10^{-3} m^2/(V $\cdot$ s)], calculate the density of free electrons in copper at room temperature.

SOLUTION

From Equation 15.4,

$$n = \frac{\sigma}{q\mu}$$

$$= \frac{58.00 \times 10^6 \Omega^{-1} \cdot m^{-1}}{0.16 \times 10^{-18} \text{ C} \times 3.5 \times 10^{-3} \text{ m}^2/(\text{V} \cdot \text{s})}$$

$$= 104 \times 10^{27} \text{ m}^{-3}$$

SAMPLE PROBLEM 15.3

Compare the density of free electrons in copper from Sample Problem 15.2 with the density of atoms.

SOLUTION

From Appendix 1,

$\rho_{Cu} = 8.93$ g $\cdot$ cm^{-3} with an atomic mass $= 63.55$ amu

$$\rho = 8.93 \, \frac{g}{cm^3} \times 10^6 \, \frac{cm^3}{m^3} \times \frac{1 \, g \cdot atom}{63.55 \, g} \times 0.6023 \times 10^{24} \, \frac{atoms}{g \cdot atom}$$

$$= 84.6 \times 10^{27} \text{ atoms/m}^3$$

This compares with 104×10^{27} electrons/m^3 from Sample Problem 15.2, that is,

$$\frac{\text{free electrons}}{\text{atom}} = \frac{104 \times 10^{27} \text{ m}^{-3}}{84.6 \times 10^{27} \text{ m}^{-3}} = 1.23$$

In other words, the conductivity of copper is high because each atom contributes roughly one free (conducting) electron. We shall see in Sample Problem 15.13 that, in semiconductors, the number of conducting electrons contributed per atom is considerably smaller.

SAMPLE PROBLEM 15.4

Calculate the drift velocity of the free electrons in copper for an electric field strength of 0.5 V/m.

SOLUTION

From Equation 15.5,

$$\bar{v} = \mu E$$
$$= [3.5 \times 10^{-3} \text{ m}^2/(\text{V} \cdot \text{s})](0.5 \text{ V} \cdot \text{m}^{-1})$$
$$= 1.75 \times 10^{-3} \text{ m/s}$$

....

PRACTICE PROBLEM 15.1

(a) The wire described in Sample Problem 15.1 shows a voltage drop of 432 mV. Calculate the voltage drop to be expected in a 0.5-mm-diameter ($\times$ 1-m-long) wire of the same alloy, also carrying a current of 10 A. (b) Repeat part (a) for a 2-mm-diameter wire.

PRACTICE PROBLEM 15.2

How many free electrons would there be in a spool of high-purity copper wire (1 mm diameter $\times$10 m long)? (See Sample Problem 15.2.)

PRACTICE PROBLEM 15.3

In Sample Problem 15.3, we compare the density of free electrons in copper with the density of atoms. How many copper atoms would be in the spool of wire described in Practice Problem 15.2?

The drift velocity of the free electrons in copper is calculated in Sample Problem 15.4. How long would a typical free electron take to move along the entire length of the spool of wire described in Practice Problem 15.2, under the voltage gradient of 0.5 V/m?

15.2 ENERGY LEVELS AND ENERGY BANDS

In Section 2.1 we saw how electron orbitals in a single atom are associated with discrete energy levels (Figure 2–3). Now, let us turn to another, similar example. Figure 15–2 shows an energy-level diagram for a single sodium atom. As indicated in Appendix 1, the electronic configuration is $1s^2 2s^2 2p^6 3s^1$. The energy-level diagram indicates that there are actually three orbitals associated with the $2p$ energy level and that each of the $1s$, $2s$, and $2p$ orbitals is occupied by two electrons. This distribution of electrons among the various orbitals is a manifestation of the **Pauli* exclusion principle**, an important concept from quantum mechanics that indicates that no two electrons can occupy precisely the same state. Each horizontal line shown in Figure 15–2 represents a different orbital (i.e., a unique set of three quantum numbers). Each such orbital can be occupied by two electrons because they are in two different states; that is, they have opposite or antiparallel electron spins (representing different values for a fourth quantum number). In general, electron spins can be parallel or antiparallel. The outer orbital ($3s$) is half-filled by a single electron. Looking at Appendix 1, we note that the next element in the periodic table (Mg) fills the $3s$ orbital with two electrons (which would, by the Pauli exclusion principle, have opposite electron spins).

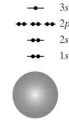

Isolated Na atom

Figure 15-2 *Energy-level diagram for an isolated sodium atom.*

Consider next a hypothetical four-atom sodium molecule, Na$_4$ (Figure 15–3). The energy diagrams for the atom core electrons ($1s^2 2s^2 2p^6$) are essentially unchanged. However, the situation for the four outer orbital electrons is affected by the Pauli exclusion principle. This is because the delocalized electrons are now being shared by all four atoms in the molecule. These electrons cannot all occupy a single orbital. The result is a "splitting" of the $3s$ energy level into four slightly different levels. This makes each level unique and satisfies the Pauli exclusion principle. It would be possible for the splitting to produce only two levels, each occupied by two electrons of opposite spin. In fact, electron pairing in a given orbital tends to be delayed until all levels of a given energy have a single electron. This is

---

* Wolfgang Pauli (1900–1958), Austrian-American physicist, was a major contributor to the development of atomic physics. To a large extent, the understanding (which the exclusion principle provides) of outer-shell electron populations allows us to understand the order of the periodic table. These outer-shell electrons play a central role in the chemical behavior of the elements.

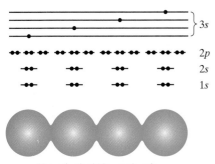

Figure 15-3 *Energy-level diagram for a hypothetical Na$_4$ molecule. The four shared, outer orbital electrons are "split" into four slightly different energy levels, as predicted by the Pauli exclusion principle.*

Hypothetical Na$_4$ molecule

referred to as **Hund's* rule**. As another example, nitrogen (element 7) has three $2p$ electrons, each in a different orbital of equal energy. Pairing of two $2p$ electrons of opposite spin in a single orbital does not occur until element 8 (oxygen). The result of this splitting is a narrow *band* of energy levels corresponding to what was a single $3s$ level in the isolated atom. An important aspect of this electronic structure is that, as in the $3s$ level of the isolated atom, the $3s$ band of the Na$_4$ molecule is only half-filled. As a result, electron mobility between adjacent atoms is quite high.

A simple extension of the effect seen in the hypothetical four-atom molecule is shown in Figure 15–4, in which a large number of sodium atoms

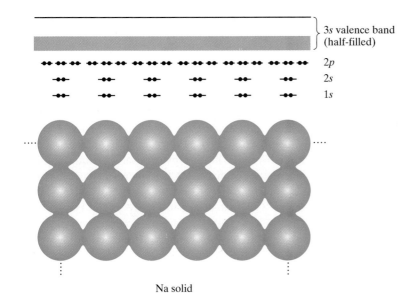

Na solid

Figure 15-4 *Energy-level diagram for solid sodium. The discrete $3s$ energy level of Figure 15–2 has given way to a pseudocontinuous energy band (half-filled). Again, the splitting of the $3s$ energy level is predicted by the Pauli exclusion principle.*

* F. Hund, Z. Physik 42, 93 (1927).

are joined by metallic bonding to produce a solid. In this metallic solid, the atom core electrons are again not directly involved in the bonding, and their energy diagrams remain essentially unchanged. However, the large number of atoms involved (e.g., on the order of Avogadro's number) produces an equally large number of energy-level splittings for the outer ($3s$) orbitals. The total range of energy values for the various $3s$ orbitals is not large. Rather, the spacing between adjacent $3s$ orbitals is extremely small. The result is a pseudocontinuous **energy band** corresponding to the $3s$ energy level of the isolated atom. As with the isolated Na atom and the hypothetical Na$_4$ molecule, the valence electron energy band in the metallic solid is only half-filled, permitting high mobility of outer orbital electrons throughout the solid. Produced from valence electrons, the energy band of Figure 15–4 is also termed the **valence band.** An important conclusion is that *metals are good electrical conductors because their valence band is only partially filled.* This statement is valid, although the detailed nature of the partially filled valence band is different in some metals. For instance, in Mg (element 12), there are two $3s$ electrons that fill the energy band, which is only half-filled in Na (element 11). However, Mg has a higher empty band that overlaps the filled one. The net result is an outer valence band only partially filled.

A more detailed picture of the nature of electrical conduction in metals is obtained by considering how the nature of the energy band varies with temperature. Figure 15–4 implied that the energy levels in the valence band are completely full up to the midpoint of the band and completely empty above. In fact, this is true only at a temperature of absolute zero (0 K). Figure 15–5 illustrates this condition. The energy of the highest filled state in the energy band (at 0 K) is known as the **Fermi* level** (E_F). The extent to which a given energy level is filled is indicated by the **Fermi function**, $f(E)$. This represents the probability that an energy level, E, is occupied by an electron and can have values between 0 and 1. At 0 K, $f(E)$ is equal to 1 up to E_F and equal to 0 above E_F. This limiting case (0 K) is not conducive to electrical conduction. Since the energy levels below E_F are full, conduction requires electrons to increase their energy to some level just above E_F (i.e., to unoccupied levels). This energy promotion requires some external energy source. One means of providing this energy is from thermal energy obtained by heating the material to some temperature above 0 K. The resulting Fermi function, $f(E)$, is shown in Figure 15–6. For $T > 0$ K, some of the electrons just below E_F are promoted to unoccupied levels just above E_F. The relationship between the Fermi function, $f(E)$, and absolute temperature, T, is

$$f(E) = \frac{1}{e^{(E-E_F)/kT} + 1} \tag{15.7}$$

---

* Enrico Fermi (1901–1954), Italian physicist, made numerous contributions to twentieth-century science, including the first nuclear reactor in 1942. His work that improved understanding of the nature of electrons in solids had come nearly 20 years earlier.

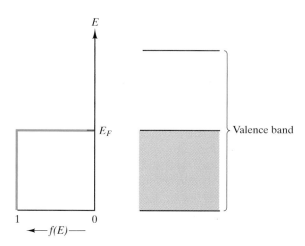

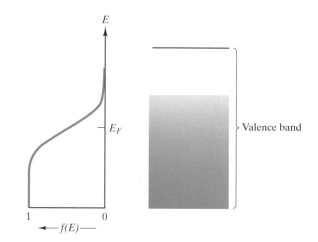

Figure 15-5 *The Fermi function, $f(E)$, describes the relative filling of energy levels. At 0 K, all energy levels are completely filled up to the Fermi level, E_F, and completely empty above E_F.*

Figure 15-6 *At $T > 0$ K, the Fermi function, $f(E)$, indicates promotion of some electrons above E_F.*

where k is the Boltzmann constant (13.8×10^{-24} J/K). In the limit of $T = 0$ K, Equation 15.7 correctly gives the step function of Figure 15–5. For $T > 0$ K, it indicates that $f(E)$ is essentially 1 far below E_F and essentially 0 far above. Near E_F, the value of $f(E)$ varies in a smooth fashion between these two extremes. At E_F, the value of $f(E)$ is precisely 0.5. As the temperature increases, the range over which $f(E)$ drops from 1 to 0 increases (Figure 15–7) and is on the order of the magnitude of kT. In summary, metals are

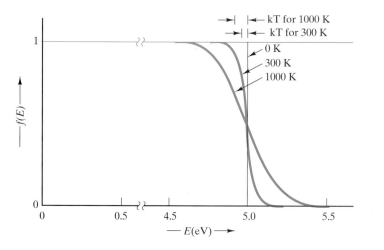

Figure 15-7 *Variation of the Fermi function, $f(E)$, with the temperature for a typical metal (with $E_F = 5$ eV). Note that the energy range over which $f(E)$ drops from 1 to 0 is equal to a few times kT.*

good electrical conductors because thermal energy is sufficient to promote electrons above the Fermi level to otherwise unoccupied energy levels. At these levels ($E > E_F$), the accessibility of unoccupied levels in adjacent atoms yields high mobility of conduction electrons known as **free electrons** through the solid.

Our discussion of energy bands to this point has focused on metals and how they are good electrical conductors. Consider now the case of a non-metallic solid, carbon in the diamond structure, which is a very poor electrical conductor. In Chapter 2 we saw that the valence electrons in this covalently bonded material are shared among adjacent atoms. The net result is that the valence band of carbon (diamond) is full. This valence band corresponds to the sp^3 hybrid energy level of an isolated carbon atom (Figure 2–3). To promote electrons to energy levels above the sp^3 level in an isolated carbon atom requires going above regions of forbidden energy. Similarly for the solid, promotion of an electron from the valence band to the **conduction band** requires going above an **energy band gap**, E_g (Figure 15–8). The concept of a Fermi level, E_F, still applies. However, E_F now falls in the center of the band gap. In Figure 15–8, the Fermi function, $f(E)$, corresponds to room temperature (298 K). One must bear in mind that the probabilities predicted by $f(E)$ can be realized only in the valence and conduction bands. Electrons are forbidden to have energy levels within the band gap. The important conclusion of Figure 15–8 is that $f(E)$ is essentially equal to 1 throughout the valence band and equal to 0 throughout the conduction band. The inability of thermal energy to promote a significant number of electrons to the conduction band gives diamond its characteristically poor electrical conductivity.

As a final example, consider silicon, element 14, residing just below carbon in the periodic table (Figure 2–2). In the same periodic table group, silicon behaves chemically in a way similar to carbon. In fact, silicon forms a covalently bonded solid with the same crystal structure as that of diamond-

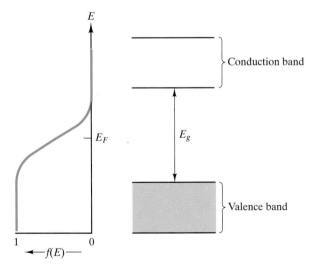

Figure 15-8 *Comparison of the Fermi function, $f(E)$, with the energy band structure for an insulator. Virtually no electrons are promoted to the conduction band [$f(E) = 0$ there] because of the magnitude of the band gap (> 2 eV).*

the diamond cubic structure discussed in Section 3.5. The energy band structure of silicon (Figure 15–9) also looks very similar to that for diamond (Figure 15–8). The primary difference is that silicon has a smaller band gap ($E_g = 1.107$ eV, compared to ~ 6 eV for diamond). The result is that, at room temperature (298 K), thermal energy promotes a small but significant number of electrons from the valence to the conduction band. Consequently, electron holes are produced in the valence band equal in number to the conduction electrons. These electron holes are positive charge carriers, as mentioned in Section 15.1. With both positive and negative charge carriers present in moderate numbers, silicon demonstrates a moderate value of electrical conductivity intermediate between that for metals and insulators (Table 15.1). This simple semiconductor is discussed further in Section 15.5.

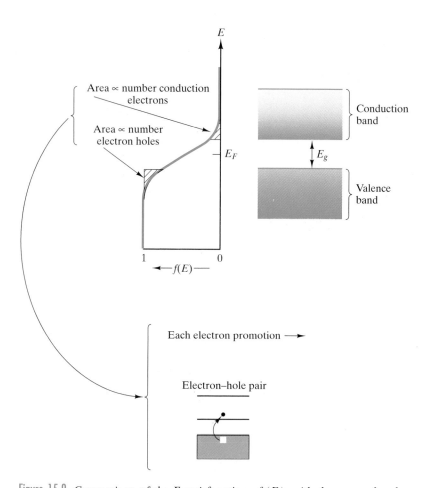

Figure 15-9 *Comparison of the Fermi function, $f(E)$, with the energy band structure for a semiconductor. A significant number of electrons is promoted to the conduction band because of a relatively small band gap (<2 eV). Each electron promotion creates a pair of charge carriers (i.e., an electron-hole pair).*

SAMPLE PROBLEM 15.5

What is the probability of an electron being thermally promoted to the conduction band in diamond ($E_g = 5.6$ eV) at room temperature (25°C)?

SOLUTION

From Figure 15–8, it is apparent that the bottom of the conduction band corresponds to

$$E - E_F = \frac{5.6}{2} \text{ eV} = 2.8 \text{ eV}$$

From Equation 15.7 and using $T = 25°C = 298$ K,

$$f(E) = \frac{1}{e^{(E-E_F)/kT} + 1}$$

$$= \frac{1}{e^{(2.8 \text{ eV})/(86.2 \times 10^{-6} \text{ eV K}^{-1})(298 \text{ K})} + 1}$$

$$= 4.58 \times 10^{-48}$$

SAMPLE PROBLEM 15.6

What is the probability of an electron being thermally promoted to the conduction band in silicon ($E_g = 1.07$ eV) at room temperature (25°C)?

SOLUTION

As in Sample Problem 15.5,

$$E - E_F = \frac{1.107}{2} \text{ eV} = 0.5535 \text{ eV}$$

and

$$f(E) = \frac{1}{e^{(E-E_F)/kT} + 1}$$

$$= \frac{1}{e^{(0.5535 \text{ eV})/(86.2 \times 10^{-6} \text{ eV K}^{-1})(298 \text{ K})} + 1}$$

$$= 4.39 \times 10^{-10}$$

While this number is small, it is 38 orders of magnitude greater than the value for diamond (Sample Problem 15.5) and is sufficient to create enough charge carriers (electron-hole pairs) to give silicon its semiconducting properties.

PRACTICE PROBLEM 15.5

What is the probability of an electron's being promoted to the conduction band in diamond at $50°C$? (See Sample Problem 15.5.)

PRACTICE PROBLEM 15.6

What is the probability of an electron's being promoted to the conduction band in silicon at $50°C$? (See Sample Problem 15.6.)

15.3 CONDUCTORS

Conductors are materials with large values of conductivity. Table 15.1 indicates that the magnitude of conductivity for typical conductors is on the order of $10 \times 10^6 \Omega^{-1} \cdot m^{-1}$. The basis for this large value was discussed in the preceding section. Recalling Equation 15.6 as the general expression for conductivity, we can write the specific form for conductors as

$$\sigma = n_e q_e \mu_e \qquad (15.8)$$

where the subscript e refers to purely **electronic conduction**, which refers to σ specifically resulting from the movement of electrons. (**Electrical conduction** refers to a measurable value of σ that can arise from the movement of any type of charge carrier.) The dominant role of the band model of the preceding section is to indicate the importance of electron mobility, μ_e, in the conductivity of metallic conductors. This is well illustrated in the effect of two variables (temperature and composition) on conductivity in metals.

The effect of temperature on conductivity in metals is illustrated in Figure 15–10. In general, an increase in temperature above room temperature results in a drop in conductivity. This drop in conductivity is predominantly due to the drop in electron mobility, μ_e, with increasing temperature. The drop in electron mobility can, in turn, be attributed to the increasing thermal agitation of the crystalline structure of the metal as temperature increases. Because of the wavelike nature of electrons, these "wave packets" can move through crystalline structure most effectively when that structure is nearly perfect. Irregularity produced by thermal vibration diminishes electron mobility.

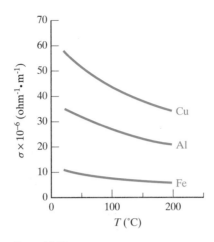

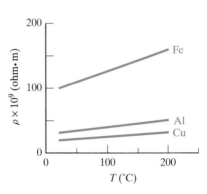

Figure 15-10 *Variation in electrical conductivity with temperature for some metals. (From J. K. Stanley,* Electrical and Magnetic Properties of Metals, *American Society for Metals, Metals Park, Ohio, 1963.)*

Figure 15-11 *Variation in electrical resistivity with temperature for the same metals shown in Figure 15–10. The linearity of these data defines the temperature coefficient of resistivity,* α.

Equation 15.3 showed that resistivity and conductivity are inversely related. Therefore, the magnitude of resistivity for typical conductors is on the order of $0.1 \times 10^{-6}\Omega \cdot$ m. Similarly, resistivity increases as temperature increases above room temperature. This relationship $[\rho(T)]$ is used more often than $\sigma(T)$ because the resistivity is found experimentally to increase quite linearly with temperature over this range; that is,

$$\rho = \rho_{rt}[1 + \alpha(T - T_{rt})] \qquad (15.9)$$

where ρ_{rt} is the room temperature value of resistivity, α the **temperature coefficient of resistivity**, T the temperature, and T_{rt} the room temperature. The data in Figure 15–10 are replotted in Figure 15–11 to illustrate Equation 15.9. Table 15.2 gives some representative values of ρ_{rt} and α for metallic conductors.

Inspection of Table 15.2 reveals that ρ_{rt} is a function of composition when forming solid solutions (e.g., $\rho_{rt, \text{ pure Fe}} < \rho_{rt, \text{ steel}}$). For small additions of impurity to a nearly pure metal, the increase in ρ is nearly linear with the amount of impurity addition (Figure 15–12). This relationship, which is reminiscent of Equation 15.9, can be expressed as

$$\rho = \rho_0(1 + \beta x) \qquad (15.10)$$

Table 15.2 *Resistivities and Temperature Coefficients of Resistivity for Some Metallic Conductors*

Material	Resistivity at 20°C ρ_{rt} ($\Omega \cdot m$)	Temperature coefficient of resistivity at 20°C α ($°C^{-1}$)
Aluminum (annealed)	28.28×10^{-9}	0.0039
Copper (annealed standard)	17.24×10^{-9}	0.00393
Gold	24.4×10^{-9}	0.0034
Iron (99.99+%)	97.1×10^{-9}	0.00651
Lead (99.73+%)	206.48×10^{-9}	0.00336
Magnesium (99.80%)	44.6×10^{-9}	0.01784
Mercury	958×10^{-9}	0.00089
Nickel (99.95% + Co)	68.4×10^{-9}	0.0069
Nichrome (66% Ni + Cr and Fe)	1000×10^{-9}	0.0004
Platinum (99.99%)	106×10^{-9}	0.003923
Silver (99.78%)	15.9×10^{-9}	0.0041
Steel (wire)	$107\text{-}175 \times 10^{-9}$	0.006–0.0036
Tungsten	55.1×10^{-9}	0.0045
Zinc	59.16×10^{-9}	0.00419

Source: Data from J. K. Stanley, *Electrical and Magnetic Properties of Metals*, American Society for Metals, Metals Park, Ohio, 1963.

where ρ_0 is the resistivity of the pure metal, β a constant for a given metal-impurity system (related to the slope of a plot such as Figure 15–12), and x the amount of impurity addition. Of course, Equation 15.10 applies to a fixed temperature. Combined variations in temperature and composition would involve the effects of both α (from Equation 15.9) and β (from Equation 15.10). It is also necessary to recall that Equation 15.10 applies for small values of x only. For large values of x, ρ becomes a nonlinear function of x.

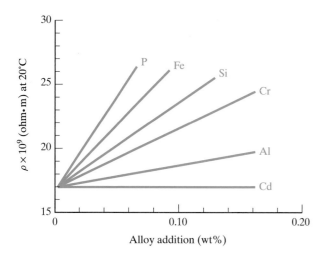

Figure 15-12 *Variation in electrical resistivity with composition for various copper alloys with small levels of elemental additions. Note that all data are at a fixed temperature (20°C). (From J. K. Stanley,* Electrical and Magnetic Properties of Metals, *American Society for Metals, Metals Park, Ohio, 1963.)*

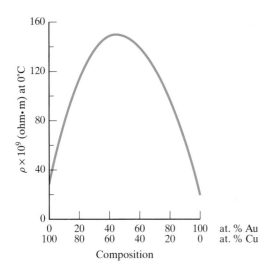

Figure 15-13 *Variation in electrical resistivity with large composition variations in the gold-copper alloy system. Resistivity increases with alloy additions for both pure metals. As a result, the maximum resistivity in the alloy system occurs at an intermediate composition (~45 at % gold, 55 at % copper). As with Figure 15–12, note that all data are at a fixed temperature (0° C). (From J. K. Stanley,* Electrical and Magnetic Properties of Metals, *American Society for Metals, Metals Park, Ohio, 1963.)*

A good example is shown in Figure 15–13 for the gold-copper alloy system. As for Figure 15–12, these data were obtained at a fixed temperature. For Figure 15–13 it is important to note that, as in Figure 15–12, pure metals (either gold or copper) have lower resistivities than alloys with impurity additions. For example, the resistivity of pure gold is less than that for gold with 10 at % copper. Similarly, the resistivity of pure copper is less than that for copper with 10 at % gold. The result of this trend is that the maximum resistivity for the gold-copper alloy system occurs at some intermediate composition (~45 at % gold, 55 at % copper). The reason that resistivity is increased by impurity additions is closely related to the reason that temperature increases resistivity. Impurity atoms diminish the degree of crystalline perfection of an otherwise pure metal.

A concept useful in visualizing the effect of crystalline imperfection on electrical conduction is the mean free path of an electron. As pointed out earlier in discussing the effect of temperature, the wavelike motion of an electron through an atomic structure is hindered by structural irregularity. The average distance that an electron wave can travel without deflection is termed its **mean free path.** Structural irregularities reduce the mean free path, which, in turn, reduces drift velocity, mobility, and finally, conductivity (see Equations 15.5 and 15.8). The nature of chemical imperfection was covered in detail in Section 4.1. For now, we only need to appreciate that any reduction in the periodicity of the metal's atomic structure hinders the movement of the periodic electron wave. For this reason, many of the structural imperfections discussed in Chapter 4 (e.g., point defects and dislocations) have been shown to cause increases in resistivity in metallic conductors.

THERMOCOUPLES

One important application of conductors is the measurement of temperature. A simple circuit, known as a **thermocouple**, which involves two metal wires for making such measurement, is shown in Figure 15–14. The effectiveness of the thermocouple can ultimately be traced to the temperature sensitivity of the Fermi function (e.g., Figure 15–7). For a given metal wire (e.g., metal A in Figure 15–14) connected between two different temperatures-T_1 (hot) and T_2 (cold)-more electrons are excited to higher energies at the hot end than at the cold end. This leads to a driving force for electron transport from the hot to the cold end. The cold end is then negatively charged and the hot end is positively charged with a voltage, V_A, between the ends of the wire. A useful feature of this phenomenon is that V_A depends only on the temperature difference $T_1 - T_2$, not on the temperature distribution along the wire. However, a useful voltage measurement requires a second wire (metal B in Figure 15–14) containing a voltmeter. If metal B is of the same material as metal A, there will also be a voltage V_A induced in metal B and the meter would read the net voltage ($= V_A - V_A = 0$ V). However, different metals will tend to develop different voltages between a given temperature difference $(T_1 - T_2)$. In general, for a metal B different from metal A, the voltmeter in Figure 15–14 will indicate a net voltage, $V_{12} = V_A - V_B$. The magnitude of V_{12} will increase with increasing temperature difference, $T_1 - T_2$. The induced voltage, V_{12}, is called the **Seebeck* potential** and the overall phenomenon illustrated by Figure 15–14 is called the **Seebeck effect.** The utility of the simple circuit for temperature measurement is apparent. By choosing a convenient reference temperature for T_2 (usually a fixed ambient temperature or an ice-water bath at $0°C$), the measured voltage, V_{12}, is a nearly linear function of T_1. The exact dependence of V_{12} on temperature is tabulated for several common thermocouple systems, such as those listed in Table 15.3. A plot of V_{12} with temperature for those common systems is given in Figure 15–15.

In Chapter 17 we shall find numerous examples of semiconductors competing with more traditional electronic materials. In the area of temperature measurement, semiconductors typically have a much more pronounced Seebeck effect than do metals. This is associated with the exponential (Arrhenius) nature of conductivity as a function of temperature in semiconductors. As a result, temperature-measuring semiconductors, or *thermistors*, are capable of measuring extremely small changes in temperature (as low as $10^{-6°}C$). However, due to a limited temperature operating range, thermistors have not displaced traditional thermocouples for general temperature-measuring applications.

SUPERCONDUCTORS

Figure 15–10 illustrated how the conductivity of metals rises gradually as temperature is decreased. This trend continues as temperature is decreased

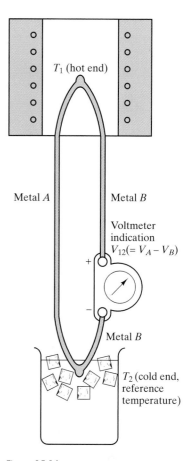

Figure 15-14 *Schematic illustration of a thermocouple. The measured voltage, V_{12}, is a function of the temperature difference, $T_1 - T_2$. The overall phenomenon is termed the "Seebeck effect."*

---

* Thomas Johann Seebeck (1770–1831), Russian-German physicist, in 1821 observed the famous effect that still bears his name. His treatment of closely related problems in thermoelectricity (interconversion of heat and electricity) was less successful, and others were to be associated with those (the Peltier and Thomson effects).

Table 15.3 *Common Thermocouple Systems*

Type	Common name	Positive element[a]	Negative element[a]	Recommended service environment(s)	Maximum service temp. (°C)
B	Platinum-rhodium/ platinum-rhodium	70 Pt–30 Rh	94 Pt–6 Rh	Oxidizing Vacuum Inert	1700
E	Chromel/constantan	90 Ni–9 Cr	44 Ni–55 Cu	Oxidizing	870
J	Iron/constantan	Fe	44 Ni–55 Cu	Oxidizing Reducing	760
K	Chromel/alumel	90 Ni–9 Cr	94 Ni-Al, Mn, Fe, Si, Co	Oxidizing	1260
R	Platinum/platinum-rhodium	87 Pt–13 Rh	Pt	Oxidizing Inert	1480
S	Platinum/platinum-rhodium	90 Pt–10 Rh	Pt	Oxidizing Inert	1480
T	Copper/constantan	Cu	44 Ni–55 Cu	Oxidizing Reducing	370

Source: Data from *Metals Handbook*, 9th ed., Vol. 3, American Society for Metals, Metals Park, Ohio, 1980.
[a] Alloy compositions expressed as weight percents.

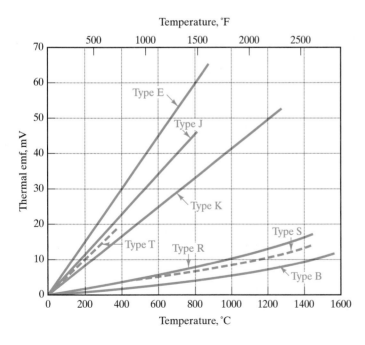

Figure 15-15 *Plot of thermocouple electromotive force (= V_{12} in Figure 15–14) as a function of temperature for some common thermocouple systems listed in Table 15.3. (From* Metals Handbook, *9th ed., Vol. 3, American Society for Metals, Metals Park, Ohio, 1980.)*

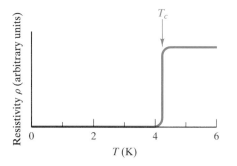

well below room temperature. But even at extremely low temperatures (such as a few degrees Kelvin), typical metals still exhibit a finite conductivity (i.e., a nonzero resistivity). A few materials are dramatic exceptions. Figure 15–16 illustrates such a case. At a critical temperature (T_c), the resistivity of mercury drops suddenly to zero-it becomes a **superconductor.** Mercury was the first material known to exhibit this behavior. In 1911, H. Kamerlingh Onnes first reported the results illustrated in Figure 15–16 as a byproduct of his research on the liquefaction and solidification of helium. Numerous other materials have since been found. (Niobium, vanadium, lead, and their alloys are good examples.) Several empirical facts about superconductivity were known following the early studies. The effect was reversible. It was generally exhibited by metals that were relatively poor conductors at room temperature. The drop in resistivity at T_c is sharp for pure metals but may occur over a 1-to 2-K span for alloys. For a given superconductor, the transition temperature is reduced by increasing current density or magnetic field strength. Until the 1980's, attention was focused on metals and alloys (especially Nb systems), and T_c was below 25 K. In fact, the development of higher-T_c materials followed a nearly straight line on a time scale from 4.12 K in 1911 (for Hg) to 23.3 K in 1975 (for Nb_3Ge). As Figure 15–17 illustrates, a dramatic leap in T_c began in 1986 with the discovery that an $(La, Ba)_2CuO_4$ ceramic exhibited superconductivity at 35 K. In 1987, $YBa_2Cu_3O_7$ was found to have a T_c of 95 K, a major milestone in that the material is superconducting well above the temperature of liquid nitrogen (77 K), a relatively economical cryogenic level. By 1988, a Tl-Ba-Ca-Cu-O ceramic exhibited a T_c of 127 K. In spite of intense research activity involving a wide range of ceramic compounds, the 127 K T_c record was not broken for five years when, in 1993, the substitution of Hg for Tl produced a T_c of 133 K. Recent research indicates that, under extremely high pressures (e.g., 235,000 atm), the T_c of this material can be increased to as high as 150 K. The impracticality of this environmental pressure and the toxicity of Tl and Hg contribute to $YBa_2Cu_3O_7$ continuing to be the most fully studied of the high-T_c materials. Considerable work continues in the effort to gradually increase T_c, with some hope that another breakthrough could occur to accelerate the approach to an ultimate goal of a room-temperature superconductor.

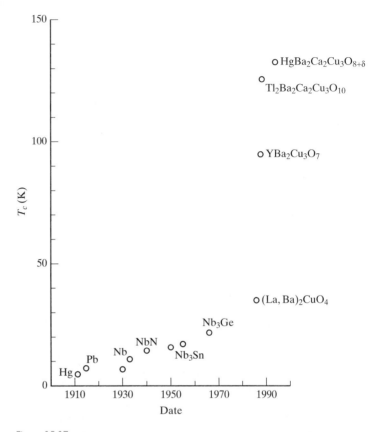

Figure 15-17 *The highest value of T_c increased steadily with time until the development of ceramic oxide superconductors in 1986.*

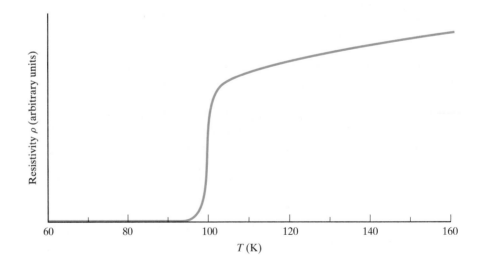

Figure 15-18 *The resistivity of $YBa_2Cu_3O_7$ as a function of temperature, indicating a $T_c \approx 95$ K.*

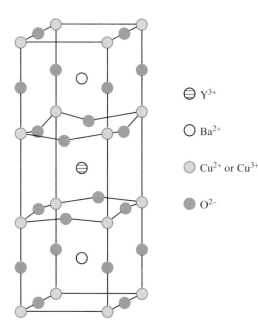

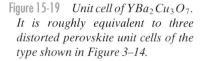

Figure 15-19   *Unit cell of $YBa_2Cu_3O_7$. It is roughly equivalent to three distorted perovskite unit cells of the type shown in Figure 3–14.*

$\ominus$ Y^{3+}

$\bigcirc$ Ba^{2+}

$\bullet$ Cu^{2+} or Cu^{3+}

$\bullet$ O^{2-}

The resistivity of a $YBa_2Cu_3O_7$ ceramic superconductor is shown in Figure 15–18. Extending the characteristic of metallic superconductors just described, we note that the drop in resistivity occurs over a wider temperature range (≈ 5 K) for this material with a relatively high T_c. Also, as poorly conductive metals exhibit superconductivity, the even more poorly conductive ceramic oxides are capable of exhibiting superconductivity to even higher temperatures.

The unit cell of $YBa_2Cu_3O_7$ is shown in Figure 15–19. This material is frequently referred to as the **1–2–3 superconductor** due to the three metal ion subscripts. Although the structure of this superconductor appears relatively complex, it is closely related to the perovskite structure of Figure 3–14. In simple perovskite, there is a ratio of two metal ions to three oxygen ions. The chemistry of the 1–2–3 superconductor has six metal ions to only seven oxygen ions, a deficiency of two oxygen ions accommodated by slight distortion of the perovskite arrangement. In fact, the unit cell in Figure 15–19 can be thought of as being equivalent to three distorted perovskite unit cells, with a Ba^{2+} ion centered in the top and bottom cells and a Y^{3+} centered in the middle cell. The boundaries between the perovskitelike subcells are distorted layers of copper and oxygen ions. A careful analysis of the charge balance between the cations and anions in the unit cell of Figure 15–19 indicates that, to preserve charge neutrality, one of the three copper ions must have the unusual valence of 3+, while the other two have the common value of 2+. The unit cell is orthorhombic. A chemically equivalent material with a tetragonal unit cell is not superconducting. Although the structure of Figure 15–19 is among the most complex considered in this text, it is still slightly idealized. The 1–2–3 superconductor, in fact, has a slight nonstoichiometry, $YBa_2Cu_3O_{7-x}$, with the value of $x \approx 0.1$.

In recent decades, substantial progress has been made in the theoretical modeling of superconductivity. Ironically, lattice vibrations, which are the source of resistivity for normal conductors, are the basis of superconductivity in metals. At sufficiently low temperatures, an ordering effect occurs between lattice atoms and electrons. Specifically, the ordering effect is a synchronization between lattice atom vibrations and the wavelike motion of conductive electrons (associated in pairs of opposite spin). This cooperative motion results in the complete loss of resistivity. The delicate nature of the lattice-electron ordering accounts for the traditionally low values of T_c in metals. Although superconductivity in high-T_c superconductors also involves paired electrons, the nature of the conduction mechanism is not as fully understood. Specifically, electron pairing does not seem to result from the same type of synchronization with lattice vibrations. What has become apparent is that the copper-oxygen planes in Figure 15–19 are the pathways for the supercurrent. In the 1–2–3 superconductor, that current is carried by electron holes. Other ceramic oxides have been developed in which the current is carried by electrons, not holes.

Whether the current is carried by electrons or by electron holes, the promise of superconductors produced by the dramatic increase in T_c has encountered an obstacle in the form of another important material parameter, namely, the **critical current density**, which is defined as the current flow at which the material stops being superconducting. Metallic superconductors used in applications such as magnets in large-scale particle accelerators have critical current densities on the order of 10^{10} A/m^2. Such magnitudes have been produced in thin films of ceramic superconductors, but bulk samples give values about one-hundredth that size. Ironically, the limitation in current density becomes more severe for increasing values of T_c. The problem is caused by the penetration of a surrounding magnetic field into the material, creating an effective resistance due to the interaction between the current and mobile magnetic flux lines. This problem does not exist in metallic superconductors because the magnetic flux lines are not mobile at such low temperatures. In the important temperature range above 77 K, this effect becomes significant and is increasingly important with increasing temperature. The result appears to be little advantage to T_c values much greater than that found in the 1–2–3 material and, even there, the immobilization of magnetic flux lines may require a thin-film configuration or some special form of microstructural control.

If the limitation of critical density can be considered a materials science challenge, a materials engineering challenge exists in the need to fabricate these relatively complex and inherently brittle ceramic compounds into useable product shapes. The issues raised in Section 6.1 in discussing the nature of ceramics as structural materials come into play here also. As with the current density limitation, materials processing challenges are directing efforts in the commercialization of high-T_c superconductors to thin-film device applications and the production of small-diameter wire for cable and solenoid applications. Wire production generally involves the addition of metallic silver to 1–2–3 superconductor particles. The resulting composite has adequate mechanical performance without a significant sacrifice of superconducting properties.

Among the most promising applications of superconductors is the use of thin films as filters for cellular telephone base stations. Compared to conventional copper metal technology, the superconductor filters can enhance the range of the the base stations, reduce channel interference, and decrease the number of dropped calls. Other applications of superconductors, both metallic and ceramic, are generally associated with their magnetic behavior and will be discussed further in Sections 18.4 and 18.5.

Whether or not high-T_c superconductors will create a technological revolution on the scale provided by semiconductors, the breakthrough in developing the new family of high-T_c materials in the late 1980s still stands as one of the most exciting developments in materials science and engineering since the development of the transistor.

SAMPLE PROBLEM 15.7

Calculate the conductivity of gold at 200°C.

SOLUTION

From Equation 15.9 and Table 15.2,

$$\rho = \rho_{rt}[1 + \alpha(T - T_{rt})]$$
$$= (24.4 \times 10^{-9}\Omega \cdot m)[1 + 0.0034°C^{-1}(200 - 20)°C]$$
$$= 39.3 \times 10^{-9}\Omega \cdot m$$

From Equation 15.3,

$$\sigma = \frac{1}{\rho}$$
$$= \frac{1}{39.3 \times 10^{-9}\Omega \cdot m}$$
$$= 25.4 \times 10^{6}\Omega^{-1} \cdot m^{-1}$$

SAMPLE PROBLEM 15.8

Estimate the resistivity of a copper–0.1 wt % silicon alloy at 100°C.

SOLUTION

Assuming that the effects of temperature and composition are independent and that the temperature coefficient of resistivity of pure copper is a good approximation to that for Cu–0.1 wt % Si, we can write

$$\rho_{100°C, \text{Cu}—0.1 \text{Si}} = \rho_{20°C, \text{Cu}—0.1 \text{Si}}[1 + \alpha(T - T_{\text{rt}})]$$

From Figure 15–12,

$$\rho_{20°C, \text{Cu}—0.1 \text{Si}} \simeq 23.6 \times 10^{-9} \Omega \cdot \text{m}$$

Then

$$\rho_{100°C, \text{Cu}—0.1 \text{Si}} = (23.6 \times 10^{-9} \Omega \cdot \text{m})[1 + 0.00393°\text{C}^{-1}(100 - 20)°\text{C}]$$

$$= 31.0 \times 10^{-9} \Omega \cdot \text{m}$$

Note. The assumption that the temperature coefficient of resistivity for the alloy was the same as for the pure metal is generally valid only for small alloy additions.

SAMPLE PROBLEM 15.9

A chromel/constantan thermocouple is used to monitor the temperature of a heat treatment furnace. The output relative to an ice-water bath is 60 mV.

(a) What is the temperature in the furnace?

(b) What would be the output relative to an ice-water bath for a chromel/alumel thermocouple?

SOLUTION

(a) Table 15.3 shows that the chromel/constantan thermocouple is "type E." Figure 15–15 shows that the type E thermocouple has an output of 60 mV at 800°C.

(b) Table 15.3 shows that the chromel/alumel thermocouple is "type K." Figure 15–15 shows that the type K thermocouple at 800°C has an output of 33 mV.

SAMPLE PROBLEM 15.10

A $YBa_2Cu_3O_7$ superconductor is fabricated in a thin film strip with dimensions 1 μm thick $\times$ 1 mm wide $\times$ 10 mm long. At 77 K, superconductivity is lost when the current along the long dimension reaches a value of 17 A. What is the critical current density for this thin film configuration?

SOLUTION

The current per cross-sectional area is

$$\text{critical current density} = \frac{17 \text{ A}}{(1 \times 10^{-6} \text{ m})(1 \times 10^{-3} \text{ m})}$$

$$= 1.7 \times 10^{10} \text{ A/m}^2$$

....

PRACTICE PROBLEM 15.7

Calculate the conductivity at 200°C of **(a)** copper (annealed standard) and **(b)** tungsten. (See Sample Problem 15.7.)

PRACTICE PROBLEM 15.8

Estimate the resistivity of a copper–0.06 wt % phosphorus alloy at 200°C. (See Sample Problem 15.8.)

PRACTICE PROBLEM 15.9

In Sample Problem 15.9, we find the output from a type K thermocouple at 800°C. What would be the output from a Pt/90 Pt–10 Rh thermocouple?

PRACTICE PROBLEM 15.10

When the 1–2–3 superconductor in Sample Problem 15.10 is fabricated in a bulk specimen with dimensions 5 mm $\times$ 5 mm $\times$ 20 mm, the current in the long dimension at which superconductivity is lost is found to be 3.25×10^3 A. What is the critical current density for this configuration?

15.4 INSULATORS

Insulators are materials with low conductivity. Table 15.1 gives magnitudes for conductivity in typical insulators from approximately 10^{-10} to $10^{-16}\Omega^{-1} \cdot m^{-1}$. This drop in conductivity of roughly 20 orders of magnitude (compared with typical metals) is the result of energy band gaps greater than 2 eV (compared to zero for metals). It is important to note that these low-conductivity materials are an important part of the electronics industry. For example, roughly 80% of the industrial ceramics market worldwide is from this category, with the structural ceramics introduced in Chapter 12 representing only the remaining 20%. As will be seen in Section 18.5, the dominant industrial use of electronic ceramics includes their applications based on the closely related magnetic behavior.

It is not a simple matter to rewrite Equation 15.6 to produce a specific conductivity equation for insulators comparable with Equation 15.8 for metals. Clearly, the density of electron carriers, n_e, is extremely small because of the large band gap. In many cases, the small degree of conductivity for insulators is not the result of thermal promotion of electrons across the band gap. Instead, it may be due to electrons associated with impurities in the material. It can also result from ionic transport (e.g., Na^+ in NaCl). Therefore, the specific form of Equation 15.6 depends on the specific charge carriers involved.

Figure 15–20 shows the nature of charge buildup in a typical insulator, or **dielectric**, application, a parallel-plate **capacitor.** On the atomic scale, the charge buildup corresponds to the alignment of electrical dipoles within the dielectric. This concept is explored in detail in conjunction with the discussion of ferroelectrics and piezoelectrics. A **charge density**, D (in units of C/m^2), is produced and is directly proportional to the electrical field strength, E (in units of V/m),

$$D = \epsilon E \tag{15.11}$$

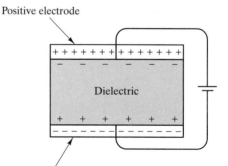

Positive electrode

Dielectric

Negative electrode

Figure 15-20 *A parallel-plate capacitor involves an insulator, or dielectric, between two metal electrodes. The charge density buildup at the capacitor surface is related to the dielectric constant of the material, as indicated by Equation 15.13.*

where the proportionality constant, ϵ, is termed the **electric permittivity** of the dielectric and has units of C/(V · m). For the case of a vacuum between the plates in Figure 15–20, the charge density is

$$D = \epsilon_0 E \qquad (15.12)$$

where ϵ_0 is the electric permittivity of a vacuum, which has a value of 8.854×10^{-12} C/(V· m). For the general dielectric, Equation 15.11 can be rewritten as

$$D = \epsilon_0 \kappa E \qquad (15.13)$$

where κ is a dimensionless material constant called the relative permittivity, relative dielectric constant, or, more commonly, the **dielectric constant.** It represents the factor by which the capacitance of the system in Figure 15–20 is increased by inserting the dielectric in place of the vacuum. For a given dielectric, there is a limiting voltage gradient, termed the **dielectric strength**, at which an appreciable current flow (or breakdown) occurs, and the dielectric fails. Table 15.4 gives representative values of dielectric constant and dielectric strength for various insulators.

FERROELECTRICS

We now turn our attention to insulators with some unique and useful electrical properties. For this discussion, let us concentrate on a representative ceramic material, barium titanate ($BaTiO_3$). The crystal structure is of the perovskite type, shown (for $CaTiO_3$) in Figure 3–14. For $BaTiO_3$, the cubic structure shown in Figure 3–14 is found above 120°C. Upon cooling just

Table 15.4 *Dielectric Constant and Dielectric Strength for Some Insulators*

Material	Dielectric constant,[a] κ	Dielectric strength (kV/mm)
Al_2O_3 (99.9%)	10.1	9.1[b]
Al_2O_3 (99.5%)	9.8	9.5[b]
BeO (99.5%)	6.7	10.2[b]
Cordierite	4.1–5.3	2.4–7.9[b]
Nylon 66-reinforced with 33% glass fibers (dry-as-molded)	3.7	20.5
Nylon 66-reinforced with 33% glass fibers (50% relative humidity)	7.8	17.3
Acetal (50% relative humidity)	3.7	19.7
Polyester	3.6	21.7

Source: Data from *Ceramic Source '86*, American Ceramic Society, Columbus, Ohio, 1985, and *Design Handbook for Du Pont Engineering Plastics.*
[a] At 10^3 Hz.
[b] Average RMS (root-mean-square) values at 60 Hz.

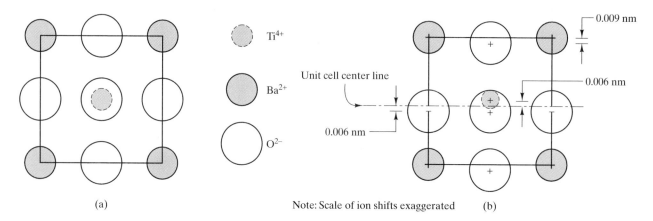

(a)

Note: Scale of ion shifts exaggerated   (b)

Figure 15-21 *(a) Front view of the cubic BaTiO₃ structure. This can be compared with Figure 3–14. (b) Below 120°C, a tetragonal modification of the structure occurs. The net result is an upward shift of cations and a downward shift of anions.*

below 120°C, BaTiO₃ undergoes a phase transformation to a tetragonal modification (Figure 15–21). The transformation temperature (120°C) is referred to as a critical temperature, T_c, in a manner reminiscent of the term for superconductivity. BaTiO₃ is said to be **ferroelectric** below T_c-that is, it can undergo spontaneous polarization. To understand the meaning of this condition, we must note that the tetragonal room-temperature structure of BaTiO₃ (Figure 15–21b) is asymmetrical. As a result, the overall center of positive charge for the distribution of cations within the unit cell is separate from the overall center of negative charge for the anion distribution. This is equivalent to a permanent electrical dipole in the tetragonal BaTiO₃ unit cell (Figure 15–22). Figure 15–23 shows that, in contrast to a cubic material, the dipole structure of the tetragonal unit cell allows for a large polarization of the material in response to an applied electrical field. This is shown as a microstructural, as well as a crystallographic, effect.

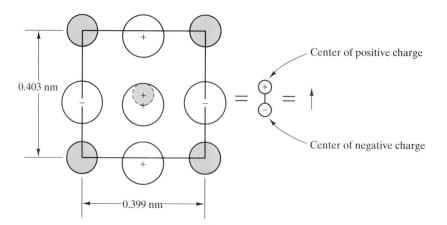

Figure 15-22 *The tetragonal unit cell shown in Figure 15–21b is equivalent to an electrical dipole (with magnitude equal to charge times distance of separation).*

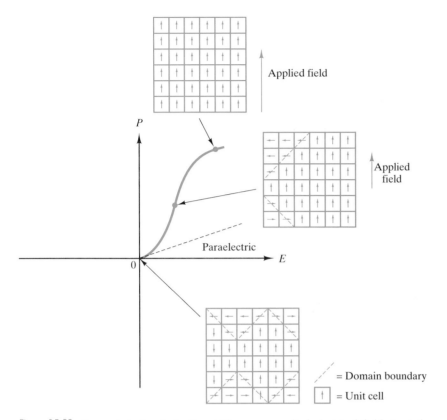

Figure 15-23 *On a plot of polarization (P) versus applied electrical field strength (E), a paraelectric material exhibits only a modest level of polarization with applied fields. In contrast, a ferroelectric material exhibits spontaneous polarization in which domains of similarly oriented unit cells grow under increasing fields of similar orientation.*

The ferroelectric material can have zero polarization under zero applied field due to a random orientation of microscopic-scale **domains**, regions in which the c axes of adjacent unit cells have a common direction. Under an applied field, unit cell dipole orientations roughly parallel to the applied field direction are favored. In this case, domains with such orientations "grow" at the expense of other, less favorably oriented ones. The specific mechanism of domain wall motion is simply the small shift of ion positions within unit cells, resulting in the net change of orientation of the tetragonal c axis. Such domain wall motion results in **spontaneous polarization.** In contrast, the symmetrical unit cell material is **paraelectric**, and only a small polarization is possible as the applied electric field causes a small induced dipole (cations drawn slightly toward the negative electrode and anions toward the positive electrode).

Figure 15–24 summarizes the **hysteresis loop** that results when the electrical field is repeatedly cycled (i.e., an alternating current is applied). Clearly, the plot of polarization versus field does not retrace itself. Several

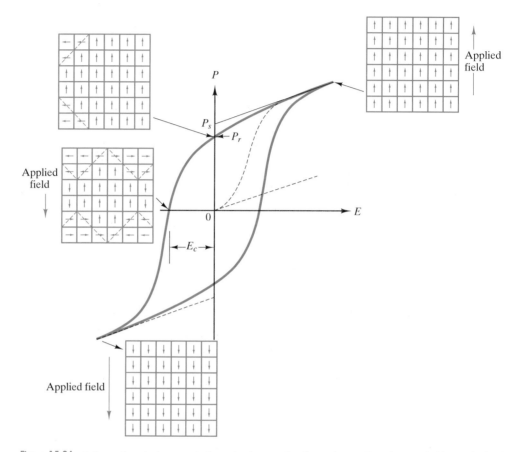

Figure 15-24 *A ferroelectric hysteresis loop is the result of an alternating electric field. A dashed line indicates the initial spontaneous polarization illustrated in Figure 15–23. Saturation polarization (P_s) is the result of maximum domain growth (extrapolated back to zero field). Upon actual field removal, some remanent polarization (P_r) remains. A coercive field (E_c) is required to reach zero polarization (equal volumes of opposing domains).*

key parameters quantify the hysteresis loop. The **saturation polarization**, P_s, is the polarization due to maximum domain growth. Note that P_s is extrapolated to zero field ($E = 0$) to correct for the induced polarization not due to domain reorientation. The **remanent polarization**, P_r, is that remaining upon actual field removal. As shown in Figure 15–24, reduction of E to zero does not return the domain structure to equal volumes of opposing polarization. It is necessary to reverse the field to a level E_c (the **coercive field**) to achieve this result. It is the characteristic hysteresis loop that gives ferroelectricity its name. "Ferro-," of course, is a prefix associated with iron-containing materials. But the nature of the P-E curve in Figure 15–24 is remarkably similar to the induction (B)-magnetic field (H) plots for ferromagnetic materials (e.g., Figure 18–5). Ferromagnetic materials generally contain iron. Ferroelectrics are named after the similar hysteresis loop and rarely contain iron as a significant constituent.

PIEZOELECTRICS

Although ferroelectricity is an intriguing phenomenon, it does not have the practical importance that ferromagnetic materials display (in areas such as magnetic storage of information). The most common applications of ferroelectrics stem from a closely related phenomenon, **piezoelectricity.** The prefix "piezo-" comes from the Greek word for pressure. Piezoelectric materials give an electrical response to mechanical pressure application. Conversely, electrical signals can make them pressure generators. This ability to convert electrical to mechanical energy and vice versa is a good example of a **transducer**, which, in general, is a device for converting one form of energy to another form. Figure 15–25 illustrates the functions of a piezoelectric transducer. Figure 15–25a specifically shows the **piezoelectric effect**, in which the application of stress produces a measurable voltage change across the piezoelectric material. Figure 15–25b illustrates the **reverse piezoelectric effect**, in which an applied voltage changes the magnitude of the polarization in the piezoelectric material and, consequently, its thickness. By constraining the thickness change (e.g., by pressing the piezoelectric against a block of solid material), the applied voltage produces a mechanical stress. It is apparent from Figure 15–25 that the functioning of a piezoelectric transducer is dependent on a common orientation of the polarization of adjacent unit cells. A straightforward means of ensuring this is to use a single-crystal transducer. Single-crystal quartz (SiO_2) is a common example and

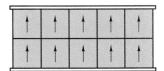

Negative electrode

Unstressed piezoelectric crystal

Positive electrode

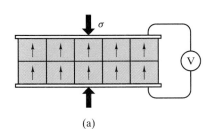

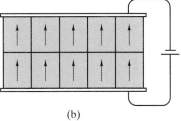

(a)   (b)

Figure 15-25 *Using schematic illustrations of piezoelectric transducers, we see that (a) the dimensions of the unit cells in a piezoelectric crystal are changed by an applied stress, thereby changing their electrical dipoles. The result is a measurable voltage change. This is the piezoelectric effect. (b) Conversely, an applied voltage changes the dipoles and, thereby, produces a measurable dimensional change. This is the reverse piezoelectric effect.*

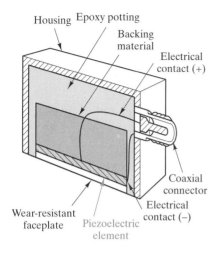

Housing   Epoxy potting
Backing material
Electrical contact (+)
Coaxial connector
Electrical contact (−)
Wear-resistant faceplate   Piezoelectric element

Figure 15-26   *A common application of piezoelectric materials is in ultrasonic transducers. In this cutaway view, the piezoelectric crystal (or "element") is encased in a convenient housing. The constraint of the backing material causes the reverse piezoelectric effect (Figure 15–25b) to generate a pressure when the faceplate is pressed against a solid material to be inspected. When the transducer is operated in this way (as an ultrasonic transmitter), an ac electrical signal (usually in the megahertz range) produces an ultrasonic (elastic wave) signal of the same frequency. When the transducer is operated as an ultrasonic receiver, the piezoelectric effect (Figure 15–25a) is employed. In that case, the high-frequency elastic wave striking the faceplate generates a measurable voltage oscillation of the same frequency. (From* Metals Handbook, *8th ed., Vol. 11, American Society for Metals, Metals Park, Ohio, 1976.)*

was widely used shortly after World War II. $BaTiO_3$ has a higher **piezoelectric coupling coefficient**, k (= fraction of mechanical energy converted to electrical energy), than does SiO_2. The k for $BaTiO_3$ is approximately 0.5 compared to 0.1 for quartz. However, $BaTiO_3$ cannot be conveniently manufactured in single-crystal form. To utilize $BaTiO_3$ as a piezoelectric transducer, the material is fabricated into a pseudo-single-crystal configuration. In this process, the particles of a fine powder of $BaTiO_3$ are aligned in a single crystallographic orientation under a strong electrical field. The powder is subsequently consolidated into a dense solid by sintering. The resulting polycrystalline material with a single, crystalline orientation is said to be **electrically poled.** During the 1950s, this technology allowed $BaTiO_3$ to become the predominant piezoelectric transducer material. While $BaTiO_3$ is still widely used, a solid solution of $PbTiO_3$ and $PbZrO_3$ [$Pb(Ti, Zr)O_3$ or **PZT**] has been more common since the 1960s. A primary reason for this is a substantially higher critical temperature, T_c. As noted earlier, the T_c for $BaTiO_3$ is 120°C. For various $PbTiO_3/PbZrO_3$ solutions, it is possible to have T_c values in excess of 200°C.

Finally, Figure 15–26 shows a typical design for a piezoelectric transducer used as an ultrasonic transmitter and/or receiver. In this common application, electrical signals (voltage oscillations) in the megahertz range produce or sense ultrasonic waves of that frequency.

SAMPLE PROBLEM 15.11

We can quantify the nature of polarization in $BaTiO_3$ by use of the concept of a *dipole moment*, which is defined as the product of charge, Q, and separation distance, d. Calculate the total dipole moment for

(a) the tetragonal $BaTiO_3$ unit cell, and

(b) the cubic $BaTiO_3$ unit cell.

SOLUTION

(a) Using Figure 15–21b, we can calculate the sum of all dipole moments relative to the midplane of the unit cell (indicated by the center line in the figure). A straightforward way of calculating the $\sum Qd$ would be to calculate the Qd product for each ion or fraction thereof relative to the midplane and sum. However, we can simplify matters by noting that the nature of such a summation will be to have a net valued associated with the relative ion shifts. For instance, the Ba^{2+} ions need not be considered because they are symmetrically located within the unit cell. The Ti^{4+} ion is shifted upward 0.006 nm, giving

$$Ti^{4+}\text{moment} = (+4q)(+0.006 \text{ nm})$$

The value of q was defined in Chapter 2 as the unit charge $(= 0.16 \times 10^{-18} \text{ C})$, giving

$$Ti^{4+} \text{ moment} = (1 \text{ ion})(+4 \times 0.16 \times 10^{-18} \text{ C/ion})$$
$$(+6 \times 10^{-3} \text{ nm})(10^{-9} \text{ m/nm})$$
$$= +3.84 \times 10^{-30} \text{ C} \cdot \text{m}$$

Inspection of the perovskite unit cell in Figure 3–14 helps us visualize that two-thirds of the O^{2-} ions in $BaTiO_3$ are associated with the midplane positions, giving (for a downward shift of 0.006 nm)

$$O^{2-} \text{ (midplane) moment} = (2 \text{ ions})(-2 \times 0.16 \times 10^{-18} \text{ C/ion})$$
$$(-6 \times 10^{-3} \text{ nm})(10^{-9} \text{ m/nm})$$
$$= +3.84 \times 10^{-30} \text{ C} \cdot \text{m}$$

The remaining O^{2-} ion is associated with a basal face position that is shifted downward by 0.009 nm, giving

$$O^{2-} \text{ (base) moment} = (1 \text{ ion})(-2 \times 0.16 \times 10^{-18} \text{ C/ion})$$
$$(-9 \times 10^{-3} \text{ nm})(10^{-9} \text{ m/nm})$$
$$= +2.88 \times 10^{-30} \text{ C} \cdot \text{m}$$

Therefore,

$$\sum Qd = (3.84 + 3.84 + 2.88) \times 10^{-30} \text{ C} \cdot \text{m}$$
$$= 10.56 \times 10^{-30} \text{ C} \cdot \text{m}$$

(b) For cubic $BaTiO_3$ (Figure 15–21a), there are no net shifts and, by definition,

$$\sum Qd = 0$$

SAMPLE PROBLEM 15.12

The polarization for a ferroelectric is defined as the density of dipole moments. Calculate the polarization for tetragonal $BaTiO_3$.

SOLUTION

Using the results of Sample Problem 15.11a and the unit cell geometry of Figure 15–22, we obtain

$$P = \frac{\sum Qd}{V}$$

$$= \frac{10.56 \times 10^{-30} \text{ C} \cdot \text{m}}{(0.403 \times 10^{-9} \text{ m})(0.399 \times 10^{-9} \text{ m})^2}$$

$$= 0.165 \text{ C/m}^2$$

....

PRACTICE PROBLEM 15.11

Using the result of Sample Problem 15.11, calculate the total dipole moment for a 2-mm-thick × 2-cm-diameter disk of $BaTiO_3$, to be used as an ultrasonic transducer.

PRACTICE PROBLEM 15.12

The inherent polarization of the unit cell of $BaTiO_3$ is calculated in Sample Problem 15.12. Under an applied electrical field, the polarization of the unit cell is increased to 0.180 C/m^2. Calculate the unit cell geometry under this condition and use sketches similar to Figures 15–21b and 15–22 to illustrate your results.

15.5   SEMICONDUCTORS

Semiconductors are materials with conductivities intermediate between those of conductors and insulators. The magnitudes of conductivity in the semiconductors in Table 15.1 fall within the range 10^{-4} to $10^{+4}\Omega^{-1} \cdot \text{m}^{-1}$. This intermediate range corresponds to band gaps of less than 2 eV. As shown in Figure 15–9, both conduction electrons and electron holes are charge carriers in a simple semiconductor. For the pure silicon example of Figure 15–9, the number of conduction electrons is equal to the number of electron holes. Pure, elemental semiconductors of this type are called *intrinsic semiconductors*. This is the only case we shall deal with in this chapter. In Chapter 17, the important role of impurities in semiconductor technology will be demonstrated in our discussion of *extrinsic semiconductors*, semiconductors with carefully controlled, small amounts of impurities. For now, we can transform the general conductivity expression (Equation 15.6) into a specific form for intrinsic semiconductors,

$$\sigma = nq(\mu_e + \mu_h) \tag{15.14}$$

where n is the density of conduction electrons ($=$ density of electron holes), q the magnitude of electron charge ($=$ magnitude of hole charge $= 0.16 \times 10^{-18}$ C), μ_e the mobility of a conduction electron, and μ_h the mobility of an electron hole. Table 15.5 gives some representative values of μ_e and μ_h together with E_g, the energy band gap and the carrier density at room temperature. Inspection of the mobility data indicates that μ_e is consistently higher than μ_h, sometimes dramatically so. The conduction of electron holes in the valence band is a relative concept. In fact, electron holes exist only in relation to the valence electrons; that is, an electron hole is a missing valence electron. The movement of an electron hole in a given direction is simply a representation that valence electrons have moved in the opposite direction (Figure 15–27). The cooperative motion of the valence electrons (represented by μ_h) is an inherently slower process than the motion of the conduction electron (represented by μ_e).

Table 15.5 *Properties of Some Common Semiconductors at Room Temperature (300 K)*

Material	Energy gap, E_g (eV)	Electron mobility, $\mu_e [\text{m}^2/(\text{V} \cdot \text{s})]$	Hole mobility, $\mu_h [\text{m}^2/(\text{V} \cdot \text{s})]$	Carrier density, $n_e (= n_h)$ (m^{-3})
Si	1.107	0.140	0.038	14×10^{15}
Ge	0.66	0.364	0.190	23×10^{18}
CdS	2.59[a]	0.034	0.0018	—
GaAs	1.47	0.720	0.020	1.4×10^{12}
InSb	0.17	8.00	0.045	13.5×10^{21}

Source: Data from C. A. Harper, Ed., *Handbook of Materials and Processes for Electronics*, McGraw-Hill Book Company, New York, 1970.

[a] This value is above our upper limit of 2 eV used to define a semiconductor. Such a limit is somewhat arbitrary. In addition, most commercial devices involve impurity levels that substantially change the nature of the band gap (see Chapter 17).

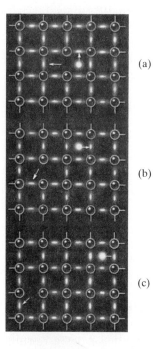

(a)

(b)

(c)

Figure 15-27 *Creation and motion of a conduction electron and an electron hole in a semiconductor. (a) An electron breaks away from the covalent bond, leaving a vacant bonding state, or a hole. The electron is now free to move in an electric field. In terms of the band model, the electron has gone from the valence band to the conduction band, leaving a hole in the valence band. The electron is shown moving upward, and the hole to the left. (b) The conduction electron will now move to the right, the hole down to the left. (c) The motions of (b) have been completed; the hole and electron continue to move outward. (From R. M. Rose, L. A. Shepard, and J. Wulff,* The Structures and Properties of Materials, *Vol. 4:* Electronic Properties, *John Wiley & Sons, Inc., New York, 1966.)*

SAMPLE PROBLEM 15.13

Calculate the fraction of Si atoms that provides a conduction electron at room temperature.

SOLUTION

As in Sample Problem 15.3, the atomic density can be calculated from data in Appendix 1:

$\rho_{Si} = 2.33 \text{ g} \cdot \text{cm}^{-3}$ with an atomic mass $= 28.09$ amu

$$\rho = 2.33 \frac{\text{g}}{\text{cm}^3} \times 10^6 \frac{\text{cm}^3}{\text{m}^3} \times \frac{1 \text{ g} \cdot \text{atom}}{28.09 \text{ g}} \times 0.6023 \times 10^{24} \frac{\text{atoms}}{\text{g} \cdot \text{atom}}$$

$$= 50.0 \times 10^{27} \text{ atoms/m}^3$$

Table 15.5 indicates that

$$n_e = 14 \times 10^{15} \text{ m}^{-3}$$

Then the fraction of atoms providing conduction electrons is

$$\text{fraction} = \frac{14 \times 10^{15} \text{ m}^{-3}}{50 \times 10^{27} \text{ m}^{-3}} = 2.8 \times 10^{-13}$$

Note. This result can be compared with the roughly $1 : 1$ ratio of conducting electrons to atoms in copper (Sample Problem 15.3).

PRACTICE PROBLEM 15.13

Using the data in Table 15.5, calculate **(a)** the total conductivity and **(b)** the resistivity of Si at room temperature. (See Sample Problem 15.13.)

15.6 COMPOSITES

There is no particular magnitude of conductivity characteristic of composites. As discussed in Chapter 14, composites are defined in terms of combinations of the four fundamental material types. A composite of two or more metals will be a conductor. A composite of two or more insulators will be an insulator. However, a composite containing both a metal and an insulator could have a conductivity characteristic of either extreme or some intermediate value, depending on the geometrical distribution of the conducting and nonconducting phases. We found in Section 14.3 that many properties of composites, including electrical conductivity, are geometry-sensitive (e.g., Equations 14.9 and 14.20).

SAMPLE PROBLEM 15.14

Calculate the electrical conductivity parallel to reinforcing fibers for an aluminum loaded with 50 vol % Al_2O_3 fibers.

SOLUTION

Using Equation 14.9 and the data in Table 15.1 (using the midrange value for Al_2O_3), we have

$$\sigma_c = v_m \sigma_m + v_f \sigma_f$$
$$= (0.5)(35.36 \times 10^6 \Omega^{-1} \cdot m^{-1}) + (0.5)(10^{-11} \Omega^{-1} \cdot m^{-1})$$
$$= 17.68 \times 10^6 \Omega^{-1} \cdot m^{-1}$$

PRACTICE PROBLEM 15.14

In Sample Problem 15.14, we calculate the electrical conductivity of an Al/Al_2O_3 composite parallel to the reinforcing fibers. Calculate the conductivity of this composite perpendicular to the reinforcing fibers.

15.7   ELECTRICAL CLASSIFICATION OF MATERIALS

We are now ready to summarize the classification system implied by the data of Table 15.1. Figure 15–28 shows these data along a log scale. The four fundamental materials categories defined by atomic bonding in Chapter 2 are now sorted based on their relative ability to conduct electricity. Metals are good conductors. Semiconductors are best defined by their intermediate values of σ caused by a small but measurable energy barrier to electronic conduction (the band gap). Ceramics and glasses and polymers are insulators characterized by a large barrier to electronic conduction. We should note, however, that certain materials such as ZnO can be a semiconductor in an electrical classification *or* a ceramic in a bonding (Chapter 2) classification. Also, in Section 15.3, we noted that certain oxides have been found to be superconducting. However, on the whole, ceramics, as discussed in Chapter 12, are usually insulators. Composites can be found anywhere along the conductivity scale depending on the nature of their components and the geometrical distribution of those components.

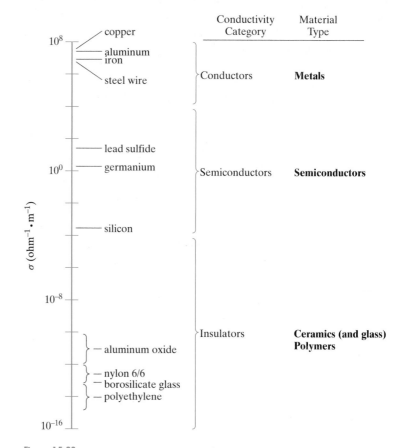

Figure 15-28 *Plot of the electrical conductivity data from Table 15.1. The conductivity ranges correspond to the four fundamental types of engineering materials.*

SUMMARY

Electrical conduction, like atomic bonding, provides a basis for classifying engineering materials. The magnitude of electrical conductivity depends on both the number of charge carriers available and the relative mobility of those carriers. Various charged species can serve as carriers, but our primary interest is in the electron. In a solid, energy bands exist corresponding to discrete energy levels in isolated atoms. Metals are termed conductors because of their high values of electrical conductivity. This is the result of an unfilled valence band. Thermal energy, even at room temperature, is sufficient to promote a large number of electrons above the Fermi level into the upper half of the valence band. Temperature increase or impurity addition causes the conductivity of metals to decrease (and the resistivity to increase). Any such decrease in the perfection of crystal structure decreases the ability of electron "waves" to pass through the metal. Important examples of conductors include the thermocouples and the superconductors.

Ceramics and glasses and polymers are termed insulators because their electrical conductivity is typically 20 orders of magnitude lower than for metallic conductors. This is because there is a large energy gap (greater than 2 eV) between their filled valence bands and their conduction bands so that thermal energy is insufficient to promote a significant number of electrons above the Fermi level into a conduction band. Important examples of insulators are the ferroelectrics and piezoelectrics. (A dramatic exception is the ability of certain oxide ceramics to exhibit superconductivity at relatively high temperatures.)

Semiconductors with intermediate values of conductivity are best defined by the nature of this conductivity. Their energy gap is sufficiently small (generally less than 2 eV) so that a small but significant number of electrons are promoted above the Fermi level into the conduction band at room temperature. The charge carriers, in this case, are both the conduction electrons and the electron holes created in the valence band by the electron promotion. Composites can have conductivity values anywhere from the conductor to the insulator range depending on the components and the geometrical distribution of those components.

KEY TERMS

capacitor (572)
charge carrier (547)
charge density (572)
coercive field (576)
conduction band (556)
conductivity (548)
conductor (559)
critical current density (568)
current (547)
dielectric (572)

dielectric constant (573)
dielectric strength (573)
domain (575)
drift velocity (548)
electric permittivity (573)
electrical conduction (559)
electrical field strength (548)
electrically poled (578)
electron hole (547)
electron-hole pair (557)

electronic conduction (559)
energy band (554)
energy band gap (556)
Fermi function (554)
Fermi level (554)
ferroelectric (574)
free electron (556)
Hund's rule (553)
hysteresis loop (575)
insulator (572)

REFERENCES

Harper, C. A., and **R. N. Sampson**, *Electronic Materials and Processes Handbook*, 2nd ed., McGraw-Hill, New York, 1994.

Kittel, C., *Introduction to Solid State Physics*, 7th ed., John Wiley & Sons, Inc., New York, 1996. Although this text is at a more advanced level, it is a classic source for information on the properties of solids.

Mayer, J. W., and **S. S. Lau**, *Electronic Materials Science: For Integrated Circuits in Si and GaAs*, Macmillan Publishing Company, New York, 1990.

Tu, K. N., **J. W. Mayer**, and **L. C. Feldman**, *Electronic Thin Film Science*, Macmillan Publishing Company, New York, 1992.

PROBLEMS

15.1 • Charge Carriers and Conduction

15.1. (a) Assume that the circuit in Figure 15–1 contains, as a sample, a cylindrical steel bar 1 cm diameter ×10 cm long with a conductivity of $7.00 \times 10^6 \Omega^{-1} \cdot m^{-1}$. What would be the current in this bar due to a voltage of 10 mV? (b) Repeat part (a) for a bar of high-purity silicon of the same dimensions. (See Table 15.1.) (c) Repeat part (a) for a bar of borosilicate glass of the same dimensions. (Again, see Table 15.1.)

15.2. A light bulb operates with a line voltage of 110 V. If the filament resistance is 200 Ω, calculate the number of electrons per second traveling through the filament.

15.3. A semiconductor wafer is 0.6 mm thick. A potential of 100 mV is applied across this thickness. (a) What is the electron drift velocity if their mobility is 0.2 $m^2/(V \cdot s)$? (b) How much time is required for an electron to move across this thickness?

D **15.4.** A 1-mm-diameter wire is required to carry a current of 10 A, but the wire must not have a power dissipation (I^2R) greater than 10 W/m of wire. Of the materials listed in Table 15.1, which are suitable for this wire application?

15.5. A strip of aluminum metallization on a solid-state device is 1 mm long with a thickness of 1 μm and a width of 6 μm. What is the resistance of this strip?

15.6. For a current of 10 mA along the aluminum strip in Problem 15.5, calculate (a) the voltage along the length of the strip and (b) the power dissipated (I^2R).

D **15.7.** A structural design involves a steel wire 2 mm in diameter that will carry an electrical current. If the resistance of the wire must be less than 25 Ω, calculate the maximum length of the wire, given the data in Table 15.1.

D **15.8.** For the design discussed in Problem 15.7, calculate the allowable wire length if a 3-mm diameter is permitted.

15.2 • Energy Levels and Energy Bands

15.9. At what temperature will the 5.60-eV energy level for electrons in silver be 25% filled? (The Fermi level for silver is 5.48 eV.)

15.10. Generate a plot comparable to Figure 15–7 at a temperature of 1000 K for copper, which has a Fermi level of 7.04 eV.

15.11. What is the probability of an electron's being promoted to the conduction band in indium antimonide, InSb, at **(a)** 25°C and **(b)** 50°C? (The band gap of InSb is 0.17 eV.)

15.12. At what temperature will diamond have the same probability for an electron's being promoted to the conduction band as silicon has at 25°C? (The answer to this question indicates the temperature range in which diamond can be properly thought of as a semiconductor rather than as an insulator.)

15.13. Gallium forms semiconducting compounds with various group VA elements. The band gap systematically drops with increasing atomic number of the VA elements. For example, the band gaps for the III-V semiconductors GaP, GaAs, and GaSb are 2.25 eV, 1.47 eV, and 0.68 eV, respectively. Calculate the probability of an electron's being promoted to the conduction band in each of these semiconductors at 25°C.

15.14. The trend discussed in Problem 15.13 is a general one. Calculate the probability of an electron's being promoted to the conduction band at 25°C in the II-VI semiconductors CdS and CdTe, which have band gaps of 2.59 eV and 1.50 eV, respectively.

15.3 • Conductors

15.15. A strip of copper metallization on a solid-state device is 1 mm long with a thickness of 1 μm and a width of 6 μm. If a voltage of 0.1 V is applied along the long dimension, what is the resulting current?

15.16. A metal wire 1 mm in diameter × 15 m long carries a current of 0.1 A. If the metal is pure copper at 30°C, what is the voltage drop along this wire?

15.17. Repeat Problem 15.16, assuming that the wire is a Cu–0.1 wt % Al alloy at 30°C.

15.18. A type K thermocouple is operated with a reference temperature of 100°C (established by the boiling of distilled water). What is the temperature in a crucible for which a thermocouple voltage of 30 mV is obtained?

15.19. Repeat Problem 15.18 for the case of a chromel/constantan thermocouple.

15.20. A furnace for oxidizing silicon is operated at 1000°C. What would be the output (relative to an ice-water bath) for **(a)** a type S, **(b)** a type K, and **(c)** a type J thermocouple?

15.21. An important application of metal conductors in the field of materials processing is in the form of metal wire for resistance-heated furnace elements. Some of the alloys used as thermocouples also serve as furnace elements. For example, consider the use of a 1-mm-diameter chromel wire to produce a 1-kW furnace coil in a laboratory furnace operated at 110 V. What length of wire is required for this furnace design? (*Note:* The power of the resistance-heated wire is equal to I^2R, and the resistivity of the chromel wire is $1.08 \times 10^6 \Omega \cdot m$.)

15.22. Given the information in Problem 15.21, calculate the power requirement of a furnace constructed of a chromel wire 6 m long and 1 mm diameter operating at 110 V.

15.23. What would be the power requirement for the furnace in Problem 15.22 if it is operated at 208 V?

15.24. A tungsten light bulb filament is 9 mm long and 100 μm diameter. What is the current in the filament when operating at 1000°C with a line voltage of 110 V?

15.25. What is the power dissipation (I^2R) in the filament of Problem 15.24?

15.26. For a bulk 1–2–3 superconductor with a critical current density of 1×10^8 A/m^2, what is the maximum supercurrent that could be carried in a 1-mm-diameter wire of this material?

15.27. If progress in increasing T_c for superconductors had continued at the linear rate followed through 1975, by what year would a T_c of 95 K be achieved?

15.28. Verify the comment regarding the presence of one Cu^{3+} valence in the $YBa_2Cu_3O_7$ unit cell in the discussion of superconductors in Section 15.3.

15.29. Verify the chemical formula for $YBa_2Cu_3O_7$ using the unit cell geometry of Figure 15–19.

• **15.30.** Describe the similarities and differences between the perovskite unit cell of Figure 3–14 and **(a)** the upper and lower thirds and **(b)** the middle third of the $YBa_2Cu_3O_7$ unit cell of Figure 15–19.

15.4 • Insulators

15.31. Calculate the charge density on a 3 mm-thick capacitor made of 99.5% Al_2O_3 under an applied voltage of 1 kV.

15.32. Repeat Problem 15.31 for the same material at its breakdown voltage gradient (= dielectric strength).

15.33. Calculate the charge density on a capacitor made of cordierite at its breakdown dielectric strength of 3 kV/mm. The dielectric constant is 4.5.

15.34. By improved processing, a new cordierite capacitor can be made with properties superior to the one described in Problem 15.33. If the dielectric constant is increased to 5.0, calculate the charge density on a capacitor operated at a voltage gradient of 3 kV/mm (which is now below the breakdown strength).

15.35. An alternate definition of polarization (introduced in Sample Problems 15.11 and 15.12) is

$$P = (\kappa - 1)\epsilon_0 E$$

where κ, ϵ_0, and E were defined relative to Equations 15.11 and 15.13. Calculate the polarization for 99.9% Al_2O_3 under a field strength of 5 kV/mm. (You might note the magnitude of your answer in comparison to the inherent polarization of tetragonal $BaTiO_3$ in Sample Problem 15.12.)

15.36. Calculate the polarization of the acetal engineering polymer in Table 15.4 at its breakdown voltage gradient (= dielectric strength). (See Problem 15.35.)

15.37. As in Problem 15.36, consider the polarization at the breakdown voltage gradient. By how much does this value increase for the nylon polymer in Table 15.4 in the humid environment, as compared to the dry condition?

15.38. By heating $BaTiO_3$ to 100°C, the unit cell dimensions change to $a = 0.400$ nm and $c = 0.402$ nm (compared to the values in Figure 15–21). In addition, the ion shifts shown in Figure 15–20b are reduced by half. Calculate **(a)** the dipole moment and **(b)** the polarization of the $BaTiO_3$ unit cell at 100°C.

15.39. If the elastic modulus of $BaTiO_3$ in the c-direction is 109×10^3 MPa, what stress is necessary to reduce its polarization by 0.1%?

• **15.40.** A central part of the appreciation of the mechanisms of ferroelectricity and piezoelectricity is the visualization of the material's crystal structure. For the case of the tetragonal modification of the perovskite structure (Figure 15–21b), sketch the atomic arrangements in the **(a)** (100), **(b)** (001), **(c)** (110), **(d)** (101), **(e)** (200), and **(f)** (002) planes.

• **15.41.** As in Problem 15.40, sketch the atomic arrangements in the **(a)** (100), **(b)** (001), **(c)** (110), and **(d)** (101) planes in the *cubic* perovskite structure (Figure 15–21a).

• **15.42.** As in Problem 15.40, sketch the atomic arrangements in the **(a)** (200), **(b)** (002), and **(c)** (111) planes in the *cubic* perovskite structure (Figure 15–21a).

15.5 • Semiconductors

15.43. Calculate the fraction of Ge atoms that provides a conduction electron at room-temperature.

15.44. What fraction of the conductivity of intrinsic silicon at room-temperature is due to **(a)** electrons and **(b)** electron holes?

15.45. What fraction of the conductivity at room temperature for **(a)** germanium and **(b)** CdS is contributed by (i) electrons and (ii) electron holes?

15.46. Using the data in Table 15.5, calculate the room temperature conductivity of intrinsic gallium arsenide.

15.47. Using the data in Table 15.5, calculate the room temperature conductivity of intrinsic InSb.

15.48. What fraction of the conductivity calculated in Problem 15.47 is contributed by **(a)** electrons and **(b)** electron holes?

15.6 • Composites

15.49. Calculate the conductivity at 20°C **(a)** parallel to and **(b)** perpendicular to the W filaments in the Cu-matrix composite in Table 14.12.

15.50. Using the form of Equation 14.21 as a guide, estimate the electrical conductivity of the dispersion-strengthened aluminum in Table 14.12. (Assume the exponent, n, in Equation 14.21 to be $\frac{1}{2}$.)

15.51. Calculate the conductivity at 20°C for the composite in Table 14.12 in which W particles are dispersed in a copper matrix. (Use the same assumptions as in Problem 15.50.)

15.52. Plot the conductivity at 20°C of a series of composites composed of W filaments in a Cu matrix. Show the extreme cases of conductivity **(a)** parallel to and **(b)** perpendicular to the filaments. As in Figure 14–13, allow the volume fraction of filaments to vary from 0 to 1.0.

CHAPTER 16
Optical Behavior

Optical behavior is often a critical factor in materials selection. An example is this automobile headlamp bezel constructed of a metal-coated polybutylene terephthalate (PBT) polymer. The bezel is placed between the reflector and the lens to enhance the aesthetics of the headlamp. The PBT polymer was chosen due to its combination of good optical reflectivity (after metal-coating), processing ability, and low warpage. (Courtesy of DuPont Automotive.)

For some materials, their optical behavior—the way in which they reflect, absorb, or transmit *visible light*—is more important than their mechanical behavior (Chapter 6). Optical behavior is intimately related to electrical
behavior (Chapter 15). The glasses of Chapter 12 are classic examples of the
primary role of optical behavior in the application of an important structural
material. The explosive growth of the communications industry at the beginning of the new millennium involves numerous applications of sophisticated
new forms of optical behavior.

Along with the x-rays introduced in Chapter 3, visible light is part
of the electromagnetic radiation spectrum. The wide use of glasses and
certain crystalline ceramics and organic polymers for optical applications
requires focusing on a number of optical properties. The refractive index is
a fundamentally important property with implications about the nature of
light reflection at the material surface and transmission through the bulk.
The transparency of a given material is limited by the nature of any second-
phase microstructure (porosity or a solid phase with an index of refraction
different from the matrix). The coloration of light-transmitting materials
results from the absorption of certain light wavelengths by species such as
ionic Fe^{2+} and Co^{2+}. A wide range of modern optical applications involves
materials capable of luminescence, the absorbing of energy followed by the
emission of visible light. The characteristic reflectivity and opacity of metals
is a direct consequence of the high density of conductive electrons in these
materials.

Some of the most important systems and devices in modern technology
are useful to us because of their optical behavior. Examples are lasers, optical
fibers, liquid crystal displays, and photoconductors.

16.1   VISIBLE LIGHT

To appreciate the nature of optical behavior, we must return to the electromagnetic radiation spectrum introduced in Section 3.7. Figure 16–1 is nearly

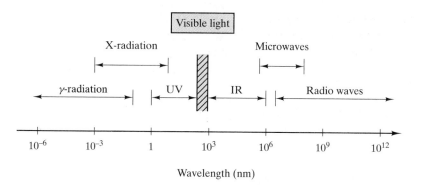

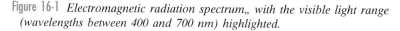

Figure 16-1 *Electromagnetic radiation spectrum,, with the visible light range
(wavelengths between 400 and 700 nm) highlighted.*

Figure 16-2   *The wavelike nature of an electromagnetic wave, such as light. Both the electric field (E) and magnetic field (H) components are sinusoidal, and the oscillations of E and H occur in perpendicular planes. The wavelength, λ, and the speed of light, c, are noted.*

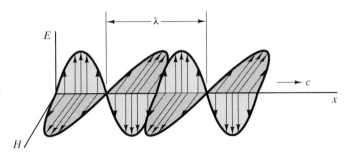

equivalent to Figure 3–34 except that **visible light**, rather than x-radiation, is highlighted. Visible light is that portion of the electromagnetic spectrum that can be perceived by the human eye. In general, this corresponds to the wavelength range of 400 to 700 nm. The wavelike nature of light is shown in Figure 16–2, which indicates the periodic variation of electric and magnetic field components along the direction of propagation. In vacuum, the speed of light, c, is $0.2998 \times 10^9 m/s$. An elegant demonstration of the relationship of light to electrical and magnetic properties is the fact that the speed of light is given precisely in terms of two fundamental constants

$$c = \frac{1}{\sqrt{\epsilon_0 \mu_0}}$$ (16.1)

where ϵ_0 is the electric permittivity of a vacuum and μ_0 is the magnetic permeability of a vacuum. As with any waveform, the frequency, v, is related to the wavelength, λ, through the velocity of the wave. In the case of light waves,

$$v = c/\lambda$$ (16.2)

In Section 2.1, we considered electrons as examples of the wave-particle duality, exhibiting both wavelike and particle like behavior. Electromagnetic radiation can be viewed in a similar way, with its particle like behavior consisting of packets of energy called **photons**. The energy, E, of a given photon is expressed by

$$E = hv = h(c/\lambda)$$ (16.3)

where h is Planck's* constant ($= 0.6626 \times 10^{-33}$ J·s).

---

* Max Karl Ernst Ludwig Planck (1858–1947), German physicist. His lifetime spanned the transition between the nineteenth and twentieth centuries. This is symbolic of his contribution in bridging classical (nineteenth century) and modern (twentieth century) physics. He introduced Equation 16.3 and the term *quantum* in 1900 while developing a successful model of the energy spectrum from a "blackbody radiator." The prestigious Max Planck Institutes in Germany bear his name.

The Material World:
A Flat Panel Display of a Medical X-Ray

A major application of modern optical materials is the increasing use of flat panel displays (FPD's) to replace bulky, television-like CRT (cathode-ray tube) displays. A major use of liquid crystal displays discussed later in this chapter is the FPD in laptop computers. A state-of-the-art application of the FPD is to provide images of medical x-rays equivalent in quality to conventional film.

The figure shows an ultra-high-resolution, 0.48 m (19-inch) diagonal flat-panel display that produces digital x-ray images with filmlike clarity, making it easier for doctors and medical professionals to view and analyze patient x-rays and medical records on line. Previously, online collaboration in hospitals and medical centers had been limited by the resolution and clarity constraints of CRTs. This flat-panel display, with a resolution of 5.2 million pixels (each with 256 gray levels), can produce sharper, clearer, distortion-free images. This system combines the advantages of digital archiving and retrieval with the conventional "feel" of a back-lighted film viewing box.

Digital radiographs are prime examples of the emerging field of tele-medicine. In addition to making vital patient information more accessible, digital radiography eliminates the time, expense, and materials consumption associated with film processing. This particular FPD is optimized for large, grayscale image files, including chest x-rays and mammograms. As shown in the figure (on the right), it is also well suited for viewing multiple images in ultrasound (such as fetal monitoring). The FPD is also effective for displaying computer-aided tomography (CAT) scans and the closely related magnetic resonance imaging (MRI) scans. In these latter cases, an entire series of images can be viewed at full resolution and clarity. The FPD can also support full-motion video for real-time fluoroscopic procedures.

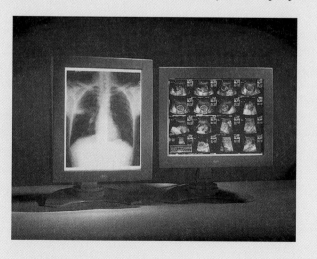

(Courtesy of dpiX, Incorporated)

SAMPLE PROBLEM 16.1

Calculate the energy of a single photon from the short wavelength (blue) end of the visible spectrum (at 400 nm).

SOLUTION

Equations 16.3 gives

$$E = \frac{hc}{\lambda}$$

$$= \frac{(0.6626 \times 10^{-33}\,\text{J·s})(0.2998 \times 10^9 m/s)}{400 \times 10^{-9}\,\text{m}}$$

$$\times \frac{6.242 \times 10^{18}\,\text{eV}}{J}$$

$$= 3.1\,\text{eV}$$

Due to the inverse relationship between E and λ, this will be the most energetic visible light photon.

....

PRACTICE PROBLEM 16.1

Calculate the energy of a single photon from the long wavelength (red) end of the visible spectrum (at 700 nm). (See Sample Problem 16.1.)

16.2 OPTICAL PROPERTIES

We shall initially concentrate on the various ways in which visible light interacts with the most common of optical materials, the oxide glasses of Chapter 12. These various optical properties also apply to many other nonmetallic materials, such as light-transmitting crystalline ceramics (Chapter 12) and organic polymers (Chapter 13). Finally, we shall note that metals (Chapter 11) do not generally transmit visible light but do have characteristic reflectivities and colors.

REFRACTIVE INDEX

When light travels from air into a transparent material at an angle, the light is refracted or "bent" (its path is changed). One of the most fundamental optical properties is the **refractive index**, n, defined as

$$n = \frac{v_{\text{vac}}}{v} = \frac{\sin \theta_i}{\sin \theta_r} \qquad (16.4)$$

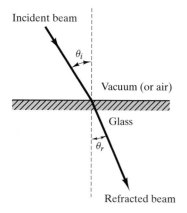

Figure 16-3   *Refraction of light as it passes from vacuum (or air) into a transparent material.*

where v_{vac} is the speed of light in vacuum (essentially equal to that in air), v is the speed of light in a transparent material, and θ_i and θ_r are angles of incidence and refraction, respectively, as defined by Figure 16–3. Typical values of n for ceramics and glasses run from 1.5 to 2.5 and for polymers from 1.4 to 1.6, meaning that the speed of light is considerably less in the solid than in vacuum. Table 16.1 gives values of n for several ceramics and glasses. Most silicate glasses have a value close to $n = 1.5$. Table 16.2 gives values of n for various polymers.

One implication of the magnitude of the refractive index, n, is characteristic appearance. The distinctive "sparkle" associated with diamonds and art glass pieces is the result of a high value of n, which allows multiple internal reflections of light. Additions of lead ($n = 2.60$) to silicate glasses raise the refractive index, giving the distinctive appearance (and associated expense) of fine "crystal" glassware.

Table 16.1   *Refractive Index for Various Ceramics and Glasses*

Material	Average refractive index
Quartz (SiO_2)	1.55
Mullite ($3Al_2O_3 \cdot 2SiO_2$)	1.64
Orthoclase ($KAlSi_3O_8$)	1.525
Albite ($NaAlSi_3O_8$)	1.529
Corundum (Al_2O_3)	1.76
Periclase (MgO)	1.74
Spinel ($MgO \cdot Al_2O_3$)	1.72
Silica glass (SiO_2)	1.458
Borosilicate glass	1.47
Soda-lime-silica glass	1.51–1.52
Glass from orthoclase	1.51
Glass from albite	1.49

Source: W. D. Kingery, H. K. Bowen, and D. R. Uhlmann, *Introduction to Ceramics*, 2nd ed., John Wiley & Sons, Inc., New York, 1976.

Table 16.2 *Refractive Index for Various Polymers*

Polymer	Average refractive index
Thermoplastic polymers	
Polyethylene	
High-density	1.545
Low-density	1.51
Polyvinyl chloride	1.54–1.55
Polypropylene	1.47
Polystyrene	1.59
Cellulosics	1.46–1.50
Polyamides (nylon 66)	1.53
Polytetrafluoroethylene (Teflon)	1.35–1.38
Thermosetting polymers	
Phenolics (phenol-formaldehyde)	1.47–1.50
Urethanes	1.5–1.6
Epoxies	1.55–1.60
Elastomers	
Polybutadiene/polystyrene copolymer	1.53
Polyisoprene (natural rubber)	1.52
Polychloroprene	1.55–1.56

Source: Data from J. Brandrup and E. H. Immergut, eds., *Polymer Handbook,* 2nd Ed., John Wiley & Sons, Inc., New York, 1975.

REFLECTANCE

Not all light striking a transparent material is refracted as described previously. Some is reflected at the surface as shown in Figure 16–4. The angle of reflection equals the angle of incidence. The **reflectance**, R, is defined as the fraction of light reflected at such an interface and is related to the index of refraction by **Fresnel's* formula:**

$$R = \left(\frac{n-1}{n+1}\right)^2 \tag{16.5}$$

Equation 16.5 is strictly valid for normal incidence ($\theta_i = 0$) but is a good approximation over a wide range of θ_i. It is apparent that high-n materials are also highly reflective. In some cases, as in glossy enamel coatings, this is desirable. In other cases, such as lens applications, high reflectivity produces undesirable light losses. Special coatings are frequently used to minimize this problem (Figure 16–5).

---

* Augustin Jean Fresnel (1788–1827), French physicist, is remembered for many contributions to the theory of light. His major advance was to identify the transverse mode of light-wave propagation.

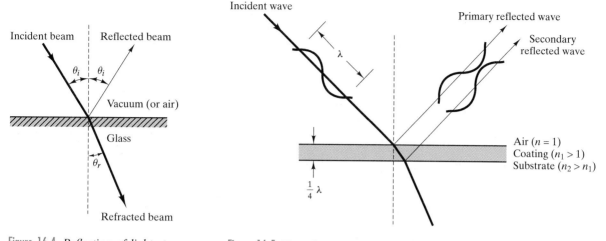

Figure 16-4 *Reflection of light at the surface of a transparent material occurs along with refraction.*

Figure 16-5 *Use of a "one-quarter-wavelength" thick coating minimizes surface reflectivity. The coating has an intermediate index of refraction and the primary reflected wave is just canceled by the secondary reflected wave of equal magnitude and opposite phase. Such coatings are commonly used on microscope lenses.*

The general appearance of a given material is strongly affected by the relative amounts of specular and diffuse reflection. **Specular reflection** is defined by Figure 16–6 as reflection relative to the "average" surface, while **diffuse reflection** is reflection due to surface roughness, where, locally, the true surface is not parallel to the average surface. The net balance between spectral and diffuse reflection for a given surface is best illustrated by **polar diagrams**. Such diagrams indicate the intensity of reflection in a

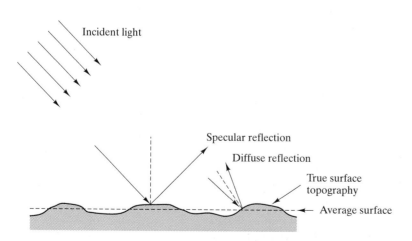

Figure 16-6 *Specular reflection occurs relative to the "average" surface, and diffuse reflection occurs relative to locally nonparallel surface elements.*

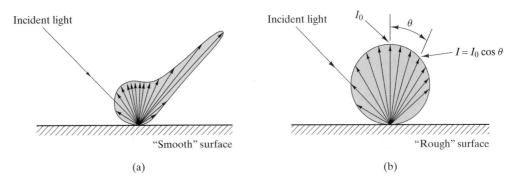

(a) (b)

Figure 16-7 *Polar diagrams illustrate the directional intensity of reflection from (a) a "smooth" surface with predominantly specular reflection and (b) a "rough" surface with completely diffuse reflection.*

given direction by the relative length of a vector. Figure 16–7 shows two polar diagrams, distinguishing (a) a "smooth" or "mirror-like" surface with predominantly specular reflection and (b) a "rough" surface with completely diffuse reflection. The perfectly circular polar diagram for Figure 16–7b is an example of the **cosine law** of scattering. The relative intensity of reflection varies as the cosine of the angle, θ, defined in Figure 16–7b:

$$I_\theta = I_0 \cos \theta \tag{16.6}$$

I_0 is the intensity of scattering at $\theta = 0°$. Since any area segment A_θ will be foreshortened when viewed at the angle θ, the brightness of the diffuse surface of Figure 16–7b will be a constant independent of viewing angle:

$$\text{brightness} = \frac{I_\theta}{A_\theta} = \frac{I_0 \cos \theta}{A_0 \cos \theta} = \text{constant} \tag{16.7}$$

We can now summarize that glasses and glassy coatings (glazes and enamels) will have high **surface gloss** due to a large index of refraction (Equation 16.5) and a smooth surface (Figure 16–7a).

TRANSPARENCY, TRANSLUCENCY, AND OPACITY

Many ceramics, glasses and polymers are effective media for transmitting light. The degree of transmission is indicated by the terms transparency, translucency, and opacity. **Transparency** simply means the ability to transmit a clear image. In Chapter 1 we saw that the elimination of porosity made polycrystalline Al_2O_3 a nearly transparent material. Translucency and opacity are somewhat more subjective terms for materials that are not transparent. In general, **translucency** means that a diffuse image is transmitted, and **opacity** means total loss of image transmission. The case of translucency is illustrated by Figures 1–20c and d and 16–8. The microscopic

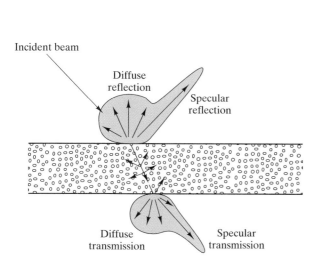

Figure 16-8 *Polar diagrams illustrate reflection and transmission of light through a translucent plate of glass. (From W. D. Kingery, H. K. Bowen, and D. R. Uhlmann,* Introduction to Ceramics, *2nd ed., John Wiley & Sons, Inc., New York, 1976.)*

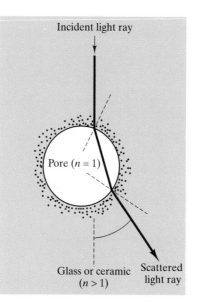

Figure 16-9 *Light scattering is the result of local refraction at interfaces of second-phase particles or pores. The case for scattering by a pore is illustrated here.*

mechanism of scattering is, as implied previously, the scattering of light by small second-phase pores or particles. Figure 16–9 illustrates how scattering can occur at a single pore by refraction. When porosity produces opacity, the refraction is due to the different indices of refraction with $n = 1$ for the pore and $n > 1$ for the solid. Many glasses and glazes contain "opacifiers," which are second-phase particles such as SnO_2 with an index of refraction ($n = 2.0$) greater than that of the glass ($n \simeq 1.5$). The degree of opacification caused by the pores or particles depends on their average size and concentration as well as mismatch of indices of refraction. If individual pores or particles are significantly smaller than the wavelength of light (400 to 700 nm), they are ineffective scattering centers. The scattering effect is maximized by pore or particle sizes within the range of 400 to 700 nm. Pore-free polymers are relatively easy to produce. In polymers, opacity is frequently due to the presence of inert additives (see Section 13.5).

COLOR

The opacity of ceramics, glasses, and polymers was just seen to be based on a scattering mechanism. By contrast, the opacity of metals is the result of an absorption mechanism intimately associated with their electrical conductivity. The conduction electrons absorb photons in the visible-light range,

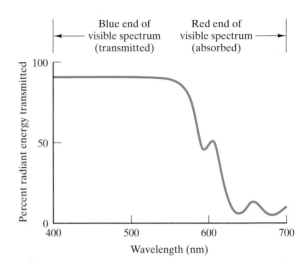

Figure 16-10 *Absorption curve for a silicate glass containing about 1% cobalt oxide. The characteristic blue color of this material is due to the absorption of much of the red end of the visible-light spectrum.*

giving a characteristic opacity to all metals. The absence of conduction electrons in ceramics and glasses accounts for their transparency. However, there is an absorption mechanism for these materials that leads to an important optical property, **color**.

In ceramics and glasses, coloration is produced by the selective absorption of certain wavelength ranges within the visible spectrum due to electron transitions in transition metal ions. As an example, Figure 16–10 shows an absorption curve for a silicate glass containing about 1% cobalt oxide (with the cobalt in the form of Co^{2+} ions). While much of the visible spectrum is efficiently transmitted, much of the red, or long-wavelength, end of the spectrum is absorbed. With the red end of the spectrum subtracted out, the net glass color is blue. The colors provided by various metal ions are summarized in Table 16.3. A single ion (such as Co^{2+}) can give different coloring in different glasses. The reason is that the ion has different coordination numbers in the different glasses. The magnitude of the energy transition for a photon-absorbing electron is affected by ionic coordination. Hence the absorption curve varies and, with it, the net color.

The general problem of light transmission is a critical one to the glass fibers used in modern fiber-optics telecommunications systems (See Section 16.3). Single fibers several kilometers long must be manufactured with a minimum of scattering centers and light-absorbing impurity ions.

For polymers, the additives introduced in Section 13.5 include **colorants**, inert *pigments* such as titanium oxide that produce opaque colors. Transparent color is provided by *dyes* that dissolve in the polymer, eliminating the mechanism of light scattering. The specific mechanism of color production in dyes is similar to that for pigments (and ceramics); that is, part of the visible light spectrum is absorbed. No simple table of color sources is

Table 16.3 *Colors Provided by Various Metal Ions in Silicate Glasses*

	In glass network		In modifier position	
Ion	Coordination number	Color	Coordination number	Color
Cr^{2+}				Blue
Cr^{3+}			6	Green
Cr^{6+}	4	Yellow		
Cu^{2+}	4		6	Blue-green
Cu^+			8	Colorless
Co^{2+}	4	Blue-purple	6–8	Pink
Ni^{2+}		Purple	6–8	Yellow-green
Mn^{2+}		Colorless	8	Weak orange
Mn^{3+}		Purple	6	
Fe^{2+}			6–8	Blue-green
Fe^{3+}		Deep brown	6	Weak yellow
U^{6+}		Orange	6–10	Weak yellow
V^{3+}			6	Green
V^{4+}			6	Blue
V^{5+}	4	Colorless		

Source: F. H. Norton, *Elements of Ceramics,* 2nd ed., Addison-Wesley
        Publishing Co., Inc., Reading, Mass., 1974.

available for dyes, in contrast to Table 16.3 for color in silicate glasses. While the mechanism of light absorption is the same, color formation with dyes is a complex function of molecular chemistry and geometry.

LUMINESCENCE

We have just seen that color is a result of the absorption of some of the photons within the visible light spectrum. Another byproduct is **luminescence**, in which photon absorption is accompanied by the reemission of some photons of visible light. The term *luminescence* is also used to describe the emission of visible light accompanying the absorption of other forms of energy (thermal, mechanical, and chemical) or particles (e.g., high-energy electrons). In fact, any emission of light from a substance for any reason other than a rise in its temperature can be termed luminescence. In general, atoms of a material emit photons of electromagnetic energy when they return to the ground state after having been in an exited state due to the absorption of energy. (Figure 16–11).

Time is a factor in distinguishing two types of luminescence. If reemission occurs rapidly (in less than about 10 nanoseconds), the phenomenon is generally termed **fluorescence**. For longer times, the phenomenon is termed **phosphorescence**. A wide variety of materials exhibit these phenomena, including many ceramic sulfides and oxides. Ordinarily, the phenomena are produced by the controlled addition of impurities.

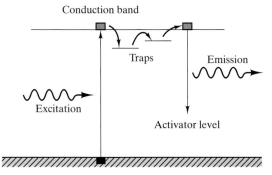

Figure 16-11   *Schematic illustration of a mechanism for luminescence. Various trap and activator energy levels within the energy band gap are produced by impurity additions to the insulating material. Following the excitation of an electron from the valence band to the conduction band, the electron moves among the traps without emitting radiation, is thermally promoted back to the conduction band, and eventually decays to the activator level with the emission of a photon of light. (After R. M. Rose, L. A. Shepard, and J. Wulff, The Structure and Properties of Materials, Vol. 4: Electronic Properties, John Wiley & Sons, Inc., New York, 1966.)*

A common example of luminescence is the fluorescent lamp in which the glass housing is coated on the inside with a tungstate or silicate film. Ultraviolet light generated in the tube from a mercury glow discharge causes the coating to fluoresce and emit white light. Similarly, the inside of a television screen is coated with a material that fluoresces as an electron beam is scanned rapidly back and forth.

Additional terminology can be assigned, based on the specific source of the energy that leads to the excited state for the atoms. For example, the term for a source of photons is **photoluminescence**, and, for a source of electrons, the term is **electroluminescence**.

REFLECTIVITY AND OPACITY OF METALS

In the discussion of color, we noted that the opacity of metals is the result of the absorption of the entire visible light spectrum by the metal's conduction electrons. Metal films greater than about 100 nm are totally absorbing. The entire range of visible light wavelength is absorbed because of the continuously available empty electron states, represented by the unfilled valance band of Figure 15–6. Between 90% and 95% of the light absorbed at the outer surface of the metal is reemitted from the surface in the form of visible light of the same wavelength. The remaining 10% to 5% of the energy is dissipated as heat. The reflectivity of the metal surface is illustrated by Figure 16–12.

The distinctive color of certain metals is the result of a wavelength dependency for the reflectivity. Copper (red-orange) and gold (yellow) have a lower reemission of the short wavelength (blue) end of the visible spectrum. The bright, silvery appearance of aluminum and silver is the result of the uniform reemission of wavelengths across the entire spectrum (i.e., white light).

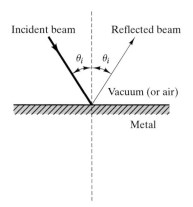

Figure 16-12 *Reflection of light at the surface of an opaque metal occurs without refraction.*

SAMPLE PROBLEM 16.2

When light passes from a high index of refraction medium to a low index of refraction medium, there is a critical angle of incidence, θ_c, beyond which no light passes the interface. This θ_c is defined at $\theta_{\text{refraction}} = 90°$. What is θ_c for light passing from silica glass to air?

SOLUTION

This is the inverse of the case shown in Figure 16–3. Here θ_i is measured in the glass and θ_r is measured in air; that is,

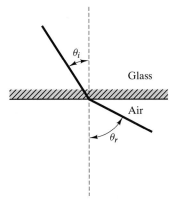

Correspondingly, Equation 16.4 has the form

$$\frac{\sin \theta_i}{\sin \theta_r} = \frac{v_{\text{glass}}}{v_{\text{air}}} = \frac{1}{n}$$

At the critical condition,

$$\frac{\sin \theta_c}{\sin 90°} = \frac{1}{n}$$

or

$$\theta_c = \arcsin \frac{1}{n}$$

From the value of n for silica glass in Table 16.1,

$$\theta_c = \arcsin \frac{1}{1.458} = 43.3°$$

Note. This is the basis of the excellent efficiency of silica glass fibers for light transmission. Light in small-diameter fibers travels along a path nearly parallel to the glass/air surface and at a θ_i well above 43.3°. As a result, light can travel along such fibers for several kilometers with only modest transmission losses. There is total internal reflection and no losses due to refraction to the surrounding environment.

SAMPLE PROBLEM 16.3

Using Fresnel's formula, calculate the reflectance, R, of a sheet of polystyrene.

SOLUTION

Using Equation 16.5, we have

$$R = \left(\frac{n-1}{n+1}\right)^2$$

Using the value of n from Table 16.2 gives us

$$R = \left(\frac{1.59-1}{1.59+1}\right)^2$$

$$= 0.0519$$

SAMPLE PROBLEM 16.4

Compare the reflectance of silica glass with that for pure PbO ($n = 2.60$).

SOLUTION

This is an application of Equation 16.5:

$$R = \left(\frac{n-1}{n+1}\right)^2$$

Using n for silica glass from Table 16.1 gives us

$$R_{\text{SiO}_2,\,\text{gl}} = \left(\frac{1.458-1}{1.458+1}\right)^2 = 0.035$$

For PbO,

$$R_{\text{PbO}} = \left(\frac{2.60-1}{2.60+1}\right)^2 = 0.198$$

or

$$\frac{R_{\text{PbO}}}{R_{\text{SiO}_2,\,\text{gl}}} = \frac{0.198}{0.035} = 5.7$$

SAMPLE PROBLEM 16.5

Calculate the range of magnitude of energy transitions that are involved in the absorption of visible light by transition metal ions.

SOLUTION

In each case, the absorption mechanism involves a photon being consumed by giving its energy as given by Equation 16.3 ($E = hc/\lambda$) to an electron that is promoted to a higher energy level. The ΔE of the electron is equal in magnitude to the E of the photon.

The wavelength range of visible light is 400 to 700 nm. Therefore,

$$\Delta E_{blue\ end} = E_{400\ nm} = \frac{hc}{400\ nm}$$

$$= \frac{(0.663 \times 10^{-33}\ \text{J} \cdot \text{s})(3.00 \times 10^{8}\ \text{m/s})}{400 \times 10^{-9}\ \text{m}}$$

$$= 4.97 \times 10^{-19}\ \text{J} \times 6.242 \times 10^{18}\ \text{eV/J} = 4.88\ \text{eV}$$

$$\Delta E_{red\ end} = E_{700\ nm} = \frac{hc}{700\ nm}$$

$$= \frac{(0.663 \times 10^{-33}\ \text{J} \cdot \text{s})(3.00 \times 10^{8}\ \text{m/s})}{700 \times 10^{-9}\ \text{m}}$$

$$= 2.84 \times 10^{-19}\ \text{J} \times 6.242 \times 10^{18}\ \text{eV/J} = 1.77\ \text{eV}$$

$$\Delta E \text{ range: } 2.84 \times 10^{-19} \text{ to } 4.97 \times 10^{-19}\ \text{J} (= 1.77 \text{ to } 4.88\ \text{eV})$$

PRACTICE PROBLEM 16.2

In Sample Problem 16.2 a critical angle of incidence is calculated for light refraction from silica glass to air. What would be the critical angle if the air were to be replaced by a water environment (with $n = 1.333$)?

PRACTICE PROBLEM 16.3

Using Fresnel's formula, calculate the reflectance, R, of **(a)** a sheet of polypropylene and **(b)** a sheet of polytetrafluoroethylene (with an average refractive index of 1.35). (See Sample Problem 16.3.)

PRACTICE PROBLEM 16.4

What is the reflectance of single-crystal sapphire, which is widely used as an optical and electronic material? (Sapphire is nearly pure Al_2O_3.) (See Sample Problem 16.4.)

The relationship between photon energy and wavelength is discussed in Sample Problem 16.5. A useful rule of thumb is that E (in electron volts) $= K\lambda$, where λ is expressed in nanometers. What is the value of K?

16.3   OPTICAL SYSTEMS AND DEVICES

We shall now focus on a few systems and devices that have, in the past few decades, gone from being research discoveries to becoming major components of modern technology. Ongoing, active research on new optical materials promises to provide dramatic new developments in the decades ahead.

LASERS

In Section 16.2, the common fluorescent light was given as an example of luminescence. This traditional light source is said to be **incoherent**, because the electron transitions that produce the light waves occur randomly so that the light waves are out of phase with each other. An important discovery in the late 1950s[*] was light amplification by stimulated emission of radiation, known now simply by the acronym **laser**, which provides a **coherent** light source in which the light waves are in phase.

Several types of lasers have been developed, but the principle of operation can be demonstrated by a common solid-state design. Single-crystal Al_2O_3 is sometimes called sapphire when relatively pure and ruby when it contains enough Cr_2O_3 in solid solution to provide a characteristic red color due to the Cr^{3+} ions. (Recall the discussion of color in Section 16.2.) The ruby laser is illustrated in Figure 16–13. The ruby is illuminated by 560-nm-wavelength photons from the surrounding xenon flash lap, exciting electrons in Cr^{3+} ions from their ground states to an excited state (Figure 16–14). Although some electrons can decay directly back to the ground state, others decay into a metastable, intermediate state, as shown in Figure 16–14, where they may reside for up to 3 ms before decaying to the ground state. The time span of 3 ms is sufficiently long that several Cr^{3+} ions can reside in this metastable excited state simultaneously. Then an initial, spontaneous decay to the ground state by a few of these electrons produces photon emissions that trigger an avalanche of emissions from the remaining electrons in the metastable state.

---

[*] C. H. Townes and A. L. Schawlow, U.S. Patent 2,929,922, March 22, 1960.

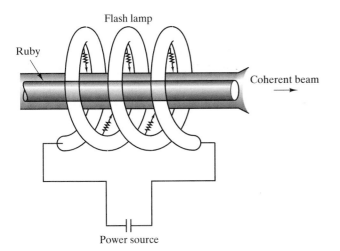

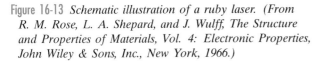

Figure 16-13 *Schematic illustration of a ruby laser. (From R. M. Rose, L. A. Shepard, and J. Wulff, The Structure and Properties of Materials, Vol. 4: Electronic Properties, John Wiley & Sons, Inc., New York, 1966.)*

A schematic of the overall sequence of stimulated emission and light amplification is shown in Figure 16–15. Note that one end of the cylindrical ruby crystal is fully silvered and the other end is partially silvered. Photons can be emitted in all directions, but those traveling nearly parallel to the long axis of the ruby crystal are the ones that contribute to the laser effect. Over-

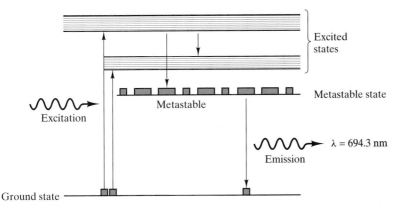

Figure 16-14 *Schematic illustration of the mechanism for the excitation and decay of electrons of a Cr^{3+} ion in a ruby laser. Although a ground-state electron can be promoted to various activated states, only the final decay from the metastable to the ground state produces laser photons of wavelength 694.3 nm. A residence time of up to 3 ms in the metastable state allows a large number of Cr^{3+} ions to emit together, producing a large light pulse. (After R. M. Rose, L. A. Shepard, and J. Wulff, The Structure and Properties of Materials, Vol. 4: Electronic Properties, John Wiley & Sons, Inc., New York, 1966.)*

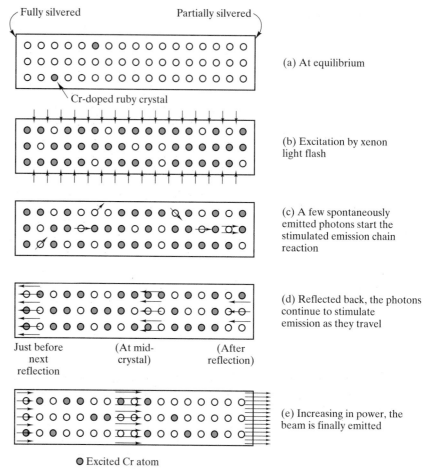

Figure 16-15 *Schematic illustration of stimulated emission and light amplification in a ruby laser. (From R. M. Rose, L. A. Shepard, and J. Wulff, The Structure and Properties of Materials, Vol. 4: Electronic Properties, John Wiley & Sons, Inc., New York, 1966.)*

all, the light beam is reflected back and forth along the rod, and its intensity increases as more emissions are stimulated. The net result is a high-intensity, coherent, and highly collimated laser beam pulse transmitted through the partially silvered end. The resulting single value (**monochromatic**) wavelength of 694.3 nm is in the red end of the visible spectrum.

Semiconductors such as gallium arsenide can also make useful lasers. In this case, the visible light photons are produced by the recombination of electrons and electron holes across the band gap, following the production of electron-hole pairs by a voltage application (recall Figure 15–9). The band gap, E_g, must be in an appropriate range if the resulting photon wavelength

Table 16.4 *Some Important Commercial Lasers*

Laser	Wavelengths (μm)	Output Type and Power
Gas		
He–Ne	0.5435–3.39	CW,[a] 1–50 mW
Liquid		
Dye	0.37–1.0	CW,[a] to a few watts
Dye	0.32–1.0	Pulsed, to tens of watts
Glass		
Nd–silicate	1.061	Pulsed, to 100 W
Solid-state		
Ruby	0.694	Pulsed, to a few watts
Semiconductor		
InGaAsP	1.2–1.6	CW,[a] to 100 mW

Source: Data from J. Hecht, in *Electrical Engineering Handbook,* R. Dorf, Ed., CRC Press, Boca Raton, Fl, 1993.
[a] CW = continuous wave

is to be in the visible range of 400 nm to 700 nm. The photon wavelength can be determined by rearranging Equation 16.3, giving

$$\lambda = hc/E_g \qquad (16.8)$$

Table 16.4 summarizes various types of commercial lasers. Lasers have been constructed from hundreds of materials giving emissions at several thousand different wavelengths in laboratories around the world. Widely used commercial lasers tend to be those that are relatively easy to operate, high in power, and energetically efficient (where a 1% conversion of the input energy into light is considered good). The laser medium can be a gas, liquid, glass, or crystalline solid (insulator or semiconductor). The power in a continuous beam, commercial laser can range from microwatts to 25 kW, and more than a megawatt in defense applications. Pulsed lasers deliver much higher peak power but comparable levels to the continuous lasers when averaged over time. Most gas and solid-state lasers emit beams with a low divergence angle of approximately one milliradian, although semiconductor lasers typically produce beams that spread out over an angle of 20 to 40 degrees. The laser principle is not confined to the visible spectrum, and some lasers emit radiation in the infrared or the ultraviolet. Lasers are light sources for various optical communication systems. Because of the highly coherent nature of the beam, lasers can be used for high-precision distance measurements. Focused laser beams can provide local heating for cutting, welding, and even surgical procedures.

OPTICAL FIBERS

As discussed in Chapters 11 through 14, the replacement of metals by non-metals has become a central issue for structural materials. A similar phenomenon has occurred in the area of telecommunications, although for quite

Figure 16-16 *The small cable on the right contains 144 glass fibers and can carry more than three times as many telephone conversations as the traditional (and much larger) copper wire cable on the left. (Courtesy of the San Francisco Examiner.)*

different reasons. A major revolution in this field has occurred with the transition from traditional metal cable to optical glass fibers (Figure 16–16). Although Alexander Graham Bell had transmitted speech several hundred meters over a beam of light shortly after his invention of the telephone, technology did not permit the practical, large-scale application of this concept for nearly a century. The key to the rebirth of this approach was the invention of the laser in 1960. By 1970, researchers at Corning Glass Works had developed an **optical fiber** with a loss as low as 20 dB/km at a wavelength of 630 nm (within the visible range). By the mid–1980s, silica fibers had been developed with losses as low as 0.2 dB/km at 1.6 μm (in the infrared range). As a result, telephone conversations and any other form of digital data can be transmitted as laser light pulses rather than the electrical signals used in copper cables. Glass fibers are excellent examples of **photonic materials** in which signal transmission is by photons rather than by the electrons of electronic materials.

Glass fiber bundles of the type illustrated in Figure 16–16 were put into commercial use by the Bell Systems in the mid–1970s. The reduced expense and size, combined with an enormous capacity for data transmission, have led to a rapid growth in the construction of optical communication systems. Now, virtually all telecommunications are transmitted in this way. Ten billion digital bits can be transmitted per second along an optical fiber in a contemporary system carrying tens of thousands of telephone calls.

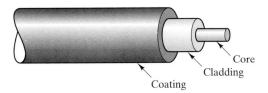

Figure 16-17 *Schematic illustration of the coaxial design of commercial optical fibers.*

As noted in Sample Problem 16.2, light can pass along a glass fiber with great efficiency because of total internal reflection and no losses due to refraction to the surrounding environment. This was seen to be a direct consequence of the critical angle of incidence for light passing from a high index of refraction medium to a low index of refraction medium. As a practical matter, commercial optical fibers are not bare glass. The glass is a core embedded in a cladding that is, in turn, covered by a coating (Figure 16–17). The light pulse signals pass along the core. The cladding is a lower index of refraction glass, providing the phenomenon of total internal reflection similar to that in Sample Problem 16.2, although the critical angle will be larger because the cladding index of refraction is much closer to that of the core than to air. The coating protects the core and cladding from environmental damage.

The core is made from a high-purity silica glass ranging in diameter from 5 μm to 100 μm. Figure 16–18 shows three common configurations

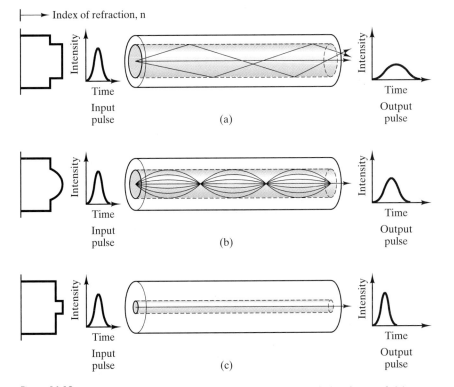

Figure 16-18 *Schematic illustration of (a) step-index, (b) graded-index, and (c) single-mode optical fiber designs.*

for commercial optical fibers. Figure 16–18a shows the **step-index fiber**, which has a large core (up to 100 μm and a sharp step-down in index of refraction at the core/cladding interface. Those light rays following various zigzag paths along the core take different times to traverse the entire fiber; this phenomenon leads to a broadening of the light pulse and limits the number of pulses that can be transmitted per second. This type of fiber is most appropriate for relatively small transmission distances such as in medical endoscopes.

Figure 16–18b shows the **graded-index fiber**, in which a parabolic variation in index of refraction within the core (produced by systematic additions of other glass formers such as B_2O_3 and GeO_2 to the silica glass) causes helical rather than zigzag paths. The various paths arrive at a receiver at nearly the same time as the beam going along the fiber axis. (Note that the fiber axis path is shorter but slower due to the higher index of refraction along the center). The digital pulse is then less distorted, allowing a higher density of information transmission. These fibers are widely used for local area networks.

Figure 16–18c shows the **single-mode fiber** with a narrow core (5μm $\leq 8\mu$m) in which light travels largely parallel to the fiber axis with little distortion of the digital pulse. Single-mode refers to the essentially singular, axial path and correspondingly undistorted pulse shape. By contrast, the step-index and graded-index designs are termed multimode. Single-mode fibers are used in millions of kilometers of fiber in telephone and cable television networks.

LIQUID CRYSTAL DISPLAYS

Liquid crystal polymers, as with the metallic quasicrystals, fall outside the conventional definitions of crystalline and noncrystalline structures. Figure 16–19 illustrates their distinctive character. Composed of extended and rigid rod-shaped molecules, the liquid crystal molecules are highly aligned even in the melt. Upon solidification, the molecular alignment is retained along with domain structures having characteristic intermolecular spacings. The principal use of these unique polymers is in **liquid crystal displays (LCDs)**, familiar features of various digital displays, especially on digital watches and laptop computers. In this system, a room-temperature liquid crystal melt is sandwiched between two sheets of glass (coated with a transparent, electrically conductive film). Character-forming number/letter elements are etched on the display face. A voltage applied over the character-forming regions disrupts the orientation of the liquid crystal molecules, creating a darkening of the material and producing a visible character.

PHOTOCONDUCTORS

In Figure 15–9, we saw that thermal energy can promote electrons in semi-conductors from the valence band to the conduction band, creating both a

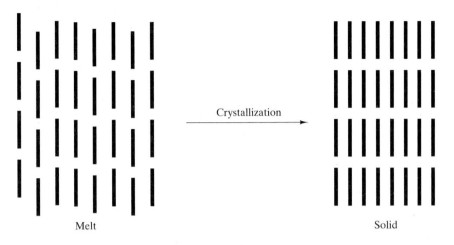

Figure 16-19 *Schematic illustration of the unique structural configuration of liquid crystal polymers.*

conduction electron and an electron hole in the valence band. Other forms of energy input can also provide the electron promotion. When a semiconductor is bombarded by light with photon energies equal to or greater than the band gap, one electron-hole pair can be produced by each photon (Figure 16–20). These **photoconductors** are best illustrated by photographic light meters. The photon-induced current is a direct function of the incident light intensity (number of photons) striking the meter per unit time. (The semiconducting lasers such as GaAs discussed earlier in this section used the electron promotion principle of photoconduction but depended on the recombination of the electron-hole pairs to produce the laser photons.) Cadmium sulfide, CdS, is widely used in photographic light meters.

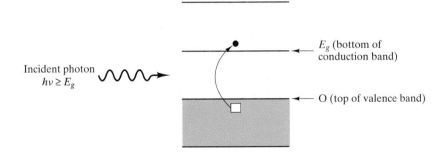

Figure 16-20 *Schematic illustration of photoconduction.*

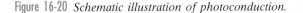

SAMPLE PROBLEM 16.6

Relative to a GaAs semiconductor laser, calculate the photon wavelength corresponding to the GaAs band gap given in Table 15.5.

SOLUTION

Using Equation 16.8 and the GaAs band gap from Table 15.5, we have

$$\lambda = hc/E_g$$

$$= \left([0.663 \times 10^{-33} \text{ J·s}][0.300 \times 10^9 \text{ m/s}]/[1.47 \text{ eV}]\right)$$

$$\times (6.242 \times 10^{18} \text{ eV/J})$$

$$= 844 \times 10^{-9} \text{ m} = 844 \text{ nm}$$

Note.   This wavelength is in the infrared region of the electromagnetic spectrum. Variations in the temperature and chemical composition of the semiconductor can be used to increase the band gap, thereby reducing the wavelength into the visible light region.

SAMPLE PROBLEM 16.7

Referring to Sample Problem 16.2, calculate the critical angle of incidence, θ_c, in a step-index fiber design for a light ray going from a glass fiber core (with index of refraction, $n = 1.470$) to the cladding (with $n = 1.460$).

SOLUTION

In this case, Equation 16.4 has the form

$$\frac{\sin \theta_i}{\sin \theta_r} = \frac{v_{\text{core}}}{v_{\text{cladding}}} = \frac{n_{\text{cladding}}}{n_{\text{core}}}$$

At the critical condition,

$$\frac{\sin \theta_c}{\sin 90°} = \frac{n_{\text{cladding}}}{n_{\text{core}}}$$

or

$$\theta_c = \arcsin \frac{n_{\text{cladding}}}{n_{\text{core}}}$$

$$= \arcsin \left(\frac{1.460}{1.479}\right) = 83.3°$$

SAMPLE PROBLEM 16.8

Calculate the maximum photon wavelength necessary to produce an electron-hole pair in the photoconductor, CdS, using the energy band gap given in Table 15.5.

SOLUTION

As in Sample Problem 16.6, we can relate the band gap and wavelength using Equation 16.8. Combining with the band gap from Table 15.5, we have

$$\lambda = hc/E_g$$

$$= \left([0.663 \times 10^{-33}\,\text{J·s}][0.300 \times 10^9\,\text{m/s}]/[2.59\,\text{eV}]\right)$$

$$\times (6.242 \times 10^{18}\,\text{eV/J})$$

$$= 479 \times 10^{-9}\,\text{m} = 479\,\text{nm}$$

PRACTICE PROBLEM 16.6

As noted in Sample Problem 16.6, the band gaps of semiconductors are a function of composition. By adding some GaP to the GaAs, the band gap can be increased to 1.78 eV. Calculate the laser photon wavelength corresponding to this larger band gap.

PRACTICE PROBLEM 16.7

In Sample Problem 16.7, we find the critical angle of incidence for a light ray going from a glass fiber core into its cladding. Calculate the critical angle of incidence for a single-mode fiber design in which the core index of refraction is $n = 1.460$ and the cladding index of refraction is very slightly smaller, $n = 1.458$.

PRACTICE PROBLEM 16.8

Repeat the wavelength calculation of Sample Problem 16.8 for a ZnSe photoconductor with a band gap of 2.67 eV.

SUMMARY

The engineering applications of many materials depend on their optical behavior, either in the visible light portion of the electromagnetic spectrum (wavelength from 400 to 700 nm) or in the adjacent ultraviolet or infrared regions. Many optical properties are most easily illustrated by the way in which visible light interacts with common glass, although these properties generally apply equally well to various crystalline ceramics and organic polymers. The refractive index (n) has a strong effect on the subjective appearance of a transparent solid or coating, and n is related to the reflectance of the surface through Fresnel's formula. Surface roughness determines the relative amounts of specular and diffuse reflection. Transparency is limited by the nature of second-phase pores or particles, which serve as scattering centers. Color is produced by the selective absorption of certain wavelength ranges in the visible light spectrum. Luminescence, like color, is a byproduct of photon absorption, in this case involving the reemission of photons of visible light. Rapid reemission is termed fluorescence, and, for longer times, the phenomenon is termed phosphorescence. The electrical conductivity of metals leads to their characteristic opacity and reflectivity. Some wavelength dependency for a metal's reflectivity can give characteristic color.

In recent decades, various optical materials have gone from being laboratory curiosities to becoming major components of modern technology. A dramatics example is the laser, which can provide a high-intensity, coherent, monochromatic, and highly collimated light beam. Lasers are widely used in products ranging from optical communication systems to surgical tools. Optical fibers, providing high-efficiency transmission of digital light signals, are excellent examples of photonic materials. Various fiber designs are used in a range of applications, including virtually all modern telecommunications. The liquid crystal displays widely used in digital watches and laptop computers are the result of novel polymeric systems that fall outside the conventional definitions of crystalline and noncrystalline materials. Photoconductors, such as photographic light meters, are semiconductors in which conductive electron-hole pairs are produced by exposure to photons of light.

KEY TERMS

coherent (606)
color (600)
colorant (600)
cosine law (598)
diffuse reflection (597)
dye (600)
electroluminescence (602)
fluorescence (601)
Fresnel's formula (596)
graded-index fiber (612)
incoherent (606)
laser (606)

liquid crystal display (LCD) (612)
luminescence (601)
monochromatic (608)
opacity (598)
optical fiber (610)
optical property (594)
phosphorescence (601)
photoconductor (613)
photoluminescence (602)
photon (592)
photonic material (610)
pigment (600)

polar diagram (597)
reflectance (596)
refractive index (594)
single-mode fiber (612)
specular reflection (597)
step-index fiber (612)
surface gloss (598)
translucency (598)
transparency (598)
visible light (592)

REFERENCES

Doremus, R. H., *Glass Science,* 2nd ed., John Wiley & Sons, Inc., New York, 1994.

Dorf, R. C., *Electrical Engineering Handbook*, CRC Press, Boca Raton, Florida, 1993.

Kingery, W. D., H. K. Bowen, and **D. R. Uhlmann,** *Intro-duction to Ceramics,* 2nd ed., John Wiley & Sons, Inc., New York, 1976.

Mark, H. F., et al., Eds., *Encyclopedia of Polymer Science and Engineering,* 2nd ed., Vols. 1–7, Index Vol., Supplementary Vol., John Wiley & Sons, Inc., New York, 1985–1989.

PROBLEMS

16.1 • Visible Light

16.1. A ruby laser produces a beam of monochromatic photons with a wavelength of 694.3 nm. Calculate the corresponding **(a)** photon frequency and **(b)** photon energy.

16.2. Given that the magnetic permeability of a vacuum is $4\pi \times 10^{-7}$N/A^2, use the constants and conversion factors of Appendix 3 to prove the quality of Equation 16.1.

16.2 • Optical Properties

16.3. By what percentage is the critical angle of incidence different for a lead oxide-containing "crystal" glass (with $n = 1.7$) compared with plain silica glass?

16.4. What would be the critical angle of incidence for light transmission from single-crystal Al_2O_3 into a coating of silica glass?

• 16.5. (a) Consider a translucent orthoclase ceramic with a thin orthoclase glass coating (glaze). What is the maximum angle of incidence at the ceramic-glaze interface to ensure that an observer can see any visible light transmitted through the product (into an air atmosphere)? (Consider only specular transmission through the ceramic.) **(b)** Would the answer to part (a) change if the glaze coating were eliminated? Briefly explain. **(c)** Would the answer to part (a) change if the product was an orthoclase glass with a thin translucent coating of crystallized orthoclase? Briefly explain.

16.6. By what percentage is the reflectance different for a lead oxide-containing "crystal" glass (with $n = 1.7$) compared with plain silica glass?

16.7. Silica glass is frequently and incorrectly referred to as "quartz." This is the result of shortening the traditional term fused quartz, which described the original technique of making silica glass by melting quartz powder. What is the percentage error in calculating the reflectance of silica glass by using the index of refraction of quartz?

16.8. For a single crystal Al_2O_3 disk used in a precision optical device, calculate **(a)** the angle of refraction for a light beam incident from a vacuum at $\theta_i = 30°$, and **(b)** the fraction of light transmitted into the Al_2O_3 at that angle.

16.9. Calculate the critical angle of incidence for the air-nylon 66 interface.

16.10. Calculate the critical angle if water ($n = 1.333$) replaced the air in Problem 16.9.

16.11. We shall find in Chapter 19 that polymers are susceptible to ultraviolet-light damage. Calculate the wavelength of ultraviolet light necessary to break the C-C single bond. (Bond energies are given in Table 2.2.)

16.12. Repeat Problem 16.11 for the $C = C$ double bond.

16.13. The highest-energy visible-light photon is the shortest wavelength light at the blue end of the spectrum. How many times greater is the energy of the CuK_α x-ray photon used in Section 3.7 for the analysis of crystal structure? (Note Sample Problem 16.5.)

16.14. The lowest-energy visible-light photon is the longest wavelength light at the red end of the spectrum. How many times greater is this energy than that of a typical microwave photon with a wavelength of 10^6 nm?

16.15. By what factor is the energy of a photon of red visible light ($\lambda = 700$ nm) greater than that of a photon of infrared light with $\lambda = 4\mu$m?

16.16. What range of photon energies are absorbed in the blue glass of Figure 16–10?

16.3 • Optical Systems and Devices

16.17. As noted in Sample Problem 16.6, the band gap of semiconductors is a function of temperature. Calculate the laser photon wavelength corresponding to the GaAs band gap at liquid nitrogen temperature (77K), $E_g = 1.508$ eV.

16.18. In Practice Problem 16.6, changes in chemical composition were seen to change emitted laser photon wavelengths. As a practical matter, GaAs and GaP are soluble in all proportions, and the band gap of the alloy increases nearly linearly with the molar additions of GaP. The band gap of pure GaP is 2.25 eV. Calculate the molar fraction of GaP required to produce the 1.78 eV band gap described in Practice Problem 16.6.

16.19. For a single-mode fiber with a core index of refraction of $n = 1.460$, how long will it take for a single light pulse to travel a length of 1 kilometer along a cable television line?

16.20. For a step-index fiber with a core index of refraction of $n = 1.470$ how long will it take for a single light pulse to travel a length of 1 meter along an endscope used to inspect a medical patient's stomach lining?

16.21. For what range of visible light wavelengths would ZnTe, with a band gap of 2.26 eV, be a photoconductor?

16.22. Repeat Problem 16.21 for CdTe, which has a band gap of 1.50 eV.

CHAPTER 17
Semiconductor Materials

The modern microprocessor is a state-of-the-art application of semiconductor materials encased in a package of conventional structural materials. This 500 MHz processor contains 10 million transistors on a single silicon chip. (Courtesy of Intel.)

In Part II of this book we discussed the four categories of structural materials. In Chapter 15, we identified a fifth type of engineering materials, semiconductors, which are important for their electrical, rather than mechanical, properties. The crystal structures of common semiconductors were introduced in Section 3.5. In Chapter 15, the nature of semiconduction was introduced by discussion of *intrinsic, elemental semiconductors*, such as impurity-free silicon. In this chapter we extend the discussion to include a variety of other semiconducting materials. We shall look at *extrinsic, elemental semiconductors*, such as silicon "doped" with a small amount of boron. The nature of the added impurity determines the nature of the semiconduction; either *n*-type (with predominantly negative charge carriers) or *p*-type (with predominantly positive charge carriers). The key elemental semiconductors come from group IVA of the periodic table. *Compound semiconductors* are ceramic like compounds formed from combinations of elements clustered around group IVA. Whereas most semiconductors are crystals of high quality, some *amorphous semiconductors* are commercially available. The processing of semiconductors has been refined to an exceptional level of structural and chemical control. This is a text on materials, not electronics, but in order to better appreciate the applications of semiconductors, a brief discussion of devices will be presented.

17.1   INTRINSIC, ELEMENTAL SEMICONDUCTORS

A few elements from the periodic table have intermediate values of conductivity in comparison with high-conductivity metals and low-conductivity insulators (ceramics, glasses, and polymers). The principles of this inherent semiconduction, or **intrinsic semiconduction**, in elemental solids (essentially free of impurities) were given in Chapter 15. To summarize, the conductivity (σ) of a solid is the sum of contributions from negative and positive charge carriers,

$$\sigma = n_n q_n \mu_n + n_p q_p \mu_p \tag{15.6}$$

where n is charge density, q is the charge of a single carrier, and μ is carrier mobility. The subscripts n and p refer to negative and positive carriers, respectively. For a solid such as elemental silicon, conduction results from the thermal promotion of electrons from a filled valence band to an empty conduction band. There, the electrons are negative charge carriers. The removal of electrons from the valence band produces electron holes, which are positive charge carriers. Because the density of conduction electrons (n_n) is identical to the density of electron holes (n_p), we may rewrite Equation 15.6 as

$$\sigma = nq(\mu_e + \mu_h) \tag{15.14}$$

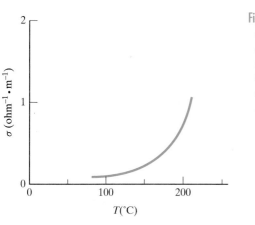

Figure 17-1 *Variation in electrical conductivity with temperature for semiconductor silicon. Contrast with the behavior shown for the metals in Figure 15–10. (This plot is based on the data in Table 17.1 using Equations 15.14 and 17.2.)*

where n is now the density of conduction electrons and the subscripts e and h refer to electrons and holes, respectively. This overall conduction scheme is possible because of the relatively small energy band gap between the valence and conduction bands in silicon (see Figure 15–9).

In Section 15.3, we found that the conductivity of metallic conductors dropped with increasing temperature. By contrast, the conductivity of semiconductors increases with increasing temperature (Figure 17–1). The reason for this opposite trend can be appreciated by inspection of Figure 15–9. The number of charge carriers depends on the overlap of the "tails" of the Fermi function curve with the valence and conduction bands. Figure 17–2 illustrates how increasing temperature extends the Fermi function,

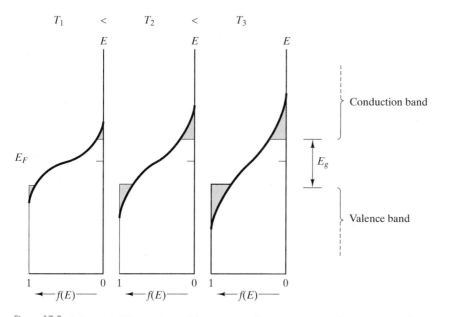

Figure 17-2 *Schematic illustration of how increasing temperature increases overlap of the Fermi function, $f(E)$, with the conduction and valence bands giving increasing numbers of charge carriers. (Note also Figure 15–9.)*

giving more overlap (i.e., more charge carriers). The precise nature of $\sigma(T)$ for semiconductors follows from the nature of charge carrier production in Figures 15–9 and 17–2—thermal activation. For this mechanism, the density of carriers increases exponentially with temperature; that is,

$$n \propto e^{-E_g/2kT} \tag{17.1}$$

where E_g is the band gap, k the Boltzmann constant, and T the absolute temperature. This is another occurrence of Arrhenius behavior (Section 5.1). Note that Equation 17.1 differs slightly from the general Arrhenius form given by Equation 5.1. There is a factor of 2 in the exponent of Equation 17.1 arising from the fact that each thermal promotion of an electron produces two charge carriers-an electron-hole pair.

Returning to Equation 15.14, we see that $\sigma(T)$ is determined by the temperature dependence of μ_e and μ_h as well as n. As with metallic conductors, μ_e and μ_h drop off slightly with increasing temperature. However, the exponential rise in n dominates the overall temperature dependence of σ, allowing us to write

$$\sigma = \sigma_0 e^{-E_g/2kT} \tag{17.2}$$

where σ_0 is a preexponential constant of the type associated with Arrhenius equations (Section 5.1). By taking the logarithm of each side of Equation 17.2, we obtain

$$\ln \sigma = \ln \sigma_0 - \frac{E_g}{2k} \frac{1}{T} \tag{17.3}$$

which indicates that a semilog plot of $\ln \sigma$ versus T^{-1} gives a straight line with a slope of $-E_g/2k$. Figure 17–3 demonstrates this linearity with the data of Figure 17–1 replotted in an Arrhenius plot.

The intrinsic, elemental semiconductors of group IVA of the periodic table are silicon (Si), germanium (Ge), and tin (Sn). This is a remarkably simple list. No impurity levels need be specified, as these materials are purposely prepared to an exceptionally high degree of purity. The role of impurities is treated in the next section. The three materials are all members of group IVA of the periodic table and have small values of E_g. Although this list is small, it has enormous importance for modern technology. Silicon is to the electronics industry what steel is to the automotive and construction industries. Pittsburgh is the "Steel City," and the Santa Clara Valley of California, once known for agriculture, is now known as "Silicon Valley" due to the major development of solid-state technology in the region.

Some elements in the neighborhood of group IVA (e.g., B from group IIIA and Te from group VIA) are also semiconductors. However, the predominant examples of commercial importance are Si and Ge from group IVA. Gray tin (Sn) transforms to white tin at 13°C. The transformation from the diamond cubic to a tetragonal structure near ambient temperature prevents gray tin from having any useful device application.

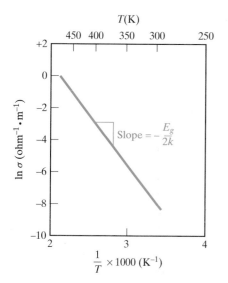

Figure 17-3 *Arrhenius plot of the electrical conductivity data for silicon given in Figure 17–1. The slope of the plot is $-E_g/2k$.*

Table 17.1 gives values of band gap (E_g), electron mobility (μ_e), hole mobility (μ_h), and conduction electron density at room temperature (n) for two intrinsic, elemental semiconductors. Because mobilities are not a strong function of composition, these values will also apply to the slightly impure, extrinsic semiconductors discussed in the next section.

Table 17.1 *Electrical Properties for Some Intrinsic, Elemental Semiconductors at Room Temperature (300 K)*

Group	Semiconductor	E_g (eV)	μ_e [m²/(V · s)]	μ_h [m²/(V · s)]	$n_e(= n_h)$ (m⁻³)
IVA	Si	1.107	0.140	0.038	14×10^{15}
	Ge	0.66	0.364	0.190	23×10^{18}

Source: Data from W. R. Runyan and S. B. Watelski, in *Handbook of Materials and Processes for Electronics*, C. A. Harper, Ed., McGraw-Hill Book Company, New York, 1970.

SAMPLE PROBLEM 17.1

For every 10^{14} atoms in pure silicon, 28 Si atoms provide a conduction electron. What is n, the density of conduction electrons?

SOLUTION

First, we calculate atomic density using data from Appendix 1:

$$\rho_{Si} = 2.33 \text{ g} \cdot \text{cm}^{-3} \text{ with an atomic mass} = 28.09 \text{ amu}$$

or

$$\rho = 2.33 \frac{\text{g}}{\text{cm}^3} \times 10^6 \frac{\text{cm}^3}{\text{m}^3} \times \frac{1 \text{ g} \cdot \text{atom}}{28.09 \text{ g}} \times 0.6023 \times 10^{24} \frac{\text{atoms}}{\text{g} \cdot \text{atom}}$$

$$= 50.0 \times 10^{27} \text{ atoms/m}^3$$

Then

$$n = \frac{28 \text{ conduction electrons}}{10^{14} \text{ atoms}} \times 50.0 \times 10^{27} \text{ atoms/m}^3$$

$$= 14 \times 10^{15} \text{ m}^{-3}$$

Note.   This is a reversal of Sample Problem 15.13. The result for n appears both in Tables 15.5 and 17.1.

SAMPLE PROBLEM 17.2

Calculate the conductivity of germanium at 200°C.

SOLUTION

From Equation 15.14,

$$\sigma = nq(\mu_e + \mu_h)$$

Using the data of Table 15.5, we have

$$\sigma_{300 \text{ K}} = (23 \times 10^{18} \text{ m}^{-3})(0.16 \times 10^{-18} \text{ C})(0.364 + 0.190) \text{ m}^2/(\text{V} \cdot \text{s})$$

$$= 2.04 \ \Omega^{-1} \cdot \text{m}^{-1}$$

From Equation 17.2

$$\sigma = \sigma_0 e^{-E_g/2kT}$$

To obtain σ_0,

$$\sigma_0 = \sigma e^{+E_g/2kT}$$

Again using data from Table 15.5, we obtain

$$\sigma_0 = (2.04 \ \Omega^{-1} \cdot m^{-1})e^{+(0.66 \ eV)/2(86.2 \times 10^{-6} \ eV/K)(300 \ K)}$$

$$= 7.11 \times 10^5 \ \Omega^{-1} \cdot m^{-1}$$

Then

$$\sigma_{200°C} = (7.11 \times 10^5 \ \Omega^{-1} \cdot m^{-1})e^{-(0.66 \ eV)/2(86.2 \times 10^{-6} \ eV/K)(473 \ K)}$$

$$= 217 \ \Omega^{-1} \cdot m^{-1}$$

SAMPLE PROBLEM 17.3

You are asked to characterize a new semiconductor. If its conductivity at 20°C is $250 \Omega^{-1} \cdot m^{-1}$ and at 100°C is $1100 \Omega^{-1} \cdot m^{-1}$, what is its band gap, E_g?

SOLUTION

From Equation 17.3,

$$\ln \sigma_{T_1} = \ln \sigma_0 - \frac{E_g}{2k} \frac{1}{T_1} \quad (a)$$

$$\ln \sigma_{T_2} = \ln \sigma_0 - \frac{E_g}{2k} \frac{1}{T_2} \quad (b)$$

Subtracting (b) from (a) yields

$$\ln \sigma_{T_1} - \ln \sigma_{T_2} = \ln \frac{\sigma_{T_1}}{\sigma_{T_2}}$$

$$= -\frac{E_g}{2k} \left(\frac{1}{T_1} - \frac{1}{T_2} \right)$$

Then

$$-\frac{E_g}{2k} = \frac{\ln(\sigma_{T_1}/\sigma_{T_2})}{1/T_1 - 1/T_2}$$

or

$$E_g = \frac{(2k)\ln(\sigma_{T_2}/\sigma_{T_1})}{1/T_1 - 1/T_2}$$

Taking $T_1 = 20°C \, (= 293 \text{ K})$ and $T_2 = 100°C \, (= 373 \text{ K})$ gives us

$$E_g = \frac{(2 \times 86.2 \times 10^{-6} \text{ eV/K}) \ln(1100/250)}{\frac{1}{373} \text{ K}^{-1} - \frac{1}{293} \text{ K}^{-1}}$$

$$= 0.349 \text{ eV}$$

PRACTICE PROBLEM 17.1

In Sample Problem 17.1, we calculate the density of conduction electrons in silicon. The result is in agreement with data in Tables 15.5 and 17.5. In Problem 15.43, the fraction of germanium atoms contributing conduction electrons at room temperature was calculated. Make a similar calculation for germanium at 150°C. (Ignore the effect of thermal expansion of germanium.)

PRACTICE PROBLEM 17.2

(a) Calculate the conductivity of germanium at 100°C, and **(b)** plot the conductivity over the range of 27 to 200°C as an Arrhenius-type plot similar to Figure 17–3. (See Sample Problem 17.2.)

PRACTICE PROBLEM 17.3

In characterizing a semiconductor in Sample Problem 17.3, we calculate its band gap. Using that result, calculate its conductivity at 50°C.

17.2   EXTRINSIC, ELEMENTAL SEMICONDUCTORS

Intrinsic semiconduction is a property of the pure material. **Extrinsic semiconduction** results from impurity additions known as **dopants**, and the process of adding these components is called *doping*. The term *impurity* has a different sense here compared to its use in previous chapters. For instance, many of the impurities in metal alloys and engineering ceramics were components "dragged along" from raw material sources. The impurities in semiconductors are added carefully after the intrinsic material has been prepared to a high degree of chemical purity. The conductivity of metallic conductors was shown to be sensitive to alloy composition in Chapter 15. Now we shall explore the composition effect for semiconductors. There are two distinct types of extrinsic semiconduction: (1) *n-type*, in which negative charge carriers dominate, and (2) *p-type*, in which positive charge carriers dominate. The philosophy behind producing each type is illustrated in Figure 17–4, which shows a small portion of the periodic table in the vicinity of silicon. The intrinsic semiconductor silicon has four valence (outer-shell) electrons.

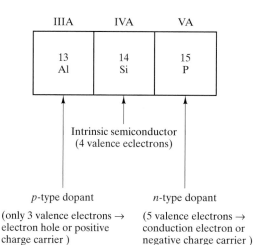

Figure 17-4 *Small section of the periodic table of elements. Silicon, in group IVA, is an intrinsic semiconductor. Adding a small amount of phosphorus from group VA provides extra electrons (not needed for bonding to Si atoms). As a result, phosphorus is an n-type dopant (i.e., an addition producing negative charge carriers). Similarly, aluminum, from group IIIA, is a p-type dopant in that it has a deficiency of valence electrons leading to positive charge carriers (electron holes).*

Phosphorus is an *n*-type dopant because it has five valence electrons. The one extra electron can easily become a conduction electron (i.e., a negative charge carrier). On the other hand, aluminum is a *p*-type dopant because of its three valence electrons. This deficiency of one electron (compared to silicon's four valence electrons) can easily produce an electron hole (a positive charge carrier).

n-TYPE SEMICONDUCTORS

The addition of a group VA atom, such as phosphorus, into solid solution in a crystal of group IVA silicon affects the energy band structure of the semiconductor. Four of the five valence electrons in the phosphorus atom are needed for bonding to four adjacent silicon atoms in the four-coordinated diamond cubic structure (see Section 3.5). Figure 17–5 shows that the extra electron that is not needed for bonding is relatively unstable and produces a **donor level** (E_d) near the conduction band. As a result, the energy barrier to forming a conduction electron ($E_g - E_d$) is substantially less than in the intrinsic material (E_g). The relative position of the Fermi function is shown in Figure 17–6. Due to the extra electrons from the doping process, the Fermi level (E_F) is shifted upward.

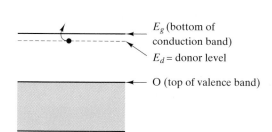

E_g (bottom of conduction band)

E_d = donor level

O (top of valence band)

Figure 17-5 *Energy band structure of an n-type semiconductor. The extra electron from the group VA dopant produces a donor level (E_d) near the conduction band. This provides relatively easy production of conduction electrons. This figure can be contrasted with the energy band structure of an intrinsic semiconductor in Figure 15–9.*

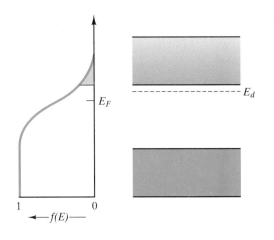

Figure 17-6 *Comparison of the Fermi function, $f(E)$, with the energy band structure for an n-type semiconductor. The extra electrons shift the Fermi level (E_F) upward compared to Figure 15–9 (where it was in the middle of the band gap for an intrinsic semiconductor).*

Since the conduction electrons provided by the group VA atoms are by far the most numerous charge carriers, the conductivity equation takes the form

$$\sigma = nq\mu_e \tag{17.4}$$

where all terms were previously defined in Equation 15.14, with n the number of electrons due to the dopant atoms. A schematic of the production of a conduction electron from a group VA dopant is shown in Figure 17–7. This can be contrasted with the schematic of an intrinsic semiconductor shown in Figure 15–27.

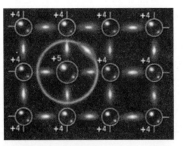

(a)

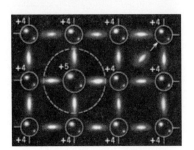

(b)

Figure 17-7 *Schematic of the production of a conduction electron in an n-type semiconductor. (a) The extra electron associated with the group VA atom can (b) easily break away, becoming a conduction electron and leaving behind an empty donor state associated with the impurity atom. This can be contrasted with the similar figure for intrinsic material in Figure 15–27. (From R. M. Rose, L. A. Shepard, and J. Wulff,* The Structure and Properties of Materials, *Vol. 4:* Electronic Properties, *John Wiley & Sons, Inc., New York, 1966.)*

Extrinsic semiconduction is another thermally activated process that follows the Arrhenius behavior. For n-type semiconductors, we can write

$$\sigma = \sigma_0 e^{-(E_g - E_d)/kT} \qquad (17.5)$$

where the various terms from Equation 17.2 again apply and E_d is defined by Figure 17–4. One should note that there is no factor of 2 in the exponent of Equation 17.5 as there was for Equation 17.2. In extrinsic semiconduction, thermal activation produces a single charge carrier as opposed to the two carriers produced in intrinsic semiconduction. An Arrhenius plot of Equation 17.5 is shown in Figure 17–8.

The temperature range for n-type extrinsic semiconduction is limited. The conduction electrons provided by group VA atoms are much easier to produce thermally than are conduction electrons from the intrinsic process. However, the number of extrinsic conduction electrons can be no greater than the number of dopant atoms (i.e., one conduction electron per dopant atom). As a result, the Arrhenius plot of Figure 17–8 has an upper limit corresponding To the temperature at which all possible extrinsic electrons have been promoted to the conduction band. Figure 17–9 illustrates this.

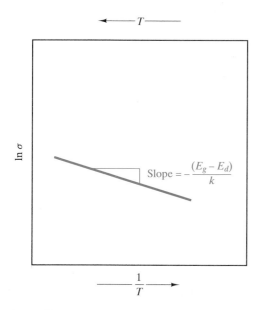

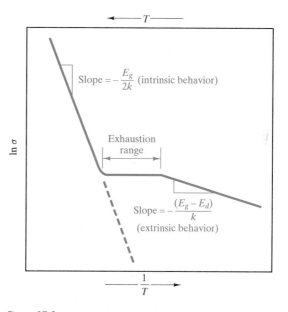

Figure 17-8 *Arrhenius plot of electrical conductivity for an n-type semiconductor. This can be contrasted with the similar plot for intrinsic material in Figure 17–3.*

Figure 17-9 *Arrhenius plot of electrical conductivity for an n-type semiconductor over a wider temperature range than shown in Figure 17–8. At low temperatures (high $1/T$), the material is extrinsic. At high temperatures (low $1/T$), the material is intrinsic. In between is the exhaustion range, in which all "extra electrons" have been promoted to the conduction band.*

Figure 17–9 represents plots of both Equation 17.5 for extrinsic behavior and Equation 17.2 for intrinsic behavior. Note that the value of σ_0 for each region will be different. At low temperatures (large $1/T$ values), extrinsic behavior (Equation 17.5) dominates. The **exhaustion range** is a nearly horizontal plateau in which the number of charge carriers is fixed ($=$ number of dopant atoms). As a practical matter, the conductivity does drop off slightly with increasing temperature (decreasing $1/T$) due to increasing thermal agitation. This is comparable to the behavior of metals in which the number of conduction electrons is fixed but mobility drops slightly with temperature (Section 15.3). As temperature continues to rise, the conductivity due to the intrinsic material (pure silicon) eventually is greater than that due to the extrinsic charge carriers (Figure 17–9). The exhaustion range is a useful concept for engineers wishing to minimize the need for temperature compensation in electrical circuits. In this range, conductivity is nearly constant with temperature.

p-TYPE SEMICONDUCTORS

When a group IIIA atom such as aluminum goes into solid solution in silicon, its three valence electrons leave it one short of that needed for bonding to the four adjacent silicon atoms. Figure 17–10 shows that the result for the energy band structure of silicon is an **acceptor level** near the valence band. A silicon valence electron can be easily promoted to this level generating an electron hole (i.e., a positive charge carrier). As with n-type material, the energy barrier to forming a charge carrier (E_a) is substantially less than in the intrinsic material (E_g). The relative position of the Fermi function is shifted downward in p-type material (Figure 17–11). The appropriate conductivity equation is

$$\sigma = nq\mu_h \qquad (17.6)$$

where n is the density of electron holes. A schematic of the production of an electron hole from a group IIIA dopant is shown in Figure 17–12.

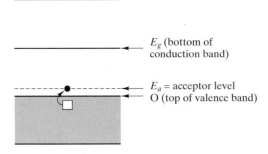

Figure 17-10   *Energy band structure of a p-type semiconductor. The deficiency of valence electrons in the group IIIA dopant produces an acceptor level (E_a) near the valence band. Electron holes are produced as a result of thermal promotion over this relatively small energy barrier.*

E_g (bottom of conduction band)

E_a = acceptor level
O (top of valence band)

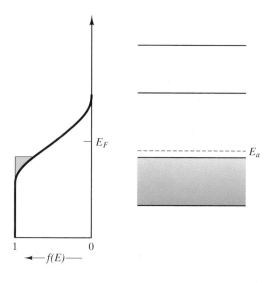

Figure 17-11 *Comparison of the Fermi function with the energy band structure for a p-type semiconductor. This electron deficiency shifts the Fermi level downward compared to Figure 15–9.*

The Arrhenius equation for p-type semiconductors is

$$\sigma = \sigma_0 e^{-E_a/kT} \qquad (17.7)$$

where the terms from Equation 17.5 again apply and E_a is defined by Figure 17–10. As with Equation 17.5, there is no factor of 2 in the exponent due to the single (positive) charge carrier involved. Figure 17–13 shows the Arrhenius plot of $\ln \sigma$ versus $1/T$ for a p-type material. This is quite similar to Figure 17–9 for n-type semiconductors. The plateau in conductivity between the extrinsic and intrinsic regions is termed the **saturation range**

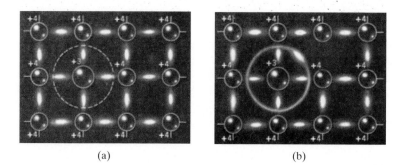

(a) (b)

Figure 17-12 *Schematic of the production of an electron hole in a p-type semiconductor. (a) The deficiency in valence electrons for the group IIIA atom creates an empty state, or electron hole, orbiting about the acceptor atom. (b) The electron hole becomes a positive charge carrier as it leaves the acceptor atom behind with a filled acceptor state. (The motion of electron holes, of course, is due to the cooperative motion of electrons.)* (From R. M. Rose, L. A. Shepard, and J. Wulff, The Structure and Properties of Materials, *Vol. 4:* Electronic Properties, *John Wiley & Sons, Inc., New York, 1966.)*

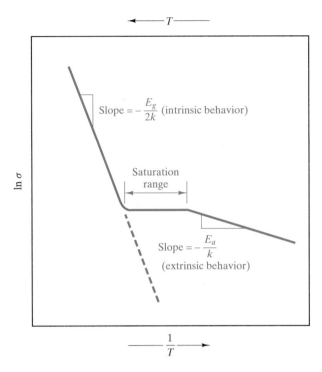

Figure 17-13 *Arrhenius plot of electrical conductivity for a p-type semiconductor over a wide temperature range. This is quite similar to the behavior shown in Figure 17–9. The region between intrinsic and extrinsic behavior is termed the saturation range corresponding to all acceptor levels being "saturated" or occupied with electrons.*

for *p*-type behavior rather than exhaustion range. Saturation occurs when all acceptor levels (= number of group IIIA atoms) have become occupied with electrons.

The similarity between Figures 17–9 and 17–13 raises an obvious question as to how you can tell whether a given semiconductor is *n* type or *p* type. The distinction can be conveniently made in a classic experiment known as a **Hall* effect** measurement, illustrated in Figure 17–14. This is one manifestation of the intimate relationship between electrical and magnetic behavior. Specifically, a magnetic field applied at right angles to a flowing current causes a sideways deflection of the charge carriers and a subsequent voltage buildup across the conductor. For negative charge carriers (e.g., electrons in metals or *n*-type semiconductors), the Hall voltage (V_H) is positive

---

*   Edwin Herbert Hall (1855–1938), American physicist. Hall's most famous discovery, the effect that still bears his name, was the basis of his Ph.D. thesis in 1880. He continued to be a productive researcher during his career as a professor of physics and even developed a popular set of high school physics experiments designed to help students prepare for the entrance examination of Harvard, where he taught.

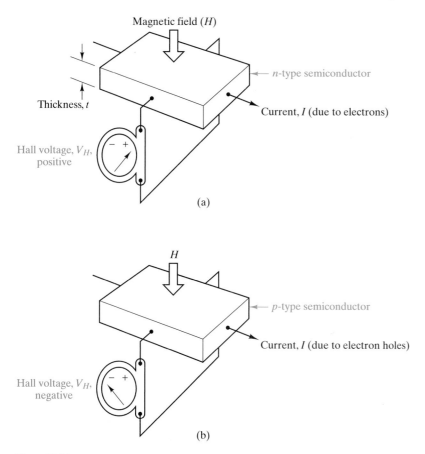

Figure 17-14 *The application of a magnetic field (with field strength, H), per-pendicular to a current, I, causes a sideways deflection of charge carriers and a resulting voltage, V_H. This phenomenon is known as the Hall effect. The Hall voltage is given by Equation 17.8. For (a) an n-type semicon-ductor, the Hall voltage is positive. For (b) a p-type semiconductor, the Hall voltage is negative.*

(Figure 17–14a). For positive charge carriers (e.g., electron holes in a p-type semiconductor), the Hall voltage is negative (Figure 17–14b). The Hall voltage is given by

$$V_H = \frac{R_H I H}{t} \qquad (17.8)$$

where R_H is the Hall coefficient (indicative of the magnitude and sign of the Hall effect), I the current, H the magnetic field strength, and t the sample thickness. Some of the common extrinsic, elemental semiconductor systems used in solid-state technology are listed in Table 17.2. Table 17.3 gives values of donor level relative to band gaps ($E_g - E_d$) and acceptor level (E_a) for various n-type and p-type donors, respectively.

Table 17.2 *Some Extrinsic, Elemental Semiconductors*

Element	Dopant	Periodic table group of dopant	Maximum solid solubility of dopant (atoms/m^3)
Si	B	IIIA	600×10^{24}
	Al	IIIA	20×10^{24}
	Ga	IIIA	40×10^{24}
	P	VA	1000×10^{24}
	As	VA	2000×10^{24}
	Sb	VA	70×10^{24}
Ge	Al	IIIA	400×10^{24}
	Ga	IIIA	500×10^{24}
	In	IIIA	4×10^{24}
	As	VA	80×10^{24}
	Sb	VA	10×10^{24}

Source: Data from W. R. Runyan and S. B. Watelski, in *Handbook of Materials and Processes for Electronics*, C. A. Harper, Ed., McGraw-Hill Book Company, New York, 1970.

Table 17.3 *Impurity Energy Levels for Extrinsic Semiconductors*

Semiconductor	Dopant	$E_g - E_d$ (eV)	E_a (eV)
Si	P	0.044	—
	As	0.049	—
	Sb	0.039	—
	Bi	0.069	—
	B	—	0.045
	Al	—	0.057
	Ga	—	0.065
	In	—	0.16
	Tl	—	0.26
Ge	P	0.012	—
	As	0.013	—
	Sb	0.096	—
	B	—	0.01
	Al	—	0.01
	Ga	—	0.01
	In	—	0.011
	Tl	—	0.01
GaAs	Se	0.005	—
	Te	0.003	—
	Zn	—	0.024
	Cd	—	0.021

Source: Data from W. R. Runyan and S. B. Watelski, in *Handbook of Materials and Processes for Electronics*, C. A. Harper, Ed., McGraw-Hill Book Company, New York, 1970.

A final note of comparison between semiconductors and metals is in order. The composition and temperature effects for semiconductors are opposite to those for metals. For metals, small impurity additions decreased conductivity [see Figure 15–12, which showed $\rho(= 1/\sigma)$ increasing with addition levels]. Similarly, increases in temperature decreased conductivity (see Figure 15–10). Both effects were due to reductions in electron mobility resulting from reductions in crystalline order. We have seen that for semiconductors, appropriate impurities and increasing temperature increase conductivity. Both effects are described by the energy band model and Arrhenius behavior.

SAMPLE PROBLEM 17.4

An extrinsic silicon contains 100 parts per billion (ppb) Al by weight. What is the atomic % Al?

SOLUTION

For 100 g of doped silicon, there will be

$$\frac{100}{10^9} \times 100 \text{ g Al} = 1 \times 10^{-5} \text{ g Al}$$

Using the data of Appendix 1, we can calculate

$$\text{no. g} \cdot \text{atoms Al} = \frac{1 \times 10^{-5} \text{ g Al}}{26.98 \text{ g/g} \cdot \text{atom}} = 3.71 \times 10^{-7} \text{ g} \cdot \text{atom}$$

$$\text{no. g} \cdot \text{atoms Si} = \frac{(100 - 1 \times 10^{-5}) \text{ g Si}}{28.09 \text{ g/g} \cdot \text{atom}} = 3.56 \text{ g} \cdot \text{atoms}$$

This gives

$$\text{atomic\% Al} = \frac{3.71 \times 10^{-7} \text{ g} \cdot \text{atom}}{(3.56 + 3.7 \times 10^{-7}) \text{ g} \cdot \text{atom}} \times 100$$

$$= 10.4 \times 10^{-6} \text{ atomic\%}$$

SAMPLE PROBLEM 17.5

In a phosphorus-doped (n-type) silicon, the Fermi level (E_F) is shifted upward 0.1 eV. What is the probability of an electron's being thermally promoted to the conduction band in silicon ($E_g = 1.107$ eV) at room temperature (25°C)?

SOLUTION

From Figure 17–6 and Equation 15.7, it is apparent that

$$E - E_F = \frac{1.107}{2}\ \text{eV} - 0.1\ \text{eV} = 0.4535\ \text{eV}$$

and then

$$f(E) = \frac{1}{e^{(E-E_F)/kT} + 1}$$

$$= \frac{1}{e^{(0.4535\ \text{eV})/(86.2\times10^{-6}\ \text{eV·K}^{-1})(298\ \text{K})} + 1}$$

$$= 2.20 \times 10^{-8}$$

This is a small number, but it is roughly two orders of magnitude higher than the value for intrinsic silicon as calculated in Sample Problem 15.6.

SAMPLE PROBLEM 17.6

For a hypothetical semiconductor with n-type doping and $E_g = 1$ eV while $E_d = 0.9$ eV, the conductivity at room temperature (25°C) is 100 $\Omega^{-1} \cdot \text{m}^{-1}$. Calculate the conductivity at 30°C.

SOLUTION

Assuming that extrinsic behavior extends to 30°C, we can apply Equation 17.5:

$$\sigma = \sigma_0 e^{-(E_g - E_d)/kT}$$

At 25°C,

$$\sigma_0 = \sigma e^{+(E_g - E_d)/kT}$$

$$= (100\ \Omega^{-1} \cdot \text{m}^{-1})e^{+(1.0-0.9)\ \text{eV}/(86.2\times10^{-6}\ \text{eV·K}^{-1})(298\ \text{K})}$$

$$= 4.91 \times 10^3\ \Omega^{-1} \cdot \text{m}^{-1}$$

At 30°C, then,

$$\sigma = (4.91 \times 10^3\ \Omega^{-1} \cdot \text{m}^{-1})e^{-(0.1\ \text{eV})/(86.2\times10^{-6}\ \text{eV·K}^{-1})(303\ \text{K})}$$

$$= 107\ \Omega^{-1} \cdot \text{m}^{-1}$$

SAMPLE PROBLEM 17.7

For a phosphorus-doped germanium semiconductor, the upper temperature limit of the extrinsic behavior is 100°C. The extrinsic conductivity at this point is 60 $\Omega^{-1} \cdot m^{-1}$. Calculate the level of phosphorus doping in parts per billion (ppb) by weight.

SOLUTION

In this case, all dopant atoms have provided a donor electron. As a result, the density of donor electrons equals the density of impurity phosphorus. The density of donor electrons is given by rearrangement of Equation 17.4:

$$n = \frac{\sigma}{q\mu_e}$$

Using data from Table 17.5, we obtain

$$n = \frac{60 \ \Omega^{-1} \cdot m^{-1}}{(0.16 \times 10^{-18} \ C)[0.364 \ m^2/(V \cdot s)]} = 1.03 \times 10^{21} \ m^{-3}$$

Important Note. As pointed out in the text, the carrier mobilities of Tables 17.1 and 17.5 apply to extrinsic as well as intrinsic materials. For example, conduction electron mobility in germanium does not change significantly with impurity addition as long as the addition levels are not great.

Using data from Appendix 1 gives us

$$[P] = 1.03 \times 10^{21} \ \frac{\text{atoms P}}{m^3} \times \frac{30.97 \ \text{g P}}{0.6023 \times 10^{24} \ \text{atoms P}}$$

$$\times \frac{1 \ cm^3 \ \text{Ge}}{5.32 \ \text{g Ge}} \times \frac{1 \ m^3}{10^6 \ cm^3}$$

$$= 9.96 \times 10^{-9} \ \frac{\text{g P}}{\text{g Ge}} = \frac{9.96 \ \text{g P}}{10^9 \ \text{g Ge}} = \frac{9.96 \ \text{g P}}{\text{billion g Ge}}$$

$$= 9.96 \ \text{ppb P}$$

SAMPLE PROBLEM 17.8

For the semiconductor in Sample Problem 17.7:

(a) Calculate the upper temperature for the exhaustion range.

(b) Determine the extrinsic conductivity at 300 K.

SOLUTION

(a) The upper temperature for the exhaustion range (see Figure 17–9) corresponds to the point where intrinsic conductivity equals the maximum extrinsic conductivity. Using Equations 15.14 and 17.2 together with data from Table 17.5, we obtain

$$\sigma_{300\text{ K}} = (23 \times 10^{18}\text{ m}^{-3})(0.16 \times 10^{-18}\text{ C})(0.364 + 0.190)\text{ m}^2/(\text{V} \cdot \text{s})$$

$$= 2.04\ \Omega^{-1} \cdot \text{m}^{-1}$$

and

$$\sigma = \sigma_0 e^{-E_g/2kT} \text{ or } \sigma_0 = \sigma e^{+E_g/2kT}$$

giving

$$\sigma_0 = (2.04\ \Omega^{-1} \cdot \text{m}^{-1})e^{+(0.66\text{ eV})/2(86.2 \times 10^{-6}\text{ eV/K})(300\text{ K})}$$

$$= 7.11 \times 10^5\ \Omega^{-1} \cdot \text{m}^{-1}$$

Then (using the value from Sample Problem 17.7)

$$60\ \Omega^{-1} \cdot \text{m}^{-1} = (7.11 \times 10^5\ \Omega^{-1} \cdot m^{-1})e^{-(0.66\text{ eV})/2(86.2 \times 10^{-6}\text{ eV/K})T}$$

giving

$$T = 408\text{ K} = 135°\text{C}$$

(b) Calculating extrinsic conductivity for this n-type semiconductor requires using Equation 17.5:

$$\sigma = \sigma_0 e^{-(E_g - E_d)/kT}$$

In Sample Problem 17.7, we were given that $\sigma = 60\ \Omega^{-1} \cdot \text{m}^{-1}$ at 100°C. In Table 17.3, we see that $E_g - E_d$ is 0.012 eV for P-doped Ge. As a result,

$$\sigma_0 = \sigma e^{+(E_g - E_d)/kT}$$

$$= (60\ \Omega^{-1} \cdot \text{m}^{-1})e^{+(0.012\text{ eV})/(86.2 \times 10^{-6}\text{ eV/K})(373\text{ K})}$$

$$= 87.1\ \Omega^{-1} \cdot \text{m}^{-1}$$

At 300 K,

$$\sigma = (87.1\ \Omega^{-1} \cdot \text{m}^{-1})e^{-(0.012\text{ eV})/(86.2 \times 10^{-6}\text{ eV/K})(300\text{ K})}$$

$$= 54.8\ \Omega^{-1} \cdot \text{m}^{-1}$$

SAMPLE PROBLEM 17.9

Plot the conductivity of the phosphorus-doped germanium of Sample Problems 17.7 and 17.8 in a manner similar to Figure 17–9.

SOLUTION

The key data are
Extrinsic:

$$\sigma_{100°C} = 60 \ \Omega^{-1} \cdot m^{-1} \quad \text{or} \quad \ln \sigma = 4.09 \ \Omega^{-1} \cdot m^{-1}$$
$$\text{at } T = 100°C = 373 \ K \quad \text{or} \quad 1/T = 2.68 \times 10^{-3} \ K^{-1}$$

and

$$\sigma_{300 \ K} = 54.8 \ \Omega^{-1} \cdot m^{-1} \quad \text{or} \quad \ln \sigma = 4.00 \ \Omega^{-1} \cdot m^{-1}$$
$$\text{at } T = 300 \ K \quad \text{or} \quad 1/T = 3.33 \times 10^{-3} \ K^{-1}$$

Intrinsic:

$$\sigma_{408 \ K} = 60 \ \Omega^{-1} \cdot m^{-1} \quad \text{or} \quad \ln \sigma = 4.09 \ \Omega^{-1} \cdot m^{-1}$$
$$\text{at } T = 408 \ K \quad \text{or} \quad 1/T = 2.45 \times 10^{-3} \ K^{-1}$$

and

$$\sigma_{300 \ K} = 2.04 \ \Omega^{-1} \cdot m^{-1} \quad \text{or} \quad \ln \sigma = 0.713 \ \Omega^{-1} \cdot m^{-1}$$
$$\text{at } T = 300 \ K \quad \text{or} \quad 1/T = 3.33 \times 10^{-3} \ K^{-1}$$

giving

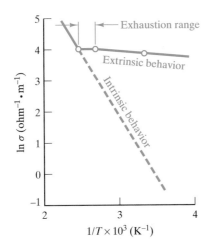

Note. The importance of plots like this to help "visualize" these calculations cannot be overemphasized.

SAMPLE PROBLEM 17.10

(a) Calculate the photon wavelength (in nanometers) necessary to promote an electron to the conduction band in intrinsic silicon.

(b) Calculate the photon wavelength (in nanometers) necessary to promote a donor electron to the conduction band in arsenic-doped silicon.

SOLUTION

(a) Substituting the band gap, E_g, for intrinsic silicon given in Table 17.1 into Equation 16.3 gives

$$E_g = 1.107 \text{ eV} = E = \frac{hc}{\lambda}$$

or

$$\lambda = \frac{hc}{1.107 \text{ eV}}$$

$$= \frac{(0.663 \times 10^{-33} \text{ J} \cdot \text{s})(3.00 \times 10^8 \text{ m/s})}{(1.107 \text{ eV}) \times 0.16 \times 10^{-18} \text{ J/eV}} \times 10^9 \frac{\text{nm}}{\text{m}}$$

$$= 1120 \text{ nm}$$

(b) Electron promotion requires only overcoming $E_g - E_d$, which is given in Table 17.3:

$$E_g - E_d = 0.049 \text{ eV} = E = \frac{hc}{\lambda}$$

or

$$\lambda = \frac{(0.663 \times 10^{-33} \text{ J} \cdot \text{s})(3.00 \times 10^8 \text{ m/s})}{(0.049 \text{ eV}) \times 0.16 \times 10^{-18} \text{ J/eV}} \times 10^9 \frac{\text{nm}}{\text{m}}$$

$$= 25{,}400 \text{ nm}$$

PRACTICE PROBLEM 17.4

A 100-ppb doping of Al, in Sample Problem 17.4, is found to represent a 10.4×10^{-6} mol % addition. What is the atomic density of Al atoms in this extrinsic semiconductor? (Compare your answer with the maximum solid solubility level given in Table 17.2.)

PRACTICE PROBLEM 17.5

In Sample Problem 17.5, we calculate the probability of an electron being thermally promoted to the conduction band in a P-doped silicon at 25°C. What is the probability at 50°C?

PRACTICE PROBLEM 17.6

The conductivity of an n-type semiconductor at 25°C and 30°C can be found in Sample Problem 17.6. **(a)** Make a similar calculation at 50°C and **(b)** plot the conductivity over the range of 25 to 50°C as an Arrhenius-type plot similar to Figure 17–8. **(c)** What important assumption underlies the validity of your results in parts (a) and (b)?

PRACTICE PROBLEM 17.7

In Sample Problems 17.7–17.9, detailed calculations about a P-doped Ge semiconductor are made. Assume now that the upper temperature limit of extrinsic behavior for an aluminum-doped germanium is also 100°C with an extrinsic conductivity at that point again being 60 $\Omega^{-1} \cdot m^{-1}$. Calculate **(a)** the level of aluminum doping in parts per billion (ppb) by weight, **(b)** the upper temperature for the saturation range, and **(c)** the extrinsic conductivity at 300 K; and **(d)** make a plot of the results similar to that in Sample Problem 17.9 and Figure 17–13.

PRACTICE PROBLEM 17.8

As in Sample Problem 17.10, calculate **(a)** the photon wavelength (in nm) necessary to promote an electron to the conduction band in intrinsic germanium and **(b)** the wavelength necessary to promote a donor electron to the conduction band in arsenic-doped germanium.

17.3   COMPOUND SEMICONDUCTORS

A large number of compounds formed from elements near group IVA in the periodic table are semiconductors. As pointed out in Chapter 3, these **compound semiconductors** generally look like group IVA elements "on the average." Many compounds have the zinc blende structure (Figure 3–24), which is the diamond cubic structure with cations and anions alternating on adjacent atom sites. Electronically, these compounds average to group IVA character. The **III-V compounds** are MX compositions, with M being a 3+ valence element and X being a 5+ valence element. The average of 4+ matches the valence of the group IVA elements and, more important, leads to a band structure comparable to Figure 15–9. Similarly, **II-VI compounds** combine a 2+ valence element with a 6+ valence element. An average of four valence electrons per atom is a good rule of thumb for semiconducting compounds. But like all such rules, there are exceptions. Some IV-VI compounds (such as GeTe) are examples. Fe_3O_4 ($= FeO \cdot Fe_2O_3$) is another. Large electron mobilities are associated with Fe^{2+}–Fe^{3+} interchanges.

Pure III-V and II-VI compounds are intrinsic semiconductors. They can be made into extrinsic semiconductors by doping in a fashion comparable to that done with elemental semiconductors (see the preceding section). Some typical compound semiconductors are given in Table 17.4. Table 17.5 gives values of band gap (E_g), electron mobility (μ_e), hole mobility (μ_h), and conduction electron density at room temperature (n) for various intrinsic, compound semiconductors. As with elemental semiconductors, the mobility values also apply to extrinsic, compound semiconductors.

Table 17.5 *Electrical Properties for Some Intrinsic, Compound Semiconductors at Room Temperature (300 K)*

Group	Semiconductor	E_g (eV)	μ_e [m²/(V·s)]	μ_h [m²/(V·s)]	$n_e (= n_h)$ (m⁻³)
III–V	AlSb	1.6	0.090	0.040	—
	GaP	2.25	0.030	0.015	—
	GaAs	1.47	0.720	0.020	1.4×10^{12}
	GaSb	0.68	0.500	0.100	—
	InP	1.27	0.460	0.010	—
	InAs	0.36	3.300	0.045	—
	InSb	0.17	8.000	0.045	13.5×10^{21}
II–VI	ZnSe	2.67	0.053	0.002	—
	ZnTe	2.26	0.053	0.090	—
	CdS	2.59	0.034	0.002	—
	CdTe	1.50	0.070	0.007	—
	HgTe	0.025	2.200	0.016	—

Source: Data from W. R. Runyan and S. B. Watelski, in *Handbook of Materials and Processes for Electronics*, C. A. Harper, Ed., McGraw-Hill Book Company, New York, 1970.

Table 17.4 *Some Compound Semiconductors*

Group	Compound	Group	Compound
III–V	BP	II–VI	ZnS
	AlSb		ZnSe
	GaP		ZnTe
	GaAs		CdS
	GaSb		CdSe
	InP		CdTe
	InAs		HgSe
	InSb		HgTe

SAMPLE PROBLEM 17.11

An intrinsic GaAs (n-type) semiconductor contains 100 ppb Se by weight. What is the mole percent Se?

SOLUTION

Considering 100 g of doped GaAs and following the method of Sample Problem 17.4, we find

$$\frac{100}{10^9} \times 100 \text{ g Se} = 1 \times 10^{-5} \text{ g Se}$$

Using data from Appendix 1, we obtain

$$\text{no. g} \cdot \text{atom Se} = \frac{1 \times 10^{-5} \text{ g Se}}{78.96 \text{ g/g} \cdot \text{atom}} = 1.27 \times 10^{-7} \text{ g} \cdot \text{atom}$$

$$\text{no. moles GaAs} = \frac{(100 - 1 \times 10^{-5}) \text{ g GaAs}}{(69.72 + 74.92) \text{ g/mol}} = 0.691 \text{ mol}$$

Finally,

$$\text{mol \% Se} = \frac{1.27 \times 10^{-7} \text{ g} \cdot \text{atom}}{(0.691 + 1.27 \times 10^{-7}) \text{ mol}} \times 100$$

$$= 18.4 \times 10^{-6} \text{ mol\%}$$

SAMPLE PROBLEM 17.12

Calculate the intrinsic conductivity of GaAs at 50°C.

SOLUTION

From Equation 15.14,

$$\sigma = nq(\mu_e + \mu_h)$$

Using the data of Table 17.5, we obtain

$$\sigma_{300 \text{ K}} = (1.4 \times 10^{12} \text{ m}^{-3})(0.16 \times 10^{-18} \text{ C})(0.720 + 0.020) \text{ m}^2/(\text{V} \cdot \text{s})$$

$$= 1.66 \times 10^{-7} \ \Omega^{-1} \cdot \text{m}^{-1}$$

From Equation 17.2,

$$\sigma = \sigma_0 e^{-E_g/2kT}$$

or

$$\sigma_0 = \sigma e^{+E_g/2kT}$$

Again using data from Table 17.5, we obtain

$$\sigma_0 = (1.66 \times 10^{-7} \ \Omega^{-1} \cdot m^{-1}) e^{+(1.47 \ eV)/2(86.2 \times 10^{-6} \ eV/K)(300 \ K)}$$

$$= 3.66 \times 10^5 \ \Omega^{-1} \cdot m^{-1}$$

Then

$$\sigma_{50°C} = (3.66 \times 10^5 \ \Omega^{-1} \cdot m^{-1}) e^{-(1.47 \ eV)/2(86.2 \times 10^{-6} \ eV/K)(323 \ K)}$$

$$= 1.26 \times 10^{-6} \ \Omega^{-1} \cdot m^{-1}$$

SAMPLE PROBLEM 17.13

In intrinsic semiconductor CdTe, what fraction of the current is carried by electrons and what fraction is carried by holes?

SOLUTION

By using Equation 15.14,

$$\sigma = nq(\mu_e + \mu_h)$$

it is apparent that

$$\text{fraction due to electrons} = \frac{\mu_e}{\mu_e + \mu_h}$$

$$\text{fraction due to holes} = \frac{\mu_h}{\mu_e + \mu_h}$$

Using the data of Table 17.5 yields

$$\text{fraction due to electrons} = \frac{0.070}{0.070 + 0.007} = 0.909$$

$$\text{fraction due to holes} = \frac{0.007}{0.070 + 0.007} = 0.091$$

PRACTICE PROBLEM 17.9

Sample Problem 17.11 describes a GaAs semiconductor with 100-ppb Se doping. What is the atomic density of Se atoms in this extrinsic semiconductor? (The density of GaAs is 5.32 Mg/m^3.)

PRACTICE PROBLEM 17.10

Calculate the intrinsic conductivity of InSb at 50°C. (See Sample Problem 17.12.)

PRACTICE PROBLEM 17.11

For intrinsic InSb, calculate the fraction of the current carried by electrons and the fraction carried by electron holes. (See Sample Problem 17.13.)

17.4   AMORPHOUS SEMICONDUCTORS

In Section 4.5 the economic advantage of **amorphous** (noncrystalline) **semiconductors** was pointed out. However, these materials have yet to replace traditional, crystalline semiconductors on a large scale. The technology of these amorphous materials is somewhat behind that of their crystalline counterparts, and the scientific understanding of semiconduction in noncrystalline solids is much less developed. However, commercial development of amorphous semiconductors appears to be on the threshold of a wide market. Already, these materials account for approximately one-quarter of the photovoltaic (solar cell) market, largely for portable consumer products. With unit energy costs approaching those from conventional fossil and nuclear fuels, amorphous semiconductors could eventually account for a practical energy source in Third World countries, where long-distance power grids are lacking. Relatively small solar power plants could provide nondepletable, pollution-free energy to small clusters of towns and villages. Table 17.6 lists some examples of amorphous semiconductors. One should note that amorphous silicon is often prepared by the decomposition of silane (SiH_4). This process is frequently incomplete and "amorphous silicon" is then, in reality, a silicon-hydrogen alloy. Table 17.6 includes a number of chalcogenides (S, Se, and Te, together with their compounds). Amorphous selenium has played a central role in the xerography process (as a photoconductive coating that permits the formation of a charged image).

Table 17.6 *Some Amorphous Semiconductors*

Group	Semiconductor	Group	Semiconductor
IVA	Si	III–V	GaAs
	Ge		
VIA	S	IV–VI	GeSe
	Se		GeTe
	Te	V–VI	As_2Se_3

SAMPLE PROBLEM 17.14

Consider an amorphous silicon that contains 20 atomic % hydrogen. To a first approximation, the hydrogen atoms are present in interstitial solid solution. If the density of pure amorphous silicon is 2.3 g · cm^{-3}, what is the effect on density of adding the hydrogen?

SOLUTION

Consider 100 g of the Si-H "alloy." This will contain x g of H and $(100 - x)$ g of Si. Or we can say that it will contain (using Appendix 1 data)

$$\frac{x \text{ g}}{1.008 \text{ g}} \text{ g} \cdot \text{atom H}$$

$$\frac{(100 - x) \text{ g}}{28.09 \text{ g}} \text{ g} \cdot \text{atom Si}$$

But

$$\frac{x/1.008}{(100 - x)/28.09} = \frac{0.2}{0.8}$$

or

$$x = 0.889 \text{ g H}$$

$$100 - x = 99.11 \text{ g Si}$$

The volume occupied by the silicon will be

$$V = \frac{99.11 \text{ g Si}}{2.3 \text{ g}} \text{ cm}^3 = 43.09 \text{ cm}^3$$

Therefore, the density of the "alloy" will be

$$\rho = \frac{100 \text{ g}}{43.09 \text{ cm}^3} = 2.32 \text{ g} \cdot \text{cm}^{-3}$$

which is an increase of

$$\frac{2.32 - 2.30}{2.30} \times 100 = 0.90\%$$

Note.   This minor difference is one reason why those people working with this material did not realize, initially, that "amorphous silicon" generally contained large quantities of hydrogen. In addition, hydrogen is an easy element to "miss" in routine chemical analyses.

PRACTICE PROBLEM 17.12

In Sample Problem 17.14, we find that 20 mol % hydrogen has a minor effect on the final density of an amorphous silicon. Suppose that we make an amorphous silicon by the decomposition of silicon tetrachloride, $SiCl_4$, rather than silane, SiH_4. Using similar assumptions, calculate the effect of 20 mol % Cl on the final density of an amorphous silicon.

17.5   PROCESSING OF SEMICONDUCTORS

The most striking feature of semiconductor processing is the ability to produce materials of unparalleled structural and chemical perfection. Table 17.7 summarizes some of the major crystal-growing techniques used for producing high-quality single crystals of semiconducting materials by growth from the melt. Following the production of the crystals, they are sliced into thin *wafers*, usually by a diamond-impregnated blade or wire. After grinding and polishing, the wafers are ready for the complex sequence of steps necessary to build a microcircuit. This processing sequence will be illustrated in Section 17.6. Structural perfection of the original semiconductor crystal is the result of the highly developed technology of crystal growing. The chemical perfection is due to a special heat treatment prior to the crystal-growing step. This treatment is a creative use of phase diagrams (Chapter 9). A bar of material containing a modest level of impurities is purified by the process of **zone refining.** Figure 17–15 shows that an induction coil produces a local molten "zone" along the length of the bar. The phase diagram illustrates that the impurity content in the liquid is substantially greater than in the solid. This allows us to define a **segregation coefficient,** K:

$$K = \frac{C_s}{C_l} \tag{17.9}$$

Table 17.7 *Some Major Crystal Growing Techniques for Semiconductors*

Description		Used for
Czochralski or Teal–Little		Si, Ge, InSb, GaAs
Float zone		Si
Zone leveling		Ge, GaAs, InSb, InAs
Verneuil		Refractory oxides
Bridgeman		Metals, some II–VI compounds
Temperature T_1, gradient T_2		SiC, diamond

Source: W. R. Runyan and S. B. Watelski, in *Handbook of Materials and Processes for Electronics*, C. A. Harper, Ed., McGraw-Hill Book Company, New York, 1970.

where C_s and C_l are the impurity concentrations in solid and liquid, respectively. Of course, K is much less than 1 for the case shown in Figure 17–15, and near the edge of the phase diagram, the solidus and liquidus lines are fairly straight, giving a constant value of K over a range of temperatures. This phenomenon allows a single pass of the heating coil along the

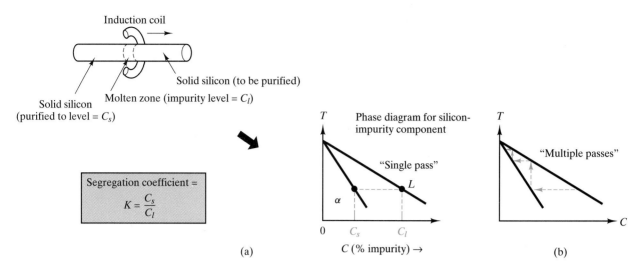

Figure 17-15 *In zone refining, (a) a single pass of the molten "zone" through the bar leads to the concentration of impurities in the liquid. This is illustrated by the nature of the phase diagram. (b) Multiple passes of the molten zone lead to increasing purification of the solid.*

bar to "sweep" the impurities along with the liquid zone to one end. Multiple sweeps lead to substantial purification (Figure 17–15b). Eventually, substantial levels of contamination will be swept to one end of the bar. That end is simply sawed off and discarded. Impurity levels in the part per billion (ppb) range are practical and, in fact, were necessary to allow the development of solid-state electronics as we know it today.

Structural defects, such as dislocations, adversely affect the performance of the silicon-based devices discussed in the next section. Such defects are a common byproduct of oxygen solubility in the silicon. A process known as **gettering** is used to "getter" (or *capture*) the oxygen, removing it from the region of the silicon where the device circuitry is developed. Ironically, the dislocation-producing oxygen is often removed by introducing dislocations on the "back" side of wafers, where the "front" side is defined as that where the circuitry is produced. Mechanical damage (by abrasion, laser impact, etc.) produces dislocations that serve as gettering sites at which SiO_2 precipitates are formed. This approach is known as **extrinsic gettering**. A more subtle approach is to heat treat the silicon wafer so that SiO_2 precipitates form within the wafer but sufficiently far below the front face to prevent interference with circuit development. This latter approach is known as **intrinsic gettering** and may involve as many as three separate annealing steps between 600 and 1250°C extending over a period of several hours. Both extrinsic and intrinsic gettering are commonly used in semiconductor processing.

The rapid development of new processes has become commonplace in the fabrication of semiconductor devices. Some examples will be introduced in Section 17.6. In addition, many modern electronic devices are based on the buildup of thin-film layers of one semiconductor on another while

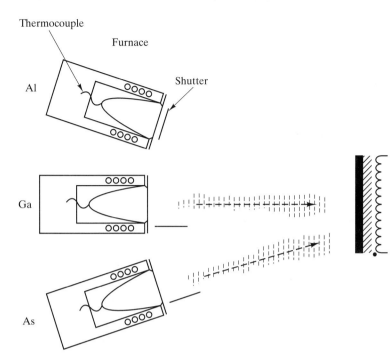

Thermocouple

Furnace

Al

Shutter

Ga

As

Figure 17-16 *Schematic illustration of the molecular-beam epitaxy technique. Resistance-heated source furnaces (also called effusion or Knudsen cells) provide the atomic or molecular beams (approximately 10-mm radius). Shutters control the deposition of each beam onto the heated substrate. (From J. W. Mayer and S. S. Lau,* Electronic Materials Science: For Integrated Circuits in Si and GaAs, *Macmillan Publishing Company, New York, 1990.)*

maintaining some particular crystallographic relationship between the layer and the substrate. This **vapor deposition** technique is called **epitaxy.** **Homoepitaxy** involves the deposition of a thin film of essentially the same material as the substrate (for example, Si on Si). **Heteroepitaxy** involves two materials of significantly different compositions (for example, $Al_x Ga_{1-x} As$ on GaAs). Advantages of epitaxial growth include careful compositional control and reduced concentrations of unwanted defects and impurities. Figure 17–16 illustrates the process of **molecular-beam epitaxy** (MBE), a highly controlled ultrahigh-vacuum deposition process. The epitaxial layers are grown by impinging heated beams of the appropriate atoms or molecules on a heated substrate. The **effusion cells**, or **Knudsen** * **cells**, provide a flux, F, of atoms (or molecules) per second given by

$$F = \frac{pA}{\sqrt{2\pi mkT}} \tag{17.10}$$

where p is the pressure in the effusion cell, A is the area of the aperture, m is the mass of a single atomic or molecular species, k is the Boltzmann's constant, and T is the absolute temperature.

---

* Martin Hans Christian Knudsen (1871–1949), Danish physicist. His brilliant career at the University of Copenhagen centered on many pioneering studies of the nature of gases at low pressure. He also developed a parallel interest in hydrography, establishing methods to define the various properties of seawater.

SAMPLE PROBLEM 17.15

We can use the Al-Si phase diagram (Figure 9–13) to illustrate the principle of zone refining. Assuming that we have a bar of silicon with aluminum as its only impurity, **(a)** calculate the segregation coefficient, K, in the Si-rich region and **(b)** calculate the purity of a 99 wt % Si bar after a single pass of the molten zone. (Note that the solidus can be taken as a straight line between a composition of 99.985 wt % Si at 1190°C and 100 wt % Si at 1414°C.)

SOLUTION

(a) Close inspection of Figure 9–13 indicates that the liquidus curve crosses the 90% Si composition line at a temperature of 1360°C.

The solidus line can be expressed in the form

$$y = mx + b$$

with y being the temperature and x being the silicon composition (in wt %). For the conditions stated,

$$1190 = m(99.985) + b$$

and

$$1414 = m(100) + b$$

Solving gives

$$m = 1.493 \times 10^4 \quad \text{and} \quad b = -1.492 \times 10^6$$

At 1360°C (where the liquidus composition is 90% Si), the solid composition is given by

$$1360 = 1.493 \times 10^4 x - 1.492 \times 10^6$$

or

$$x = \frac{1360 + 1.492 \times 10^6}{1.493 \times 10^4}$$

$$= 99.99638$$

The segregation coefficient is calculated in terms of impurity levels, that is,

$$c_s = 100 - 99.99638 = 0.00362 \text{ wt \% Al}$$

and

$$c_l = 100 - 90 = 10 \text{ wt } \% \text{ Al}$$

yielding

$$K = \frac{c_s}{c_l} = \frac{0.00362}{10} = 3.62 \times 10^{-4}$$

(b) For the liquidus line, a similar straight-line expression takes on the values

$$1360 = m(90) + b$$

and

$$1414 = m(100) + b$$

yielding

$$m = 5.40 \quad \text{and} \quad b = 874$$

A 99 wt % Si bar will have a liquidus temperature

$$T = 5.40(99) + 874 = 1408.6°C$$

The corresponding solidus composition is given by

$$1408.6 = 1.493 \times 10^4 x - 1.492 \times 10^6$$

or

$$x = \frac{1408.6 + 1.492 \times 10^6}{4924} = 99.999638 \text{wt } \% \text{ Si}$$

An alternate composition expression is

$$\frac{(100 - 99.999638)\% \text{ Al}}{100\%} = 3.62 \times 10^{-6} \text{Al}$$

or 3.62 parts per million Al.

Note. These calculations are susceptible to round-off errors. Values for m and b in the solidus line equation must be carried to several places.

PRACTICE PROBLEM 17.13

The purity of a 99 wt % Si bar after one zone refining pass is found in Sample Problem 17.15. What would be the purity after two passes?

17.6 SEMICONDUCTOR DEVICES

Little space in this book is devoted to details of the final applications of engineering materials. Our focus is on the nature of the material itself. For structural and optical applications, detailed descriptions are often unnecessary. The structural steel and glass windows of modern buildings are familiar to us all, but the miniaturized applications of semiconductor materials are generally less so. In this section we look briefly at some solid-state devices.

Miniaturized electrical circuits are the result of the creative combination of p-type and n-type semiconducting materials. An especially simple example is the **rectifier**, or **diode**, shown in Figure 17–17. This contains a

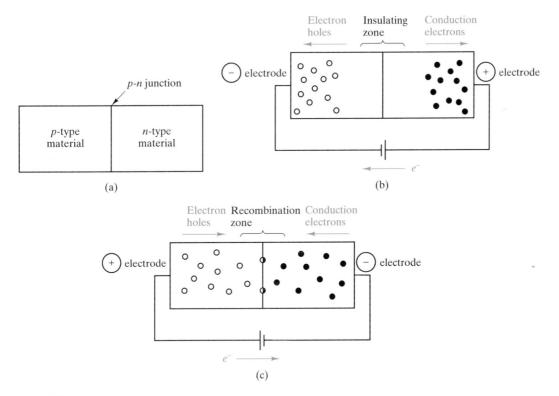

Figure 17-17 (a) A solid-state rectifier, or diode, contains a single p–n junction. (b) In reverse bias, polarization occurs and little current flows. (c) In forward bias, majority carriers in each region flow toward the junction, where they are continuously recombined.

single **p-n junction;** that is, a boundary between adjacent regions of p-type and n-type materials. This junction can be produced by physically joining two pieces of material, one p-type and one n-type. Later, we shall see more subtle ways of forming such junctions by diffusing different dopants (p-type and n-type) into adjacent regions of an (initially) intrinsic material. When voltage is applied to the device, as shown in Figure 17–17b, the charge carriers are driven away from the junction (the positive holes toward the negative electrode, and the negative electrons toward the positive electrode). This **reverse bias** quickly leads to polarization of the rectifier. Majority charge carriers in each region are driven to the adjacent electrodes, and only a minimal current (due to intrinsic charge carriers) can flow. Reversing the voltage produces a **forward bias**, as shown in Figure 17–17c. In this case, the majority charge carriers in each region flow toward the junction where they are continuously recombined (each electron filling an electron hole). This allows a continuous flow of current in the overall circuit. This continuous flow is aided at the electrodes. Electron flow in the external circuit provides a fresh supply of electron holes (i.e., removal of electrons) at the positive electrode and a fresh supply of electrons at the negative electrode. Figure 17–18a shows the current flow as a function of voltage in an ideal rectifier, while Figure 17–18b shows current flow in an actual device. The ideal rectifier allows zero current to pass in reverse bias and has zero resistivity in forward bias. The actual device has some small current in reverse bias (from minority carriers) and a small resistivity in forward bias. This simple solid-state device replaced the relatively bulky vacuum-tube rectifier (Figure 17–19). The solid-state devices to replace the various vacuum tubes allowed a substantial miniaturization of electrical circuitry in the 1950s.

Perhaps the most heralded component in this solid-state revolution was the **transistor** shown in Figure 17–20. This device consists of a pair of nearby p-n junctions. Note that the three regions of the transistor are termed **emitter, base**, and **collector.** Junction 1 (between the emitter and base) is forward biased. As such, it looks identical to the rectifier of Figure 17–17c.

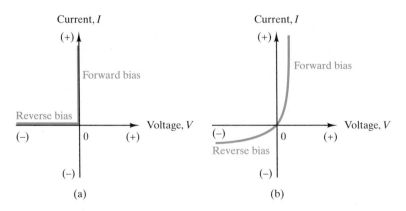

Figure 17-18 *Current flow as a function of voltage in (a) an ideal rectifier and (b) an actual device such as that shown in Figure 17–17.*

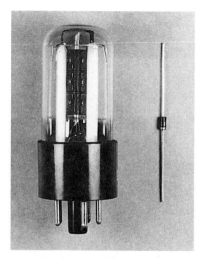

holes

p
(emitter)

n
(base)

p
(collector)

External load

V_e

V_c

Figure 17-19 *Comparison of a vacuum-tube rectifier with a solid-state counterpart. Such components allowed substantial miniaturization in the early days of solid-state technology. (Courtesy of R. S. Wortman.)*

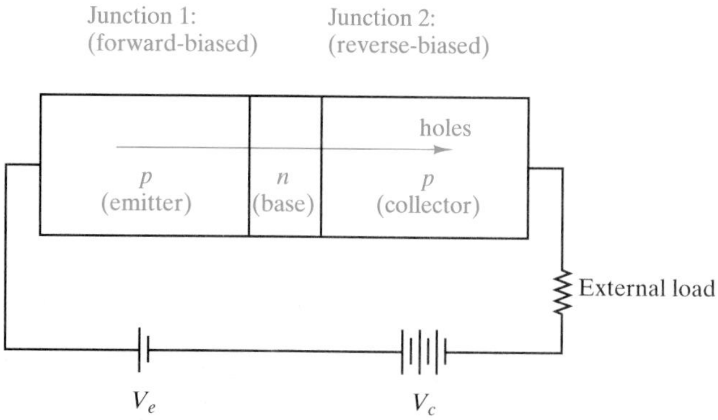

Figure 17-20 *Schematic of a transistor (a p–n–p sandwich). The overshoot of electron holes across the base (n-type region) is an exponential function of the emitter voltage, V_e. Because the collector current (I_c) is similarly an exponential function of V_e, this device serves as an amplifier. An n–p–n transistor functions similarly except that electrons rather than holes are the overall current source.*

However, the function of the transistor requires behavior not considered in our description of the rectifier. Specifically, the recombination of electrons and holes shown in Figure 17–17 does not occur immediately. In fact, many of the charge carriers move well beyond the junction. If the base (n-type) region is narrow enough, a large number of the holes (excess charge carriers) pass across junction 2. A typical base region is less than 1 μm wide. Once in the collector, the holes again move freely (as majority charge carriers). The extent of "overshoot" of holes beyond junction 1 is an exponential function of the emitter voltage, V_e. Because of this, the current in the collector, I_c, is an exponential function of V_e,

$$I_c = I_0 e^{V_e/B} \qquad (17.11)$$

where I_0 and B are constants. The transistor is an **amplifier**, since slight increases in emitter voltage can produce dramatic increases in collector current. The p-n-p "sandwich" in Figure 17–20 is not a unique transistor design. An n-p-n system functions in a similar way, with electrons, rather than holes, being the overall current source. In integrated circuit technology, the configuration in Figure 17–20 is also called a **bipolar junction transistor** (BJT).

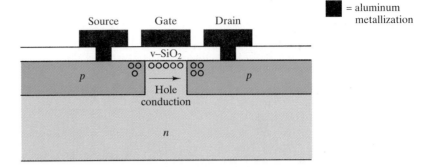

Figure 17-21 *Schematic of a field-effect transistor (FET). A negative voltage applied to the gate produces a field under the vitreous silica layer and a resulting p-type conductive channel between the source and the drain. The width of the gate is less than 1 μm in contemporary integrated circuits.*

A contemporary variation of transistor design is illustrated in Figure 17–21. The **field-effect transistor** (FET) incorporates a "channel" between a **source** and a **drain** (corresponding to the emitter and collector, respectively, in Figure 17–20). The p-channel (under an insulating layer of vitreous silica) becomes conductive upon application of a negative voltage to the **gate** (corresponding to the base in Figure 17–20). The channel's field, which results from the negative gate voltage, produces an attraction for holes from the substrate. (In effect, the n-type material just under the silica layer is distorted by the field to p-type character.) The result is the free flow of holes from the p-type source to the p-type drain. The removal of the voltage on the gate effectively stops the overall current.

An n-channel FET is comparable to Figure 17–21, but with the p- and n-type regions reversed and electrons, rather than electron holes, serving as charge carriers. The operating frequency of high-speed electronic devices is limited by the time required for an electron to move from the source to the drain across such an n-channel. A primary effort in silicon-based **integrated-circuit** (IC) technology is to reduce the gate length, with a value of $< 1\mu$m being typical and $\approx 0.1\mu$m being the current minimum. An alternative is to go to a semiconductor with a higher electron mobility, such as GaAs. (Note the relative values of μ_e for Si and GaAs in Tables 17.1 and 17.5.) The use of GaAs must be balanced against its higher cost and more difficult processing technology.

Modern technology has moved with great speed, but nowhere has progress moved at quite such a dizzy pace as in solid-state technology. Solid-state circuit elements such as diodes and transistors allowed substantial miniaturization when they replaced vacuum tubes. Further increases in miniaturization have occurred by eliminating separate solid-state elements. A sophisticated electrical **microcircuit** such as that shown in Figure 1–17 can be produced by application of precise patterns of diffusible n-type and p-type dopants to give numerous elements within a single **chip** of

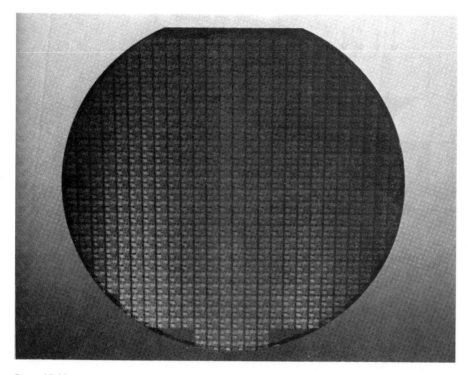

Figure 17-22 *A silicon wafer (1.5 mm thick × 150 mm diameter) containing numerous chips of the type illustrated in Figure 1–17. (Courtesy of R. D. Pashley, Intel Corporation)*

single-crystal silicon. Figure 17–22 shows an array of many such chips that have been produced on a single silicon **wafer**, a thin slice from a cylindrical single crystal of high-purity silicon. A typical wafer is 150 mm (6 in.) or 200 mm (8 in.) in diameter and 250 μm thick, with chips being 5 to 10 mm on an edge. Individual circuit element patterns are produced by **lithography**, originally a print-making technique involving patterns of ink on a porous stone (giving the prefix "litho" from the Greek word *lithos*, meaning "stone"). The sequence of steps used to produce a vitreous SiO_2 pattern on silicon is shown in Figure 17–23. The original uniform SiO_2 layer is produced by thermal oxidation of the Si between 900 and 1100°C. The key to the IC lithography process is the use of a polymeric **photoresist.** In Figure 17–23, a "positive" photoresist is used in which the material is depolymerized by exposure to ultraviolet radiation. A solvent is used to remove the *exposed* photoresist. A "negative" photoresist is used in the metallization process of Figure 17–24, in which the ultraviolet radiation leads to cross-linking of the polymer, allowing the solvent to remove the *unexposed* material. The production of a controlled region of doped material is illustrated in Figure 17–25, showing the two-step process of ion implantation through a vitreous SiO_2 mask, followed by diffusion of the dopant in a temperature range of 950 to 1050°C.

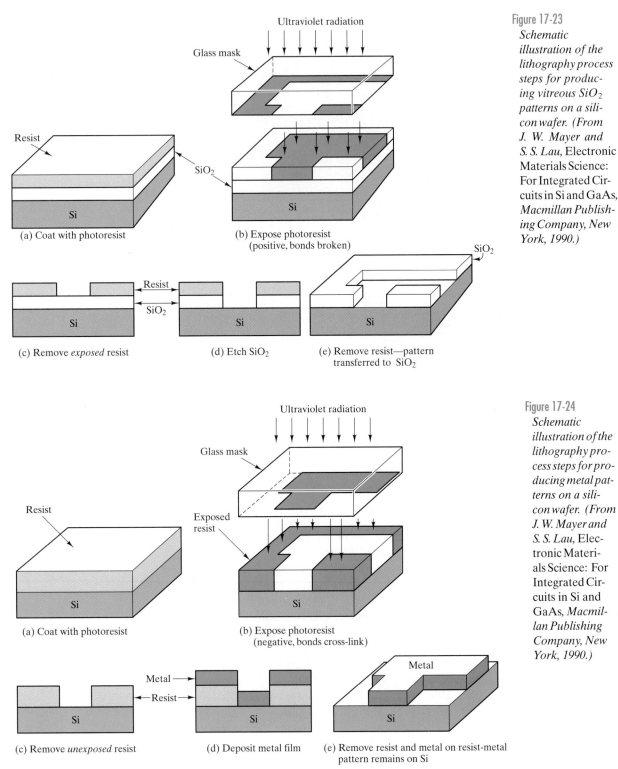

Figure 17-23

Schematic illustration of the lithography process steps for producing vitreous SiO$_2$ patterns on a silicon wafer. (From J. W. Mayer and S. S. Lau, Electronic Materials Science: For Integrated Circuits in Si and GaAs, Macmillan Publishing Company, New York, 1990.)

Ultraviolet radiation

Glass mask

Resist

SiO$_2$

Si

(a) Coat with photoresist

(b) Expose photoresist (positive, bonds broken)

SiO$_2$

Si

Resist

SiO$_2$

Si

(c) Remove *exposed* resist

Si

(d) Etch SiO$_2$

SiO$_2$

Si

(e) Remove resist—pattern transferred to SiO$_2$

Figure 17-24

Schematic illustration of the lithography process steps for producing metal patterns on a silicon wafer. (From J. W. Mayer and S. S. Lau, Electronic Materials Science: For Integrated Circuits in Si and GaAs, Macmillan Publishing Company, New York, 1990.)

Ultraviolet radiation

Glass mask

Resist

Si

(a) Coat with photoresist

Exposed resist

Si

(b) Expose photoresist (negative, bonds cross-link)

Metal

Resist

Si

(c) Remove *unexposed* resist

Metal

Resist

Si

(d) Deposit metal film

Metal

Si

(e) Remove resist and metal on resist-metal pattern remains on Si

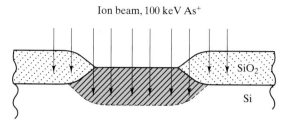

Ion beam, 100 keV As⁺

SiO₂

Si

(a) Ion implantation of dopants (As).

Figure 17-25 *Schematic illustration of the two-step doping of a silicon wafer with arsenic producing an n-type region beneath the vitreous SiO₂ mask. (From J. W. Mayer and S. S. Lau,* Electronic Materials Science: For Integrated Circuits in Si and GaAs, *Macmillan Publishing Company, New York, 1990.)*

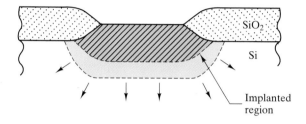

SiO₂

Si

Implanted region

(b) Drive-in diffusion, 950-1050°C

During the later stages of circuit fabrication, oxide and nitride layers are often deposited on silicon to serve as insulating films between metal lines or as insulating, protective covers. These layers are generally deposited by **chemical vapor deposition (CVD)**. SiO_2 films can be produced between 250 and 450°C by the reaction of silane and oxygen:

$$SiH_4 + 0_2 \rightarrow SiO_2 + 2\,H_2 \qquad (17.12)$$

Silicon nitride films involve the reaction of silane and ammonia:

$$3\,SiH_4 + 4\,NH_3 \rightarrow Si_3N_4 + 12\,H_2 \qquad (17.13)$$

Ultimately, the sub-micron-scale IC patterns must be connected to the macroscopic electronic "package" by relatively large-scale metal wires with diameters of 25 to 75 μm (Figure 17–26).

An active area of development at the current time is the production of **quantum wells**; that is, thin layers of semiconducting material in which the wavelike electrons are confined within the layer thickness, a dimension as small as 2 nm. Advanced processing techniques are being used to develop such confined regions in two dimensions (**quantum wires**) or in three dimensions (**quantum dots**, again as small as 2 nm on a side). These small dimensions permit electron transit times of less than a picosecond and correspondingly high device operating speeds.

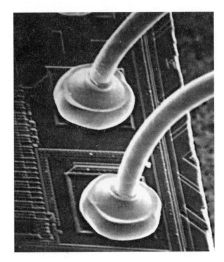

Figure 17-26   *Typical metal wire bond to an integrated circuit. (From C. Woychik and R. Senger, in* Principles of Electronic Packaging, *D. P. Seraphim, R. C. Lasky, and C.-Y. Li, Eds., McGraw-Hill Book Company, New York, 1989.)*

In summary, semiconductor devices have revolutionized modern life by providing for the miniaturization of electronic circuits. The replacement of traditional, large-scale elements such as diodes and transistors with separate solid-state counterparts started a revolution. The development of integrated microcircuits accelerated this revolution. The rapid pace at which chips replaced separate elements is indicated in Figure 17–27. Miniaturization is continuing by reducing the size of microcircuit elements. At the outset, miniaturization of electronics was driven by the aerospace industry, for which computers and circuitry needed to be small and low in power consumption. The steady and dramatic reduction in cost that has accompanied these developments has led to the increasing crossover of applications into industrial production and control, as well as consumer electronics.

The dramatic changes in our technology are nowhere more evident than in the computer field. In addition to the amplification of electrical signals discussed previously, transistors and diodes can also serve as switching devices. This application is the basis of the computational and information storage functions of computers. The elements within the microcircuit represent the two states ("off" or "on") of the binary arithmetic of the digital circuit. In this application, miniaturization has led to the trend away from main-frames to personal computers and, further, to portable (laptop) computers. Simultaneous with this trend has been a steady reduction in cost and an increase in computational power.

Figure 17–28 illustrates the dramatic progress in the miniaturization of computer chips. The number of transistors produced in the microcircuitry of a single chip has, over roughly three decades, gone from a few thousand to 10 million. This has generally represented a doubling in number every two years. This steady pace of miniaturization has become widely known as

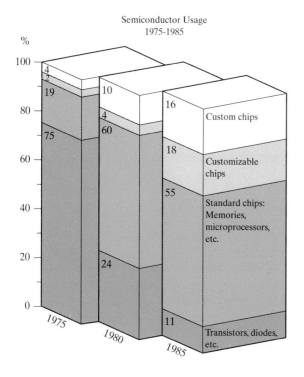

Semiconductor Usage
1975-1985

Figure 17-27 *Although separate solid-state elements such as transistors and diodes (e.g., Figure 17–19) provide miniaturization compared with vacuum tubes, microcircuits (e.g., Figure 1–17) allow substantially greater size reduction. The trend in which industry moved to microcircuit chips is shown here. Custom chips are those designed for specific applications. Standard chips represent more general-purpose circuit designs. Customizable chips are produced partway like standard chips but, in final stages, are prepared for specific circuit applications. There can be a fivefold difference in cost between a fully custom chip and a standard one. (Courtesy of the* San Francisco Examiner, *based on data provided by the Digital Equipment Corporation)*

Moore's law after Gordon Moore, cofounder of Intel Corporation, who predicted this capability in the early days of integrated circuit technology. While current lithography techniques can produce features on the order of a few tenths of a μm (a few hundred nm) in width, intense research is being conducted on the use of short wavelength ultraviolet and x-ray litography. As we can see in Figure 16–1, UV and x-ray wavelengths are much shorter that those of visible light and can potentially allow production of litographic features on the order of 0.1 μm (100 nm). In this way, microcircuit technology could continue to follow Moore's law into a level of one billion transistors per chip.

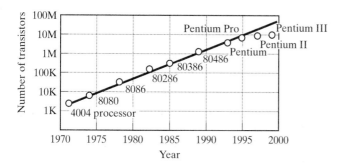

Figure 17-28 *The rapid and steady growth in the number of transistors contained on a single microcircuit chip has generally followed Moore's law, which states that the number doubles roughly every two years. (After data from Intel Corporation)*

The Material World:
A Brief History of the Electron

Given the ubiquitous role of electronics in modern life, it is perhaps surprising that the electron itself was "discovered" only toward the end of the nineteenth century. In 1897, Professor J. J. Thomson of Cambridge University in England showed that the cathode rays in a device equivalent to a primitive version of the television tube were streams of negatively charged particles. He called these particles corpuscles. By varying the magnitudes of electric and magnetic fields through which the corpuscles traveled, he was able to measure the ratio of mass m to charge q. Thomson furthermore made the bold claim that his corpuscles were a basic constituent of all matter and that they were more than a 1000 times lighter than the lightest known atom, hydrogen. He was right on both counts. (The electron's mass is 1/1836.15 that of the hydrogen atom.) Thomson's m/q measurement combined with his bold and accurate claims have caused him to be remembered as the discoverer of the electron.

Fifty years later, John Bardeen, Walter Brattain, and William Shockley at Bell Labs in New Jersey discovered that a small single crystal of germanium provided electronic signal amplification. This first transistor ushered in the era of solid-state electronics. In 1955, Dr. Shockley left Bell Labs and started the Shockley Semiconductor Laboratory of Beckman Instruments in Mountain View, California. In 1963, he moved on to Stanford University, where he was a Professor of Engineering Science for many years. Shockley's experience at the Shockley Semiconductor

THE

LONDON, EDINBURGH, AND DUBLIN

PHILOSOPHICAL MAGAZINE

AND

JOURNAL OF SCIENCE.

[FIFTH SERIES.]

OCTOBER 1897.

XL. *Cathode Rays.* By J. J. THOMSON, *M.A., F.R.S., Cavendish Professor of Experimental Physics, Cambridge*.

THE experiments† discussed in this paper were undertaken in the hope of gaining some information as to the nature of the Cathode Rays. The most diverse opinions are held as to these rays; according to the almost unanimous opinion of German physicists they are due to some process in the æther to which—inasmuch as in a uniform magnetic field their course is circular and not rectilinear—no phenomenon hitherto observed is analogous: another view of these rays is that, so far from being wholly ætherial, they are in fact

(Courtesy of the Philosophical Magazine)

Laboratory soured in 1957 when a group of young engineers he dubbed the "traitorous eight" left to form their own company, Fairchild Semiconductor, with the backing of the Fairchild Camera and Instrument Company. Among these young rebels were Gordon Moore and Robert Noyce.

By 1968, Moore and Noyce had become disenchanted with Fairchild. They were not alone. Many of the engineers were leaving with the feeling that technology was being supplanted by the politics of the workplace. Moore and Noyce left Fairchild to form a new company called Intel. They were joined in this venture by another former Fairchild employee named Andrew Grove. At this time, there were 30,000 computers in the world. Most were mainframes big enough to fill a room; the rest were minicomputers roughly the size of a refrigerator. Computer codes were entered mechanically using punch cards. The young Intel company had reasonably good success making computer memory but in 1971 took a bold step in developing for a client, Busicom of Japan, a radically new product called a microprocessor. The 4004 product took nine months to develop and contained 2,300 transistors on a single chip of silicon. Modest by today's standards, the 4004 had as much computing power as the pioneering ENIAC computer invented in 1946, which weighed 27 Mg and contained 18,000 vacuum tubes. Realizing it had a product with substantial potential, Intel bought the design and marketing rights to the 4004 microprocessor back from Busicom for $60,000. Shortly after this, Busicom went bankrupt. The rest, as they say, is history.

Andrew Grove, Robert Noyce, and Gordon Moore in 1975. (Courtesy of Intel Corporation)

SAMPLE PROBLEM 17.16

A given transistor has a collector current of 5 mA when the emitter voltage is 5 mV. Increasing the emitter voltage to 25 mV (a factor of 5) increases the collector current to 50 mA (a factor of 10). Calculate the collector current produced by further increasing the emitter voltage to 50 mV.

SOLUTION

Using Equation 17.11,

$$I_c = I_0 e^{V_e/B}$$

we are given

$$I_c = 5 \text{ mA} \quad \text{when } V_e = 5 \text{ mV}$$

and

$$I_c = 50 \text{ mA} \quad \text{when } V_e = 25 \text{ mV}$$

Then

$$\frac{50 \text{ mA}}{5 \text{ mA}} = e^{25 \text{ mV}/B - 5 \text{ mV}/B}$$

giving

$$B = 8.69 \text{ mV}$$

and

$$I_0 = 5 \text{ mA} e^{-(5 \text{ mV})/(8.69 \text{ mV})}$$

$$= 2.81 \text{ mA}$$

Therefore,

$$I_{c,50 \text{ mV}} = (2.81 \text{ mA}) e^{50 \text{ mV}/8.69 \text{ mV}}$$

$$= 886 \text{ mA}$$

PRACTICE PROBLEM 17.14

In Sample Problem 17.16, we calculate, for a given transistor, the collector current produced by increasing the emitter voltage to 50 mV. Make a continuous plot of collector current versus emitter voltage for this device over the range of 5 to 50 mV.

SUMMARY

Following the discussion of intrinsic, elemental semiconductors in Chapter 15, we note that the Fermi function indicates that the number of charge carriers increases exponentially with temperature. This effect so dominates the conductivity of semiconductors that conductivity also follows an exponential increase with temperature (an example of an Arrhenius equation). This increase is in sharp contrast to the behavior of metals.

We consider the effect of impurities in extrinsic, elemental semiconductors. Doping a group IVA material, such as Si, with a group VA impurity, such as P, produces an n-type semiconductor in which negative charge carriers (conduction electrons) dominate. The "extra" electron from the group VA addition produces a donor level in the energy band structure of the semiconductor. As with intrinsic semiconductors, extrinsic semiconduction exhibits Arrhenius behavior. In n-type material, the temperature span between the regions of extrinsic and intrinsic behavior is called the exhaustion range. A p-type semiconductor is produced by doping a group IVA material with a group IIIA impurity, such as Al. The group IIIA element has a "missing" electron, producing an acceptor level in the band structure and leading to formation of positive charge carriers (electron holes). The region between extrinsic and intrinsic behavior for p-type semiconductors is called the saturation range. Hall effect measurements can distinguish between n-type and p-type conduction.

Compound semiconductors usually have an MX composition with an average of four valence electrons per atom. The III-V and II-VI compounds are the common examples. Amorphous semiconductors are the noncrystalline materials with semiconducting behavior. Elemental and compound materials are both found in this category. Chalcogenides are important members of this group.

Semiconductor processing is unique in the production of commercial materials in that exceptionally high-quality crystalline structure is required together with chemical impurities in the parts per billion level. Structural perfection is approached with various crystal-growing techniques. Chemical perfection is approached by the process of zone refining. Vapor deposition techniques, such as molecular-beam epitaxy, are used to produce thin films for advanced electronic devices.

To appreciate the applications of semiconductors, we review a few of the solid-state devices that have been developed in the past few decades. The solid-state rectifier, or diode, contains a single p-n junction. Current flows readily when this junction is forward biased but is almost completely choked off when reverse biased. The transistor is a device consisting of a pair of nearby p-n junctions. The net result is a solid-state amplifier. Replacing vacuum tubes with solid-state elements such as these produced substantial miniaturization of electrical circuits. Further miniaturization has resulted by the production of microcircuits consisting of precise patterns of n-type and p-type regions on a single-crystal chip. An increasingly finer degree of miniaturization is an ongoing goal of this integrated-circuit technology.

The major electrical properties needed to specify an intrinsic semiconductor are band gap, electron mobility, hole mobility, and conduction electron density (= electron-hole density) at room temperature. For extrinsic semiconductors, one needs to specify either the donor level (for n-type material) or the acceptor level (for p-type material).

KEY TERMS

acceptor level (630)

amorphous semiconductor (645)

amplifier (655)

base (654)

bipolar junction transistor (BJT) (655)

chemical vapor deposition (CVD) (659)

chip (656)

collector (654)

compound semiconductor (642)

conduction electron (620)

device (653)

diode (653)

donor level (627)

dopant (626)

drain (656)

effusion cell (650)

emitter (654)

epitaxy (650)

exhaustion range (630)

extrinsic gettering (649)

extrinsic semiconductor (626)

field-effect transistor (FET) (656)

forward bias (654)

gate (656)

gettering (649)

Hall effect (632)

heteroepitaxy (650)

homoepitaxy (650)

integrated circuit (IC) (656)

intrinsic gettering (649)

intrinsic semiconductor (620)

Knudsen cell (650)

lithography (657)

microcircuit (656)

molecular-beam epitaxy (MBE) (650)

Moore's law (661)

n-type semiconductor (627)

p-n junction (654)

p-type semiconductor (630)

photoresist (657)

quantum dot (659)

quantum well (659)

quantum wire (659)

rectifier (653)

reverse bias (654)

saturation range (631)

segregation coefficient (647)

source (656)

III-V compound (642)

II-VI compound (642)

transistor (654)

vapor deposition (650)

wafer (657)

zone refining (647)

REFERENCES

Harper, C. A., and **R. N. Sampson**, *Electronic Materials and Processes Handbook*, 2nd ed., McGraw-Hill, New York, 1994.

Kittel, C., *Introduction to Solid State Physics*, 7th ed., John Wiley & Sons, Inc., New York, 1996.

Mayer, J. W., and **S. S. Lau**, *Electronic Materials Science: For Integrated Circuits in Si and GaAs*, Macmillan Publishing Company, New York, 1990.

Tu, K. N., J. W. Mayer, and **L. C. Feldman**, *Electronic Thin Film Science*, Macmillan Publishing Company, New York, 1992.

PROBLEMS

17.1 • Intrinsic, Elemental Semiconductors

17.1. In a 150-mm diameter × 0.5-mm-thick wafer of pure silicon at room temperature, **(a)** how many conduction electrons would be present, and **(b)** how many electron holes would be present?

17.2. In a 60-mm diameter × 0.5-mm-thick wafer of pure germanium at room temperature, **(a)** how many conduction electrons would be present, and **(b)** how many electron holes would be present?

17.3. Using data from Table 17.1, make a plot similar to Figure 17–3 showing both intrinsic silicon and intrinsic germanium over the temperature range of 27 to 200°C.

17.4. Superimpose a plot of the intrinsic conductivity of GaAs on the result of Problem 17.3.

17.5. Starting from an ambient temperature of 300 K, what temperature increase is necessary to double the conductivity of pure silicon?

17.6. Starting from an ambient temperature of 300 K, what temperature increase is necessary to double the conductivity of pure germanium?

17.7. There is a slight temperature dependence for the band gap of a semiconductor. For silicon, this dependence can be expressed as

$$E_g(T) = 1.152 \text{ eV} - \frac{AT^2}{T + B}$$

where $A = 4.73 \times 10^{-4}$ eV/K, $B = 636$ K, and T is in Kelvin. What is the percentage error in taking the band gap at 200°C to be the same as that at room temperature?

17.8. Repeat Problem 17.7 for GaAs, in which

$$E_g(T) = 1.567 \text{ eV} - \frac{AT^2}{T + B}$$

where $A = 5.405 \times 10^{-4}$ eV/K and $B = 204$ K.

17.2 • Extrinsic, Elemental Semiconductors

17.9. An n-type semiconductor consists of 100 ppb of P doping, by weight, in silicon. What is **(a)** the mole percentage P and **(b)** the atomic density of P atoms? Compare your answer in part (b) with the maximum solid solubility level given in Table 17.2.

17.10. An As-doped silicon has a conductivity of 2.00×10^{-2} $\Omega^{-1} \cdot$ m^{-1} at room temperature. **(a)** What is the predominant charge carrier in this material? **(b)** What is the density of these charge carriers? **(c)** What is the drift velocity of these carriers under an electrical field strength of 200 V/m? (The μ_e and μ_h values given in Table 15.5 also apply for an extrinsic material with low impurity levels.)

17.11. Repeat Problem 17.10 for the case of a Ga-doped silicon with a conductivity of 2.00×10^{-2} $\Omega^{-1} \cdot$ m^{-1} at room temperature.

17.12. Calculate the conductivity for the saturation range of silicon doped with 10-ppb boron.

17.13. Calculate the conductivity for the saturation range of silicon doped with 20-ppb boron. (Note Problem 17.12)

17.14. Calculate the conductivity for the exhaustion range of silicon doped with 10-ppb antimony.

17.15. Calculate the upper temperature limit of the saturation range for silicon doped with 10-ppb boron. (Note Problem 17.12.)

17.16. Calculate the upper temperature limit of the exhaustion range for silicon doped with 10-ppb antimony. (Note Problem 17.14.)

17.17. If the lower temperature limit of the saturation range for silicon doped with 10-ppb boron is 110°C, calculate the extrinsic conductivity at 300 K. (Note Problems 17.12 and 17.15.)

17.18. Plot the conductivity of the B-doped Si of Problem 17.17 in a manner similar to Figure 17–13.

17.19. If the lower temperature limit of the exhaustion range for silicon doped with 10-ppb antimony is 80°C, calculate the extrinsic conductivity at 300 K. (Note Problems 17.14 and 17.16.)

17.20. Plot the conductivity of the Sb-doped Si of Problem 17.19 in a manner similar to Figure 17–9.

D **17.21.** In designing a solid-state device using B-doped Si, it is important that the conductivity not increase more than 10% (relative to the value at room temperature) during the operating lifetime. For this factor alone, what is the maximum operating temperature to be specified for this design?

D **17.22.** In designing a solid-state device using As-doped Si, it is important that the conductivity not increase more than 10% (relative to the value at room temperature) during the operating lifetime. For this factor alone, what is the maximum operating temperature to be specified for this design?

• **17.23.** **(a)** It was pointed out in Section 15.3 that the temperature sensitivity of conductivity in semiconductors makes them superior to traditional thermocouples for certain high-precision temperature measurements. Such devices are referred to as thermistors. As a simple example, consider a wire 0.5 mm in diameter × 10 mm long made of intrinsic silicon. If the resistance of the wire can be measured to within 10^{-3} Ω, calculate the temperature sensitivity of this device at 300 K. (*hint:* The very small differences here may make you want to develop an expression for $d\sigma/dT$.) **(b)** Repeat the calculation for an intrinsic germanium wire of the same dimensions. **(c)** For comparison with the temperature sensitivity of a metallic conductor, repeat the calculation for a copper (annealed standard) wire of the same dimensions. (The necessary data for this case can be found in Table 15.2.)

17.24. An application of semiconductors of great use to materials engineers is the "lithium-drifted silicon," Si(Li), solid-state photon detector. This is the basis for the detection of microstructural-scale elemental distributions as illustrated in Figure 4–39. A characteristic x-ray photon striking the Si(Li) promotes a number of electrons (N) to the conduction band creating a current pulse, where

$$N = \frac{\text{photon energy}}{\text{band gap}}$$

For an Si(Li) detector operating at liquid nitrogen temperature (77 K), the band gap is 3.8 eV. What would be the size of a current pulse (N) created by **(a)** a copper K_α characteristic x-ray photon ($\lambda = 0.1542$ nm) and **(b)** an iron K_α characteristic x-ray photon ($\lambda = 0.1938$ nm)? (By the way, "lithium-drifted" refers to the Li dopant being diffused into the Si under an electrical potential. The result is a highly uniform distribution of the dopant.)

• **17.25.** Using the information from Problem 17.24, sketch a "spectrum" produced by chemically analyzing a stainless steel with characteristic x-rays. Assume the yield of x-rays is proportional to the atomic fraction of elements in the sample being bombarded by an electron beam. The spectrum itself consists of sharp spikes of height proportional to the x-ray yield (number of photons). The spikes are located along a "current pulse (N)" axis. For an "18–8" stainless steel (18 wt % Cr, 8 wt % Ni, bal. Fe), the following spikes are observed:

$$FeK_\alpha(\lambda = 0.1938\text{nm})$$

$$FeK_\beta(\lambda = 0.1757\text{nm})$$

$$CrK_\alpha(\lambda = 0.2291\text{nm})$$

$$CrK_\beta(\lambda = 0.2085\text{nm})$$

$$NiK_\alpha(\lambda = 0.1659\text{nm})$$

$$NiK_\beta(\lambda = 0.1500\text{nm})$$

(K_β photon production is less probable than K_α production. Take the height of a K_β spike to be only 10% of that of the K_α spike for the same element.)

• **17.26.** Using the information from Problems 17.24 and 17.25, sketch a spectrum produced by chemically analyzing a specialty alloy (75 wt % Ni, 25 wt % Cr) using characteristic x-rays.

17.3 • Compound Semiconductors

17.27. Calculate the atomic density of Cd in a 100-ppb doping of GaAs.

17.28. The band gap of intrinsic InSb is 0.17 eV. What temperature increase (relative to room temperature = 25°C) is necessary to increase its conductivity by **(a)** 10%, **(b)** 50%, and **(c)** 100%?

17.29. Illustrate the results of Problem 17.28 on an Arrhenius-type plot.

17.30. The band gap of intrinsic ZnSe is 2.67 eV. What temperature increase (relative to room temperature = 25°C) is necessary to increase its conductivity by **(a)** 10%, **(b)** 50%, and **(c)** 100%?

17.31. Illustrate the results of Problem 17.30 on an Arrhenius-type plot.

17.32. Starting from an ambient temperature of 300 K, what temperature increase is necessary to double the conductivity of intrinsic InSb?

17.33. Starting from an ambient temperature of 300 K, what temperature increase is necessary to double the conductivity of intrinsic GaAs?

17.34. Starting from an ambient temperature of 300 K, what temperature increase is necessary to double the conductivity of intrinsic CdS?

17.35. What temperature increase (relative to room temperature) is necessary to increase the conductivity of intrinsic GaAs by 1%?

17.36. What temperature increase (relative to room temperature) is necessary to increase by 1% the conductivity of **(a)** Se-doped GaAs and **(b)** Cd-doped GaAs?

17.37. In intrinsic semiconductor GaAs, what fraction of the current is carried by electrons and what fraction is carried by holes?

17.38. What fraction of the current is carried by electrons and what fraction is carried by holes in **(a)** Se-doped GaAs and **(b)** Cd-doped GaAs, in the extrinsic behavior range?

17.4 • Amorphous Semiconductors

17.39. Estimate the atomic packing factor of amorphous germanium if its density is reduced by 1% relative to the crystalline state. (See Practice Problem 4.7.)

17.40. Ion implantation treatment of crystalline silicon can lead to the formation of an amorphous surface layer extending from the outer surface to the ion penetration depth. This is considered a structural defect for the crystalline device. What is the appropriate processing treatment to eliminate this defect?

17.5 • Processing of Semiconductors

17.41. When the aluminum impurity level in a silicon bar has reached 1 part per billion, what would have been the purity of the liquid on the previous pass?

17.42. Suppose you have a bar of 99 wt % Sn, with the impurity being Pb. Determine the impurity level after one zone refining pass. (Recall the phase diagram for the Pb-Sn system in Figure 9–16.)

17.43. For the bar in Problem 17.42, what would be the impurity level after **(a)** two passes or **(b)** three passes?

17.44. **(a)** Calculate the flux of gallium atoms out of an MBE effusion cell at a pressure of 2.9×10^{-6} atm and a temperature of 970°C with an aperture area of 500 mm^2. **(b)** If the atomic flux in part (a) is projected onto an area of 45,000 mm^2 on the substrate side of the growth chamber, how much time is required to build up a monolayer of gallium atoms? (Assume, for simplicity, a square grid of adjacent Ga atoms.)

17.6 • Semiconductor Devices

17.45. The high-frequency operation of solid-state devices can be limited by the transit time of an electron across the gate between the source and drain of an FET. For a device to operate at 1 gigahertz (10^9 s^{-1}), a transit time of 10^{-9} s is required. **(a)** What electron velocity is required to achieve this transit time across a 1-μm gate? **(b)** What electric field strength is required to achieve this electron velocity in silicon? **(c)** For the same gate width and electric field strength, what operating frequency would be achieved with GaAs, a semiconductor with a higher electron mobility?

17.46. Make a schematic illustration of an n-p-n transistor analogous to the p-n-p case shown in Figure 17–20.

17.47. Make a schematic illustration of an n-channel FET analogous to the p-channel FET shown in Figure 17–21.

17.48. Figure 17–26 illustrates the relatively large metal connection needed to communicate with an integrated circuit. A limit to the increasing scale of integration in integrated circuits (IC) is the density of interconnection. A useful empirical equation to estimate the number of signal input/output (I/O) pins in a device package is $P = KG^{\alpha}$ where K and α are empirical constants and G is the number of gates of the IC. (One gate is equal to approximately four transistors.) For K and α values of 7 and 0.2, respectively, calculate the number of pins for devices with **(a)** 1000, **(b)** 10,000, and **(c)** 100,000 gates.

CHAPTER 18
Magnetic Materials

Each of these steel billets is being heated uniformly in a cost-effective way by two adjacent coils which utilize an oscillating electrical current to produce an oscillating magnetic flux and the resulting heating due to the "hysteresis" of a ferromagnetic core material. (Courtesy of CoreFlux Heating Systems.)

After concentrating on the major electronic and optical materials, we finish Part III of the book with a discussion of magnetic materials. This requires a brief discussion of *magnetism*, an important branch of physics that is intimately associated with the electronic phenomena treated in the preceding three chapters. The introduction is somewhat detailed because the basic vocabulary associated with magnetic terms and units is not as familiar as some others treated earlier in the book (e.g., basic chemical and mechanical concepts).

Although materials exhibit a variety of magnetic behavior, one of the most important types is *ferromagnetism* associated, as the name implies, with iron-containing metal alloys. The microscopic-scale *domain structure* of these materials allows for their distinctive response to magnetic fields. *Ferrimagnetism* is a subtle variation of ferromagnetic behavior found in a number of ceramic compounds. *Metallic magnets* are traditionally characterized as "soft" or "hard" in terms of their relative ferromagnetic behavior. This nomenclature is conveniently associated with mechanical hardness. The primary examples of *soft magnets* are the iron-silicon alloys that are routinely used in electrical power applications, where the magnetization of the alloy (e.g., in a transformer core) is easily reversed. In contrast, certain alloys are *hard magnets* that are useful as *permanent magnets*. *Ceramic magnets* are widely used and best illustrated by the many *ferrite* compounds based on the *inverse spinel* crystal structure. Substantial interest is currently focused on magnets with the unique property of *superconductivity*. Ceramic superconductors, rather than their metallic counterparts, are proving to have relatively high operating temperatures and, thereby, vastly wider potential applications. In order to appreciate more fully the nature of these various magnetic materials, we now turn to a brief discussion of the fundamentals of magnetic behavior.

18.1   MAGNETISM

A simple example of **magnetism**, the physical phenomenon associated with the attraction of certain materials, is shown in Figure 18–1. An electrical current loop generates a region of physical attraction, or **magnetic field**, represented by a set of **magnetic flux lines.** The magnitude and direction of the magnetic field at any given point near the current loop is given by **H**, a vector quantity. Some materials are inherently magnetic; that is, they can generate a magnetic field without a macroscopic electrical current. A bar magnet shown in Figure 18–2 is a simple example. It exhibits an identifiable *dipole* (north-south) orientation. Much of the utility of magnetism is, of course, the force of attraction that it can provide. Figure 18–3 illustrates this with the attraction of two adjacent bar magnets. Note the orientation of the two bar magnets. The "pairing" of dipoles is symbolic of the interaction of electron orbitals that occurs on the atomic scale in magnetic materials. We shall return to this point when defining ferromagnetism in the next section.

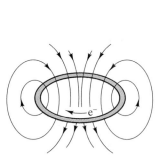

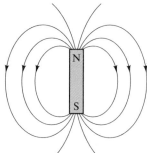

Figure 18-1 *A simple illustration of magnetism shows the magnetic field (seen as magnetic flux lines) generated around an electrical current loop.*

Figure 18-2 *A magnetic material can generate a magnetic field without an electrical current. This simple bar magnet is an example.*

Figure 18-3 *Attraction of two adjacent bar magnets.*

For the free space surrounding a magnetic field source, we can define an **induction, B**, whose magnitude is the **flux density.** The induction is related to the **magnetic field strength, H**, by

$$\mathbf{B} = \mu_0 \mathbf{H} \tag{18.1}$$

where μ_0 is the **permeability** of vacuum. If a solid is inserted in the magnetic field, the magnitude of induction will change but can still be expressed in a similar form:

$$\mathbf{B} = \mu \mathbf{H} \tag{18.2}$$

where μ is the permeability of the solid. It is useful to note that this basic equation for magnetic behavior is a direct analog of the more commonly expressed relationship for electronic behavior, Ohm's law. If we take our basic statement of Ohm's law from Chapter 15,

$$V = IR \tag{15.1}$$

and combine with it the definitions of resistivity and conductivity (also from Chapter 15),

$$\rho = \frac{RA}{l} \tag{15.2}$$

and

$$\sigma = \frac{1}{\rho} \tag{15.3}$$

we obtain an alternative form for Ohm's law:

$$\frac{I}{A} = \sigma \frac{V}{l} \qquad (18.3)$$

Here I/A is the current density and V/l is the voltage gradient. We then see that the magnetic induction (**B**) is analogous to current density and the magnetic field strength (**H**) is analogous to voltage gradient (electric field strength), with permeability (μ) corresponding to conductivity. The presence of the solid has changed the induction. The separate contribution of the solid is illustrated by the expression

$$\mathbf{B} = \mu\mathbf{H} = \mu_0(\mathbf{H} + \mathbf{M}) \qquad (18.4)$$

where **M** is called the **magnetization** of the solid and the term $\mu_0\mathbf{M}$ represents the "extra" magnetic induction field associated with the solid. The magnetization, **M**, is the volume density of magnetic dipole moments associated with the electronic structure of the solid.

The units for these various magnetic terms are webers[*] $/m^2$ for B (the magnitude of **B**), webers/ampere-meter or henries[†] $/m$ for μ, and amperes/m for H and M. The magnitude of μ_0 is $4\pi \times 10^{-7}$ H/m/(=N/A^2). It is sometimes convenient to describe the magnetic behavior of a solid in terms of its **relative permeability**, μ_r, given by

$$\mu_r \equiv \frac{\mu}{\mu_0} \qquad (18.5)$$

which is, of course, dimensionless. These various units are in the meter-kilogram-second (mks) system and are consistent with those accepted in the SI system.

Some solids, such as the highly conductive metals copper and gold, have relative permeabilities of slightly less than 1 (about 0.99995 to be exact). Such materials exhibit **diamagnetism.** In effect, the material's electronic structure responds to an applied magnetic field by setting up a slight opposing field. A large number of solids have relative permeabilities slightly greater than 1 (between 1.00 and 1.01). Such materials exhibit **paramagnetism.** Their electronic structure allows them to set up a reinforcing field parallel to the applied field. The magnetic effect for both diamagnetic and paramagnetic

---

[*] Wilhelm Eduard Weber (1804–1891), German physicist, was a long-time collaborator of Gauss. He developed a logical system of units for electricity to complement the system for magnetism developed by Gauss. Also with Gauss, he constructed one of the first practical telegraphs.

[†] Joseph Henry (1797–1878), American physicist. Like Weber, Henry developed a telegraph but was not interested in financial reward. He left to Samuel Morse the opportunity to patent the idea. He succeeded in building the most powerful electromagnet of his day and later developed the concept of the electric motor.

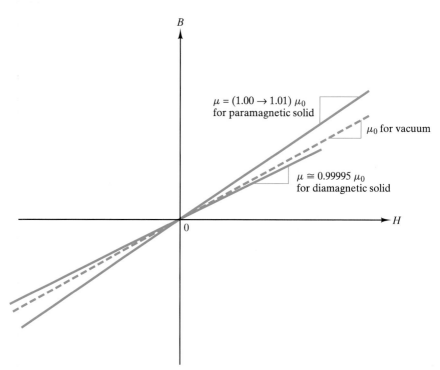

Figure 18-4 *Comparison of diamagnetism and paramagnetism on a plot of induction (B) versus magnetic field strength (H). Neither of these phenomena is of practical engineering importance due to the modest level of induction that can be generated.*

materials is small. Figure 18–4 illustrates a *B-H* plot for these two categories. For magnetic properties, the *B-H* plot is comparable to the σ-ϵ plot for mechanical behavior and will be used frequently through the remainder of this chapter. There is, however, another category of magnetic behavior in which the relative permeability is substantially greater than 1 (as much as 10^6). Such magnitudes provide important engineering applications and are the subject of the next section.

SAMPLE PROBLEM 18.1

A magnetic field strength of 2.0×10^5 amperes/m (provided by an ordinary bar magnet) is applied to a paramagnetic material with a relative permeability of 1.01. Calculate the values of $|\mathbf{B}|$ and $|\mathbf{M}|$.

SOLUTION

We can rewrite Equation 18.4 as

$$B = \mu H = \mu_0(H + M)$$

where

$$B = |\mathbf{B}| \quad \text{and} \quad M = |\mathbf{M}|$$

Using the first equality, we obtain

$$B = \mu H$$

$$= \mu_r \mu_0 H$$

$$= (1.01)(4\pi \times 10^{-7} \text{ henry/m})(2.0 \times 10^5 \text{ amperes/m})$$

$$= 0.254 \text{ henry} \cdot \text{amperes/m}^2$$

$$= 0.254 \text{ weber/m}^2 = |\mathbf{B}|$$

Using the second equality, we obtain

$$\mu H = \mu_0 (H + M)$$

and

$$\mu H - \mu_0 H = \mu_0 M$$

or

$$\frac{\mu}{\mu_0} H - H = M$$

and finally

$$M = \left(\frac{\mu}{\mu_0} - 1 \right) H = (\mu_r - 1)H$$

$$= (1.01 - 1)(2.0 \times 10^5 \text{ amperes/m})$$

$$= 2.0 \times 10^3 \text{ amperes/m} = |\mathbf{M}|$$

PRACTICE PROBLEM 18.1

In Sample Problem 18.1, we calculate the induction and magnetization of a paramagnetic material under an applied field strength of 2.0×10^5 amperes/m. Repeat this calculation for the case of another paramagnetic material that has a relative permeability of 1.005.

18.2 FERROMAGNETISM

For some materials, the induction increases dramatically with field strength. Figure 18–5 illustrates this phenomenon, **ferromagnetism**, and is in sharp contrast to the simple, linear behavior of Figure 18–4. This term *ferromagnetism* comes from the early association of the phenomenon with ferrous, or iron-containing, materials. You should recall from Section 15.4 that ferroelectric materials are so named because they exhibit a plot of polarization versus electric field similar to the *B-H* curve of Figure 18–5. In general, ferroelectric materials do not contain iron as a significant component.

The arrows in Figure 18–5 allow you to follow the induction, B, as a function of magnetic field strength, H. Initially, the sample studied was "demagnetized," with $B = 0$ in the absence of a field ($H = 0$). The initial application of the field generates a slight increase in induction in a manner

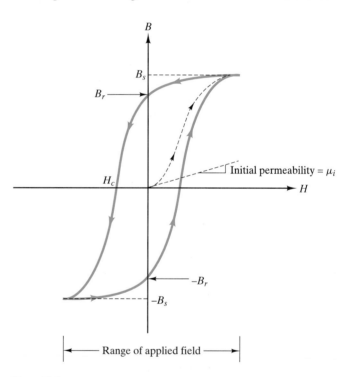

Figure 18-5 *In contrast to Figure 18–4, the B-H plot for a ferromagnetic material indicates substantial utility for engineering applications. A large rise in B occurs during initial magnetization (shown by the dashed line). The induction reaches a large, "saturation" value (B_s) upon application of sufficient field strength. Much of that induction is retained upon removal of the field (B_r = remanent induction). A coercive field (H_c) is required to reduce the induction to zero. By cycling the field strength through the range indicated, the B-H plot continuously follows the path shown as a solid line. This is known as a hysteresis loop.*

comparable with that for a paramagnetic material. However, after a modest field increase, a sharp rise in induction occurs. With further increase in field strength, the magnitude of induction levels off at a **saturation induction**, B_s. The magnetization, M, (introduced in Equation 18.4) is, in fact, the quantity that saturates. A close inspection of that equation indicates that B, which includes a $\mu_0 H$ contribution, will continue to increase with increasing H. Since the magnitude of B is much greater than $\mu_0 H$ at the point of saturation, B appears to level off and the term saturation induction is widely used. In fact, induction never truly saturates.

Equal in significance to the large value of B_s is the fact that much of that induction is retained upon the removal of the field. Following the *B-H* curve as the field is removed (arrows to the left), the induction drops to a nonzero, **remanent induction**, B_r, at a field strength of $H = 0$. In order to remove this remanent induction, the field must be reversed. In so doing, B is reduced to zero at a **coercive field** (sometimes called **coercive force**) of H_c. By continuing to increase the magnitude of the reversed field, the material can again be saturated (at an induction of $-B_s$). As before, a remanent induction $(-B_r)$ remains as the field is reduced to zero. The dashed line with arrows in Figure 18–5 represents the initial magnetization, but the solid line represents a completely reversible path that will continue to be traced out as long as the field is cycled back and forth between the extremes indicated. The solid line is known as a **hysteresis loop.**

To understand the nature of this hysteresis loop, we must explore both the atomic-scale and microscopic-scale structure of this material. As noted in Section 18.1, a current loop is the source of a magnetic field having a given orientation (see Figure 18–1). This provides a primitive model of a magnetic contribution of the orbital motion of the electrons in an atom. More important to our current discussion is the magnetic contribution of **electron spin.** This phenomenon is sometimes likened to the motion of a spinning planet, independent of its orbital motion. Though a useful concept for visualizing this contribution, electron spin is actually a relativistic effect associated with the intrinsic angular moment of the electron. In any case, the magnitude of the magnetic dipole, or **magnetic moment**, due to electron spin is the **Bohr*** **magneton**, $\mu_B (= 9.27 \times 10^{-24}$ ampere $\cdot$ m^2). It can be a positive quantity (for spin "up") or negative (for "spin down"). The orientation of spins is, of course, relative but is important in terms of the magnetic contribution of associated electrons. In a filled atomic shell, the electrons are all paired with each pair consisting of electrons of opposite spin and zero net magnetic moment ($+ \mu_B - \mu_B = 0$). The electron configuration for atoms was discussed in Section 2.1 and is given in detail for the elements in Appendix 1. It can be seen by inspecting Appendix 1 that the buildup of electrons in the orbital shells of the elements proceeds in a simple and systematic fashion from element 1 (hydrogen) to element 18 (argon). The

---

* Niels Henrik David Bohr (1885–1962), Danish physicist, was a major force in the development of atomic physics. He is perhaps best remembered for the development of the model of the simple hydrogen atom, which allowed him to explain its characteristic spectrum.

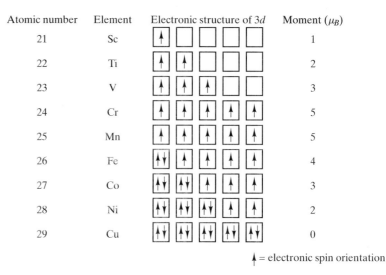

Atomic number	Element	Electronic structure of $3d$	Moment (μ_B)
21	Sc	↑	1
22	Ti	↑ ↑	2
23	V	↑ ↑ ↑	3
24	Cr	↑ ↑ ↑ ↑ ↑	5
25	Mn	↑ ↑ ↑ ↑ ↑	5
26	Fe	↑↓ ↑ ↑ ↑ ↑	4
27	Co	↑↓ ↑↓ ↑ ↑ ↑	3
28	Ni	↑↓ ↑↓ ↑↓ ↑ ↑	2
29	Cu	↑↓ ↑↓ ↑↓ ↑↓ ↑↓	0

↑ = electronic spin orientation

Figure 18-6 *The electronic structure of the 3d orbital for transition metals. Unpaired electrons contribute to the magnetic nature of these metals.*

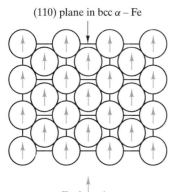

(110) plane in bcc α – Fe

Each ↑ = $4\mu_B$

Figure 18-7 *The alignment of magnetic moments for adjacent atoms leads to the large net magnetic moment (and B_s on a B-H plot) for the bulk solid. The example here is pure bcc iron at room temperature.*

pattern of adding electrons changes, however, as we move above element 18. Electrons are added first to the 4s orbital for elements 19 (potassium) and 20 (calcium), and then subsequent additions involve going back to fill in the 3d orbital. Having an unfilled inner orbital creates the possibility of unpaired electrons. This is precisely the case for elements 21 (scandium) to 28 (nickel). For these, the term **transition metals** has been given; that is, elements in the periodic table representing a gradual shift from the strongly electropositive elements of groups IA and IIA to the more electronegative elements of groups IB and IIB. Elements 21 through 28 represent, in fact, the first group of transition metals. Close inspection of Appendix 1 reveals additional groups of elements in which inner orbitals are systematically filled providing some unpaired electrons. A more detailed illustration of the electronic structure of the 3d orbital for the transition metals is given in Figure 18–6. Each unpaired electron contributes a single Bohr magneton to the "magnetic nature" of the metal. The number of Bohr magnetons per element is given in Figure 18–6. We see that iron (element 26) is one of the transition metals and has four unpaired 3d electrons, and, consequently, a contribution of $4\mu_B$. So ferromagnetism is clearly associated with iron, but we can identify several other transition metals with the same behavior.

We can now appreciate why transition metals can have high values of induction. If adjacent atoms in the crystal structure have their net magnetic moments aligned, the result is a substantial magnetic moment for the bulk crystal (Figure 18–7). The tendency of adjacent atoms to have aligned magnetic moments is a consequence of the **exchange interaction** between adjacent electron spins in the adjacent atoms. This is simply a case of an electron configuration stabilizing the system as a whole. As such, this case is analogous to the electron sharing that is the basis of the covalent bond (Section 2.3). The exchange interaction is a sensitive function of crystallography. In α (bcc) iron, the degree of interaction (and the resulting saturation induction) varies with crystallographic direction. More significantly, γ (fcc) iron is paramagnetic. This allows austenitic stainless steels (Section 11.1) to be used in designs requiring "nonmagnetic" steels.

Figure 18–7 shows how a high value of induction (B_s) is possible, but it also raises a new question as to how the induction can ever be zero. The answer to this question and the related explanation of the shape of the ferromagnetic hysteresis loop come at the microstructural level. The case of an unmagnetized iron crystal with $B = 0$ is shown in Figure 18–8. The microstructure is composed of **domains**, which have an appearance similar to polycrystalline grains. However, this illustration represents a single crystal. All of the domains have a common crystallographic orientation. Adjacent domains differ not in crystallographic orientation but in the orientation of magnetic moments. By having equal volumes of oppositely oriented moments, the net effect is zero induction. The dramatic rise in induction during initial magnetization is due to a large fraction of the individual atomic moments orienting toward a direction parallel to the direction of the applied field (Figure 18–9). In effect, domains favorably oriented with the applied field "grow" at the expense of those not favorably oriented. We can appreciate the ease with which this growth can occur by noting the magnetic

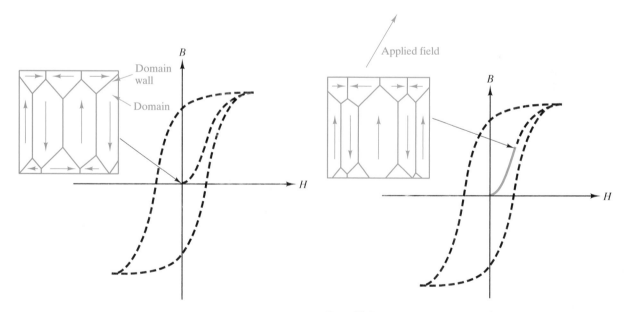

Figure 18-8 *The domain structure of an unmagnetized iron crystal gives a net $B = 0$ even though individual domains have the large magnetic moment indicated by Figure 18–7.*

Figure 18-9 *The sharp rise in B during initial magnetization is due to domain growth.*

Figure 18-10 *The domain, or Bloch, wall is a narrow region in which atomic moments change orientation by 180°. Domain wall motion (implied in Figures 18–8 and 18–9) simply involves a shift in this reorientation region. No atomic migration is required.*

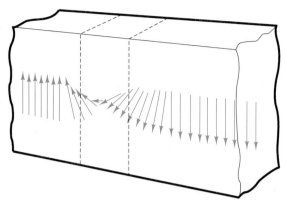

"structure" of the boundary between adjacent domains. A **Bloch* wall**, shown in Figure 18–10, is a narrow region in which the orientation of atomic moments changes systematically by 180°. During domain growth, the domain wall shifts in a direction to favor the domain more closely oriented with the applied field. Figure 18–11 monitors the domain microstructure during the course of the ferromagnetic hysteresis loop.

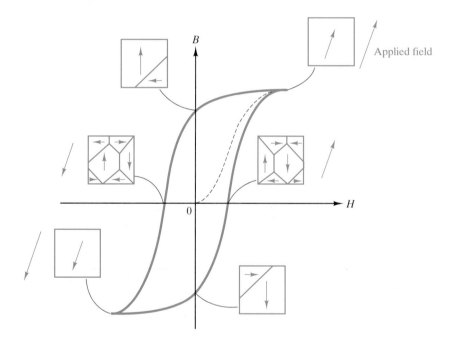

Figure 18-11 *Summary of domain microstructures during the course of a ferromagnetic hysteresis loop.*

---

* Felix Bloch (1905–1983), Swiss-American physicist. In addition to his contributions to the understanding of solid-state magnetism, Bloch's work in nuclear physics led to the development of nuclear magnetic resonance, an important addition to the fields of analytical chemistry and, more recently, medical imaging.

SAMPLE PROBLEM 18.2

The electronic structure of the $3d$ orbitals for a series of transition metals (Sc to Cu) is shown in Figure 18–6. Generate a similar illustration for the $4d$ orbitals (and resulting magnetic moments) for the series: Y to Pd.

SOLUTION

The illustration can essentially be made by inspection. The population of the $4d$ orbital is available in Appendix 1. Note that in filling the d orbitals, electron pairs form as a "last resort" when more than five electrons are involved.

Atomic number	Element	Electronic structure of $4d$	Moment (μ_B)
39	Y	↑ ▢ ▢ ▢ ▢	1
40	Zr	↑ ↑ ▢ ▢ ▢	2
41	Nb	↑ ↑ ↑ ↑ ▢	4
42	Mo	↑ ↑ ↑ ↑ ↑	5
43	Tc	↑↓ ↑ ↑ ↑ ↑	4
44	Ru	↑↓ ↑↓ ↑ ↑ ↑	3
45	Rh	↑↓ ↑↓ ↑↓ ↑ ↑	2
46	Pd	↑↓ ↑↓ ↑↓ ↑↓ ↑↓	0

SAMPLE PROBLEM 18.3

The following data are obtained for a cunife (copper-nickel-iron) alloy during the generation of a steady-state ferromagnetic hysteresis loop such as Figure 18–5:

H (amperes/m)	B (weber/m^2)
6×10^4	0.65 (saturation point)
1×10^4	0.58
0	0.56
-1×10^4	0.53
-2×10^4	0.46
-3×10^4	0.30
-4×10^4	0
-5×10^4	-0.44
-6×10^4	-0.65

(a) Plot the data.

(b) What is the remanent induction?

(c) What is the coercive field?

SOLUTION

(a) A plot of the data reveals one-half of the hysteresis loop. The remaining half can be drawn as a mirror image:

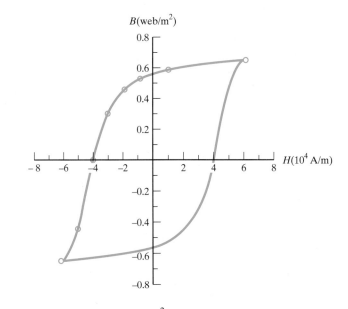

(b) $B_r = 0.56$ weber/m^2 (at $H = 0$)

(c) $H_c = -4 \times 10^4$ amperes/m (at $B = 0$)

Note. The use of the negative sign for H_c is somewhat arbitrary due to the symmetry of the hysteresis loop.

....

PRACTICE PROBLEM 18.2

In Sample Problem 18.2, we illustrate the electronic structure and resulting magnetic moments for the $4d$ orbitals of a series of transition metals. Generate a similar illustration for the $5d$ orbitals of the series Lu to Au.

PRACTICE PROBLEM 18.3

As pointed out in the beginning of Section 18.2, magnetization rather than induction is the quantity that saturates during ferromagnetic hysteresis. **(a)** For the case given in Sample Problem 18.3, what is the saturation induction? **(b)** What is the saturation magnetization at that point?

18.3   FERRIMAGNETISM

Ferromagnetism is the basis of most of the useful metallic magnets to be discussed in Section 18.4. For the ceramic magnets (Section 18.5), a slightly different mechanism is involved. The hysteresis behavior, as shown in Figure 18–5, is essentially the same. However, the crystal structure of the most common magnetic ceramics leads to some **antiparallel spin pairing** (where antiparallel is defined as parallel but of opposite direction), thereby reducing the net magnetic moment below that possible in metals. This similar phenomenon is distinguished from ferromagnetism by the slightly different spelling **ferrimagnetism.** Since the nature of domain structure and motion (associated with hysteresis, as shown in Figure 18–11) is not different, let us turn to the distinction that occurs at the atomic level. The most commercially important ceramic magnets are associated with the *spinel* ($MgAl_2O_4$) crystal structure, which was illustrated in Figure 3–15. This was one of the more complex crystal structures covered in Chapter 3. The cubic unit cell contains 56 ions, of which 32 are O^{2-} anions. Magnetic behavior is associated with the remaining 24 cation positions. Of course, spinel itself is nonmagnetic since neither Mg^{2+} nor Al^{3+} is a transition metal ion. However, some compounds containing transition metal ions do crystallize in this structure. An even larger number crystallize in the closely related *inverse spinel* structure, also discussed in Section 3.3. For the "normal" spinel structure, the divalent (M_I^{2+}) ions are tetrahedrally (fourfold) coordinated by O^{2-} ions, and the trivalent (M_{II}^{3+}) ions are octahedrally (sixfold) coordinated. This corresponds to 8 divalent and 16 trivalent ions per unit cell. For the inverse spinel, the trivalent ions occupy the tetrahedral sites and one-half of the octahedral sites. The divalent ions occupy the remaining half of the octahedral sites. This corresponds to the 16 trivalent ions being equally divided between tetrahedral and octahedral sites. All 8 divalent ions are, then, located at octahedral sites.

The historical model of a magnetic material is **magnetite** ($Fe_3O_4 = FeFe_2O_4$), with Fe^{2+} and Fe^{3+} ions being distributed in the inverse spinel configuration. Our need to do a detailed inventory of the distribution of cations among the available sites is due to an important fact: The magnetic moments of cations on tetrahedral and octahedral sites are antiparallel. Therefore, the equal distribution of trivalent ions between these two sites in the inverse spinel structure leads to cancellation of their contribution to the net magnetic moment for the crystal. The net moment is, then, provided by the divalent ions. The number of Bohr magnetons contributed by various transition metal ions is summarized in Table 18.1. The net magnetic moment for a unit cell of magnetite can be predicted by the use of this table to be eight times the magnetic moment of divalent $Fe^{2+}(= 8 \times 4 \, \mu_B = 32 \, \mu_B)$, which is in good agreement with the measured value of 32.8 μ_B based on the saturation induction for magnetite.

Table 18.1 *Magnetic Moment of Various Transition Metal Ions*

Ion	Moment $(\mu_B)^a$
Mn^{2+}	5
Fe^{2+}	4
Fe^{3+}	5
Co^{2+}	3
Ni^{2+}	2
Cu^{2+}	1

[a] $\mu_B = 1$ Bohr magneton $= 9.27 \times 10^{-24}$ ampere $\cdot$ m^2.

SAMPLE PROBLEM 18.4

In the text, the calculated magnetic moment of a unit cell of magnetite ($32\ \mu_B$) was found to be close to the measured value of $32.8\ \mu_B$. Make a similar calculation for nickel ferrite, which has a measured value of $18.4\ \mu_B$.

SOLUTION

As with the case of magnetite, there is an equal number of Fe^{3+} ions on tetrahedral and octahedral (antiparallel) sites. The net magnetic moment is determined solely by the 8 divalent (Ni^{2+}) ions. Using Table 18.1, we obtain

$$\text{magnetic moment/unit cell}$$

$$= (\text{no. } Ni^{2+} / \text{ unit cell})(\text{moment } Ni^{2+})$$

$$= 8 \times 2\mu_B = 16\mu_B$$

Note. This calculation represents an error of

$$\frac{18.4 - 16}{18.4} \times 100\% = 13\%$$

This discrepancy is largely due to a lack of perfect stoichiometry for the commercial ferrite material, providing some contribution to the net magnetic moment by the Fe^{3+}.

SAMPLE PROBLEM 18.5

What would be the saturation magnetization, $|M_s|$, for the nickel ferrite described in Sample Problem 18.4? (The lattice parameter for nickel ferrite is 0.833 nm.)

SOLUTION

Magnetization is defined relative to Equation 18.4 as the volume density of magnetic dipole moments. The magnetic moment per unit cell (assuming saturation, i.e., the parallel alignment of eight Ni^{2+} moments) is given in Sample Problem 18.4 as 18.4 μ_B. Therefore, the saturation magnetization is

$$|M_s| = \frac{18.4\,\mu_B}{\text{vol. of unit cell}}$$

$$= \frac{(18.4)(9.274 \times 10^{-24} \text{ A} \cdot \text{m}^2)}{(0.833 \times 10^{-9} \text{ m})^3}$$

$$= 2.95 \times 10^5 \text{ A/m}$$

PRACTICE PROBLEM 18.4

Calculate the magnetic moment of a unit cell of copper ferrite. (See Sample Problem 18.4.)

PRACTICE PROBLEM 18.5

Calculate the saturation magnetization for the copper ferrite described in Practice Problem 18.4. (The lattice parameter for copper ferrite is 0.838 nm.) (See Sample Problem 18.5.)

18.4 METALLIC MAGNETS

The commercially important **metallic magnets** are ferromagnetic. In general, these materials are categorized as either soft or hard magnets. Ferromagnetic materials with domain walls easily moved by applied fields are termed **soft magnets.** Those with less mobile domain walls are termed **hard magnets.** The compositional and structural factors that lead to magnetic hardness are generally the same ones that produce mechanical hardness (see Section 6.4). The relative appearance of hysteresis loops for soft and hard materials is shown in Figure 18–12. Until the recent development of ceramic superconductors, the best examples of superconducting magnets were certain metals, such as Nb and its alloys.

Figure 18-12 *Comparison of typical hysteresis loops for "soft" and "hard" magnets.*

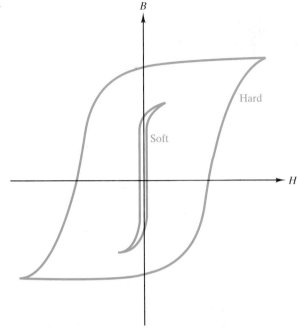

SOFT MAGNETS

The largest use of magnetic materials is in power generation. A common example is the ferromagnetic core in a transformer. This application calls for a soft magnet. The area in a ferromagnetic hysteresis loop represents the energy consumed in traversing the loop. For alternating current (ac) power applications, the loop can be traversed at frequencies of 50 to 60 Hz (hertz,* or cycles per second) and greater. As a result, the small area hysteresis loop of a soft magnet (Figure 18–12) provides a minimum source of **energy loss.** Of course, small loop area is important, but a high saturation induction (B_s) is equally desirable for minimizing the size of the transformer core.

A second source of energy loss in ac applications is the generation of fluctuating electrical currents **(eddy currents)** induced by the fluctuating magnetic field. The energy loss comes directly from **Joule**[†] **heating**

---

* Heinrich Rudolf Hertz (1857–1894), German physicist. Although he died at an early age, Hertz was one of the leading scientists of the nineteenth century. A primary achievement was the experimental demonstration of the nature of electromagnetic waves.

† James Prescott Joule (1818–1889), English physicist. The discovery of the quantity of heating caused by an electrical current was an early achievement by Joule (in his early twenties). His determined efforts to perfect the measurement of the mechanical equivalent of heat led to the fundamental unit of energy being named in his honor. Joule became one of England's most renowned scientists in the latter part of his life. His recognition was hindered for some time because his scientific research was done in conjunction with his primary role as manager of his family business, a brewery.

$(= I^2R$, where I is current and R is resistance). This loss can be reduced by increasing the resistivity of the material. At first glance, increasing resistivity would seem to increase the I^2R term because resistivity is proportional to resistance, R. But the reduction in current, I, which is a squared term, more than compensates for the increased magnitude of R (i.e., $I^2R = [V^2/R^2]R = V^2/R$). For this reason, higher-resistance iron-silicon alloys have replaced plain carbon steels in low-frequency power applications.* The silicon addition also increases magnetic permeability and, consequently, B_s. Further improvement of magnetic properties is produced by cold-rolling sheets of the silicon steel. This takes advantage of the greater permeability along certain crystallographic directions. The production of such a **preferred-orientation**, or **textured, microstructure** is shown in Figure 18–13. Figure 18–14 gives a comparison of initial magnetization for three materials: a plain carbon cast iron (3 wt % C), a random texture Fe–3.25 wt % Si alloy, and a (100) [001] texture Fe–3.25 wt % Si alloy. Some typical magnetic properties of various soft magnetic metals are given in Table 18.2.

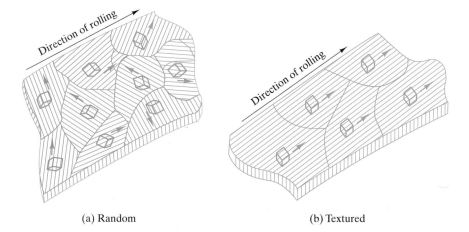

(a) Random   (b) Textured

Figure 18-13 *A comparison of (a) random and (b) textured (with preferred orientation) microstructures in polycrystalline iron-silicon alloy sheets. The preferred orientation is the result of cold rolling. The small cubes represent the orientation (but not the size) of unit cells in each grain's crystal structure. The preferred orientation (b) is termed (100)[001], corresponding to the plane and direction of the unit cells relative to the sheet geometry. Figure 18–14 shows how the textured microstructure takes advantage of the crystallographic anisotropy of magnetic properties. (From R. M. Rose, L. A. Shepard, and J. Wulff,* The Structure and Properties of Materials, *Vol. 4:* Electronic Properties, *John Wiley & Sons, Inc., New York, 1966.)*

---

* You might recall the biographical footnote for Augustin Charpy, inventor of the Charpy impact test. The development of the silicon steels for power applications was one of his many achievements.

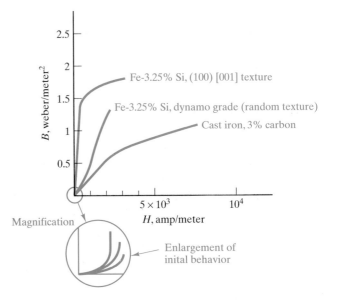

Figure 18-14 *A comparison of initial magnetization for three ferrous alloys. The silicon addition increases magnetic permeability and, consequently, B_s. Preferred orientation, or texturing, increases initial magnetization substantially (see Figure 18–13). (From R. M. Rose, L. A. Shepard, and J. Wulff,* The Structure and Properties of Materials, *Vol. 4:* Electronic Properties, *John Wiley & Sons, Inc., New York, 1966.)*

Table 18.2 *Typical Magnetic Properties of Various Soft Magnetic Metals*

Material	Initial relative permeability (μ_r at $B \sim 0$)	Hysteresis loss (J/m^3 per cycle)	Saturation induction (Wb/m^2)
Commercial iron ingot	250	500	2.16
Fe–4% Si, random	500	50–150	1.95
Fe–3% Si, oriented	15,000	35–140	2.0
45 Permalloy (45% Ni–55% Fe)	2,700	120	1.6
Mumetal (75% Ni–5% Cu–2% Cr–18% Fe)	30,000	20	0.8
Supermalloy (79% Ni–15% Fe–5% Mo)	100,000	2	0.79
Amorphous ferrous alloys (80% Fe–20% B)	–	25	1.56
(82% Fe–10% B–8% Si)	–	15	1.63

Source: R. M. Rose, L. A. Shepard, and J. Wulff, *The Structure and Properties of Materials*, Vol. 4: *Electronic Properties*, John Wiley & Sons, Inc., New York, 1966, and J. J. Gilman, "Ferrous Metallic Glasses," *Metal Progress*, July 1979.

The iron-nickel alloys listed provide higher permeability in weak fields. This provides for superior performance in high-fidelity communication equipment. This fidelity is possible by the sacrifice of saturation induction.

One of the early commercial applications of amorphous metals (see Sections 4.5 and 11.1) has been as ribbons for soft magnet applications. These ferrous alloys are chemically distinct from conventional steels in that boron rather than carbon is the primary alloying element. The absence of grain boundaries in this material apparently accounts for the easy motion of domain walls. This is coupled with relatively high resistivity to make these materials attractive for applications such as transformer cores. Table 18.2 includes data for amorphous ferrous alloys. A design case study involving the use of an amorphous metal for a transformer core in electric power distribution is given in Section 20.3.

HARD MAGNETS

Although unsuited for ac power applications, the hard magnets (Figure 18–12) are ideal as **permanent magnets.** The large area contained within the hysteresis loop, which implies large ac losses, simultaneously defines the "power" of a permanent magnet. Specifically, the product of B and H during the demagnetization portion of the hysteresis loop leads to a maximum value, $(BH)_{max}$, which is a convenient measure of this power (Figure 18–15). Table 18.3 gives values of $(BH)_{max}$ for various hard magnets. The alnico alloys are especially important commercially.

SUPERCONDUCTING MAGNETS

In Section 15.3 the interesting property of superconductivity was introduced. Although still largely in the development stage, **superconducting magnets** are demonstrating exciting potential applications. Metallic superconducting magnets are already making possible high-field solenoids with no steady-state power consumption and high-switching-speed computer circuitry. The major barrier to wide application has been the relatively low critical temperature, T_c, above which the superconducting behavior is lost. Progress in producing higher T_c values with the development of oxide superconductors raises the possibility of a revolution in the application of magnetic materials. The development of the high-T_c superconductors is outlined in Section 15.3. Superconducting ceramic magnets are discussed further in Section 18.5.

Table 18.3 *Maximum BH Products for Various Hard Magnetic Metals*

Alloy	$(BH)_{max}(A \cdot Wb/m^3)$
Samarium-cobalt	120,000
Platinum-cobalt	70,000
Alnico	36,000

Source: Data from R. A. Flinn and P. K. Trojan, *Engineering Materials and Their Applications*, 2nd ed., Houghton Mifflin Company, Boston, 1981.

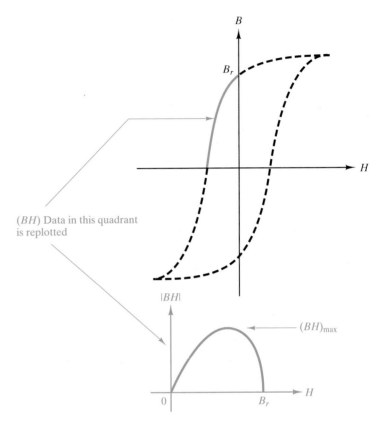

(BH) Data in this quadrant is replotted

Figure 18-15 *Replotting data in the "demagnetization quadrant" of the hysteresis loop demonstrates a maximum value of the $|BH|$ product, $(BH)_{max}$. This quantity is a convenient measure of the "power" of permanent magnets. Table 18.3 gives $(BH)_{max}$ for various hard magnets.*

SAMPLE PROBLEM 18.6

The data presented in Sample Problem 18.3 are representative of a hard magnet. Calculate the energy loss of that magnet (i.e., the area within the loop).

SOLUTION

A careful measurement of the area of the plot in Sample Problem 18.3 gives

$$\text{area} = 8.9 \times 10^4 \ (\text{amperes/m})(\text{webers/m}^2)$$

$$= 8.9 \times 10^4 \ \frac{\text{amperes} \cdot \text{webers}}{\text{m}^3}$$

One ampere · weber is equal to 1 joule. The area is, then, a volume density of energy, or

$$\text{energy loss} = 8.9 \times 10^4 \text{ J/m}^3$$

$$= 89 \text{ kJ/m}^3 (\text{per cycle})$$

Note.   A comparison of this result with the values given in Table 18.2 for soft magnetic materials indicates the disadvantage of a hard magnet for ac applications.

SAMPLE PROBLEM 18.7

For the hard magnet discussed in Sample Problem 18.6, calculate the power of the magnet [i.e., the $(BH)_{max}$ value].

SOLUTION

Replotting the data of Sample Problem 18.3 in the manner of Figure 18–15, we obtain

B (webers/m^2)	H (A/m)	$|BH|$(weber · A/m^3 = J/m^3)
0	-4×10^4	0
0.3	-3×10^4	9×10^3
0.46	-2×10^4	9.2×10^3
0.53	-1×10^4	5.3×10^3
0.56	0	0

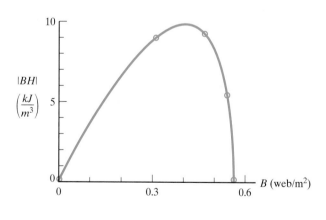

or

$$(BH)_{max} \approx 10 \times 10^3 \text{ J/m}^3$$

Note.   Although the large energy loss calculated for this material in Sample Problem 18.6 was a disadvantage, the large result here is the basis of this alloy's selection as a permanent magnet.

··

PRACTICE PROBLEM 18.6

In Sample Problem 18.6, we analyze data for a hard magnet (cunife). Use the similar data for a soft magnet (armco iron) given in Problem 18.7 to calculate the energy loss.

PRACTICE PROBLEM 18.7

For the soft magnet referred to in Practice Problem 18.6, calculate the power of the magnet, as done in Sample Problem 18.7.

18.5   CERAMIC MAGNETS

Ceramic magnets can be divided into two categories. The traditional ones have the low conductivity characteristic of most ceramics. As pointed out in Section 15.4, roughly 80% of the industrial ceramics worldwide are used for their electronic or magnetic behavior. As discussed in Section 18.4, the most dramatic superconducting magnets are members of a new family of ceramic oxides.

LOW-CONDUCTIVITY MAGNETS

The traditional, commercially important **ceramic magnets** are ferrimagnetic. The dominant examples are the **ferrites**, based on the inverse spinel crystal structure discussed in some detail in Section 18.3. We noted in the preceding section that steel alloys for transformer cores are selected for maximum resistivity to minimize eddy-current losses. For high-frequency applications, no metallic alloy has a sufficiently large resistivity to prevent substantial eddy-current losses. The characteristically high resistivities of ceramics (see Section 15.4) make ferrites the appropriate material for such applications. Numerous transformers in the communication industry are made of ferrites. The deflection transformers used to form electronic images on television screens are routine examples. A list of representative commercial ferrites is given in Table 18.4. Although the term *ferrite* is sometimes used interchangeably with the term *magnetic ceramic*, the ferrites are only one group of ceramic crystal structures exhibiting ferrimagnetic behavior. Another is the **garnets**, which have a relatively complex crystal structure similar to that of the natural gem garnet, $Al_2Mg_3Si_3O_{12}$. This structure has three types of crystal environments for cations. The Si^{4+} cation is tetrahedrally coordinated, Al^{3+} is octahedrally coordinated, and Mg^{2+} is in a dodecahedral (eightfold) site. Ferrimagnetic garnets contain Fe^{3+} ions. For example, yttrium iron garnet (YIG) has the formula $Fe_2^{3+}[Y_3^{3+}Fe_3^{3+}]O_{12}^{2-}$. The first two Fe^{3+} in the formula are in octahedral sites. The second three Fe^{3+} ions are in tetrahedral sites. The three Y^{3+} ions are in dodecahedral sites. Garnets are the primary materials used for waveguide components in mi-

Table 18.4 *Some Commercial Ferrite Compositions*

Name	Composition	Comments
Magnesium ferrite	$MgFe_2O_4$	
Magnesium-zinc ferrite	$Mg_xZn_{1-x}Fe_2O_4$	$0 < x < 1.$
Manganese ferrite	$MnFe_2O_4$	
Manganese-iron ferrite	$Mn_xFe_{3-x}O_4$	$0 < x < 3.$
Manganese-zinc ferrite	$Mn_xZn_{1-x}Fe_2O_4$	$0 < x < 1.$
Nickel ferrite	$NiFe_2O_4$	
Lithium ferrite	$Li_{0.5}Fe_{2.5}O_4$	Li^+ occurs in combination with Fe^{3+}, producing a $(Li_{0.5}Fe_{0.5})^{2+}Fe_2^{3+}O_4$ configuration.

crowave communication. A list of commercial garnet compositions is given in Table 18.5.

Both ferrites and garnets are soft magnets. Some ceramics have been developed that are magnetically hard. Important examples are the **magnetoplumbites.** These materials have a hexagonal crystal structure with a chemical composition $MO \cdot 6Fe_2O_3$ (M = divalent cation) similar to that of the mineral magnetoplumbite. As with the garnets, the crystal structure of the magnetoplumbites is substantially more complex than that of the ferrite spinels. There are, in fact, five different types of coordination environments for cations (as opposed to two for the inverse spinels and three for garnets). The three divalent cations of greatest commercial interest are strontium (Sr^{2+}), barium (Ba^{2+}), and lead (Pb^{2+}). Permanent magnets fabricated from these materials are characterized by high coercive fields and low cost. They find applications in small dc motors, radio loudspeakers, and magnetic door latches.

One of our more common examples of a ceramic magnet is recording tape. This consists of fine particles of γ-Fe_2O_3 (see Figure 3–13 for a comparable crystal structure) oriented on a plastic tape. The resulting thin film of Fe_2O_3 has a "hard" hysteresis loop. High-fidelity recording is the result of the residual magnetization of the film being proportional to the

Table 18.5 *Some Commercial Garnet Compositions*

Name	Composition	Comments
Yttrium iron garnet (YIG)	$Y_3Fe_5O_{12}$	
Aluminum substituted YIG	$Y_3Al_xFe_{5-x}O_{12}$	Al^{3+} prefers tetrahedral sites.
Chromium substituted YIG	$Y_3Cr_xFe_{5-x}O_{12}$	Cr^{3+} prefers octahedral sites.
Lanthanum iron garnet (LaIG)	$La_3Fe_5O_{12}$	
Praseodymium iron garnet (PrIG)	$Pr_3Fe_5O_{12}$	Many of the pure garnets are not commercially prepared; all form at least limited solid solutions with each other (e.g., $Pr_xY_{3-x}Fe_5O_{12}$ with $x_{max} = 1.5$).

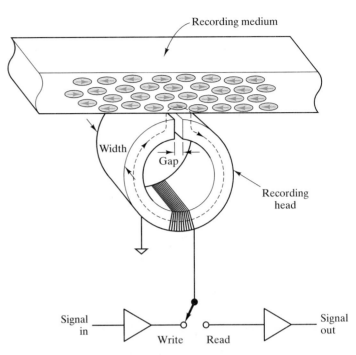

Figure 18-16 *Schematic illustration of magnetic storage and retrieval using a recording medium composed of needlelike particles of γ-Fe_2O_3. (From J. U. Lemke, MRS Bulletin, 15, 31 [1990].)*

electrical signal produced by sound. The same concept has been applied in the production of both flexible ("floppy") and rigid computer disks. The floppy disks are composed of a similar iron-oxide coating on a flexible plastic substrate. The rigid disks usually have an exceptionally flat and smooth aluminum alloy substrate.

The general principle of magnetic recording is illustrated in Figure 18–16, which shows how aligned, needlelike particles of γ-Fe_2O_3 store binary information as zeros and ones by means of their orientation (pointing right or left in the illustration). Data are "written" to the tape or disk by the electrical signal in the coil, which generates a magnetic field across the gap in the recording head. This field magnetizes a small region of the recording medium near the gap. Once the region of the medium passes beyond the recording head, the magnetization remains, and the signal has been stored. The same head can also "read" the stored data as also shown in Figure 18–16.

Recording heads are made of ceramic ferrites such as nickel zinc ferrite and manganese zinc ferrite. For early recording heads, those ceramics were sufficient. High-density recording applications, however, require higher coercive fields and have led to the use of thin films of metallic alloys within the gap. A "metal-in-gap" or MiG recording head design is shown in Figure 18–17. Permalloy (NiFe) and Sendust (FeAlSi) alloys are used for this purpose.

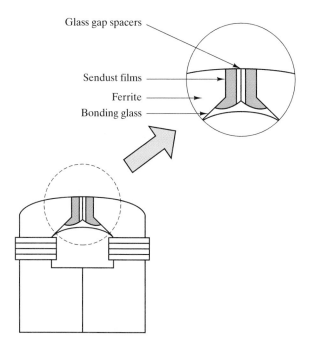

Glass gap spacers

Sendust films

Ferrite

Bonding glass

Figure 18-17 *Schematic illustration of a thin film of a
metal alloy, Sendust (FeAlSi), on the pole faces of the
ferrite recording head in a metal-in-gap or MiG design. (From A. S. Hoagland and J. E. Monson, Digital Magnetic Recording, 2nd ed., Wiley-Interscience,
New York, 1991.)*

SUPERCONDUCTING MAGNETS

As introduced in Section 15.3, a relatively new family of ceramic materials
has exhibited the property of superconductivity. As noted in that introduction, superconductivity in metals had been associated with a limited set
of elements and alloys that, above T_c, were relatively poor conductors. In
a similar way, a specific family of oxide ceramics, traditionally included in
the category of insulators, was found to exhibit superconductivity with substantially higher T_c values than was possible with the best of the metallic
superconductors. In Section 15.3, the encouraging rise in critical temperature, T_c, was contrasted with the discouraging limitation to critical current
density. There is a third, major property for superconductors, namely, the
critical magnetic field, above which the material stops being superconducting. Unlike the critical current density, the critical magnetic field increases
along with critical temperature for the high-T_c materials, Figure 18–18. Unfortunately, the limited current density (due to magnetic field penetration)
remains as a primary obstacle, especially for large-scale applications such as
power transmission and transportation. As pointed out in Section 15.3, some
of the most promising applications of these new materials appear to be in thin

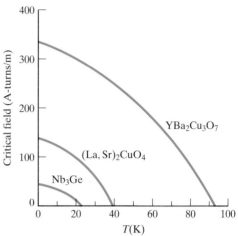

Figure 18-18 *Comparison of the critical magnetic field versus temperature for a metallic superconductor (Nb_3Ge) and two ceramic superconductors.*

film devices, such as **Josephson* junctions**, which consist of a thin layer of insulator between superconducting layers. These devices switch voltages at very high frequencies while consuming much less energy than conventional devices. Resulting applications may include more compact computers and ultrasensitive magnetic field detectors. Also noted in Section 15.3 were the development of composite (metallic silver/$YBa_2Cu_3O_7$) superconducting wire with the requisite mechanical and superconducting properties necessary for solenoid coil fabrication and thin film filters for cellular telephone base stations.

SAMPLE PROBLEM 18.8

The γ-Fe_2O_3 ceramic used in magnetic recording shares the corundum structure illustrated in Figure 3–13. Using the data of Appendix 2 and the principles of Section 2.2, confirm that the Fe^{3+} ion should reside in an octahedral coordination.

SOLUTION

We find in Appendix 2 that

$$r_{Fe^{3+}} = r = 0.067 \text{ nm}$$

and

$$r_{O^{2-}} = R = 0.132 \text{ nm}$$

---

* Brian David Josephson (1940-   ), English physicist. While still a graduate student at Cambridge University, Josephson developed the theoretical concept of the layered junction, which now bears his name. Later experimental verification helped to confirm earlier theoretical models of conduction in metallic superconductors.

The resulting radius ratio is

$$\frac{r}{R} = \frac{0.067 \text{ nm}}{0.132 \text{ nm}} = 0.508$$

which is within the range of 0.414 to 0.732 given in Table 2.1 for sixfold (octahedral) coordination.

SAMPLE PROBLEM 18.9

Many commercial ferrites can be prepared in a combination of both normal and inverse spinel structures. Calculate the magnetic moment of a unit cell of manganese ferrite if it occurs in **(a)** an inverse spinel, **(b)** a normal spinel, or **(c)** a 50:50 mixture of inverse and normal spinel structures.

SOLUTION

(a) For the inverse spinel case, we can follow the example of Sample Problem 18.4 and note from Table 18.1 that the moment for Mn^{2+} is $5\mu_B$:

magnetic moment/unit cell

$$= 8 \times 5\mu_B = 40\mu_B$$

(b) For the normal spinel case, we would have all Mn^{2+} ions in tetrahedral coordination and all Fe^{3+} ions in octahedral coordination. As with the inverse spinel structure, the magnetic moments associated with the tetrahedral and octahedral sites are antiparallel. Arbitrarily taking those associated with tetrahedral sites as negative and those with octahedral sites as positive, we obtain

magnetic moment/unit cell

$$= -(\text{no. } Mn^{2+}/\text{unit cell})(\text{moment of } Mn^{2+})$$

$$+ (\text{no. } Fe^{3+}/\text{unit cell})(\text{moment of } Fe^{3+})$$

$$= -(8)(5\mu_B) + (16)(5\mu_B) = 40\mu_B$$

(c) A 50:50 mixture will give

magnetic moment/unit cell

$$= (0.5)(40\mu_B) + (0.5)(40\mu_B) = 40\mu_B$$

Note. In this particular case, the total moment is the same for any combination of the two structures. In the more general case, when the moment of the divalent ion is not $5\mu_B$, different values are obtained. By the way, the structure of manganese ferrite is generally 80% normal spinel and 20% inverse spinel.

PRACTICE PROBLEM 18.8

In Sample Problem 18.8, we use a radius ratio calculation to confirm the octahedral coordination of Fe^{3+} in γ-Fe_2O_3. Do similar calculations for Ni^{2+} and Fe^{3+} in the inverse spinel, nickel ferrite, introduced in Sample Problem 18.4.

PRACTICE PROBLEM 18.9

(a) Calculate the magnetic moment of a unit cell of $MgFe_2O_4$ in an inverse spinel structure. **(b)** Repeat part (a) for the case of a normal spinel structure. **(c)** Given that the experimental value of the unit cell moment for $MgFe_2O_4$ is 8.8 μ_B, estimate the fraction of the ferrite in the inverse spinel structure. (See Sample Problem 18.9.)

SUMMARY

Our primary means for characterizing magnetic materials is the *B-H* plot, which traces the variation in induction (B) with magnetic field strength (**H**). This serves for magnetic applications as the σ-ϵ plot did for mechanical applications. Diamagnetism and paramagnetism exhibit linear *B-H* plots of small slopes and little commercial significance. A highly nonlinear *B-H* plot, called a hysteresis loop, is characteristic of ferromagnetism. The large value of saturation induction (B_s) possible with ferromagnetic materials is of substantial commercial importance. The magnitude of the coercive field (H_c) necessary to reduce the induction to zero indicates whether the material is a soft (small H_c) or hard (large H_c) magnet. As the name implies, ferromagnetism is associated with ferrous (iron-containing) alloys. However, because of their similar electronic structure, a variety of transition metals share this behavior. Transition metals have unfilled inner orbitals that allow unpaired electron spins to contribute one or more Bohr magnetons to the net magnetic moment of the atom. This electronic structure explains the magnitude of B_s, but the shape of the hysteresis loop results from a microstructural feature, domain wall motion. Iron-silicon alloys are excellent examples of soft magnets. A small area within the *B-H* hysteresis loop corresponds to small energy loss for ac applications. Increased resistivity, compared with that of plain carbon steels, reduces eddy-current losses. Permanent magnets, such as alnico alloys, are characterized by large hysteresis loop areas and $(BH)_{max}$ values. Metallic superconducting magnets have exhibited some practical applications, limited primarily by their relatively low operating temperature range.

Ferrimagnetism is a phenomenon closely related to ferromagnetism. It occurs in magnetic ceramic compounds. In these systems, transition metal ions provide magnetic moments, as transition metal atoms do in ferromagnetism. The difference is that the magnetic moments of certain cations are canceled by antiparallel spin pairing. The net saturation induction is then diminished in comparison with ferromagnetic metals. Ceramic magnets, like ferromagnetic metals, can be magnetically soft or hard. Our primary examples are the ferrites based on the inverse spinel crystal structure. Yttrium iron garnet (YIG) is another example of a ferrimagnetic ceramic, in this case, based on the crystal structure of the gem garnet. Both ferrites and garnets are soft magnets. Hexagonal ceramic compounds based on the structure of the mineral magnetoplumbite are hard magnets, with characteristically high coercive fields and low costs. Thin films of fine-particle γ-Fe_2O_3 are widely used hard magnets in applications from recording tape to computer disks. Ceramic superconducting magnets, with substantially higher operating temperatures than their metallic counterparts, show promise for expanding the application of superconductors, especially in the area of thin film devices for filters, compact computers, and ultrasensitive magnetic field detectors, plus the development of wire for solenoid coils.

KEY TERMS

antiparallel spin pairing (683)
Bloch wall (680)
Bohr magneton (677)
ceramic magnet (692)
coercive field (677)
coercive force (677)
critical magnetic field (695)
diamagnetism (673)
domain (679)
domain wall (680)
eddy current (686)
electron spin (677)
energy loss (686)
exchange interaction (678)
ferrimagnetism (683)
ferrite (692)

ferromagnetism (676)
flux density (672)
garnet (692)
hard magnet (685)
hysteresis loop (677)
induction (672)
Josephson junction (696)
Joule heating (686)
magnetic dipole (671)
magnetic field (671)
magnetic field strength (672)
magnetic flux line (671)
magnetic moment (677)
magnetism (671)
magnetite (683)
magnetization (673)

magnetoplumbite (693)
metallic magnet (685)
paramagnetism (673)
permanent magnet (689)
permeability (672)
preferred-orientation microstructure (687)
relative permeability (673)
remanent induction (677)
saturation induction (677)
soft magnet (685)
superconducting magnet (689)
textured microstructure (687)
transition metal (678)

REFERENCES

Cullity, B. D., *Introduction to Magnetic Materials*, Addison-Wesley Publishing Co., Inc., Reading, Mass., 1972.

Kittel, C., *Introduction to Solid State Physics*, 7th ed., John Wiley & Sons, Inc., New York, 1996.

Suzuki, T., *et al.*, Editors, *Magnetic Materials-Microstructure and Properties*, Materials Research Society, Pittsburgh, Pa, 1991.

PROBLEMS

18.1 • Magnetism

18.1. Calculate the induction and magnetization of a diamagnetic material (with $\mu_r = 0.99995$) under an applied field strength of 2.0×10^5 amperes/m.

18.2. Calculate the induction and magnetization of a paramagnetic material with $\mu_r = 1.001$ under an applied field strength of 5.0×10^5 amperes/m.

18.3. Plot the B versus H behavior of the paramagnetic material of Problem 18.2 over a range of $-5.0 \times 10^5 \,\text{A/m} < H < 5.0 \times 10^5$ A/m. Include a dashed-line plot of the magnetic behavior of a vacuum.

18.4. Superimpose on the plot of Problem 18.3 the behavior of the diamagnetic material of Problem 18.1.

18.5. The following data are obtained for a metal subjected to a magnetic field:

H (amperes/m)	B (weber/m^2)
0	0
4×10^5	0.50263

(a) Calculate the relative permeability for this metal.
(b) What type of magnetism is being demonstrated?

18.6. The following data are obtained for a ceramic subjected to a magnetic field:

H (amperes/m)	B (weber/m^2)
0	0
4×10^5	0.50668

(a) Calculate the relative permeability for this ceramic. (b) What type of magnetism is being demonstrated?

18.2 • Ferromagnetism

18.7. The following data are obtained for an armco iron alloy during the generation of steady-state ferromagnetic hysteresis loop:

H (amperes/m)	B (weber/m^2)
56	0.50
30	0.46
10	0.40
0	0.36
-10	0.28
-20	0.12
-25	0
-40	-0.28
-56	-0.50

(a) Plot the data. (b) What is the remanent induction? (c) What is the coercive field?

18.8. For the armco iron of Problem 18.7, determine (a) the saturation induction and (b) the saturation magnetization.

18.9. The following data are obtained for a nickel-iron alloy during the generation of a steady-state ferromagnetic hysteresis loop:

H (amperes/m)	B (weber/m^2)
50	0.95
25	0.94
0	0.92
-10	0.90
-15	0.75
-20	-0.55
-25	-0.87
-50	-0.95

(a) Plot the data. (b) What is the remanent induction? (c) What is the coercive field?

18.10. For the nickel-iron alloy of Problem 18.9, determine (a) the saturation induction and (b) the saturation magnetization.

18.11. Illustrate the electronic structure and resulting magnetic moments for the heavy elements No and Lw, which involve an unfilled $6d$ orbital.

• **18.12.** Let us explore further the difference between induction, which does not truly saturate, and magnetization, which does. **(a)** For the magnet treated in Practice Problem 18.3, what would be the induction at a field strength of 60×10^4 amperes/m, 10 times greater than that associated with saturation induction? **(b)** Sketch quantitatively the hysteresis loop for the case of cycling the magnetic field strength between -60×10^4 and $+60 \times 10^4$ amperes/m.

18.3 • Ferrimagnetism

18.13. **(a)** Calculate the magnetic moment of a unit cell of manganese ferrite. **(b)** Calculate the corresponding saturation magnetization, given a lattice parameter of 0.850 nm.

18.14. Make a photocopy of Figure 3–15. Relabel the ions so that the unit cell represents the structure of an inverse spinel, $CoFe_2O_4$. (Do not try to label each site.)

18.15. Calculate the magnetic moment of the unit cell generated in Problem 18.14.

18.16. Estimate the saturation magnetization of the unit cell generated in Problem 18.14.

18.17. A key aspect of the ferrite crystal structure based on spinel, $MgAl_2O_4$ (Figure 3–15), is the tendency toward tetrahedral or octahedral coordination of the metal ions by O^{2-}. Calculate the radius ratio for **(a)** Mg^{2+} and **(b)** Al^{3+}. In each case, comment on the corresponding coordination number. (Recall that the relationship of radius ratio to coordination number was introduced in Section 2.2.)

18.18. In regard to the discussion of Problem 18.17, calculate the radius ratio for **(a)** Fe^{2+} and **(b)** Fe^{3+}. In each case, comment on the corresponding coordination number in the inverse spinel structure of magnetite, Fe_3O_4.

18.4 • Metallic Magnets

18.19. Assuming that the hysteresis loop in Practice Problem 18.6 is traversed at a frequency of 60 Hz, calculate the rate of energy loss (i.e., power loss) for this magnet.

18.20. Repeat Problem 18.19 for the hard magnet of Sample Problems 18.3 and 18.6.

18.21. Calculate the energy loss (i.e., area within the loop) for the nickel-iron alloy of Problem 18.9.

18.22. Given the result of Problem 18.21, comment on whether the nickel-iron alloy is a "soft" or "hard" magnet.

18.23. Assuming the hysteresis loop in Problem 18.9 is traversed at a frequency of 60 Hz, calculate the rate of energy loss (i.e., power loss) for this magnet.

18.24. Assuming the hysteresis loop in Problem 18.9 is traversed at a frequency of 1 kHz, calculate the rate of energy loss (i.e., power loss) for this magnet.

18.25. The hysteresis loss for soft magnets is generally given in units of W/m^3. Calculate the loss in these units for the Fe-B amorphous metal in Table 18.2 at a frequency of 60 Hz.

18.26. Repeat Problem 18.25 for the Fe-B-Si amorphous metal in Table 18.2.

• **18.27.** Many of the highest T_c and H_c metal superconductors (such as Nb_3Sn with $T_c = 18.5$ K) have the A_3B "β-tungsten" structure, with A atoms at tetrahedral sites [(0 1/2 1/4)-type positions] in a bcc unit cell of B atoms. Sketch the unit cell of such a material. (*Hint:* Only two of four such A sites are occupied in a given unit cell face, and the A atoms would form, with adjacent unit cells, three orthogonal chains.)

• **18.28.** Verify that the composition of the β-tungsten structure described in Problem 18.27 is A_3B.

18.5 • Ceramic Magnets

18.29. Characterize the ionic coordination of the cations in yttrium iron garnet using radius ratio calculations.

18.30. Characterize the ionic coordination of the cations in aluminum-substituted YIG using radius ratio calculations.

18.31. Characterize the ionic coordination of the cations in chromium-substituted YIG using radius ratio calculations.

D • **18.32.** **(a)** Derive a general expression for the magnetic moment of a unit cell of a ferrite with a divalent ion moment of $n\mu_B$ and y being the fraction of inverse spinel structure. (Assume $[1 = y]$ to be the fraction of normal spinel structure.) **(b)** Use the expression derived in part (a) to calculate the moment of a copper ferrite given a heat treatment that produces 25% normal spinel and 75% inverse spinel structure.

18.33. The plot of H_c versus T for a metallic compound, such as Nb_3Ge in Figure 18–18, can be approximated by the equation for a parabola, namely

$$H_c = H_0 \left[1 - \left(\frac{T^2}{T_c^2} \right) \right]$$

where H_0 is the critical field at 0 K. Given that H_c for Nb_3Ge is 22×10^4 A-turns/m at 15 K, **(a)** calculate the value of H_0 using the foregoing equation and **(b)** calculate the percentage error of your result in comparison to the experimental value of $H_0 = 44 \times 10^4$ A-turns/m.

18.34. Repeat Problem 18.33 in order to evaluate the utility of the parabolic equation to describe the relationship between H_c and T for the 1–2–3 superconductor shown in Figure 18–18. For the particular sample illustrated, T_c is 93 K, H_0 is 328×10^4 A-turns/m, and H_c is 174×10^4 A-turns/m at 60 K.

18.35. A commercial ceramic magnet, Ferroxcube A, has a hysteresis loss per cycle of 40 J/m^3. **(a)** Assuming the hysteresis loop is traversed at a frequency of 60 Hz, calculate the power loss for this magnet. **(b)** Is this a soft or hard magnet?

18.36. A commercial ceramic magnet, Ferroxdur, has a hysteresis loss per cycle of 180 kJ/m^3. **(a)** Assuming the hysteresis loop is traversed at a frequency of 60 Hz, calculate the power loss for this magnet. **(b)** Is this a soft or hard magnet?

CHAPTER **19**

Environmental Degradation

This heat exchanger for the chemical processing industry contains over 17 kilometers of zirconium tubing. Zirconium is the material selected for many such applications due to its strong corrosion resistance in a variety of acidic and basic chemical environments. (Courtesy of Teledyne Wah Chang, Albany, Oregon.)

19.1 *Oxidation-Direct Atmospheric Attack*

19.2 *Aqueous Corrosion-Electrochemical Attack*

19.3 *Galvanic Two-Metal Corrosion*

19.4 *Corrosion by Gaseous Reduction*

19.5 *Effect of Mechanical Stress on Corrosion*

19.6 *Methods of Corrosion Prevention*

19.7 *Polarization Curves*

19.8 *Chemical Degradation of Ceramics and Polymers*

19.9 *Radiation Damage*

19.10 *Wear*

19.11 *Surface Analysis*

In the first three parts of the book, we surveyed the scientific principles that apply to the different categories of engineering materials. For each category, we found that atomic-and microscopic-scale structures were the basis of the important material properties. In this final part of the book, we turn to general considerations for the practical application of these various materials in engineering design. In this chapter we must dwell on an essentially negative concept—that all materials, to some degree, are susceptible to environmental degradation. In the final chapter we return to a more positive note in identifying the best candidate material for a specific application. An important part of that process is to incorporate our understanding of why given materials tend to be stable (or unstable) in given environments. Environmental degradation will be classified into one of four mechanisms: chemical, electrochemical, radiation induced, or wear related.

The first part of this chapter concentrates on the oxidation and corrosion of metals. *Oxidation* represents the direct chemical reaction between the metal and atmospheric oxygen (O_2). There are various mechanisms for the buildup of oxide scale on metals. Each is distinguished by a specific type of diffusion through the scale. For some metals, the oxide coating is tenacious and provides protection against further environmental attack. For others, the coating tends to crack and is not protective. Oxygen is not the only atmospheric gas that can be responsible for direct chemical attack. Similar problems occur, for example, with nitrogen and sulfur. *Aqueous corrosion* is a common form of electrochemical attack. A variation in metal ion concentration in aqueous solution above two different regions of a metal surface leads to an electrical current through the metal. The low ionic concentration region corrodes (i.e., loses material to the solution). *Galvanic corrosion* results when a more active metal is in contact with a more noble metal in an aqueous environment. The active metal is anodic and corroded. The relative performance of metals and alloys as "active" or "noble" depends on the specific aqueous environment. In the absence of ionic concentration differences or galvanic couples, corrosion can still occur by *gaseous reduction.* In these cases, the gaseous reduction reaction establishes a cathodic region. Several practical examples include rusting and corrosion beneath scale and dirt films. Corrosion can be enhanced by the presence of *mechanical stress.* This is true of both applied stresses and internal stresses associated with microstructure (e.g., grain boundaries). Corrosion can be prevented by careful *materials selection, design selection, protective coatings, galvanic protection (sacrificial anodes* or *impressed voltage),* and chemical *inhibitors.*

The use of nonmetallic coatings to prevent corrosion indicates the superior performance of ceramics and polymers against environmental attack. Their low electrical conductivity precludes corrosion, which is an electrochemical process. Of course, no material is totally inert. Silicates exhibit significant reactions with atmospheric moisture. Polymers, being organic compounds, are susceptible to attack by various solvents.

All categories of materials can be damaged by *radiation.* The nature of the damage varies with the category and the type of radiation. *Wear* is another form of material degradation that is generally physical, rather than chemical, in nature. An important type of materials characterization relevant to this chapter is *surface analysis* by means of *Auger electron* spectroscopy and related techniques.

19.1   OXIDATION-DIRECT ATMOSPHERIC ATTACK

In general, metals and alloys form stable oxide compounds under exposure to air at elevated temperatures. A few notable exceptions such as gold are highly prized. The stability of metal oxides is demonstrated by their relatively high melting points compared to the pure metal. For example, Al melts at $660°C$, whereas Al_2O_3 melts at $2054°C$. Even at room temperature, thin surface layers of oxide can form on some metals. For some metals, the reactivity with atmospheric oxygen, or **oxidation,** can be a primary limitation to their engineering application. For others, surface oxide films can protect the metal from more serious environmental attack.

There are four mechanisms commonly identified with metal oxidation as illustrated in Figure 19–1. The oxidation of a given metal or alloy can usually be characterized by one of these four diffusional processes, including (a) an "unprotective," porous oxide film through which molecular oxygen (O_2) can continuously pass and react at the metal-oxide interface; (b) a nonporous film through which cations diffuse in order to react with oxygen at the outer (air-oxide) interface; (c) a nonporous film through which O^{2-} ions diffuse in order to react with the metal at the metal-oxide interface; and (d) a nonporous film in which both cations and O^{2-} anions diffuse at roughly the same rate, causing the oxidation reaction to occur within the oxide film rather than at an interface. At this point, you may wish to review the discussion

y = oxide thickness

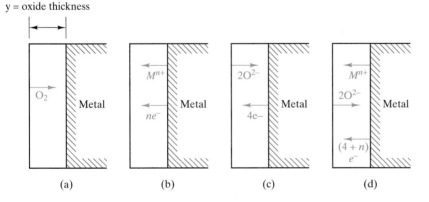

(a)   (b)   (c)   (d)

Figure 19-1 *Four possible metal oxidation mechanisms. (a) "Unprotective" film is sufficiently porous to allow continuous access of molecular O_2 to the metal surface. Mechanisms (b)-(d) represent nonporous films that are "protective" against O_2 permeation. In (b), cations diffuse through the film reacting with oxygen at the outer surface. In (c), O^{2-} ions diffuse to the metal surface. In (d), both cations and anions diffuse at nearly equal rates, leading to the oxidation reaction occurring within the oxide film.*

of ionic diffusion in Section 5.3. You should also note that charge neutrality requires cooperative migration of electrons with the ions in mechanisms (b)-(d). It may not be obvious that the diffusing ions in mechanisms (b)-(d) are the result of two distinct mechanisms: $M \rightarrow M^{n+} + ne^-$ at the metal-oxide interface and $O_2 + 4e^- \rightarrow 2O^{2-}$ at the air-oxide interface.

The rate at which oxidation occurs is, of course, a primary concern for the engineer responsible for a given materials selection. For an unprotective oxide (Figure 19–1a), oxygen gas is available to the metal surface (by way of the porous coating) at an essentially constant rate. As a result, the rate of oxide film growth is given by

$$\frac{dy}{dt} = c_1 \qquad (19.1)$$

where y is the thickness of the oxide film, t the time, and c_1 a constant. Integration of Equation 19.1 gives

$$y = c_1 t + c_2 \qquad (19.2)$$

where c_2 is a constant representing film thickness at $t = 0$. This time dependence is appropriately termed a **linear growth rate law.**

For film growth that is limited by ionic diffusion (Figure 19–1b-d), the growth rate diminishes as the film thickness grows. As illustrated in Figure 19–2, a uniform drop in O^{2-} concentration is present at a given instant during the oxidation process. From Fick's first law (Equation 5.8), it is apparent that the rate of film growth is inversely proportional to film thickness:

$$\frac{dy}{dt} = c_3 \frac{1}{y} \qquad (19.3)$$

where c_3 is a constant different, in general, from those in Equations 19.1 and 19.2. Integration gives

$$y^2 = c_4 t + c_5 \qquad (19.4)$$

where c_4 and c_5 are two additional constants. It is straightforward to show that $c_4 = 2c_3$ and that c_5 is the square of the film thickness at $t = 0$. The time dependence in Equation 19.4 is termed a **parabolic growth rate law.** This name is consistent with the plot of y versus t shown in Figure 19–3, which compares the parabolic with the linear rate Law (Equation 19.2). Because the oxide coating is of generally uniform density, a plot of weight gain during oxidation has the same general appearance as Figure 19–3. The weight gain is sometimes easier to measure than the rather small film thickness. Figure 19–3 illustrates a sharp contrast in behavior between a totally unprotective coating and one that diminishes further oxidation as its thickness increases.

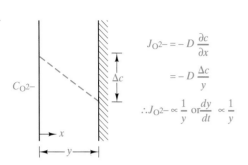

$$J_{O^{2-}} = -D\frac{\partial c}{\partial x}$$

$$= -D\frac{\Delta c}{y}$$

$$\therefore J_{O^{2-}} \propto \frac{1}{y} \text{ or} \frac{dy}{dt} \propto \frac{1}{y}$$

Figure 19-2 *A linear drop in oxygen concentration across the oxide film thickness leads to a film growth rate that is inversely proportional to film thickness.*

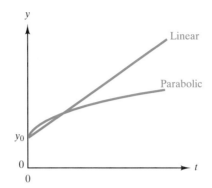

Figure 19-3 *Comparison of film growth kinetics for linear and parabolic growth laws. The diminishing growth rate with time for parabolic growth leads to protection against further oxidation.*

For thin oxide layers (less than 100 nm) at relatively low temperatures (a few hundred °C), some metals and alloys follow a **logarithmic growth rate law** in which

$$y = c_6 \ln(c_7 t + 1) \tag{19.5}$$

where c_6 and c_7 are constants different from those appearing in Equation 19.1 through 19.4. In Figure 19–4, the slow build-up of film thickness during logarithmic film growth in superimposed on the plots of linear and parabolic growth from Figure 19–3. The mechanism for logarithmic film growth is the subject of ongoing research but may be associated with electric fields across the film due to very low levels of electronic conductivity. In any case, the resulting thin films are seldom of consequence in selecting materials for corrosion resistance. For these oxide systems, thicker films and higher temperatures tend to produce parabolic growth kinetics.

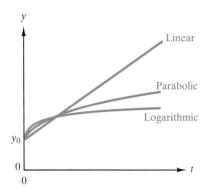

Figure 19-4 *A logarithmic growth rate law is superimposed on the linear and parabolic growth law plots of Figure 19–3. The logarithmic kinetics applies primarily to thin oxide films at low temperatures and provides little benefit for corrosion resistance.*

Table 19.1 *Pilling-Bedworth Ratios for Various Metal Oxides*

Protective oxides	Nonprotective oxides
Be—1.59	Li—0.57
Cu—1.68	Na—0.57
Al—1.28	K—0.45
Si—2.27	Ag—1.59
Cr—1.99	Cd—1.21
Mn—1.79	Ti—1.95
Fe—1.77	Mo—3.40
Co—1.99	Hf—2.61
Ni—1.52	Sb—2.35
Pd—1.60	W—3.40
Pb—1.40	Ta—2.33
Ce—1.16	U—3.05
	V—3.18

Source: B. Chalmers, *Physical Metallurgy,* John Wiley & Sons, Inc., New York, 1959.

The tendency of a metal to form a protective oxide coating is indicated by an especially simple parameter known as the **Pilling-Bedworth* ratio,** R, given as

$$R = \frac{Md}{amD} \tag{19.6}$$

where M is the molecular weight of the oxide (with formula M_aO_b and density D) and m is the atomic weight of the metal (with density d). Careful analysis of Equation 19.6 reveals that R is simply the ratio of the oxide volume produced to the metal volume consumed. For $R < 1$, the oxide volume tends to be insufficient to cover the metal substrate. The resulting oxide coating tends to be porous and unprotective. For R equal to or slightly greater than 1, the oxide tends to be protective. For R greater than 2, there are likely to be large compressive stresses in the oxide that cause the coating to buckle and flake off, a process known as **spalling.** The general utility of the Pilling-Bedworth ratio to predict the protective nature of an oxide coating is illustrated in Table 19.1. The protective oxides generally have R values between 1 and 2. The nonprotective oxides generally have R values less than 1 or greater than 2. There are exceptions, such as Ag and Cd. A number of factors, in addition to R, must be favorable to produce a protective coating. Similar coefficients of thermal expansion and good adherence are such additional factors.

---

* N. B. Pilling and R. E. Bedworth, *J. Inst. Met. 29*, 529 (1923).

Two of our most familiar protective oxide coatings in everyday applications are those on anodized aluminum and stainless steel. **Anodized aluminum** represents a broad family of aluminum alloys with Al_2O_3 as the protective oxide. However, the oxide coating is produced in an acid bath rather than by routine atmospheric oxidation. The stainless steels were introduced in Section 11.1. The critical alloy addition is chromium and the protective coating is an iron-chromium oxide. We shall refer to these coatings again in Section 19.6 as various forms of corrosion protection are reviewed.

Before leaving the subject of oxidation, we must note that oxygen is not the only chemically reactive component of environments to which engineering materials are subjected. Under certain conditions, atmospheric nitrogen can react to form nitride layers. A more common problem is the reaction of sulfur from hydrogen sulfide and other sulfur-bearing gases from various industrial processes. In jet engines, even nickel-based superalloys show rapid reaction with sulfur-containing combustion products. Cobalt-based superalloys are alternatives, although cobalt sources are limited. An especially insidious example of atmospheric attack is **hydrogen embrittlement,** in which hydrogen gas, also commonly found in a variety of industrial processes, permeates into a metal such as titanium, creating substantial internal pressure and even reacting to form brittle hydride compounds. The result, in either case, is a general loss of ductility.

SAMPLE PROBLEM 19.1

A nickel-based alloy has a 100-nm-thick oxide coating at time (t) equals zero, upon being placed in an oxidizing furnace at 600°C. After 1 hour, the coating has grown to 200 nm in thickness. What will be the thickness after 1 day, assuming a parabolic growth rate law?

SOLUTION

Equation 19.4 is appropriate:

$$y^2 = c_4 t + c_5$$

For $t = 0$, $y = 100$ nm, or

$$(100 \text{ nm})^2 = c_4(0) + c_5$$

giving

$$c_5 = 10^4 \text{ nm}^2$$

For $t = 1$ h, $y = 200$ nm, or

$$(200 \text{ nm})^2 = c_4(1 \text{ h}) + 10^4 \text{ nm}^2$$

Solving, we obtain

$$c_4 = 3 \times 10^4 \text{ nm}^2/\text{h}$$

Then, at $t = 24$ h,

$$y^2 = 3 \times 10^4 \text{ nm}^2/\text{h}(24 \text{ h}) + 10^4 \text{ nm}^2$$

$$= 73 \times 10^4 \text{ nm}^2$$

or $y = 854$ nm ($= 0.854 \mu$m).

SAMPLE PROBLEM 19.2

Given that the density of Cu_2O is 6.00 Mg/m^3, calculate the Pilling-Bedworth ratio for copper.

SOLUTION

The expression for the Pilling-Bedworth ratio is given by Equation 19.6:

$$R = \frac{Md}{amD}$$

Additional data, other than the density of Cu_2O, are available in Appendix 1:

$$R = \frac{[2(63.55) + 16.00](8.93)}{(2)(63.55)(6.00)}$$

$$= 1.68$$

(which is the value in Table 19.1).

PRACTICE PROBLEM 19.1

In Sample Problem 19.1, we calculate the thickness of an oxide coating after 1 day in an oxidizing atmosphere. What would be the thickness of the coating if the same measurements apply but the oxide grows by a linear growth rate law?

PRACTICE PROBLEM 19.2

(a) The Pilling-Bedworth ratio for copper is calculated in Sample Problem 19.2. In this case, we assume that cuprous oxide, Cu_2O, is formed. Calculate the Pilling-Bedworth ratio for the alternate possibility that cupric oxide, CuO, is formed. (The density of CuO is 6.40 Mg/m^3.) (b) Do you expect CuO to be a protective coating? Briefly explain.

19.2   AQUEOUS CORROSION-ELECTROCHEMICAL ATTACK

Corrosion is the dissolution of a metal into an aqueous environment. The metal atoms dissolve as ions. A simple model of this **aqueous corrosion** is given in Figure 19–5. This is an **electrochemical cell** in which chemical change (such as the corrosion of the anodic iron) is accompanied by an electrical current. In the next few sections of this chapter, various types of electrochemical cells will be described. The specific type introduced in Figure 19–5 is termed a **concentration cell** because corrosion and the associated electrical current are due to a difference in ionic concentration. The metal bar on the left side of the electrochemical cell is an **anode,** that is, a metal that supplies electrons to the external circuit and dissolves, or corrodes. The **anodic reaction** can be given as

$$Fe^0 \rightarrow Fe^{2+} + 2e^- \tag{19.7}$$

This reaction is driven by the tendency to equilibrate the ionic concentration in both sides of the overall cell. The porous membrane allows the transport of Fe^{2+} ions between the two halves of the cell (thereby completing the total electrical circuit) while maintaining a distinct difference in concentration levels. A metal bar on the right side of the electrochemical cell is a **cathode,**

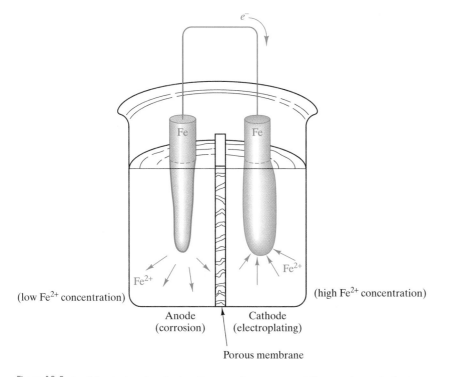

Figure 19-5 *In this electrochemical cell, corrosion occurs at the anode, and electroplating occurs at the cathode. The driving force for the two "half-cell" reactions is a difference in ionic concentration.*

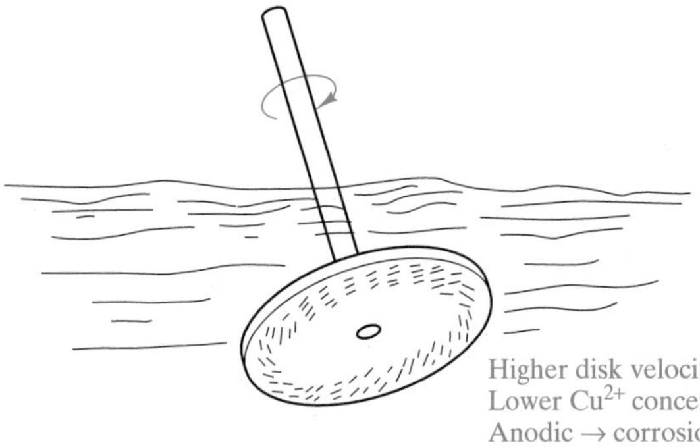

Figure 19-6 *A rotating brass disk in an aqueous solution containing Cu^{2+} ions produces a gradient in ionic concentration near its surface. The Cu^{2+} concentration is lower next to the more rapidly moving surface near the disk edge. As a result, that area is anodic and corrodes. This problem is analogous to the ionic concentration cell in Figure 19–5.*

Higher disk velocity →
Lower Cu^{2+} concentration →
Anodic → corrosion pits result

a metal that accepts the electrons from the external circuit and neutralizes ions in a **cathodic reaction:**

$$Fe^{2+} + 2e^- \rightarrow Fe^0 \tag{19.8}$$

At the cathode, then, metal builds up as opposed to dissolving. This process is known as **electroplating.** Each side of the electrochemical cell is appropriately termed a **half cell,** and Equations 19.7 and 19.8 are each **half-cell reactions.** An example of an actual corrosion problem due to an ionic concentration cell is shown in Figure 19–6. We can now turn to a variety of other corrosion examples involving somewhat different types of electrochemical cells.

SAMPLE PROBLEM 19.3

In a laboratory demonstration of an ionic concentration corrosion cell (such as Figure 19–5), an electrical current of 10 mA is measured. How many times per second does the reaction shown in Equation 19.7 occur?

SOLUTION

The current indicates a flow rate of electrons:

$$I = 10 \times 10^{-3} \text{ A} = 10 \times 10^{-3} \frac{C}{s} \times \frac{1 \text{ electron}}{0.16 \times 10^{-18} \text{ C}}$$

$$= 6.25 \times 10^{16} \text{ electrons/s}$$

As the oxidation of each iron atom (Equation 19.7) generates two electrons,

$$\text{reaction rate} = (6.25 \times 10^{16} \text{ electrons/s})(1 \text{ reaction/2 electrons})$$

$$= 3.13 \times 10^{16} \text{ reactions/s}$$

PRACTICE PROBLEM 19.3

For the experiment described in Sample Problem 19.3, how many times per second does the reduction (electroplating) reaction given in Equation 19.8 occur?

19.3   GALVANIC TWO-METAL CORROSION

In the preceding section, we created an electrochemical cell by allowing different ionic concentrations adjacent to a given type of metal. Figure 19–7 shows that a cell can be generated by two different metals even though each is surrounded by an equal concentration of its ions and aqueous solution. In this **galvanic*** **cell,** the iron bar, surrounded by a 1 molar solution of Fe^{2+}, is

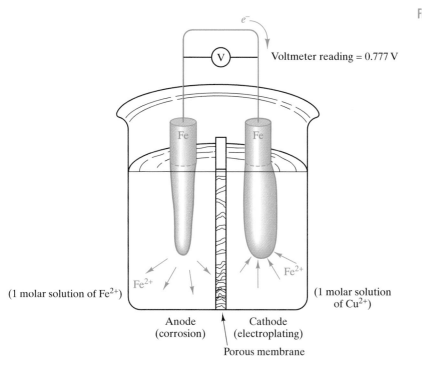

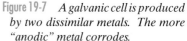
Voltmeter reading = 0.777 V

(1 molar solution of Fe^{2+})

Fe^{2+}

Fe^{2+}

(1 molar solution of Cu^{2+})

Anode
(corrosion)

Cathode
(electroplating)

Porous membrane

Figure 19-7   *A galvanic cell is produced by two dissimilar metals. The more "anodic" metal corrodes.*

---

* Luigi Galvani (1737–1798), Italian anatomist. In Section 5.3 we saw that a major contribution to materials science came from the medical sciences (the diffusional laws of Adolf Fick). In a similar way, we owe much of our basic understanding of electricity to Galvani, a professor of anatomy at the University of Bologna. He used the twitching of frog leg muscles to monitor electrical current. To duplicate Benjamin Franklin's lightning experiment, Galvani laid the frog leg muscles on brass hooks near an iron lattice work. During a thunderstorm, the muscles did indeed twitch, demonstrating again the electrical nature of lightning. But Galvani noticed that they also twitched whenever the muscles simultaneously touched the brass and iron. He thus identified the "Galvanic cell." When an instrument was developed to measure electrical current in 1820, Ampère

the anode and is corroded. (Recall that a 1 molar solution contains 1 gram atomic weight of ions in 1 liter of solution.) The copper bar, surrounded by a 1 molar solution of Cu^{2+}, is the cathode and Cu^0 plates out on the bar. The anodic reaction is equivalent to Equation 19.7, and the cathodic reaction is

$$Cu^{2+} + 2e^- \rightarrow Cu^0 \tag{19.9}$$

The driving force for the overall cell of Figure 19–7 is the relative tendency for each metal to ionize. The net flow of electrons from the iron bar to the copper bar is a result of the stronger tendency of iron to ionize. A voltage of 0.777 V is associated with the overall electrochemical process. Because of the common occurrence of galvanic cells, a systematic collection of voltages associated with half-cell reactions has been made. This **electromotive force (emf) series** is given in Table 19.2. Of course, half-cells exist only as pairs. All emf values in Table 19.2 are defined relative to a reference electrode that, by convention, is taken as the ionization of H_2 gas over a platinum surface. The metals toward the bottom of the emf series are said to be more

Table 19.2 *Electromotive Force Series*

	Metal-metal ion equilibrium (unit activity)	Electrode potential versus normal hydrogen electrode at 25°C (V)
	Au-Au^{3+}	+1.498
↑	Pt-Pt^{2+}	+1.2
Noble or	Pd-Pd^{2+}	+0.987
cathodic	Ag-Ag^+	+0.799
	Hg-Hg_2^{2+}	+0.788
	Cu-Cu^{2+}	+0.337
	H_2-H^+	0.000
	Pb-Pb^{2+}	−0.126
	Sn-Sn^{2+}	−0.136
	Ni-Ni^{2+}	−0.250
	Co-Co^{2+}	−0.277
	Cd-Cd^{2+}	−0.403
	Fe-Fe^{2+}	−0.440
	Cr-Cr^{3+}	−0.744
	Zn-Zn^{2+}	−0.763
Active or	Al-Al^{3+}	−1.662
anodic	Mg-Mg^{2+}	−2.363
↓	Na-Na^+	−2.714
	K-K^+	−2.925

Source: After A. J. de Bethune and N. A. S. Loud, as summarized in M. G. Fontana and N. D. Greene, *Corrosion Engineering,* 2nd ed., John Wiley & Sons, Inc., New York, 1978.

(see Section 15.1) suggested that it be known as a galvanometer.

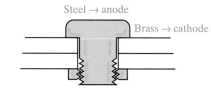

Steel → anode

Brass → cathode

Figure 19-8   *A steel bolt in a brass plate creates a galvanic cell analogous to the model system in Figure 19–7.*

active (i.e., anodic). Those toward the top are more *noble* (i.e., cathodic). The total voltage measured in Figure 19–7 is the difference of two half-cell potentials $[+0.337V - (-0.440V) = 0.777V]$.

While a useful guide to galvanic corrosion tendencies, Table 19.2 is somewhat idealistic. Engineering designs seldom involve pure metals in standard concentration solutions. Instead, they involve commercial alloys in various aqueous environments. For such cases, a table of standard voltages is less useful than a simple, qualitative ranking of alloys by their relative tendencies to be active or noble. As an example, the **galvanic series** in sea water is given in Table 19.3. This listing is a useful guide for the design engineer in predicting the relative behavior of adjacent materials in marine applications. Close inspection of this galvanic series indicates that alloy composition can radically affect the tendency toward corrosion. For example, (plain carbon) steel is near the active end of the series, whereas some passive stainless steels are among the most noble alloys. Steel and brass are fairly widely separated, and Figure 19–8 shows a classic example of galvanic corrosion in which a steel bolt is unwisely selected for securing a brass plate in a marine environment. Figure 19–9 reminds us that two-phase microstructures can provide a small-scale galvanic cell, leading to corrosion even in the absence of a separate electrode on the macroscopic scale.

Cu-Zn phase diagram

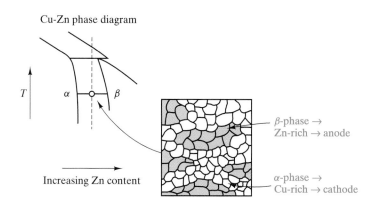

T α β

Increasing Zn content

β-phase →
Zn-rich → anode

α-phase →
Cu-rich → cathode

Figure 19-9   *A galvanic cell can be produced on the microscopic scale. Here β-brass (bcc structure) is zinc-rich and anodic relative to α-brass (fcc structure), which is copper-rich.*

Table 19.3 *Galvanic Series in Seawater*[a]

↑ Noble or cathodic	Platinum Gold Graphite Titanium Silver
	⌈ Chlorimet 3 (62 Ni, 18 Cr, 18 Mo) ⌊ Hastelloy C (62 Ni, 17 Cr, 15 Mo)
	⌈ 18–8 Mo stainless steel (passive) ∣ 18–8 stainless steel (passive) ⌊ Chromium stainless steel 11–30% Cr (passive)
	⌈ Inconel (passive) (80 Ni, 13 Cr, 7 Fe) ⌊ Nickel (passive)
	Silver solder
	⌈ Monel (70 Ni, 30 Cu) ∣ Cupronickels (60–90 Cu, 40–10 Ni) ∣ Bronzes (Cu-Sn) ∣ Copper ⌊ Brasses (Cu-Zn)
	⌈ Chlorimet 2 (66 Ni, 32 Mo, 1 Fe) ⌊ Hastelloy B (60 Ni, 30 Mo, 6 Fe, 1 Mn)
	⌈ Inconel (active) ⌊ Nickel (active)
	Tin Lead Lead-tin solders
	⌈ 18–8 Mo stainless steel (active) ⌊ 18–8 stainless steel (active)
	Ni-resist (high-nickel cast iron) Chromium stainless steel, 13% Cr (active)
	⌈ Cast iron ⌊ Steel or iron
Active or anodic ↓	2024 aluminum (4.5 Cu, 1.5 Mg, 0.6 Mn) Cadmium Commercially pure aluminum (1100) Zinc Magnesium and magnesium alloys

Source: From tests conducted by International Nickel Company and summarized in M. G. Fontana and N. D. Greene, *Corrosion Engineering,* 2nd ed., John Wiley & Sons, Inc., New York, 1978.

[a] (Active) and (passive) designations indicate whether or not a passive, oxide film has been developed on the alloy surface.

SAMPLE PROBLEM 19.4

Suppose that you set up a laboratory demonstration of a galvanic cell similar to Figure 19–7 but are forced to use zinc and iron for the electrodes. **(a)** Which electrode will be corroded? **(b)** If the electrodes are immersed in 1 molar solutions of their respective ions, what will be the measured voltage between the electrodes?

SOLUTION

(a) Inspection of Table 19.2 indicates that zinc is anodic relative to iron. Therefore, zinc will be corroded.

(b) Again using Table 19.2, the voltage will be

$$\text{voltage} = (-0.440\text{V}) - (-0.763\text{V})$$
$$= 0.323\text{V}$$

....

PRACTICE PROBLEM 19.4

In Sample Problem 19.4, we analyze a simple galvanic cell composed of zinc and iron electrodes. Make a similar analysis of a galvanic cell composed of copper and zinc electrodes immersed in 1 molar ionic solutions.

19.4   CORROSION BY GASEOUS REDUCTION

So far, our examples of aqueous corrosion have involved corrosion at the anode and electroplating at the cathode. You may have recalled, from your own experience, various examples of corrosion in which no plating process was apparent. Such cases are, in fact, quite common (e.g., rusting). Electroplating is not the only process that will serve as a cathodic reaction. Any chemical reduction process that consumes electrons will serve the purpose. Figure 19–10 illustrates a model electrochemical cell based on **gaseous reduction.** The anodic reaction is again given by Equation 19.7. Now the cathodic reaction is

$$O_2 + 2H_2O + 4e^- \rightarrow 4OH^- \qquad (19.10)$$

where two water molecules are consumed together with four electrons from the external circuit to reduce a molecule of oxygen to four hydroxyl ions. The iron at the cathode serves only as a source of electrons. It is not, in this

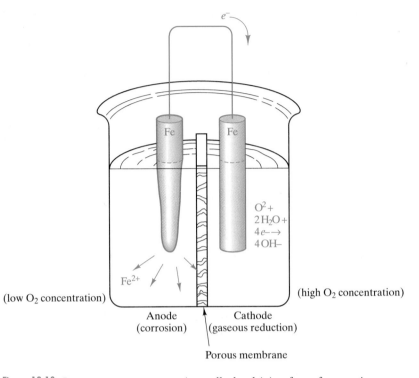

Figure 19-10 *In an oxygen concentration cell, the driving force for reaction is the difference in oxygen concentration. Corrosion occurs at the oxygen-deficient anode. The cathodic reaction is gaseous reduction.*

case, a substrate for electroplating. Figure 19–10 can be labeled an **oxygen concentration cell,** in contrast to Figure 19–5, which was an **ionic concentration cell.** Some practical oxidation cells are illustrated in Figure 19–11. This is an especially troublesome source of corrosion damage. As an example, a surface crack (Figure 19–11a) is a "stagnant" region of the aqueous environment with a relatively low oxygen concentration. As a result, corrosion occurs at the crack tip, leading to crack growth. This increase in crack depth increases the degree of oxygen depletion, which, in turn, enhances the corrosion mechanism further. Our most familiar corrosion problem, the **rusting** of ferrous alloys, is another example of oxygen reduction as a cathodic reaction. The overall process is illustrated in Figure 19–12. Rust is the reaction product, $Fe(OH)_3$, that precipitates onto the iron surface. While we have dwelled on oxygen reduction as an example, many gaseous reactions are available to serve as cathodes. For metals immersed in acids, the cathodic reaction can be

$$2H^+ + 2e^- \rightarrow H_2 \uparrow \qquad (19.11)$$

in which some of the high concentration of hydrogen ions is reduced to hydrogen gas, which is then evolved from the aqueous solution.

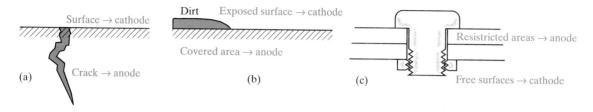

Figure 19-11 *Various practical examples of corrosion due to oxygen concentration cells. In each case, metal corrodes next to oxygen-deficient regions of an aqueous environment.*

In the past few sections, we have seen a variety of electrochemical cells associated with corrosion. We have not generally been specific about concentrations other than referring to standard, 1 molar (1 M) solutions relative to galvanic corrosion (Table 19.2). The effect of variations in the concentration of solutions is given by the **Nernst* equation**, which states that the cell voltage, V, is given by

$$V = V^0 - (RT/nF)\ln K \qquad (19.12)$$

where V^0 is the voltage under standard state conditions. R is the universal gas constant, T is the absolute temperature, n is the number of electrons transferred in the corrosion equation, F is the Faraday's[†] constant, and K

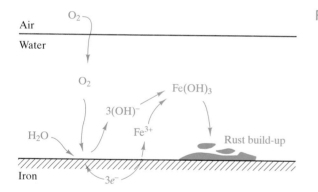

Figure 19-12 *The rusting of ferrous alloys is another corrosion reaction associated with gaseous reduction.*

---

* Hermann Walther Nernst (1864–1941), German chemist. Originally trained as a physicist, Nernst demonstrated an unparalleled command of the theoretical foundations of physical chemistry. He made numerous contributions to electrochemistry, including the famous equation that bears his name.

[†] Michael Faraday (1791–1867), English chemist and physicist. Born in poverty, Faraday was apprenticed to a bookbinder at the age of 14. He seized this opportunity to read books and attend related lectures in London. Demonstrating tremendous intellectual gifts, he was eventually offered, in 1812, the position laboratory assistant at the Royal Institution. He eventually succeeded his mentor, Humphry Davy, as the most famous and influential scientist in England. He made numerous contributions to chemistry and the theory of electricity, including their convergence in the subject of electrochemistry.

is a reaction quotient. The Faraday's constant is the quantity of electrical charge on 1 mole of electrons (= 96,500 C/mol). The reaction quotient is defined for a given chemical reaction such as

$$aA + bB \rightarrow cC + dD \tag{19.13}$$

as

$$K = \frac{[C]^c[D]^d}{[A]^a[B]^b} \tag{19.14}$$

where $[A]$, etc. represent the "activity" of the given chemical species. For ionic solutions, the activity is the molar concentration. For gases, the activity is the pressure in atm. The activity of a pure, solid metal is taken as 1.0.

For simplicity, the term $(RT/nF) \ln K$ in Equation 19.2 is generally calculated at 25°C, and the natural logarithm is replaced by the common (base 10) logarithm, giving the form

$$V = V^0 - (0.059/n) \log_{10} K \tag{19.15}$$

As an example of the application of the Nernst equation, consider the combination of Equation 19.7 and 19.11:

$$
\begin{array}{rcl}
Fe^0 & \rightarrow & Fe^{2+} + 2e^- \\
+ \quad 2H^+ + 2e^- & \rightarrow & H_2 \uparrow \\
\hline
Fe^0 + 2H^+ & \rightarrow & Fe^2 + H_2 \uparrow
\end{array} \tag{19.16}
$$

For this overall cell reaction, Equation 19.15 becomes

$$V = V^0 - (0.059/2) \log_{10} \frac{[Fe^{2+}][H_2]}{[Fe^0][H^+]^2} \tag{19.17}$$

Equation 19.17 indicates that the cell voltage is a sensitive function of the pH ($\equiv -\log_{10}[H^+]$). In fact, the common pH meter is a specially designed electrochemical cell with a voltmeter calibrated to read pH directly.

SAMPLE PROBLEM 19.5

In the oxygen concentration cell of Figure 19–10, what volume of oxygen gas (at STP) must be consumed at the cathode to produce the corrosion of 100 g of iron? [STP stands for "standard temperature and pressure," which are 0°C (= 273 K) and 1 atm.]

SOLUTION

The common link between the corrosion reaction (Equation 19.7) and the gaseous reduction reaction (Equation 19.10) is the production (and consumption) of electrons:

$$Fe^0 \rightarrow Fe^{2+} + 2e^- \qquad \text{(Eq. 19.7)}$$

and

$$O_2 + 2H_2O + 4e^- \rightarrow 4OH^- \qquad \text{(Eq. 19.10)}$$

One mole of iron produces 2 moles of electrons, but only $\frac{1}{2}$ mole of O_2 gas is needed to consume 2 moles of electrons. Using data from Appendix 1, we can write

$$\text{moles of } O_2 \text{ gas} = \frac{100 \text{ g Fe}}{(55.85 \text{ g Fe/g} \cdot \text{atom Fe})} \times \frac{\frac{1}{2} \text{ mole } O_2}{1 \text{ mole Fe}}$$

$$= 0.895 \text{ mole } O_2$$

Using the ideal gas law, we obtain

$$pV = nRT \quad \text{or} \quad V = \frac{nRT}{p}$$

At STP,

$$V = \frac{(0.895 \text{ mole})(8.314 \text{ J/mol} \cdot \text{K})(273 \text{ K})}{(1 \text{ atm})(1 \text{ Pa}/9.869 \times 10^{-6} \text{ atm})}$$

$$= 0.0201 \frac{\text{J}}{\text{Pa}} = 0.0201 \frac{\text{N} \cdot \text{m}}{\text{N/m}^2}$$

$$= 0.0201 \text{ m}^3$$

SAMPLE PROBLEM 19.6

Equation 19.16 represents an Fe/H_2 cell. Consider a similar Zn/H_2 cell that has a cell voltage of 0.45 V at 25°C when $[Zn^{2+}] = 1.0$ M and $p_{H_2} = 1.0$ atm. Calculate the corresponding pH.

SOLUTION

For this cell, we can rewrite Equation 19.16 as

$$Zn^0 + 2H^+ \rightarrow Zn^{2+} + H_2 \uparrow$$

We see from data in Table 19.2 that the standard state cell voltage is

$$V^0 = 0.000\,V - (-0.763\,V) = 0.763\,V$$

As two electrons are transferred per Zn atom, $n = 2$. For this cell, Equations 19.14 and 19.15 can be combined as

$$V = V^0 = (0.059/2) \log_{10} \frac{[Zn^{2+}]p_{H_2}}{[Zn^0][H^+]^2}$$

or

$$0.45\,V = 0.763\,V - (0.59/2) \log_{10} \frac{[1.0][1.0]}{[1.0][H^+]^2}$$

Rearranging,

$$\frac{0.45\,V - 0.763\,V}{(-0.059)/2} = \log_{10} \frac{1}{[H^+]^2} = \log_{10}[H^+]^{-2}$$

$$= -2 \log_{10}[H^+]$$

or

$$- \log_{10}[H^+] = \frac{(0.45V - 0.763V)2}{(-0.059)2} = 5.29$$

or, by definition, pH = 5.29.

···

PRACTICE PROBLEM 19.5

Calculate the volume of O_2 gas consumed (at STP) by the corrosion of 100 g of chromium. (In this case, trivalent Cr^{3+} ions are found at the anode.) (See Sample Problem 19.5.)

PRACTICE PROBLEM 19.6

If reducing the concentration of Zn^{2+} in the Zn/H$_2$ cell of Sample Problem 19.6 to 0.1 M increases the cell voltage to 0.542 V, what is the corresponding pH?

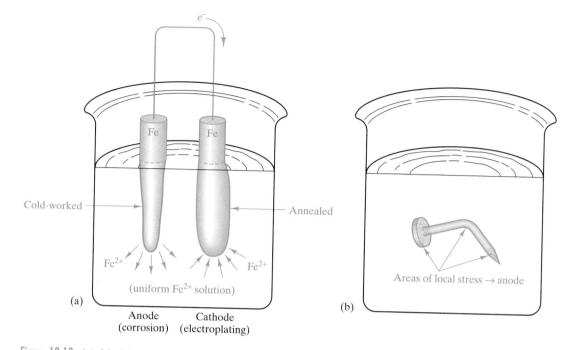

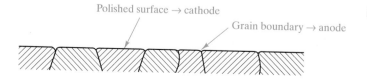

Figure 19-13 *(a) Model electrochemical stress cell. The more highly stressed electrode is anodic and corrodes. (b) Common example of a stress cell. In an aqueous environment, the regions of a nail that were stressed during fabrication or use corrode locally.*

19.5 EFFECT OF MECHANICAL STRESS ON CORROSION

In addition to the various chemical factors that lead to corrosion, mechanical stress can also contribute. High-stress regions in a given material are anodic relative to low-stress regions. In effect, the high-energy state of the stressed metal lowers the energy barrier to ionization. Figure 19–13a illustrates a model electrochemical **stress cell.** A practical example of such stress cells is given in Figure 19–13b, in which regions of a nail stressed during fabrication or use become susceptible to local corrosive attack.

Grain boundaries are microstructural regions of high energy (see Section 4.4). As a result, they are susceptible to accelerated attack in a corrosive environment (Figure 19–14). Although this can be the basis of problems such as intergranular fracture, it can also be the basis of useful processes such as the etching of polished specimens for microscopic inspection.

Polished surface → cathode

Grain boundary → anode

Figure 19-14 *On the microscopic scale, grain boundaries are regions of local stress and are susceptible to accelerated attack.*

19.6   METHODS OF CORROSION PREVENTION

Sections 19.2 to 19.5 have outlined such a broad range of corrosion mechanisms that we need not be surprised that corrosion is now an annual multibillion-dollar expense to modern society. Even thin layers of condensed atmospheric moisture are sufficient aqueous environments for metallic alloys to lead to appreciable corrosion by some of these mechanisms. A major challenge to all engineers employing metals in their designs is to prevent corrosive attack. When complete prevention is impossible, losses should be minimized. Consistent with the wide variety of corrosion problems, a wide range of preventative measures is available.

Our primary means of corrosion prevention is **materials selection.** Boating enthusiasts quickly learn to avoid steel bolts for brass hardware. A careful application of the principles of this chapter allows the materials engineer to find those alloys least susceptible to given corrosive environments. In a similar fashion, **design selection** can minimize damage. Threaded joints and similar high-stress regions are to be avoided when possible. When galvanic couples are required, a small-area anode next to a large-area cathode should be avoided. The resulting large current density at the anode accelerates corrosion.

When an alloy must be used in an aqueous environment in which corrosion could occur, additional techniques are available to prevent degradation. **Protective coatings** provide a barrier between the metal and its environment. Table 19.4 lists various examples. These are divided into three categories, corresponding to the fundamental structural materials: metals, ceramics, and polymers. Chrome plating has traditionally been used on decorative trim for automobiles. **Galvanized steel** operates on a somewhat different principle. As seen in Figure 19–15, protection is provided by a zinc coating. Because zinc is anodic relative to the steel, any break in the coating does not lead to corrosion of the steel, which is cathodic and preserved. This is in contrast to more noble coatings (e.g., tin on steel in Figure 19–15), in which a break leads to accelerated corrosion of the substrate. As discussed in Section 19.1, stable oxide coatings on a metal can be protective. The (Fe, Cr) oxide coating on stainless steel is a classic example. But Figure 19–16 illustrates a limitation for this material. Excessive heating (e.g., welding)

Table 19.4 *Protective Coatings for Corrosion Prevention*

Category	Examples
Metallic	Chrome plating
	Galvanized steel
Ceramic	Stainless steel
	Porcelain enamel
Polymeric	Paint

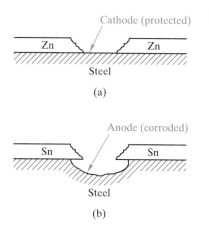

(a)

(b)

Figure 19-15 *(a) Galvanized steel consists of a zinc coating on a steel substrate. Since zinc is anodic to iron, a break in the coating does not lead to corrosion of the substrate. (b) In contrast, a more noble coating such as "tin plate" is protective only as long as the coating is free of breaks. At a break, the anodic substrate is preferentially attacked.*

can cause precipitation of chromium carbide at grain boundaries. The result is chromium depletion adjacent to the precipitates and susceptibility to corrosive attack in that area. An alternative to an oxide reaction layer is a deposited ceramic coating. **Porcelain enamels** are silicate glass coatings with thermal expansion coefficients reasonably close to those of their metal substrates. Polymeric coatings can provide similar protection, usually at a lower cost. Paint is our most common example. (We should distinguish enamel paints, which are organic polymeric coatings, from the porcelain enamels, which are silicate glasses.)

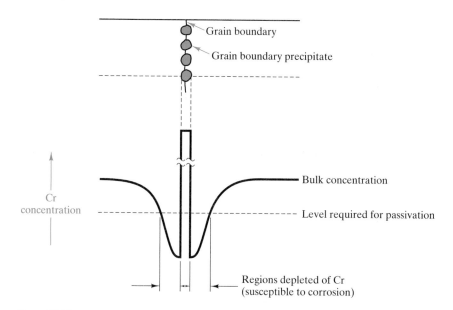

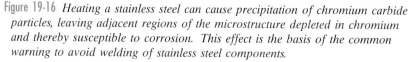

Figure 19-16 *Heating a stainless steel can cause precipitation of chromium carbide particles, leaving adjacent regions of the microstructure depleted in chromium and thereby susceptible to corrosion. This effect is the basis of the common warning to avoid welding of stainless steel components.*

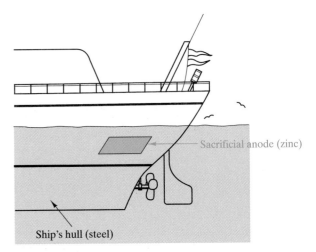

Figure 19-17 *A sacrificial anode is a simple form of galvanic protection. The galvanized steel in Figure 19–15a is a special form of this protection.*

A galvanized steel coating (see Table 19.4) is a specialized example of a **sacrificial anode.** A general, noncoating example is given in Figure 19–17. This is one type of **galvanic protection.** Another is the use of an **impressed voltage**, in which an external voltage is used to oppose the one due to the electrochemical reaction. The impressed voltage stops the flow of electrons needed for the corrosion reaction to proceed. Figure 19–18 illustrates a common example of this technique.

A final approach to corrosion prevention is the use of an **inhibitor,** defined as a substance used in small concentrations, which decreases the rate of corrosion. Most inhibitors are organic compounds that form adsorbed layers on the metal surface. This provides a system similar to the protective coatings discussed earlier. Other inhibitors affect gaseous reduction reactions associated with the cathode (see Section 19.4).

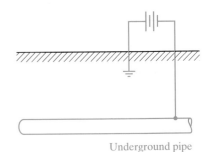

Figure 19-18 *An impressed voltage is a form of galvanic protection that counters the corroding potential.*

SAMPLE PROBLEM 19.7

A 2-kg sacrificial anode of magnesium is attached to the hull of a ship (see Figure 19–17). If the anode lasts 3 months, what is the average corrosion current during that period?

SOLUTION

Using data from Appendix 1, we can write

$$\text{current} = \frac{2 \text{ kg}}{3 \text{ months}} \times \frac{1000 \text{ g}}{1 \text{ kg}} \times \frac{0.6023 \times 10^{24} \text{ atoms}}{24.31 \text{ g}}$$

$$\times \frac{2 \text{ electrons}}{\text{atom}} \times \frac{0.16 \times 10^{-18} \text{ C}}{\text{electron}}$$

$$\times \frac{1 \text{ month}}{31 \text{ d}} \times \frac{1 \text{ d}}{24 \text{ h}} \times \frac{1 \text{ h}}{3600 \text{ s}} \times \frac{1 \text{ A}}{1 \text{ C/s}}$$

$$= 1.97 \text{ A}$$

PRACTICE PROBLEM 19.7

In Sample Problem 19.7, we calculate the average current for a sacrificial anode. Assume that the corrosion rate could be diminished by 25% by using an annealed block of magnesium. What mass of such an annealed anode would be needed to provide corrosion prevention for **(a)** 3 months and **(b)** a full year?

19.7 POLARIZATION CURVES

In Sections 19.2 through 19.6, we have seen various forms of corrosion and methods by which corrosion damage can be prevented. A common way to monitor corrosion behavior is to plot the relationship between the electrochemical potential (in volts) of a given half-cell reaction versus the resulting corrosion rate. Figure 19–19 gives such a plot for an anodic half-cell. We should note that the relationship is linear when the corrosion rate is plotted on a logarithmic scale. Figure 19–20 shows that the intersection of plots for individual cathodic and anodic half-cell reactions defines a corrosion potential, V_c. In Figure 19–20, we can focus on the anodic half-cell reaction, defining an anodic **polarization** as an **overvoltage** (η) above the corrosion potential. Physically, the anodic polarization represents a deficiency of electrons produced in a metal oxidation reaction such as Equation 19.7 as an overvoltage is applied. Similarly, the cathodic polarization, or negative overvoltage, shown in Figure 19–20 corresponds to a buildup of electrons at the metal surface for a reduction reaction such as Equation 19.8.

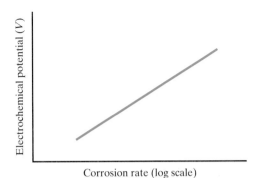

Figure 19-19 *Schematic illustration of the linear, semilogarithmic plot of electrochemical potential versus corrosion rate for an anodic half-cell.*

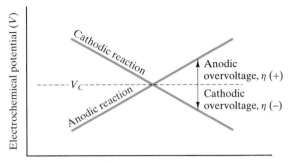

Figure 19-20 *Schematic illustration of the establishment of a corrosion potential, V_c, at the intersection of anodic and cathodic reaction plots. Anodic polarization corresponds to a positive overvoltage, η.*

For metals such as chrominun and alloys such as stainless steel, the plot of potential versus corrosion rate above the range shown in Figure 19–20 shows a sudden, sharp drop in corrosion rate above some critical potential, V_p (Figure 19–21). Despite a high level of anodic polarization above V_p, the corrosion rate drops precipitously due to the formation of a thin, protective oxide film as a barrier to the anodic dissolution reaction. Resistance to corrosion above V_p is termed **passivity**. The drop in corrosion rate above V_p can be as much as 10^3 to 10^6 times below the maximum rate in the active state. With increasing corrosion potential, the low corrosion rate remains constant until at a relatively high potential, the passive film breaks down, and the normal increase in corrosion rate resumes in a **transpassive** region.

Figure 19–22 illustrates how a given anode can exhibit either active or passive behavior depending on the specific corrosive environment. In effect, the curve of Figure 19–21 is intersected by two cathodic curves for environments A and P. Environment A produces an intersection corresponding to r_A, a relatively high corrosion rate representing active behavior. Environment P produces an intersection corresponding to r_p, a relatively low corrosion rate representing passive behavior. A specific example of Figure 19–22 is type 304 stainless steel, which is active in deaerated salt water and passive in aerated salt water.

The schematic illustrations of Figures 19–19 through 19–22 used horizontal axes of corrosion rate on a logarithmic scale. Corrosion rate, r (in mol/m$^2\cdot$ s), can be related to the corresponding current density, i (in A/m^2), by

$$r = \frac{i}{nF} \tag{19.18}$$

where n and F were defined relative to Equation 19.12.

In general, an overpotential can be related to a standard state current density, i_0, by

$$\eta = \beta \log_{10}(i/i_0) \tag{19.19}$$

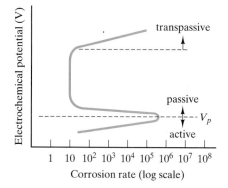

Figure 19-21 *Schematic illustration of passivity. The corrosion rate for a given metal drops sharply above an oxidizing potential of V_p.*

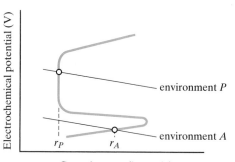

Figure 19-22 *The position of a cathodic half-cell plot can affect the nature of metallic corrosion. Environment A intersects the anodic polarization curve in the active region, and environment P intersects in the passive region.*

where β is a constant equal to the slope of the electrochemical potential plot. The value of β is positive for an anodic half-cell and negative for a cathodic half-cell. Figure 19–23 shows various plots of Equation 19.19 corresponding to the corrosion of zinc in an acid solution. The corrosion potential, V_c, and corrosion current density, i_c, are determined by the intersection of anodic and cathodic plots. The standard state potentials for the zinc and hydrogen half-cells in Figure 19–23 correspond to the data in Table 19.2.

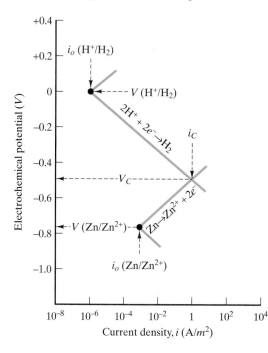

Figure 19-23 *The anodic and cathodic half-cell reactions for zinc in an acid solution show, by their intersection, the corrosion potential, V_c, and corrosion current density, i_c. (After M. G. Fontana, Corrosion Engineering, 3rd ed., McGraw-Hill, New York, 1986.)*

SAMPLE PROBLEM 19.8

For the corrosion of zinc in an acid solution, the standard state current density for the zinc half-cell is 10^{-3} A/m^2 and the slope β in Equation 19.19 is $+0.09$ V. Calculate the electrochemical potential of this half-cell at a current density of 1 A/m^2.

SOLUTION

First, we can note from Table 19.2 that the standard state potential for the zince half-cell is -0.763 V. The overpotential for the given condition will be from Equation 19.19:

$$\eta = \beta \log_{10}(i/i_0)$$
$$= (+0.09\,\text{V}) \log_{10}(1/10^{-3})$$
$$= (+0.09\,\text{V})(3.0) = 0.27\,\text{V}$$

giving an electrochemical potential at 1 A/m^2 of

$$V = -0.763\,\text{V} + 0.27\,\text{V} = -0.493\,\text{V}$$

PRACTICE PROBLEM 19.8

For the corrosion of zinc in an acid solution, the standard state current density for the hydrogen half-cell is 10^{-6} A/m^2 and the slope β in Equation 19.19 is -0.08 V. Calculate the electrochemical potential of this half-cell at a current density of 1 A/m^2. (Note Sample Problem 19.8.)

19.8 CHEMICAL DEGRADATION OF CERAMICS AND POLYMERS

The high electrical resistivity of ceramics and polymers removes them from consideration of corrosive mechanisms. The use of ceramic and polymeric protective coatings on metals leads to the general view of these nonmetallic materials as "inert." In fact, any material will undergo chemical reaction under suitable circumstances. As a practical matter, ceramics and polymers are relatively resistant to the environmental reactions associated with typical metals. Although electrochemical mechanisms are not significant, some direct chemical reactions can limit utility. A good example was the reaction of H_2O with silicates, leading to the phenomenon of static fatigue (see Section 8.3). Also, ceramic refractories are selected, as much as possible, to resist chemical reaction with molten metals that they contain in metal casting processes.

The cross-linking of polymers during vulcanization was an example of chemical reaction affecting mechanical properties in polymers (see Section 6.6). The sensitivity of the mechanical properties of nylon to atmospheric moisture was illustrated by Figure 6–17. Polymers are also reactive with various organic solvents. This is an important consideration in those industrial processes in which such solvents are part of the material's "environment."

19.9   RADIATION DAMAGE

This chapter has concentrated on chemical reactions between materials and their environments. Increasingly, materials are also subjected to radiation fields. Nuclear power generation, radiation therapy, and communication satellites are a few of the applications in which materials must withstand severe radiation environments.

Table 19.5 summarizes some common forms of radiation. For electromagnetic radiation, the energy of a given photon, E, was introduced in Chapter 16 as

$$E = h\nu \qquad (16.3)$$

where h is Planck's constant ($= 0.6626 \times 10^{-33}$ J·s) and ν is the vibrational frequency, which, in turn, is equal to

$$\nu = \frac{c}{\lambda} \qquad (16.2)$$

Table 19.5 *Common Forms of Radiation*

Category	Description
Electromagnetic[a]	
Ultraviolet	1 nm $< \lambda < 400$ nm
X ray	10^{-3} nm $< \lambda < 10$ nm
γ ray	$\lambda < 0.1$ nm
Particles	
α particle (α ray)	He^{2+} (helium nucleus = two protons + two neutrons)
β particle (β ray)	e^+ or e^- (positive or negative particle with mass of a single electron)
Neutron	$_0 n^1$

[a] See Figure 19–24. Obviously, the wavelength ranges of these high-energy photons overlap. A basic distinction is the mechanism of radiation production. Ultraviolet light is produced by outer electron orbital transitions. X rays are produced by generally higher energy, inner orbital transitions; γ rays are produced by radioactive decay (a nuclear rather than electronic process).

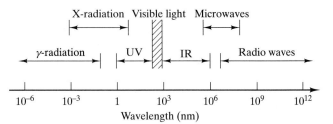

Figure 19-24 *Electromagnetic radiation spectrum. (This chart was introduced in Figure 3–34 in which x-radiation was first described as a medium for identifying crystalline structures and later in Figure 16–1 as an introduction to the optical behavior of materials.)*

where c is the speed of light ($= 0.2998 \times 10^9$ m/s) and λ is the wavelength. Figure 19–24 summarizes the wavelength range for electromagnetic radiation. It is the radiation with wavelengths shorter than those of visible light that tends to damage materials. As Equations 16.2 and 16.3 indicate, photon energy increases with decreasing wavelength.

The response of different materials to a given type of radiation varies considerably. Similarly, a given material can be affected quite differently by different types of radiation. In general, a radiation-induced atomic displacement is an inefficient process and requires a displacement energy substantially greater than a simple binding energy discussed in Chapter 2. Figure 19–25 illustrates the nature of atom displacements generated by a single

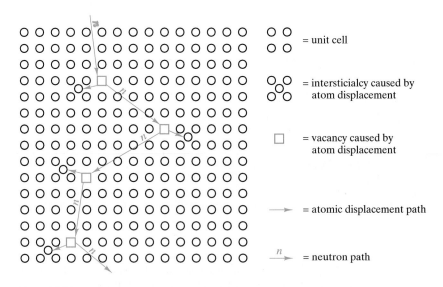

Figure 19-25 *Schematic of the sequence of atomic displacements in a metallic crystal structure caused by a single, high-energy neutron. (An electron micrograph of a neutron-damaged microstructure of a zirconium alloy is shown in Figure 4–35a.)*

neutron during the course of neutron irradiation of a metal. Figure 19–26 shows a microstructural consequence of electron irradiation of a structural ceramic, Al_2O_3. The substantial number of dislocation loops produced by the irradiation is visible in the electron microscope (see Section 4.7). Polymers are especially susceptible to ultraviolet (UV) radiation damage. A single UV photon has sufficient energy to break a single C-C bond in many linear-chain polymers. The broken bonds serve as reactive sites for oxidation reactions. One of the reasons for using carbon black as an additive to polymers is to shield the material from UV radiation.

Although we have concentrated on structural materials, radiation can also strongly affect the performance of electrical and magnetic materials. Radiation damage to semiconductors used in communication satellites can be a major limitation for their design applications.

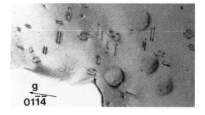

Figure 19-26 *Electron micrograph of dislocation loops produced in Al_2O_3 as a result of electron beam irradiation. (Courtesy of D. G. Howitt)*

SAMPLE PROBLEM 19.9

Electromagnetic radiation with photon energies greater than 15 eV can damage a particular semiconductor intended for use in a communications satellite. Will visible light be a source of this damage?

SOLUTION

The most energetic visible light photon will be the short wavelength (blue) end of the visible spectrum (at 400 nm). Combining Equations 16.2 and 16.3, we obtain

$$E = \frac{hc}{\lambda}$$

$$= \frac{(0.6626 \times 10^{-33} \text{ J} \cdot \text{s})(0.2998 \times 10^9 \text{ m/s})}{400 \times 10^{-9} \text{ m}}$$

$$\times \frac{6.242 \times 10^{18} \text{ eV}}{\text{J}}$$

$$= 3.1 \text{ eV}$$

As this is less than 15 eV, visible light will not be a source of this damage.

....

PRACTICE PROBLEM 19.9

In Sample Problem 19.9, we find that a given photon energy necessary for radiation damage of a semiconductor is not available in visible light. **(a)** What wavelength of radiation is represented by the photon energy of 15 eV? **(b)** What type of electromagnetic radiation has such wavelength values?

19.10 WEAR

As with radiation damage, **wear** is generally a physical (rather than chemical) form of material degradation. Specifically, wear can be defined as the removal of surface material as a result of mechanical action. The amount of wear need not be great to be relatively devastating. (A 1500-kg automobile can be "worn out" as a result of the loss of only a few grams of material from surfaces in sliding contact.) Although the systematic study of wear has largely been confined to the last few decades, several key aspects of this phenomenon are now well characterized. Four main forms of wear have been identified: (1) **Adhesive wear** occurs when two smooth surfaces slide over each other and fragments are pulled off one surface and adhere to the other. The adjective for this category comes from the strong bonding or "adhesive" forces between adjacent atoms across the intimate contact surface. A typical example of adhesive wear is shown in Figure 19–27. (2) **Abrasive wear** occurs when a rough, hard surface slides on a softer surface. The result is a series of grooves in the soft material and the resulting formation of wear particles. (3) **Surface fatigue wear** occurs during repeated sliding or rolling over a track. Surface or subsurface crack formation leads to breakup of the surface. (4) **Corrosive wear** takes place with sliding in a corrosive environment and, of course, adds chemical degradation to the physical effects of wear. The sliding action can break down passivation layers and, thereby, maintain a high corrosion rate.

Figure 19-27 *The sliding of a copper disk against a 1020 steel pin produces irregular wear particles. (Courtesy of I. F. Stowers)*

In addition to the four main types of wear, related mechanisms can occur in certain design applications. **Erosion** by a stream of sharp particles is analogous to abrasive wear. **Cavitation** involves damage to a surface caused by the collapse of a bubble in an adjacent liquid. The surface damage results from the mechanical shock associated with the sudden bubble collapse.

In addition to the qualitative description of wear just given, some progress has been made in the quantitative description. For the most common form of wear, adhesive wear,

$$V = \frac{kPx}{3H} \qquad (19.20)$$

where V is the volume of material worn away under a load P sliding over a distance x, with H being the hardness of the surface being worn away. The term k is referred to as the **wear coefficient** and represents the probability that an adhesive fragment will be formed. Like the coefficient of friction, the wear coefficient is a dimensionless constant. Table 19.6 gives values of k for a wide range of sliding combinations. (Note that k is rarely greater than 0.1).

In Section 19.1, we saw the potential damage that can result from the direct chemical reaction of a metal with environmental gases such as hydrogen. In subsequent sections, we saw a wide variety of environmental damage produced by aqueous corrosion. In this section, we see that mechanical wear by itself or in combination with aqueous corrosion provides additional forms of environmental degradation. Figure 19–28 illustrates practical examples of this wide spectrum of environmental degradation for metals, including **stress-corrosion cracking** and **corrosion-fatigue**, first introduced in Chapter 8.

Nonmetallic structural materials are frequently selected for their superior wear resistance. High-hardness ceramics generally provide excellent resistance to wear. Aluminum oxide, partially stabilized zirconia, and tungsten carbide (as a coating) are common examples. Polymers and polymer-matrix composites are increasingly replacing metals in bearings, cams, gears,

Table 19.6 *Values of the Wear Coefficient (k) for Various Sliding Combinations*

Combination	$k(\times 10^3)$
Zinc on zinc	160
Low-carbon steel on low-carbon steel	45
Copper on copper	32
Stainless steel on stainless steel	21
Copper on low-carbon steel	1.5
Low-carbon steel on copper	0.5
Phenol-formaldehyde on phenol-formaldehyde	0.02

Source: After E. Rabinowicz, *Friction and Wear of Materials*, John
　　　 Wiley & Sons, Inc., New York, 1965.

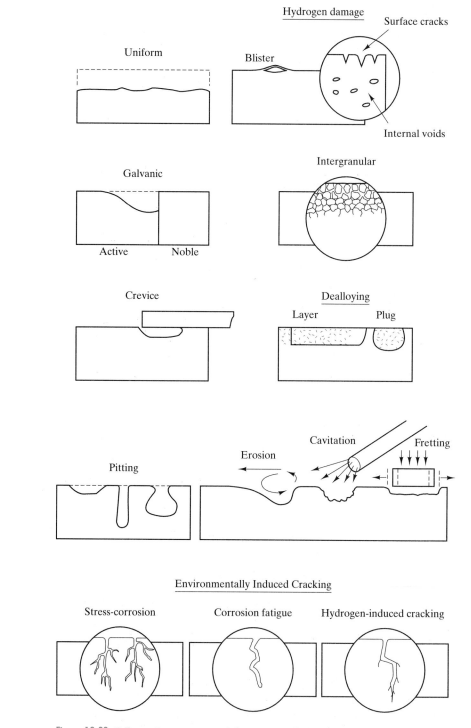

Figure 19-28 *Schematic summary of the various forms of environmental degradation for metals. (From D. A. Jones, Principles and Prevention of Corrosion, Macmillan Publishing Company, New York, 1992.)*

and other sliding components. Polytetrafluoroethylene (PTFE) is an example of a self-lubricating polymer that is widely used for its wear resistance. Fiber reinforcement of PTFE improves other mechanical properties without sacrificing the wear performance.

SAMPLE PROBLEM 19.10

Estimate the particle size of a wear fragment produced by the adhesive wear of two 1040 steel surfaces under a load of 50 kg with a sliding distance of 5 mm. (Assume the particle to be a hemisphere of diameter d.)

SOLUTION

From Table 19.6, we find a value of $k = 45 \times 10^{-3}$ for low-carbon steel on low-carbon steel. From Table 6.11, we find the hardness of 1040 steel to be 235 (in units of kg/mm^2, as noted in Sample Problem 6.9).

Using Equation 19.20, we obtain

$$V = \frac{kPx}{3H}$$

$$= \frac{(45 \times 10^{-3})(50 \text{ kg})(5 \text{ mm})}{3(235 \text{ kg/mm}^2)}$$

$$= 0.0160 \text{ mm}^3$$

As the volume of a hemisphere is $(1/12)\pi d^3$,

$$(1/12)\pi d^3 = 0.0160 \text{ mm}^3$$

or

$$d = \sqrt[3]{\frac{12(0.0160 \text{ mm}^3)}{\pi}}$$

$$= 0.394 \text{ mm} = 394 \ \mu\text{m}$$

....

PRACTICE PROBLEM 19.10

In Sample Problem 19.10, the diameter of a wear particle is calculated for the case of two steel surfaces sliding together. In a similar way, calculate the diameter of a wear particle for the same sliding combination under the same conditions, but with the 1040 steel heat treated to a hardness of 200 Brinell hardness number (BHN).

19.11   SURFACE ANALYSIS

Much of the environmental degradation associated with this chapter occurs at free surfaces or interfaces, such as grain boundaries. The characterization of this degradation frequently requires chemical analysis in the surface region. An example of microstructural chemical analysis done with a scanning electron microscope (SEM) is shown in Figure 4–39. In order to understand the basis of this type of chemical analysis, we need to return to the description of electronic energy levels in Chapter 2. Figure 19–29 reproduces the energy level diagram for a carbon atom, as shown in Figure 2–3, along with an alternate labeling system, namely, K for the innermost electron orbital shell, L for the next shell, and so on. In the case of carbon with only six orbital electrons, we deal only with the K and L shells. In heavier elements, electrons can also occupy M, N, and so on, shells. Figure 19–30 summarizes the two-part mechanism in which a carbon atom in the surface region of an SEM sample can be identified chemically. An electron from the beam being used to produce the topographical image has sufficient energy to eject a K shell electron (Figure 19–30a). The resulting unstable state for the atom leads to an electron transition from the L to K shell. By conservation of energy, the reduction in energy associated with the L to K transition is balanced by the emission of a characteristic x-ray photon with energy $|E_K - E_L|$.

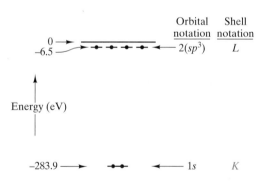

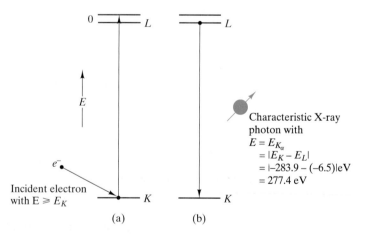

Figure 19-29 *The energy level diagram for a carbon atom (introduced in Figure 2–3) is labeled with a K for the lowest energy (innermost) electron shell and an L for the next lowest energy shell.*

Figure 19-30 *The mechanism for producing a characteristic x-ray photon for chemical analysis of an atom of the element carbon can be represented in two steps. (a) An electron with energy greater than or equal to the binding energy of a K shell electron (283.9 eV) can eject that electron from the atom. (b) The resulting unstable state is eliminated by an L to K electron transition. The reduction in electron energy produces a K_α photon with a specific energy characteristic of the carbon atom.*

The photon is labeled K_α because it was emitted as a result of the filling of a vacancy in the K electron shell by an electron from the nearest electron orbital. In heavier elements, it is possible to have a K_β photon in which an electron transition from the M to K shell fills the K shell, thereby leading to emission of a photon with energy $|E_K - E_M|$. (As a practical matter, the probability of an L to K electron transition is greater than the probability of an M to K electron transition and the yield of K_α photons is typically 10 times greater than that for K_β photons for those heavier elements.) When used in conjunction with an SEM, this technique of chemical analysis is termed **energy-dispersive x-ray spectrometry (EDX)**. EDX for chemically identifying elements using characteristic x-ray photons produced by bombarding the sample with electrons is patterned after an earlier technology known as **x-ray fluorescence (XRF)**. In that case, the mechanism is identical to that shown in Figure 19–30 except that the initial K shell electron ejection is caused by an x-ray photon with an energy greater than the binding energy of the K shell electron. A disadvantage of the x-ray fluorescence technique is that the incident x-ray beam generally cannot be focused to a micron-size spot in the way an electron beam can.

Although the SEM analysis just outlined provides a map of microstructural distributions of the elements in the surface of a sample (e.g., Figure 4–39), we must use care in defining the term *surface*. For example, an electron beam in a typical SEM (with a typical beam energy of 25 keV) can penetrate to a depth of about one micron into the surface of the sample. As a result, the characteristic photons analyzed for chemical information have escaped from a depth of one micron, a distance corresponding to a few thousand atomic layers. Unfortunately, many of the environmental reactions described in this chapter occur over a depth of just a few atomic layers. Relative to those cases, the SEM chemical analysis is insensitive. True surface analysis on the order of a few atomic layers can be done by a somewhat different mechanism, as illustrated in Figure 19–31. In this case, the characteristic x-ray photon illustrated in Figure 19–30 does not escape the vicinity of the atomic core but, instead, ejects one of the L shell electrons. The result is an **Auger* electron** with a kinetic energy characteristic of the chemical element (carbon in this illustration). As shown in Figure 19–31, the corresponding notation for the Auger electron is KLL. The key to the use of this mechanism for true surface analysis is that the Auger electron has a substantially lower escape depth from the surface than does a characteristic x-ray photon. The escape depth, or depth of sample surface analyzed, ranges from 0.5 to 5.0 nm, that is, from 1 to 10 atomic layers. A typical commercial instrument for doing Auger electron spectroscopy is shown in Figure 19–32. A typical microstructural analysis is shown in

---

* Pierre Victor Auger (1899–), French physicist, identified the nonradiating electron transition of Figure 19–31 in the 1920s during an early application of the cloud chamber method of experimental particle physics. Roughly 40 years were required before high-vacuum instrumentation and rapid data analysis systems were available to allow this principle to be the basis of routine chemical analysis.

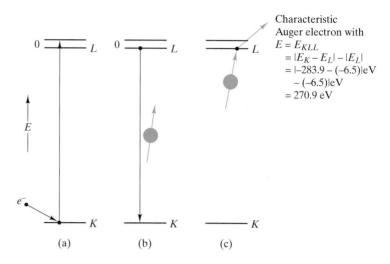

Characteristic
Auger electron with
$E = E_{KLL}$
$= |E_K - E_L| - |E_L|$
$= |{-}283.9 - (-6.5)|eV$
$- (-6.5)|eV$
$= 270.9\ eV$

(a)   (b)   (c)

Figure 19-31 *The mechanism for producing a characteristic electron for chemical analysis of a carbon atom in the first few atomic layers of a sample surface can be represented in three steps. Steps (a) and (b) are essentially identical to Figure 19–30. In step (c), the characteristic K_α photon ejects an L shell electron. The resulting kinetic energy of this Auger electron has a specific value characteristic of the carbon atom.*

Figures 19–33 and 19–34. In Figure 19–33 an SEM topographical image is shown along with an Auger spectrum of a specific point on the image. In Figure 19–34 the corresponding elemental maps of the surface chemistry are shown.

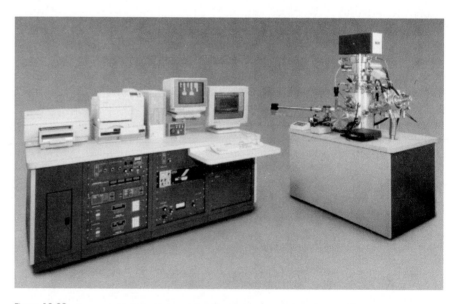

Figure 19-32 *A commercial microprobe for doing scanning Auger electron spectroscopy. (Courtesy of Perkin-Elmer, Physical Electronics Division)*

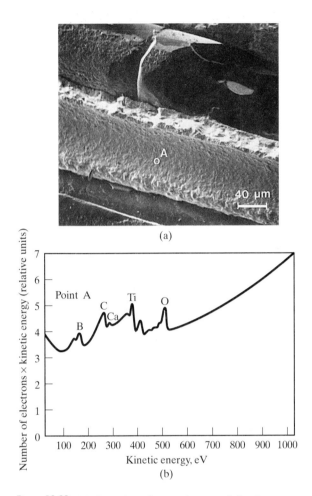

(a)

(b)

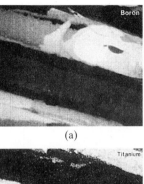

(a)

(b)

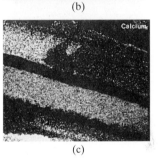

(c)

Figure 19-33 *(a) Scanning electron image of the fracture surface of a composite of boron carbide fibers in a titanium matrix. (b) Auger electron spectrum measured at point A in the image of part (a). Note the magnitude of the kinetic energy of the carbon Auger electron as calculated in Figure 19–31. Note also the presence of Ca and O impurities at the fracture interface. (Courtesy of Perkin-Elmer, Physical Electronics Division)*

Figure 19-34 *(a) Boron, (b) titanium, and (c) calcium maps of the fracture surface shown in Figure 14–33a at the same magnification. These Auger electron images can be powerful indicators of impurity concentrations [such as calcium in (c)] confined within a few atomic layers at the interface between the composite matrix and the reinforcing phase. Such interfacial segregation can play a major role in material properties, such as fracture strength. (Courtesy of Perkin-Elmer, Physical Electronics Division)*

Numerous techniques have been developed in the last few decades that share with Auger electron spectroscopy the ability to analyze true surface chemistry. The most closely related is **x-ray photoelectron spectroscopy (XPS)**, also known as **electron spectroscopy for chemical analysis (ESCA)**, in which characteristic electrons are produced by an incident x-ray photon rather than an incident electron. Figure 19–35 and Table 19.7 summarize the various, related techniques for surface chemical analysis.

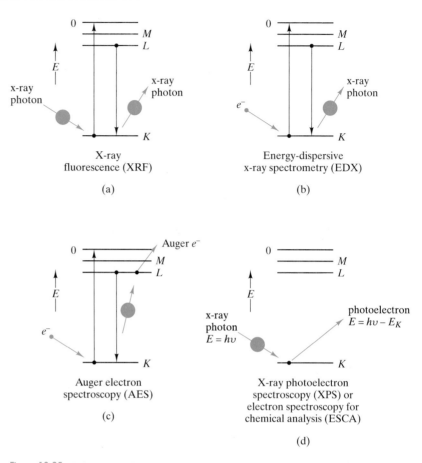

Figure 19-35 *Schematic illustration summarizing the four, related techniques for surface chemical analysis given in Table 19.7. Note that (a) x-ray fluoresence and (b) energy-dispersive x-ray spectrometry essentially involve bulk chemical analysis, whereas (c) Auger electron spectroscopy and (d) x-ray photoelectron spectroscopy (or electron spectoscropy for chemical analysis) represent true surface analysis, providing the analysis of the outermost atomic layer(s).*

Table 19.7 *Techniques for Surface Chemical Analysis*[a]

Technique	Input	Output	Depth of analysis	Diameter of spot analysis
X-ray fluorescence (XRF)	x-ray photon	x-ray photon	100 μm	1 mm
Energy-dispersive x-ray spectrometry (EDX)	electron	x-ray photon	1 μm	1μm
Auger electron spectroscopy	electron	electron	0.5–5 nm	50 nm
X-ray photoelectron spectroscopy (XPS) *or* electron spectroscopy for chemical analysis (ESCA)	x-ray photon	electron	0.5–5 nm	1 mm

[a] See Figure 19–35.

SAMPLE PROBLEM 19.11

The electron energy levels for an iron atom are $E_K = -7112$ eV, $E_L = -708$ eV, and $E_M = -53$ eV. Calculate **(a)** the K_α and **(b)** the K_β photon energies used in an SEM chemical analysis of iron. Also calculate **(c)** the KLL Auger electron energy for iron.

SOLUTION

(a) As illustrated in Figure 19–30,

$$E_{K_\alpha} = |E_K - E_L|$$
$$= |-7112 \text{ eV} - (-708 \text{ eV})| = 6404 \text{ eV}$$

(b) Similarly,

$$E_{K_\beta} = |E_K - E_M|$$
$$= |-7112 \text{ eV} - (-53 \text{ eV})| = 7059 \text{ eV}$$

(c) As illustrated in Figure 19–31,

$$E_{KLL} = |E_K - E_L| - |E_L|$$
$$= |-7112 \text{ eV} - (-708 \text{ eV})| - |-708 \text{ eV}| = 5696 \text{ eV}$$

PRACTICE PROBLEM 19.11

Characteristic photon and electron energies are calculated in Sample Problem 19.11. Using the given data, calculate **(a)** the L_α characteristic photon energy and **(b)** the LMM Auger electron energy.

SUMMARY

A wide range of environmental reactions limits the utility of engineering materials considered in this book. Oxidation is the direct chemical reaction of a metal with atmospheric oxygen. There are four mechanisms for oxidation associated with various modes of diffusion through the oxide scale. Two extreme cases are (1) the linear growth rate law for an unprotective oxide and (2) the parabolic growth rate law for a protective oxide. The tendency of a coating to be protective can be predicted by the Pilling-Bedworth ratio, R. Values of R between 1 and 2 are associated with a coating under moderate compressive stresses, and, as a result, it can be protective. Familiar examples of protective coatings are those on anodized aluminum and stainless steel. Unprotective coatings are susceptible to buckling and flaking off, a process known as spalling. Other atmospheric gases, including nitrogen, sulfur, and hydrogen, can lead to direct chemical attack of metals.

Corrosion is the dissolution of a metal into an aqueous environment. An electrochemical cell is a simple model of such aqueous corrosion. In an ionic concentration cell, the metal in the low-concentration environment is anodic and corrodes. The metal in the high-concentration environment is cathodic and experiences electroplating. A galvanic cell involves two different metals with different tendencies toward ionization. The more active, or ionizable, metal is anodic and corrodes. The more noble metal is cathodic and plates out. A list of half-cell potentials showing relative corrosion tendencies is the electromotive force series. A galvanic series is a more qualitative list for commercial alloys in a given corrosive medium such as seawater. Gaseous reduction can serve as a cathodic reaction, eliminating the need for electroplating to accompany corrosion. The oxygen concentration cell is such a process, and the rusting of ferrous alloys is a common example. For simple corrosion cells, the cell voltage can be calculated using the Nernst equation. In a stress cell, a stressed metal is anodic to the same metal in an annealed state. On the microstructural scale, grain boundaries are anodic relative to the adjacent grains. Metallic corrosion can be prevented by materials selection, design selection, protective coatings of various kinds, galvanic protection (using sacrificial anodes or impressed voltage), and chemical inhibitors. Polarization curves (plots of electrochemical potential versus corrosion rate) are helpful in monitoring corrosion and passivation.

Although nonmetals are relatively inert in comparison with corrosion-sensitive metals, direct chemical attack can affect design applications. The attack of silicates by moisture and the vulcanization of rubber are examples. All materials can be selectively damaged by certain forms of radiation. Neutron damage of metals, electron damage of ceramics, and UV degradation of polymers are examples. Wear is the removal of surface material as a result of mechanical action, such as continuous or cyclic sliding.

Many of the environmental reactions in this chapter are associated with material surfaces. Auger electron spectroscopy and related tools have become powerful methods for the chemical analysis of the first few atomic layers at a free surface or interface. The Auger electron and the photoelectron have kinetic energies characteristic of the element being analyzed.

KEY TERMS

abrasive wear (734)

adhesive wear (734)

anode (711)

anodic reaction (711)

anodized aluminum (709)

aqueous corrosion (711)

Auger electron (739)

cathode (712)

cathodic reaction (711)

cavitation (735)

concentration cell (711)

corrosion (711)

corrosion fatigue (735)

corrosive wear (734)

design selection (724)

electrochemical cell (711)

electromotive force series (714)

electron spectroscopy for chemical analysis (ESCA) (741)

electroplating (712)

energy-dispersive x-ray spectrometry (EDX) (739)

erosion (735)

galvanic cell (713)

galvanic corrosion (713)

galvanic protection (726)

REFERENCES

ASM Handbook, Vol. 13: Corrosion, ASM International, Materials Park, Ohio, 1987.

Briggs, D., and **M. P. Seah,** Eds., *Practical Surface Analysis,* 2nd ed., 2 Vols., Chichester: Wiley, New York, 1990–1992.

Jones, D. A., *Principles and Prevention of Corrosion,* 2nd

ed., Prentice-Hall, Upper Saddle River, N. J., 1996.

Kelly, B. T., *Irradiation Damage to Solids,* Pergamon Press, Inc., Elmsford, N.Y., 1966.

Rabinowicz, E., *Friction and Wear of Materials,* 2nd ed., John Wiley & Sons, Inc., New York, 1995.

PROBLEMS

19.1 • Oxidation-Direct Atmospheric Attack

19.1. The following data are collected during the oxidation of a small bar of metal alloy:

Time	Weight gain (mg)
1 min	0.40
1 hr	24.0
1 day	576

The weight gain is due to the oxide formation. Because of the experimental arrangement, you are unable to visually inspect the oxide scale. Predict whether the scale is (1) porous and discontinuous or (2) dense and tenacious. Briefly explain your answer.

19.2. The densities for three iron oxides are FeO (5.70 Mg/m^3), Fe_3O_4 (5.18 Mg/m^3), and Fe_2O_3 (5.24 Mg/m^3). Calculate the Pilling-Bedworth ratio for iron relative to each type of oxide and comment on the implications for the formation of a protective coating.

19.3. Given the density of SiO_2 (quartz) = 2.65 Mg/m^3, calculate the Pilling-Bedworth ratio for silicon and comment on the implication for the formation of a protective coating if quartz is the oxide form.

19.4. In contrast to the assumption in Problem 19.3, silicon oxidation tends to produce a vitreous silica film with density = 2.20 Mg/m^3. Semiconductor fabrication routinely involves such vitreous films. Calculate the Pilling-Bedworth ratio for this case and comment on the implication for the formation of a tenacious film.

19.5. Verify the statement in regard to Equation 19.4 that $c_4 = 2c_3$ and $c_5 = y^2$ at $t = 0$.

19.6. Verify that the Pilling-Bedworth ratio is the ratio of oxide volume produced to metal volume consumed.

19.2 • Aqueous Corrosion-Electrochemical Attack

19.7. In an ionic concentration corrosion cell involving nickel (forming Ni^{2+}), an electrical current of 5 mA is measured. How many Ni atoms per second are oxidized at the anode?

19.8. For the concentration cell described in Problem 19.7, how many Ni atoms per second are reduced at the cathode?

19.9. In an ionic concentration corrosion cell involving chromium, which forms a trivalent ion (Cr^{3+}), an electrical current of 10 mA is measured. How many atoms per second are oxidized at the anode?

19.10. For the concentration cell described in Problem 19.9, how many Cr atoms per second are reduced at the cathode?

19.3 • Galvanic Two-Metal Corrosion

19.11. (a) In a simple galvanic cell consisting of Co and Cr electrodes immersed in 1 molar ionic solutions, calculate the cell potential. (b) Which metal would be corroded in this simple cell?

19.12. (a) In a simple galvanic cell consisting of Al and Mg electrodes immersed in 1 molar ionic solutions, calculate the cell potential. (b) Which metal would be corroded in this simple cell?

19.13. Identify the anode in the following galvanic cells, including a brief discussion of each answer: (a) copper and nickel electrodes in standard solutions of their own ions, (b) a two-phase microstructure of a 50 : 50 Pb-Sn alloy, (c) a lead-tin solder on a 2024 aluminum alloy in sea water, and (d) a brass bolt in a Hastelloy C plate, also in sea water.

D **19.14.** Figure 19–9 illustrates a microstructural-scale galvanic cell. In selecting a material for the outer shell of a marine compass, use the Cu-Zn phase diagram from Chapter 9 to specify a brass composition range that would avoid this problem.

19.4 • Corrosion by Gaseous Reduction

19.15. A copper-nickel (35 wt %-65 wt %) alloy is corroded in an oxygen concentration cell using boiling water. What volume of oxygen gas (at 1 atm) must be consumed at the cathode to corrode 10 g of the alloy? (Assume only divalent ions are produced.)

19.16. Assume that iron is corroded in an acid bath, with the cathode reaction being given by Equation 19.11. Calculate the volume of H_2 gas produced at STP in order to corrode 100 gm of iron.

19.17. For the rusting mechanism illustrated in Figure 19–12, calculate the volume of O_2 gas consumed (at STP) in the production of 100 gm of rust [$Fe(OH)_3$].

19.18. In the failure analysis of an aluminum vessel, corrosion pits are found. The average pit is 0.1 mm in diameter and the vessel wall is 1 mm thick. If the pit developed over a period of 1 year, calculate (a) the corrosion current associated with each pit and (b) the corrosion current density (normalized to the area of the pit).

19.19. The Nernst equation was applied to a Zn/H_2 cell in Sample Problem 19.6. In a similar way, calculate the cell voltage at 25°C for a Zn/Fe galvanic cell when [Zn^{2+}] = 0.5 M and [Fe^{2+}] = 0.1 M.

19.20. Calculate the cell voltage at 25°C for the Zn/Fe galvanic cell introduced in Problem 19.19 if the ion concentrations are [Zn^{2+}] = 0.1 M and [Fe^{2+}] = 0.5 M.

19.21. Calculate the cell voltage at 25°C for the Zn/Fe galvanic cell introduced in Problem 19.19 if the ion concentrations are [Zn^{2+}] = 1.0 M and [Fe^{2+}] = 10^{-3} M.

19.22. Verify the equivalence of Equations 19.12 and 19.15.

19.6 • Methods of Corrosion Prevention

D **19.23.** In designing the hull of a new fishing vessel to ensure corrosion protection, you find that a sacrificial anode of zinc provides an average corrosion current of 2 A over the period of 1 year. What mass of zinc is required to give this protection?

D **19.24.** In designing an off-shore steel structure to ensure corrosion protection, you find that a sacrificial anode of magnesium provides an average corrosion current of 1.5 A over a period of 2 years. What mass of magnesium is required to give this protection?

D **19.25.** The maximum corrosion current density in a galvanized steel sheet used in the design of the new engineering laboratories on campus is found to be 5 mA/m². What thickness of the zinc layer is necessary to ensure at least (a) 1 year and (b) 5 years of rust resistance?

D **19.26.** A galvanized steel sheet used in the design of the new chemistry laboratories on campus has a zinc coating 18 μm thick. The corrosion current density is found to be 4 mA/m². What is the corresponding duration of rust resistance provided by this system?

19.7 • Polarization Curves

19.27. For the Zn/H_2 cell considered in Sample Problem 19.8 and Practice Problem 19.8, calculate the corrosion current density, i_c.

19.28. Calculate the corrosion potential, V_c, for the Zn/H_2 cell of Problem 19.27.

19.29. For the corrosion of lead in an acid solution, assume that divalent Pb^{2+} is formed and that the standard state current densities for the lead and hydrogen half-cells are 2×10^{-5} A/m^2 and 10^{-4} A/m^2, respectively. Also, take the slope β in Equation 19.19 for lead and hydrongen to be $+0.12$ V and -0.10 V, respectively. Calculate the corrosion current density, i_c.

19.30. Calculate the corrosion potential, V_c, for the Pb/H_2 cell of Problem 19.29.

19.9 • Radiation Damage

19.31. In Problem 16.11, the ultraviolet wavelengths necessary to break carbon bonds (in polymers) were calculated. Another type of radiation damage found in a variety of solids is associated with electron-positron "pair production," which can occur at a threshold photon energy of 1.02 MeV. **(a)** What is the wavelength of such a threshold photon? **(b)** Which type of electromagnetic radiation is this?

19.32. Calculate the full range of photon energies associated with **(a)** ultraviolet radiation and **(b)** x-radiation in Table 19.5.

19.10 • Wear

19.33. Calculate the diameter of a wear particle for copper sliding on a 1040 steel. Take the load to be 40 kg over a distance of 10 mm. (Take the hardness of 1040 steel to be given by Table 6.11.)

19.34. Calculate the diameter of a wear particle produced by the adhesive wear of two 410 stainless steel surfaces under the load conditions of Problem 19.33. (Note Table 6.11 for hardness data.)

19.11 • Surface Analysis

19.35. The electron energy levels for a copper atom are $E_K = -8982$ eV, $E_L = -933$ eV, and $E_M = -75$ eV. Calculate **(a)** the K_α photon energy, **(b)** the K_β photon energy, **(c)** the L_α photon energy, **(d)** the KLL Auger electron energy, and **(e)** the LMM Auger electron energy.

19.36. The K shell electron energy for nickel is $E_K = -8333$ eV and the wavelengths of the NiK_α and NiK_β photons are 0.1660 nm and 0.1500 nm, respectively. **(a)** Draw an energy-level diagram for a nickel atom. Calculate **(b)** the KLL and **(c)** the LMM Auger electron energies for nickel.

• **19.37.** Characteristic photon energies are generally measured in a so-called energy-dispersive mode in which a solid-state detector measures energy directly (see Problem 17.24). An alternate technique is the "wavelength dispersive" mode in which the photon energy is determined indirectly by measuring the x-ray wavelength by diffraction (see Section 3.7). **(a)** Calculate the diffraction angle (2θ) needed to identify the FeK_α photon using the (200) planes of an NaCl single crystal. **(b)** Sketch the experimental system for this measurement.

• **19.38.** As pointed out in this section, an alternate surface analysis technique is x-ray photoelectron spectroscopy (XPS). In this case, a "soft" x ray such as AlK_α is used to eject an inner orbital electron (from an atom in the sample) giving a characteristic kinetic energy. Calculate the specific photoelectron energy that could be used to identify an iron atom. (Note that aluminum electron energy levels are $E_K = -1560$ eV and $E_L = -72.8$ eV.)

CHAPTER **20**
Materials Selection

The manufacturer of this carbon-fiber reinforced composite bicycle uses a sophisticated software package [utilizing "finite element analysis"] to analyze how the frame will respond to stress, allowing the engineers to tailor the frame stiffness to the individual rider. (Courtesy of Algor, Inc.)

The task of selecting the right material for a given engineering design, can appear to be overwhelming. The number of materials commercially available to the design engineer is finite but nonetheless quite large. The job of sorting through the options to arrive at the optimum selection requires a systematic approach based on an understanding of the nature of materials science and engineering. Development of this understanding has been the primary goal of the previous 19 chapters.

A giant step in establishing a systematic approach to materials selection is the use of the five materials categories. In this chapter we discuss examples of materials selection relative to either a competition among the five categories or within a single category. Our examples are illustrative and relatively brief. At this introductory level, we cannot cover all of the detailed procedures required to produce materials specifications for a given product in a given industry. Although these pragmatic details are beyond the scope of this book, we can nonetheless outline the general philosophy of materials selection.

First, we must acknowledge that each of the material properties discussed in the previous chapters can be translated into design parameters used by the engineer to specify quantitatively the material requirements of that design. Second, we shall recall that the processing of materials can have a strong effect on their design parameters. Examples of selection within each category of structural materials include a discussion of the competition among the four different types. Then, the selection of electronic, optical, and magnetic materials including semiconductors, will be discussed. Finally, we must be aware of the effects of engineering materials on our environment.

20.1   MATERIAL PROPERTIES—ENGINEERING DESIGN PARAMETERS

In looking at fundamentals in Part I of this book and material categories in Parts II and III, we have defined dozens of the basic properties of engineering materials. For materials *science,* the nature of these properties is an end in itself. They serve as the basis of our understanding of the solid state. For materials *engineering,* the properties assume a new role. They are the **design parameters,** which are the basis for selecting a given material for a given application. A graphic example of this engineering perspective is shown in Figure 20–1, in which some of the basic mechanical properties defined in Chapter 6 appear as parameters in an engineering handbook. A broader view of the role of material properties as a link between materials science and materials engineering is given in Figure 20–2. Also, Figure 20–3 illustrates the integral relationship among materials (and their properties), the processing of the materials, and the effective use of the materials in engineering design.

A convenient format for facilitating materials selection is to have tabular data listed in order of increasing property values. For example, Table 20.1 gives the tensile strength of various tool steels listed in order of increasing strength. Similarly, Table 20.2 gives the ductility as % elongation for the same set of tool steels listed in order of increasing ductility.

Figure 20-1   *The basic mechanical properties obtained from the tensile test introduced in Chapter 6 lead to a list of engineering design parameters for a given alloy. (The parameters are reproduced from a list in* ASM Handbook, *Vol. 2, ASM International, Materials Park, Ohio, 1990.)*

Specifications
UNS number. A92036

Chemical Comosition
Composition limits. 0.50 max Si; 0.50 max Fe; 2.2 max Cu; 0.10 to 0.40 Mn; 0.30 to 0.6 Mg; 0.10 max Cr; 0.25 max Zn; 0.15 max Ti; 0.05 max others (each); 0.15 max others (total); bal Al

Applications
Typical uses. Sheet for auto body panels

Mechanical Properties

Tensile properties. Typical, for 0.64 to 3.18 mm (0.025 to 0.125 in.) flat sheet, T4 temper; tensile strength, 340 MPa (49 ksi); yield strength, 195 MPa (28 ksi); elongation, 24% in 50 mm or 2 in. Minimum, for 0.64 to 3.18 mm flat sheet, T4 temper; tensile strength, 290 MPa (42 ksi); yield strength, 160 MPa (23 ksi); elongation, 20% in 50 mm or 2 in.

Hardness. Typical, T4 temper: 80 HR15T strain-hardening exponent, 0.23

Elastic modulus. Tension, 70.3 GPa (10.2×10^6 ksi); compression, 71.7 GPa (10.4×10^6 ksi)

Fatigue strength. Typical, T4 temper: 124 MPa (18 ksi) at 10^7 cycles for flat sheet tested in reversed flexure

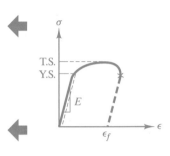

Figure 20-2   *Schematic illustration of the central role played by properties in the selection of materials. Properties are a link between the fundamental issues of materials science and the practical challenges of materials engineering. (From G. E. Dieter, in ASM Handbook, Vol. 20: Materials Selection and Design, ASM International, Materials Park, Ohio, 1997, p. 245.)*

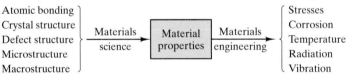

Figure 20-3   *Schematic illustration of the integral relationship among materials, the processing of those materials, and engineering design. (From G. E. Dieter, in ASM Handbook, Vol. 20: Materials Selection and Design, ASM International, Materials Park, Ohio, 1997, p. 243.)*

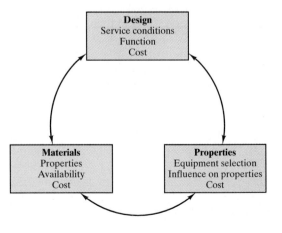

Table 20.1 *Selecting Tensile Strength of Tool Steels*

Type	Condition	Tensile strength (MPa)
S7	Annealed	640
L6	Annealed	655
S1	Annealed	690
L2	Annealed	710
S5	Annealed	725
L2	Oil quenched from 855°C and single tempered at 650°C	930
L6	Oil quenched from 845°C and single tempered at 650°C	965
S5	Oil quenched from 870°C and single tempered at 650°C	1035
S7	Fan cooled from 940°C and single tempered at 650°C	1240
L2	Oil quenched from 855°C and single tempered at 540°C	1275
L6	Oil quenched from 845°C and single tempered at 540°C	1345
S5	Oil quenched from 870°C and single tempered at 540°C	1520
L2	Oil quenched from 855°C and single tempered at 425°C	1550
L6	Oil quenched from 845°C and single tempered at 425°C	1585
L2	Oil quenched from 855°C and single tempered at 315°C	1790
S7	Fan cooled from 940°C and single tempered at 540°C	1820
S5	Oil quenched from 870°C and single tempered at 425°C	1895
S7	Fan cooled from 940°C and single tempered at 425°C	1895
S7	Fan cooled from 940°C and single tempered at 315°C	1965
L2	Oil quenched from 855°C and single tempered at 205°C	2000
L6	Oil quenched from 845°C and single tempered at 315°C	2000
S7	Fan cooled from 940°C and single tempered at 205°C	2170
S5	Oil quenched from 870°C and single tempered at 315°C	2240
S5	Oil quenched from 870°C and single tempered at 205°C	2345

Table 20.2 *Selecting Elongation of Tool Steels*

Type	Condition	Elongation (%)
L6	Oil quenched from 845°C and single tempered at 315°C	4
L2	Oil quenched from 855°C and single tempered at 205°C	5
S5	Oil quenched from 870°C and single tempered at 205°C	5
S5	Oil quenched from 870°C and single tempered at 315°C	7
S7	Fan cooled from 940°C and single tempered at 205°C	7
L6	Oil quenched from 845°C and single tempered at 425°C	8
S5	Oil quenched from 870°C and single tempered at 425°C	9
S7	Fan cooled from 940°C and single tempered at 315°C	9
L2	Oil quenched from 855°C and single tempered at 315°C	10
S5	Oil quenched from 870°C and single tempered at 540°C	10
S7	Fan cooled from 940°C and single tempered at 425°C	10
S7	Fan cooled from 940°C and single tempered at 540°C	10
L2	Oil quenched from 855°C and single tempered at 425°C	12
L6	Oil quenched from 845°C and single tempered at 540°C	12
S7	Fan cooled from 940°C and single tempered at 650°C	14
L2	Oil quenched from 855°C and single tempered at 540°C	15
S5	Oil quenched from 870°C and single tempered at 650°C	15
L6	Oil quenched from 845°C and single tempered at 650°C	20
S1	Annealed	24
L2	Annealed	25
L2	Oil quenched from 855°C and single tempered at 650°C	25
L6	Annealed	25
S5	Annealed	25
S7	Annealed	25

Source: (Tables 20.1 and 20.2) Data from *ASM Metals Reference Book, Second Edition,* American Society for
Metals, Metals Park, Ohio, 1984, as re-configured in J. F. Shackelford, W. Alexander, and J. S. Park,
CRC Practical Handbook of Materials Selection, CRC Press, Boca Raton, Florida, 1995.

A global view of the relative behavior of different classes of engineering materials is given by charts such as Figure 20–4, in which pairs of materials properties are plotted against each other. In Figure 20–4. elastic modulus,

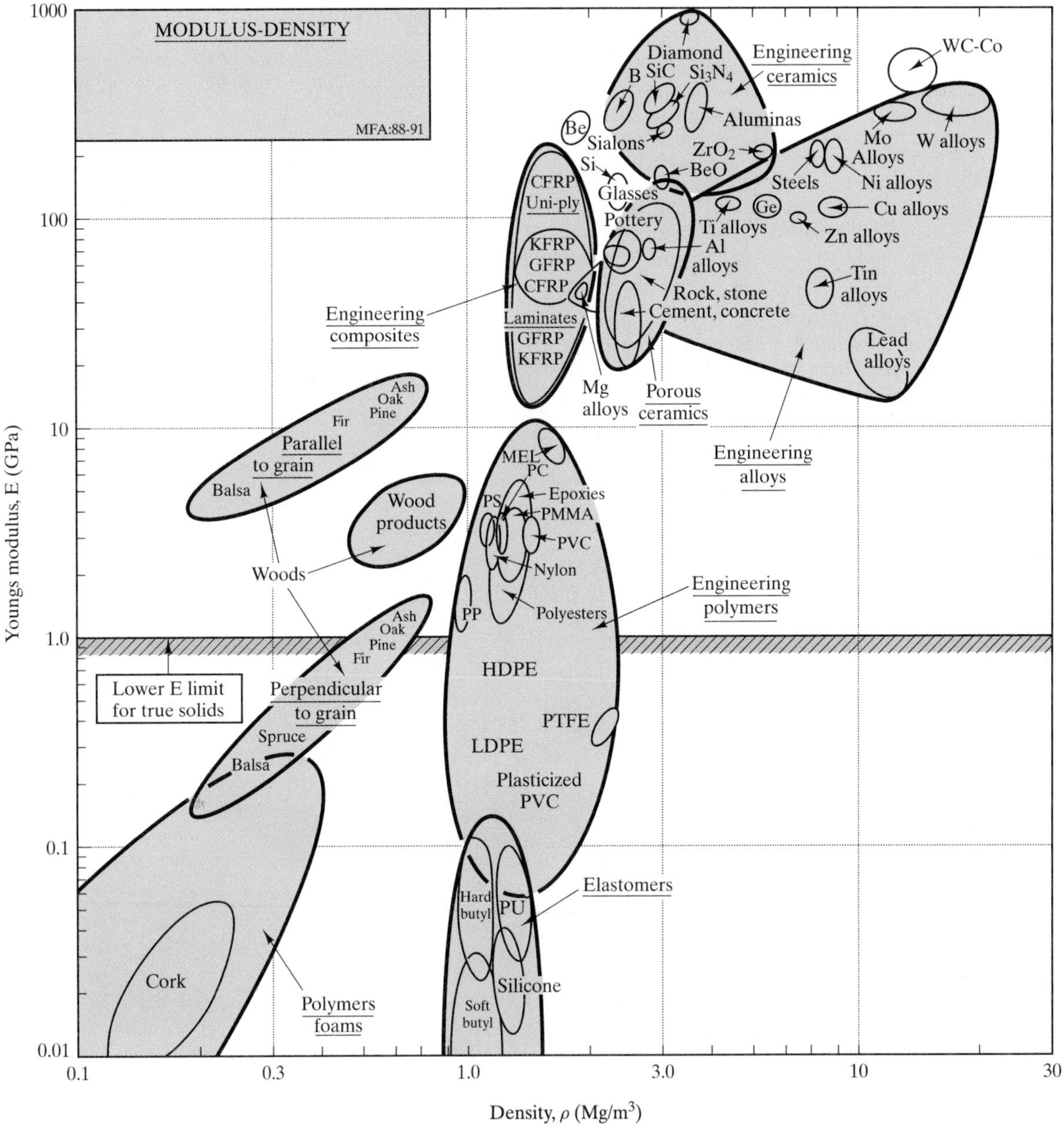

Figure 20-4 *A materials property chart with a global view of relative materials performance. In this case, plots of elastic modulus and density data (on logarithmic scales) for various materials indicate that members of the different categories of structural materials tend to group together. (After M. F. Ashby, Materials Selection in Engineering Design, Pergamon Press, Inc., Elmsford, N.Y., 1992.)*

E, is plotted against density, ρ, on logarithmic scales. Clearly, the various categories of structural materials tend to group together with, for example, the modulus-density combination for metal alloys generally being distinct from ceramics and glasses, polymers, and composites.

D SAMPLE PROBLEM 20.1

In selecting a tool steel for a machining operation, the design specification calls for a material with a tensile strength of $\geq$ 1500 MPa and a % elongation of $\geq$ 10%. Which specific alloys would meet this specification?

SOLUTION

By inspection of Tables 20.1 and 20.2, we see that the following alloy/heat treatment combinations would meet these specifications:

Type	Condition
S5	Oil quenched from 870°C and single tempered at 540°C
L2	Oil quenched from 855°C and single tempered at 425°C
L2	Oil quenched from 855°C and single tempered at 315°C
S7	Fan cooled from 940°C and single tempered at 540°C
S7	Fan cooled from 940°C and single tempered at 425°C

D PRACTICE PROBLEM 20.1

In reordering tool steel for the machining operation discussed in Sample Problem 20.1, you notice that the design specification has been updated. Which of the alloys would meet the new criteria of a tensile strength of $\geq$ 1800 MPa and a % elongation of $\geq$ 10%?

20.2   SELECTION OF STRUCTURAL MATERIALS—CASE STUDIES

In Chapter 1 we introduced the concept of materials selection by outlining the steps leading to the choice of a metal alloy for a high-pressure cylinder (see Figure 1–24). The selection process begins with the choice of a category of structural material (metal, ceramic, polymer, or composite). Once the metals category is decided upon, it is necessary to choose the optimal alloy for the given application. With the wide range of commercial materials and properties (parameters) introduced in Parts I and II, we now have a clearer idea of our range of selection. The foundation for the fine-tuning of structural design parameters was laid in Chapters 9 and 10. In general, we look for an optimal balance of strength and ductility for a given application. Some of the common examples of composite materials in Chapter 14 provided this balance by the combination of a strong (but brittle) dispersed phase with a ductile (but weak) matrix. Many of the polymers of Chapter 13 provide adequate mechanical properties in combination with both formability and modest cost. Materials selection, however, can be more than an objective consideration of design parameters. The subjective factor of consumer appeal can play an equal part. In 1939 nylon 66 stockings were introduced, with 64 million pairs being sold in the first year alone.

When ductility is not essential, traditional, brittle ceramics (Chapter 12) can be used for their other attributes, such as high-temperature resistance or chemical durability. Many glasses (Chapter 12) and polymers (Chapter 13) are selected for their optical properties, such as transparency and color. Also, these traditional considerations are being modified by the development of new materials, such as high fracture toughness ceramics.

No material selection is complete until the issues of failure prevention and environmental degradation are taken into account (see Chapters 8 and 19). Consideration of the design configuration is important, (e.g., whether stress-concentrating geometries or fatigue-producing cyclic loads are specified.) Chemical reaction, electrochemical corrosion, radiation damage, and wear can eliminate an otherwise attractive material.

The design selection process can be illustrated by some specific case studies. We shall look in some detail at the materials selection for windsurfer masts and more briefly at a variety of other case studies.

MATERIALS FOR WINDSURFER MASTS

Windsurfing is a relatively young sport that has become widely popular due to a successful merger of mechanical and materials engineering in the overall design process. Figure 20–5 illustrates the basic components of a windsurfer. Although the idea of a "freesail" was put forward in the mid–1960s, a practical device awaited the development of a universal joint that allowed the mast to swivel freely about the surfboard. The mast is central to the design, and the selection of its specific shape and material of construction follows from a careful consideration of design criteria. The mast influences the dynamics of the sail and must flex under wind pressure. The resulting need to specify stiffness must be done in conjunction with a limit on outer diameter in order to reduce its influence on air flow. For stability, the mass of the mast must be minimized. The design criteria for windsurfer masts are summarized in Table 20.3.

The length L, of a typical windsurfer mast is 4.6 m. The mast stiffness, S, is defined by

$$S = \frac{P}{\delta} \tag{20.1}$$

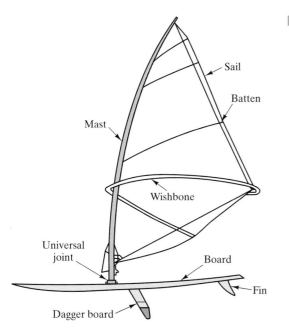

Figure 20-5 *Components of a windsurfer design. The stiffness of the mast controls the sail shape, and the pivoting of the mast about the universal joint controls the response of the craft. (After M. F. Ashby, "Performance Indices," in ASM Handbook, Vol. 20: Materials Selection and Design, ASM International, Materials Park, Ohio, 1997, pp. 281–290.)*

where P is the load created by hanging a weight at the midpoint of the mast supported horizontally at its ends and δ is the resulting deflection. As a practical matter, the stiffness of masts is characterized by the International Mast Check System (IMCS) number defined as

$$\text{IMCS number} = \frac{L}{\delta} \qquad (20.2)$$

with the deflection, δ, produced by a standard weight of 30 kg. The IMCS number ranges between 20 (for a soft mast) and 32 (for a hard one).

The limit on the outer radius of the mast, r_{max}, can be stated in a formal way. The outer radius of the mast, r, must be

$$r \leq r_{max} \qquad (20.3)$$

Table 20.3 *Design Criteria for Winsurfer Masts*[a]

Criterion	Constraint	Basis of constraint
Stiffness	Specify	Sailing characteristics
Outer diameter	Limit	Reduce influence on air flow
Mass	Minimize	Stability

[a] After M. F. Ashby, "Performance Indices," in *ASM Handbook, Vol. 20: Materials Selection and Design*, ASM International, Materials Park, Ohio, 1997, pp. 281–290.

For lightness, the mast is hollow, and the mass, m, which is minimized in the design is given by

$$m = AL\rho \tag{20.4}$$

where A is the cross-sectional area of the hollow mast, ρ is the density, and L was defined relative to Equation 20.2 A typical mast weighs between 1.8 and 3.0 kg.

Mast design is a specific application of a thin-walled tube, for which we can define a shape factor, ϕ, as

$$\phi = \frac{r}{t} \tag{20.5}$$

where t is the wall thickness and r was defined relative to Equation 20.3. An analysis of the mechanics of the thin-walled tube shows that the mass is related to the shape factor and the material properties elastic modulus, E, and density by

$$m = B\left(\frac{\rho}{[\phi E]^{1/2}}\right) \tag{20.6}$$

where B is a constant specific to design loading conditions and includes the bending stiffness. Equation 20.6 indicates that the mass can be minimized by maximizing a performance index, M, defined as

$$M = \frac{(\phi E)^{1/2}}{\rho} \tag{20.7}$$

Table 20.4 summarizes the values of the shape factor, ϕ, performance index, M, and minimum mass, m, for a range of candidate mast materials for a midrange IMCS number of 26 and a maximum radius of $r_{max} = 20$ mm. The optimal materials tend to be those with the largest values of M. Furthermore, we can modify Equation 20.7 so that

$$M = \frac{(1/\phi)(\phi E)^{1/2}}{(1/\phi)\rho} = \frac{(E/\phi)^{1/2}}{(\rho/\phi)} \tag{20.8}$$

Then the E/ϕ term can be given by a logarithmic expression by noting that

$$M^2(\rho/\phi)^2 = E/\phi \tag{20.9}$$

or

$$\ln(E/\phi) = 2\ln(\rho/\phi) + 2\ln M \tag{20.10}$$

Table 20.4 *Design Results for Windurfer Mast Materials*[a]

Material	Shape factor $\phi(= r/t)$	Performance index M (GPa$^{1/2}$/[Mg/m^3])	Mass[b] m (kg)
Carbon-fiber reinforced polymer (CFRP)	14.3	22.9	2.0
Wood (spruce)	1.7	9.0	5.0
Aluminum	8.0	8.5	5.3
Glass-fiber reinforced polymer (GFRP)	4.3	6.2	7.3

[a] After M. F. Ashby, "Performance Indices," in *ASM Handbook, Vol. 20: Materials Selection and Design,* ASM International, Materials Park, Ohio, 1997, pp. 281–290.

[b] For $r_{max} = 20$ mm and International Mast Check System (IMCS) number $= 26$.

The practical consequence of Equation 20.10 is that the behavior of a specific design geometry (the thin-walled tube) can be superimposed on the materials property chart of Figure 20–4 normalized by the shape factor, ϕ, as shown in Figure 20–6. For example, the carbon fiber-reinforced polymer (CFRP) location for the bulk material in Figure 20–4 corresponds to a shape factor of $\phi = 1$. In Figure 20–6, that position is connected to the tubing position corresponding to E/ϕ and ρ/ϕ, where $\phi = 14.3$ as given in Table 20.4.

Finally, we should acknowledge that windsurfer design blurs the boundary between art and science. For example, the mast is routinely "tuned" by varying the stiffness along its length in order to match the weight of the surfer as well as the type of windsurfing (slalom, race, or wave).

METAL SUBSTITUTION WITH A POLYMER

The increasing trend of replacement of metal parts by engineering polymers has been emphasized repeatedly in Chapters 1 and 13. An example is given in Figure 20–7, which shows a motocross (racing motorcycle) drive sprocket made of a dispersion-toughened nylon. The nylon product has become widely used due largely to reduced chain breakage. Tensile stresses on the drive chain can reach 65 MPa (9.4 ksi) and greater (during impact loading). Improved performance is related to a combination of high toughness and impact strength.

The drive sprocket is machined from an injection-molded disk, 13.7 mm thick with a diameter that may range from 130 to 330 mm. Additional attractive features include increased resistance to corrosion and attack by most solvents and lubricants, as well as resistance to wear. (Chain wear is similarly reduced.) A 0.34-kg nylon sprocket replaces a 0.45-kg aluminum alloy or a 0.90-kg steel. The cost of the nylon product is comparable to the aluminum but approximately one-third less expensive than the steel.

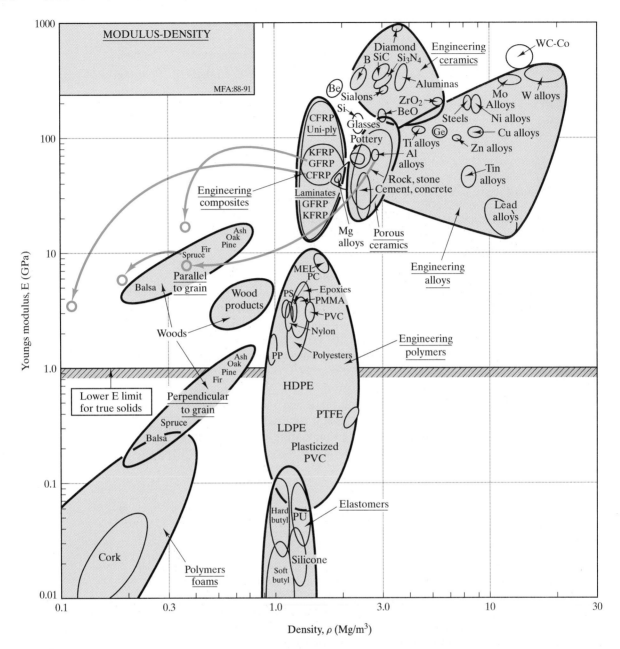

Figure 20-6 *The behavior of the windsurfer mast materials in Table 20.4 are superimposed on the* ln *E versus* ln ρ *chart of Figure 20–4 normalized by the shape factor, ϕ, of a thin-walled tube. For example, the CFRP mast with a shape factor of $\phi = 14.3$ is shown at a position of $(E/14.3, \rho/14.3)$ relative to the (E, ρ) position of the bulk material for which $\phi = 1$. (After M. F. Ashby, "Performance Indices," in ASM Handbook, Vol. 20: Materials Selection and Design, ASM International, Materials Park, Ohio, 1997, pp. 281–290.)*

Figure 20-7 *A drive sprocket made from dispersion-toughened nylon has replaced aluminum and steel parts in many motocross racing designs. (Courtesy of the Du Pont Company, Engineering Polymers Division)*

METAL SUBSTITUTION WITH COMPOSITES

A key example of the driving force for replacing metals with lower density composites is in the commercial aircraft industry. Manufacturers had developed parts of fiberglass for improved dynamics and cost savings by the early 1970s. In the mid–1970s, the "oil crisis" led to a rapid rise in fuel costs, from 18% of direct operating costs to 60% within a few years. (One kilogram of "dead weight" on a commercial jet aircraft can consume 830 liters of fuel per year.) An early response to the need for materials substitution for fuel savings was the use of over 1100 kg of Kevlar-reinforced composites in the Lockheed L–1011–500 long-range aircraft. The result was a net 366-kg weight saving on the secondary exterior structure. Similar substitutions were made later on all L–1011 models. An excellent example of this effort is the design of the Boeing 767. Figure 20–8 illustrates this case. A significant fraction of the exterior surface consists of advanced composites, primarily with Kevlar and graphite reinforcements. The resulting weight savings using advanced composites is 570 kg.

HONEYCOMB STRUCTURE

The honeycomb made by bees has inspired a structural configuration widely used in modern engineering design. The **honeycomb structure** illustrated in Figure 20–9 originated in the 1940s for aircraft sandwich panel construction.

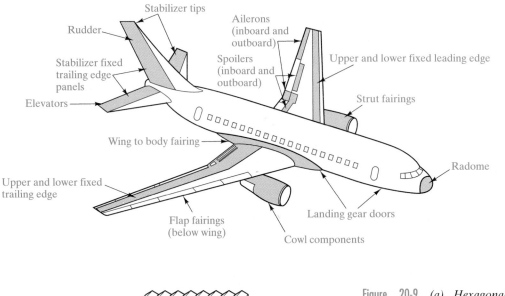

Figure 20-8 *Schematic illustration of the composite structural applications for the exterior surface of a Boeing 767 aircraft. (After data from the Boeing Airplane Company.)*

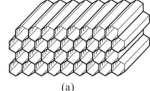

(a)

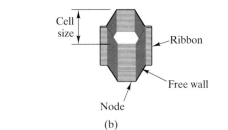

(b)

Figure 20-9 *(a) Hexagonal cell honeycomb is composed of (b) individual cells composed of adhesively bonded layers which are (c) subsequently bonded to face sheets to form the overall sandwich panel. (After J. Corden, "Honeycomb Structure," in Engineered Materials Handbook, Vol. 1, Composites, ASM International, Metals Park, Ohio, 1987, p. 721.)*

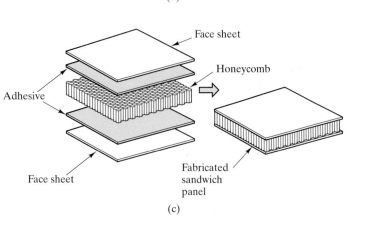

(c)

The adhesively bonded hexagonal core is subsequently bonded to face sheets to form the sandwich panel. The honeycomb structure is extremely light weight and exhibits high levels of stiffness and strength-to-weight ratios. This popular aircraft design is sometimes referred to as a **structural composite**, although it falls outside the formal definition of *composite* used in Chapter 14, which required the different material components to be distributed on a microscopic scale.

Sandwich structures are designed to meet a range of structural criteria as summarized in Figure 20–10. Cost effectiveness and durability in a given service environment are additional criteria.

In addition to sandwich panels for aircraft and architectural applications, honeycomb structure is used in various engineering designs for energy absorption, radio frequency shielding, light diffusion, and the directing of air flows. The honeycomb is manufactured from a wide variety of materials. The Chinese made paper honeycomb about 2000 years ago. Paper and other nonmetallic materials (including graphite, aramid polymer, and fiberglass) are still used. Common metallic core materials are aluminum, corrosion-resistant steel, titanium, and nickel-based alloys.

CERAMIC TILE FOR THE SPACE SHUTTLE

One of the most demanding applications for any structural material is the thermal protection system for the National Aeronautics and Space Administration (NASA) Space Transportation System (STS), more commonly known as the Space Shuttle Orbiter. The Space Shuttle is a rocket-launched, reusable space vehicle that carries a wide variety of cargo, from scientific experiments to commercial satellites. At the end of an orbital mission, the space craft reenters the atmosphere and eventually lands in a manner similar to a normal aircraft. Since the maiden flight of the *Columbia* in April 1981, numerous missions have been carried out by a number of Shuttles. Upon reentry from space, the Shuttle experiences enormous frictional heating in the earth's atmosphere.

The successful development of a fully reusable outer skin to serve as a thermal protection system was a major part of the overall Shuttle design. High-performance thermal insulating materials previously available in the aerospace industry proved inadequate for the Space Shuttle's design specifications because they were either not reusable or were too dense. Skin temperatures can range from $-110°C$ to $+1500°C$. The inner airframe cannot rise above $175°C$. In addition, the insulation system should be reusable for 100 missions with a maximum turnaround time of 160 hours, provide an aerodynamically smooth outer surface, resist severe thermomechanical loads, and resist moisture and other atmospheric contaminants between missions. Finally, the thermal protection system must be attached to an aluminum alloy airframe.

A variety of specific materials has been used to provide the appropriate thermal insulation, depending on the local, maximum skin temperature. For relatively low temperature areas in which the temperature will not rise above

1. The facings should be thick enough to withstand the tensile, compressive, and shear stresses induced by the design load.

2. The core should have sufficient strength to withstand the shear stresses induced by the design loads. Adhesive must have sufficient strength to carry shear stress into core.

3. The core should be thick enough and have sufficient shear modulus to prevent overall buckling of the sandwich under load, and to prevent crimping.

4. Compressive modulus of the core and the compressive modulus of the facings should be sufficient to prevent wrinkling of the faces under design load.

5. The core cells should be small enough to prevent intracell dimpling of the facings under design load.

6. The core should have sufficient compressive strength to resist crushing by design loads acting normal to the panel facings or by compressive stresses induced through flexure.

7. The sandwich structure should have sufficient flexural and shear rigidity to prevent excessive deflections under design load.

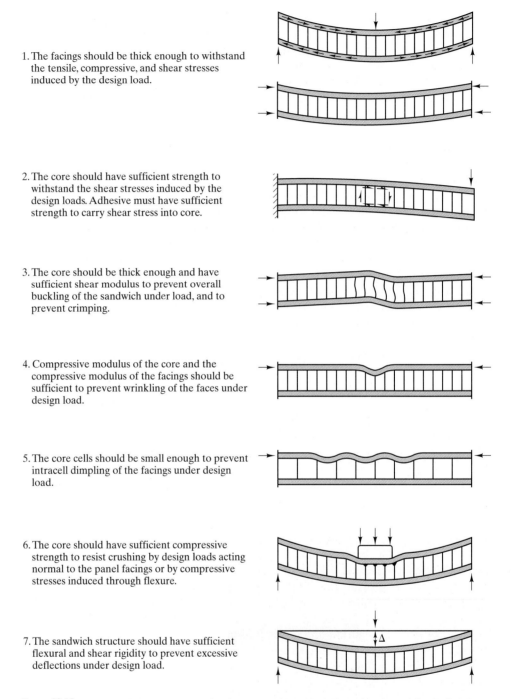

Figure 20-10 *Structural design criteria for honeycomb structural sandwich panels. (After J. Corden, "Honeycomb Structure," in Engineered Materials Handbook, Vol. 1, Composites, ASM International, Metals Park, Ohio, 1987, p. 727.)*

400°C, a felt reusable surface insulation (FRSI) is used and is composed of a nylon felt blanket in which the nylon fibers are coated with a silicone rubber. The nose and wing leading edges of the Shuttle experience extreme temperatures (above 1260°C), for which a reinforced carbon-carbon (RCC) composite is used. A carbon matrix is reinforced with graphite fibers, and the overall surface is coated with a thin layer of SiC to protect against oxidation.

Approximately 70% of the Shuttle surface, however, must protect against temperatures between 400°C and 1260°C. For this major portion of the thermal protection system, ceramic tiles are used. For the temperature range of 400°C to 650°C, a low-temperature reusable surface insulation (LRSI) is used. LRSI tiles are generally composed of high-purity vitreous silica fibers with diameters between 1 and 4 μm and fiber lengths of approximately 3000 μm. Loose packings of these fibers are sintered together to form a highly porous and lightweight material, as shown in Figure 20–11. Ceramic and glass materials are inherently good thermal insulators, and combining the extremely high porosity (approximately 93% by volume) of the microstructure in Figure 20–11 results in exceptionally low thermal conductivity values. LRSI tiles are coated with a 300-μm-thick layer of borosilicate glass. These characteristic white tiles effectively reflect the sun's rays while in orbit, thereby maintaining a moderate temperature for the Orbiter crew and cargo.

For the vehicle underbody and tail leading and trailing edges, a temperature range of 650°C to 1260°C is experienced, requiring a high-temperature reusable surface insulation (HRSI). HRSI tiles are also composed of sintered, high-purity vitreous silica fibers, but they are coated with a layer composed of both borosilicate glass and silicon tetraboride, SiB_4. The additional tetraboride gives a characteristic black color. The tile's high optical emittance causes 90% of the heat generated upon reentry to be radiated away from the tile surface. Figure 20–12 summarizes the distribution of the various components of the thermal protection system.

Figure 20-11 *A scanning electron micrograph of sintered silica fibers in a Space Shuttle Orbiter ceramic tile. (Courtesy of Daniel Leiser, National Aeronautics and Space Administration [NASA])*

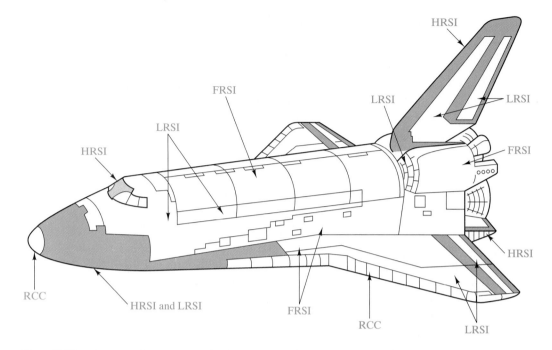

Figure 20-12 *Schematic illustration of the distribution of the components of the thermal protection system for the Space Shuttle Orbiter: felt reusable surface insulation (FRSI), low-temperature reusable surface insulation (LRSI), high-temperature reusable surface insulation (HRSI), and reinforced carbon-carbon composite (RCC). (After L. J. Korb, et al.,* Bull. Am. Ceram. Soc. 61, *1189 [1981].)*

Each ceramic tile must be custom-fitted to the local configuration of the Shuttle exterior and is shaped by diamond-machining. Figure 20–13 shows typical tiles approximately 200 mm × 200 mm × 50 mm in dimension. The Space Shuttle Orbiter has over 30,000 of these ceramic tiles. It is important to note that the tile design includes a strain isolator pad (SIP) to cushion this

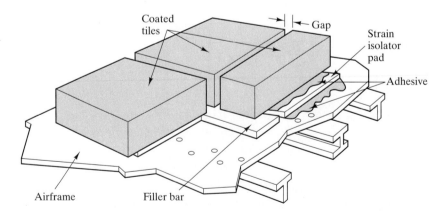

Figure 20-13 *Schematic of a typical ceramic tile configuration in the thermal protection system for the Space Shuttle Orbiter. (After L. J. Korb, et al.,* Bull. Am. Ceram. Soc. 61, *1189 [1981].)*

brittle ceramic from thermomechanical loads. The filler bar is composed of nylon felt and seals the SIP from water or plasma penetration. Room-temperature vulcanizing (RTV) silicone adhesive is used to bond the tile to the SIP and, in turn, the SIP and filler bar to the airframe.

Finally, it is worth noting that we refer to these tiles as "ceramic" even though their central component is generally a glass (vitreous silica fibers). This is because of the common tendency in industry to include "glass" as a subset of "ceramics," as well as the fact that some tiles use alumino-borosilicate fibers, which can devitrify to become a true crystalline ceramic.

MATERIALS FOR HIP JOINT REPLACEMENT

Some of the most dramatic developments in the applications of advanced materials have come in the field of medicine. One of the most successful has been the artificial hip **prosthesis** (a device for replacing a missing body part). Figure 20–14 illustrates the surgical procedure involved in replacing a damaged or diseased hip joint and replacing it with a prosthesis. Figure 20–15 shows a typical example, a cobalt-chrome alloy (e.g., 50 wt % Co, 20 wt % Cr, 15 wt % W, 10 wt % Ni, and 5 wt % other elements) constituting the main stem and head, with an ultrahigh-molecular-weight polyethylene (with a molecular weight of 1 to 4×10^6 atomic mass units) cup completing the ball and socket system. The term total hip replacement (THR) refers to the

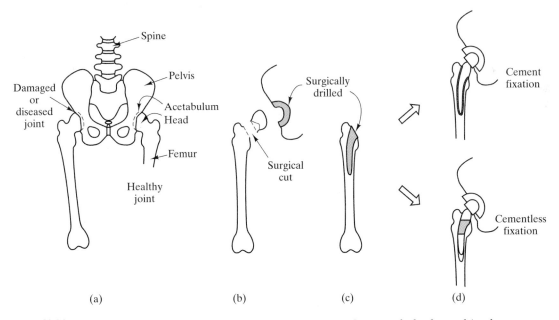

Figure 20-14 *Schematic of the total hip replacement (THR) surgery. In general, the femoral implant stem is anchored to the bone by either a thin layer (a few mm thick) of polymethylmethacrylate (PMMA) cement or a cementless system involving a snug fit of the stem in the femoral shaft. In typical cementless fixation, the upper one-third of the stem is covered with a porous coating of sintered metal alloy beads. Bone growth into the porous surface provides a mechanical anchoring.*

Figure 20-15 *A cobalt–chrome stem and ball, with a polyethylene cup, form a ball and socket system for an artificial hip joint. (Courtesy of DePuy, a Division of Boehringer Mannheim Corporation.)*

simultaneous replacement of both the ball and socket with engineered materials. The orthopedic surgeon removes the degenerative hip joint and drills out a cavity in the femoral bone to accommodate the stem. The stem is either anchored to the skeletal system with a polymethylmethacrylate (PMMA) cement or by bone ingrowth into a porous surface coating ("cementless fixation"). The cup is generally attached to a metal backing, which, in turn, is attached to the hip by metallic bone screws. The titanium alloy Ti–6Al–4V (Table 6.1) is generally preferred for the stem in cementless applications. The elastic modulus of Ti–6Al–4V (1.10×10^5 MPa) is closer to that of bone (1.4×10^4 MPA) than is cobalt-chrome (2.42×10^5 MPa) and, therefore, creates less stress on the bone due to modulus mismatch. (Cobalt-chrome is preferred for cemented implants for a similar reason. The lower elastic modulus of Ti–6Al–4V leads to excessive load on the interfacial cement.)

The metal/polymer interface provides a low-friction contact surface, and each material (the metal and the polymer) has good resistance to degradation by highly corrosive body fluids. Early artificial hip designs with cement fixation had a typical lifetime of 5 years (limited primarily by mechanical loosening), adequate for elderly patients but requiring painful replacement surgeries for younger people. The cementless fixation can extend the implant lifetime by a factor of three. More than 250,000 THR surgeries are performed in the United States each year, with a similar number in Europe.

The metal alloys and polymers involved in the total hip replacement are examples of **biomaterials,** which can be defined as engineered materials created for applications in biology and medicine. Biomaterials can be contrasted with the naturally occurring bone, which would be an example of a **biological material.** Both of these new terms are reminiscent of the biomimetic materials introduced in Chapter 12. Some of those engineered materials being produced by low-temperature, liquid-phase processing routes are prime candidates to be the next generation of biomaterials.

For roughly three decades, ceramics and glasses have been the focus of substantial research on their potential applications as biomaterials. The compressive load on the THR ball makes a high-density structural ceramic, such as Al_2O_3, a good candidate for that application. An attractive feature of this substitution would be the typically low surface wear of structural ceramics. In a similar way, the THR cup can be fabricated from Al_2O_3. As with many potential applications for ceramics in engineering designs, the inherent brittleness and low fracture toughness of these materials have limited their use in the THR.

In the past decade, however, a significant biomedical application has been found for one ceramic system. Although the PMMA cement for fixation does not have a serious ceramic competitor, a ceramic substitute has been found for the traditional porous surfaces of cementless designs, viz., hydroxyapatite coatings. It is ironic that such an obvious candidate as hydroxyapatite has only recently come into fashion as a biomaterial. Hydroxyapatite, $Ca_{10}(PO_4)_6(OH)_2$, is the primary mineral content of bone, representing 43% by weight. It has the distinct physiochemical advantages of being stable, inert, and biocompatible. The most successful application of hydroxyapatite in biomedicine has been in the form of a thin coating on a prosthetic implant, as shown in Figure 20–16. These coatings have been plasma-sprayed on both Co-Cr and Ti–6 Al–4V alloys. Optimal performance has come from coating thicknesses on the order of 25–30 μm. Interfacial strengths between the implant and bone are as much as 5 to 7 times as great as with the uncoated specimens. The enhanced interfacial development corresponds to the mineralization of bone directly onto the hydroxyapatite surface with no signs of intermediate, fibrous tissue layers. Unlike the porous,

Figure 20-16 *The Omnifit® HA Hip Stem consists of hydroxyapatite coating on a hip replacement prosthesis for the purpose of improved adhesion between the prosthesis and bone. Hydroxyapatite is the predominant mineral phase in natural bone. (Courtesy of Osteonics, Allendale, New Jersey.)*

metallic coatings that they replace, this ceramic substitute does not have to be porous. The bone adhesion can occur directly on a smooth hydroxyapatite surface. The substantial success of this coating system has led to its widespread use in THR surgery.

SAMPLE PROBLEM 20.2

Estimate the annual fuel savings due to the weight reduction provided by composites in a fleet of 50 767 aircraft owned by a commercial airline.

SOLUTION

Using the information from the discussion of the case study in Section 20.2, we have

$$\text{fuel savings} = (\text{wt savings/aircraft}) \times \frac{(\text{fuel/year})}{(\text{wt savings})} \times 50 \text{ aircraft}$$

$$= (570 \text{ kg}) \times (830 \text{ l/yr})/\text{kg} \times 50$$

$$= 23.7 \times 10^6 \text{ l}$$

PRACTICE PROBLEM 20.2

An annual fuel savings is calculated in Sample Problem 20.2. For this commercial airline, estimate the fuel savings that would have been provided by a fleet of 50 L–1011 aircraft. (See the case study in Section 20.2.)

The Material World: The Future Car

The 1997 FutureCar Challenge was a contest in which twelve schools were selected to compete against each other while matching government and industry efforts to create a midsize family sedan that attains 2.9 liters/100 km (80 mpg) fuel consumption and maintains the performance, utility, and cost of a conventional car. Shown here is Joule, the first place winner from the University of California, Davis. Joule is a modified 1996 Ford Taurus converted to a hybrid electric vehicle by students of the UC Davis Future-Car Project under the supervision of Professor Andrew Frank. Joule is driven using an electric motor and/or an internal combustion engine. The two power sources are combined in a parallel hybrid powertrain. The electric motor and the engine drive a common input shaft to the manual transmission. An onboard computer dictates the use of each power source.

The electric motor is used exclusively below 24 km/h (15 mph). Above this speed, the engine provides the primary power, but the electric motor augments engine power during hard accelerations and at speeds above 120 km/h (75 mph). The power switching is transparent to the driver. The overall power strategy is designed to maximize energy efficiency while minimizing emissions.

Joule's fuel consumption was 3.7 liters/100 km (63 mpg) on the highway and 5.6 liters/100 km (42 mpg) in the city. Although below the ultimate goal, this represents a doubling of the stock vehicle's fuel economy while maintaining comparable performance. For example, Joule can accelerate from 0 to 100 km/h in 12.0 seconds compared to 12.5 seconds for the stock vehicle.

Materials selection is playing a central role in the automobile industry for improving fuel economy by decreasing vehicle weight. Joule's success was aided by this philosophy. The stock body was substantially reshaped by the use of carbon fiber-reinforced polymer in place of stamped steel panels. In addition to reducing weight, the more streamlined shape improved air flow. Many other lightweight materials were used in Joule. The headlights and turn signals were covered with polycarbonate lenses flush with the body. Besides being light in weight, the electric battery boxes had to be strong, compact, and protect occupants in case of an accident. The boxes were constructed of 5052 and 6061-T6 aluminum alloys and lined with Phenolic G10, a glass fiber-reinforced polymer. The overall curb weight of Joule was 1150 kg compared to 1500 kg for the stock vehicle

(Courtesy of the University of California, Davis)

20.3 SELECTION OF ELECTRONIC, OPTICAL, AND MAGNETIC MATERIALS — CASE STUDIES

In Chapter 15 we saw that solid-state materials are naturally sorted into one of the three categories: conductor, insulator, or semiconductor. Conductor selection is frequently determined by formability and cost as much as by specific conductivity values. The selection of an insulator can also be dominated by these same factors. A ceramic substrate may be limited by its ability to bond to a metallic conductor, or a polymeric insulation for conductive wire may be chosen due to its low cost. In some cases, new concepts in engineering design can make possible radically new materials choices (e.g., optical glass fibers in Chapter 16).

In Chapter 17, we found that design and materials selection are combined in an especially synergistic way in the semiconductor industry. Materials engineers and electronics engineers must work together effectively to ensure that the complex pattern of n-and p-type semiconducting regions in a chip provides a microcircuit of optimal utility.

The outline of magnetic materials in Chapter 18 was a nearly chronological unfolding of expanding choices for the design engineer. The iron-silicon alloys largely replaced plain-carbon steels as soft magnets due to reduced Joule heating effects. Various alloys with large hysteresis loops were developed as hard magnets. The high resistivity of ceramics makes materials such as ferrites the appropriate choices for high-frequency applications. Developments in the area of superconducting magnets promise new options for engineering design applications.

As with the structural materials, we can obtain a fuller appreciation of the process of selecting electronic, optical, and magnetic materials by looking at a few representative case studies. We shall look in some detail at the use of amorphous metals for electric-power distribution and more briefly at a variety of other case studies.

AMORPHOUS METAL FOR ELECTRIC-POWER DISTRIBUTION

As noted in Section 18.4, the development of amorphous metals in recent decades has provided an attractive new choice for transformer cores. A key to the competitiveness of amorphous metals is the absence of grain boundaries, allowing for easier domain wall motion. High resistivity (which damps eddy currents) and the absence of crystal anisotropy also contribute to domain wall mobility. Ferrous "glasses" are among the most easily magnetized of all ferromagnetic materials. Figure 20–17 illustrates a transformer core application using an amorphous ferrous alloy. As seen in Table 18.2, these ferrous alloys have especially low core losses. As manufacturing costs for the amorphous ribbons and wires were reduced, the energy conservation due to low core losses led to commercial applications. Fe-B-Si cores were put in commercial service as early as 1982, providing electrical power to homes and industrial facilities. The replacement of grain-oriented silicon steel with amorphous metals in transformer cores reduced core losses by 75%.

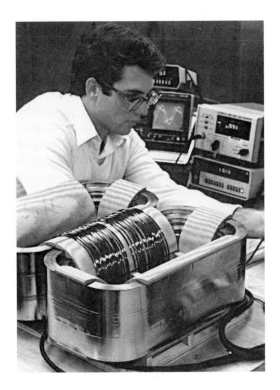

Figure 20-17   *Transformer core winding using an amorphous ferrous alloy wire. (Courtesy of Allied-Signal, Inc.)*

Given the large number of units in service and the continuous magnetization and demagnetization of the cores at line frequencies, transformers account for the largest portion of the energy losses in electric power distribution systems. It is estimated that more than 50×10^9 kW·h of electrical energy is dissipated annually in the United States in the form of core losses. At an average generating cost of $0.035/kW·h, these core losses are valued at more than $1.5 billion.

Since 1982, well over one million amorphous metal distribution transformers have been installed worldwide, providing substantial improvement in the efficiency of electric-power distribution. The substitution of Fe-based amorphous metals for grain-oriented silicon steel is not, however, without design challenges. The amorphous metal tends to be thinner, harder, and more fragile than the silicon steel. New amorphous metal transformer installations must be compatible with existing electric power distribution systems and must survive 30 years of continuous service. These fundamental requirements have affected the transformer manufacturing process and mandated laboratory and field testing to ensure that potential energy savings could be provided reliably at a practical cost.

An early observation of alloy compositions that could, at reasonable quench rates, form amorphous structures was that they tended to involve a metallic element alloyed with a metalloid. Furthermore, such systems tend to form a "deep eutectic" at a composition of approximately 20 atomic %

metalloid, where "deep eutectic" is defined as a low eutectic temperature compared to the melting points of the pure metal and metalloid. The classic example was an $Au_{80}Si_{20}$ alloy, which has a eutectic temperature of 363°C, in contrast to melting points of Au and Si of 1064°C and 1414°C, respectively. In effect, at a relatively low temperature these eutectic compositions provide liquid metal, which can more easily be "frozen" in the amorphous state. The association of amorphous metals with rapid solidification and eutectic compositions continues to be the basis of alloy design. Note that commercial amorphous metal transformer cores are generally 80 atomic % Fe, with the balance largely metalloids such as B, P, and C.

An amorphous $Fe_{80}P_{13}C_7$ alloy was produced in the laboratory in the late 1960s and the first commercial product, $Fe_{40}Ni_{40}P_{14}B_6$, known as METGLAS* 2826, was available in the early 1970s. Although the properties of this first product were optimized by annealing, relatively low saturation induction and critical temperature (above which it is nonmagnetic) limited its use to low-power, high-frequency applications. The development of amorphous alloys for electric power distribution have since focused primarily on Fe-based alloys. Table 20.5 represents distinct families of amorphous ferrous alloys in comparison to grain-oriented silicon steel. $Fe_{80}P_{13}C_7$ alloy offers the lowest raw materials cost by the use of P and C. $Fe_{80}B_{20}$ and $Fe_{86}B_8C_6$ alloys offer higher saturation induction by replacing P and C by B and using a higher Fe content. $Fe_{80}B_{11}Si_9$ alloy offers the best thermal stability (highest critical temperature). Thermal stability has proven to be a primary design consideration in the commercialization of amorphous alloys for electric-power distribution. The primary concern in this regard is the crystallization of the alloy either during processing or in service leading to diminished performance. As a result, $Fe_{80}B_{11}Si_9$ is the most commonly used amorphous alloy in power applications. Although its saturation induction is only 80% of that for the grain-oriented silicon steel, the amorphous alloy generates only 30% of the core loss.

Table 20.5 *Characteristics of Traditional Electrical Steel and Fe-Based Amorphous Metals*[a]

Material	Saturation induction B_s (Wb/m^2)	Critical temperature T_c (K)	Coercive force H_c (A/m)	Core loss @ 60 Hz, 1.4 Wb/m^2 CL (W/kg)
Grain-oriented Fe–3.2% Si	2.01	1019	24	0.7
Amorphous $Fe_{80}P_{13}C_7$	1.40	587	5	—
Amorphous $Fe_{80}P_{20}$	1.60	647	3	0.3
Amorphous $Fe_{86}P_8C_6$	1.75	< 600	4	0.4 (est.)
Amorphous $Fe_{80}P_{11}Si_9$	1.59	665	2	0.2

[a] After N. DeCristofaro, "Amorphous Metals in Electric-Power Distribution Applications," *MRS Bulletin, 23*, 50 (1998).

---

* METGLAS is a registered trademark of Allied Signal, Inc. (formerly Allied Chemical).

The processing of amorphous alloys has also provided numerous engineering challenges. To achieve the necessary $10^{5\circ}$ C/s quench rates to form these alloys by rapid solidification, at least one dimension must be small. The original laboratory demonstration of feasibility involved the splat-quenching of liquid droplets against a copper plate. Subsequent development led to the use of the chill block melt spinning technique, available since the 1870s for producing solder wire, in which a continuous stream of molten alloy is projected against the outer surface of a rotating drum. Refinements of this method led to the production of the initial commercial product ($Fe_{40}Ni_{40}P_{14}B_6$ alloy) in a continuous ribbon 50 μm thick and 1.7 mm wide. Furthermore refinements of this processing technology could only produce ribbons up to 5 mm in width. The production of transformer cores was significantly enhanced by the availability of wide sheets of alloy rather than narrow ribbon. Such sheets became possible with the development of planar flow casting, in which the molten alloy is forced through a slotted nozzle in close proximity ($\approx$ 0.5 mm) to the surface of a moving substrate. The melt puddle is then constrained between the nozzle and substrate, producing a stable, rectangular cross section. The planar flow casting technique has produced amorphous metal sheets up to 300 mm in width, with 210-mm widths commercially available.

The last quarter of the twentieth century has seen major changes in the global electric-power industry. After the 1973 oil embargo, electric utilities developed an understandable interest in higher-efficiency transformers, even as energy supplies and prices stabilized in the 1980s. In the United States, the Electric Power Research Institute (EPRI) focused attention on distribution transformers, typically mounted on utility poles or concrete pads (Figure 20–18). Distribution transformers step down electrical voltages from the 5–14

Figure 20-18   *Pole-mounted amorphous metal distribution transformer. (Courtesy of Allied Signal, Inc.)*

Table 20.6 *Environmental Impact of Amorphous Metal Transformers*[a]

Benefit	Country or Region				
	U.S.	Europe	Japan	China	India
Energy savings (10^9 kW·h)	40	25	11	9	2
Oil (10^6 barrels)	70	45	20	15	4
CO_2 (10^9 kg)	32	18	9	11	3
NO_x (10^6 kg)	100	63	27	82	20
SO_2 (10^6 kg)	240	150	68	190	47

[a] After N. DeCristofaro, "Amorphous Metals in Electric-Power Distribution Applications," *MRS Bulletin, 23*, 50 (1998).

kV range used for local transmission to the 120–240 V used in homes and businesses. Given the thinner geometry and more brittle mechanical behavior of amorphous alloys, the specific design of the transformer core has been modified somewhat in comparison to the traditional grain-oriented silicon steel. To similate the geometry of silicon steel sheet, thin amorphous sheets are "prespooled" in multiple layered packages. The resulting thick package makes handling and installation of the core substantially more practical.

Global political and economic forces are motivating electric power companies to reduce costs and improve service. These changes must be balanced, however, against substantial environmental concerns. The Third Conference of the Parties of the United Nations Framework Convention on Climate Change, held in Kyoto, Japan in December 1997, adopted an agreement known as the Kyoto Protocol to cut greenhouse gas emissions by 5% from 1990 levels. Table 20.6 indicates the potential benefits, in this regard, associated with amorphous metal distribution transformers. Note that in the United States alone an energy equivalent of 70 million barrels of oil could be saved along with an attendant reduction in CO_2, NO_x, and SO_2 gases.

Amorphous metal transformers are often more expensive than silicon steel models. The amorphous metal units can, however, be more cost-effective in many electric power systems. Utility engineers commonly use a "loss-evaluation" method, which includes economic factors such as transformer loading patterns, energy cost, inflation, and interest rates. Their goal to combine the initial transformer cost with the cost of operation creating a total owning cost (TOC) defined by

$$TOC = BP + (F_{CL} \times CL) + (F_{LL} \times LL) \qquad (20.11)$$

where BP is the bid price, F_{CL} is the core loss factor, CL is the core loss, F_{LL} is the load loss factor, and LL is the load loss (defined as the energy loss by the system other than the transformer core). Table 20.7 shows that for this case the amorphous metal transformer costs 15% more than its traditional counterpart but provides an overall 3% reduction in TOC.

Table 20.7 *Economic Comparison of Traditional Electrical Steel and Fe-Based Amorphous Metal Transformers*[a]

Distribution transformer @ 60 Hz, 500 kW (15 kV/480–277 V)	Amorphous metal core	Grain-oriented silicon steel core
1. Core loss (W)	230	610
2. Core loss factor ($/W)	$5.50	$5.50
3. Load loss (W)	3192	3153
4. Load loss factor ($/W)	$1.50	$1.50
5. Efficiency (%)	99.6	99.4
6. Bid price	$11,500	$10,000
7. Core loss value (1×2)	$1,265	$3,355
8. Load loss value (3×4)	$4,788	$4,730
9. Total owing cost $(6 + 7 + 8)$	$17,558	$18,085

[a] After N. DeCristofaro, "Amorphous Metals in Electric-Power Distribution Applications," *MRS Bulletin, 23,* 50 (1998).

REPLACEMENT OF A THERMOSETTING POLYMER WITH A THERMOPLASTIC

Although the emergence of engineering polymers has primarily been discussed as a challenge to traditional structural metals, one of the predominant traditional dielectrics is the group of thermosetting phenolics, such as phenol-formaldehyde (see Table 13.2). Since the introduction of these phenolics in 1905, they have been the routine material of choice for housings, terminal blocks, connectors, and the myriad other dielectric parts required by the electronics industry. Certain thermoplastics, however, have been developed with sufficiently competitive properties to provide the designer with an option. Figure 20–19 illustrates an application of polyethylene terephtha-

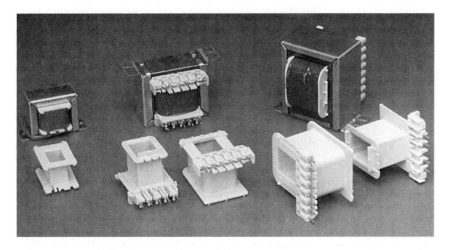

Figure 20-19 *Small transformer bobbins molded from a polyester thermoplastic are shown in the foreground. Wound, fully assembled transformers are in the background. (Courtesy of the Du Pont Company, Engineering Polymers Division.)*

late (PET), a polyester thermoplastic (see Table 13.1). This bobbin would traditionally have been fabricated from a phenolic thermoset. The polyester thermoplastic provides properties comparable to the thermoset (resistance to high-voltage arcing and hot-wire ignition, ability to withstand soldering heat, and strength to withstand the stress imposed by winding the conductive coil wire). Preference for the thermoplastic in this case is largely an economical one. Although the phenolic has a lower unit price than the polyester, the polyester allows a substantial saving in fabrication costs because of greater flexibility of thermoplastic processing (recall the discussions in Section 13.4).

METAL ALLOY SOLDER FOR FLIP-CHIP TECHNOLOGY

In Section 17.6, we were introduced to some common solid-state device technologies, including the use of wire bonds to connect the device to the outside world. An alternative technique using Pb-Sn solder balls allows a greater number of output leads in the device package. In wire bonding, the connecting pads are located along the edge of the chip. In "flip-chip" technology, solder balls are placed in an array over the face of the chip, allowing over 100 interconnect pads. Figure 20–20 illustrates how solder is melted into a ball (approximately 125 μm in diameter) at a space in the aluminum metallization. The chip is then flipped over and mated with a metallized ceramic substrate with a matching pattern of input/output (I/O) terminals.

Wire bonding involves solid-state joining mechanisms. Solder-ball technology, on the other hand, involves the melting of the alloy. (Recall

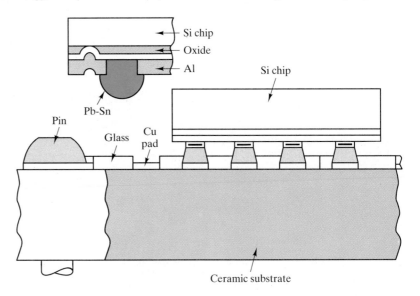

Figure 20-20 *A schematic illustration of a flip-chip solder bonded to a ceramic substrate. The enlarged view shows Pb-Sn solder prior to bonding. (From J. W. Mayer and S. S. Lau,* Electronic Materials Science: For Integrated Circuits in Si and GaAs, *Macmillan Publishing Company, New York, 1990.)*

that the eutectic temperature for the Pb-Sn system is 183°C.) The solder connection is to a Cu pad, which, of course, has a much higher melting point. A glass dam limits the flow of liquid solder.

Heat dissipation in the flip-chip configuration is through the solder joints to the substrate. This is in contrast to the case of a chip fully bonded to the substrate. The amount of heat dissipation depends on the number and size of solder joints.

LIGHT-EMITTING DIODE

In Section 16.2, we were introduced to a variety of mechanisms for luminescence, defined as light emission accompanying various forms of energy absorption. Electroluminescence was the term for electron-induced light emission. An important form of electroluminescence results when a forward-biased potential is applied across the p–n junction introduced in Section 17.6. Within a recombination region near the junction, electrons and holes can annihilate each other and emit photons of visible light. The emitted wavelength was given by Equation 16.8:

$$\lambda = hc/E_g \qquad (20.12)$$

The E_g and corresponding wavelength emitted are functions of the semiconductor composition with a variety of common examples given in Table 20.8. As a practical matter, the electron transition is from a small range of energies at the lower end of the conduction band to a small range of energies at the upper end of the valence band (Figure 20–21). The resulting range of values for E_g can produce a **spectral width** of light on the order of a few nm. Figure 20–22 shows both surface-and edge-emitting configurations for these **light-emitting diodes (LEDs)**. The construction of the ubiquitous digital display based on LEDs is illustrated in Figure 20–23. Many of the lasers introduced in Section 16.3 provide the carrier waves for most optical fiber networks and, in fact, are important examples of LEDs.

Table 20.8 *Common Light-Emitting Diode Compounds and Wavelengths*[a]

Compound	Wavelength (nm)	Color
GaP	565	Green
GaAsP	590	Yellow
GaAsP	632	Orange
GaAsP	649	Red
GaAlAs	850	Near IR
GaAs	940	Near IR
InGaAs	1060	Near IR
InGaAsP	1300	Near IR
InGaAsP	1550	Near IR

[a] (From R. C. Dorf, *Engineering Handbook*, CRC Press, Boca Raton, Fl, 1993, p. 751.)

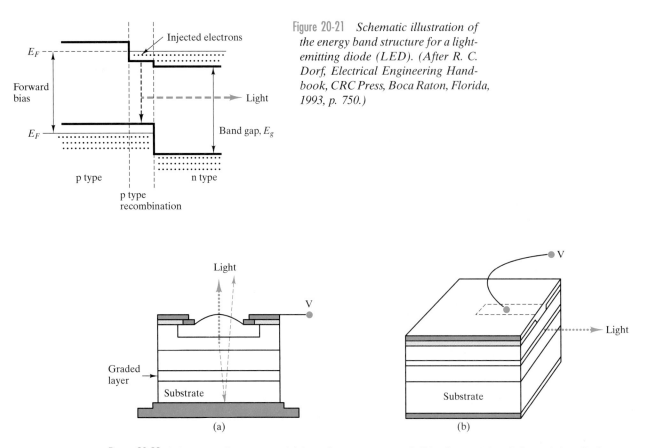

Figure 20-21 *Schematic illustration of the energy band structure for a light-emitting diode (LED). (After R. C. Dorf, Electrical Engineering Handbook, CRC Press, Boca Raton, Florida, 1993, p. 750.)*

Figure 20-22 *Schematic illustration of (a) surface emitting and (b) edge emitting light-emitting diodes (LEDs). (After R. C. Dorf, Electrical Engineering Handbook, CRC Press, Boca Raton, Florida, 1993, p. 750.)*

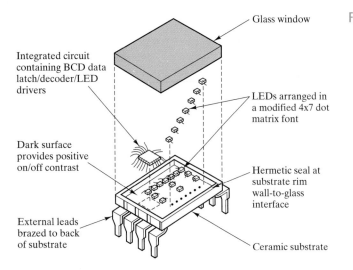

Figure 20-23 *Schematic illustration of a digital display employing an array of light-emitting diodes (LED). (From S. Gage et al., Optoelectronics/Fiber-Optics Applications Manual, 2nd ed., Hewlett-Packard/McGraw-Hill, New York, 1981.)*

PIEZOELECTRIC ACTUATOR AS A SMART MATERIAL

Among the most fascinating developments in materials science and engineering in the past decade has been the emergence of so-called **smart materials,** which have the ability to perform both sensing and responding functions. We traditionally associate this ability only with living systems, as opposed to inanimate matter. The distinguishing feature of smart materials is, then, their ability to imitate this fundamental aspect of life.

Advances in this active field of research are combining sophisticated sensors and responsive components with traditional materials, often blurring the interface between mechanism and material. An important example under active research is the use of embedded sensors in the structural members of bridges and aircraft. Sensing crack growth would be followed by crack-arresting responses.

An already well-established application of smart materials is illustrated in Figure 20–24. In this case, a so-called ceramic actuator is the central component of an impact dot-matrix printer head. In regard to smart materials performing sensing and response functions, this example concentrates on the response, and **actuator** can be defined as a device that carries out a function in response to an input signal. In this case, of course, the signal is coming in the form of a computer text or image to be printed. Each character formed in a typical printer is a 24 × 24 dot matrix. A printing ribbon is impacted by the multiwire array. The printing element is composed of a multilayer piezoelectric device involving 100 thin ceramic sheets. Each sheet is 100 μm thick. Many of the emerging smart material systems involve the piezoelectric ceramics introduced in Section 15.4. The hinged lever provides a

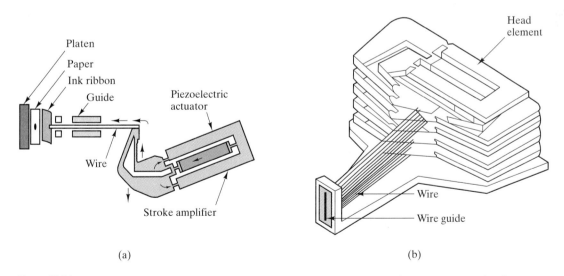

(a) (b)

Figure 20-24 *A schematic illustration of a ceramic actuator as an example of a smart material. The specific application is an impact dot-matrix printer. (a) Overall structure of the printer head. (b) Close-up of the multilayer piezoelectric printer-head element. (From K. Uchino,* MRS Bulletin, 18, *42 [1993].)*

displacement magnification of 30 times, resulting in a net motion at the ink ribbon of 0.5 mm with an energy transfer efficiency of $> 50\%$.

The advantages of the piezoelectric design application compared with conventional electromagnetic types are order-of-magnitude higher printing speed, order-of-magnitude lower energy consumption, and reduced printing noise. The reason for the last advantage is that the low level of heat generation by the piezoelectric allows greater sound shielding.

POLYMER AS AN ELECTRICAL CONDUCTOR

In Chapters 13 and 15 as well as previously in this section, we have seen the conventional role of polymers as electrical insulators. The discovery in the late 1970s, however, that doped polyacetylene had a relatively high conductivity opened up the possibility of a radically different role for polymers in electrical applications. In the following two decades, polymeric conductors went from being scientific curiosities to applications in electronic devices.

A key feature of **electronic polymers** is a chain backbone consisting of alternating single and double bonds. The "extra" electrons associated with the double bonds can then move along the polymeric backbone relatively easily. The corresponding, small energy gap can lead to semiconductor levels of conduction and even metallic behavior (Figure 20–25). The term *synthetic metals* has come into use for this latter case.

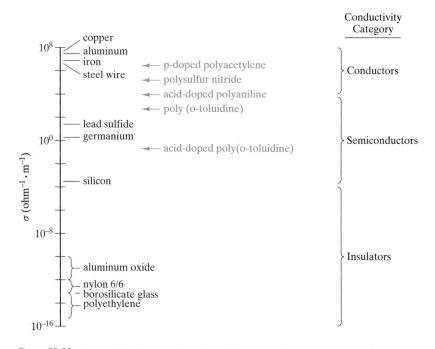

Figure 20-25 *Plot of the electrical conductivity of various electronic polymers, which challenge the conventional classifications given in Figure 15–28. (After A. J. Epstein, MRS Bulletin, 22, 19 [1997].)*

For several years after their discovery, electronic polymers presented a special challenge due to their instability in air. Significant progress has since been made, and these materials can now be processed under a broad range of environments, including organic and inorganic solvents and aqueous media. Some electronic polymers are even melt-processable.

The intensive research on these materials has helped to specify the conduction mechanisms and identify initial applications. Semiconducting polymers can be used as light-emitting devices. Numerous applications are dependent on the polymer's ability to change optical or charge-transport properties on exposure to various environments. Examples include drug-delivery systems, photovoltaics, and anticorrosion coatings for ferrous metal alloys. Other applications include gas-separation media and antistatic coatings for photographic film.

SAMPLE PROBLEM 20.3

Given the following data, indicate the economic advantage of a thermo-plastic polyester (unit cost $4.30/kg) over a phenolic thermoset ($1.21/kg). Each part being fabricated weighs 2.9 g. (Assume that the machinery costs are the same for each material and that operator labor costs are $10/hour.)

	Phenolic	Polyester
Fabrication yield rate	70%	95%
Fabrication cycle time/machine	35 s	20 s
Number parts formed/cycle	4	4
Number machines operated by single operator	1	5

SOLUTION

First, the true unit material costs (correcting for yield) would be

$$\text{phenolic: } \frac{\$1.21/kg}{0.70} = \$1.73/kg$$

$$\text{polyester: } \frac{\$4.30/kg}{0.95} = \$4.53/kg$$

Then the net materials cost per part is

phenolic: $1.73/kg × 2.9 g/part × 1 kg/1000 g = $0.005/part

$$= 0.5 \text{ cents/part}$$

polyester: $4.53/kg × 2.9 g/part × 1kg/1000 g = $0.013\/part

$$= 1.3 \text{ cents/part}$$

(The greater yield rate for polyester cannot, in itself, overcome the higher inherent material cost.)

However, the net labor costs are

phenolic: $\dfrac{\$10/\text{hour}}{\text{operator}} \times 1\ \text{operator} \times \dfrac{35\ \text{s/cycle}}{4\ \text{parts/cycle}} \times \dfrac{1\ \text{hour}}{3600\ \text{s}} = 0.024/\text{part}$

$= 2.4\ \text{cents/part}$

polyester: $\dfrac{\$10/\text{hour}}{\text{operator}} \times \dfrac{1}{5}\ \text{operator} \times \dfrac{20\ \text{s/cycle}}{4\ \text{parts/cycle}} \times \dfrac{1\ \text{hour}}{3600\ \text{s}} = \$0.003/\text{part}$

$= 0.3\ \text{cents/part}$

The total cost (materials + labor) is then

phenolic: $(0.5\text{cents} + 2.4\text{cents})/\text{part} = 2.9\text{cents/part}$

polyester: $(1.3\text{cents} + 0.3\text{cents})/\text{part} = 1.6\text{cents/part}$

The greatly reduced labor costs have given a net economic advantage to the polyester.

PRACTICE PROBLEM 20.3

We calculate the cost savings due to more economical processing of a thermoplastic in Sample Problem 20.3. The largest single factor is the ability of a single operator to work with multiple fabrication machines for the thermoplastics. By what factor would this "machine operator" parameter have to be increased for thermosets before the two materials would be exactly equal in price?

20.4 MATERIALS AND OUR ENVIRONMENT

In Chapter 19, we look closely at the various ways in which the application of engineering materials can be limited by reaction with their environment. As engineers, we are increasingly responsible for ensuring that, in turn, engineering materials do not adversely affect the environment in which all of us must live. Recycling is one effective way to limit the impact of materials on the environment.

ENVIRONMENTAL ASPECTS OF DESIGN

In Section 20.3, we looked in some detail at a case study in which conventional silicon steel was replaced by amorphous metals in transformer cores. The driving forces for this materials substitution were seen to be economic and environmental. The reduced core loss for amorphous metals led to substantial reductions in both energy costs and pollution emissions.

Public policy is increasingly dictating reduced pollution emissions whether or not this reduction correlates with economic benefits to industry. In the United States, the Environmental Protection Agency (EPA) has numerous regulations regarding environmental performance. Major environmental legislation of the past three decades is summarized in Table 20.9. Enforcement of these regulations by the EPA falls under the criminal code of law. Similar regulations also exist at the state level. Incentives for environmental regulation compliance are sufficiently great that systematic strategies have evolved for ensuring that the engineering design process incorporates regulatory compliance. Specifically, **design for the environment (DFE)** includes carrying out a **life cycle assessment (LCA)**, a "cradle-to-grave" evaluation of environmental and energy impacts of a given product design. An **environmental impact assessment (EIA)** is related to the LCA and is illustrated by Figure 20–26. The EIA is intended to structure a

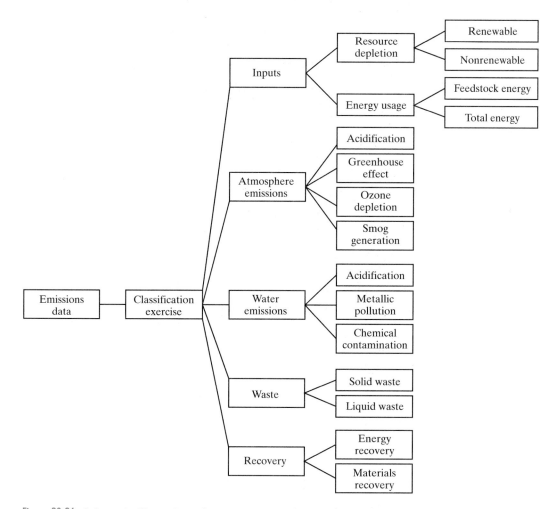

Figure 20-26 *Schematic illustration of an environmental impact assessment (EIA) of emissions data. (After L. Holloway et al., Materials and Design, 15, 259 [1994].)*

Table 20.9 *Major Environmental Legislation in the United States*[a]

Clean Air Act, 1970 (CAA)	Controlled chemicals list, allowed EPA to assess risk and act to prevent harm; EPA allowed to enforce law without proof of harm; in 1990 establish Maximum Achievable Control Technology (MACT) instead of previous public health standards; emission levels set by EPA or states (Reauthorized 1996, amendments 1977, 1990, 1995)
Clean Water Act, 1972 (CWA)	Gave EPA authority to control industrial discharge to water by imposing discharge requirements to industry, placing special controls on toxic discharge and requiring a variety of safety and construction measures to reduce spills to waterways
Toxic Substances Control Act, 1976 (TSCA)	All new (toxic) substances or new uses of substances entering the marketplace evaluated for health and environmental effects
Resource Conservation and Recovery Act, 1976 (RCRA)	Clarified waste disposal issues and established cradle-to grave control of hazardous waste
Comprehensive Environmental Response, Compensation, and Liability, 1980 (CERCLA)	Reauthorized and amended RCRA, clarified responsibility and liability of parties involved in hazardous material management, established Superfund to pay for remediation (by 1984 the EPA identifies 378,000 sites requiring corrective action)
Hazardous and Solid Waste Amendments, 1984 (HSWA)	Step up national efforts to improve hazardous waste management
Emergency Planning and Community Right to Know Act, 1986 (EPCRA), also known as Superfund Amendments and Reauthorization Act (SARA)	Section 313 of Title III of this act includes the Toxics Release Inventory (TRI), which makes hazardous waste and toxins a matter of public record. This act sets fines for violations.
Polution Prevention Act, 1990 (PPA)	Establishes hierarchy of Pollution Prevention (PP or P^2), places source reduction at the head of the list for pollution prevention; firms that must report under TRI must also report level of pollution prevention accomplishments
Examples of Nonlegislative EPA Measures/Actions	
Environmental auditing, 1986	The EPA attempts to formalize procedures for environmental auditing by developing a generic protocol for environmental auditing.
Industrial Toxics Project (33/50 Program), 1991	Companies with large releases of the 17 chemicals of the TRI (Toxic Release Inventory) reported with largest volume releases asked to target 33% reduction by 1992, 50% reduction by 1995. This is a voluntary measure. There are hopes to foster a pollution prevention ethic in business.

[a] After S. T. Fleischmann, "Environmental Aspects of Design," in *ASM Handbook, Vol. 20: Materials Selection and Design*, ASM International, Materials Park, Ohio, 1997, pp. 131–138.

study of the impact of emissions so that potential major problems are identified. Clearly, environmental regulations have spawned a substantial, new vocabulary. Some have labeled this overall exercise of environmentally sensitive design as simply **"green engineering."**

Representative of specific challenges to materials engineers is the National Emission Standard for Hazardous Air Pollutants (NESHAPS), which is section 112 under section 40 of the Clean Air Act (see Table 20.9). NESHAPS regulates the emissions of the chemicals listed in Table 20.10, various combinations of which can be byproducts of many of the common materials processing technologies outlined earlier in this book.

Airborne particles known as atmospheric **aerosols** are an increasing focus of environmental regulators. The deadly potential of such pollution has been well known for some time. Especially infamous was a week-long episode in London during 1952 in which thousands of children and elderly people died as a result of high levels of soot and sulfur dioxide. Since the Clean Air Act of 1970, the EPA has limited the levels of atmospheric particles. Originally, the rules covered particle diameters up to 50 μm. Health studies showed, however, that such coarse particles tend to be safely expelled from the body's upper airways. Since 1987, the EPA has only restricted particulate matter less than 10 μm in diameter, referred to simply as **PM10**. Several health studies in the 1990s have suggested that the greatest risk comes from an even finer scale of aerosols-viz, those with diameters less than 2.5 μm (**PM2.5**). A detailed example of the distribution of fine metal

Table 20.10 *Chemicals Included under the National Emission Standard for Hazardous Air Pollutants (NESHAPS)*[a]

Asbestos
Benzene
Beryllium
Coke oven emissions
Mercury
Vinyl chloride
Inorganic arsenic
Rn–222
Radionuclides
Copper
Nickel
Phenol
Zinc and zinc oxide

[a] After S. T. Fleischmann, "Environmental Aspects of Design," in *ASM Handbook, Vol. 20: Materials Selection and Design*, ASM International, Materials Park, Ohio, 1997, pp. 131–138.

Table 20.11 *Distribution of Metallic $\leq 2.5\mu m$ Particulate Matter (PM2.5) from an Urban Setting*[a]

| | Concentration (ng/m³) for size range: | | | | |
	0.069–0.24 μm	0.24–0.34 μm	0.34–0.56 μm	0.56–1.15 μm	1.15–2.5 μm
Vanadium	2.5	6.1	10.5	12.2	8.6
Nickel	1.3	4.4	7.7	4.5	0.5
Zinc	17.6	46.3	140.4	189.4	39
Selenium	< 0.3	0.32	3.00	1.40	0.65
Lead	71.4	47.6	59.9	69.9	25.4

[a] Averaged data collected over a 6-day period in November 1987 in Long Beach, California, by T. A. Cahill et al., University of California, Davis.

particles within a PM2.5 sample from an industrialized urban setting is summarized in Table 20.11. The chemical identification of these particles was done by x-ray fluorescence (XRF) introduced in Section 19.11 and a closely related technique known as **proton-induced x-ray emission (PIXE)**, in which a beam of high-energy protons rather than x-ray photons is used to stimulate the emission of characteristic x-rays. As such, PIXE is also quite similar to electron-dispersive x-ray spectrometry (EDX) introduced in Section 19.11. XRF, EDX, and PIXE differ only in the stimulus for characteristic x-ray emission (x-ray photons, electrons, and protons, respectively).

RECYCLING

Materials that are highly inert in their environmental surroundings are candidates for **recycling**. The ubiquitous aluminum beverage can is a prime example. The opposite case would be a material that is completely biodegradable. An example would be the paper napkin, which might be a temporary visual pollutant but, after a relatively short time, would deteriorate. There can be substantial energy savings and an attendant reduction in pollution emissions when materials such as aluminum are recycled.

Table 20.12 summarizes the recyclability of the five categories of engineering materials discussed in this book. Among the metals, aluminum alloys are excellent examples of recycling. The energy requirement for aluminum recycling is small in comparison to that for initial production. Many commercial aluminum alloys have been designed to accommodate impurity contamination. Many other alloys, both ferrous and nonferrous, are recyclable. Difficult-to-recycle metals are often the result of design configurations more than the nature of an isolated alloy. Examples include the galvanized (zinc-coated) steel introduced in Section 19.6 and rivets fastened to nonmetallic components. In general, dissimilar material combinations are a challenge for recycling.

Glass containers are as well known for recycling as are aluminum cans. Glass container manufacturing routinely incorporates a combination of recycled glass (termed "cullet") and raw materials (sand and carbonates of

Table 20.12 *Recycling Characteristics of the Five Materials Categories*

Category	Characteristics
Metals	Many commercial alloys are recyclable.
	Aluminum cans are prime examples.
	Difficulty associated with dissimilar material design configurations.
Ceramics and glasses	Crystalline ceramics generally not recycled.
	Glass containers widely recycled.
	Greater difficulty associated with combinations of dissimilar materials.
Polymers	Recycling code system in place for thermoplastic polymers.
	Thermosetting polymers generally not recycled.
	Elastomers, other than thermoplastic variety, are generally difficult to recycle.
Composites	The inherent fine-scale combination of dissimilar materials makes recycling impractical.
Semiconductors	Wafer reclaiming (stripping away surface circuitry) is widely practiced.
	Associated recycling of other solid and chemical waste is done at fabrication laboratories.

sodium and calcium). This use of recycled glass provides increased production rates and reduced pollution emissions. On the other hand, care is required to sort the recycled glass by type (container versus plate) and color (clear versus various colors). Also, glass containers with polymeric coatings are additional examples of dissimilar material combinations that make recycling more difficult. In contrast to glasses, crystalline ceramic products are, in general, not recycled.

An elaborate recycling code system is now routinely noted on a variety of polymeric products, especially food and beverage containers. As a practical matter, recycling of polymers is practical only for thermoplastic materials, such as polyethylene terephthalate (PET), polyethylene (PE), and polypropylene (PP). The presence of fillers that are difficult to separate out presents a challenge to recycling. Thermosetting polymers are in general not amenable to recycling. Rubber, once vulcanized, behaves as a thermosetting polymer and is not generally recyclable. Rubber tires, in fact, pose a major challenge to landfills. A major benefit of thermoplastic elastomers is, as noted in Section 13.3, their recyclability.

Composite materials, by definition, are a fine-scale combination of different material components and are impractical to recycle.

The substantial demands and related expense in producing high-quality silicon wafers for the semiconductor industry have led to a substantial market for wafer reclaiming. The surface circuitry produced by the elaborate techniques described in Section 17.6 can be stripped away by mechanical and chemical means to leave behind a high-quality, albeit thinner wafer suitable for test wafer applications. In addition, semiconductor fabrication operations are scrutinized by regulatory agencies to maximize both solid and chemical waste recycling. Large volumes of polymeric packaging materials are routinely recycled at semiconductor "fab labs."

SAMPLE PROBLEM 20.4

Using the data in Sample Problem 19.11, calculate the K_α photon energy that would be used in a PIXE analysis of iron-containing PM2.5 particles.

SOLUTION

As noted in the definition of PIXE, it differs from XRF and EDX only in the source of photon stimulation, not in the nature of the resulting, characteristic x-ray photon. So again,

$$E_{K_\alpha} = |E_K - E_L|$$
$$= |-7112 \text{ eV} - (-708 \text{ eV})| = 6404 \text{ eV}$$

PRACTICE PROBLEM 20.4

As in Sample Problem 20.4, calculate the K_β photon energy that would be used in a PIXE analysis of iron-containing PM2.5 particles.

SUMMARY

The many properties of materials defined in the first 19 chapters of this text become the design parameters to guide engineers in the selection of materials for a given engineering design. These engineering design parameters are often dependent on the processing of the material.

In selecting structural materials, we are faced first with a competition among the four types outlined in Part II of this book. Once a given category is chosen, a specific material must be identified as an optimal choice. In general, a balance must be reached between strength and ductility. Relatively brittle ceramics and glasses can still find structural applications based on properties such as temperature resistance and chemical durability. Improved toughness is expanding their design options further. Glasses and polymers may be selected as structural materials because of optical properties. Final design issues for any structural material are environmental degradation and failure prevention.

The selection of electronic materials begins with defining the need for one of three categories: conductor, insulator, or semiconductor. The selection of semiconductors is part of a complex engineering design process leading to increasingly complex and miniaturized electrical circuits. The selection of optical materials is often done in conjunction with their structural and electronic applications. The discussion of metallic and ceramic magnetic materials in Chapter 18 provided a clear outline of the material selections appropriate for specific magnetic applications. For both structural and electronic, optical, and magnetic categories, materials selection is further illustrated by specific case studies.

As engineers, we must be increasingly aware of our responsibility to the environment. Governmental entities such as the Environmental Protection Agency oversee a vast array of regulations. Design for the environment (DFE) has become an integral part of overall professional practice. The definition of hazardous air pollutants is being broadened to include increasingly fine-scale particles, byproducts of many common materials-processing technologies. Substantial energy savings and attendant reductions in pollution emissions are possible with the recycling of many engineering materials.

KEY TERMS

actuator (779)

aerosol (785)

biological material (766)

biomaterial (766)

design for the environment (DFE) (783)

design parameter (749)

electronic polymer (780)

environmental impact assessment (EIA) (783)

green engineering (785)

honeycomb structure (759)

life cycling assessment (LCA) (783)

life-emitting diode (LED) (777)

PM2.5 (785)

PM10 (785)

prosthesis (765)

proton-induced x-ray emission (PIXE) (786)

recycling (786)

spectral width (777)

smart material (779)

structural composite (761)

REFERENCES

ASM Handbook, Vol. 20: Materials Selection and Design, ASM International, Materials Park, Ohio, 1997.

Ashby, M. F., *Materials Selection in Engineering Design,* Pergamon Press, Inc., Elmsford, N.Y., 1992.

Ashby, M. F., and **D. R. H. Jones**, *Engineering Materials 1— An Introduction to Their Properties and Applications,* 2nd ed., Butterworth-Heinemann, Oxford, 1996.

Ashby, M. F., and **D. R. H. Jones**, *Engineering Materials 2— An Introduction to Microstructures, Processing, and Design,*

2nd ed., Butterworth-Heinemann, Oxford, 1998.

Dorf, R. C., *Electrical Engineering Handbook,* 2nd ed., CRC Press, Boca Raton, Fl, 1997.

Shackelford, J. F., W. Alexander, and **J. S. Park**, *CRC Materials Science and Engineering Handbook,* 2nd ed., CRC Press, Boca Raton, Fl, 1994.

Shackelford, J. F., W. Alexander, and **J. S. Park**, *CRC Practical Handbook of Materials Selection,* CRC Press, Boca Raton, Fl, 1995.

PROBLEMS

At the beginning of this chapter, we noted that materials selection is a complex topic and our introductory examples are somewhat idealistic. As in previous chapters, we shall continue to avoid subjective homework problems and instead deal with problems in the objective style. The many subjective problems dealing with materials selection will be left to more advanced courses or your own experiences as a practicing engineer.

20.1 • Material Properties-Engineering Design Parameters

20.1. A government contractor requires the use of tool steels with a tensile strength of $\geq$ 1000 MPa and a % elongation of $\geq$ 15%. Which alloys in Tables 20.1 and 20.2 would meet this specification?

20.2. Which additional alloys could be specified if the government contractor in Problem 20.1 would allow the selection of tool steels with a tensile strength of $\geq$ 1000 MPa and a % elongation of $\geq$ 10%?

D **20.3.** In selecting a ductile cast iron for a high-strength/low-ductility application, you find the following two tables in a standard reference:

Tensile Strength of Ductile Irons		Elongation of Ductile Irons	
Grade or class	T.S. (MPa)	Grade or class	Elongation (%)
Class C	345	120–90–02	2
Class B	379	100–70–03	3
Class A	414	80–55–06	6
65–45–12	448	Class B	7
80–55–06	552	65–45–12	12
100–70–03	689	Class A	15
120–90–02	827	Class C	20

If specifications require a ductile iron with a tensile strength of ≥ 550 MPa and a % elongation of ≥ 5%, which alloys in these tables would be suitable?

D **20.4.** In selecting a ductile cast iron for a low-strength/high-ductility application, which of the alloys in Problem 20.3 would meet the specifications of a tensile strength of ≥ 350 MPa and a % elongation of ≥ 15%?

D **20.5.** In selecting a polymer for an electronic packaging application, you find the following two tables in a standard reference:

Volume Resistivity of Polymers	
Polymer	Resistivity ($\Omega \cdot$ m)
epoxy	1×10^5
phenolic	1×10^9
cellulose acetate	1×10^{10}
polyester	1×10^{10}
polyvinyl chloride	1×10^{12}
nylon 6/6	5×10^{12}
acrylic	5×10^{12}
polyethylene	5×10^{13}
polystyrene	2×10^{14}
polycarbonate	2×10^{14}
polypropylene	2×10^{15}
PTFE	2×10^{16}

Thermal Conductivity of Polymers	
Polymer	Conductivity (J/s $\cdot$ m $\cdot$ K)
polystyrene	0.12
polyvinyl chloride	0.14
polycarbonate	0.19
polyester	0.19
acrylic	0.21
phenolic	0.22
PTFE	0.24
cellulose acetate	0.26
polyethylene	0.33
epoxy	0.52
polypropylene	2.2
nylon 6/6	2.9

If specifications require a polymer with a volume resistivity of ≥ $10^{13} \Omega \cdot$ m and a thermal conductivity of ≥ 0.25 J/s $\cdot$ m $\cdot$ K, which polymers in these tables would be suitable?

D **20.6.** In reviewing the performance of polymers chosen in Problem 20.5, concerns over product performance lead to the establishment of more stringent specifications. Which polymers would have a volume resistivity of ≥ $10^{14} \Omega \cdot$ m and a thermal conductivity of ≥ 0.35 J/s $\cdot$ m $\cdot$ K?

20.7. On a photocopy of Figure 20–4, superimpose modulus-density data for (a) 1040 carbon steel, (b) 2048 plate aluminum, and (c) Ti–5Al–2.5Sn. Take the modulus data from Table 6.2 and approximate the alloy densities by using the values for pure Fe, Al, and Ti in Appendix 1.

20.8. On a photocopy of Figure 20–4, superimpose modulus-density data for (a) the sintered alumina from Table 6.5, (b) the high-density polyethylene from Table 6.7, and (c) the E-glass/epoxy fiberglass from Table 14.12. Take the densities to be 3.8, 1.0, and 1.8 Mg/m^3, respectively.

20.2 • Selection of Structural Materials-Case Studies

20.9. Use Equation 20.4 to calculate the mass of a CFRP windsurfer mast with a length of 4.6 m, outer radius of 18 mm, shape factor of $\phi = 5$, and density of 1.5 Mg/m^3.

20.10. Repeat Problem 20.9 for a GFRP windsurfer mast with a length of 4.6 m, outer radius of 17 mm, shape factor of $\phi = 4$, and density of 1.8 Mg/m^3.

20.11. The Boeing 767 aircraft uses a Kevlar-reinforced composite for its cargo liner. The structure weighs 125 kg. **(a)** What weight savings does this represent compared to an aluminum structure of the same volume? (For simplicity, use the density of pure aluminum. A calculation of density for a Kevlar composite was made in Problem 14.4.) **(b)** What annual fuel savings would this one material substitution represent?

20.12. What is the annual fuel savings resulting from the Al-Li alloy substitution described in Problem 11.7?

20.13. Consider an inspection program for Ti–6Al–4V THR femoral stems. Would the ability to detect a flaw size of 1 mm be adequate to prevent fast fracture? Use the data in Table 8.3 and assume an extreme loading of five times body weight for an athletic 200 lb$_f$ patient. The cross-sectional area of the stem is 650 mm^2.

20.14. Consider an inspection program for a more traditional application of the Ti–6Al–4V alloy discussed in Problem 20.13. Is the ability to detect a 1-mm flaw size adequate in an aerospace structural member loaded to 90% of the yield strength given in Table 6.2?

20.15. Consider the loading of a simple, cantilever beam:

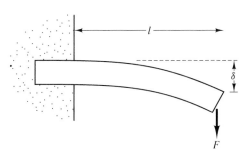

From basic mechanics, it can be shown that the mass of the beam subjected to a deflection, δ, by a force, F, is given by

$$M = (4l^5 F/\delta)^{1/2}(\rho^2/E)^{1/2}$$

where ρ is the density, E is the elastic modulus, and the other terms are defined by the figure. Clearly, the mass of this structural member is minimized for a given loading by minimizing ρ^2/E. Given the data in Problem 14.49, which of those materials would be the optimal choice for this type of structural application? (Moduli can generally be obtained from Tables 6.2 and 14.12. The modulus of reinforced concrete can be taken as 47×10^3 MPa.)

20.16. How would the material selection in Problem 20.15 be modified if cost were included in the minimization?

20.3 • Selection of Electronic, Optical, and Magnetic Materials-Case Studies

20.17. Verify the total owning cost calculation for the amorphous metal core in Table 20.7.

20.18. Repeat Problem 20.17 for the grain-oriented silicon steel core in Table 20.7.

D 20.19. In comparing two polyester thermoplastics for use in the design of fuseholders (mass = 5 g) in a new automobile, the following data are available:

	Polyester 1	Polyester 2
Cost/kg	$4.25	$4.50
Yield rate	95%	92%
Cycle time	25 s	20 s

Given that all other production factors are comparable and equal to the parameters for the polyester in Practice Problem 20.3, carry out an economic comparison and recommend a choice between these two engineering polymers.

20.20. Given that the thermal conductivity, k, of a 95 Pb: 5 Sn solder used in flip-chip technology is 63 J/(s · m · K), calculate the heat flow rate across an array of 100 solder balls with an average diameter of 100 μm and thickness of 75 μm. The temperature of the chip can be taken as 80°C and the substrate as 25°C. (Recall the discussion of heat transfer in Section 7.3.)

D 20.21. Consider the following design considerations for a typical solid-state device, a memory cell in a field-effect transistor (FET). The cell contains a thin film of SiO$_2$, which serves as a small capacitor. You want the capacitor to be as small as possible, but big enough so that an α radiation particle will not cause errors with your 5 V operating signal. To ensure this result, design specifications require a capacitor with a capacitance (=[charge density x area]/voltage) of 50×10^{-15} coul/volt. How much area is required for the capacitor if the SiO_2 thickness is 1 μm? (Note that the dielectric constant of SiO_2 is 3.9, and recall the discussion of capacitance in Section 15.4.)

Element	E_K (keV)
V	5.465
Ni	8.333
Zn	9.665
Se	12.66
Pb	88.02

D 20.22. If the specifications for the device introduced in Problem 20.21 are changed so that a capacitance of 70×10^{-15} coul/volt is required but the area is fixed by the overall circuit design, calculate the appropriate SiO_2 film thickness for the modified design.

20.23. Calculate the band gap, E_g, corresponding to the LEDs in Table 20.8 made of (a) GaP and (b) GaAsP.

20.24. Repeat Problem 20.23 for the LEDs in Table 20.8 made of (a) GaAs and (b) InGaAs.

20.4 • Materials and Our Environment

20.25. Given the following data for the K shell binding energies of the particulate metals listed in Table 20.11, calculate the maximum x-ray wavelengths necessary to create K_α radiation for each of the elements. (Recall the mechanism illustrated by Figure 19–30.)

20.26. Given the data in Problem, 20.25, calculate the minimum proton kinetic energy necessary to create K_α radiation for each of the elements listed in Table 20.11.

20.27. Would you consider the following materials systems to be easy or difficult to recycle: (a) steel casters fixed to polymeric chair bases and (b) clear frozen food bags made of low-density polyethylene? Briefly explain your answers.

20.28. Repeat Problem 20.27 for (a) polymeric trays used to carry electronic devices locally in a fabrication plant and (b) polymeric bottles with well-adhered paper labels.

APPENDIX ONE

Physical and Chemical Data for the Elements

Atomic number	Element	Symbol	1s	2s	2p	3s	3p	3d	4s	4p	4d	4f	5s	5p	5d	5f	6s	6p	6d	7s	Atomic mass (amu)	Density of solid (at 20°C) (Mg/m³ = g/cm³)	Crystal structure (at 20°C)	Melting point (°C)	Atomic number
1	Hydrogen	H	1																		1.008			−259.34 (T.P.)	1
2	Helium	He	2																		4.003			−271.69	2
3	Lithium	Li		1																	6.941	0.533	bcc	180.6	3
4	Berylium	Be		2																	9.012	1.85	hcp	1289	4
5	Boron	B		2	1																10.81	2.47		2092	5
6	Carbon	C		2	2																12.01	2.27	hex.	3826 (S.P.)	6
7	Nitrogen	N		2	3																14.01			−210.0042 (T.P.)	7
8	Oxygen	O		2	4																16.00			−218.789 (T.P.)	8
9	Flourine	F		2	5																19.00			−219.67 (T.P.)	9
10	Neon	Ne		2	6																20.18			−248.587 (T.P.)	10
11	Sodium	Na				1															22.99	0.966	bcc	97.8	11
12	Magnesium	Mg				2															24.31	1.74	hcp	650	12
13	Aluminum	Al				2	1														26.98	2.70	fcc	660.452	13
14	Silicon	Si				2	2														28.09	2.33	dia. cub.	1414	14
15	Phosphorus	P				2	3														30.97	1.82 (white)	ortho.	44.14 (white)	15
16	Sulfer	S				2	4														32.06	2.09	ortho.	115.22	16
17	Chlorine	Cl				2	5														35.45			−100.97 (T.P.)	17
18	Argon	Ar				2	6														39.95			−189.352 (T.P.)	18
19	Potassium	K							1												39.10	0.862	bcc	63.71	19
20	Calcium	Ca							2												40.08	1.53	fcc	842	20
21	Scandium	Sc						1	2												44.96	2.99	fcc	1541	21
22	Titanium	Ti						2	2												47.90	4.51	hcp	1670	22
23	Vanadium	V						3	2												50.94	6.09	bcc	1910	23
24	Chromium	Cr						5	1												52.00	7.19	bcc	1863	24
25	Manganese	Mn						5	2												54.94	7.47	cubic	1246	25
26	Iron	Fe						6	2												55.85	7.87	bcc	1538	26
27	Cobalt	Co						7	2												58.93	8.8	hcp	1495	27
28	Nickel	Ni						8	2												58.71	8.91	fcc	1455	28
29	Copper	Cu						10	1												63.55	8.93	fcc	1084.87	29
30	Zinc	Zn						10	2												65.38	7.13	hcp	419.58	30
31	Gallium	Ga						10	2	1											69.72	5.91	ortho.	29.7741 (T.P.)	31
32	Germanium	Ge						10	2	2											72.59	5.32	dia. cub.	938.3	32
33	Arsenic	As						10	2	3											74.92	5.78	rhomb.	603 (S.P.)	33

Electronic configuration[a] (number of electrons in each group)

Helium core · Neon core · Argon core

Atomic number	Element	Symbol	1s	2s	2p	3s	3p	3d	4s	4p	4d	4f	5s	5p	5d	5f	6s	6p	6d	7s	Atomic mass[a] (amu)	Density of solid[b] (at 20°C) ($Mg/m^3 = g/cm^3$)	Crystal structure[c] (at 20°C)	Melting point[d] (°C)	Atomic number
							Electronic configuration[a] (number of electrons in each group)																		
34	Selenium	Se						10	2	4											78.96	4.81	hex.	221	34
35	Bromine	Br						10	2	5											79.90			−7.25 (T.P.)	35
36	Krypton	Kr						10	2	6											83.80			−157.385	36
37	Rubidium	Rb				Krypton core							1								85.47	1.53	bcc	39.48	37
38	Strontium	Sr											2								87.62	2.58	fcc	769	38
39	Yttrium	Y									1		2								88.91	4.48	hcp	1522	39
40	Zirconium	Zr									2		2								91.22	6.51	hcp	1855	40
41	Niobium	Nb									4		1								92.91	8.58	bcc	2469	41
42	Molybdenum	Mo									5		1								95.94	10.22	bcc	2623	42
43	Technetium	Tc									6		1								98.91	11.50	hcp	2204	43
44	Ruthenium	Ru									7		1								101.07	12.36	hcp	2334	44
45	Rhodium	Rh									8		1								102.91	12.42	fcc	1963	45
46	Palladium	Pd									10										106.4	12.00	fcc	1555	46
47	Silver	Ag									10		1								107.87	10.50	fcc	961.93	47
48	Cadmium	Cd									10		2								112.4	8.65	hcp	321.108	48
49	Indium	In									10		2	1							114.82	7.29	fct	156.634	49
50	Tin	Sn									10		2	2							118.69	7.29	bct	231.9681	50
51	Antimony	Sb									10		2	3							121.75	6.69	rhomb.	630.755	51
52	Tellurium	Te									10		2	4							127.60	6.25	hex.	449.57	52
53	Iodine	I									10		2	5							126.90	4.95	ortho.	113.6 (T.P)	53
54	Xenon	Xe									10		2	6							131.30			−111.7582 (T.P.)	54
55	Cesium	Cs					Xenon core										1				132.91	1.91 (−10°)	bcc	28.39	55
56	Barium	Ba															2				137.33	3.59	bcc	729	56
57	Lanthanum	La													1		2				138.91	6.17	hex.	918	57
58	Cerium	Ce										2					2				140.12	6.77	fcc	798	58
59	Praseodymium	Pr										3					2				149.91	6.78	hex.	931	59
60	Neodymium	Nd										4					2				144.24	7.00	hex.	1021	60
61	Promethium	Pm										5					2				(145)		hex.	1042	61
62	Samarium	Sm										6					2				150.4	7.54	rhomb.	1074	62
63	Europium	Eu										7					2				151.96	5.25	bcc	822	63
64	Gadolinium	Gd										7			1		2				157.25	7.87	hcp	1313	64
65	Terbium	Tb										9					2				158.93	8.27	hcp	1356	65
66	Dysprosium	Dy										10					2				162.50	8.53	hcp	1412	66
67	Holmium	Ho										11					2				164.93	8.80	hcp	1474	67
68	Erbium	Er										12					2				167.26	9.04	hcp	1529	68
69	Thulium	Tm										13					2				168.93	9.33	hcp	1545	69

Atomic number	Element	Symbol	1s	2s	2p	3s	3p	3d	4s	4p	4d	4f	5s	5p	5d	5f	6s	6p	6d	7s	Atomic mass[a] (amu)	Density of solid[b] (at 20°C) (Mg/m³ = g/cm³)	Crystal structure[c] (at 20°C)	Melting point[d] (°C)	Atomic number
70	Ytterbium	Yb	\<— Xenon core —\>									14					2				173.04	6.97	fcc	819	70
71	Lutetium	Lu										14			1		2				174.97	9.84	hcp	1663	71
72	Hafnium	Hf										14			2		2				178.49	13.28	hcp	2231	72
73	Tantalum	Ta										14			3		2				180.95	16.67	bcc	3020	73
74	Tungsten	W										14			4		2				183.85	19.25	bcc	3422	74
75	Rhenium	Re										14			5		2				186.2	21.02	hcp	3186	75
76	Osmium	Os										14			6		2				190.2	22.58	hcp	3033	76
77	Iridium	Ir										14			9						192.22	22.55	fcc	2447	77
78	Platinum	Pt										14			9		1				195.09	21.44	fcc	1769.0	78
79	Gold	Au										14			10		1				196.97	19.28	fcc	1064.43	79
80	Mercury	Hg										14			10		2				200.59			−38.836	80
81	Thallium	Tl										14			10		2	1			204.37	11.87	hcp	304	81
82	Lead	Pb										14			10		2	2			207.2	11.34	fcc	327.502	82
83	Bismuth	Bi										14			10		2	3			208.98	9.80	rhomb.	271.442	83
84	Polonium	Po										14			10		2	4			(~210)	9.2	monoclinic	254	84
85	Astatine	At										14			10		2	5			(210)			≈302	85
86	Radon	Rn										14			10		2	6			(222)			−71	86
87	Francium	Fr	\<— Radon core —\>																	1	(223)		bcc	≈27	87
88	Radium	Ra																		2	226.03		bct	700	88
89	Actinium	Ac																	1	2	(227)		fcc	1051	89
90	Thorium	Th																	2	2	232.04	11.72	fcc	1755	90
91	Protoactinium	Pa														2			1	2	231.04		bct	1572	91
92	Uranium	U														3			1	2	238.03	19.05	ortho.	1135	92
93	Neptunium	Np														4			1	2	237.05		ortho.	639	93
94	Plutonium	Pu														6				2	(244)	19.82	monoclinic	640	94
95	Americium	Am														7				2	(243)		hex.	1176	95
96	Curium	Cm														7			1	2	(247)		hex.	1345	96
97	Berkelium	Bk														9				2	(247)		hex.	1050	97
98	Californium	Cf														10				2	(251)			900	98
99	Einsteinium	Es														11				2	(254)			860	99
100	Fermium	Fm														12				2	(257)			≈1527	100
101	Medelevium	Md														13				2	(258)			≈827	101
102	Nobelium	No														14				2	(259)			≈827	102
103	Lawrencium	Lw														14			1	2	(260)			≈1627	103

SOURCES: Data from

[a] *Handbook of Chemistry and Physics,* 58th ed., R. C. Weast, Ed., CRC Press, Boca Raton, Fla., 1977.

[b] X-ray diffraction measurements are tabulated in B. D. Cullity, *Elements of X-Ray Diffraction,* 2nd ed., Addison-Wesley Publishing Co., Inc., Reading, Mass., 1978.

[c] R. W. G. Wyckoff, *Crystal Structure,* 2nd ed., Vol. 1, Interscience Publishers, New York, 1963; and *Metals Handbook,* 9th ed., Vol. 2, American Society for Metals, Metals Park, Ohio, 1979.

[d] *Binary Alloy Phase Diagrams,* Vols. 1 and 2, T. B. Massalski, ed., American Society for Metals, Metals Park, Ohio, 1986. T.P. = triple point. S.P. = sublimation point at atmospheric pressure.

Atomic and Ionic Radii of the Elements

Atomic number	Symbol	Atomic radius (nm)	Ion	Ionic radius (nm)
1	H	0.046	H^-	0.154
2	He	—	—	—
3	Li	0.152	Li^+	0.078
4	Be	0.114	Be^{2+}	0.054
5	B	0.097	B^{3+}	0.02
6	C	0.077	C^{4+}	<0.02
7	N	0.071	N^{5+}	0.01–0.02
8	O	0.060	O^{2-}	0.132
9	F	—	F^-	0.133
10	Ne	0.160	—	—
11	Na	0.186	Na^+	0.098
12	Mg	0.160	Mg^{2+}	0.078
13	Al	0.143	Al^{3+}	0.057
14	Si	0.117	Si^{4-}	0.198
			Si^{4+}	0.039
15	P	0.109	P^{5+}	0.03–0.04
16	S	0.106	S^{2-}	0.174
			S^{6+}	0.034
17	Cl	0.107	Cl^-	0.181
18	Ar	0.192	—	—
19	K	0.231	K^+	0.133
20	Ca	0.197	Ca^{2+}	0.106
21	Sc	0.160	Sc^{2+}	0.083
22	Ti	0.147	Ti^{2+}	0.076
			Ti^{3+}	0.069
			Ti^{4+}	0.064
23	V	0.132	V^{3+}	0.065
			V^{4+}	0.061
			V^{5+}	~ 0.04
24	Cr	0.125	Cr^{3+}	0.064
			Cr^{6+}	0.03–0.04

Source: After a tabulation by R. A. Flinn and P. K. Trojan, *Engineering Materials and Their Applications*, Houghton Mifflin Company, Boston, 1975. The ionic radii are based on the calculations of V. M. Goldschmidt, who assigned radii based on known interatomic distances in various ionic crystals.

Atomic number	Symbol	Atomic radius (nm)	Ion	Ionic radius (nm)
25	Mn	0.112	Mn^{2+}	0.091
			Mn^{3+}	0.070
			Mn^{4+}	0.052
26	Fe	0.124	Fe^{2+}	0.087
			Fe^{3+}	0.067
27	Co	0.125	Co^{2+}	0.082
			Co^{3+}	0.065
28	Ni	0.125	Ni^{2+}	0.078
29	Cu	0.128	Cu^{+}	0.096
			Cu^{2+}	0.072
30	Zn	0.133	Zn^{2+}	0.083
31	Ga	0.135	Ga^{3+}	0.062
32	Ge	0.122	Ge^{4+}	0.044
33	As	0.125	As^{3+}	0.069
			As^{5+}	~ 0.04
34	Se	0.116	Se^{2-}	0.191
			Se^{6+}	0.03–0.04
35	Br	0.119	Br^{-}	0.196
36	Kr	0.197	—	—
37	Rb	0.251	Rb^{+}	0.149
38	Sr	0.215	Sr^{2+}	0.127
39	Y	0.181	Y^{3+}	0.106
40	Zr	0.158	Zr^{4+}	0.087
41	Nb	0.143	Nb^{4+}	0.074
			Nb^{5+}	0.069
42	Mo	0.136	Mo^{4+}	0.068
			Mo^{6+}	0.065
43	Tc	—	—	—
44	Ru	0.134	Ru^{4+}	0.065
45	Rh	0.134	Rh^{3+}	0.068
			Rh^{4+}	0.065
46	Pd	0.137	Pd^{2+}	0.050
47	Ag	0.144	Ag^{+}	0.113
48	Cd	0.150	Cd^{2+}	0.103
49	In	0.157	In^{3+}	0.092
50	Sn	0.158	Sn^{4-}	0.215
			Sn^{4+}	0.074
51	Sb	0.161	Sb^{3+}	0.090
52	Te	0.143	Te^{2-}	0.211
			Te^{4+}	0.089
53	I	0.136	I^{-}	0.220
			I^{5+}	0.094
54	Xe	0.218	—	—
55	Cs	0.265	Cs^{+}	0.165
56	Ba	0.217	Ba^{2+}	0.143
57	La	0.187	La^{3+}	0.122
58	Ce	0.182	Ce^{3+}	0.118
			Ce^{4+}	0.102

Atomic number	Symbol	Atomic radius (nm)	Ion	Ionic radius (nm)
59	Pr	0.183	Pr^{3+}	0.116
			Pr^{4+}	0.100
60	Nd	0.182	Nd^{3+}	0.115
61	Pm	—	Pm^{3+}	0.106
62	Sm	0.181	Sm^{3+}	0.113
63	Eu	0.204	Eu^{3+}	0.113
64	Gd	0.180	Gd^{3+}	0.111
65	Tb	0.177	Tb^{3+}	0.109
			Tb^{4+}	0.089
66	Dy	0.177	Dy^{3+}	0.107
67	Ho	0.176	Ho^{3+}	0.105
68	Er	0.175	Er^{3+}	0.104
69	Tm	0.174	Tm^{3+}	0.104
70	Yb	0.193	Yb^{3+}	0.100
71	Lu	0.173	Lu^{3+}	0.099
72	Hf	0.159	Hf^{4+}	0.084
73	Ta	0.147	Ta^{5+}	0.068
74	W	0.137	W^{4+}	0.068
			W^{6+}	0.065
75	Re	0.138	Re^{4+}	0.072
76	Os	0.135	Os^{4+}	0.067
77	Ir	0.135	Ir^{4+}	0.066
78	Pt	0.138	Pt^{2+}	0.052
			Pt^{4+}	0.055
79	Au	0.144	Au^{+}	0.137
80	Hg	0.150	Hg^{2+}	0.112
81	Tl	0.171	Tl^{+}	0.149
			Tl^{3+}	0.106
82	Pb	0.175	Pb^{4-}	0.215
			Pb^{2+}	0.132
			Pb^{4+}	0.084
83	Bi	0.182	Bi^{3+}	0.120
84	Po	0.140	Po^{6+}	0.067
85	At	—	At^{7+}	0.062
86	Rn	—	—	—
87	Fr	—	Fr^{+}	0.180
88	Ra	—	Ra^{+}	0.152
89	Ac	—	Ac^{3+}	0.118
90	Th	0.180	Th^{4+}	0.110
91	Pa	—	—	—
92	U	0.138	U^{4+}	0.105

APPENDIX
THREE

Constants and Conversion Factors

Constants

Avogadro's number, N_A	0.6023×10^{24} mol^{-1}
Atomic mass unit, amu	1.661×10^{-24} g
Electric permittivity of a vacuum, ϵ_0	8.854×10^{-12} C/(V $\cdot$ m)
Electron mass	0.9110×10^{-27} g
Elementary charge, e	0.1602×10^{-18} C
Gas constant, R	8.314 J/(mol $\cdot$ K)
	1.987 cal/(mol $\cdot$ K)
Boltzmann's constant, k	13.81×10^{-24} J/K
	86.20×10^{-6} eV/K
Planck's constant, h	0.6626×10^{-33} J $\cdot$ s
Speed of light (in vacuum), c	0.2998×10^9 m/s
Bohr magneton, μ_B	9.274×10^{-24} A $\cdot$ m^2
Faraday's constant, F	$96,500$ C/mol

SI prefixes

giga, G	10^9
mega, M	10^6
kilo, k	10^3
milli, m	10^{-3}
micro, μ	10^{-6}
nano, n	10^{-9}
pico, p	10^{-12}

Conversion factors

Length	1 meter $= 10^{10}$ Å
	$= 3.281$ ft
	$= 39.37$ in.
Mass	1 kilogram $= 2.205$ lb$_m$
Force	1 newton $= 0.2248$ lb$_f$
Pressure	1 pascal $= 1$ N/m^2
	$= 0.1019 \times 10^{-6}$ kg$_f$/mm^2
	$= 9.869 \times 10^{-6}$ atm
	$= 0.1450 \times 10^{-3}$ lb$_f$/in.^{2}
Viscosity	1 Pa $\cdot$ s $= 10$ poise
Energy	1 joule $= 1$ W $\cdot$ s
	$= 1$ N $\cdot$ m
	$= 1$ V $\cdot$ C
	$= 0.2389$ cal
	$= 6.242 \times 10^{18}$ eV
	$= 0.7377$ ft lb$_f$
Temperature	°C $=$ K $- 273$
	$= ($°F $- 32)/1.8$
Current	1 ampere $= 1$ C/s
	$= 1$ V/Ω

FOUR

Properties of the Structural Materials

The following tables are provided as convenient compilations of the key properties of the structural materials covered in Part II of the book. Reference is frequently made to specific tables located throughout the book.

Appendix 4A   Physical Properties of Selected Materials

Material	Density [Mg/m^3]	Melting temperature [°C]	Glass transition temperature [°C]
Metals[a]			
Aluminum	2.70	660	
Copper	8.93	1085	
Gold	19.28	1064	
Iron	7.87	1538	
Lead	11.34	328	
Nickel	8.91	1455	
Silver	10.50	962	
Titanium	4.51	1670	
Tungsten	19.25	3422	
Ceramics and glasses[b]			
Al_2O_3	3.97	2054	
MgO	3.58	2800	
SiO_2	2.26–2.66	1726	
Mullite ($3Al_2O_3 \cdot 2SiO_2$)	3.16	1890	
ZrO_2	5.89	2700	
Silica glass	2.2		1100
Soda-lime-silica glass	2.5		450
Polymers[c]			
Epoxy (mineral filled)	1.22		400
Nylon 66	1.13–1.15		150
Phenolic	1.32–1.46		375
Polyethylene (high-density)	0.94–0.97		
Polypropylene	0.90–0.91		
Polytetrafluoroethylene (PTFE)	2.1–2.3		

Source: Data from [a] Appendix 1, [b] D. R. Lide, *Handbook of Chemistry and Physics*, 71st ed., CRC Press, Boca Raton, Florida, 1990, ceramic phase diagrams in Chapter 9, and the Corning Glass Works, and [c] J. F. Shackelford, W. Alexander, and J. S. Park, *The CRC Materials Science and Engineering Handbook*, 2nd ed., CRC Press, Boca Raton, Florida, 1994 and Figure 6–51.

Appendix 4B   Tensile and Bend Test Data for Selected Engineering Materials

Material	E [GPa (psi)]	E_{flex} [MPa (ksi)]	E_{Dyn} [MPa (ksi)]	Y.S. [MPa (ksi)]	T.S. [MPa (ksi)]	Flexural strength [MPa (ksi)]	Compressive strength [MPa (ksi)]	Percent elongation at failure
Metal alloys[a]								
1040 carbon steel	200(29 × 10^6)			600(87)	750(109)			17
8630 low-alloy steel	193(28 × 10^6)			680(99)	800(116)			22
304 stainless steel	200(29 × 10^6)			205(30)	515(75)			40
410 stainless steel	200(29 × 10^6)			700(102)	800(116)			22
L2 tool steel				1380(200)	1550(225)			12
Ferrous superalloy (410)				700(102)	800(116)			22
Ductile iron, quench	165(24 × 10^6)			580(84)	750(108)			9.4
Ductile iron, 60–40–18	169(24.5 × 10^6)			329(48)	461(67)			15
3003-H14 aluminum	70(10.2 × 10^6)			145(21)	150(22)			8–16
2048, plate aluminum	70.3(10.2 × 10^6)			416(60)	457(66)			8
AZ31B magnesium	45(6.5 × 10^6)			220(32)	290(42)			15
AM100A casting magnesium	45(6.5 × 10^6)			83(12)	150(22)			2
Ti–5Al–2.5Sn	107 – 110(15.5 – 16 × 10^6)			827(120)	862(125)			15
Ti–6Al–4V	110(16 × 10^6)			825(120)	895(130)			10
Aluminum bronze, 9% (copper alloy)	110(16.1 × 10^6)			320(46.4)	652(94.5)			34
Monel 400 (nickel alloy)	179(26 × 10^6)			283(41)	579(84)			39.5
AC41A zinc					328(47.6)			7
50:50 solder (lead alloy)				33(4.8)	42(6.0)			60
Nb–1 Zr(refractory metal)				138(20)	241(35)			20
Dental gold alloy (precious metal)	68.9(10 × 10^6)				310–380(45–55)			20–35
Ceramics and glasses[b]								
Mullite (aluminosilicate) porcelain	69(10 × 10^6)					69(10)		
Steatite (magnesia aluminosilicate) porcelain	60(10 × 10^6)					140(20)		
Superduty fireclay (aluminosilicate) brick	97(14 × 10^6)					5.2(0.75)		
Alumina (Al$_2$O$_3$) crystals	380(55 × 10^6)					340–1000(49–145)		
Sintered alumina (~ 5% porosity)	370(54 × 10^6)					210–340(30–49)		
Alumina porcelain (90–95% alumina)	370(54 × 10^6)					340(49)		
Sintered magnesia (~ 5% porosity)	210(30 × 10^6)					100(15)		
Magnesite (magnesia) brick	170(25 × 10^6)					28(4)		
Sintered spinel (magnesia aluminate) (~ 5% porosity)	238(35 × 10^6)					90(13)		
Sintered stabilized zirconia (~ 5% porosity)	150(22 × 10^6)					83(12)		
Sintered beryllia (~ 5% porosity)	310(45 × 10^6)					140–280(20–41)		
Dense silicon carbide (~ 5% porosity)	470(68 × 10^6)					170(25)		
Bonded silicon carbide (~ 20% porosity)	340(49 × 10^6)					14(2)		
Hot-pressed boron carbide (~ 5% porosity)	290(42 × 10^6)					340(49)		
Hot-pressed boron nitride (~ 5% porosity)	83(12 × 10^6)					48–100(7–15)		
Silica gass	72.4(10.5 × 10^6)					107(16)		
Borosilicate glass	69(10 × 10^6)					69(10)		

Appendix 4B *Continued*

Material	E [GPa (psi)]	E_{flex} [MPa (ksi)]	E_{Dyn} [MPa (ksi)]	Y.S. [MPa (ksi)]	T.S. [MPa (ksi)]	Flexural strength [MPa (ksi)]	Compressive strength [MPa (ksi)]	Percent elongation at failure
Polymers^c								
Polyethylene								
High-density	$0.830(0.12 \times 10^6)$				28(4)			15–100
Low-density	$0.170(0.025 \times 10^6)$				14(2)			90–800
Polyvinylchloride	$2.80(0.40 \times 10^6)$				41(6)			2–30
Polypropylene	$1.40(0.20 \times 10^6)$				34(5)			10–700
Polystyrene	$3.10(.045 \times 10^6)$				48(7)			1–2
Polyesters		8960(1230)			158(22.9)			2.7
Acrylics (Lucite)	$2.90(0.42 \times 10^6)$				55(8)			5
Polyamides (nylon 66)	$2.80(0.41 \times 10^6)$	2830(410)			82.7(12.0)			60
Cellulosics	$3.40 - 28.0(0.50 - 4.0 \times 10^6)$				14–55(2–8)			5–40
ABS	$2.10(0.30 \times 10^6)$				28–48(4–7)			20–80
Polycarbonates	$2.40(0.35 \times 10^6)$				62(9)			110
Acetals	$3.10(0.45 \times 10^6)$	2830(410)			69(10)			50
Polytetrafluoroethylene (Teflon)	$0.41(0.060 \times 10^6)$				17(2.5)			100–350
Polyester-type thermoplastic elastomers		585(85)			46(6.7)			400
Phenolics (phenolformaldehyde)	$6.90(1.0 \times 10^6)$				52(7.5)			0
Urethanes					34(5)			—
Urea-melamine	$10.0(1.5 \times 10^6)$				48(7)			0
Polyesters	$6.90(1.0 \times 10^6)$				28(4)			0
Epoxies	$6.90(1.0 \times 10^6)$				69(10)			0
Polybutadiene/polystyrene copolymer								
Vulcanized	$0.0016(0.23 \times 10^3)$		0.8(0.12)		1.4–3.0(0.20–0.44)			440–600
Vulcanized with 33% carbon black	$0.003 - 0.006(0.4 - 0.9 \times 10^3)$		8.7(1.3)		17–28(2.5–4.1)			400–600
Polyisoprene								
Vulcanized	$0.0013(0.19 \times 10^3)$		0.4(0.06)		17–25(2.5–3.6)			750–850
Vulcanized with 33% carbon black	$0.003 - 0.008(0.44 - 1.2 \times 10^3)$		6.2(0.90)		25–35(3.6–5.1)			550–650
Polychloroprene								
Vulcanized	$0.0016(0.23 \times 10^3)$		0.7(0.10)		25–38(3.6–5.5)			800–1000
Vulcanized with 33% carbon black	$0.003 - 0.005(0.4 - 0.7 \times 10^3)$		2.8(0.41)		21–30(3.0–4.4)			500–600
Polyisobutene/polyisoprene copolymer								
Vulcanized	$0.0010(0.15 \times 10^3)$		0.4(0.06)		18–21(2.6–3.0)			750–950
Vulcanized with 33% carbon black	$0.003 - 0.004(0.4 - 0.6 \times 10^3)$		3.6(0.52)		18–21(2.6–3.0)			650–850
Silicones					7(1)			4000
Vinylidene fluoride/hexafluoropropylene					12.4(1.8)			

Appendix 4B *Continued*

Material	E [GPa (psi)]	E_{flex} [MPa (ksi)]	E_{Dyn} [MPa (ksi)]	Y.S. [MPa (ksi)]	T.S. [MPa (ksi)]	Flexural strength [MPa (ksi)]	Compressive strength [MPa (ksi)]	Percent elongation at failure
Composites[d]								
E-glass (73.3 vol %) in epoxy (parallel loading of continuous fibers)	56(8.1 × 10⁶)				1640(238)	—	—	2.9
Al₂O₃ whiskers (14 vol %) in epoxy	41(6 × 10⁶)				779(113)	—	—	—
C (67 vol %) in epoxy (parallel loading)	221(32 × 10⁶)				1206(175)	—	—	—
Kevlar (82 vol %) in epoxy (parallel loading)	86(12 × 10⁶)				1517(220)	—	—	—
B (70 vol %) in epoxy (parallel loading of continuous filaments)	210 − 280(30 − 40 × 10⁶)				1400–2100(200–300)	—	—	—
Al₂O₃ (10 vol %) dispersion-strengthened aluminum	—				330(48)	—	—	—
W (50 vol %) in copper (parallel loading of continuous filaments)	260(38 × 10⁶)				1100(160)	—	—	—
W particles (50 vol %) in copper	190(27 × 10⁶)				380(55)	—	—	—
SiC whiskers in Al₂O₃	—				—	800(116)	—	—
SiC fibers in SiC	—				—	750(109)	—	—
SiC whiskers in reaction-bonded Si₃N₄	—				—	900(131)	—	—
Douglas fir, kiln-dried at 12% moisture (loaded parallel to grain)	13.4(1.95 × 10⁶)				85.5(12.4)	—	49.9(7.24)	—
Douglas fir, kiln-dried at 12% moisture (loaded perpendicular to grain)	—				—	—	5.5(0.80)	—
Standard concrete, water/cement ration of 4 (after 28 days)	—				—	—	41(6.0)	—
Standard concrete, water/cement ration of 4 (after 28 days) with air entrainer	—				—	—	33(4.8)	—

[a] From Table 6.2
[b] From Table 6.5
[c] From Tables 6.7 and 6.8
[d] From Table 14.12

Appendix 4C Miscellaneous Mechanical Properties Data for Selected Engineering Materials

	Poisson's ratio ν	Brinell hardness number	Rockwell hardness R scale	Charpy Impact Energy [J(ft·lb)]	Izod Impact Energy [J(ft·lb)]	K_{IC} (MPa $\sqrt{m}$)	Fatigue limit [MPa (ksi)]
Metals and alloys[a]							
1040 carbon steel	0.3	235		180(133)			280(41)
Mild steel						140	
Medium-carbon steel							
8630 low-alloy steel	0.3	220		51(41)		51	400(58)
304 stainless steel	0.29						170(25)
410 stainless steel		250		34(25)			
L2 tool steel				26(19)			
Rotor steels (A533; Discalloy)						204–214	
Pressure-vessel steels (HY130)						170	
High-strength steels (HSS)						50–154	
Ductile iron	0.29	167		9(7)		6–20	
Cast iron						100–350	
Pure ductile metals (e.g., Cu, Ni, Ag, Al)							
Be (brittle hcp metal)						4	
3003-H14 aluminum	0.33	40					62(9)
2048, plate aluminum				10.3(7.6)		23–45	
Aluminum alloys (high strength-low strength)							
AZ31B magnesium	0.35	73		4.3(3.2)			69(10)
AM100A casting magnesium	0.35	53		0.8(0.6)			
Ti–5Al–2.5Sn	0.35	335		23(17)			410(59)
Ti–6Al–4V	0.33						
Titanium alloys						55–115	
Aluminum bronze, 9% (copper alloy)	0.33	165		48(35)			200(29)
Monel 400 (nickel alloy)	0.32	110–150		298(220)			290(42)
AC41A zinc		91					56(8)
50:50 solder (lead alloy)		14.5		21.6(15.9)			
Nb–1 Zr (refractory metal)				174(128)			
Dental gold alloy (precious metal)		80–90					

Appendix 4C *Continued*

	Poisson's ratio ν	Brinell hardness number	Rockwell hardness R scale	Charpy Impact Energy [J(ft·lb)]	Izod Impact Energy [J(ft·lb)]	K_{IC} (MPa $\sqrt{m}$)	Fatigue limit [MPa (ksi)]
Ceramics and glasses[b]							
Al$_2$O$_3$	0.26					3–5	
BeO	0.26						
CeO$_2$	0.27–0.31						
MgO						3	
Cordierite (2MgO·2Al$_2$O$_3$·5SiO$_2$)	0.31						
Mullite (3Al$_2$O$_3$·2SiO$_2$)	0.25						
SiC	0.19					3	
Si$_3$N$_4$	0.24					4–5	
TaC	0.24						
TiC	0.19						
TiO$_2$	0.28						
Partially stabilized ZrO$_2$	0.23					9	
Fully stabilized ZrO$_2$	0.23–0.32						
Glass-ceramic (MgO-Al$_2$O$_3$-SiO$_2$)	0.24						
Electrical porcelain						1	
Cement/concrete, unreinforced						0.2	
Soda glass (Na$_2$O-SiO$_2$)						0.7–0.8	
Borosilicate glass	0.2						
Glass from cordierite	0.26						
Polymers[c]							
Polythylene							
High-density			40		1.4–16(1–12)	2	
Low-density			10		22(16)	1	
Polyvinylchloride			110		1.4(1)	—	
Polypropylene			90		1.4–15(1–11)	3	
Polystyrene			75		0.4(0.3)	2	
Polyesters			120		1.4(1)	0.5	40.7(5.9)
Acrylics (Lucite)			130		0.7(0.5)	—	
Polyamides (nylon 66)	0.41		121		1.4(1)	3	
Cellulosics			50 to 115		3–11(2–8)	—	
ABS			95		1.4–14(1–10)	4	
Polycarbonates			118		19(14)	1.0–2.6	
Acetals	0.35		120		3(2)	—	31(4.5)
Polytetrafluoroethylene (Teflon)			70		5(4)	—	
Polyester-type thermoplastic elastomers					—	—	
Phenolics (phenolformaldehyde)			125		0.4(0.3)	—	
Urethanes			—		—	—	
Urea-melamine			115		0.4(0.3)	—	
Polyesters			100		0.5(0.4)	—	
Epoxies			90		1.1(0.8)	0.3–0.5	

Appendix 4C *Continued*

	Poisson's ratio ν	Brinell hardness number	Rockwell hardness R scale	Charpy Impact Energy [J(ft·lb)]	Izod Impact Energy [J(ft·lb)]	K_{IC} (MPa $\sqrt{m}$)	Fatigue limit [MPa (ksi)]
Composites[d]							
E-glass (73.3 vol %) in epoxy (parallel loading of continuous fibers)						42–60	
B (70 vol %) in epoxy (parallel loading of continuous filaments)						46	
SiC whiskers in Al_2O_3						8.7	
SiC fibers in SiC						25.0	
SiC whiskers in reaction-bonded Si_3N_4						20.0	
Douglas fir, kiln-dried at 12% moisture (loaded parallel to grain)						11–13	
Douglas fir, kiln-dried at 12% moisture (loaded perpendicular to grain)						0.5–1	
Concrete							
Standard concrete, water/cement ratio of 4 (after 28 days)						0.2	

[a] From Tables 6.4, 6.11, 8.1, 8.3, and 8.4
[b] From Tables 6.6 and 8.3
[c] From Tables 6.7, 6.12, 8.2, and 8.3
[d] From Table 14.12

807

Appendix 4D Thermal Properties Data for Selected Materials

	Specific heat[a] c_p [J/kg · K]	Linear coefficient of thermal expansion[b] α[mm/(mm ·° C) × 10^6]			Thermal conductivity[c] k[J/(s · m · K)]			
		27° C	527° C	0–1000° C	27° C	100° C	527° C	1000° C
Metals[a]								
Aluminum	900	23.2	33.8		237		220	
Copper	385	16.8	20.0		398		371	
Gold	129	14.1	16.5		315		292	
Iron (α)	444				80		43	
Lead	159							
Nickel	444	12.7	16.8		91		67	
Silver	237	19.2	23.4		427		389	
Titanium	523				22		20	
Tungsten	133	4.5	4.8		178		128	
Ceramics and glasses[b]								
Mullite ($3Al_2O_3 \cdot 2SiO_2$)				5.3		5.9		3.8
Porcelain				6.0		1.7		1.9
Fireclay refractory				5.5		1.1		1.5
Al_2O_3	160			8.8		30		6.3
Spinel ($MgO \cdot Al_2O_3$)				7.6		15		5.9
MgO	457			13.5		38		7.1
UO_2				10.0				
ZrO_2 (stabilized)				10.0		2.0		2.3
SiC	344			4.7				
TiC						25		5.9
Carbon (diamond)	519							
Carbon (graphite)	711							
Silica glass				0.5		2.0		2.5
Soda-lime-silica glass				9.0		1.7		
Polymers[c]								
Nylon 66	1260–2090	30–31			2.9			
Phenolic	1460–1670	30–45			0.17–0.52			
Polyethylene (high-density)	1920–2300	149–301			0.33			
Polypropylene	1880	68–104			2.1–2.4			
Polytetrafluoroethylene (PTFE)	1050	99			0.24			

[a] From Table 7.1
[b] From Table 7.2
[c] From Table 7.4

Properties of the Electronic, Optical, and Magnetic Materials

Appendix 5A. Electrical Conductivities of Selected Materials at Room Temperature

Material	Conductivity, σ ($\Omega^{-1} \cdot m^{-1}$)
Metals and alloys[a]	
Aluminum (annealed)	35.36×10^6
Copper (annealed standard)	58.00×10^6
Gold	41.0×10^6
Iron (99.9+%)	10.3×10^6
Lead (99.73+%)	4.84×10^6
Magnesium (99.80%)	22.4×10^6
Mercury	1.04×10^6
Nickel (99.95% + Co)	14.6×10^6
Nichrome (66% Ni + Cr and Fe)	1.00×10^6
Platinum (99.99%)	9.43×10^6
Silver (99.78%)	62.9×10^6
Steel (wire)	$5.71 - 9.35 \times 10^6$
Tungsten	18.1×10^6
Zinc	16.90×10^6
Semiconductors[b]	
Silicon (high purity)	0.40×10^{-3}
Germanium (high purity)	2.0
Gallium arsenide (high purity)	0.17×10^{-6}
Indium antimonide (high purity)	17×10^3
Lead sulfide (high purity)	38.4
Ceramics, glasses, and polymers[c]	
Aluminum oxide	$10^{-10} - 10^{-12}$
Borosilicate glass	10^{-13}
Polyethylene	$10^{-13} - 10^{-15}$
Nylon 66	$10^{-12} - 10^{-13}$

[a] From Tables 15.1 and 15.2
[b] From Tables 15.1 and 15.5
[c] From Table 15.1

Appendix 5B.   Properties of some Common Semiconductors at Room Temperature

Material	Energy gap, E_g (eV)	Electron mobility, $\mu_e [m^2/(V \cdot s)]$	Hole mobility, $\mu_h [m^2/(V \cdot s)]$	Carrier density, $n_e (= n_h)$ (m^{-3})
Elements[a]				
Si	1.107	0.140	0.038	14×10^{15}
Ge	0.66	0.364	0.190	23×10^{18}
III-V Compounds[b]				
AlSb	1.6	0.090	0.040	—
GaP	2.25	0.030	0.015	—
GaAs	1.47	0.720	0.020	1.4×10^{12}
GaSb	0.68	0.500	0.100	—
InP	1.27	0.460	0.010	—
InAs	0.36	3.300	0.045	—
InSb	0.17	8.000	0.045	13.5×10^{21}
II-VI Compounds[b]				
ZnSe	2.67	0.053	0.002	—
ZnTe	2.26	0.053	0.090	—
CdS	2.59	0.034	0.002	—
CdTe	1.50	0.070	0.007	—
HgTe	0.025	2.200	0.016	—

[a] From Table 15.5
[b] From Table 17.5

Appendix 5C.   Dielectric Constant and Dielectric Strength for Selected Insulators

Material	Dielectric constant,[a] κ	Dielectric strength,[a] (kV/mm)
Al_2O_3 (99.9%)	10.1	9.1
Al_2O_3 (99.5%)	9.8	9.5
BeO (99.5%)	6.7	10.2
Cordierite	4.1–5.3	2.4–7.9
Nylon 66—reinforced with 33% glass fibers (dry-as-molded)	3.7	20.5
Nylon 66—reinforced with 33% glass fibers (50% relative humidity)	7.8	17.3
Acetal (50% relative humidity)	3.7	19.7
Polyester	3.6	21.7

[a] From Table 15.4

Appendix 5D. Refractive Index for Selected Optical Materials

Material	Average refractive index
Ceramics and glasses[a]	
Quartz (SiO_2)	1.55
Mullite ($3Al_2O_3 \cdot 2SiO_2$)	1.64
Orthoclase ($KAlSi_3O_8$)	1.525
Albite ($NaAlSi_3O_8$)	1.529
Corundum (Al_2O_3)	1.76
Periclase (MgO)	1.74
Spinel ($MgO \cdot Al_2O_3$)	1.72
Silica glass (SiO_2)	1.458
Borosilicate glass	1.47
Soda-lime-silica glass	1.51–1.52
Glass from orthoclase	1.51
Glass from albite	1.49
Polymers[b]	
Polythylene	
High-density	1.545
Low-density	1.51
Poly(vinyl chloride)	1.54–1.55
Polypropylene	1.47
Polystyrene	1.59
Cellulosics	1.46–1.50
Polyamides (nylon 66)	1.53
Polytetrafluoroethylene (Teflon)	1.35–1.38
Phenolics (phenol-formaldehyde)	1.47–1.50
Urethanes	1.5–1.6
Epoxies	1.55–1.60
Polybutadiene/polystyrene copolymer	1.53
Polyisoprene (natural rubber)	1.52
Polychloroprene	1.55–1.56

[a] From Table 16.1
[b] From Table 16.2

Appendix 5E. Magnetic Properties Data for Selected Materials

Material	Initial relative permeability (μ_r at $B \sim 0$)	Hysteresis loss (J/m^3 per cycle)	Saturation induction (Wb/m^2)	Maximum BH product $(BH)_{max}$(A·Wb/m^3)
Metals and alloys[a]				
Commercial iron ingot	250	500	2.16	
Fe–4% Si, random	500	50–150	1.95	
Fe–3% Si, oriented	15,000	35–140	2.0	
45 Permalloy (45% Ni–55% Fe)	2,700	120	1.6	
Mumetal (75% Ni–5% Cu–2% Cr–18% Fe)	30,000	20	0.8	
Supermalloy (79% Ni–15% Fe–5% Mo)	100,000	2	0.79	
Amorphous ferrous alloys				
(80% Fe–20% B)	—	25	1.56	
(82% Fe–10% B–8% Si)	—	15	1.63	
Samarium-cobalt				120,000
Platinum-cobalt				70,000
Alnico				36,000
Ceramics[b]				
Ferroxcube III (Mn Fe$_2$O$_4$-Zn Fe$_2$O$_4$)	1,000	—	0.25	—
Vectolite (30% Fe$_2$O$_3$-44% Fe$_3$O$_4$-26% Cr$_2$O$_3$)	—	—	—	6,000

[a] From Tables 18.2 and 18.3

[b] From J. F. Shackelford, "Properties of Materials," in *The Electronics Handbook*, J. C. Whitaker, Ed., CRC Press, Boca Raton, Florida, 1996, pp. 2326–51.

Materials Characterization Locator

The key experimental tools for characterizing the structure and chemistry of engineering materials have been introduced throughout the text. This table provides a compact summary of the location of those discussions.

Materials characterization technique	Location
X-ray diffraction	Section 3.7, pp. 101–109
Optical microscopy	Section 4.7, pp. 144–145
Electron microcopy (scanning, transmission, and atomic-resolution)	Section 4.7, pp. 144–151
Scanning tunneling microscope (STM) and atomic force microscope (AFM)	Section 4.7, pp. 148–150
Surface analysis [x-ray fluorescence (XRF), Energy-dispersive x-ray spectrometry (EDX), Auger electron spectroscopy (AES) and electron spectroscopy for chemical analysis (ESCA) or x-ray photoelectron spectroscopy (XPS)]	Section 19.11, pp. 000–000
Nondestructive testing (especially x-radiography and ultrasonic testing)	Section 8.4, pp. 289–295

Glossary

Following are definitions of the key words that appeared at the end of each chapter.

abrasive wear   Wear occurring when a rough, hard surface slides on a softer surface. Grooves and wear particles form on the softer surface.

acceptor level   Energy level near the valence band in a *p*-type semiconductor. (See Figure 17–10.)

acoustic-emission testing   A type of nondestructive testing in which defects are monitored as a result of ultrasonic emissions caused by stressing the flawed material.

activation energy   Energy barrier that must be overcome by atoms in a given process or reaction.

active   Tending to be oxidized in an electrochemical cell.

actuator   A device that carries out a function in response to an input signal.

addition polymerization   See *Chain growth*.

additive   Material added to a polymer to provide specific characteristics.

adhesive wear   Wear occurring when two smooth surfaces slide over each other. Fragments are pulled off one surface and adhere to the other.

admixture   Addition to cement to provide desirable features, for example, coloring agent.

advanced composites   Synthetic fiber-reinforced composites with relatively high modulus fibers. The fiber modulus is generally higher than that of E-glass.

aerosol   Airborn particle.

age hardening   See *Precipitation hardening*.

aggregate   Nonfibrous dispersed phase in a composite. Specifically refers to sand and gravel dispersed in concrete.

aggregate composite   Material reinforced with a dispersed particulate (rather than fibrous) phase.

alloy   Metal composed of more than one element.

aluminum alloy   Metal alloy composed of predominantly aluminum.

amorphous metal   Metal lacking long-range crystalline structure.

amorphous semiconductor   Semiconductor lacking long-range crystalline structure.

amplifier   Electronic device for increasing current.

anion   Negatively charged ion.

anisotropic   Having properties that vary with direction.

annealing   Heat treatment for the general purpose of softening or stress-relieving a material.

annealing point   Temperature at which a glass has a viscosity of $10^{13.4}$ P and internal stresses can be relieved in about 15 minutes.

anode   The electrode in an electrochemical cell that oxidizes, giving up electrons to an external circuit.

anodic reaction   The oxidation reaction occurring at the anode in an electrochemical cell.

anodized aluminum   Aluminum alloy with a protective layer of Al_2O_3.

antiparallel spin pairing   The alignment of atomic magnetic moments in opposite directions in ferrimagnetic materials.

aqueous corrosion   Dissolution of a metal into a water-based environment.

Arrhenius behavior   Having a property (such as diffusivity) that follows the Arrhenius equation.

Arrhenius equation   General expression for a thermally activated process, such as diffusion. (See Equation 5.1.)

Arrhenius plot   A semilog plot of ln (rate) versus the reciprocal of absolute temperature $(1/T)$. The slope of the straight-line plot gives the activation energy of the rate mechanism.

atactic   Irregular alteration of side groups along a polymeric molecule. (See Figure 13–11.)

atomic force microscope (AFM)   A derivative of the scanning tunneling microscope in which a small force, rather than an electrical current, is used to trace out an image of the structure of the surface of a material.

atomic mass   Mass of an individual atom expressed in atomic mass units.

atomic mass unit   Equal to 1.66×10^{-24} g. Approximately equal to the mass of a proton or neutron.

atomic number   Number of protons in the nucleus of an atom.

atomic packing factor (APF)   Fraction of unit cell volume occupied by atoms.

atomic radius   Distance from atomic nucleus to outermost electron orbital.

atomic resolution electron microscope   A refined version of the transmission electron microscope in which the packing arrangements of atoms in a thin sample can be imaged.

atomic-scale architecture   Structural arrangement of atoms in an engineering material.

Auger electron   Secondary electron with a characteristic energy, providing a basis for chemical identification. (See Figure 19–31.)

austempering   Heat treatment of a steel involving holding just above the martensitic transformation range long enough to completely form bainite. (See Figure 10–20.)

austenite   Face-centered cubic (γ) phase of iron or steel.

austenitic stainless steel   Corrosion-resistant ferrous alloy with a predominant face-centered cubic (γ) phase.

Avogadro's number   Equal to the number of atomic mass units per gram of material (0.6023×10^{24}).

bainite   Extremely fine needlelike microstructure of ferrite and cementite. (See Figure 10–9.)

base   Intermediate region between emitter and collector in a transistor.

Bernal model   Representation of the atomic structure of an amorphous metal as a connected set of polyhedra. (See Figure 4–24.)

bifunctional   Polymer with two reaction sites for each mer, resulting in a linear molecular structure.

binary diagram   Two-component phase diagram.

biological material   A naturally occurring structural material, such as bone.

biomaterial   An engineered material created for a biological or medical application.

biomimetic processing   A ceramic fabrication technique which imitates natural processes, such as sea shell formation.

bipolar junction transistor (BJT)   A "sandwich" configuration such as the *p-n-p* transistor shown in Figure 17–20.

blend   Molecular-scale polymeric mixture. (See Figure 13–4.)

bloch wall   Narrow region of magnetic moment orientation change separating adjacent domains. (See Figure 18–10.)

block copolymer   Combination of polymeric components in "blocks" along a single molecular chain. (See Figure 13–3.)

blow molding   Processing technique for thermoplastic polymers.

body-centered cubic   Common atomic arrangement for metals, as illustrated in Figure 3–4.

Bohr magneton   Unit of magnetic moment ($= 9.27 \times 10^{-24}$ ampere $\cdot$ m^2).

bond angle   Angle formed by three adjacent, directionally bonded atoms.

bond energy   Net energy of attraction (or repulsion) as a function of separation distance between two atoms or ions.

bond force   Net force of attraction (or repulsion) as a function of separation distance between two atoms or ions.

bond length   Center-to-center separation distance between two adjacent, bonded atoms or ions.

borosilicate glass   High-durability, commercial glassware composed of primarily silica, with a significant component of B_2O_3.

Bragg angle   Angle relative to a crystal plane from which x-ray diffraction occurs. (See Figure 3–35.)

Bragg's law   The relationship defining the condition for x-ray diffraction by a given crystal plane (Equation 3.5).

branching   The addition of a polymeric molecule to the side of a main molecular chain. (See Figure 13–12.)

Brinell hardness number   Parameter obtained from indentation test, as defined in Table 6.10.

brittle   Lacking in deformability.

brittle fracture   Failure of a material following mechanical deformation with the absence of significant ductility.

bronze age   The period of time from roughly 2000 B.C. to 1000 B.C. representing the foundation of metallurgy.

buckminsterfullerene   Carbon molecule named in honor of R. Buckminster Fuller, inventor of the geodesic dome, which resembled the initial fullerene, C_{60}.

buckyball   Nickname for the buckminsterfullerene molecule, C_{60}.

buckytube   Cylindrical buckminsterfullerene molecule composed of hexagonal carbon rings.

bulk diffusion   See *Volume diffusion.*

Burgers vector   Displacement vector necessary to close a stepwise loop around a dislocation.

capacitor   Electrical device involving two electrodes separated by a dielectric.

carbon-carbon composite   An advanced composite system with especially high modulus and strength.

carbon steel   Ferrous alloy with nominal impurity level and carbon as the primary compositional variable.

carburization   The diffusion of carbon atoms into the surface of steel for the purpose of hardening the alloy.

carrier mobility   Drift velocity per unit electric field strength for a given charge carrier in a conducting material.

cast iron   Ferrous alloy with greater than 2 wt % carbon.

casting   Material processing technique involving pouring of molten liquid into a mold, followed by solidification of the liquid.

cathode   The electrode in an electrochemical cell that accepts electrons from an external circuit.

cathodic reaction   The reduction reaction occurring at the cathode in an electrochemical cell.

cation   Positively charged ion.

cavitation   A form of wear damage associated with the collapse of a bubble in an adjacent liquid.

cement   Matrix material (usually a calcium aluminosilicate) in concrete, an aggregate composite.

ceramic   Nonmetallic, inorganic engineering material.

ceramic magnet   Ceramic material with an engineering application predominantly based on its magnetic properties.

ceramic-matrix composite   Composite material in which the reinforcing phase is dispersed in a ceramic.

cermet   A ceramic-metal composite.

cesium chloride   Simple compound crystal structure as illustrated in Figure 3–8.

chain growth   Polymerization process involving a rapid, "chain reaction" of chemically activated monomers.

chalcogenide   Compound containing S, Se, or Te.

charge   Quantity of positive or negative carriers.

charge carrier   Atomic-scale species by which electricity is conducted in materials.

charge density   Number of charge carriers per unit volume.

charge neutrality   Absence of net positive or negative charge. A common "ground rule" in ionic compound formation.

Charpy test   Method for measuring impact energy, as illustrated by Figure 8–1.

chemical vapor deposition   The production of thin layers of materials in the fabrication of integrated circuits by means of specific chemical reactions.

chemically strengthened glass   A fracture-resistant glass produced by the compressive stressing of the silicate network as a result of the chemical exchange of larger radius K^+ ions for the Na^+ ions in the glass surface.

chip   A thin slice of crystalline semiconductor upon which electrical circuitry is produced by controlled diffusion.

clay   Fine-grained soil, composed chiefly of hydrous aluminosilicate minerals.

coercive field   Magnitude of a reverse electric field necessary to return a polarized ferroelectric to zero polarization. Also, the magnitude of a reverse magnetic field necessary to return a magnetized ferromagnetic material to zero induction.

coercive force   Alternate term for the coercive field in a ferromagnetic material.

coherent   Light source in which the light waves are in phase.

coherent interface   Interface between a matrix and precipitate at which the crystallographic structures maintain registry.

coincident site lattice (CSL)   Array of atomic sites common to both crystal lattice orientations of the grains adjacent to a given grain boundary.

cold work   Mechanical deformation of a metal at relatively low temperatures.

collector   Region in a transistor that receives charge carriers.

color   Visual sensation associated with various portions of the electromagnetic spectrum with wavelengths between 400 and 700 nm.

colorant   Additive for a polymer for the purpose of providing color.

complete solid solution   Binary phase diagram representing two components that can dissolve in all proportions. (See Figure 9–5.)

complex failure   Material fracture involving the sequential operation of two distinct mechanisms.

component   Distinct chemical substance, for example, Al or Al_2O_3.

composite   Material composed of a microscopic-scale combination of individual materials from the categories of metals, ceramics (and glasses), and polymers.

compound semiconductor   Semiconductor consisting of a chemical compound, rather than a single element.

compression molding   Processing technique for thermosetting polymers.

concentration cell   See *Ionic concentration cell.*

concentration gradient   Change in concentration of a given diffusing species with distance.

concrete   Composite composed of aggregate, cement, water, and, in some cases, admixtures.

condensation polymerization   See *Step growth*.

conduction band   A range of electron energies in a solid associated with an unoccupied energy level in an isolated atom. An electron in a semiconductor becomes a charge carrier upon promotion to this band.

conduction electron   A negative charge carrier in a semiconductor. (See also *Conduction band*.)

conductivity   Reciprocal of electrical resistivity.

conductor   Material with a substantial level of electrical conduction, for example, a conductivity greater than 10^{+4} $\Omega^{-1} \cdot m^{-1}$.

congruent melting   The case in which the liquid formed upon melting has the same composition as the solid from which it was formed.

continuous coolng transformation (CCT) diagram   A plot of percentage phase transformation under nonisothermal conditions using axes of temperature and time.

continuous fiber   Composite reinforcing fiber without a break within the dimensions of the matrix.

controlled devitrification   Processing technique for glass-ceramics in which a glass is transformed to a fine-grained, crystalline ceramic.

coordination number   Number of adjacent ions (or atoms) surrounding a reference ion (or atom).

copolymer   Alloylike result of polymerization of an intimate solution of different types of monomers.

copper alloy   Metal alloy composed of predominantly copper.

cored structure   Microstructure in which concentration gradients occur in individual grains. (See Figure 11–5.)

corrode   Lose material to a surrounding solution in an electrochemical process.

corrosion   The dissolution of a metal in an aqueous environment.

corrosion-fatigue failure   Metal fracture due to the combined actions of a cyclic stress and a corrosive environment.

corrosive wear   Wear that takes place with sliding in a corrosive environment.

corundum   Compound crystal structure as illustrated in Figure 3–13.

cosine law   Expression describing the scattering of light from an ideally "rough" surface, as defined by Equation 16.6.

coulombic attraction   Tendency toward bonding between oppositely charged species.

covalent bond   Primary, chemical bond involving electron sharing between atoms.

creep   Plastic (permanent) deformation occurring at a relatively high temperature under constant load over a long time period.

creep curve   Characteristic plot of strain versus time for a material undergoing creep deformation. (See Figure 6–31.)

creep-rupture failure   Material fracture following plastic deformation at a relatively high temperature (under constant load over a long time period).

cristobalite   Compound crystal structure as illustrated in Figure 3–11.

critical current density   Current flow at which a material stops being superconducting.

critical magnetic field   Field above which a material stops being superconducting.

critical resolved shear stress   Stress operating on a slip system and great enough to produce slip by dislocation motion. (See Equation 6.15.)

cross-linking   Joining of adjacent, linear polymeric molecules by chemical bonding. See, for example, the vulcanization of rubber in Figure 6–46.

crystal (Bravais) lattice   The 14 possible arrangements of points (with equivalent environments) in three-dimensional space.

crystal growing   Processing techniques for forming single crystals, as summarized in Table 17.7.

crystal system   The seven, unique unit cell shapes that can be stacked together to fill three-dimensional space.

crystalline   Having constituent atoms stacked together in a regular, repeating pattern.

crystalline ceramic   A ceramic material with a predominantly crystalline atomic structure.

cubic   Simplest of the seven crystal systems. (See Table 3.1.)

cubic close packed   See *Face-centered cubic.*

current   Flow of charge carriers in an electrical circuit.

Debye temperature   The temperature above which the value of the heat capacity at constant volume, C_v, levels off at approximately $3R$, where R is the universal gas constant.

degree of polymerization   Average number of mers in a polymeric molecule.

degrees of freedom   Number of independent variables available in specifying an equilibrium microstructure.

delocalized electron   An electron equally probable to be associated with any of a large number of adjacent atoms.

dendritic structure   A nonequilibrium microstructure from casting. (See Figure 11–6.)

design for the environment (DFE)   The inclusion of environmental concerns in the process of engineering design.

design parameter   Material property that serves as the basis for selecting a given engineering material for a given application.

design selection   Method for corrosion prevention, for example, avoid a small-area anode next to a large-area cathode.

device   A functional electronic design, for example, the transistor.

devitrified   Crystallized. (In reference to a material initially in a glassy, or vitreous, state.)

diamagnetism   Magnetic behavior associated with a relative permeability of slightly less than one.

diamond cubic   Important crystal structure for covalently bonded, elemental solids, as illustrated in Figure 3–23.

dielectric   Electrically insulating material.

dielectric constant   The factor by which the capacitance of a parallel-plate capacitor is increased by inserting a dielectric in place of a vacuum.

dielectric strength   The voltage gradient at which a dielectric "breaks down" and becomes conductive.

diffraction angle Twice the Bragg angle. (See Figure 3–36.)

diffraction geometry Geometry associated with Bragg's law, as illustrated in Figure 3–35.

diffractometer An electromechanical scanning device for obtaining an x-ray diffraction pattern of a powder sample.

diffuse reflection Light reflection due to surface roughness. (See Figure 16–6.)

diffusion The movement of atoms or molecules from an area of higher concentration to an area of lower concentration.

diffusion coefficient (diffusivity) Proportionality constant in the relationship between flux and concentration gradient, as defined in Equation 5.8.

diffusional transformation Phase transformation with strong time dependence, due to a mechanism of atomic diffusion.

diffusionless transformation Phase transformation that is essentially time independent, due to the absence of a diffusional mechanism.

diode A simple electronic device that limits current flow to an application of positive voltage (forward bias).

dipole Asymmetrical distribution of positive and negative charge, associated with secondary bonding.

dipole moment Product of charge and separation distance between centers of positive and negative charge in a dipole.

discrete (chopped) fiber Composite reinforcing fiber broken into segments.

dislocation Linear defect in a crystalline solid.

dislocation climb A mechanism for creep deformation, in which the dislocation moves to an adjacent slip plane by diffusion.

dispersion-strengthened metal An aggregate-type composite in which the metal contains less than 15 vol % oxide particles (0.01 to 0.1 micron in diameter).

domain Microscopic region of common electrical dipole alignment (in a ferroelectric material) or common magnetic moment alignment (in a ferromagnetic material).

domain structure Microstructural arrangement of magnetic domains.

domain wall See *Bloch wall.*

donor level Energy level near the conduction band in an *n*-type semiconductor. (See Figure 17–5.)

dopant Purposeful impurity addition in an extrinsic semiconductor.

double bond Covalent sharing of two pairs of valence electrons.

drain Region in a field-effect transistor that receives charge carriers.

drift velocity Average velocity of the charge carrier in an electrically conducting material.

ductile Deformable.

ductile fracture Metal failure occurring after the material is taken beyond its elastic limit.

ductile iron A form of cast iron that is relatively ductile due to spheroidal graphite precipitates, rather than flakes.

ductile-to-brittle transition temperature   Narrow temperature region in which the fracture of bcc alloys changes from brittle (at lower temperatures) to ductile (at higher temperatures).

ductility   Deformability. (Percent elongation at failure is a quantitative measure.)

dye   Soluble, organic colorant for polymers.

dynamic modulus of elasticity   Parameter representing the stiffness of a polymeric material under an oscillating load.

E-glass   Most generally used glass fiber composition for composite applications. (See Table 14.1.)

eddy current   Fluctuating electrical current in a conductor. A source of energy loss in ac applications of magnets.

eddy-current testing   A type of nondestructive testing in which defects are monitored as a result of changes in the impedance of an inspection coil, as the coil carries an alternating current and passes through the vicinity of the flawed material.

edge dislocation   Linear defect with the Burgers vector perpendicular to the dislocation line.

effusion cell   Resistance-heated furnace providing a source of atoms (or molecules) for a deposition process.

elastic deformation   Temporary deformation associated with the stretching of atomic bonds. (See Figure 6–18.)

elastomer   Polymer with a pronounced rubbery plateau in its plot of modulus of elasticity versus temperature.

electric permittivity   Proportionality constant in the relationship between charge density and electrical field strength, as defined in Equation 15.11.

electrical conduction   Having a measurable conductivity due to the movement of any type of charge carrier.

electrical field strength   Voltage per unit distance.

electrically poled   Polycrystalline material having a single, crystalline orientation due to the alignment of the starting powder under a strong electrical field.

electrochemical cell   System providing for connected anodic and cathodic electrode reactions.

electroluminescence   Luminescence caused by electrons.

electromotive force series   Systematic listing of half-cell reaction voltages, as summarized in Table 19.2.

electron   Negatively charged subatomic particle located in an orbital about a positively charged nucleus.

electron cloud   Assembly of delocalized electrons in a metallic solid.

electron density   Concentration of negative charge in an electron orbital.

electron gas   See *Electron cloud*.

electron hole   Missing electron in an electron cloud. A charge carrier with an effective positive charge.

electron-hole pair   Two charge carriers produced when an electron is promoted to the conduction band, leaving behind an electron hole in the valence band.

electron orbital   Location of negative charge about a positive nucleus in an atom.

electron sharing   Basis of the covalent bond.

electron spectroscopy for chemical analysis (ESCA)   See *X-ray photoelectron spectroscopy (XPS)*.

electron spin   Relativistic effect associated with the intrinsic angular momentum of an electron. For simplicity, can be likened to the motion of a spinning planet.

electron transfer   Basis of the ionic bond.

electronegativity   The ability of an atom to attract electrons to itself.

electronic ceramic   Ceramic material with an engineering application predominantly based on its electronic properties.

electronic conduction   Having a measurable conductivity due specifically to the movement of electrons.

electronic, optical, and magnetic materials   Those engineering materials used primarily for their electronic, optical, or magnetic properties.

electronic polymer   A linear polymer with alternating single and double bonds. The "extra" electrons associated with the double bonds are relatively conductive.

electroplating   Metal buildup at the cathode of an electrochemical cell.

element   Fundamental chemical species, as summarized in the periodic table.

emitter   Region in a transistor that serves as the source of charge carriers.

enamel   Glass coating on a metal substrate.

energy band   A range of electron energies in a solid associated with an energy level in an isolated atom.

energy band gap   Range of electron energies above the valence band and below the conduction band.

energy-dispersive x-ray spectrometry (EDX)   Chemical analysis by the use of characteristic x-ray photons produced by exposure to high-energy electrons.

energy level   Fixed binding energy between an electron and its nucleus.

energy loss   Area contained within the hysteresis loop of a ferromagnetic material.

energy trough   See *Energy well*.

energy well   The region around the energy minimum in a bonding energy curve such as Figure 2–18.

engineering polymer   Polymer with sufficient strength and stiffness to be a candidate for a structural application previously reserved for a metal alloy.

engineering strain   Increase in sample length at a given load divided by the original (stress-free) length.

engineering stress   Load on a sample divided by the original (stress-free) area.

environmental impact assessment (EIA)   An evaluation of the potential impact of emissions associated with an engineering process for the purpose of identifying major problems.

epitaxy   A vapor deposition technique involving the buildup of thin-film layers of one semiconductor on another while maintaining some particular crystallographic relationship between the layer and the substrate.

erosion   Wear caused by a stream of sharp particles and analogous to abrasive wear.

eutectic composition   Composition associated with the minimum temperature at which a binary system is fully melted. (See Figure 9–11.)

eutectic diagram Binary phase diagram with the characteristic eutectic reaction.

eutectic reaction The transformation of a liquid to two solid phases upon cooling, as summarized by Equation 9.3.

eutectic temperature Minimum temperature at which a binary system is fully melted. (See Figure 9–11.)

eutectoid diagram Binary phase diagram with the eutectoid reaction (Equation 9.4). (See Figure 9–18.)

exchange interaction Phenomenon among adjacent electron spins in adjacent atoms that leads to aligned magnetic moments.

exhaustion range Temperature range over which conductivity in an n-type semiconductor is relatively constant due to the fact that impurity-donated electrons have all been promoted to the conduction band.

extended length Length of a polymeric molecule that is extended as straight as possible. (See Figure 13–10.)

extrinsic gettering The capture of oxygen in a silicon device by using mechanical damage to produce dislocations as "gettering sites."

extrinsic semiconductor Semiconducting material with a purposeful impurity addition that, over a certain temperature range, establishes the level of conductivity.

extrusion molding Processing technique for thermoplastic polymers.

face-centered cubic Common atomic arrangement for metals, as illustrated in Figure 3–5.

failure analysis A systematic methodology for characterizing the failure of engineering materials.

failure prevention The application of knowledge obtained by failure analysis to prevent future catastrophes.

family of directions A set of structually equivalent crystallographic directions.

family of planes A set of structurally equivalent crystallographic planes.

fast fracture See *Flaw-induced fracture*.

fatigue The general phenomenon of material failure after several cycles of loading to a stress level below the ultimate tensile stress.

fatigue curve Characteristic plot of stress versus number of cycles to failure. (See Figure 8–10.)

fatigue failure The fracture of a metal part by a mechanism of slow crack growth, with a resulting "clamshell" fracture surface.

fatigue strength (endurance-limit) Lower limit of the applied stress at which a ferrous alloy will fail by cyclic loading.

Fermi function Temperature-dependent function that indicates the extent to which a given electron energy level is filled. (See Equation 15.7.)

Fermi level Energy of an electron in the highest filled state in the valence energy band at 0 K.

ferrimagnetism Ferromagneticlike behavior that includes some antiparallel spin pairing.

ferrite Ferrous alloy based on the bcc structure of pure iron at room temperature. Also, ferrimagnetic ceramic based on the inverse spinel structure.

ferritic stainless steel   Corrosion-resistant ferrous alloy with a predominant body-centered cubic (α) phase.

ferroelectric   Material that exhibits spontaneous polarization under an applied electrical field. (See Figure 15–23.)

ferromagnetism   Phenomenon of sharp rise in induction with applied magnetic field strength.

ferrous alloy   Metal alloy composed of predominantly iron.

fiber-reinforced composite   Material reinforced with a fibrous phase.

fiberglass   Composite system composed of a polymeric matrix reinforced with glass fibers.

Fibonacci series   Mathematical series in which each term is the sum of the two previous terms. The limiting value of the ratio of two consecutive terms is the golden ratio (1.618).

Fick's first and second laws   The basic, mathematical descriptions of diffusional flow. (See Equations 5.8 and 5.9.)

field-effect transistor   Solid-state amplifier, as illustrated in Figure 17–21.

filler   Relatively inert additive for a polymer, providing dimensional stability and reduced cost.

fired ceramic   Ceramic material after processing at a sufficiently high temperature to drive off any volatiles and to allow necessary strengthening mechanisms, such as sintering.

firing   The processing of a ceramic by heating raw materials to a high temperature, typically above 1000°C.

fivefold symmetry   Property of a structure that is equivalent after a 360°/5 rotation. This characteristic of quasicrystals is not consistent with traditional crystallography.

flame retardant   Additive used to reduce the inherent combustibility of certain polymers.

flaw-induced fracture   The rapid failure of a material due to the stress-concentration associated with a pre-existing flaw.

flexural modulus   Stiffness of a material, as measured in bending. (See Equation 6.12.)

flexural strength   Failure stress of a material, as measured in bending. (See Equation 6.10.)

fluorescence   Luminescence in which photon emission occurs rapidly (in less than about ten nanoseconds).

fluorite   Compound crystal structure, as illustrated in Figure 3–10.

flux density   Magnitude of the magnetic induction.

forward bias   Orientation of electrical potential to provide a significant flow of charge carriers in a rectifier.

Fourier's law   Relationship between rate of heat transfer and temperature gradient, as expressed in Equation 7.5.

fracture mechanics   Analysis of failure of structural materials with preexisting flaws.

fracture toughness   Critical value of the stress-intensity factor at a crack tip necessary to produce catastrophic failure.

free electron   Conducting electron in a metal.

free radical   Reactive atom or group of atoms containing an unpaired electron.

Frenkel defect   Vacancy-interstitialcy defect combination, as illustrated in Figure 4–9.

Fresnel's formula   Fraction of light reflected at an interface as a function of the index of refraction, as expressed by Equation 16.5.

fullerene   See *Buckminsterfullerene*.

fusion casting   Ceramic processing technique similar to metal casting.

gage length   Region of minimum cross-sectional area in a mechanical test specimen.

galvanic cell   Electrochemical cell in which the corrosion and associated electrical current are due to the contact of two dissimilar metals.

galvanic corrosion   Corrosion produced by the electromotive force associated with two dissimilar metals.

galvanic protection   Design configuration in which the structural component to be protected is made to be the cathode and, thereby, protected from corrosion.

galvanic series   Systematic listing of relative corrosion behavior of metal alloys in an aqueous environment, such as seawater.

galvanization   Production of a zinc coating on a ferrous alloy for the purpose of corrosion protection.

galvanized steel   Steel with a zinc coating for the purpose of corrosion protection.

garnet   Ferrimagnetic ceramic with a crystal structure similar to that of the natural gem garnet.

gaseous reduction   A cathodic reaction that can lead to the corrosion of an adjacent metal.

gate   Intermediate region in a field-effect transistor. (See Figure 17–21.)

Gaussian error function   Mathematical function based on the integration of the "bell-shaped" curve. It appears in the solution to many diffusion-related problems.

general diagram   A binary phase diagram containing more than one of the simple types of reactions described in Section 9.2.

general yielding   The slow "failure" of a material due to plastic deformation occurring at the yield strength.

gettering   A process used to capture oxygen and remove it from a region of a silicon wafer where device circuitry is developed.

Gibbs phase rule   General relationship between microstructure and state variables, as expressed in Equation 9.1.

glass   Noncrystalline solid, unless otherwise noted, with a chemical composition comparable to a crystalline ceramic.

glass-ceramic   Fine-grained, crystalline ceramic produced by the controlled devitrification of a glass.

glass forming   Processing technique for a glass.

glass transition temperature   The temperature range, above which a glass becomes a supercooled liquid, and below which, it is a true, rigid solid.

glaze   Glass coating applied to a ceramic such as clayware pottery.

golden ratio   The irrational number $(\sqrt{5}+1)/2 = 1.618$, which plays a fundamental role in numerous shapes in the natural world, including the recently discovered quasi-crystals.

graded-index fiber   Optical fiber with a parabolic variation in index of refraction within the core. (See Figure 16–18.)

graft copolymer   Combination of polymeric components in which one or more components is grafted onto a main polymeric chain.

grain   Individual crystallite in a polycrystalline microstructure.

grain boundary   Region of mismatch between two adjacent grains in a polycrystalline microstructure.

grain boundary diffusion   Enhanced atomic flow along the relatively open structure of the grain boundary region.

grain boundary dislocation (GBD)   Linear defect within a grain boundary, separating regions of good correspondence.

grain growth   Increase in average grain size of a polycrystalline microstructure due to solid-state diffusion.

grain-size number   Index for characterizing the average grain size in a microstructure, as defined by Equation 4.1.

gram-atom   Avogadro's number of atoms of a given element.

gray cast iron   A form of cast iron containing sharp graphite flakes that contribute to a characteristic brittleness.

green engineering   Environmentally-sensitive engineering design.

Griffith crack model   Prediction of stress intensification at the tip of a crack in a brittle material.

group   Chemical elements in a vertical column of the periodic table.

Guinier-Preston zone   Structure developed in the early stages of precipitation of an Al-Cu alloy. (See Figure 10–28.)

half-cell reaction   Chemical reaction associated with either the anodic or the cathodic half of an electrochemical cell.

Hall effect   Sideways deflection of charge carriers (and resulting voltage) associated with the application of a magnetic field perpendicular to an electrical current.

hard magnet   Magnet with relatively immobile domain walls.

hard sphere   Atomic (or ionic) model of an atom as a spherical particle with a fixed radius.

hardenability   Relative ability of a steel to be hardened by quenching.

hardness   Resistance of a material to indentation.

hardwood   Relatively high-strength wood from deciduous trees with covered seeds.

heat capacity   The amount of heat necessary to raise the temperature of one gram-atom (for elements) or one mole (for compounds) by 1 K ($= 1°$C). (See also *Specific heat.*)

heat treatment   Temperature versus time history necessary to generate a desired microstructure.

hemicellulose   One component of the matrix of the wood microstructure.

heteroepitaxy   Deposition of a thin film of a composition significantly different from the substrate.

heterogeneous nucleation   The precipitation of a new phase occurs at some structural imperfection such as a foreign surface.

hexagonal   One of the seven crystal systems, as illustrated in Table 3.1.

hexagonal close packed   Common atomic arrangement for metals, as illustrated in Figure 3–6.

high-alloy steel   Ferrous alloy with more than 5 wt % noncarbon additions.

high-strength, low-alloy steel   Steel with relatively high strength but significantly less than 5 wt % noncarbon additions.

high T_c superconductor   Ceramic material such as $YBa_2Cu_3O_7$, which is superconducting at a temperature higher than is possible with traditional metal superconductors, for example, above 30 K.

Hirth-Pound model   Atomistic model of the ledge-like structure of the surface of a crystalline material, as illustrated in Figure 4–17.

homoepitaxy   Deposition of a thin film of essentially the same material as the substrate.

homogeneous nucleation   The precipitation of a new phase occurs within a completely homogeneous medium.

honeycomb structure   A structural configuration illustrated by Figure 20–9.

Hooke's law   The linear relationship between stress and strain during elastic deformation. (See Equation 6.3.)

hot isostatic pressing (HIP)   Powder metallurgical technique combining high temperature and isostatic forming pressure.

Hume-Rothery rules   Four criteria for complete miscibility in metallic solid solutions.

Hund's rule   Electron pairing in a given orbital tends to be delayed until all levels of a given energy have a single electron.

hybrid   Woven fabric with two or more types of reinforcing fibers for use in a single composite.

hybridization   Formation of four equivalent electron energy levels (sp^3-type) from initially different levels (s-type and p-type).

hydrogen bridge   Secondary bond formed between two permanent dipoles in adjacent water molecules.

hydrogen embrittlement   Form of environmental degradation in which hydrogen gas permeates into a metal and forms brittle hydride compounds.

hypereutectic   Composition greater than that of the eutectic.

hypereutectoid   Composition greater than that of the eutectoid.

hypoeutectic   Composition less than that of the eutectic.

hypoeutectoid   Composition less than that of the eutectoid.

hysteresis   Characteristic behavior, such as ferromagnetism, in which a material property plot follows a closed loop.

hysteresis loop   Graph in which a plot of a material property (such as induction) does not retrace itself upon the reversal of an independent variable (such as magnetic field strength).

icosahedral phases   Materials with fivefold, threefold, and twofold symmetries corresponding to the structure of the icosahedron.

icosahedron   Polyhedron composed of 20 identical equilateral triangular faces.

impact energy   Energy necessary to fracture a standard test piece with an impact load.

impressed voltage   A method of corrosion protection in which an external voltage is used to oppose the one due to an electrochemical reaction.

impurity   A chemically foreign constituent, sometimes present due to raw material sources and, sometimes, purposefully added (as in extrinsic semiconductors).

incoherent   Light source in which the light waves are out of phase.

incongruent melting   The case in which the liquid formed upon melting has a different composition than the solid from which it was formed.

induced dipole   A separation of positive and negative centers of charge in an atom due to the coulombic attraction of an adjacent atom.

induction   Parameter representing the degree of magnetism due to a given field strength.

inhibitor   A substance, used in small concentrations, that decreases the rate of corrosion in a given environment.

initiator   Chemical species that triggers a chain growth polymerization mechanism.

injection molding   Processing technique for thermoplastic polymers.

insulator   Material with a low level of electrical conduction (e.g., a conductivity less than $10^{-4}\Omega^{-1}m^{-1}$).

integrated circuit (IC)   A sophisticated electrical circuit produced by the application of precise patterns of diffusable n-type and p-type dopants to give numerous elements within a single-crystal chip.

interfacial strength   Strength of the bonding between a composite matrix and its reinforcing phase.

intermediate   An oxide whose structural role in a glass is between that of a network former and a network modifier.

intermediate compound   A chemical compound formed between two components in a binary system.

interplanar spacing   Distance between the centers of atoms in two adjacent crystal planes.

interstitial solid solution   An atomic-scale combination of more than one kind of atom, with a solute atom located in an interstice of the solvent crystal structure.

interstitialcy   Atom occupying an interstitial site not normally occupied by an atom in the perfect crystal structure *or* an extra atom inserted into the perfect crystal such that two atoms occupy positions close to a singly occupied atomic site in the perfect structure.

intrinsic gettering   The capture of oxygen in a silicon device by heat treating the silicon wafer to form SiO_2 precipitates below the surface region where the integrated circuit is being developed.

intrinsic semiconductor   Material with semiconducting behavior independent of any impurity additions.

invariant point   Point in a phase diagram that has zero degrees of freedom.

inverse spinel   Compound crystal structure associated with ferrimagnetic ceramics and a variation of the spinel structure.

ion   Charged species due to an electron(s) added to, or removed from, a neutral atom.

ionic bond   Primary, chemical bond involving electron transfer between atoms.

ionic concentration cell   Electrochemical cell in which the corrosion and associated electrical current are due to a difference in ionic concentration.

ionic packing factor (IPF)   The fraction of the unit cell volume for a ceramic occupied by the various cations and anions.

ionic radius   See *Atomic radius*. (Ionic radius is associated, of course, with an ion rather than a neutral atom.)

iron age   The period of time from roughly 1000 B.C. to 1 B.C. during which iron alloys largely replaced bronze for tool and weapon making in Europe.

isostrain   Loading condition for a composite in which the strain on the matrix and dispersed phase are the same.

isostress   Loading condition for a composite in which the stress on the matrix and dispersed phase are the same.

isotactic   Polymeric structure in which side groups are along one side of the molecule. (See Figure 13–11.)

isothermal transformation diagram   See *TTT diagram*.

isotope   Any of two or more forms of a chemical element with the same number of protons but different numbers of neutrons.

isotropic   Having properties that do not vary with direction.

Izod test   Impact test typically used for polymers.

Jominy end-quench test   Standardized experiment for comparing the hardenability of different steels.

Josephson junction   Device consisting of a thin layer of insulator between superconducting layers.

Joule heating   Heating of a material due to the resistance to electrical current flow. A source of energy loss in ferromagnetic materials.

kaolinite   Silicate crystal structure as illustrated in Figure 3–16.

kinetics   Science of time-dependent phase transformations.

Knudsen cell   See *Effusion cell*.

laminate   Fiber-reinforced composite structure in which woven fabric is layered with the matrix.

laser   Coherent light source based on light amplification by stimulated emission of radiation.

lattice constant   Length of unit cell edge and/or angle between crystallographic axes.

lattice direction   Direction in a crystallographic lattice. (See Figure 3–28 for standard notation.)

lattice parameter   See *Lattice constant*.

lattice plane   Plane in a crystallographic lattice. (See Figure 3–30 for standard notation.)

lattice point   One of a set of theoretical points that are distributed in a periodic fashion in three-dimensional space.

lattice position   Standard notation for a point in a crystallographic lattice, as illustrated in Figure 3–26.

lattice translation   Vector connecting equivalent positions in adjacent unit cells.

Laue camera Device for obtaining an x-ray diffraction pattern of a single crystal. (See Figure 3–38.)

lead alloy Metal alloy composed of predominantly lead.

leathery Mechanical behavior associated with a polymer near its glass transition temperature. (See Figure 6–44.)

lever rule Mechanical analog for the mass balance with which one can calculate the amount of each phase present in a two-phase microstructure. (See Equations 9.9 and 9.10.)

life cycle assessment (LCA) A "cradle-to-grave" evaluation of environmental and energy impacts of a given product design.

light-emitting diode (LED) An electro-optical device illustrated by Figure 20–22.

lignin One component of the matrix of the wood microstructure. A phenol-propane network polymer. (See also *Hemicellulose.*)

linear coefficient of thermal expansion Material parameter indicating dimensional change as a function of increasing temperature. (See Equation 7.4.)

linear defect One-dimensional disorder in a crystalline structure, associated primarily with mechanical deformation. (See also *Dislocation.*)

linear density The number of atoms per unit length along a given direction in a crystal structure.

linear growth rate law An expression for the buildup of an unprotective oxide coating, as given by Equation 19.2.

linear molecular structure Polymeric structure associated with a bifunctional mer and illustrated by Figure 2–15.

liquid crystal display (LCD) An optical device in which a liquid crystal (composed of rod-shaped polymeric molecules) is used to create visible characters by voltage-produced disruption of the molecular orientation.

liquid-erosion failure A special form of wear damage in which a liquid is responsible for the removal of material.

liquid-metal embrittlement A form of degradation in which a material loses some ductility or fractures below its yield stress in conjunction with surface wetting by a lower-melting-point liquid metal.

liquid-penetrant testing A type of nondestructive testing in which surface defects are observed by the presence of a high-visibility liquid that has previously penetrated into the defects and is subsequently drawn out by the capillary action of a fine powder.

liquidus In a phase diagram, the line above which a single liquid phase will be present.

lithography Print-making technique applied to the processing of integrated circuits.

logarithmic growth rate law An expression for the buildup of a thin oxide film, as given by Equation 19.5.

long-range order A structural characteristic of crystals (and not glasses).

longitudinal cell Tubelike microstructural feature of wood aligned with the vertical axis of the tree.

low-alloy steel Ferrous alloy with less than 5 wt % noncarbon additions.

lower-yield point The onset of general plastic deformation in a low-carbon steel. (See Figure 6–10.)

luminescence   The reemission of photons of visible light in association with photon absorption.

magnesium alloy   Metal alloy composed of predominantly magnesium.

magnetic ceramic   See *Ceramic magnet.*

magnetic dipole   North-south orientation of a magnet.

magnetic field   Region of physical attraction produced by an electrical current. (See Figure 18–1.)

magnetic field strength   Intensity of the magnetic field.

magnetic flux line   Representation of the magnetic field.

magnetic moment   Magnetic dipole associated with electron spin.

magnetic-particle testing   A type of nondestructive testing in which defects are observed by the presence of a fine powder of magnetic particles attracted to the magnetic leakage flux around the surface or near-surface discontinuities.

magnetism   The physical phenomenon associated with the attraction of certain materials, such as ferromagnets.

magnetite   The historically important ferrous compound (Fe_3O_4) with magnetic behavior.

magnetization   Parameter associated with the induction of a solid. (See Equation 18.4.)

magnetoplumbite   Ceramic that is magnetically "hard." Its hexagonal crystal structure and chemical composition are similar to those of the mineral of the same name.

malleable iron   A traditional form of cast iron with modest ductility. It is first cast as white iron and then heat-treated to produce nodular graphite precipitates.

martempering   Heat treatment of a steel involving a slow cool through the martensitic transformation range to reduce stresses associated with that crystallographic change.

martensite   Iron-carbon solid solution phase with an acicular, or needlelike, microstructure produced by a diffusionless transformation associated with the quenching of austenite.

martensitic stainless steel   Corrosion-resistant ferrous alloy with a predominant martensitic phase.

martensitic transformation   Diffusionless transformation most commonly associated with the formation of martensite by the quenching of austenite.

mass balance   Method for calculating the relative amounts of the two phases in a binary microstructure. (See also *Lever rule.*)

materials science and engineering   Label for the general branch of engineering dealing with materials.

materials selection   Decision that is a critical component of the overall engineering design process.

matrix   The portion of a composite material in which a reinforcing, dispersed phase is embedded.

Maxwell-Boltzmann distribution   Description of the relative distribution of molecular energies in a gas.

mean free path   The average distance that an electron wave can travel without deflection.

mechanical stress A source of corrosion in metals. (See Figure 19–13.)

medium-range order Structural ordering occurring over the range of a few nanometers in an otherwise noncrystalline material.

melting point Temperature at which a solid-to-liquid transformation occurs upon heating.

melting range Temperature range over which the viscosity of a glass is between 50 and 500 P.

mer Building block of a long-chain or network (polymeric) molecule.

metal Electrically conducting solid with characteristic metallic bonding.

metal matrix composite Composite material in which the reinforcing phase is dispersed in a metal.

metallic bond Primary, chemical bond involving the nondirectional sharing of delocalized electrons.

metallic magnet Metal alloy with an engineering application predominantly based on its magnetic properties.

metastable A state that is stable with time, although it does not represent true equilibrium.

microcircuit Microscopic-scale electrical circuit produced on a semiconductor substrate by controlled diffusion.

microscopic-scale architecture Structural arrangement of the various phases in an engineering material.

microstructural development Changes in the composition and distribution of phases in a material's microstructure as a result of thermal history.

Miller indices Set of integers used to characterize a crystalline plane.

Miller-Bravais indices Four-digit set of integers used to characterize a crystalline plane in the hexagonal system.

mixed-bond character Having more than one type of atomic bonding, for example, covalent and secondary bonding in polyethylene.

mixed dislocation Dislocation with both edge and screw character. (See Figure 4–13.)

modulus of elasticity Slope of the stress-strain curve in the elastic region.

modulus of elasticity in bending See *Flexural modulus.*

modulus of rigidity See *Shear modulus.*

modulus of rupture See *Flexural strength.*

mole Avogadro's number of atoms or ions in the compositional unit of a compound, for example, mole of Al_2O_3 contains 2 moles of Al^{3+} ions and 3 moles of O^{2-} ions.

molecular-beam epitaxy (MBE) A highly controlled ultrahigh-vacuum deposition process.

molecular length Length of a polymeric molecule expressed in one of two ways. (See *Extended length* and *Root-mean-square length.*)

molecular weight Number of atomic mass units for a given molecule.

molecule Group of atoms joined by primary bonding (usually covalent).

monochromatic A single wavelength of radiation.

monomer   Individual molecule that combines with similar molecules to form a polymeric molecule.

Moore's law   During much of the history of the integrated circuit, the number of transistors produced in a single chip has roughly doubled every two years. (See Figure 17–28.)

n-type semiconductor   Extrinsic semiconductor in which the electrical conductivity is dominated by negative charge carriers.

near net-shape processing   Material processing with the goal of minimizing any final shaping operation.

negative charge carrier   Charge carrier with a negative electrical charge.

Nernst equation   Expression for the voltage of an electrochemical cell as a function of solution concentrations. (See Equation 19.12.)

net-shape processing   Material processing that does not require a subsequent shaping operation.

network copolymer   Alloylike combination of polymers with an overall network, rather than linear, structure.

network former   Oxides that form oxide polyhedra, leading to network structure formation in a glass.

network modifier   Oxides that do not form oxide polyhedra and, therefore, break up the network structure in a glass.

network molecular structure   Polymeric structure associated with a polyfunctional mer and illustrated by Figure 13–7.

neutron   Subatomic particle without a net charge and located in the atomic nucleus.

nickel alloy   Metal alloy composed of predominantly nickel.

nickel-aluminum superalloy   Nonferrous alloy with exceptional high-temperature strength and corrosion resistance.

noble   Tending to be reduced in an electrochemical cell.

noncrystalline   Atomic arrangement lacking in long-range order.

noncrystalline solid   Solid lacking in long-range stuctural order.

nondestructive testing   The evaluation of engineering materials without impairing their usefulness.

nonferrous alloy   Metal alloy composed predominantly of an element(s) other than iron.

nonoxide ceramic   Ceramic material composed predominantly of a compound(s) other than an oxide.

nonprimitive   Crystal structure having atoms at unit cell positions in addition to the unit cell corners.

nonprimitive unit cell   See *Nonprimitive.*

nonsilicate glass   Glass composed predominantly of a compound(s) other than silica.

nonsilicate oxide ceramic   Ceramic material composed predominantly of an oxide compound(s) other than silica.

nonstoichiometric compound   Chemical compound in which variations in ionic charge lead to variations in the ratio of chemical elements, for example, $Fe_{1-x}O$.

nuclear ceramic Ceramic material with a primary engineering application in the nuclear industry.

nucleate Initiate a phase transformation.

nucleation First stage of a phase transformation, such as precipitiation. (See Figure 10–2.)

nucleus Central core of atomic structure, about which electrons orbit.

nylon 66 Poly (hexamethylene adipamide), an important engineering polymer. (See Figure 3–22 for the unit cell structure.)

octahedral position Site in a crystallographic structure at which an atom or ion would be surrounded by six neighboring atoms or ions.

Ohm's law Relationship between voltage, current, and resistance in an electrical circuit. (See Equation 15.1.)

1–2–3 superconductor The material $YBa_2Cu_3O_7$, which is the most commonly studied ceramic superconductor and whose name is derived from the three metal ion subscripts.

opacity Total loss of image transmission.

optical fiber A small-diameter glass fiber in which digital light pulses can be transmitted with low losses. (See Figure 16–17.)

optical microscope Instrument using visible light to produce images of the structure of materials at a scale greater than possible with the unaided eye. (See Figure 4–33a.)

optical property Material characteristic relative to the nature of its interaction with light.

orbital See *Electron orbital.*

orbital shell Set of electrons in a given orbital.

ordered solid solution Solid solution in which the solute atoms are arranged in a regular pattern. (See Figure 4–3.)

orientational order Parallel or antiparallel alignment of structural units.

overaging Continuation of the age-hardening process so long that the precipitates coalesce into a coarse dispersion, becoming a less effective dislocation barrier and leading to a drop in hardness.

overvoltage A positive or negative change in electrochemical potential relative to a corrosion potential. (See Figure 19–20.)

oxidation Reaction of a metal with atmospheric oxygen.

oxide Compound between an elemental metal(s) and oxygen.

oxide glass Noncrystalline solid in which the predominant components(s) is (are) an oxide(s).

oxygen concentration cell Electrochemical cell in which the corrosion and associated electrical current are due to a difference in gaseous oxygen concentrations.

PM10 Airborne particulate matter less than 10 μm in diameter.

PM2.5 Airborne particulate matter less than 2.5 μm in diameter.

p-n junction Boundary between adjacent regions of p-type and n-type material in a solid-state electronic device.

p-type semiconductor Extrinsic semiconductor in which the electrical conductivity is dominated by positive charge carriers.

PZT   Lead zirconate-titanate ceramic used as a piezoelectric transducer.

parabolic growth rate law   An expression for the buildup of a protective oxide coating in which growth is limited by ionic diffusion. (See Equation 19.4.)

paraelectric   Having a modest polarization with applied electrical field. (See dashed line in Figure 15–23.)

paramagnetism   Magnetic behavior in which a modest increase in induction (compared to that for a vacuum) occurs with applied magnetic field. (See Figure 18–4.)

partially stabilized zirconia (PSZ) ZrO_2 ceramic with a modest second component addition (e.g., CaO) producing a two-phase microstructure. Retention of some ZrO_2-rich phase in PSZ allows for the mechanism of transformation toughening.

particulate composite   Composite material with relatively large dispersed particles (at least several microns in diameter) and in a concentration greater than 25 vol %.

passivity   The resistance to corrosion due to the formation of a thin, protective oxide film.

Pauli exclusion principle   Quantum mechanical concept that no two electrons can occupy precisely the same state.

pearlite   Two-phase eutectoid microstructure of iron and iron carbide. (See Figure 9–2.)

Penrose tilings   Patterns that fill two-dimensional space with a resulting fivefold symmetry.

periodic table   Systematic graphical arrangement of the elements indicating chemically similar groups. (See Figure 2–2.)

peritectic diagram   Binary phase diagram with the peritectic reaction (Equation 9.5). (See Figure 9–22.)

peritectic reaction   The transformation of a solid to a liquid and a solid of a different composition upon heating, as summarized by Equation 9.5.

permanent dipole   Molecular structure with an inherent separation of centers of positive and negative charge.

permanent magnet   Magnet, typically of a hard steel, that retains its magnetization once magnetized.

permeability   Proportionality constant between induction and magnetic field strength. (See Equations 18.1 and 18.2.)

perovskite   Compound crystal structure as illustrated in Figure 3–14.

phase   Chemically homogeneous portion of a microstructure.

phase diagram   Graphical representation of the state variables associated with microstructures.

phase field   Region of a phase diagram corresponding to the existence of a given phase.

phosphorescence   Luminescence in which photon emission occurs after more than about ten nanoseconds. (See also *fluorescence*.)

photoconductor   Semiconductor in which electron-hole pairs are produced by exposure to photons.

photoluminescence   Luminescence caused by photons.

photon   The particle-like packet of energy corresponding to a given wavelength of electromagnetic radiation.

photonic material   Optical material in which signal transmission is by photons rather than by the electrons of electronic materials.

photoresist   Polymeric material used in the lithography process.

piezoelectric coupling coefficient   Fraction of mechanical energy converted to electrical energy by a piezoelectric transducer.

piezoelectric effect   Production of a measurable voltage change across a material as a result of an applied stress.

piezoelectricity   An electrical response to mechanical pressure application.

pigment   Insoluble, colored additive for polymers.

Pilling-Bedworth ratio   Ratio of oxide volume produced to the metal volume consumed in oxidation. (See Equation 19.6.)

planar defect   Two-dimensional disorder in a crystalline structure, for example, a grain boundary.

planar density   The number of atoms per unit area in a given plane in a crystal structure.

plastic   See *Polymer*.

plastic deformation   Permanent deformation associated with the distortion and reformation of atomic bonds.

plasticizer   Additive for the purpose of softening a polymer.

point defect   Zero-dimensional disorder in a crystalline structure, associated primarily with solid-state diffusion.

point lattice   See *Crystal (Bravais) lattice*.

Poisson's ratio   Mechanical property indicating the contraction perpendicular to the extension caused by a tensile stress. (See Equation 6.5.)

polar diagram   Plot of intensity of reflected light from a surface. (See Figure 16–7.)

polar molecule   Molecule with a permanent dipole moment.

polarization   Development of an overvoltage relative to a corrosion potential. (See Figure 19–20.)

polyethylene   The most widely used polymeric material. See Figure 3–20 for the unit cell structure.

polyfunctional   Polymer with more than two reaction sites for each mer, resulting in a network molecular structure.

polymer   Engineering material composed of long-chain or network molecules.

polymer-matrix composite   Composite material in which the reinforcing phase is dispersed in a polymer.

polymeric molecule   A long-chain or network molecule composed of many building blocks (mers).

polymerization   Chemical process in which individual molecules (monomers) are converted to large molecular weight molecules (polymers).

porcelain enamel   Silicate glass coating on a metal substrate for the purpose of shielding the metal from a corrosive environment.

portland cement   Calcium aluminosilicate used as a matrix for aggregate in concrete.

positive charge carrier   Charge carrier with a positive electrical charge.

pottery   Burned clayware ceramic.

powder metallurgy   Processing technique for metals involving the solid-state bonding of a fine-grained powder into a polycrystalline product.

precious metal   Generally corrosion-resistant metal or alloy, such as gold, platinum, and their alloys.

precipitation hardening   Development of obstacles to dislocation motion (and, thereby, increased hardness) by the controlled precipitation of a second phase.

precipitation-hardened stainless steel   Corrosion-resistant ferrous alloy that has been strengthened by precipitation hardening.

preexponential constant   Temperature-independent term that appears before the exponential term of the Arrhenius equation. (See Equation 5.1.)

preferred orientation   Alignment of a given crystallographic direction in adjacent grains of a microstructure as the result of cold rolling.

preferred-orientation microstructure   Grain structure in a polycrystalline material in which the grains tend to have a common crystallographic orientation.

prestressed concrete   An aggregate composite (concrete) in which steel bars have been embedded prior to the setting of the cement. Release of tension on the bars after the concrete hardens places the material under a residual, compressive stress. The material is more crack-resistant as a result.

primary bond   Relatively strong bond between adjacent atoms resulting from the transfer or sharing of outer orbital electrons.

primitive   Crystal structure having atoms located at unit cell corners only.

primitive unit cell   See *Primitive*.

processing   Production of a material into a form convenient for engineering applications.

proeutectic   Phase that forms by precipitation in a temperature range above the eutectic temperature.

proeutectoid   Phase that forms by solid-state precipitation in a temperature range above the eutectoid temperature.

property   Observable characteristic of a material.

property averaging   Determination of the overall property (such as elastic modulus) of a composite material as the geometrical average of the properties of the individual phases.

prosthesis   Device for replacing a missing body part.

protective coating   A barrier between a metal and its corrosive environment.

proton   Positively charged subatomic particle located in the atomic nucleus.

proton-induced x-ray emission (PIXE)   Chemical analysis by the use of characteristic x-ray photons produced by exposure to high-energy protons.

pure oxide   Ceramic compound with relatively low impurity level (typically less than 1 wt %).

quantum dot   Semiconducting material with three thin dimensions of the scale associated with a quantum well.

quantum well   Thin layer of semiconducting material in which the wavelike electrons are confined within the layer thickness.

quantum wire Semiconducting material with two thin dimensions of the scale associated with a quantum well.

quasicrystal Material with a structural state intermediate between traditional crystalline and noncrystalline solids.

radial cell Wood cell extending radially out from the center of the tree trunk.

radiation Various photons and atomic-scale particles that can be a source of environmental damage to materials.

radius ratio The radius of a smaller ion divided by the radius of a larger one. This ratio establishes the number of larger ions that can be adjacent to the smaller one.

random network theory Statement that a simple oxide glass can be described as the random linkage of "building blocks" (such as the silica tetrahedron).

random solid solution Solid solution in which the solute atoms are arranged in an irregular fashion. (See Figure 4–3.)

random walk Atomistic migration in which the direction of each step is randomly selected from among all possible orientations. (See Figure 5–6.)

rapid solidification Processing technique for cooling a melt below its melting point at a high quench rate (e.g., $10^6°C/s$) resulting in the possibility of forming an amorphous structure or metastable crystalline phases.

rapidly solidified alloy Metal alloy formed by a rapid solidification process.

rate-limiting step Slowest step in a process involving sequential steps. The overall process rate is, thereby, established by that one mechanism.

recovery Initial state of annealing in which atomic mobility is sufficient to allow some softening of the material without a significant microstructural change.

recrystallization Nucleation and growth of a new stress-free microstructure from a cold-worked microstructure. (See Figures 10–30a-d.)

recrystallization temperature The temperature at which atomic mobility is sufficient to affect mechanical properties as a result of recrystallization. The temperature is approximately one-third to one-half times the absolute melting point.

rectifier See *Diode.*

recycling The reprocessing of a relatively inert engineering material.

reflectance Fraction of light reflected at an interface. (See Equation 16.5.)

reflection rules Summary of which crystal planes in a given structure cause x-ray diffraction. (See Table 3.4.)

refractive index Fundamental optical property defined by Equation 16.4.

refractory High-temperature-resistant material, such as many of the common ceramic oxides.

refractory metal Metals and alloys (such as molybdenum) that are resistant to high temperatures.

reinforcement Additive (such as glass fibers) providing a polymer with increased strength and stiffness.

relative permeability The magnetic permeability of a solid divided by the permeability of a vacuum.

relaxation time The time necessary for the stress on a polymer to fall to $0.37 \, (= 1/e)$ of the initial applied stress.

remanent induction   The induction (of a ferromagnetic material) remaining after the applied magnetic field is removed.

remanent polarization   The polarization (of a ferroelectric material) remaining after the applied electrical field is removed.

repulsive force   Force due to the like-charge repulsion of both (negative) electron orbitals and (positive) nuclei of adjacent atoms.

residual stress   The stress remaining within a structural material after all applied loads are removed.

resistance   Property of a material by which it opposes the flow of an electrical current. (See Equation 15.1.)

resistivity   The material property for electrical resistance normalized for sample geometry.

resolved shear stress   Stress operating on a slip system. (See Equation 6.14.)

reverse bias   Orientation of electrical potential to provide a minimal flow of charge carriers in a rectifier.

reverse piezoelectric effect   Production of a thickness change in a material as a result of an applied voltage.

rigid   Mechanical behavior associated with a polymer below its glass transition temperature. (See Figure 6–44.)

Rockwell hardness   Common mechanical parameter, as defined in Table 6.10.

root-mean-square length   Separation distance between the ends of a randomly coiled polymeric molecule. (See Equation 13.4 and Figure 13–9.)

rubbery   Mechanical behavior associated with a polymer just above its glass transition temperature. (See Figure 6–44.)

rusting   A common corrosion process for ferrous alloys. (See Figure 19–12.)

sacrificial anode   Use of a less noble material to protect a structural metal from corrosion. (See Figure 19–17.)

saturation induction   The apparent maximum value of induction for a ferromagnetic material under the maximum applied field. (See Figure 18–5.)

saturation polarization   Polarization of a ferroelectric material due to maximum domain growth. (See Figure 15–24.)

saturation range   Temperature range over which conductivity in a p-type semiconductor is relatively constant due to the fact that all acceptor levels are "saturated" with electrons.

scanning electron microscope (SEM)   Instrument for obtaining microstructural images using a scanning electron beam. (See Figure 4–36.)

scanning tunneling microscope (STM)   An instrument capable of providing direct images of atomic packing patterns by monitoring the quantum-mechanical tunneling of electrons near the sample surface.

Schottky defect   A pair of oppositely charged ion vacancies. (See Figure 4–9.)

screw dislocation   Linear defect with the Burgers vector parallel to the dislocation line.

secondary bond   Atomic bond without electron transfer or sharing.

Seebeck effect   The development of an induced voltage in a simple electrical circuit as the result of a temperature differential.

Seebeck potential   Induced voltage due to a dissimilar metal thermocouple between two different temperatures. (See Figure 15–14.)

segregation coefficient   Ratio of the saturation impurity concentrations for the solid and liquid solution phases, as defined by Equation 17.9.

self-diffusion   Atomic-scale migration of a species in its own phase.

self-propagating high temperature synthesis (SHS)   Material processing involving the heat evolved by certain chemical reactions to sustain the reaction and produce the final product.

semiconductor   Material with a level of electrical conductivity intermediate between that for an insulator and a conductor, for example, a conductivity between $10^{-4}\ \Omega^{-1} \cdot m^{-1}$ and $10^{+4}\ \Omega^{-1} \cdot m^{-1}$.

shear modulus   Elastic modulus under pure shear loading. (See Equation 6.8.)

shear strain   Elastic displacement produced by pure shear loading. (See Equation 6.7.)

shear stress   Load per unit area (parallel to the applied load). See Equation 6.6.

short-range order   Local "building block" structure of a glass (comparable to the structural unit in a crystal of the same composition).

silica   Silicon dioxide. One of various, stable crystal structures is illustrated in Figure 3–11.

silicate   Ceramic compound with SiO_2 as a major constituent.

silicate glass   Noncrystalline solid with SiO_2 as a major constituent.

silicon   Element 14 and an important semiconductor.

single-mode fiber   Optical fiber with a narrow core in which light travels largely parallel to the fiber axis. (See Figure 16–18.)

sintering   Bonding of powder particles by solid-state diffusion.

slip casting   Processing technique for ceramics in which a powder-water mixture (slip) is poured into a porous mold.

slip system   A combination of families of crystallographic planes and directions corresponding to dislocation motion.

smart material   An engineered material with the ability to perform both sensing and responding functions.

soda-lime-silica glass   Noncrystalline solid composed of sodium, calcium, and silicon oxides. The majority of windows and glass containers are in this category.

sodium chloride   Simple compound crystal structure as illustrated in Figure 3–9.

soft magnet   Magnet with relatively mobile domain walls.

soft sphere   Atomic (or ionic) model that acknowledges that the outer orbital electron density does not terminate at a fixed radius.

softening point   Temperature at which a glass has a viscosity of $10^{7.6}$ P, corresponding to the lower end of the working range.

softwood   Relatively low-strength wood from "evergreen" trees.

sol-gel processing   Technique for forming ceramics and glasses of high density at a relatively low temperature by means of an organometallic solution.

solid solution   Atomic-scale intermixing of more than one atomic species in the solid state.

solidus   In a phase diagram, the line below which only a solid phase(s) is (are) present.

solute   Species dissolving in a solvent to form a solution.

solution hardening   Mechanical strengthening of a material associated with the restriction of plastic deformation due to solid solution formation.

solution treatment   Heating of a two-phase microstructure to a single-phase region.

solvent   Species into which a solute is dissolved in order to form a solution.

source   Region in a field-effect transistor that provides charge carriers.

spalling   Buckling and flaking off of an oxide coating on a metal, due to large compressive stresses.

specific heat   The amount of heat necessary to raise the temperature of one unit mass of material by 1 K ($= 1°C$). (See also *Heat capacity*.)

specific strength   Strength-per-unit density.

spectral width   A range of wavelengths.

specular reflection   Light reflection relative to the "average" surface. (See Figure 16–6.)

spinel   Compound crystal structure as illustrated in Figure 3–15.

spontaneous polarization   A sharp rise in polarization in a ferroelectric material, due to a modest field application. (See Figure 15–23.)

stabilizer   Additive for a polymer for the purpose of reducing degradation.

stainless steel   Ferrous alloy resistant to rusting and staining, due primarily to the addition of chromium.

state   Condition for a material, typically defined in terms of a specific temperature and composition.

state point   A pair of temperature and composition values that define a given state.

state variables   Material properties, such as temperature and composition, that are used to define a state.

static fatigue   For certain ceramics and glasses, a degradation in strength that occurs without cyclic loading.

static modulus of elasticity   Modulus of elasticity for a polymer obtained from the slope of the initial, incremental load-deflection plot of an overall measurement of dynamic elastic properties.

steady-state diffusion   Mass transport that is unchanging with time.

steel   A ferrous alloy with up to approximately 2.0 wt % carbon.

step growth   Polymerization process involving individual chemical reactions between pairs of reactive monomers.

step-index fiber   Optical fiber with a sharp step down in index of refraction at the core/cladding interface. (See Figure 16–18.)

stone age   The time when our human ancestors, or hominids, chipped stones to form weapons for hunting. This age has been traced back as far as 2.5 million years ago.

strain hardening   The strengthening of a metal alloy by deformation (due to the increasing difficulty for dislocation motion through the increasingly dense array of dislocations).

strain-hardening exponent   The slope of a log-log plot of true stress versus true strain between the onset of plastic deformation of a metal alloy and the onset of necking. This parameter is an indicator of the alloy's ability to be deformed.

strength-to-weight ratio   See *Specific strength*.

stress cell   An electrochemical cell in which corrosion can occur due to the presence of variations in the degree of mechanical stress within a metal sample.

stress-corrosion cracking   A combined mechanical and chemical failure mechanism in which a noncyclic tensile stress (below the yield strength) leads to the initiation and propagation of fracture in a relatively mild chemical environment.

stress-intensity factor   Parameter indicating the greater degree of mechanical stress at the tip of a pre-existing crack in a material under a mechanical load.

stress relaxation   Mechanical phenomenon in certain polymers in which the stress on the material drops exponentially with time under a constant strain. (See Equation 6.17.)

stress-rupture failure   See *Creep-rupture failure*.

structural clay product   Traditional engineering ceramic, such as a brick, tile, or sewer pipe.

structural composite   See *Honeycomb structure*.

structural material   Engineering material used primarily for its mechanical properties relative to a structural application.

substitutional solid solution   An atomic-scale combination of more than one kind of atom, with a solute atom substituting for a solvent atom at an atomic lattice site.

superalloy   Broad class of metals with especially high strength at elevated temperatures.

superconducting magnet   Magnet made from a superconducting material.

superconductor   Material that is generally a poor conductor at elevated temperatures but, upon cooling below a critical temperature, has zero resistivity.

superplastic forming   Technique for forming complex-shaped metal parts from certain fine-grained alloys at elevated temperatures.

surface   Exterior planar boundary of a solid that can be considered a defect structure, for example, as illustrated by Figure 4–17.

surface analysis   Technique such as Auger electron spectroscopy in which the first few atomic layers of the material's surface are chemically analyzed.

surface diffusion   Enhanced atomic flow along the relatively open structure of a material's surface.

surface fatigue wear   Wear occurring during repeated sliding or rolling of a material over a track.

surface gloss   A condition of specular, rather than diffuse, reflection from a given surface.

syndiotactic   Regular alteration of side groups along a polymeric molecule. (See Figure 13–11.)

TTT diagram   A plot of the time necessary to reach a given percent transformation at a given temperature. (See Figure 10–6.)

temperature coefficient of resistivity   The coefficient that indicates the dependence of a metal's resistivity on temperature. (See Equation 15.9.)

tempered glass   A strengthened glass involving a heat treatment that serves to place the exterior surface in a residual compressive state.

tempered martensite   An $\alpha + Fe_3C$ microstructure produced by heating the more brittle martensite phase.

tempering   A thermal history for steel, as illustrated in Figure 10–17, in which martensite is reheated.

tensile strength   The maximum engineering stress experienced by a material during a tensile test.

terminator   Chemical species that ends a chain growth polymerization mechanism.

tetrahedral   Involving fourfold coordination.

tetrahedral position   Site in a crystallographic structure at which an atom or ion would be surrounded by four neighboring atoms or ions.

textured microstructure   Microstructure associated with preferred orientation.

theoretical critical shear stress   High stress level associated with the sliding of one plane of atoms over an adjacent plane in a defect-free crystal.

thermal activation   Atomic-scale process in which an energy barrier is overcome by thermal energy.

thermal conductivity   Proportionality constant in the relationship between heat transfer rate and temperature gradient, as defined in Equation 7.5.

thermal shock   The fracture (partial or complete) of a material as the result of a temperature change (usually a sudden cooling).

thermal vibration   Periodic oscillation of atoms in a solid at a temperature above absolute zero.

thermocouple   Simple electrical circuit for the purpose of temperature measurement. (See Figure 15–14.)

thermoplastic   Polymer that becomes soft and deformable upon heating.

thermoplastic elastomer   Compositelike polymer with rigid, elastomeric domains in a relatively soft matrix of a crystalline, thermoplastic polymer.

thermoplastic polymer   See *Thermoplastic.*

thermosetting   Polymer that becomes hard and rigid upon heating.

thermosetting polymer   See *Thermosetting.*

III-V compound   Chemical compound between a metallic element in group III and a nonmetallic element in group V of the periodic table. Many of these compounds are semiconducting.

tie line   Horizontal line (corresponding to constant temperature) connecting two phase compositions at the boundaries of a two-phase region of a phase diagram. (See Figure 9–6.)

tilt boundary   Grain boundary associated with the tilting of a common crystallographic direction in two adjacent grains. (See Figure 4–19.)

titanium alloy   Metal alloy composed of predominantly titanium.

tool steel   Ferrous alloy used for cutting, forming, or otherwise shaping another material.

toughness   Total area under the stress-strain curve.

transducer   A device for converting one form of energy to another form.

transfer molding   Processing technique for thermosetting polymers.

transformation toughening   Mechanism for enhanced fracture toughness in a partially stabilized zirconia ceramic involving a stress-induced phase transformation of tetragonal grains to the monoclinic structure. (See Figure 8–7.)

transistor   A solid-state amplifier.

transition metal   Element in a region of the periodic table associated with a gradual shift from the strongly electropositive elements of groups IA and IIA to the more electronegative elements of groups IB and IIB.

transition metal ion   Charged species formed from a transition metal atom.

translucency   Transmission of a diffuse image.

transmission electron microscope (TEM)   Instrument for obtaining microstructual images using electron transmission in a design similar to a conventional optical microscope. (See Figure 4–33.)

transpassive   The increase in corrosion rate at a relatively high potential due to the breakdown of the passive surface film. (See Figure 19–21.)

twin boundary   A planar defect separating two crystalline regions that are, structually, mirror images of each other. (See Figure 4–15.)

II-VI compound   Chemical compound between a metallic element in group II and a nonmetallic element in group VI of the periodic table. Many of these compounds are semiconducting.

ultrasonic testing   A type of nondestructive testing in which defects are detected using high-frequency acoustical waves.

unit cell   Structural unit that is repeated by translation in forming a crystalline structure.

upper yield point   Distinct break from the elastic region in the stress-strain curve for a low-carbon steel. (See Figure 6–10.)

vacancy   Unoccupied atom site in a crystal structure.

vacancy migration   Movement of vacancies in the course of atomic diffusion without significant crystal structure distortion. (See Figure 5–5.)

valence   Electronic charge of an ion.

valence band   A range of electron energies in a solid associated with the valence electrons of an isolated atom.

valence electron   Outer orbital electron that takes part in atomic bonding. In a semiconductor, an electron in the valence band.

van der Waals bond   See *Secondary bond.*

vapor deposition   Processing technique for semiconductor devices involving the buildup of material from the vapor phase.

viscoelastic deformation   Mechanical behavior involving both fluidlike (viscous) and solidlike (elastic) characteristics.

viscosity   Proportionality constant in the relationship between shearing force and velocity gradient, as defined in Equation 6.19.

viscous   Mechanical behavior associated with a polymer near its melting point. (See Figure 6–44.)

viscous deformation   Liquidlike mechanical behavior associated with glasses and polymers above their glass transition temperatures.

visible light   The portion of the electromagnetic radiation spectrum that can be perceived by the human eye (the wavelength range of 400 to 700 nm).

vitreous silica   Commercial glass that is nearly pure SiO_2.

voltage   A difference in electrical potential.

volume (bulk) diffusion   Atomic flow within a material's crystal structure by means of some defect mechanism.

vulcanization   The transformation of a polymer with a linear structure into one with a network structure by means of cross-linking.

wafer   Thin slice from a cylindrical single crystal of high-purity material, usually silicon.

wear   Removal of surface material as a result of mechanical action.

wear coefficient   Mechanical property representing the probability that an adhesive wear fragment will be formed. (See Equation 19.20.)

wear failure   Surface-related damage phenomena, such as wear debris on sliding contact surfaces.

welding   Joining of metal parts by local melting in the vicinity of the join.

whisker   Small, single-crystal fiber with a nearly perfect crystalline structure that serves as a high-strength composite reinforcing phase.

white cast iron   A hard, brittle form of cast iron with a characteristic white, crystalline fracture surface.

whiteware   Commercial fired ceramic with a typically white and fine-grained microstructure. Examples include tile, china, and pottery.

wood   A natural, fiber-reinforced composite.

working range   Temperature range in which glass product shapes are formed (corresponding to a viscosity range of 10^4 to 10^8 P).

woven fabric   Composite reinforcing fiber configuration as illustrated in Figure 14–3.

wrought alloy   Metal alloy that has been rolled or forged into a final, relatively simple shape following an initial casting operation.

wrought process   Rolling or forging of an alloy into a final, relatively simple shape following an initial casting step. (See Figure 11–2 for examples for steel products.)

wurtzite   Compound crystal structure as illustrated in Figure 3–25.

x-radiation   The portion of the electromagnetic spectrum with a wavelength on the order of 1 nanometer. X-ray photons are produced by inner orbital electron transitions.

x-radiography   A type of nondestructive testing in which defects are inspected using the attenuation of x-rays.

x-ray diffraction   The reinforced scattering of x-ray photons by an atomic structure. Bragg's law indicates the structural information available from this phenomenon.

x-ray fluorescence (XRF)   Chemical analysis by the use of characteristic x-ray photons produced by exposure to x-ray photons.

x-ray photoelectron spectroscopy (XPS)   Chemical analysis by the use of characteristic energy photoelectrons produced by exposure to x-ray photons.

YIG   Yttrium iron garnet, a ferrimagnetic ceramic.

yield point   See *Upper yield point.*

yield strength   The strength of a material associated with the approximate upper limit of Hooke's law behavior, as illustrated in Figure 6–4.

Young's modulus   See *Modulus of elasticity*.

Zachariasen model   Visual definition of the random network theory as illustrated in Figure 4–23b.

zinc alloy   Metal alloy composed of predominantly zinc.

zinc blende   Compound crystal structure as illustrated in Figure 3–24.

zone refining   Technique for purifying materials by passing an induction coil along a bar of the material and using principles of phase equilibria, as illustrated in Figure 17–15.

Answers to Practice Problems (PP) and Odd-Numbered Problems

Chapter 2

PP 2.1 **(a)** 3.38×10^{10} atoms, **(b)** 2.59×10^{10} atoms

PP 2.2 3.60 kg

PP 2.3 **(a)** 19.23 mm, **(b)** 26.34 mm

PP 2.4 **(b)** Mg: $1s^2 2s^2 2p^6 3s^2$, Mg^{2+} : $1s^2 2s^2 2p^6$; O: $1s^2 2s^2 2p^4$, O^{2-} : $1s^2 2s^2 2p^6$ **(c)** Ne for both cases of Mg^{2+} and O^{2-}

PP 2.5 $F_c = 20.9 \times 10^{-9}$ N, $F_R = -F_c = -20.9 \times 10^{-9}$ N

PP 2.6 **(a)** $\sin(109.5°/2) = R/(r + R)$ or $r/R = 0.225$ **(b)** $\sin 45° = R/(r + R)$ or $r/R = 0.414$

PP 2.7 CN = 6 for both cases

PP 2.8

PP 2.9 Same as for Practice Problem 2.8 except $R = C_6H_5$

PP 2.10 **(a)** 60 kJ/mol, **(b)** 60 kJ/mol

PP 2.11 794

PP 2.12 A greater degree of covalency in Si–Si bond provides even stronger directionality and lower coordination number

2.1 10.4×10^{21} atoms

2.3 3.31×10^{15} atoms Si and 6.62×10^{15} atoms O

2.5 0.921×10^{12} atoms Al and 1.38×10^{12} atoms O

2.7 **(a)** 0.372×10^{24} molecules O_2, **(b)** 0.617 mol O_2

2.9 **(a)** 1.41 g, **(b)** 6.50×10^{-4} g

2.11 4.47 nm

2.15 -1.49×10^{-9} N

2.17 -8.13×10^{-9} N

2.19 -10.4×10^{-9} N

2.23 Pink

2.25 22.1×10^{-9} N

2.27 **(b)** 1.645

2.29 229 kJ

2.31 **(b)** 60 kJ/mol, **(c)** 678 kJ

2.33 335 kJ/mol

2.35 60,068 amu

2.37 12.0 kJ

2.39 **(b)** 60 kJ/mol, **(c)** 32,020 amu

2.49 392 m^2/kg

2.51 2.77×10^{23} atoms/(m^3 atm)

Chapter 3

PP 3.2 **(a)** $a = (4/\sqrt{3})r$, **(b)** $a = 2r$

PP 3.3 7.90 g/cm^3

PP 3.4 **(a)** 0.542, **(b)** 0.590, **(c)** 0.627

PP 3.5 3.45 g/cm^3

PP 3.6 5.70×10^{24}

PP 3.7 Directionality of covalent bonding dominates over efficient packing of spheres.

PP 3.8 5.39 g/cm^3

PP 3.9 **(a)** Body-centered position: $\frac{1}{2}\frac{1}{2}\frac{1}{2}$ **(b)** same, **(c)** same

PP 3.10 **(a)** 000, $\frac{1}{2}\frac{1}{2}\frac{1}{2}$, 111 **(b)** same, **(c)** same

PP 3.12 **(a)**
$$\langle 100 \rangle = [100], [010], [001]$$
$$[\bar{1}00], [0\bar{1}0], [00\bar{1}]$$

PP 3.13 (a) $45°$, (b) $54.7°$

PP 3.16 (a) 4.03 atoms/nm, (b) 1.63 atoms/nm

PP 3.17 (a) 7.04 atoms/nm^2, (b) 18.5 atoms/nm^2

PP 3.18 $(1.21\ Ca^{2+} + 1.21\ O^{2-})/nm$

PP 3.19 10.2 $(Ca^{2+}$ or $O^{2-})/nm^2$

PP 3.20 2.05 atoms/nm

PP 3.21 7.27 atoms/nm^2

PP 3.22 0.483 nm

PP 3.23 $78.5°, 82.8°, 99.5°, 113°, 117°$

3.7 1.74 g/cm^3

3.13 Diameter of opening in center of unit cell $= 0.21$ nm

3.17 3.75 g/cm^3

3.19 0.317

3.21 1.24 eV

3.23 0.12

3.25 3.09 g/cm^3

3.27 3.10 g/cm^3

3.29 (a) 000, 100, 010, 001, 110, 101, 011, 111 (b) same

3.31 (a) $\frac{1}{2}\frac{1}{2}\frac{1}{2}$ (b) $\frac{1}{2}\frac{1}{2}0, \frac{1}{2}\frac{1}{2}1$

3.33 Octahedron

3.35 $[\bar{1}10], [1\bar{1}0], [\bar{1}01], [10\bar{1}], [01\bar{1}],$ and $[0\bar{1}1]$

3.37 $[110], [\bar{1}\bar{1}0], [101], [\bar{1}0\bar{1}], [0\bar{1}1],$ and $[01\bar{1}]$

3.39 $[\bar{1}11], [\bar{1}\bar{1}1], [11\bar{1}],$ and $[1\bar{1}\bar{1}]$

3.41 $[111], [\bar{1}\bar{1}\bar{1}], [1\bar{1}1],$ and $[\bar{1}1\bar{1}]$

3.43 $(01\bar{1}0), (0\bar{1}10), (10\bar{1}0), (\bar{1}010), (1\bar{1}00),$ and $(\bar{1}100)$

3.45 (a) $[100], [010], [\bar{1}00], [0\bar{1}0]$ (b) $[100], [\bar{1}00]$

3.47 (a) 000, $\frac{1}{2}\frac{1}{2}1$, 112

3.49 (a) 000, 112, 224

3.51 $[\bar{1}10]$ or $[1\bar{1}0]$

3.55 (a) $< 112 > =$ $\begin{array}{l} [112], [121], [211], [\bar{1}\bar{1}2], [\bar{1}\bar{2}1], [\bar{2}\bar{1}1] \\ [11\bar{2}], [12\bar{1}], [21\bar{1}], [\bar{1}12], [\bar{1}21], [\bar{2}11] \\ [1\bar{1}2], [1\bar{2}1], [2\bar{1}1], [\bar{1}1\bar{2}], [\bar{1}2\bar{1}], [\bar{2}1\bar{1}] \\ [\bar{1}12], [\bar{1}21], [\bar{2}11], [1\bar{1}\bar{2}], [1\bar{2}\bar{1}], [2\bar{1}\bar{1}] \end{array}$

3.57 000, $\frac{2}{3}\frac{1}{3}\frac{1}{2}$

3.59 000, $\frac{1}{2}\frac{1}{2}0$

3.63 Eight tetrahedral sites and four octahedral sites

3.65 $(1.05 U^{4+} + 2.11\ O^{2-})/nm$

3.67 Cl^- at 000, $\frac{1}{2}\frac{1}{2}0, \frac{1}{2}0\frac{1}{2}, 0\frac{1}{2}\frac{1}{2}$ and Na^+ at $00\frac{1}{2}, \frac{1}{2}\frac{1}{2}\frac{1}{2}, \frac{1}{2}01, 0\frac{1}{2}1$

3.69 $(3.76\ Ca^{2+} + 11.3 O^{2-})/nm^2$

3.73 $6.56(Zn^{2+}$ or $S^{2-})/nm^2$

3.75 Zn^{2+} at 000, $\frac{1}{2}\frac{1}{2}0, \frac{1}{2}0\frac{1}{2}, 0\frac{1}{2}\frac{1}{2}$ and S^{2-} at $\frac{1}{4}\frac{1}{4}\frac{1}{4}, \frac{3}{4}\frac{3}{4}\frac{1}{4}, \frac{3}{4}\frac{1}{4}\frac{3}{4}, \frac{1}{4}\frac{3}{4}\frac{3}{4}$

3.77 0.468

3.79 $44.8°, 65.3°, 82.5°$

3.81 (220), (310), (222)

3.85 (100), (002), (101)

3.87 $48.8°, 52.5°, 55.9°$

3.89 $56.5°, 66.1°, 101°, 129°, 142°$

3.91 Cr

Chapter 4

PP 4.1 No, % radius difference $> 15\%$

PP 4.2 (b) Roughly 50% too large

PP 4.3 $5.31 \times 10^{25}\ m^{-3}$

PP 4.4 0.320 nm

PP 4.5 (a) 16.4 nm, (b) 3.28 nm

PP 4.6 $G \cong 2$

PP 4.7 0.337

PP 4.8 The ratio of "up" to "down" pentagons $= 20/21 = 0.95$ or essentially 1:1

PP 4.9 (a) $1.05°$, (b) $1.48°$

4.1 Rule number 3 (electronegativities differ by 27%), rule number 4 (valences are different)

4.3 Rule number 2 (different crystal structures), rule number 3 (electronegativities differ by 19%), possibly rule number 4 (same valences shown in Appendix 2, although Cu^+ is also stable)

4.9 0%

4.11 $0.6023 \times 10^{24}\ Mg^{2+}$ vacancies

4.13 (a) 8.00×10^{-6} at %, (b) 7.69×10^{-6} wt %

4.15 (a) 8.00×10^{-6} at %, (b) 8.83×10^{-6} wt %

4.17 $5.00 \times 10^{21}\ m^{-3}$

4.19 $0.269 \times 10^{24}\ m^{-3}$

4.21 (a) 2.67, (b) 1.33

4.23 **(a)** 3.00, **(c)** 2.67

4.25 ≈ 5.6

4.27 170 μm, 41.3 μm

4.31 0.264 nm

4.39 $d_{\text{int}}/d = 0.90$

4.43 **(a)** 15.8 mm, **(b)** 18.3 mm, **(c)** 25.8 mm

4.45 13.9 mm

4.47 M to L transition

Chapter 5

PP 5.1 0.572 kg/(m^4·s)

PP 5.2 **(a)** 1.25×10^{-4}, **(b)** 9.00×10^{-8}, **(c)** 1.59×10^{-12}

PP 5.3 6.52×10^{19} atoms/(m^2· s)

PP 5.4 **(a)** 0.79 wt % C, **(b)** 0.34 wt % C

PP 5.5 **(a)** 0.79 wt % C, **(b)** 0.34 wt % C

PP 5.6 970° C

PP 5.7 2.88×10^{-3} kg/h

PP 5.8 21.9 mm

5.1 500 kJ/mol

5.3 290 kJ/mol

5.15 179 kJ/mol

5.17 4.10×10^{-15} m^2/s

5.19 1.65×10^{-13} m^2/s

5.21 264 kJ/mol

5.23 Al^{3+} diffusion controlled ($Q = 477$ kJ/mol)

5.25 0.324×10^{-3} kg/m^2·h

5.27 12.0×10^{12} atoms/s

5.29 86.6 kJ/mol

Chapter 6

PP 6.1 **(d)** 193×10^3 MPa (28.0×10^6 psi), **(e)** 275 MPa (39.9 ksi), **(f)** 550 MPa (79.8 ksi), **(g)** 30%

PP 6.2 **(a)** 1.46×10^{-3}, **(b)** 2.84×10^{-3}

PP 6.3 9.9932 mm

PP 6.4 **(a)** 80 MPa, **(b)** 25 MPa

PP 6.5 80.1 MPa

PP 6.6 3.57×10^{-5}

PP 6.7 **(a)** 0.2864 nm, **(b)** 0.2887 nm

PP 6.8 0.345 MPa (50.0 psi)

PP 6.9 4.08 mm

PP 6.10 **(a)** 1.47×10^{-3} % per hour, **(b)** 1.48×10^{-2} % per hour, **(c)** 9.98×10^{-2} % per hour

PP 6.11 **(a)** $\approx 550°$C, **(b)** $\approx 615°$C

PP 6.12 **(a)** 83.2 days, **(b)** 56.1 days

PP 6.13 511 to 537°C

6.1 108 GPa

6.3 **(a)** 4.00 GPa

6.5 3.14×10^5 N (7.06×10^4 lb$_f$)

6.7 **(a)** 2.96 MPa, **(b)** 2.39×10^{-5}

6.9 **(b)** Ti$-$5Al$-$2.5Sn, **(c)** 1040 carbon steel

6.13 **(a)** 78.5°, **(b)** 31.3 GPa

6.15 **(a)** 29.7 MPa, **(b)** 5.10×10^4 N

6.17 4.69 μm

6.19 8970 MPa

6.21 **(a)** 1.11 GPa, **(b)** 1.54 GPa

6.23 97.5 MPa

6.25 $\left(\dfrac{dF}{da}\right)_{a_0} = -42\dfrac{K_A}{a_0^8} + 156\dfrac{K_R}{a_0^{14}}$

6.27 0.136 MPa

6.29 20.4 MPa

6.31 $(1\bar{1}1)[110]$, $(\bar{1}11)[110]$, $(11\bar{1})[101]$, $(\bar{1}11)[101]$, $(11\bar{1})[011]$, $(1\bar{1}1)[011]$

6.37 $(211)[\bar{1}11]$, $(121)[1\bar{1}1]$, $(112)[11\bar{1}]$
$(\bar{2}11)[111]$, $(1\bar{2}1)[111]$, $(11\bar{2})[111]$
$(2\bar{1}1)[11\bar{1}]$, $(\bar{1}21)[11\bar{1}]$, $(\bar{1}12)[1\bar{1}1]$
$(21\bar{1})[1\bar{1}1]$, $(12\bar{1})[\bar{1}11]$, $(1\bar{1}2)[\bar{1}11]$

6.39 400 ± 140 MPa

6.41 26.7 MPa

6.43 Rockwell B99

6.45 Y.S. = 400 MPa, T.S. $\cong$ 550 MPa

6.47 347 BHN

6.49 **(a)** 252 kJ/mol, **(b)** 1.75×10^{-5} % per hour

6.51 1.79 hours

6.53 **(a)** 30.3 hr, **(b)** 303 hr

6.55 **(a)** 8.13×10^{-8} mm/mm/hr, **(b)** 14 years

6.57 **(a)** 247 days, **(b):** **(i)** 0.612 MPa, **(ii)** 0.333 MPa, **(iii)** 0.171 MPa

6.59 1.37 days

6.61 (a) 405 kJ/mol, (b) 759 to 1010°C, (c) 1120 to 1218°C

6.63 15°C

Chapter 7

PP 7.1 392 J/kg·K $\approx$ 385 J/kg·K

PP 7.2 0.517 mm

PP 7.3 1.86×10^6 J/m²·s

PP 7.4 670°C

PP 7.5 $\approx$ 250°C to 1000°C ($\approx$ 700°C midrange)

7.1 (a) 66.6 kJ, (b) 107 kJ, (c) 282 kJ

7.3 2940

7.5 10.07 mm

7.7 49.4°C

7.9 -11.7 kW

7.11 -43.7 kW

7.13 357 MPa (compressive)

7.15 35.3 MPa (compressive)

7.17 -277 MPa

7.19 No

Chapter 8

PP 8.1 $\leq 0.20\%$

PP 8.2 (a) 12.9 mm, (b) 2.55 mm

PP 8.3 (a) 169 MPa, (b) 508 MPa

PP 8.4 77.5 MPa

PP 8.5 (a) 213 s, (b) 11.6 s

PP 8.6 (a) 0.289, (b) 1.19×10^{-27}

PP 8.7 0.707

8.1 1040 carbon steel, 8630 low-alloy steel, L2 tool steel, ductile iron, Nb–1Zr

8.3 1.8 wt % Mn

8.5 ≥ 11.3 mm

8.9 Yes

8.11 57 μm $\leq a \leq$ 88 μm

8.13 286 μm

8.19 226 MPa

8.21 18.3 MPa

8.23 52.8%

8.25 91°C

8.27 (a) 1.75×10^6 per year, (b) 17.5×10^6 over 10 years

8.29 32.4 hr

8.31 6.67 MPa

8.33 $+0.14$ mm, -0.13 mm

8.35 0.43 mm to 0.68 mm

8.37 (a) 33.7 μs, (b) 36.9 μs, (c) 40.0 μs

8.39 0.046%

Chapter 9

PP 9.1 (a) 2, (b) 1, (c) 0

PP 9.2 The first solid to precipitate is β; at the peritectic temperature, the remaining liquid solidifies, leaving a two-phase microstructure of solid solutions β and γ

PP 9.3 $m_L = 952$ g, $m_{SS} = 48$ g

PP 9.4 $m_\alpha = 831$ g, $m_{Fe_3C} = 169$ g

PP 9.5 38.5 mol % monoclinic and 61.5 mol % cubic

PP 9.6 (a) 667 g, (b) 0.50

PP 9.7 60.8 g

PP 9.9 (a) $\approx$ 680°C, (b) solid solution β with a composition of $\approx$ 100 wt % Si, (c) 577°C, (d) 84.7 g, (e) 13.0 g Si in eutectic α, 102.0 g Si in eutectic β, 85.0 g Si in proeutectic β

PP 9.11 (a) For 200°C, (i) liquid only, (ii) L is 60 wt % Sn, (iii) 100 wt % L; for 100°C, (i) α and β, (ii) α is $\approx$ 5 wt % Sn, β is $\approx$ 99 wt % Sn, (iii) 41.5 wt % α, 58.5 wt % β

(b) For 200°C, (i) α and liquid, (ii) α is $\approx$ 18 wt % Sn, L is $\approx$ 54 wt % Sn, (iii) 38.9 wt % α, 61.1 wt % L; for 100°C, (i) α and β, (ii) α is $\approx$ 5 wt % Sn, β is $\approx$ 99 wt % Sn, (iii) 62.8 wt % α, 37.2 wt % β

PP 9.12 0 mol % Al_2O_3 = 0 wt % Al_2O_3, 60 mol % Al_2O_3 = 71.8 wt % Al_2O_3, 44.5 mol % SiO_2 = 36.1 wt % SiO_2, 55.5 mol % mullite = 63.9 wt % mullite

9.3 (a) 2, (b) 1, (c) 2

9.17 (a) $m_L = 1$ kg, $m_\alpha = 0$ kg; (b) $m_L = 667$ g, $m_\alpha = 333$ g; (c) $m_L = 0$ kg, $m_\alpha = 1$ kg

9.19 (a) $m_L = 1$ kg; (b) $m_L = 611$ g, $m_{\alpha-Pb} = 389$ g; (c) $m_{\alpha-Pb} = 628$ g, $m_{\beta-Sn} = 372$ g; (d) $m_{\alpha-Pb} = 606$ g, $m_{\alpha-Sn} = 394$ g

9.21 (a) $m_L = 50$ kg; (b) $m_\alpha = 20.9$ kg, $m_\beta = 29.1$ kg; (c) $m_\alpha = 41.8$ kg, $m_\beta = 8.2$ kg; (d) $m_\alpha = 50$ kg; (e) $m_\alpha = 39.9$ kg, $m_\beta = 10.1$ kg; (f) $m_\alpha = 33.2$ kg, $m_\beta = 16.8$ kg

9.23 760 g

9.25 (a) 33.8%, (b) 6.31%

9.27 16.7 mol % SiO_2 and 83.3 mol % mullite

9.29 92.8 wt % kaolinite and 7.2 wt % silica

9.31 6.9 wt % CaO

9.33 (a) $m_L = 1$ kg; (b) $m_L = 549$ g, $m_{alumina} = 451$ g; (c) $m_{mullite} = 549$ g, $m_{alumina} = 451$ g

9.35 mol. frac. cubic = 0.608, mol. frac. monoclinic = 0.392

9.37 (a) 0.258, (b) 0.453

9.41 4.34 wt % Si

9.43 59.4 kg

9.53 97.7 kg

9.57 (a) $\approx 950°$C, (b) solid solution α with a composition of $\approx$ 26 wt % Zn, (c) = 920°C, (d) $\approx$ 920°C to $\approx 40°$C.

9.63 ≈ 1 to 17.1 wt % Mg, ≈ 42 to ≈ 57 wt % Mg, ≈ 57 to 59.8 wt % Mg, 87.4 to ≈ 99 wt % Mg

9.65 $m_\alpha = 58.2$ g, $m_\beta = 58.8$ g

Chapter 10

PP 10.1 582°C

PP 10.2 (a) ≈ 1 s (at 600°C), ≈ 80 s (at 300°C); (b) ≈ 7 s (at 600°C), ≈ 1500 s = 25 min (at 300°C)

PP 10.3 (a) $\approx 10\%$ fine pearlite + 90% γ, (b) $\approx 10\%$ fine pearlite + 90% bainite, (c) $\approx 10\%$ fine pearlite + 90% martensite (including a small amount of retained γ)

PP 10.4 > 90% for 0.5 wt % C, $\approx 20\%$ for 0.77 wt % C, 0% for 1.13 wt % C

PP 10.5 (a) ≈ 15 s, (b) $\approx 2\frac{1}{2}$ min. (c) ≈ 1 hour

PP 10.6 (a) $\approx 7°$C/s (at 700°C), (b) $\approx 2.5°$C/s (at 700°C)

PP 10.7 (a) Rockwell C38, (b) Rockwell C25, (c) Rockwell C21.5

PP 10.8 (a) 7.55%, (b) 7.55%

PP 10.10 62 wt %

10.1 8.43×10^{12}

10.3 32.9 kJ/mol

10.5 $-2\sigma/\Delta G_v$

10.9 (a) 100% bainite, (b) austempering

10.11 (a) 100% bainite, (b) > 90% martensite, balance retained austenite

10.13 (b) $\approx 710°$C, (c) coarse pearlite

10.15 (b) $\approx 225°$C, (c) martensite

10.19 100% fine pearlite

10.23 (a) $> \approx 10°$C/s (at 700°C), (b) $< \approx 10°$C/s (at 700°C)

10.25 84.6% increase

10.27 Rockwell C53

10.31 Alloys 4340 and 9840

10.33 (a) 23.5%, (b) β phase

10.35 $\approx 400°$C

10.37 98.0 kJ/mol

10.39 No

10.41 31% CW

10.43 185 kJ/mol

10.45 (a) 503°C, (b) 448°C, (c) 521°C

10.47 475 kJ/mol

10.49 62.6 kJ/mol

Chapter 11

PP 11.1 $N_C = 13{,}851$, $N_{Mn} = 668$, $N_{Si} = 3225$, $N_P = 189$, $N_S = 183$

PP 11.2 2.75 Mg/m^3 (3003 Al), 2.91 Mg/m^3 (2048 Al)

PP 11.3 (a) 34 < % Ni < 79, (b) 34% Ni alloy

PP 11.4 (a) 63%, (b) 86%

11.1 (a) 7.85 $g/Mg/m^3$, (b) 99.7%

11.3 9.82 $g/Mg/m^3$

11.5 6.72 $g/Mg/m^3$

11.7 (a) 5.94%, (b) 4158 kg

11.9 332 kg

11.11 10.1 $g/Mg/m^3$

11.13 (a) 450 MPa, (b) 8%

Chapter 12

PP 12.1 0.717

PP 12.2 14.8 wt % Na_2O, 9.4 wt % CaO, 1.1 wt % Al_2O_3, 74.7 wt % SiO_2

PP 12.3 77.1 mol % SiO_2, 8.4 mol % Li_2O, 9.8 mol % Al_2O_3, 4.7 mol % TiO_2

PP 12.4 845 kg

PP 12.5 876°C to 934°C

12.1 72.4 wt % Al_2O_3, 27.6 wt % SiO_2

12.3 36.1 wt % SiO_2, 63.9 wt % mullite

12.5 0.346 kg

12.7 13.7 wt % Na_2O, 9.9 wt % CaO, 76.4 wt % SiO_2

12.9 42.2 kg

12.11 1.30 nm

12.13 0.0425 μm

12.15 4.0 vol %

12.17 0.100 μm

12.19 1587°C

12.21 512°C

12.23 22.6 g

12.25 sintered beryllia

Chapter 13

PP 13.1 400

PP 13.2 **(a)** 0.243 wt %, **(b)** 0.121 wt %

PP 13.3 69.0 mol % ethylene, 31.0 mol % vinyl chloride

PP 13.4 833

PP 13.5 **(a)** 15.5%, **(b)** 33.3%

PP 13.6 3.76×10^{23}

PP 13.7 25.1 wt % A, 25.6 wt % B, 49.3 wt % S

PP 13.8 84,100 amu

PP 13.9 3.37 g

PP 13.10 0.701 vinylidene fluoride, 0.299 hexafluoropropylene

PP 13.11 1.49 Mg/m^3

PP 13.12 866

13.1 21,040 amu

13.3 1.35 g

13.5 35.8 wt %

13.7 34,060 amu

13.9 612

13.11 0.552

13.13 101 nm

13.15 152 nm

13.17 **(a)** 50,010 amu, **(b)** 4.87 nm, **(c)** 126 nm

13.19 393

13.21 **(b)** 226.3 amu

13.23 296 kg

13.25 3.61×10^{25} amu

13.27 0.0426

13.29 **(a)** 49.7 wt %, **(b)** 1.53 Mg/m^3

13.31 43.3%

13.33 0.131 mm

13.35 23.0 wt %

13.37 12.8×10^6 kJ

13.39 acrylics, polyamides, polycarbonates

Chapter 14

PP 14.1 **(a)** 1.82 Mg/m^3, **(b)** 2.18 Mg/m^3

PP 14.2 **(a)** 24,320 g/mol, **(b)** 40,540 g/mol

PP 14.3 89.0 wt %

PP 14.4 14.1 Mg/m^3

PP 14.5 39.7×10^3 MPa

PP 14.6 0.57 W/(m·K)

PP 14.7 $E_c = 12.6 \times 10^3$ MPa, $k_c = 0.29$ W/(m·K)

PP 14.8 419×10^3 MPa

PP 14.9 $3.2 - 38\%$ error

PP 14.10 6.40×10^6 mm (epoxy), 77.8×10^6 mm (composite)

PP 14.11 26.8%

14.1 1.50 Mg/m^3

14.3 119,100 amu

14.5 370 kg

14.7 586

14.9 88.7 wt %

14.11 2.81 Mg/m^3

14.13 $8.43 \ \text{Mg/m}^3$

14.15 $289 \times 10^3 \ \text{MPa}$

14.17 $286 \times 10^3 \ \text{MPa}$

14.21 $49.2 \times 10^3 \ \text{MPa}$

14.23 $119 \times 10^3 \ \text{MPa}$

14.31 $430 \times 10^3 \ \text{MPa}$

14.35 0.38% error

14.37 $4.1 - 16\%$ error

14.39 $n \approx 0$

14.41 0.783

14.43 $(69.3 \ \text{to} \ 104) \times 10^6 \ \text{mm}$

14.45 $7.96 \times 10^6 \ \text{mm}$

14.47 $20.8 \times 10^6 \ \text{mm}$

14.51 2600 to 3400 psi

14.53 $0.341 \le v_{\text{clay}} \le 0.504$

Chapter 15

PP 15.1 **(a)** 1.73 V, **(b)** 108 mV

PP 15.2 8.17×10^{23} electrons

PP 15.3 6.65×10^{23} atoms

PP 15.4 1.59 hr

PP 15.5 2.11×10^{-44}

PP 15.6 2.32×10^{-9}

PP 15.7 **(a)** $34.0 \times 10^6 \Omega^{-1} \cdot \text{m}^{-1}$, **(b)** $10.0 \times 10^6 \Omega^{-1} \cdot \text{m}^{-1}$

PP 15.8 $43.2 \times 10^{-9} \Omega \cdot \text{m}$

PP 15.9 7 mV

PP 15.10 $1.30 \times 10^{-8} \ \text{A/m}^2$

PP 15.11 $1.034 \times 10^{-7} \ \text{C·m}$

PP 15.13 **(a)** $3.99 \times 10^{-4} \Omega^{-1} \cdot \text{m}^{-1}$, **(b)** $2.50 \times 10^3 \Omega \cdot \text{m}$

PP 15.14 $2.0 \times 10^{-11} \Omega^{-1} \cdot \text{m}^{-1}$

15.1 **(a)** 55.0 A, **(b)** 3.14×10^{-9} A, **(c)** 7.85×10^{-19} A

15.3 **(a)** 33.3 m/s, **(b)** 18.0 μs

15.5 4.71 Ω

15.7 734 m

15.9 994°C

15.11 **(a)** 0.0353, **(b)** 0.0451

15.13 3.75×10^{-13} (for GaP), 1.79×10^{-6} (for GaSb)

15.15 34.8 mV

15.17 36.9 mV

15.19 $\approx 500°$C

15.21 8.80 m

15.23 5.24 kW

15.25 1.73×10^{-7} W

15.27 ≈ 2210

15.31 $2.89 \times 10^{-5} \ \text{C/m}^2$

15.33 $1.20 \times 10^{-4} \ \text{C/m}^2$

15.35 $4.03 \times 10^{-4} \ \text{C/m}^2$

15.37 112% increase

15.39 109 MPa

15.43 5.2×10^{-10}

15.45 **(a)** 0.657 (electrons), 0.343 (holes); **(b)** 0.950 (electrons), 0.050 (holes)

15.47 $1.74 \times 10^4 \Omega^{-1} \cdot \text{m}^{-1}$

15.49 **(a)** $38.1 \times 10^6 \Omega^{-1} \cdot \text{m}^{-1}$, **(b)** $27.6 \times 10^6 \Omega^{-1} \cdot \text{m}^{-1}$

15.51 $3.53 \times 10^7 \Omega^{-1} \cdot \text{m}^{-1}$

Chapter 16

PP 16.1 1.77 eV

PP 16.2 66.1°

PP 16.3 **(a)** 0.0362, **(b)** 0.0222

PP 16.4 0.0758

PP 16.5 1240

PP 16.6 697 nm

PP 16.7 87.0°

PP 16.8 464 nm

16.1 **(a)** $3.318 \times 10^{14} \ \text{s}^{-1}$ **(b)** 1.786 eV

16.3 16.9% smaller

16.5 **(a)** 40.98°, **(c)** 41.47°

16.7 34.0%

16.9 40.8°

16.11 323 nm

16.13 2,594

16.15 5.71

16.17 822 nm

16.19 4.87 μs

16.21 400 nm $\le \lambda \le$ 549 nm

Chapter 17

PP 17.1 2.1×10^{-8}

PP 17.2 **(a)** $24.8\Omega^{-1} \cdot m^{-1}$

PP 17.3 $474\Omega^{-1} \cdot m^{-1}$

PP 17.4 5.2×10^{21} atoms/m^3 ($\ll 20 \times 10^{24}$ atoms/m^3)

PP 17.5 8.44×10^{-8}

PP 17.6 **(a)** $135\Omega^{-1} \cdot m^{-1}$

PP 17.7 **(a)** 16.6 ppb, **(b)** 135°C, **(c)** $55.6\Omega^{-1} \cdot m^{-1}$

PP 17.8 **(a)** 1880 nm, **(b)** 95,600 nm

PP 17.9 4.07×10^{21} atoms/m^3

PP 17.10 $2.20 \times 10^4\Omega^{-1} \cdot m^{-1}$

PP 17.11 0.9944 (electrons), 0.0056 (holes)

PP 17.12 31.6% increase

PP 17.13 1.31 parts per billion (ppb) Al

17.1 **(a)** 1.24×10^{11}, **(b)** 1.24×10^{11}

17.5 10 K

17.7 4.77%

17.9 **(a)** 9.07×10^{-6} mol %, **(b)** 4.54×10^{21} atoms/m^3 ($\ll 1000 \times 10^{24}$ atoms/m^3)

17.11 **(a)** Electron hole, **(b)** $3.29 \times 10^{18} m^{-3}$, **(c)** 7.6 m/s

17.13 $15.8\Omega^{-1} \cdot m^{-1}$

17.15 285°C

17.17 $5.41\Omega^{-1} \cdot m^{-1}$

17.19 $2.07\Omega^{-1} \cdot m^{-1}$

17.21 317 K ($= 44°$C), taking room temperature as 300 K

17.23 **(a)** 1.11×10^{-10} K, **(b)** 9.31×10^{-7} K, **(c)** 289 K

17.27 2.85×10^{21} atoms/m^3

17.33 7.5 K

17.35 0.1°C

17.37 0.973 (electrons), 0.027 (holes)

17.39 0.337

17.41 2.76 ppm

17.43 **(a)** 55.1 ppm Pb, **(b)** 4.1 ppm Pb

17.45 **(a)** 10^3 m/s, **(b)** 7.14×10^3 V/m, **(c)** 5.14 gigahertz

Chapter 18

PP 18.1 $B = 0.253$ weber/m^2, $M = 1.0 \times 10^3$ amperes/m

PP 18.3 **(a)** 0.65 weber/m^2, **(b)** 4.57×10^5 amperes/m

PP 18.4 $8\mu_\beta$

PP 18.5 1.26×10^5 A/m

PP 18.6 44 J/m^3

PP 18.7 ≈ 3.3 J/m^3

PP 18.8 0.508 (Fe^{3+}), 0.591 (Ni^{2+})

PP 18.9 **(a)** 0, **(b)** 80 μ_β, **(c)** 0.89

18.1 $B = 0.251$ weber/m^2, $M = 10$ amperes/m

18.5 **(a)** 0.99995, **(b)** diamagnetism

18.7 **(b)** 0.36 weber/m^2, **(c)** -25 A/m

18.9 **(b)** 0.92 weber/m^2, **(c)** -18 A/m

18.13 **(a)** $40\mu_B$, **(b)** 6.04×10^5 A/m

18.15 24 μ_B

18.17 **(a)** 0.591, **(b)** 0.432

18.19 2.64 kW/m^3

18.21 67 J/m^3

18.23 4.02 kW/m^3

18.25 1.50 kW/m^3

18.33 **(a)** 38×10^4 A-turns/m, **(b)** 14% error

18.35 **(a)** 2.4 kW/m^3, **(b)** soft

Chapter 19

PP 19.1 2.5 μm

PP 19.2 **(a)** 1.75, **(b)** yes

PP 19.3 $3.13 \times 10^{16} s^{-1}$

PP 19.4 **(a)** Zinc corroded, **(b)** 1.100 V

PP 19.5 0.0323 m^3

PP 19.6 4.25

PP 19.7 **(a)** 1.5 kg, **(b)** 6.0 kg

PP 19.8 -0.48 V

PP 19.9 **(a)** 82.7 nm, **(b)** ultraviolet

PP 19.10 415 μm

PP 19.11 $E(L_\alpha) = 655$ eV, $E(LMM) = 602$ eV

19.3 1.88

19.7 1.56×10^{16} s^{-1}

19.9 2.08×10^{16} s^{-1}

19.11 **(a)** 0.467 V, **(b)** Cr

19.13 **(a)** nickel, **(b)** tin-rich phase, **(c)** 2024 aluminum, **(d)** brass

19.15 2.54×10^{-3} m^3

19.17 0.0157 m^3

19.19 0.302 V

19.21 0.235 V

19.23 21.8 kg

19.25 **(a)** 7.50 μm, **(b)** 37.5 μm

19.27 1.19 A/m^2

19.29 1.55×10^{-4} A/m^2

19.31 **(a)** 1.22×10^{-3} nm, **(b)** x-ray or γ-ray

19.33 148 μm

19.35 **(a)** 8049 eV, **(b)** 8907 eV, **(c)** 858 eV, **(d)** 7116 eV, **(e)** 783 eV

19.37 **(a)** 40.6°

Chapter 20

PP 20.1 S7 Fan cooled from 940°C and single-tempered at 540°C
S7 Fan cooled from 940°C and single-tempered at 425°C

PP 20.2 15.2×10^6 l

PP 20.3 2.18

PP 20.4 7059 eV

20.1 S5 Oil-quenched from 870°C and single-tempered at 650°C
L2 Oil-quenched from 855°C and single-tempered at 540°C

20.3 80–55–06

20.5 Polyethylene and polypropylene

20.9 2.53 kg

20.11 **(a)** 120 kg, **(b)** 9.96×10^4l (per aircraft)

20.13 Adequate

20.15 Carbon fiber-reinforced polymer

20.19 Polyester 1

20.21 1.45×10^{-9} m^2

20.23 **(a)** 2.19 eV, **(b)** 1.91 to 2.10 eV

20.25 0.2269 nm(V), 0.1488 nm(Ni), 0.1283 nm(Zn), 0.09794 nm(Se), 0.01409 nm(Pb)

20.27 **(a)** difficult, **(b)** easy

Index

859

PHYSICAL AND CHEMICAL DATA FOR SELECTED ELEMENTS[a]

Atomic number	Element	Symbol	Atomic mass (amu)	Density of solid (at 20°C) (Mg/m³ = g/cm³)	Crystal structure (at 20°C)	Melting point (°C)	Atomic number
1	Hydrogen	H	1.008			−259.34 (T.P.)	1
2	Helium	He	4.003			−271.69	2
3	Lithium	Li	6.941	0.533	bcc	180.6	3
4	Beryllium	Be	9.012	1.85	hcp	1289	4
5	Boron	B	10.81	2.47		2092	5
6	Carbon	C	12.01	2.27	hex.	3826 (S.P.)	6
7	Nitrogen	N	14.01			−210.0042 (T.P.)	7
8	Oxygen	O	16.00			−218.789 (T.P.)	8
9	Fluorine	F	19.00			−219.67 (T.P.)	9
10	Neon	Ne	20.18			−248.587 (T.P.)	10
11	Sodium	Na	22.99	0.966	bcc	97.8	11
12	Magnesium	Mg	24.31	1.74	hcp	650	12
13	Aluminum	Al	26.98	2.70	fcc	660.452	13
14	Silicon	Si	28.09	2.33	dia. cub.	1414	14
15	Phosphorus	P	30.97	1.82 (white)	ortho.	44.14 (white)	15
16	Sulfur	S	32.06	2.09	ortho.	115.22	16
17	Chlorine	Cl	35.45			−100.97 (T.P.)	17
18	Argon	Ar	39.95			−189.352 (T.P.)	18
19	Potassium	K	39.10	0.862	bcc	63.71	19
20	Calcium	Ca	40.08	1.53	fcc	842	20
21	Scandium	Sc	44.96	2.99	fcc	1541	21
22	Titanium	Ti	47.90	4.51	hcp	1670	22
23	Vanadium	V	50.94	6.09	bcc	1910	23
24	Chromium	Cr	52.00	7.19	bcc	1863	24
25	Manganese	Mn	54.94	7.47	cubic	1246	25
26	Iron	Fe	55.85	7.87	bcc	1538	26
27	Cobalt	Co	58.93	8.8	hcp	1495	27
28	Nickel	Ni	58.71	8.91	fcc	1455	28
29	Copper	Cu	63.55	8.93	fcc	1084.87	29
30	Zinc	Zn	65.38	7.13	hcp	419.58	30
31	Gallium	Ga	69.72	5.91	ortho.	29.7741 (T.P.)	31
32	Germanium	Ge	72.59	5.32	dia. cub.	938.3	32
33	Arsenic	As	74.92	5.78	rhomb.	603 (S.P.)	33
34	Selenium	Se	78.96	4.81	hex.	221	34
35	Bromine	Br	79.90			−7.25 (T.P.)	35
36	Krypton	Kr	83.80			−157.385	36
37	Rubidium	Rb	85.47	1.53	bcc	39.48	37
38	Strontium	Sr	87.62	2.58	fcc	769	38
39	Yttrium	Y	88.91	4.48	hcp	1522	39
40	Zirconium	Zr	91.22	6.51	hcp	1855	40
41	Niobium	Nb	92.91	8.58	bcc	2469	41
42	Molybdenum	Mo	95.94	10.22	bcc	2623	42

[a]For a complete listing, see Appendix 1.